FROM THE LIBRARY OF
JBG-0006X © Antioch Publishing Company

AFTER THE FIRST THREE MINUTES

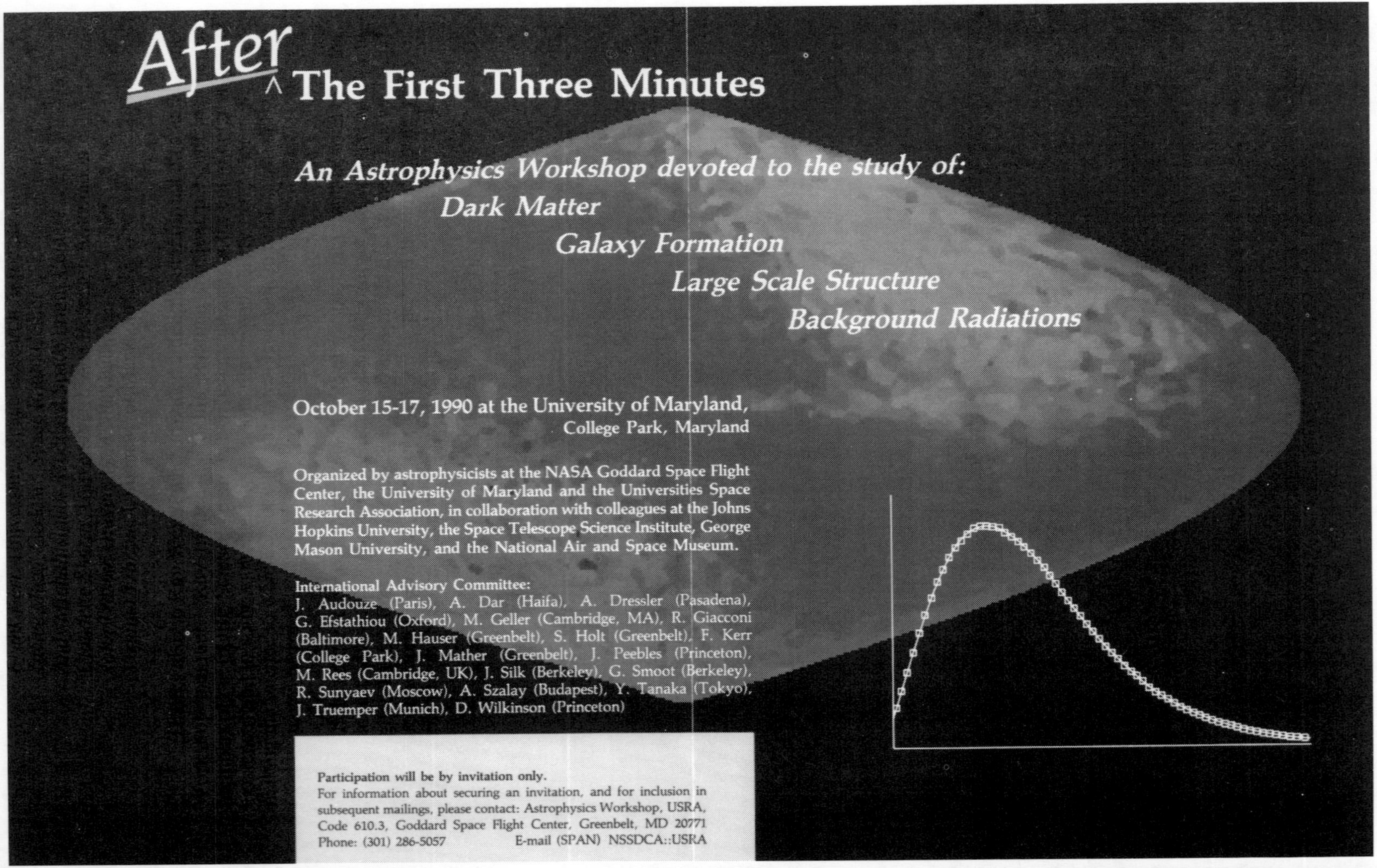
After
The First Three Minutes
An Astrophysics Workshop devoted to the study of:
Dark Matter
Galaxy Formation
Large Scale Structure
Background Radiations
October 15-17, 1990 at the University of Maryland, College Park, Maryland
Organized by astrophysicists at the NASA Goddard Space Flight Center, the University of Maryland and the Universities Space Research Association, in collaboration with colleagues at the Johns Hopkins University, the Space Telescope Science Institute, George Mason University, and the National Air and Space Museum.
International Advisory Committee:
J. Audouze (Paris), A. Dar (Haifa), A. Dressler (Pasadena), G. Efstathiou (Oxford), M. Geller (Cambridge, MA), R. Giacconi (Baltimore), M. Hauser (Greenbelt), S. Holt (Greenbelt), F. Kerr (College Park), J. Mather (Greenbelt), J. Peebles (Princeton), M. Rees (Cambridge, UK), J. Silk (Berkeley), G. Smoot (Berkeley), R. Sunyaev (Moscow), A. Szalay (Budapest), Y. Tanaka (Tokyo), J. Truemper (Munich), D. Wilkinson (Princeton)
Participation will be by invitation only.
For information about securing an invitation, and for inclusion in subsequent mailings, please contact: Astrophysics Workshop, USRA, Code 610.3, Goddard Space Flight Center, Greenbelt, MD 20771
Phone: (301) 286-5057
E-mail (SPAN) NSSDCA::USRA

AIP CONFERENCE PROCEEDINGS 222

AFTER THE FIRST THREE MINUTES

COLLEGE PARK, MD 1990

EDITORS:

STEPHEN S. HOLT
CHARLES L. BENNETT
NASA GODDARD SPACE FLIGHT CENTER
VIRGINIA TRIMBLE
UNIVERSITY OF MARYLAND

AIP

American Institute of Physics New York

L.C. Catalog Card No. 91-55214
ISBN 0-88318-828-7
DOE CONF-9010268

Printed in the United States of America.

Table of Contents

Preface

These Proceedings are the published record of the first in what is anticipated to be a new series of annual October topical astrophysics workshops.

A few years ago, Menas Kafatos organized a similar series of workshops at George Mason University (jointly sponsored by the Goddard Space Flight Center). Although George Mason University is maximally inconvenient for those of us who are permanent residents of Maryland working at the Goddard Space Flight Center or the University of Maryland (it is almost exactly 180° around the Washington Beltway), the four-year series was so well-conceived and well-organized that many of us attended it regularly in spite of the inconvenience. The series lapsed only because it became burdensome for its lone organizer.

There are a number of other astronomical institutions of excellence in the greater Washington D.C. area, including the Naval Observatory, the Naval Research Laboratory, the National Air and Space Museum, the Carnegie Institution, and the Applied Physics Laboratory of the Johns Hopkins University. A long-standing Washington Area Astronomers group meets twice each year, which also includes attendees from the Space Telescope Science Institute and the Johns Hopkins University in Baltimore, as well as some from as far away as the University of Virginia at Charlottesville. These meetings are semi-social gatherings that feature informal papers by some of the attendees, and are therefore neither topical (i.e. devoted to a single theme) nor do they attract expert visitors from outside the Washington area.

This year, some of us had the determination to reinstate the idea of a local annual October topical astronomy workshop. The Goddard Space Flight Center and the University of Maryland have a long tradition of scientific cooperation, and leaders of the scientific staffs of these two organizations took it upon themselves to form a permanent executive committee that will oversee the continuing operation of such a workshop series. The committee members are the Chiefs of the three space science laboratories at Goddard (Dan Baker, Mike Hauser and Jonathan Ormes), the University of Maryland Program Chairs of Astronomy (Roger Bell) and Physics (originally Chuan Sheng Liu, now replaced by his successor, Derek Boyd), and two other key individuals. One is Menas Kafatos, the Chairman of Physics and Astronomy at George Mason University and the originator of the previous incarnation of this series. The other is Frank Kerr, the Chief Scientist of the Universities Space Research Association (which administers a visiting scientist program at Goddard, and which provided considerable support to the workshop this year); Frank has formerly been both Astronomy Program Chair and Provost of the University of Maryland. As the Director of Space Sciences at Goddard, I serve as the chair of this committee.

The excitement generated by the new results from the COBE spacecraft made "observational cosmology" the natural choice for this year's topic. We formed a Scientific Organizing Committee consisting (largely) of local experts in relevant subdisciplines, including members from several Washington area institutions. This committee made two key decisions regarding the content of this workshop, to distinguish it from a plethora of workshops with apparently similar subjects: one was to attempt to balance the theory that otherwise dominates the content of many such conferences with a strong emphasis on the best new observational data, and the second was to concentrate on the remnant radiations and structures rather than on big-bang physics. The latter consideration rather naturally gave rise to a title for our workshop: "*After* the First Three Minutes."

We developed a list of potential invitees, which was sent to an International Advisory Committee of well-known astrophysics generalists, for their approval. We were gratified that a significant fraction of the most important contributors to the field were able to support the meeting with their attendance. In fact, we found that we had an embarrassment of riches, insofar as we could not possibly maintain a "workshop" atmosphere if we asked all those who were deserving to give invited papers. Instead, we asked some experts to give summary invited talks, and some others to serve as session chairs. The session chairs were given the authority to design their sessions from both invited talks and poster papers that could be "promoted" to short oral contributions. In retrospect, we were at least as successful in selecting the session chairs as we were in selecting invited speakers; the session chairs exhibited their own individual styles in constructing their sessions, but uniformly provided the energy and leadership necessary to make them work.

Neta Bahcall, Princeton University
Chuck Bennett, Goddard Space Flight Center
Roger Blandford, California Institute of Technology
Elihu Boldt, Goddard Space Flight Center
Dave Burstein, Arizona State University
Kris Gorski, Los Alamos National Laboratory
Lawrence Krauss, Yale University
Bruce Partridge, Haverford College
Simon White, Steward Observatory, University of Arizona
Ned Wright, University of California, Los Angeles

Several local institutions were invited to participate in the Local Organizing Committee, chaired by John Trasco, which consisted largely of astrophysicists whose specialties were generally less closely related to the subject material than were those of the members of the Scientific Organizing Committee, and who unselfishly gave of their time to provide the ideas and hard work that made the workshop as comfortable as it was informative. I think that all of the attendees will agree that the arrangements for posters, hotels, transportation, coffee, meals, etc. were handled with exceptional efficiency.

These Proceedings are organized in general accordance with the actual chronology of the meeting (see Appendix A). Each session was designed to run for no more than about 2 hours, and poster papers that appeared to be relevant to that session were displayed with a parallel numbering scheme. Without the necessity of meeting session time constraints in these Proceedings, the editors have decided to utilize functional divisions for the subject material, rather than to adhere strictly to the original chronology of the meeting. This is a mixed blessing, insofar as there are many papers that can just as easily be grouped within one functional category as they can another, but we hope that, on the whole, the regrouping makes it easier for the reader to locate papers relevant to a particular subject that might otherwise be missed. We have utilized camera-ready manuscripts, and so we have not attempted to correct typographical errors; we trust that the provision of a Subject Index is at least partial recompense to the reader for occasional typos.

Another useful aspect of these Proceedings is the inclusion of a variety of independent overviews, as a natural result of numerous expert invited contributions. The introductory paper by Jim Peebles (the "keynote" paper of the workshop) not only states "The Problem" (the fact that "standard" Cold Dark Matter (CDM), which was the most-fashionable key ingredient for large scale structure in the universe prior to this workshop, cannot successfully confront the new data) but also suggests the revitalization of a cosmological constant as a possible modification to CDM. The final paper presented at the workshop was Dick Bond's valuable *rapporteur* contribution, which deftly reviews all the major elements of the workshop contributions and discussions without the benefit of anything but his own notes and recollections. The concluding sections of these Proceedings contain no less than four purportedly objective summaries in four different styles, ranging from deadly serious to utterly whimsical -- the reader will get no help from the editors in determining which adjectives should be attached to each.

In the final analysis, the workshop succeeded because a few people worked very hard to do whatever needed doing, from the development of the first version of the scientific programme to the unplugging of the last overhead projector. At the risk of slighting some people who worked almost as hard, the following are my "top ten" (in alphabetical order) to whom special thanks are due:

Chuck Bennett, Goddard Space Flight Center
Arnon Dar, Goddard Space Flight Center
David Holdridge, Universities Space Research Association
Demos Kazanas, Goddard Space Flight Center
Frank Kerr, Universities Space Research Association
Jaylee Mead, Goddard Space Flight Center
Andy Michalitsianos, Goddard Space Flight Center
Claudette Sharps, Universities Space Research Association
John Trasco, University of Maryland
Virginia Trimble, University of Maryland

Thanks also to Ray Weiss, who provided a delightful informal discourse at the Workshop banquet. Ray treated us to his personal musical personifications of members of the COBE Science Team (which he has chaired for the entire duration of its existence) and to some valuable advice about the responsibility that all of us have for making the science to which we devote our careers relevant to our less scientific friends (and taxpayer-patrons).

In the future, the most significant change in our procedure will be to involve the International Advisory Committee earlier, so that they might actively participate in selecting workshop topics, invited speakers and session chairs. Otherwise, we intend to follow the model described here for what we hope will be an ongoing series of memorable October topical astrophysics workshops.

Steve Holt
December 15, 1990

Introductory Overviews

GENERAL INTRODUCTION

P. J. E. Peebles
Institute for Advanced Study
and
Princeton University
Princeton NJ 08540

ABSTRACT

What will be the character of research in physical cosmology in the 1990s? We may find a greater emphasis on semiempirical models whose development is informed by a more nearly equal balance of theory and observation than has been feasible up to now. I present an example of how this might go in a model for galaxy formation in a low density universe.

INTRODUCTION

There are shifting patterns of research in physical science, as in any human endeavor. The tenor of the discussion at this conference is quite different from what you would have observed in a meeting on cosmology in the early 1980s, when inflation and non-baryonic dark matter were young, and different again from what you would have observed in the early 1970s.

The shifts in directions of research in physical cosmology have been triggered by observational developments and by attractive new ideas. In the middle 1960s the microwave cosmic background radiation (the CBR) was discovered and interpreted as the primeval fireball left over from the dense very early universe. The detection of this radiation lent credence to the idea that the universe really did expand from a dense state, and it led to discussions of the role of the CBR in the physics of the evolving universe. Models tended to be semiempirical, attempting to extrapolate from what is observed at the present epoch to what things might have been like in the past. That is what one does in the computation of light element production, and in the 1970s it was the theme of discussions of the evolution of irregularities in the distributions of matter and radiation in an initially hot expanding universe. I would include in this tradition the discussions of explosion pictures by Ostriker and Cowie (1981) and Ikeuchi (1982). But during the 1980s the focus was more on ideas that motivated *ab initio* theories that were compared to observations. Inflation offered the possibility of predicting the character of the primeval density fluctuations that grew into the large-scale structure we observe. That led to the influential cold dark matter theory. Also influential was the idea of topological defects, particularly cosmic strings. This was a healthy phase, and I certainly hope the flow of ideas will continue. But, to put it bluntly, the *ab initio* approaches have not yet given us anything that is believable according to the standard rules of evidence in physical science. That is one reason why I am speculating that the 1990s may see a renewed interest in semiempirical models. Another very good reason is of course the dramatic increase in the the store of observational results that might be used in semiempirical analyses.

The prefix "semi" is meant to indicate that model building has to be informed by our understanding of physics, for there is no way we could piece together from observations alone an understanding of how the universe has been behaving even at modest expansion factors. How does one combine and weigh the theory and the observational and laboratory results? That of course is a matter of taste; my inclinations are to try the most straightforward way first, as will be seen in the example below.

Such a semiempirical phase of research in cosmology, if successful, would tend to seed its replacement: it would show us what really happened at modest expansion factors, and that surely would trigger a new phase of invention of *ab initio* theories of the very early universe.

To illustrate how I imagine semiempirical model building might go, I present here a picture for galaxy formation. Since it seems to me that the straightforward interpretation of the observations is that the mean mass density is well below the critical Einstein-de Sitter value, I develop the model for galaxy formation in a low density model universe. Let us begin by recalling the arguments for this case.

THE CASE FOR A LOW DENSITY UNIVERSE

If optically selected galaxies trace the distribution of mass then dynamical studies on scales $\lesssim 10h^{-1}$ Mpc ($H = 100h$ km s^{-1} Mpc^{-1}) pretty consistently indicate that the density parameter is $\Omega \sim 0.1$, rather than the critical value $\Omega = 1$ for an Einstein-de Sitter model (Peebles 1986). It has been argued that this result is not necessarily significant, because it assumes galaxies trace mass. That is the truth, but not the whole truth. For example, Figure 1 shows the rms relative line of sight velocity $\Delta v(r)$ of pairs of galaxies as a function of the galaxy separation r (Davis and Peebles 1983, Peebles 1984a). It will be noted that the horizontal axis spans three decades in separation, while the vertical axis is linear. One sees that Δv increases very slowly with increasing galaxy separation r. If the galaxy-mass cross correlation function, $\xi_{g\rho}$, is equal to the galaxy two-point correlation function, $\xi_{gg} \propto r^{-\gamma}$, with $\gamma = 1.77 \pm 0.04$, and the same is true of the three point functions, then the rms relative velocity is predicted to scale with separation as

$$\Delta v \propto r^{(2-\gamma)/2} \sim r^{0.1} \ . \tag{1}$$

That is, the slow increase of Δv with r seen in Figure 1 is just the behavior expected if galaxies trace mass. However, the size of Δv is consistent with a mean mass density ~ 10 % of the critical Einstein-de Sitter value.

It was the result in Figure 1, that the form of $\Delta v(r)$ is consistent with the assumption that galaxies trace mass but that the amplitude is wrong for the Einstein-de Sitter case, that first caused me seriously to consider the possibility that we live in a low density universe.

It is of course possible that the coincidence of shapes of $\xi_{g\rho}$ and ξ_{gg} is only an accident, and that there is a considerable amount of mass not clustered with galaxies on small scales. That is the case in the biasing concept pioneered by Kaiser (1986), Bardeen (1986), and the team of Davis, Efstathiou Frenk and White (1985). This has been an exceedingly influential and stimulating idea, and it may even be right. However, there is not a strong empirical case for it.

To my mind the most serious problem for the biasing concept is that it would predict that galaxies in low density regions (voids) are irregular and/or dwarfs, as has been cogently argued by Dekel and Silk (1986) and many others. The point simply is that if most seeds in the voids failed to make it to visible galaxies then the few that did surely would show the stigmata of an exceedingly

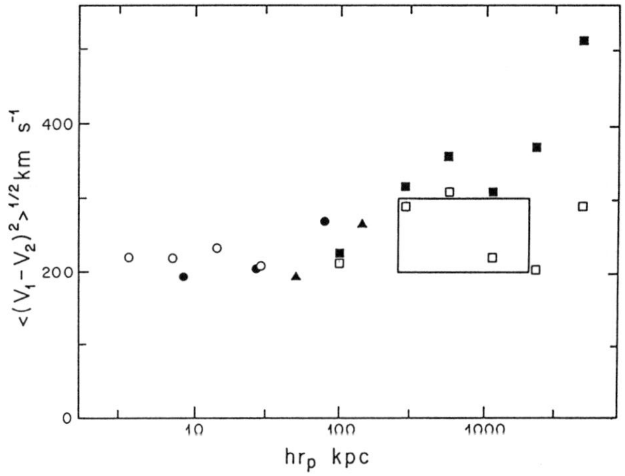

Fig. 1. The rms relative velocity of pairs of galaxies as a function of separation. For a list of the catalogs used in these estimates see Peebles (1984a).

unhealthy early environment. However, this is not supported by the observations (eg. Binggeli 1989; Peebles 1989).

My impression is that, by the standard rules of evidence in a physical science, biasing ought to be treated as an interesting idea but one to be taken with caution until we find some evidence that the straightforward interpretation of the observations is wrong.

LEARNING TO LIVE WITH LAMBDA

If the density parameter is well below unity, general relativity theory requires that the universe has non-zero space curvature and/or a non-zero cosmological constant. The economical assumption is that only one is significant. The one unambiguous prediction of inflation is that space curvature is negligibly small; if we accept that and a low density in moveable matter we must learn to live with lambda. (Perhaps a more elegant picture would be that the large effective Λ present during inflation dropped precipitously at reheating, but not to zero, and has since since been rolling to zero but so slowly that at late times it comes to dominate again. A model of this sort is discussed by Peebles and Ratra 1988.)

A second argument for low space curvature in a low density universe is that it makes the growth of density fluctuations in linear perturbation theory insensitive to Ω (Peebles 1984b). If space curvature dominates, the late time growth of density fluctuations in linear perturbation theory is more strongly suppressed, while the contrast in bound systems of course continues to grow as the background density drops. In this case it might be surprising that the mean run of galaxy number density with radius in a cluster is fairly close to a power law, $n(r) \propto r^{-2}$, from the dense inner parts to the low contrast halo. Arguments of this sort are what led me to argue for an Einstein-de Sitter model (Peebles 1974); but maybe a Λ-dominated universe would do about as well.

There are other more direct observational hints. First, if the product $H_o t_o$

of the Hubble constant H_o and the time t_o since the Big Bang is greater than unity, then Λ must be positive. There is a prominent case for a value of the Hubble constant near 50 km s^{-1} Mpc^{-1} (Sandage and Tammann 1990), but one also encounters a considerable number of recent papers that indicate a larger value. Several recent estimates of the distance to the Virgo cluster, from the luminosities of planetary nebulae (Jacoby *et al.* 1990), surface brightness fluctuations (Tonry *et al.* 1989), and the Tully Fisher relation (van den Bergh 1989) give the "short distance," $\sim 14.5 \pm 1$ Mpc. Since the ratios of distances to the Virgo cluster and more distant clusters are fairly well known one can use the distance to Virgo to find the distances to clusters whose redshifts are thought to be dominated by the expansion of the universe. For example, Dressler *et al.* (1987) find from their $D_n - \sigma$ relation that the ratio of distances to the Coma and Virgo clusters is 5.37. Taking the redshift of the Coma cluster to be 7100 km s^{-1}, this says the cosmological redshift of the Virgo cluster is $cz = 1300$ km s^{-1}. From a more thorough survey of cluster distance ratios, Sandage and Tammann (1990) find $cz = 1140$ km s^{-1}. This with the short distance to the Virgo cluster gives $H_o = 1140/14.5 = 80$ km s^{-1} Mpc^{-1}. Application of the B-band Tully Fisher relation directly to Coma cluster galaxies gives a similar result (Fukugita *et al.* (1990a). If this high value for H_o is right, the Einstein-de Sitter age is $t_o \sim 8$ by, and the maximum age of a universe with $\Lambda = 0$ is $t_o \lesssim 12$ by. The former is unacceptable, the latter dangerously low. If the large value of H_o and the conventional estimates $t_o \sim 15$ billion years should hold up, we would have to resort to a Λ-like term.

Koo (1989) emphasized that the high counts of faint galaxies might be most simply understood if we lived in a low density universe, where the comoving volume per unit increment of redshift is larger than in the Einstein-de Sitter model. The volume is larger for zero space curvature than for zero Λ. In two recent analyses of deep galaxy counts, by Fukugita *et al.* (1990b) and Lilly *et al.* (1990), the preliminary results are that a low density cosmologically flat model universe gives a not unreasonable fit to the data, while the Einstein-de Sitter model seems problematic.

Another possible benefit of low Ω is the interpretation of the frequency of detection of damped Lyman-α Wolfe clouds. In a cosmologically flat universe containing objects with time independent mean cross section σ and fixed comoving number density n_o, the probability per unit redshift that a line of sight intersects an object is

$$\frac{dP}{dz} = \frac{\sigma n_o c}{H_o} \frac{(1+z)^2}{[\Omega(1+z)^3 + 1 - \Omega]^{1/2}} . \qquad (2)$$

In the Einstein-de Sitter case, dP/dz for Wolfe clouds at $z \sim 2.5$ is roughly five times what would be expected from the abundance of low redshift galaxies (Wolfe 1989). One can think of many possible explanations, to be sure. But it is interesting that the discrepancy is appreciably reduced by using a lower value of Ω in equation (2).

Turner (1990) has pointed out that the frequency of gravitational lensing of high redshift quasars by galaxies could provide a sensitive test for Λ. In the approximation of the lensing galaxies as isothermal gas spheres with negligible core radius (mass density run $\rho \propto r^{-2}$), the probability that a quasar at redshift

z appears as a double image due to an intervening galaxy is

$$P = \frac{2\pi^3}{15}\left(\frac{c}{H_o}\right)^3 (H_o a_o x(z)/c)^3 \sum n_i (v_i/c)^4 \ , \tag{3}$$

where the coordinate distance $\int dt/a$ to the quasar is

$$H_o a_o x(z)/c = \int_{1/(1+z)}^{1} \frac{dy}{[\Omega y + (1-\Omega)y^4]^{1/2}} \ , \tag{4}$$

if space curvature can be neglected. The sum is over the comoving number density n_i of galaxies with circular velocity v_i. An informal estimate of this sum by Turner *et al.* (1984) gave

$$\sum n_i \left(\frac{v_i}{200 \text{ km s}^{-1}}\right)^4 \sim 0.2h^3 \text{ Mpc}^{-3} \ . \tag{5}$$

Turner (1990) points out that with this value for the sum the large coordinate distance $x(z)$ in a low Ω model makes the lensing probability P uncomfortably high. However, my impression is that the sum in equation (5) may be a little high. (At conventional abundances and circular velocities, the contribution of spirals to the sum would be perhaps 20 % of eq. [5]. Giant ellipticals have larger velocity dispersions, but they are considerably less abundant.) If the quantity in equation (5) is high, then the lensing probability in the Einstein-de Sitter case may be uncomfortably low.

The main point is that, as Turner (1990) notes, the lensing probability is quite sensitive to the cosmology, so it will be of considerable interest to have detailed surveys of the frequency distribution of v_i^4, as well as the corrections for core size and absorption in the lensing galaxies and for the efficiency of detection of the lensing events.

GALAXY FORMATION IN A LOW DENSITY UNIVERSE

My conclusion from the preceeding two sections is that we ought to give some thought to the possibility that we live in a low density cosmologically flat universe. Let us consider how we might build a semiempirical model for galaxy formation in such a universe. This partially repeats the discussion in Peebles (1987, 1988, 1990a, and 1990b).

The goal is to find a path of least resistance through the observations. As Schramm describes in these proceedings, homogeneous light element production in the hot Big Bang works if the density parameter in baryons is

$$\Omega = \Omega_{\text{baryonic}} \sim 0.03 \ . \tag{6}$$

Since we have seen that a low density universe has some observationally attractive features, a reasonably straight path through the observations leads us to a cosmologically flat universe now dominated by baryons with the density parameter in equation (6). (Perhaps an even straighter path would be to $\Omega \sim 0.1$, which is closer to the bulk of the dynamical estimates, but let us concentrate on one possibility, that requires some "antibiasing.")

To get a reasonable time scale we need a large Hubble constant. The choice

$$H_o = 100 \text{ km s}^{-1} \text{ Mpc}^{-1} \tag{7}$$

gives

$$t_o = 16 \text{ by} . \tag{8}$$

As indicated above, these values are within the ranges now under discussion.

The background radiation drag on ionized matter became unimportant at redshift $z \sim 100$. Let us consider the situation a little later, at expansion factor $1 + z = 30$. At this epoch Hubble's constant is

$$H = 2800 \text{ km s}^{-1} \text{ Mpc}^{-1} , \tag{9}$$

and the age of the universe is

$$t \sim 3 \times 10^8 \text{ y} . \tag{10}$$

The present mean luminosity density in galaxies is $\sim 2 \times 10^8 \text{ L}_\odot \text{ Mpc}^{-3}$, and a characteristic luminosity for a large galaxy is $\sim 10^{10} \text{ L}_\odot$, so a characteristic present number density of bright galaxies is $n_o \sim 0.02 \text{ Mpc}^{-3}$. The characteristic mean distance between bright galaxies scaled back to $1 + z = 30$ is then

$$n^{-1/3} = l \sim 100 \text{ kpc} , \tag{11}$$

so at this redshift the typical relative velocity of neighboring galaxies is

$$Hl \sim 300 \text{ km s}^{-1} . \tag{12}$$

If each of these galaxies had a dark halo, with circular velocity 200 km s^{-1}, the halo radius needed to produce the assumed mean mass density in equation (6) is

$$R_{\text{max}} \sim 50 \text{ kpc} . \tag{13}$$

The point of these numbers is that protogalaxies with mass structures and comoving number density comparable to present day giant galaxies could have existed at $1 + z \sim 30$, but not much earlier. A straight path through the observations and theory thus leads to the assumption that protogalaxies were forming at $z \sim 30$.

It will be noted that the characteristic relative velocity in equation (12) is about right for the motions within the massive halo of a galaxy, and the collapse factor is about right, so it is sensible to assume that massive halos are forming at $z \sim 30$. We have to assume that much of the accreted material is fragmenting into massive comets, or "Jupiters," or low mass stars, or other dark things, to keep the halo dark. It might be relevant that halos are not entirely dark, so we would want some production of stars with masses near that of the sun. And it is reasonable to expect that this activity added heavy elements to the remaining gas. The gas that falls into protogalaxies toward the end of the period of heavy accretion might tend to end up in discs. The collapse factor, from radii $\sim l \sim 100$ kpc to a disc radius ~ 5 kpc, is on the order of what is thought to be needed to spin up a disc.

The collapse to the disc requires dissipation in the motion of gas clouds accreted along with the bulk of the halo mass from radii ~ 100 kpc. The dissipation time might be longer than the orbit time in equation (10), but it is easy to imagine that discs formed in ten orbit times, say, which would put disc formation at $z \sim 5$. This accretion picture for the formation of discs is a variant of the Ostriker-Thuan (1975) model of enrichment in the spheroid prior to collapse to the disc.

Early galaxy formation by gravitational instability has to be associated with large perturbations to the CBR, but we can imagine that the perturbations are erased by scattering by free electrons in the young galaxies. To see this, note that the Hubble length at $z \sim 30$ is

$$L = ct \sim \frac{2}{3}\frac{c}{H} \sim 70 \text{ Mpc} . \tag{14}$$

This is about the present distance to the Coma cluster, so if observers had been present they would only have needed modest telescopes to observe galaxies at redshifts near unity. However, if much of the mass at $z \sim 30$ were still in optically thin plasma, perhaps concentrated in and around the protogalaxies, the optical depth for electron scattering at the Hubble length at $1 + z = 30$ would be

$$\tau = \sigma n L \sim 1 . \tag{15}$$

That is, galaxies at redshifts near unity would be obscured by free electron scattering. The same effect would erase primeval irregularities in the CBR on scales much smaller than the Hubble length L in equation (14). In our low density cosmology this smoothing length subtends an angle

$$\theta_s \sim 6^\circ , \tag{16}$$

at the present epoch.

On angular scales $\gtrsim \theta_s$ the fluctuations in the CBR are unaffected by the scattering. But this probes the primeval mass fluctuations on comoving scales

$$R \gtrsim (1 + z)L \sim 2000 \text{ Mpc} , \tag{16}$$

considerably larger than the scales on which we have to build structures. The key point here is that in our semiempirical approach we are free to assume that the mass fluctuations on the scales of equation (16) are as small as necessary to aviod any conflict with the CBR measurements, because the observed distribution of galaxy counts only give us upper bounds on the fluctuations on these very large scales.

As discussed by Juszkiewicz in these proceedings, coherent motions of the gas within protogalaxies reinsert anisotropy in the CBR, but at levels that need not be easy to detect.

The postulated scattering of the CBR need not have much effect on its spectrum. Gas within protogalaxies might be moving at velocities $v \sim 300$ km s^{-1} at $z \sim 30$. If the mean optical depth for scattering by free electrons in the protogalaxies were $\tau \sim 100$, say, the effective Compton-Thomson parameter would be $y \sim \tau(v/c)^2 \sim 10^{-4}$, which is $\sim$ 10 % of the present upper bound. It will be noted also that in this model the gas might be expected to pass through the thermal bremsstrahlung phase at fairly high redshifts, which would considerably suppress the perturbation to the long wavelength part of the CBR spectrum (A. Loeb, private communication).

To my mind it is attractive that this model makes protogalaxies with modest (and I think reasonable) collapse factors for halo and disc, and that the model easily accounts for the observed existence of quasars and the intergalactic medium at expansion factor $1 + z \sim 6$. In this model the primeval perturbations to the CBR are large, but that can be hidden by free electron scattering. (It is helpful that the Hubble length at high redshift is increased by lowering Ω, so protogalaxies can already have formed before the universe becomes transparent to ionized matter.) Early galaxy formation also leaves traces in the CBR, but that more subtle effect will require a more careful analysis, and of course if detectable would be an important test of the model.

CONCLUDING REMARKS

Modern cosmology originated from a wonderfully successful *ansatz*, Einstein's assumption that the universe is homogeneous and isotropic. People inevitably are going to try to repeat this brilliant coup. But with the large and growing fund of observational evidence on what the universe is like now and in the past it would be crazy to ignore the possibility of proceeding in a semiempirical way, as in the example presented here. Perhaps this will be a significant element of cosmology in the 1990s.

The results in the present example are unsatisfactory in the sense that we have invoked primeval density fluctuations *ad hoc*, with no justification other than the hope that it may lead to a picture that works. Perhaps further considerably more careful discussion will convince us that this really is roughly the way it happened, or will lead us to some other convincing semiempirical picture. If so it would considerable accomplishment, and perhaps a key hint to the fundamental theory of what the very early universe was like.

This work was supported in part at Princeton University by the US National Science Foundation, and at the Institute for Advanced Study by the Ambrose Monell Foundation.

REFERENCES

Bardeen, J. M. 1986, in *Inner Space/Outer Space* (Chicago: Univ. of Chicago Press), eds. E. W. Kolb *et al.*, p. 212.

Binggeli, B. 1989, in *Large-Scale Structures and Motions in the Universe*, eds. M. Mezzetti *et al.* (Dordrecht: Kluwer Academic Publishers) p. 47.

Davis, M., Efstathiou, G., Frenk, C. S. and White, S. D. M. 1985, *Ap. J.*, **292**, 394.

Davis, M. and Peebles, P. J. E. 1983, *Ap. J.*, **267**, 465.

Dekel, A. and Silk, J. 1986, *Ap. J.*, **303**, 39.

Dressler, A. *et al.* 1987, *Ap. J.*, **313**, 42.

Fukugita, M., Okamura, S., Tarusawa, K., Rood, H. J. and Williams, B. A. 1990a, preprint (Kyoto: YITP/K-874).

Fukugita, M., Takahara, F., Yamashita, K. and Yoshii, Y. 1990b, *Ap. J.*, **361**, L1.

Ikeuchi, S. 1982, *Publ. Astron. Soc. J.*, **33**, 211.

Jacoby, G. H., Ciardullo, R. and Ford, H. C. 1990, *Ap. J.*, **356**, 332.
Kaiser, N. 1986, in *Inner Space/Outer Space* (Chicago: Univ. of Chicago Press), eds. E. W. Kolb *et al.*, p. 258.
Koo, D. C. 1989, in *The Epoch of Galaxy Formation* (eds. Frenk, C. S. *et al.*) Kluwer Publishers, p. 71.
Lilly, S. J., Cowie, L. L. and Gardner, J. P. 1990, Ap. J., in press.
Ostriker, J. P. and Cowie, L. L. 1981, *Ap. J.*, **243**, L127.
Ostriker, J. P. and Thuan, T. X. 1975, *Ap. J.*, **202**, 353.
Peebles, P. J. E. 1974, *Ap. J.*, **189**, L51.
Peebles, P. J. E. 1984a, *Science*, **224**, 1385.
Peebles, P. J. E. 1984b, *Ap. J.*, **284**, 439.
Peebles, P. J. E. 1986, *Nature*, **321**, 27.
Peebles, P. J. E. 1987, *Ap. J.*, **315**, L73.
Peebles, P. J. E. 1988, in *The Early Universe*, eds. W. G. Unruh and G. W. Semenoff (Dordrecht: Reidel) p. 203.
Peebles, P. J. E. 1989, *J. R. Astron. Soc. Canada*, **83**, 363.
Peebles, P. J. E. 1990a, in *The Galactic and Extragalactic Background Radiation*, eds. S. Bowyer and C. Leinert (The Netherlands: 1990 IAU) p. 295.
Peebles, P. J. E. 1990b, in *II Rencontres de Moriond: Physical Cosmology*, ed. J. Tran Thanh Van.
Peebles, P. J. E. and Ratra, B. 1988, *Ap. J.*, **325**, L17.
Sandage, A. and Tammann, G. A. 1990, *Ap. J.*, in press.
Tonry, J. L., Ajhar, E. A. and Luppino, G. A. 1989, *Ap. J.*, **346** L59.
Turner, E. L. 1990, preprint.
Turner, E. L., Ostriker, J. P. and Gott, J. R. 1984, *Ap. J.*, **284**, 1.
van den Bergh, S. 1989, *Astron. Ap. Rev.*, **1**, 111.
Wolfe, A. 1989, in *The Epoch of Galaxy Formation*, eds C. S. Frenk *et al.* (Dordrecht: Kluwer).

THE FIRST THREE MINUTES: 1990 VERSION

David N. Schramm
The University of Chicago,
5640 S. Ellis Avenue, Chicago, IL 60637
and
NASA/Fermilab Astrophysics Center,
Fermi National Accelerator Laboratory
Box 500, Batavia, IL 60510-0500

ABSTRACT

Early universe cosmology has come a long way since Steve Weinberg's famous book, *The First Three Minutes.* This talk will summarize some of the key areas: inflation, phase transitions, dark matter, and nucleosynthesis. The summary will focus on what comes out of the universe at the end of three minutes and thus serves as the initial conditions for later studies of the universe. Particular emphasis will be made on those aspects of the early universe that are confirmed by experiment or observation and those aspects that are relatively insensitive to the model-dependent assumptions.

INTRODUCTION

The study of the early universe has undergone a tremendous explosion of activity since Steve Weinberg published his famous book, *The First Three Minutes*, in 1977.[1] While Weinberg's book focused mainly on big bang nucleosynthesis occurring from cosmological times of 1 sec to $\sim$ 3 minutes, modern early universe cosmology pushes back to times as early as $\sim 10^{-43}$ sec. Even as to big bang nucleosynthesis, the situation is somewhat changed in that Weinberg's book mainly focused on the ^4He synthesis, whereas now we have confidence in the synthesis with the other light isotopes as well, ^2H, ^3He and ^7Li and the implied constraint they give to the density of baryons in the universe. Furthermore, we now have particle-collider experiments verifying the predictions of big bang nucleosynthesis regarding the number of neutrino families, N_ν. This latter point, not only gives us greater confidence in the big bang nucleosynthesis arguements, but also illustrates the close symbiotic relationship that has now developed between cosmology and particle physics. This relationship was only just beginning in the late 1970's.

This 1990's version of the first three minutes, will focus on those things coming out of the early universe that lead to potentially observable consequences. Furthermore, rather than discuss very model-dependent arguments, this paper will try to emphasize those facets that are model-independent or at least generic to broad classes of models. For example, the first of these restrictions means that quantum gravity and string theory are not yet appropriate for inclusion. Similarly, the latter restriction towards generic results mean that inflation will be treated in a very broad brush manner.

Because of the length restrictions, this paper will, for the most part, refer to recent comprehensive reviews rather than attempt to be a full scholarly review in its own right.

The epochs of the 1990's version of the first three minutes are shown in Table 1. This paper will go through the first two rapidly, emphasizing the generic observables. It will then go into some detail on the latter three. Since these latter three all can affect the observable results of big bang nucleosynthesis (hereafter denoted as BBN), these topics will be integrated together into an extensive discussion of that subject.

THE T.O.E. EPOCH

T.O.E., or the Theory of Everything, is the name given to the real zone of mystery where physicists hope an ultimate unified theory can be found. The current front runner for such a theory is Superstring Theory.[2] This theory appears to have all the nice mathematical properties one would wish for an ultimate theory. However, the best versions seem to work in 10 or 26 dimensions rather than our three spatial and one time. Thus, some compactification of the excess dimensions or inflation of the preferred dimensions is needed. Such theories hope eventually to answer such fundamental questions as the origin of space and time and the generation of an expanding universe out of nothing. While there have been many nobel attempts at such problems, none has yet reached the point of

TABLE 1

EPOCHS IN THE EARLY UNIVERSE

Time	Temperature	Epoch Ending	Observables
$\sim 10^{-43} sec$	$\sim 10^{19} GeV$	T.O.E. Quantum Gravity (Superstrings?)	Space-Time Expansion Λ?
$10^{-34} - 10^{-38} sec$	$10^{14} - 10^{16} GeV$	GUT	Inflation, B2aryons, Fluctuations and/or Defects?
$\sim 10^{-10} sec$	$\sim 10^{2} GeV$	Electroweak	?
$\sim 10^{-4} - 10^{-6} sec$	$100 MeV - 1 GeV$	QCD Confinement (Quark-Hadron Transition)	Hadrons Fluctuations?
$\sim 1 sec$	$1 MeV$	Weak Freezeout	Neutrinos n/p ratio Dark Matter?
$\sim 10^{3} sec$	$30 keV$	Nucleosynthesis	^4He, ^3He, ^2H, ^7Li (Ω_b, N_ν)

anything beyond a "postdiction" since we already know the universe is here, has 3 + 1 dimensions, and is expanding.

One potential observation in quantum gravity theories is the vacuum energy in the absence of matter usually labeled as Λ (the cosmological constraint). Much recent controversy surrounds the attempts of Coleman and others to get Λ to go to zero via "wormholes" in the early universe.[3] At present, it can safely be said that there is no complete agreement on how to get $\Lambda = 0$; however, most physicists would agree that it is probably zero or at least negligibly small. The argument is as follows: Since the microphysics that affects Λ can come from the quantum gravity epoch, then the energy scale is $M_{Pl(anck)} \sim 10^{19} Gev$. Expressing Λ as an energy density means $\Lambda \propto M_{Pl}^4$ in its natural units. But modern observations of the dynamics of systems tell us that $\Lambda \lesssim 10^{-121} M_{Pl}^4$. In fact, if one desires Λ to have dynamic effects on the universe during the present epoch,[1] then that demands $\Lambda \simeq 10\ ^{121} M_{Pl}^4$ exactly! From a microphysics point of view, this is tremendous fine tuning. It is one thing to have a process that overshoots and thus drives Λ towards zero. It is quite another to have it explicitly pick out a value accurate to 121 decimal places. Thus, barring such a remarkable fine tuning, most physicists expect $\Lambda = 0$. However, as mentioned before, the detailed mechanism to drive it towared zero is still under debate.

THE GUT EPOCH, INFLATION, BARYOSYNTHESIS, ETC.

One of the intellectually richest arenas in the early universe has come from Grand Unified Theories or GUTs. These are the theories which attempt to unify the strong interaction with the already unified electroweak. These theories have given us models[5] ("baryosynthesis") to enable an understanding of the observed[6] baryon-antibaryon asymmetry. They have also given us a mechanism[7] known as "inflation" to solve the traditional initial condition problems of the standard big bang. These problems include the "horizon problem" and the "flatness problem."

In the course of solving these problems, inflation makes two predictions: (1) $\Omega = 1$ to high accuracy; and (2) the universe has a primordial density fluctuation spectrum of the Zeldovich-Harrison variety, that is, equal power on all scales. The first prediction is inherent in all reasonable inflation scenarios; the second is slightly more model dependent. In particular, some models are able to produce special, "designer," fluctuation spectra. Furthermore, the amplitude of the spectrum produced is somewhat arbitrary and instead of tuning it to the specific value that it takes eventually to produce some of the observed structure, it is certainly possible to overshoot on the amplitude and get it arbitrarily close to zero, analogous to Λ, in which case inflation could satisfy the horizon and flatness problems and require $\Omega = 1$ but not influence the galaxy formation scenario any further.

Rather than go into the details of baryosynthesis or inflation here, let the reader seek more detailed reviews such as Kolb's and Turner's book on the early universe[8] or Linde's book[9] or Olive's extensive review of inflation.[10] In general, it should be noted that inflation is fairly easy to achieve in most unified models since, as Linde showed, any scalar field can lead to inflation. One further point to note is that inflation is not possible unless baryons can be synthesized at the end of (or after) the inflation, since otherwise the universe would be empty. This presumably limits inflation to the GUT or TOE transitions, although recent ideas about electroweak baryosynthesis may relax this point.

Before leaving the GUT epoch, let us also note that in addition to making random gaussian fluctuation at the end of a vacuum phase transition, vacuum phase transitions can also make topological defects. For example, phase transitions in three dimensions can produce:

(1) monopoles (point defects),
(2) strings (1D defects).
(3) walls (2D defects), or
(4) textures (3D transient phenomena).

Thus, coming out of the GUT and/or TOE epoch, the universe could have density fluctuations and/or topological defects. These could also be generated at any late phase transition as well. However, in terms of providing seeds for galaxy and structure formation, only those early transitions accompanied by the final inflationary phase (GUT or possibly TOE) can produce galaxy-size seeds. This constraint is true, unless the phase transition occurs after the horizon of the universe gets large enough to contain galaxy-size seeds. Recently, Hill, Schramm and Fry [11] have proposed such a late-time transition, but that would occur long after three minutes and so is inappropriate for further discussion here.

HISTORY OF BIG BANG NUCLEOSYNTHESIS

Big bang nucleosynthesis, BBN, has become our most quantitative probe of the early universe, and much has happened since the publication of Weinberg's *The First Three Minutes.* Because of these developments, we will first give a brief history before describing some of the scientific points in more detail. (This section will rely heavily on the recent reviews of reference 12.)

It should be noted that there is a symbiotic connection between BBN and the $3K$ background dating back to Gamow and his associates Alpher and Herman. The initial BBN calculations of Gamow and his associates[13] assumed pure neutrons as an initial condition and thus were not particularly accurate, but their inaccuracies had little effect on the group's predictions for a background radiation.

Once Hayashi recognized in 1950 the role of neutron-proton equilibration, the framework for BBN calculations themselves has not varied significantly. The work of Alpher, Follin and Herman[14] and Taylor and Hoyle[15], preceeding the discovery of the 3K background, and Peebles[16] and Wagoner, Fowler and Hoyle,[17] immediately following the discovery, and the more recent work of our group of collaborators[18,19,20,21,22] all do essentially the same basic calculation, the results of which are shown in Figure 1. As far as the calculation itself goes, solving the reaction network is relatively simple by the standards of explosive nucleosynthesis calculations in supernovae (c.f. the 1965 calculation of Truran *et al.*),[23] with the changes over the last 25 years being mainly in terms of more recent

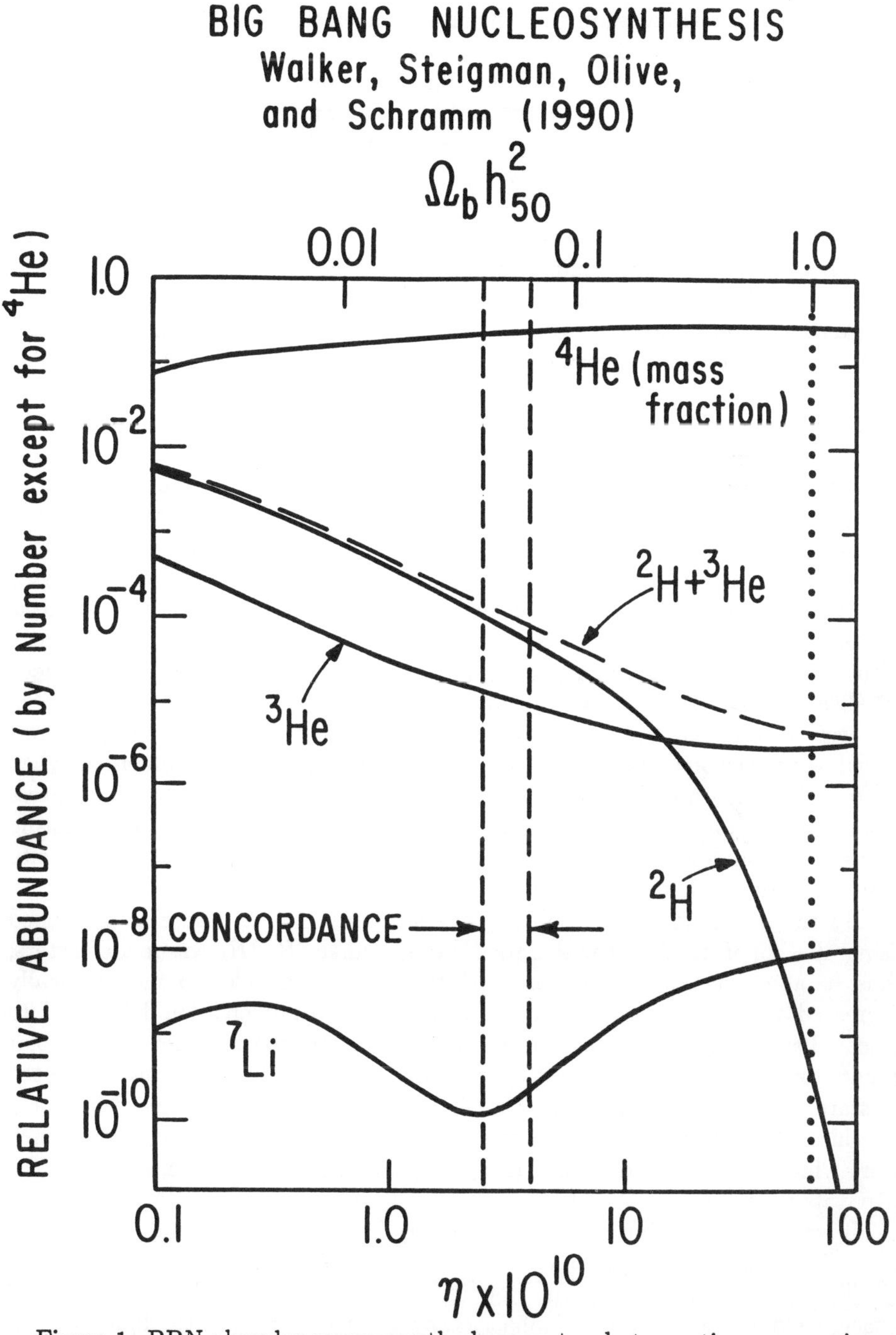

Figure 1. BBN abundances versus the baryon to photon ratio, η, or equivalently the fraction of the critical density, Ω_b.

nuclear reaction rates as input, not as any great calculational insight (although the current Kawano/Walker code[21,22] is somewhat streamlined relative to the earlier Wagoner code[17]). With the possible exception of ^7Li yields, the reaction rate changes over the past 25 years have not had any major affect.[19,21,22,23,24] The one key improved input is a better neutron lifetime determination, a point to which we will also return shortly.

With the exception of the effects of elementary particle assumptions to which we will also return, the real excitement for BBN over the last 25 years has not really been in redoing the basic calculation. Instead, the true action is focused on understanding the evolution of the light element abundances and using that information to make powerful conclusions. In particular, in the 1960's, the main focus was on ^4He which is very insensitive to the baryon density. The agreement between BBN predictions and observations helped support the basic big bang model but gave no significant information at that time with regard to density. In fact, in the mid-1960's, the other light isotopes (which are, in principle, capable of giving density information) were generally assumed to have been made during the T-Tauri phase of stellar evolution,[25] and so, were not then taken to have cosmological significance. It was during the 1970's that BBN fully developed as a tool for probing the Universe. This possibility was in part stimulated by Ryter *et al.*[26] who showed that the T-Tauri mechanism for light element synthesis failed. Furthermore, ^2H abundance determinations[27,28] improved significantly with solar wind measurements and the interstellar work from the Copernicus satellite. Reeves, Audouze, Fowler and Schramm[29] argued for cosmological ^2H and were able to place a constraint on the baryon density excluding a universe closed with baryons. Subsequently, the ^2H arguments were cemented when Epstein, Lattimer and Schramm[30] proved that no realistic astrophysical process other than the big bang could produce significant ^2H. It was also interesting that the baryon density implied by BBN was in good agreement with the density implied by the dark galactic halos.[31]

By the late 1970's, a complimentary argument to ^2H had also developed using ^3He. In particular, it was argued[32] that, unlike ^2H, ^3He was made in stars; thus, its abundance would increase with time. Since ^3He like ^2H monotonically decreased with cosmological baryon density, this argument could be used to place a lower limit on the baryon density[33] using ^3He measurements from solar wind[27] or interstellar determinations.[34] Since the bulk of the ^2H was converted in stars to ^3He, the constraint was shown to be quite restrictive.[29] Support for this point[35] also comes from the observation of ^3He in horizontal branch stars which, as processed stars still having ^3He on their surface, indicates the survivability of ^3He.

It was interesting that the lower boundary from ^3He and the upper boundary from ^2H yielded the requirement that ^7Li be near its minimum of $^7Li/H \sim 10^{-10}$, which was verified by the Pop II Li measurements of Spite and Spite,[36] hence, yielding the situation emphasized by Yang *et al.*[19] that the light element abundances are consistent over nine orders of magnitude with BBN, but only if the cosmological baryon density is constrained to be around 6% of the critical value. It is worth noting that ^7Li alone gives both an upper and a lower limit to Ω_b. However, while its derived upper limit is more than competitive with the ^2H limit, the ^7Li lower limit is not nearly as restictive as the $^2H +^3 He$ limit. Claims that big bang nucleosynthesis can yield Ω_b lower than 0.01 must necessarily neglect the $^3He +^2 H$ limit.

Another development of the 70's for BBN was the explicit calculation of Steigman, Schramm and Gunn,[37] showing that the number of neutrino generations, N_ν, had to be small to avoid overproduction of ^4He. This will subsequently be referred to as the SSG limit.(Earlier work had noted a dependency of the ^4He abundance on assumptions about the fraction of the cosmological stress-energy in exotic particles,[38,15] but had not actually made an explicit calculation probing the quantity of interest to particle physicists, N_ν.) To put this in perspective, one should remember that the mid-1970's also saw the discovery of charm, bottom and tau, so that it almost seemed as if each new detector produced new particle discoveries, and yet, cosmology was arguing against this "conventional" wisdom. Over the years the SSG limit on N_ν improved with ^4He abundance measurements, neutron lifetime measurements and with limits on the lower bound to the baryon density; hovering at $N_\nu \lesssim 4$ for most of the 1980's and dropping to slightly lower than 4[39,40,20] just before LEP and SLC turned on.

BIG BANG NUCLEOSYNTHESIS: Ω_b and N_ν

The power of Big Bang Nucleosynthesis comes from the fact that essentially all of the physics input is well determined in the terrestrial laboratory. The appropriate temperature regimes, 0.1 to $1MeV$, are well explored in nuclear physics labs. Thus, what nuclei do under such conditions is not a matter of guesswork, but is precisely known. In fact, it is known for these temperatures far better than it is for the centers of stars like our sun. The center of the sun is only a little over $1keV$, thus, below the energy where nuclear reaction rates yield significant results in laboratory experiments, and only the long times and higher densities available in stars enable anything to take place.

To calculate what happens in the big bang, all one has to do is follow what a gas of baryons with density ρ_b does as the universe expands and cools. As far as nuclear reactions are concerned, the only relevant region is from a little above $1MeV$ ($\sim 10^{10}K$) down to a little below $100keV$ ($\sim 10^9K$). At higher temperatures, no complex nuclei other than free single neutrons and protons can exist, and the ratio of neutrons to protons, n/p, is just determined by

$$n/p = e^{-Q/T},$$

where $Q = (m_n - m_p)c^2 \sim 1.3MeV$.

Equilibrium applies because the weak interaction rates are much faster than the expansion of the universe at temperatures much above $10^{10}K$. At temperatures much below 10^9K, the electrostatic repulsion of nuclei prevents nuclear reactions from proceeding as fast as the cosmological expansion separates the particles.

Because of the equilibrium existing for temperatures much above $10^{10}K$, we don't have to worry about what went on in the universe at higher temperatures. Thus, we can start our calculation at $10MeV$ and not worry about speculative physics like T.O.E. or GUTs as long as a gas of neutrons and protons exists in thermal equilibriuim by the time the universe has cooled to $\sim 10MeV$.

After the weak interaction drops out of equilibrium, a little above $10^{10}K$, the ratio of neutrons to protons changes more slowly due to free neutrons decaying to protons, and similar transformations of neutrons to protons via interactions with the ambient leptons. By the time the universe reaches 10^9K $(0.1MeV)$, the ratio is slightly below 1/7. For temperatures above 10^9K, no significant abundance of complex nuclei can exist due to the continued existence of gammas with greater than MeV energies. Note that the high photon to baryon ratio in the universe ($\sim 10^{10}$) enables significant population of an MeV high energy Boltzman tail until $T \lesssim 0.1\ MeV$.

Once the temperature drops to about 10^9K, nuclei can exist in statistical equilibrium through reactions such as $n + p \leftrightarrow ^2H + \gamma$ and $^2H + p \leftrightarrow ^3He + \gamma$ and $^2H + n \leftrightarrow ^3H + \gamma$, which in turn react to yield ^4He. Since ^4He is the most tightly bound nucleus in the region, the flow of reactions converts almost all the neutrons that exist at 10^9K into ^4He. The flow essentially stops there because there are no stable nuclei at either mass-5 or mass-8. Since the baryon density at big bang nucleosynthesis is relatively low (much less than $1g/cm^3$) and the time-scale short ($t \lesssim 10^2 sec$), only reactions involving two-particle collisions occur. It can be seen that combining the most abundant nuclei, protons and ^4He via two body interactions always leads to unstable mass-5. Even when one combines ^4He with rarer nuclei like ^3H or ^3He, we still get only to mass-7, which, when hit by a proton, the most abundant nucleus around, yields mass-8. (A loophole around the mass-8 gap can be found if $n/p > 1$, so that excess neutrons exist, but for the standard case $n/p < 1$). Eventually, ^3H radioactively decays to ^3He, and any mass-7 made radioactively decays to ^7Li. Thus, big bang nucleosynthesis makes ^4He with traces of ^2H, ^3He, and ^7Li. (Also, all the protons left over that did not capture neutrons remain as hydrogen.) For standard homogeneous BBN, all other chemical elements are made later in stars and in related processes. (Stars jump the mass-5 and -8 instability by having gravity compress the matter to sufficient densities and have much longer times available so that three-body collisions can occur.) With the possible exception of ^7Li,[19,20,21,22,24] the results are rather insensitive to the detailed nuclear reaction rates. This insensitivity was discussed in reference 19 and most recently using a Monte Carlo study by Krauss and Romanelli.[24] An n/p ratio of $\sim 1/7$ yields a ^4He primordial mass fraction,

$$Y_p \simeq \frac{2n/p}{n/p + 1} \approx \frac{1}{4}$$

The only parameter we can easily vary in such calculations is the density that corresponds to a given temperature. From the thermodynamics of an expanding universe we know that $\rho_b \propto T^3$; thus, we can relate the baryon density at $10^{10}K$

to the baryon density today, when the temperature is about $3K$. The problem is that we don't know today's ρ_b, so the calculation is carried out for a range in ρ_b. Another aspect of the density is that the cosmological expansion rate depends on the total mass-energy density associated with a given temperature. For cosmological temperatures much above $10^4 K$, the energy density of radiation exceeds the mass-energy density of the baryon gas. Thus, during big bang nucleosynthesis, we need the radiation density as well as the baryon density. The baryon density determines the density of the nuclei and thus their interaction rates, and the radiation density controls the expansion rate of the universe at those times. The density of radiation is just proportional to the number of types of radiation. Thus, the density of radiation is not a free parameter if we know how many types of relativistic particles exist when big bang nucleosynthesis occurred.

Assuming that the allowed relativistic particles at $1MeV$ are photons, e, μ, and τ neutrinos (and their antiparticles) and electrons (and positrons), Figure 1 shows the BBN yields for a range in present ρ_b, going from less than that observed in galaxies to greater than that allowed by the observed large-scale dynamics of the universe. The ^4He yield is almost independent of the baryon density, with a very slight rise in the density due to the ability of nuclei to hold together at slightly higher temperatures and at higher densities, thus enabling nucleosynthesis to start slightly earlier, when the baryon to photon ratio is higher. No matter what assumptions one makes about the baryon density, it is clear that ^4He is predicted by big bang nucleosynthesis to be around 1/4 of the mass of the universe.

THE SSG LIMIT - COSMOLOGICAL NEUTRINO COUNTING

Let us now look at the connection to N_ν. Remember that the yield of ^4He is very sensitive to the n/p ratio. The more types of relativistic particles, the greater the energy density at a given temperature, and thus, a faster cosmological expansion. A faster expansion yields the weak-interaction rates being exceeded by the cosmological expansion rate at an earlier, higher temperature; thus, the weak interaction drops out of equilibrium sooner, yielding a higher n/p ratio. It also yields less time between dropping out of equilibrium and nucleosynthesis at $10^9 K$, which gives less time for neutrons to change into protons, thus also increasing the n/p ratio. A higher n/p ratio yields more ^4He. As we will see in the next section, quark-hadron induced variations[41] in the standard model also yield higher ^4He for higher values of Ω_b. Thus, such variants still support the constraint on the number of relativistic species.[42]

In the standard calculation we allowed for photons, electrons, and the three known neutrino species (and their antiparticles). However, following SSG and doing the calculation (see Figure 2) for additional species of neutrinos, we can see when ^4He yields exceed observational limits while still yielding a density consistent with the ρ_b bounds from ^2H, ^3He, and now ^7Li. (The new ^7Li value gives approximately the same constraint on ρ_b as the others, thus strengthening

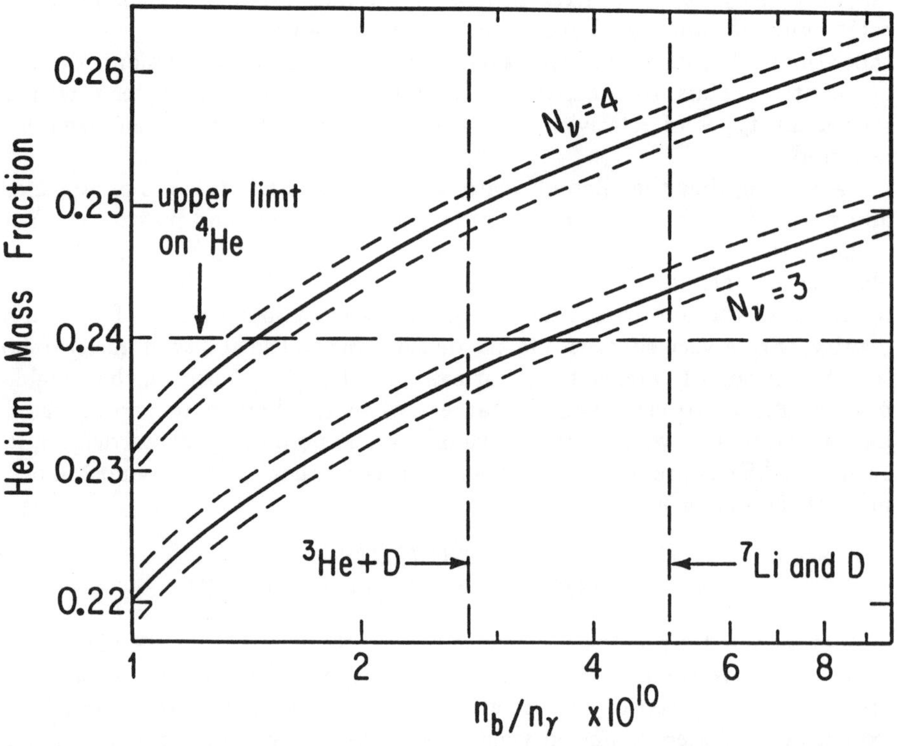

Figure 2. The SSG argument with recent parameter constraints showing the BBN helium mass fraction versus η for $N_\nu = 3$ and 4. Note that 4 is excluded.

the conclusion.) The bound on ^4He comes from observations of helium in many different objects in the universe. However, since ^4He is not only produced in the big bang but in stars as well, it is important to estimate what part of the helium in some astronomical object is primordial—from the big bang—and what part is due to stellar production after the big bang. The pioneering work of the Peimberts[43] showing that ^4He varies with oxygen has now been supplemented by examination of how ^4He varies with nitrogen and carbon. The observations have also been systematically reexamined by Pagel[44]. The conclusions of Pagel[44], Steigman *et al.*[45] and Walker *et al.*[21] all agree that the ^4He mass fraction, Y_p, extrapolated to zero heavy elements, whether using N, O, or C, is $Y_p \sim 0.23$ with an upper bound allowing for possible systematics of 0.24.

The other major uncertainty in the ^4He production used to be the neutron lifetime. However, the new world average of $\tau_n = 890 \pm 4s(\tau_{1/2} = 10.3\ min)$ is dominated by the dramatic results of Mampe *et al.*[46] using a neutron bottle. This new result is quite consistent with a new counting measurement of Byrne *et al.*[47] and within the errors of the previous world average of $896 \pm 10s$ and is also consistent with the precise C_A/C_V measurements from PERKEO[48] and others. Thus, the old ranges of $10.4 \pm 0.2\ min$, used for the half-life in calculations,[49,19] seem to have converged towards the lower side. The convergence means that, instead of the previous broad bands for each neutrino flavour, we obtain relatively narrow bands (see Figure 2). Note that $N_\nu = 4$ is excluded. In fact, the SSG limit is now$N_\nu < 3.4$.[20,21]

The recent verification of this cosmological standard model prediction by LEP, $N_\nu = 2.98 \pm 0.06$, from the average of ALEPH, DELPHI, L3 and OPAL[50] collaborations as well as the SLC[50] results, thus, experimentally confirms our confidence in BBN. (However, we should also remember that LEP and cosmology are sensitive to different things.[51] Cosmology counts all relativistic degrees of freedom for $m_x \lesssim 10MeV$, with LEP and SLC counting particles coupling to the Z^0 with $m_x \lesssim 45GeV$. While ν_e and ν_μ are obviously counted equally in both situations, a curious loophole exists for ν_τ since the current experimental limit $m_{\nu_\tau} < 35MeV$ could allow it not to contribute as a full neutrino in the cosmology argument[52]. Proposed experiments which push the m_{ν_τ} limit down to less than a few MeV should eliminate this loophole.

It might also be noted that if we assume m_{ν_τ} is light so that cosmologically $N_\nu = 3$, we can turn the argument around and use LEP to predict the primordial helium abundance ($\sim$ 24%), or even use limits on ^4He to give an upper limit on Ω_b (also $\lesssim 0.10$). Thus, LEP strengthens the argument that we need non-baryonic dark matter if $\Omega = 1$. In fact, note also that with $N_\nu = 3$, if Y_p is ever proven to be less than ~ 0.235, standard BBN is in difficulty. Similar difficulties occur if Li/H is ever found below $\sim 10^{-10}$. In other words, BBN is a falsifiable theory.

ALTERNATIVE PROPOSALS

As noted above, BBN yields all agree with observations using only one freely adjustable parameter, ρ_b. Thus, BBN can make strong statements regarding ρ_b

if the observed light element abundances cannot be fit with any alternative theory. Before exploring the implications for ρ_b, let us examine alternative proposals which have arisen to try to escape the power of the homogeneous BBN conclusions.

The two alternatives that have recently received interest are:

(1) Decaying particles;[53] and

(2) Quark-hadron transition inspired inhomogeneities.[41]

The first of these notes that if a species of massive particle ($M \gtrsim few\ GeV$) were to decay after traditional BBN, it could redo nucleosynthesis. While previous decaying particle proposals had been made, the new idea[43] emphasizes the importance of the resulting hadron cascade which, they argue, will dominate the yields. While interesting results are obtained, problems with detailed abundance determinations do result. In particular, this class of models seems to predict inevitably that $^6Li/^7Li >> 1$, whereas observations show $^7Li/^6Li \gtrsim 10$. While at first this might seem fatal, it is almost avoidable by noting that ^6Li is much more fragile than ^7Li; thus, it is easy to deplete ^6Li and obtain the observed ratios. However, Brown and Schramm[54] have pointed out that for high surface temperature Pop II stars, the convective zones do not go deep enough to destroy any primordial ^6Li. Pilachowski *et al.*[55] have now looked at those specific stars and indeed find no ^6Li, again seeing $^7Li/^6Li > 10$. Therefore, unless the Brown and Schramm convection argument can be surmounted, ^6Li seems to contrain this model seriously. Steigman, Audouze and others have noted additional problems with this model for ^3He and ^2H ratios.

Let us now look at the quark-hadron inspired inhomogeniety models.[41] While inhomogeneity models had been looked at previously (c.f. reference 19) and were found to make little difference, the quark-hadron inspired models had the added ingredient of variations in n/p ratios.

The initial claim by Applegate *et al.*, followed by a similar argument from Alcock *et al.*, that $\Omega_b \sim 1$ might be possible, created tremendous interest. Their argument was that if the quark-hadron transition was a first-order phase transition (as some preliminary lattice gauge calculations then implied), it was possible that large inhomogeneities could develop at $T \gtrsim 100 MeV$. The preferential diffusion of neutrons versus protons out of the high density regions could lead to big bang nucleosynthesis occurring under conditions with both density inhomogeneities and variable neutron/proton ratios. In the first round of calculations, it was claimed that such conditions might allow $\Omega_b \sim 1$, while fitting the observed primordial abundances of $^4He, ^2H, ^3He$ with an overproduction of ^7Li. Since ^7Li is the most recent of the cosmological abundance constraints and has a different observed abundace in Pop I stars versus the traditionally more primitive Pop II stars,[36] some argued that perhaps some special depletion process might be going on to reduce the excess ^7Li. Reeves and Audouze each argued against such processes and tried to turn the argument around and use lithium abundances to constrain the quark-hadron transition.

At first it appeared that if the lithium constraint could be surmounted, then the constraints of standard big bang nucleosynthesis might disintegrate. (Although Audouze, Reeves and Schramm emphasized that the number of parameters needed to fit the light elements was somewhat larger for these non-standard models, nonetheless, a non-trivial loophole appeared to be forming.) To further stimulate the flow through the loophole, Mullaney and Fowler showed that, in addition to looking at the diffusion of neutrons out of high density regions, one must also look at the subsequent effect of excess neutrons diffusing back into the high density regions as the nucleosynthesis goes to completion in the low density regions. (The initial calculations treated the two regions separately.) Mullaney and Fowler argued that for certain phase transition parameter values (e.g. nucleation site separations $\sim 10m$ at the time of the transition), this back diffusion could destroy much of the excess lithium. Recent work by Banerjee and Chitre (private communication) suggests that more accurate treatment of the diffusion calculation could reduce the interesting separation distance by several orders of magnitude.

However, Kurki-Suonio, Matzner, Olive and Schramm,[42] the Tokyo group,[56] and the Livermore group[57] have recently argued that in their detailed diffusion calculations, the back diffusion not only effects ^7Li, but also the other light nuclei as well. They find that for $\Omega_b \sim 1$, ^4He is also overproduced (although it does go to a minimum for similar parameter values as does the lithium). One can understand why these models might tend to overproduce ^4He and ^7Li by remembering that in standard homogeneous big bang nucleosynthesis, high baryon densities lead to excesses in these nuclei. As back diffusion evens out the effects of the initial fluctuation, the averaged result should approach the homogeneous value. Furthermore, it can be argued that any narrow range of parameters, such as those which yield relatively low lithium and helium, are unrealistic since in most realisitic phase transitions there are distributions of parameter values (distribution of nucleation sites, separations, density fluctuations, etc.). Therefore, narrow minima are washed out which would bring the ^7Li and ^4He values back up to their excesive levels for all parameter values with $\Omega \sim 1$. Furthermore, Adams and Freese[58] have argued that the boundary betwen the two phases may be fractal-like rather than smooth. The large surface area of a fractal-like boundary would allow more interaction between the regions and minimize exotic effects.

Figure 3 shows the results of Kurki-Suonio *et al.*[42] for varying spacing l with the constraints from the different light element abundances. Notice that the Li and even the ^4He constraint do not allow $\Omega_b \sim 1$. (The ^4He abundance constraint used in Kurki-Suonio *et al.* was a generous $Y_p \lesssim 0.25$; for the preferred $Y_p \lesssim 0.24$, the ^4He bound is about as tight as the Pop II Li constraint.) Note also that with the Pop II ^7Li constraint, the results for Ω_b are quite similar to the standard model with a slight excess in Ω_b possible if l is tuned to ~ 10.

Furthermore, initially it looked like quark-hadron inspired models might enable leakage[59] beyond mass-7, thus enabling ^9Be, ^{14}N, or maybe even r-process elements to become probes as whether or not the universe had such a transition (even if $\Omega_b \sim 1$). However, Tarasawa and Sato[56] have shown that when full multizone calculations of the type used by Kurki-Suonio *et al.* are

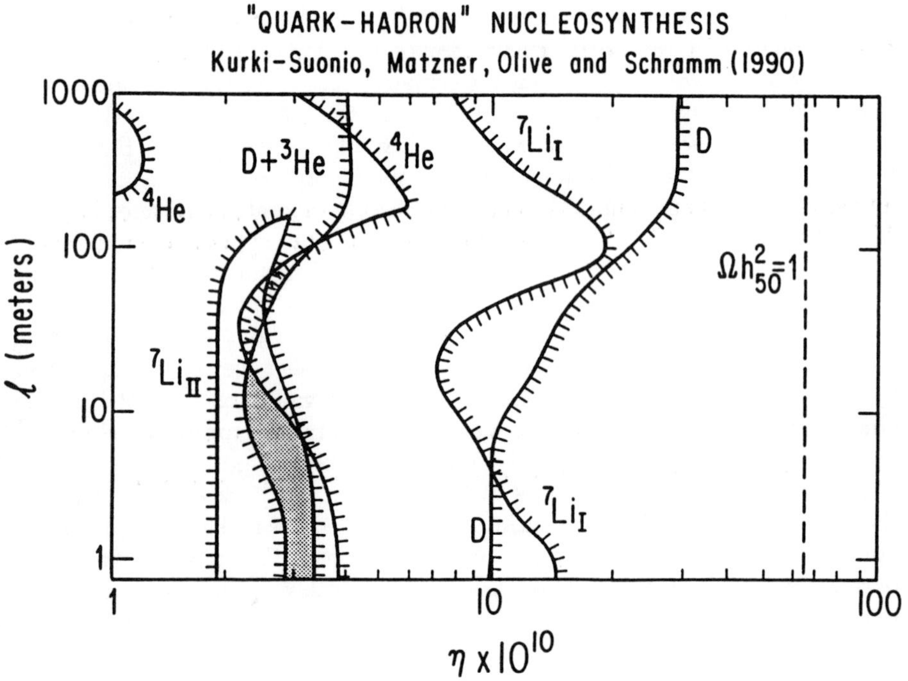

Figure 3. This shows the constraints on η of the various observed abundances in a first-order quark-hadron phase transition with nucleation sites separated by a distance l with density contrast $R \lesssim 10^3$. The Pop II lithium abundance used here is from the compilation of data given by Walker *et al.*[11] and is slightly more restrictive on η than that used in Figure 2 or used in the original Kurki-Suonio *et al.*[32] calculation from which this figure is derived. It should be noted that work by the Tokyo[46] group and by the Livermore group[47] confirms the conclusions on restricting Ω_b to values similar to the standard result even when $R \to \infty$.

utilized, then no significant leakage occurs.

One possible signature that remains for a first order quark-hadron transition is a slightly larger allowed range for Y_p that is concordant with $N_\nu = 3$ and with the other light element abundances. In particular, if ^4He were ever shown to be definitively < 0.235, it might be evidence for such a quark-hadron induced behavior since the standard homogenous case cannot accomodate such values. Of course, excessively low values for Y_p would still be unallowable.

One can conclude from the failure of the attempts to circumvent the standard BBN results that the results are amazingly robust. Even when many new free parameters are added, as in the quark-hadron case, the bottom line, when one requires concordance with the light element abundances, is essentially the same as the standard result. In other words, $\Omega_b \sim 0.06$ (although with fine-tuning the upper bound might be relaxed a bit to ~ 0.2 rather than 0.1).

LIMITS ON Ω_b AND DARK MATTER REQUIREMENTS

The narrow range in baryon density for which concordance occurs is very interesting. Let us convert it into units of the critical cosmological density for the allowed range of Hubble expansion rates. For the big bang nucleosynthesis constraints,[19,20,21,22,39,40] the dimensionless baryon density, Ω_b, that fraction of the critical density that is in baryons, is less than 0.11 and greater than 0.02 for $0.04 \lesssim h_0 \lesssim 0.7$, where h_0 is the Hubble constant in units of $100 km/sec/Mpc$. The lower bound on h_0 comes from direct observational limits and the upper bound from age of the universe constraints.[59] Note that the constraint on Ω_b means that the universe *cannot be closed with baryonic matter*. If the universe is truly at its critical density, then nonbaryonic matter is required. This argument has led to one of the major areas of research at the particle-cosmology interface, namely, the search for non-baryonic dark matter.

Another important conclusion regarding the allowed range in baryon density is that it is in very good agreement with the density implied from the dynamics of galaxies, *including their dark halos*. An early version of this argument, using only deuterium, was described over fifteen years ago.[31] As time has gone on, the argument has strengthened, and the fact remains that galaxy dynamics and nucleosynthesis agree at about 6% of the critical density. Thus, if the universe is indeed at its critical density, as many of us believe, it requires most matter not to be associated with galaxies and their halos, as well as to be nonbaryonic.

Let us put the nucleosynthetic arguments in context. The arguments requiring some sort of dark matter fall into two separate and quite distinct areas. First are the arguments using Newtonian mechanics applied to various astronomical systems that show that there is more matter present than the amount that is shining. These arguments are summarized in the first part of Table 2. It should be noted that these arguments reliably demonstrate that galactic halos seem to have a mass ~ 10 times the visible mass.

Note however that big bang nucleosynthesis requires that the bulk of the baryons in the universe are dark since $\Omega_{vis} << \Omega_b$. Thus, the dark halos could in principle be baryonic.[31] Recently arguments on very large scales[61] (bigger

TABLE 2

"OBSERVED" DENSITIES

$$\left[\Omega \equiv \rho/\rho_c \text{ where } \rho_c = 2\cdot 10^{-29} h_0^2 \text{g/cm}^3 \text{and } h_o \equiv \frac{\text{Ho}}{100 \text{ km/sec/mpc}}\right]$$

Newtonian Mechanics (cf. Faber and Gallagher[80])	
Visible	$\Omega \sim 0.007$ (factor of 2 accuracy)
Binaries Small groups Extended flat relation curves	$\Omega \sim 0.07$ (factor of 2 accuracy)
Clusters Gravitational lenses	$\Omega \sim 0.1$ to 0.3
Big Bang Nucleosynthesis (with $t_u \gtrsim 10^{10}$yrs.) (c.f. Walker *et al.*[21] and ref. therein)	$\Omega_b = 0.065 \pm 0.045$
Preliminary Large Scale Studies	
IRAS red shift study and peculiar velocities (Ref. [61])	$\Omega \gtrsim 0.3$
Density redshift counts (Loh and Spillar[81])	$\Omega \sim 1 \pm 0.6$
Inflation Paradigm (Guth[7])	$\Omega = 1$

than cluster of galaxies) hint that Ω on those scales is indeed greater than Ω_b, thus forcing us to need non-baryonic matter. However, until these arguments are confirmed, we must look to the inflation paradigm.

This is the argument that the only long-lived natural value for Ω is unity, and that inflation[7] or something like it provided the early universe with the mechanism to achieve that value and thereby solve the flatness and smoothness problems. Thus, our need for exotica is dependent on inflation and big bang nucleosythesis and not on the existence of dark galatic halos. This point is frequently forgotten, not only by some members of the popular press but occasionally by active workers in the field.

Some baryonic dark matter must exist since from the ${}^2H + {}^3He$ argument we know that the lower bound from big bang nucleosynthesis is greater than the upper limits on the amount of visible matter in the universe. However, we do not know what form this baryonic dark matter is in. It could be either in condensed objects in the halo, such as brown dwarfs and jupiters (objects with $\lesssim 0.08 M_\odot$ so they are not bright shining stars), or in black holes (which at the time of nucleosynthesis would have been baryons). Or, if the baryonic dark matter is not in the halo, it could be in hot intergalactic gas, hot enough not to show absorption lines in the Gunn-Peterson test, but not so hot as to be seen in the x-rays. Evidence for some hot gas is found in clusters of galaxies. However, the amount of gas in clusters would not be enough to make up the entire missing baryonic matter. Another possible hiding place for the dark baryons would be failed galaxies, large clumps of baryons that condense gravitationally but did not produce stars. Such clumps are predicted in galaxy formation scenarios that include large amounts of biasing where only some fraction of the clumps shine.

Hegyi and Olive[62] have argued that dark baryonic halos are unlikely. However, they do allow for the loopholes mentioned above of low mass objects or of massive black holes. It is worth noting that these loopholes are not that unlikely.[12] Furthermore, recent observational evidence,[63,64], seems to show disk formation is relatively late, occurring at red shifts $z \lesssim 1$. Thus, the first several billion years of a galaxy's life may have been spent prior to the formation of the disk. In fact, if the first large objects to form are less than galactic mass, as many scenarios imply, then mergers are necessary for eventual galaxy size objects. Mergers stimulate star formation while putting early objects into halos rather than disks. Mathews and Schramm[65] have recently developed a galactic evolution model which does just that and gives a reasonable scenario for chemical evolution. (This scenario also provides a natural explanation for the number-versus-magnitude relation of low luminosity galaxies found by Cowie.[66]) Thus, while making halos out of exotic material may be more exciting, it is certainly not impossible for the halos to be in the form of dark baryons. One application of William of Ockham's famous razor would be to have us not invoke exotic matter until we are forced to do so.

Non-baryonic matter can be divided following Bond and Szalay[67] into two major categories for cosmological purposes: hot dark matter (HDM) and cold dark matter (CDM). Hot dark matter is matter that is relativistic until just before the epoch of galaxy formation, the best example being low mass neutrinos with $m_\nu \sim 25eV$. (Remember $\Omega_\nu \sim \frac{m_\nu(eV)}{100h_0^2}$).

Cold dark matter is matter that is moving slowly at the epoch of galaxy formation. Because it is moving slowly, it can clump on very small scales, whereas HDM tends to have more difficulty in being confined on small scales. Examples of CDM could be massive neutrino-like particles with masses, M_x, greater than several GeV or the lightest super-symmetric particle which is presumed to be stable and might also have masses of several GeV. Following Michael Turner, all such weakly interacting massive particles are called "WIMPS." Axions, while very light, would also be moving very slowly[68] and, thus, would clump on small scales. Or, one could also go to non-elementary particle candidates, such as planetary mass blackholes or quark nuggets of strange quark matter, possibly produced at the quark-hadron transition.[69] Another possibility would be any sort of massive toplogical remnant left over from some early phase transition. Note that CDM would clump in halos, thus requiring the dark baryonic matter to be out between galaxies, whereas HDM would allow baryonic halos.

When thinking about dark matter candidates, one should remember the basic work of Zeldovich,[70] resurrected by Lee and Weinberg[71] and others,[72] which showed for a weakly interacting particle that one can obtain closure densities, either if the particle is very light, $\sim 25eV$, or if the particle is very massive, $\sim 3GeV$. This occurs because, if the particle is much lighter than the decoupling temperature, then its number density is the number density of photons (to within spin factors and small corrections), and so the mass density is in direct proportion to the particle mass, since the number density is fixed. However, if the mass of the particle is much greater than the decoupling temperature, then annihilations will deplete the particle number since, as the temperature of the expanding universe drops below the rest mass of the particle, Boltzmann suppression prohibits production while the number is depleted via annihilations until the annihilation reacton freezes out. For normal weakly interacting particles, decoupling occurs at a temperature of $\sim 1MeV$, so higher mass particles are depleted. It should also be noted that the curve of density versus particle mass turns over again (see Figure 4) once the mass of the WIMP exceeds the mass of the coupling boson[73,74,75] so that the annihilation cross section varies as $\frac{1}{M_x^2}$, independent of the mass of the coupling boson. In this latter case, $\Omega = 1$ can be obtained for $M_x \sim 1TeV \sim (3K \times M_{Planck})^{1/2}$, where $3K$ and M_{Planck} are the only energy scales left in the calculation (see Figure 4). A loophole to this argument occurs if there is a matter-antimatter assymmetry as in the case of baryons. However, such particles would have to be Dirac particles and we will see that they are still severely constrained.

A few years ago the preferred candidate particle was probably a few GeV mass WIMP. However, LEP's lack of discovery of any new particle coupling to the Z° with $M_x \lesssim 45GeV$, coupled with underground experiments[76] clearly eliminates that candidate[77,78]. Constraints for particles not fully coupled to

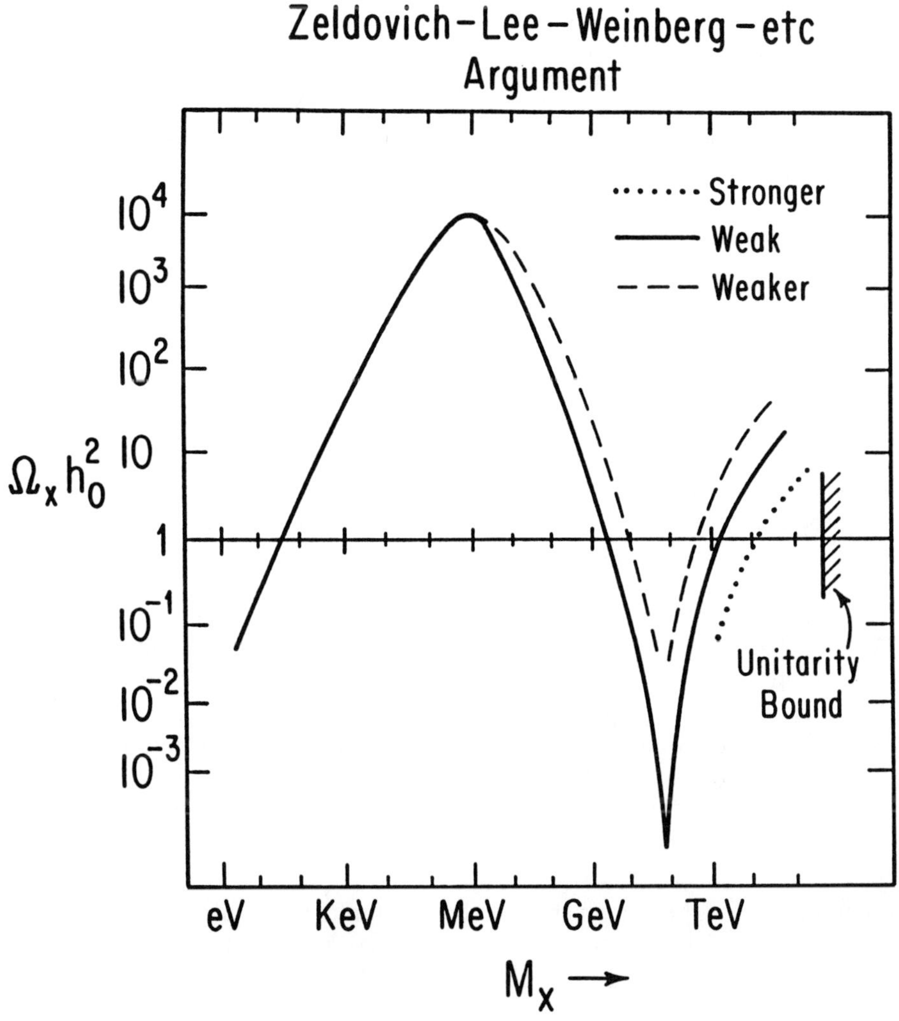

Figure 4. $\Omega_x h_0^2$ versus M_x for weakly interacting particles showing three crossings of $\Omega h_0^2 = 1$. Note also how the curve shifts at high M_x for interactions weaker or stronger than normal weak interaction (where normal weak is that of neutrino coupling through Z^0). Extreme strong couplings reach a unitarity limit at $M_x \sim 340 TeV$.

the Z^0 were discussed by Ellis, Nanopoulos, Roskowski and Schramm[78] and are updated and presented in Figures 5a and 5b. (The inclusion of the Kamiokande II results as well as the newer LEP limits yields an important update over the results of Ellis *et al.*[78] since it closes the loophole for Dirac particles near $12GeV$.) Note also that the generic constraints of Figure 5 also apply to other hypothetical particles since CDF and UA2 do not see any squarks, sleptons, W' or Z' up to masses significantly greater than M_{Z^0}. Thus, whatever the coupling boson is, it must be greater than M_{Z^0} which means the effective value for $\sin^2 \phi_Z$ is < 1.

Furthermore, as Krauss[77] has emphasized, scaler particles such as sneutrinos interact like Dirac neutrinos so that the Kamiokande II and ionization experimental limits[76] also apply. Since asymmetric candidates are all Dirac particles, the restricted part of Figure 5b constrains asymmetric candidates where $\Omega = 1$ is no longer required to follow the locus shown. Thus, it seems that whether the particle is matter-antimatter symmetric or not, it is required to have an interaction weaker than weak and/or have a mass greater than about $20GeV$. Future dark matter searches should thus focus on more massive and more weakly interacting particles.

Also, as Dimopoulos[73] has emphasized, the next appealing crossing of $\Omega = 1$ (see Figure 4) is $\gtrsim 1TeV$ (but, in any case, $\lesssim 340TeV$ from the unitarity bound[75]), which can be probed by SSC and LHC as well as by underground detectors. After the current experimental constraints are taken into account, the favoured CDM particle candidate is now either a $10^{-5}eV$ axion or a gaugino with a mass of many tens of GeV. Of course an HDM ν_τ with $m_{\nu_\tau} \sim 20 \pm 10 eV$ is still a fine candidate as long as galaxy formation proceeds by some mechanism other than adiabatic gaussian matter fluctuations.[79,82] This latter candidate becomes particularly attractive if recent hints from the gallium experiment[83] require the solution to the solar neutrino problem to have neutrino mixing with $\nu_e - \nu_\mu$ mass scales of 0.01 to 0.001 eV, making multiple eV mass scales for ν_τ quite plausible from see-saw type models where $m_{\nu_\tau} \sim m_{\nu_\mu}(\frac{M_{f_3}}{M_{f_2}})^2$ and M_{f_i} is an associated fermion mass for the ith generation. For example, if one uses the heavy quark masses $(\frac{M_t}{M_c})^2 \sim 10^4$ so that ν_τ becomes ideal HDM. Such possibilities also may help late-time phase transition models for producing structure.[11,82]

CONCLUSION

We have seen that our study of the early universe has pushed back to epochs as early as $10^{-43} sec$ and that there are many more observations, experiments and potential observables than existed when Weinberg's book was published 13 years ago. In particular, we have seen that the early universe studies of big bang nucleosynthesis are remarkably robust and have been verified by light element abundance measurements as well as accelerator neutrino counting experiments. Of particular significance for studies "after the first three minutes" is that the nucleosynthetic arguments require $\Omega_b \sim 0.06$. That this exceeds $\Omega_{visible}$ argues that many baryons are dark and not associated with shining material. However, we have also seen that inflation enables many of the big bang initial condition

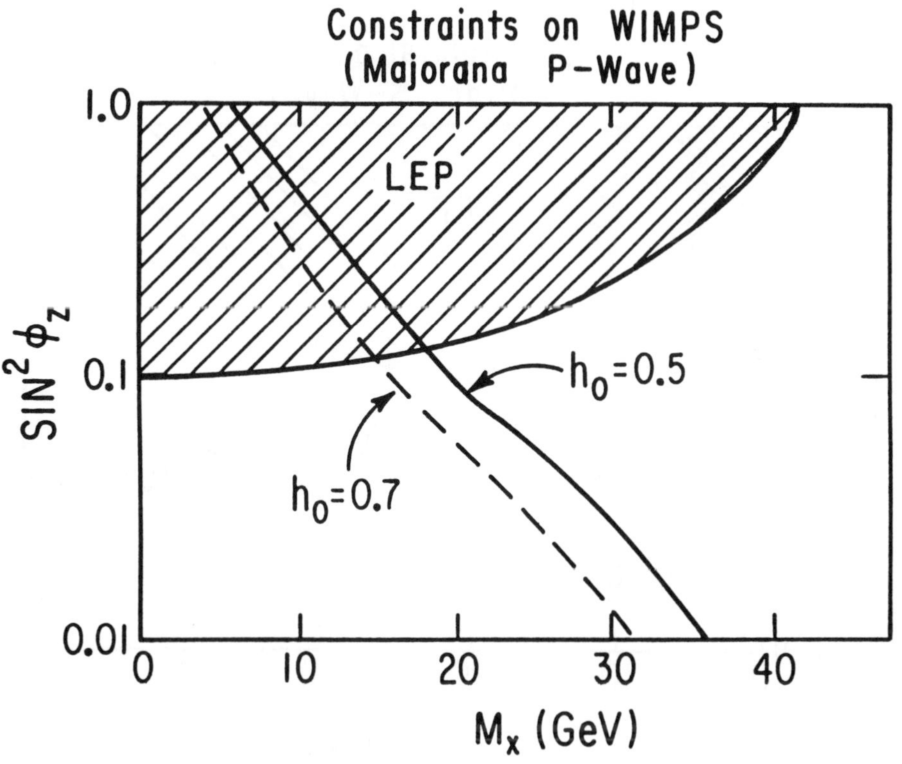

Figure 5a. Constraints on WIMPS of mass M_x versus $\sin^2 \phi_z$, the relative coupling to the Z^0. The constraints are shown assuming Majorana particles (p-wave interactions). The diagonal lines show the combinations of M_x and $\sin^2 \phi_z$ that yield $\Omega = 1$. The cross-hatched region is what is ruled out by the current LEP results. Note that $\Omega = 1$ with $h_0 = 0.5$ is possible only if $M_x \gtrsim 20 GeV$ and $\sin^2 \phi_z < 0.1$. This figure is revised from that of Ref.[68] using latest LEP results.

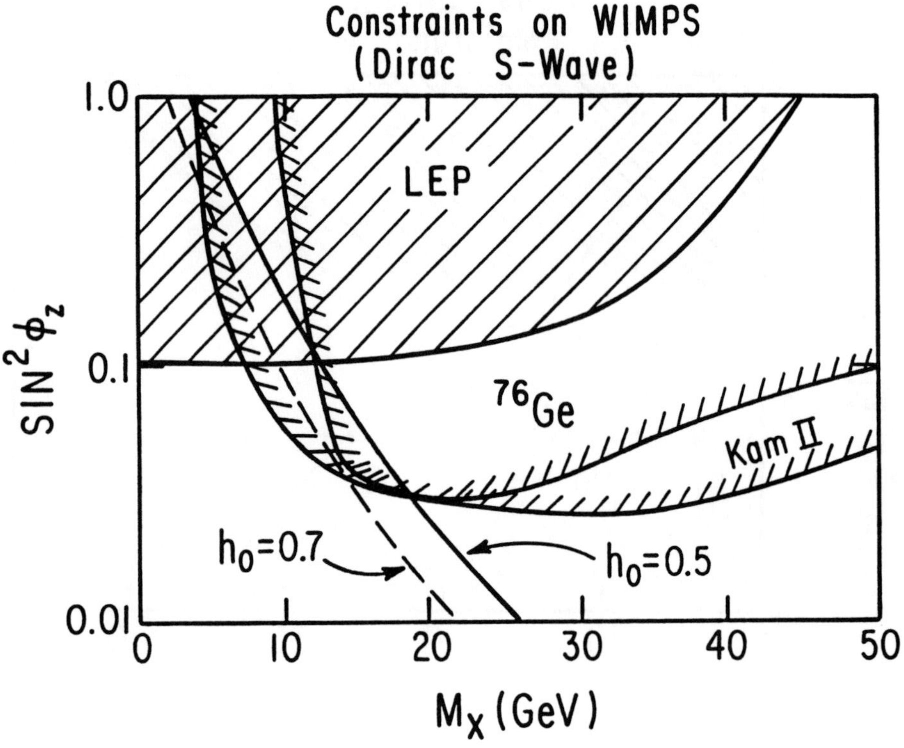

Figure 5b. This is the same as 5a but for Dirac particles (s-wave interactions). The ^{76}Ge region is that ruled out by the Caldwell *et al.* double-β decay style experiments. This figure is revised from that of Ref.[68] using latest LEP results and using new Kamiokande limits which closed a possible loophole near $M_x \sim 10 GeV$. The current results require $M_x \gtrsim 20 GeV$ and $\sin^2 \phi_z \lesssim 0.03$ for matter-antimatter symmetric particles and also exclude the entire cross hatched region for asymmetric particle candidates.

problems to be satisfied but requires $\Omega_{total} = 1.000....$ Thus, ~94% of the universe is required to be in some non-baryonic form. We have also seen that the coupling of particle physics to cosmology gives many possible candidates for this exotic dark matter. Furthermore, we've seen that vacuum phase transitions in the universe, such as the one producing inflation, can produce both adiabatic random gaussian fluctuations and topological defects. The latter can be non-gaussian and behave more like isocurvature (or isothermal) seeds. We have also seen that a non-zero Λ is difficult to produce.

The study of the early universe has become a rich and active field with a close interplay between theory and observation and experiment.

ACKNOWLEDGEMENTS

I would like to thank my recent collaborators Chris Hill, John Ellis, Savas Dimopoulos, Gary Steigman, Keith Olive, Michael Turner, Rocky Kolb and Terry Walker for many useful discussions. I would also like to acknowledge useful communications with Denys Wilkinson, Walter Mampe, Jim Byrne and Stuart Friedman on the neutron lifetime and with Bernard Pagel and Jay Gallagher on helium abundances. This work was supported in part by NSF Grant # AST 88-22595 and by NASA Grant # NAGW-1321 at the University of Chicago, and by the DoE and NASA Grant # NAGW-1340 at the NASA/Fermilab Astrophysics Center.

REFERENCES

1. S. Weinberg, *The First Three Minutes* (NY: Basic Books, 1977).
2. J. Schwartz and D. Green, *Superstrings.*
3. c.f. discussions at Fermilab meeting on wormholes, FNAL, 1989.
4. P.J.E. Peebles, in *Proc. "After the First Three Minutes" Astrophysics Workshop, University of Maryland, October 1990*, (1990).
5. A. Sakharov *Zh. Ek. Teor. Fiz.* **5**, 32 (1990).
6. G. Steigman, in *Cargesse Lectures in Physics*, ed. E. Schatzman **6**, 505 (1971).
7. A. Guth, *Phys. Rev. D* **23**, 347 (1981).
8. E. Kolb and M. Turner, *The Early Universe* (Redwood City, CA: Addison-Wesley, 1990).
9. A. Linde, *Particle Physics and Inflationary Cosmology* (Harwood Academic Publishers, 1990).
10. K. Olive, *Phys. Rep.* **190**, 309-403 (1990).
11. C. Hill, D. Schramm and J. Fry, *Comments Nucl. Part. Phys.* **19**, 25-39 (1989)
12. D. Schramm, *Proc. of the 1990 Rencontre de Physique at La Thuile, March 1990*, ed. M. Greco, in press (1990).

    D. Schramm, *Proc. 1990 Nobel Symposium, Gräftvållen, Sweden*, in press (1990).

    D. Schramm, *Proc. of the Moriond Astrophysics Meeting on Particle Astrophysics at Les Arcs, Savoie, France, March 1990*, ed. T. Piran, in press (1990).
13. R.A. Alpher, H. Bethe and G. Gamow, *Phys. Rev.* **73**, 803 (1948).
14. R.A. Alpher, J.W. Follin and R.C. Herman, *Phys. Rev.* **92**, 1347 (1953).
15. R. Taylor and F. Hoyle, *Nature* **203**, 1108 (1964).
16. P.J.E. Peebles, *Phys. Rev. Lett.* **16**, 410 (1966).
17. P. Wagoner, W.A. Fowler and F. Hoyle, *Astrophys. J.* **148**, 3 (1967).
18. D.N. Schramm and R.V. Wagoner, *Ann. Rev. of Nuc. Sci.* **27**, 37 (1977).

    K. Olive, D.N. Schramm, G. Steigman, M. Turner and J. Yang, *Astrophys. J.* **246**, 557 (1981);

    A. Boesgaard and G. Steigman, *Ann. Rev. of Astron. and Astrophys.* **23**, 319 (1985).

19. J. Yang, M. Turner, G. Steigman, D.N. Schramm and K. Olive, *Astrophys. J.* **281**, 493 (1984).

20. K. Olive, D.N. Schramm, G. Steigman and T. Walker, *Phys. Lett B.* **236**, 454 (1990).

21. T. Walker, G. Steigman, D. Schramm, H.-S. Kang and K. Olive, *Astrophys. J.*, submitted (1990).

22. L. Kawano, D.N. Schramm and G. Steigman, *Astrophys. J.* **327**, 750 (1988).

23. J.W. Truran, A.G. Cameron and A. Gilbert, *Canadian Journal of Physics* **44**, 563 (1966).

24. L. Krauss and P. Romanelli, Yale University preprint and *Ap.J.*, in press (1990).

25. W. Fowler, J. Greenstein and F. Hoyle *Geophys. J.R.A.S.* **6**, 6 (1962).

26. C. Ryter, H. Reeves, E. Gradstajn and J. Audouze, *Astron. and Astrophys.* **8**, 389 (1970).

27. J. Geiss and H. Reeves, *Astron. and Astrophys.* **18**, 126 (1971);

D. Black, *Nature* **234**, 148 (1971).

28. J. Rogerson and D. York, *Astrophys. J.* **186**, L95 (1973).

29. H. Reeves, J. Audouze, W.A. Fowler and D.N. Schramm, *Astrophys. J.* **179**, 909 (1973).

30. R. Epstein, J. Lattimer and D.N. Schramm, *Nature* **263**, 198 (1976).

31. J. R. Gott, III, J. Gunn, D.N. Schramm and B.M. Tinsley, *Astrophys. J.* **194**, 543 (1974).

32. R.T. Rood, G. Steigman and B.M. Tinsley, *Astrophys. J.* **207**, L57 (1976).

33. J. Yang, D.N. Schramm, G. Steigman and R.T. Rood, *Astrophys. J.* **227**, 697 (1979).

34. T. Wilson, R.T. Rood and T. Bania, in *Proc. of the ESO Workshop on Primordial Healing*, ed. P. Shaver and D. Knuth (Garching: European Southern Observatory, 1983).

35. M.R. Hartoog, *Ap.J.* **231**, 161 (1979).

J. Ostriker and D.N. Schramm, FNAL/Princeton preprint (1990).

36. J. Spite and F. Spite, *Astron. and Astrophys.* **115**, 357 (1982);

R. Rebolo, P. Molaro and J. Beckman, *Astron. and Astrophys.* **192**, 192 (1988);

L. Hobbs and C. Pilachowski, *Astrophys. J.* **326**, L23 (1988).

37. G. Steigman, D.N. Schramm and J. Gunn, *Phys. Lett. B* **66**, 202 (1977).

38. V.F. Schvartzman, *JETP Letters* **9**, 184 (1969);

P.J.E. Peebles, *Physical Cosmology* (Princeton University Press, 1971).

39. D.N. Schramm and L. Kawano, *Nuc. Inst. and Methods* **A284**, 84 (1989).

40. B. Pagel, *Proc. of 1989 Rencontre de Moriond*

41. R. Scherrer, J. Applegate and C. Hogan, *Phys. Rev. D* **35**, 1151 (1987);

C. Alcock, G. Fuller and G. Mathews, *Astrophys. J.* **320**, 439 (1987);

W.A. Fowler and R. Malaney, *Astrophys. J.* **333**, 14 (1988).

42. H. Kurki-Suonio, R. Matzner, K. Olive and D.N. Schramm, *Astrophys. J.* **353**, 406 (1990);

43. M. Peimbert and S. Torres-Peimbert, *Astrophys. J.* **193**, 327 (1974).

44. B. Pagel, *ESO/CERN Proc.* (1990).

45. G. Steigman, D.N. Schramm and J. Gallagher, *Comments on Astrophys.* **14**, 97 (1989).

46. W. Mampe, P. Ageron, C. Bates, J.M. Pendlebury and A. Steyerl, *Phys. Rev. Lett.* **63**, 593 (1989).

47. J. Byrne, *et al. Phys. Lett. B*, submitted (1990).

48. H. Abele, M. Arnould, H.A. Borel, J. Dohner, D. Dubbers, S. Freedman, J. Last and I. Reichert, *Proc. Grenoble Workshop on Slow Neutrons* (1989).

49. G. Steigman, K. Olive, D.N. Schramm and M. Turner, *Phys. Lett. B* **176**, 33 (1986).

50. ALEPH, DELPHI, L3, OPAL and MARK II collaboration papers in *Proc. of the Int. High Energy Conference, Singapore* (1990).

51. D.N. Schramm and G. Steigman, *Phys. Lett. B* **141**, 337 (1984).

52. E. Kolb and R. Scherrer, *Phys. Rev. D* **25**, 1481 (1982);

A. Chakravorty, E. Kolb and D.N. Schramm, in progress (1991).

53. S. Dimopoulos, R. Esmailzadeh, L. Hall and G. Starkman, *Astrophys. J.* **330**, 545 (1988).

54. L. Brown and D.N. Schramm, *Astrophys. J.* **329**, L103 (1988).

55. C. Pilachowski, L. Hobbs and D. De Young, *Astrophys. J. Lett.* 345, L39 (1989).

56. H. Reeves, K. Sato and M. Tarasawa, U. of Tokyo preprint (1989).

N. Tarasawa and K. Sato, U. of Tokyo preprint (1990).

57. C. Alcock, G. Fuller, G. Mathews, and B. Meyer, Livermore preprint (1990).

58. F. Adams and K. Freese, MIT preprint (1990)

59. L. Kawano, W. Fowler and R. Malaney, Caltech-Kellogg preprint (1990).

J. Applegate, Columbia University preprint (1989).

60. K. Freese and D.N. Schramm, *Nucl. Phys.* **B233**, 167 (1984).

61. M. Strauss, M. Davis and A. Yahil, U.C. Berkeley preprint (1989);

N. Kaiser and A. Stebbins, CITA preprint (1990);

E. Bertschinger, A. Dekel, and A Yahil, in *Proc. Blois Symposium on the Microwave Background* (1990);

M. Rowan-Robinson and A. Yahil, *Proc. Rencontres de Moriond* (1989).

62. D. Hegyi and K. Olive, *Astrophys. J.* **303** (1986) 56.

63. J. Gunn, Talk at ITP Santa Barbara (1988);

64 D. York, Talk at University of Chicago (1988).

65. G. Mathews and D.N. Schramm, *Astrophys. J.*, submitted (1990).

66. L. Cowie, in *Proc. IUPAP Conference on Primordial Nucleosynthesis and Evolution of Early Universe, Tokyo, September 1990*, in press (1990).

67. R. Bond and A. Szalay, in *Proc. Texas Relativistic Astrophysical Symposium, Austin, Texas* (1982).

68. M. Turner, F. Wilczek and A. Zee, *Phys. Lett. B* **125**, 35 (1983); **125**, 519 (1983).

69. M. Crawford and D.N. Schramm, *Nature* **298**, 538 (1982).

E. Witten, *Phys. Rev. D* **30**, 272 (1984).

See also C. Alcock and A. Olinto, *Ann. Rev. Nuc. Part. Phys.* **38**, 161 (1988).

70. Ya. Zeldovich, *Adv. Astron. and Astrophys.* **3**, 241 (1965).

71. B. Lee and S. Weinberg, *Phys. Rev. Lett.* **39**, 165 (1977).

72. H-Y. Chiu, *Phys. Rev. Lett.* **17**, 712 (1966).

C.P. Hut, *Phys. Lett. B* **69**, 85 (1977);

K. Sato and H. Koyayashi, *Prog. Theor. Phys.* **58**, 1775 (1977).

73. S. Dimopoulos, R. Esmailzadeh, L. Hall and N. Tetradis, *Nucl. Phys. B*, in press (1991).

74. D. Brahm and L. Hall, *Phys. Rev. D* **41**, 1067 (1990).

75. K. Griest, M. Kamionkowski and M. Turner, *Phys. Rev. Lett.* **41**, 3565 (1990).

    K. Olive and M. Srednicki, *Phys. Lett. B* **230**, 78 (1989).

76. D. Caldwell, *et al.*, in *Proc. La Thuile*, ed. M. Greco (1990); see also Kamiokande II Collaboration, Proc. IUPAP Sumposium on Primodial Nucleosynthesis and Evolution of Early Universe, Tokyo, September 1990, in press (1990).

77. L. Krauss, *Phys. Rev. Lett.*, in press (1990);

    K. Griest and J. Silk, U.C. Berkeley preprint (1990).

78. J. Ellis, D. Nanopoulos, L. Roskowski and D.N. Schramm, *Phys. Lett. B*, in press (1990).

79. D. N. Schramm, *Proc. 1988 Berkeley Workshop on Particle Astrophysics*, ed, E. Norman (World Scientific, 1989).

80. S.M. Faber and J.S. Gallagher *Ann. Rev. Astron. and Astrophys.* **17**, 135 (1979).

81. E. Loh and E. Spillar, *Astrophys. J.* **329**, 24 (1988).

82. D. Schramm, in *Proc. IUPAP Symposium on Primordial Nucleosynthesis and Evolution of Early Universe, Tokyo, September 1990*, in press (1990).

83. SAGE collaboration *Proc. Int. High Energy Physics Meeting, Singapore, August 1990*, in press (1990).

Spectrum of the Cosmic Microwave Background

EARLY RESULTS FROM THE FAR INFRARED ABSOLUTE SPECTROPHOTOMETER (FIRAS)

J. C. Mather, E. S. Cheng, and R. A. Shafer
Laboratory for Astronomy and Solar Physics
Goddard Space Flight Center, Greenbelt, MD 20771

R. E. Eplee and R. B. Isaacman
General Sciences Corporation, Laurel, MD

D. J. Fixsen
University Space Research Association, Greenbelt, MD

S. M. Read
STX Corporation, Lanham, MD

S. S. Meyer and R. Weiss
Massachusetts Institute of Technology, Cambridge, MA

E. L. Wright
UCLA Department of Astronomy, Los Angeles, CA

ABSTRACT

The Far Infrared Absolute Spectrophotometer *(FIRAS)* on the Cosmic Background Explorer *(COBE)* mapped 98% of the sky, 60% of it twice, before the liquid helium coolant was exhausted. The *FIRAS* covers the frequency region from 1 to 100 cm^{-1} (10 mm to 0.1 mm wavelength) with a 7° angular resolution. The spectral resolution is 0.2 cm^{-1} for frequencies less than 20 cm^{-1} and 0.8 cm^{-1} for higher frequencies. Preliminary results include: a limit on the deviations from a Planck curve of 1% of the peak brightness from 1 to 20 cm^{-1} , a temperature of 2.735 ± 0.06 K, a limit on the Comptonization parameter y of 10^{-3}, on the chemical potential parameter μ of 10^{-2}, a strong limit on the existence of a hot smooth intergalactic medium, and a confirmation that the dipole anisotropy spectrum is that of a Doppler shifted blackbody. Although there are many unresolved issues about the data processing and analysis of systematic errors, some of which are described here, the instrument is expected to achieve its goal of an accuracy and rms sensitivity for νI_ν better than 10^{-9} W/m^{2}sr, for each 7° beamwidth and each 5% spectral resolution element, for frequencies from 1 to 20 cm^{-1} .

INTRODUCTION

The cosmic microwave background radiation (CMBR) from the Big Bang, and the cosmic infrared background radiation (CIB) from the first objects to form after the Big Bang, are the subjects of NASA's first satellite designed primarily for observational cosmology, the *COBE*. The mission has been described by Mather (1982) and Gulkis *et al.* (1990), and early results have been summarized by Mather *et al.* (1990a,b).

We summarize here the key events in the expanding universe which influence these backgrounds. Processes in the early expansion of the universe such as inflation and subsequent nucleosynthesis, which take place within the first few minutes after the initial explosion, set the scale for the largest anisotropies and inhomogeneities. The number of quanta in the CMBR is essentially fixed when the time scales for photon creation and destruction exceed the expansion time scale, which occurs at a redshift $z \sim 10^6$ when photons outnumber matter particles by a factor of $\sim 10^8$. After this epoch the spectrum of the CMBR can be modified to a Bose–Einstein distribution with a chemical potential μ by such processes as black hole formation, cosmic string annihilation, matter/antimatter annihilation, or the decay of exotic particles. After a redshift of $z \sim 3 \times 10^4$, the processes which re-establish equilibrium become too slow to achieve a Bose–Einstein distribution, and any other energy release into the radiation field mediated by hot electrons produces a mixture of blackbodies at different temperatures, characterized by the parameter y (Zel'dovich and Sunyaev 1969). After the decoupling of matter and radiation at $z \sim 10^3$, the photons are free to move unimpeded, and the major features of the spectrum and angular distribution of the CMBR as now observed should have been established.

The processes leading to the formation of galaxies and large scale structures, as well as the epoch when these processes occurred, are major unknowns in current cosmology. One can bracket the epoch by noting that the earliest quasars have a $z<5$ and that galaxies at their present size would overlap at z between 20 and 100. The formation of galaxies may well be accompanied by the creation of a hot intergalactic medium, which could produce anisotropies and spectral distortions in the CMBR as well as a signature in the infrared. Furthermore, it is expected that precursors to the large scale structures should have left anisotropies and possibly spectral distortions in the CMBR and X-ray background.

Galactic and Solar System foreground radiation limit the ability of solar system based instruments to measure the cosmic background radiation. The primary such background for the *FIRAS* instrument is the interstellar dust, which absorbs starlight and reradiates at a temperature of the order of 22 K. A full sky survey has been made to find the directions where such interference is a minimum and to enable accurate extrapolations to a viewpoint outside our Galaxy.

FIRAS INSTRUMENT DESCRIPTION

The purpose of the *FIRAS* is to compare the spectrum of the CMBR with that of a precise blackbody, enabling the measurement of very small deviations from a Planckian spectrum. The *FIRAS* instrument covers two frequency ranges, a low frequency channel from 1 to 20 cm^{-1} and a high frequency channel from 20 to 100 cm^{-1} . It has a 7° diameter beamwidth, established by a non-imaging parabolic concentrator, which has a flared aperture to reduce diffractive sidelobe responses. The instrument is calibrated by a full aperture, temperature-controlled external blackbody, which can be moved into the beam by command.

The *FIRAS* is the first instrument to measure the background radiation and compare it with such an accurate external full beam calibrator in flight. The spectral resolution is obtained with a polarizing Michelson interferometer, with separated input and output beams to permit fully symmetrical differential operation. One input beam views the sky or the full aperture calibrator, while the second input beam views an internal temperature controlled reference blackbody, with its own parabolic concentrator. Both input concentrators and both calibrators are temperature controlled and can be set by command to any temperature between 2 and 25 K. The standard operating condition is for the two concentrators and the internal reference body to be commanded to match the sky temperature, thereby yielding a nearly nulled interferogram and reducing almost all instrumental gain errors to negligible values.

The instrument's apodized spectral resolution is limited by the maximum stroke length to 0.2 cm^{-1} in the low frequency channels, and by the microprocessor buffer size and telemetry bandwidth to 0.8 cm^{-1} in the high frequency channels. Beam divergence within the instrument limits the spectral resolution to ~1% and shifts the effective frequency scale by 0.4%. The rms sensitivity for frequencies from 2 to 20 cm^{-1} is $\nu I_\nu < 10^{-9}$ W/m^{2}sr per 7° field and per 5% spectral resolution element after one year of operation. There are four large area (0.5 cm^2) composite bolometer detectors, two on each output of the spectrometer. The separation of low and high frequencies is accomplished by a capacitive grid dichroic filter.

The external calibrator determines the accuracy of the instrument for broad band sources like the cosmic background radiation. It is a re-entrant cone shaped like a trumpet mute, made of Eccosorb CR-110 iron-loaded epoxy. The angles at the point and groove are 25°, so that a ray reaching the detector has undergone 7 specular reflections from the calibrator. This can be seen by approximating the circular groove by a straight groove and drawing the multiple images of each surface in the other. The calculated reflectance for this design, including diffraction and surface imperfections, is less than 10^{-4} from 2 to 20 cm^{-1} . Measurements of the reflectance of an identical calibrator in an identical antenna using coherent radiation at 1 cm^{-1} and 3 cm^{-1} frequencies confirm this calculation. The instrument is calibrated by measuring spectra with the calibrator in the sky horn while operating all other controllable sources within the instrument at a sequence of different temperatures. The overall responsivity and the emissivity of each source can be determined relative to the external calibrator by solving a set of coupled linear equations.

The first results of the *FIRAS* (Mather *et al.* 1990a, Cheng *et al.* 1990) may be summarized as follows. The intensity of the background sky radiation is consistent with a blackbody at 2.735 ± 0.06 K. Deviations from this blackbody at the spectral resolution of the instrument are less than 1% of the peak brightness. The quoted uncertainty in temperature is due to an uncertainty in the thermometer calibration; we expect to reduce this uncertainty by additional tests. The measured spectrum is shown in Figure 1, and is converted to temperature units in Figure 2, where it is compared to previous measurements. The deviations can be fitted to the Sunyaev-Zel'dovich form for Comptonization (Zel'dovich and Sunyaev 1969), giving a limit on the y parameter of $|y| < 10^{-3}$ ($3\ \sigma$). They can also be fitted to a Bose-Einstein distribution with a chemical potential, giving a limit of $|\mu| < 10^{-2}$

(3 σ). There is a strong limit on the existence of a smooth hot intergalactic medium: it can contribute less than 3% of the X-ray background radiation even at a reheating time as recent as $z = 2$. There is no evidence of a distortion of the spectrum, such as that reported by Matsumoto *et al.* (1988), and the measured temperature is consistent with previous reports and the recent rocket result of Gush *et al.* (1990).

The variation of the spectrum with position in the sky as measured by the *FIRAS* is dominated by the dipole pattern of the Doppler shift of the background temperature, plus a variation in the interstellar dust emission. A preliminary analysis of three weeks of data taken near the Galactic poles shows exactly such a variation. A determination of the spectrum of the dipole was made by calculating the average spectra in two large circular regions of the sky each of angular diameter 60°, one centered at $(\alpha, \delta) = (11.1^h, -6.3°)$ and the other at $(23.1^h, 6.3°)$, which lie along opposite ends of the dipole axis as determined by the *DMR*. The difference between these spectra can be fit extremely well by the difference of two blackbodies, and is consistent with a peak dipole amplitude of 3.3 ± 0.3 mK and the assumed dipole direction.

SYSTEMATIC ERROR ANALYSIS AND ULTIMATE ACCURACY

The *FIRAS* data set includes about 500,000 interferograms per detector, each susceptible to a wide variety of possible kinds of errors. We have characterized the possible errors and begun to develop software to deal with all of them on a routine basis, since automation is required for such a large data set. These errors, and the approaches to deal with them, are outlined below. We believe that every one can be understood and dealt with in software to the degree necessary to meet our accuracy requirements. They include:

- telemetry errors
- cosmic rays striking the detectors
- cosmic rays striking the control sensors for the mirror mechanism and stopping the scanning
- errors in thermometer calibration
- possible gradients in temperature in the calibrators
- a shift in the frequency scale
- detector nonlinearities
- stray signals from the Earth when it rises above the top of the shield and shines on the instruments
- stray light from the Moon
- Sun glints from defects in the sunshade edge
- thermal crosstalk from the *DIRBE*
- emission from the *DMR* local oscillators
- vibrations of the moving mirror
- variations of the optical response across the antenna beam and
- multipass phenomena within the interferometer.

Telemetry errors are detected in numerous ways. The data stream from the spacecraft contains words whose values are known, such as frame sync words, clocks, and frame

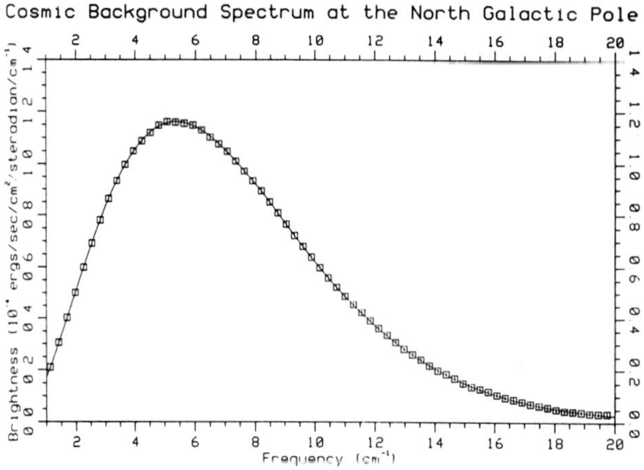

Figure 1. Preliminary spectrum of the cosmic microwave background from the *FIRAS* instrument, compared to a blackbody. Boxes are measured points and show size of assumed 1% error band.

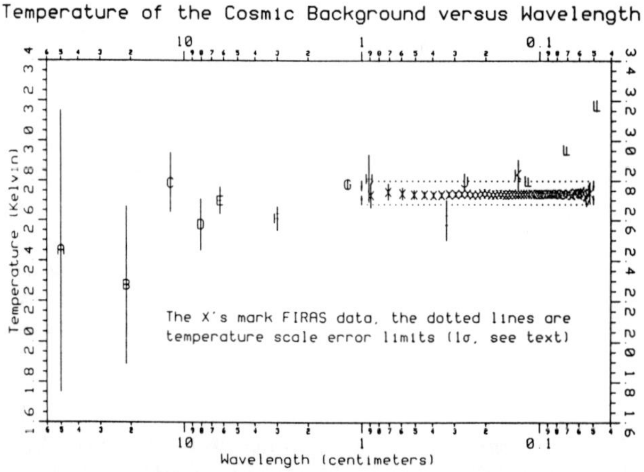

Figure 2. Composite plot of recent measurements of the temperature of the sky. A = Sironi *et al.* (1987), B = Levin *et al.* (1987), C = Sironi and Bonelli (1986), D = De Amici *et al.* (1988), E = Mandolesi *et al.* (1986), F = Kogut *et al.* (1988), G = Johnson and Wilkinson (1987), H = Smoot *et al.* (1985), I = Smoot *et al.* (1987), J = Crane *et al.* (1989), K = Meyer *et al.* (1989), Palazzi *et al.* (1990), L = Matsumoto *et al.* (1988).

counters, which must agree with expectations. Unfortunately there are no overall data checksums. The interferograms do have check sums, so most bad telemetry can be detected in this way. Interferograms are not used in sky maps or calibrations unless they are consistent with others taken under the same instrument conditions and looking in the same direction, but considerable effort is still necessary to understand the criteria for consistency.

Cosmic rays strike the detectors several times per second in normal operation, delivering impulses of heat and producing "glitches" in the interferograms. They are flagged by the on-board microprocessor but not removed; that is reserved for ground software. The ground software recognizes the glitches by comparing multiple observations of each pixel and removing the impulses that exceed a criterion for normal detector noise. This software works well for the nearly null interferograms, but glitches that occur near the peak of the interferogram are harder to discern. This is important in the calibration data, and for observations of the Galactic Plane where the interferograms are not all consistent because the Galaxy is thinner than the beamwidth. Tests of new methods are in progress.

Cosmic rays also strike the control sensors for the mirror mechanism and stop the scanning several times per day, even outside the South Atlantic Anomaly (the SAA, a region of high trapped proton flux) where the scanning is commanded off. Each such event results in a loss of data and a thermal impulse to the cryostat, since the mechanism stops by pushing against a mechanical stop, and the servo drives a high current into the motor. The loss of data is not serious but the temperature impulses cause a major loss of calibration stability and accuracy. New software is being developed to incorporate the measured effects of temperature changes of the scanning mirrors, and to delete data taken when the detector temperature is either out of range or changing too rapidly.

Errors in thermometer calibration are difficult to determine from flight data alone. The photometric calibration data are being examined for consistency, and small adjustments to the temperature scales are being tested to see if the overall calibration residuals can be reduced. Present indications are that adjustments of a few milliKelvin are sufficient at 2.7 K, but adjustments up to 1 K may be required at 20 K. A backup calibrator is being instrumented for further tests on the ground, and backup thermometers have been recalibrated to investigate the possibility of long term drifts in the calibration constants.

The flight data indicate significant gradients in the temperatures of both the internal and external calibrators, amounting to 17 and 6 mK respectively at 2.7 K. These values are the same as before launch, and are the reason why the announced cosmic background temperature uncertainty is 60 mK. The thermal design for the external calibrator predicts that it cannot have such a large gradient, so this will also be investigated in the test of the backup unit. Similar tests for the internal calibrator are also planned.

A shift in the frequency scale is possible for two reasons. First, the reference scale used to govern the sampling of the interferogram is made of glass, and shrinks during the cooldown from room temperature. Second, beam divergence within the interferometer multiplies the effective frequency by the cosine of the angle from the beam center. These effects add up to about 0.5%, an amount that is consistent with the photometric calibration.

Detector nonlinearities may be important for large signal sources, such as the Moon and some of our calibration data. In addition, the detector responsivity is strongly dependent on the operating point, as determined by the DC bias voltage and temperature. Our software models these variations, and preliminary fits to the calibration data show that preflight values for the detector constants are approximately correct.

Stray signals from the Earth when it rises above the top of the shield and shines on the instruments are very important to the *FIRAS*. The heat from the Earth at worst case during the June solstice is sufficient to raise the input horn antenna from 2.7 K to 7 K, and the cooldown time is of the order of 20 minutes. These temperature changes are a serious challenge to the calibration software, and it is not known whether good measurements of the background can be retrieved from them. Fortunately, most of the sky seen during that season was also observed early in the mission without interference from the Earth.

Stray light from the Moon is presently being investigated. It is clearly detectable in the high frequency channel interferogram peak height as far away as 20° from the Moon, approximately as expected from preflight infrared laser measurements of the antenna beam profile. At the present the software makes no attempt to model or remove the effects of the Moon except to cut out data taken too close to the Moon and not include them in the sky maps. About 3% of the observations are taken within 20° of the Moon.

Sun glints are reflections from the Sun into the top of the cryostat, presumably from some small part of the insulation blanket on the sunshield that has come loose, or from an unexpected position of the horn covers for the *DMR* after they successfully deployed. They are detected by their thermal effect on the vapor cooled shields of the cryostat, and occur once per spacecraft rotation, but not always with the same amplitude. Searches of both *DIRBE* and *FIRAS* data have revealed no effect of such glints, probably because the Sun is not a particularly bright far infrared source and because the instrument response is very small at large angles from the line of sight.

Thermal crosstalk from the *DIRBE* is not presently known to be a problem but needs further investigation. The *DIRBE* thermal effects on the *FIRAS* are much smaller than the temperature changes caused by mode changes of the mirror mechanism, cosmic ray events in the mirror mechanism electronics, and temperature controller setpoint changes for the horns and calibrators. In principle the calibration software should handle all such effects without any residual errors.

Emission from the *DMR* local oscillators would be seen as constant spectrum lines in all the sky spectra at the known oscillator frequencies or their harmonics, which would not be present in the calibration data. No such lines have been recognized to date. In addition, the emission from the oscillators might be expected to be modulated at 100 Hz, the frequency of the Dicke switches used by the *DMR*. In that case, the noise level of the *FIRAS* detectors should show a peak at 100 Hz, and there is no evidence for such a peak.

Vibrations of the moving mirror produce errors because they cause the interferograms to be sampled at incorrect locations, creating frequency modulation sidebands of spectral features. Most spacecraft vibrations are not synchronous with the scans, but some vibrations are excited at each velocity change of the moving mirror and occur on every stroke. These errors are still negligible when the interferograms are of small amplitude, as for normal sky measurements, but they are important for calibrations. We see small ripples in the instrument phase function which may be due to these vibrations.

Variations of the optical response across the antenna beam have been determined using the Moon as a source. We find that not only does the efficiency vary with the location of the Moon in the beam, but it varies differently at different frequencies, and the effective center (zero path difference point) of the interferogram also changes. These effects do not cause errors for diffuse sources but limit our ability to measure the Moon and use it for confirming our calibration.

Multipass phenomena within the interferometer were known before launch from tests of breadboard interferometers. They arise from radiation that is reflected from the bolometer or its housing and filter, which then returns to the interferometer, is modulated a second time, reflects from the optics at the input ports, returns through the interferometer and is modulated a third time, and finally reaches the detectors again. Such radiation should not be a problem for the null test, where the sky is compared directly to an external calibrator at the same temperature. However, it may limit the accuracy of the instrument at higher frequencies where Galactic emissions are of interest. Studies are beginning to determine the magnitude of these effects, and preliminary results limit the multipass energy to less than 1% of the total.

The ultimate accuracy of the instrument is limited by the reflectance of the calibrator, by leakage around the edge of the calibrator, by instrument stability between calibrations, by the detector noise, by the time devoted to calibration, and by residual statistical and systematic errors in the determination of the calibration parameters. The reflectance of the calibrator is important because the calibrator is illuminated by radiation from the instrument, including contributions from the antenna (adjusted to match the sky temperature) and from the interferometer (usually around 1.6 K). As described above, we believe the reflectance is less than 10^{-4}. There is also a possibility of leakage around the edge of the calibrator. Tests were done in orbit to investigate the change of the interferogram as a function of the position of the calibrator, but no changes were seen until the calibrator was far out of its normal position. The stability of the instrument is always difficult to analyze, but over the course of the 10 months of operation the instrument transmission function changed by less than 2%, with no evident trend. During the last month of operation, calibration and sky data were interleaved with a 50% duty cycle and a 7 day period to reduce sensitivity to potential instabilities. A total of one month of near-null calibration data were obtained, out of a total of 10 months of operation, so the calibration is potentially accurate to 10^{-10} W/m^{2}sr, or even better if averaged over frequencies.

SUMMARY AND CONCLUSIONS

The detailed analysis of the *FIRAS* data to model and remove systematic errors and local foreground sources is just beginning, but the temperature has been measured as 2.735 ± 0.06 K, and limits on the distortion parameters have been obtained, with $|y| < 10^{-3}$ and $|\mu| < 10^{-2}$ ($3\ \sigma$). The dipole anisotropy has the expected shape for a Doppler shifted blackbody as measured by *FIRAS*, and the dipole amplitude and direction are consistent with previous measurements. Systematic errors limit the data accuracy, but analysis in progress is expected to reduce them below the goal of 0.1% of the peak CMBR brightness.

ACKNOWLEGEMENTS

It is a pleasure to acknowledge the vital contributions of all those at GSFC who devoted their efforts to making this challenging mission not only possible but enjoyable as well. The National Aeronautics and Space Administration/Goddard Space Flight Center (NASA/GSFC) is responsible for the design, development, and operation of the Cosmic Background Explorer (*COBE*). GSFC is also responsible for the development of the analysis software and for the production of the mission data sets. The *COBE* program is supported by the Astrophysics Division of NASA's Office of Space Science and Applications.

REFERENCES

Cheng, E. C., Mather, J., Shafer, R., Meyer, S., Weiss, R., Wright, E., Eplee, R., Isaacman, R., and Smoot, G. 1990, *Bull. APS,* **35,** 937.

Crane, P., Hegyi, D. J., Kutner, M. L., and Mandolesi, N. 1989, *Ap. J.,* **346,** 136-142.

DeAmici, G., Smoot, G., Aymon, J., Bersanelli, M., Kogut, A., Levin, S., and Witebsky, C., 1988, *Ap. J.,* **329,** 556.

Gulkis, S., Lubin, P.M., Meyer, S.S., and Silverberg, R. F.,1990, *Sci. Am.,* **262(1),** 132-139.

Gush, H.P., Halpern, M., and Wishnow, E.,1990, *Phys. Rev. Lett.,* **65,** 537.

Johnson, D. G., and Wilkinson, D. T. 1987, *Ap. J. (Letters),* **313,** L1-L4 .

Kogut, A., Bersanelli, G., De Amici, G., Friedman, S., Griffith, M., Grossan, B., Levin, S., Smoot, G., and Witebsky, C. 1988, *Ap. J.,* **325,** 1.

Levin, S., Witebsky, C., Bensadoun, M., Bersanelli, M., De Amici, G. , Kogut, A., and Smoot, G. 1987, *Ap. J.,* **334,** 14.

Mandolesi, N., Calzolari, P., Cortiglioni., S., and Morigi, G., 1986, *Ap. J.,* **310,** 561.

Mather, J. C.,1982, *Opt. Eng.,* **21(4),** 769-774.

Mather, J.C., Cheng, E.S., Eplee, R.E., Jr., Isaacman, R. B., Meyer, S. S., Shafer, R. A., Weiss, R., Wright, E.L., Bennett, C. L, Boggess, N. W., Dwek, E., Gulkis, S., Hauser, M. G., Janssen, M., Kelsall, T., Lubin, P. M., Moseley, S. H. Jr., Murdock, T. L., Silverberg, R. F., Smoot, G. F., and Wilkinson, D. T. :1990a, *Ap. J.,* **354,** L37-L41.

Mather, J.C., *et al.* 1990b, *Proc. IAU Colloquium 123 "Observatories in Earth Orbit and Beyond",* ed. Y. Kondo, Kluwer Acad. Publ., Dordrecht.

Matsumoto, T., Hayakawa, S., Matsuo, H., Murakami, H., Sato, S., Lange, A. E., and Richards, P. L.,1988, *Ap. J.,* **329,** 567-571.

Meyer, D.M., Roth, K.C., and Hawkins, I., 1989, *Ap. J. (Letters),* **343,** L1.

Palazzi, E., Mandolesi, N., Crane, P., Kutner, M.L., Blades, J.C., and Hegyi, D.J., 1990, *Ap. J.,* **357,** 111-C3.

Sironi, G.,*et al.* 1987, *Proc. 13th Texas Symposium on Relativistic Astrophysics,* ed. M. P. Ulmer, (Singapore: World Scientific), 245.

Sironi, G., and Bonelli, G., 1986, *Ap. J.,* **311,** 418.

Smoot, G., De Amici, G., Levin,. S., and Witebsky, C. 1985, *Soc. Italiana di Fis. Conf. Proc.,* **1,** 27.

Smoot, G., Bensadoun, M., Bersanelli, M., De Amici, G., Kogut, A., Levin, S., and Witebsky, C. 1987, *Ap. J. (Letters),* **317,** L45.

Zel'dovich, Ya. B., and Sunyaev, R. A. 1969, *Ap. Space Sci.*, **4**, 301.

AUDIENCE QUESTIONS

Ofer Lahav: An upper limit of 10^{-3} on the Comptonization parameter y is not a small number, in the sense that spectral distortions due to hot IGM could still be hidden in the "perfect" blackbody spectrum. What are the prospects to get an even lower limit on y from COBE and the other experiments?

Answer: The *FIRAS* sensitivity and accuracy should be good enough to measure a y as small as 10^{-4}, since the early result was derived from only 9 minutes of data. The known ultimate limitation of the FIRAS is the external calibrator reflectance of about 10^{-4}.

Rocket Measurement of the Spectrum of the CBR

M. Halpern, H.P. Gush and E.H. Wishnow

Department of Physics, U.B.C., Vancouver, B.C. Canada V6T 2A6

Abstract

The spectrum of the cosmic background radiation (CBR) has been measured using a liquid helium cooled spectrometer carried to high altitude by a rocket. The spectrum in the frequency band 2 to 30 cm^{-1} is very well fitted by a Planck function of temperature T = 2.736 K. The scatter of the equivalent temperature in the band 3 to 16 cm^{-1} is ±10 mK, about 1/3% of the mean, whereas the estimated overall experimental accuracy of the mean temperature is ±17 mK. These results are inconsistent with previously reported infrared excess in CBR, but are in good agreement with the COBE results. They form the most precise constraints reported to date on the spectrum of the CBR.

INTRODUCTION

The cosmic background radiation (CBR) has been recognized as a very important observable in cosmology since its discovery in 1965,[1,2] being one of the oldest quantities accessible to measurement. Substantial experimental effort has been devoted in the past 25 years to investigating the properties of this radiation for a combination of reasons: the information to be gained is very important to an understanding of the early Universe and the experimental difficulties have been such that definitive measurements have not previously been obtained. We report here measurements which provide a dramatic improvement in our knowledge of the spectrum of the CBR[3].

In the simple big bang scenario the CBR is highly redshifted radiation which was in thermal equilibrium with matter as the universe cooled to approximately 4000K allowing the formation of neutral Hydrogen. The spectral distribution of the radiation is strictly Planckian and this form remains under expansion of the universe, the temperature dropping as the inverse of the relative scale factor, z.

For values of $z \geq 10^7$ electron positron pairs can be produced thermally and radiation and matter are held in tight thermal equilibrium. Once z falls below 10^6 however, radiation and matter are coupled primarily through single Compton scattering which is insufficient to thermalize radiation since it does not change the number of photons. If matter and radiation are brought out of thermal equilibrium between $z \simeq 10^6$ and $z \simeq 10^4$ kinetic, but not thermodynamic, equilibrium can be reestablished resulting in a Bose Einstein spectrum characterized by a chemical potential μ, a measure of the energy deposited relative to the radiative energy density.

The spectrum would show a dip in intensity at wavelengths longer than that of the peak intensity, the exact form of this dip depending on density and expansion rate.

After $z = 10^4$ not even kinetic equilibrium can be reestablished and energy release results in a combination of thermal spectra at several temperatures or, for lower z, a Compton distorted spectrum characterized by the parameter y, a measure of the discrepancy between the electron and radiation temperatures. In either case the resulting spectrum rises at short wavelengths compared to a Planck function. See, for example, the excellent review by Sunyaev and Zel'dovich.[4]

The measurements presented here place tight constraints on μ and y as well as on other, less orthodox distortion mechanisms,[5] such as the recent decay of low mass primordial particles or substantial thermal scattering at low redshifts.[6]

Previous measurements indicate that the CBR spectrum is roughly consistent with a Planck function at T = 2.75 K over three decades in wavelength, although statistically significant departures have been reported at both long and short wavelengths.[7,8] Experience in this field has taught us to treat such reports with caution and so we have designed a new experiment to be as free from sources of systematic error as possible. Our goal is to check the thermal character of the spectrum over a broad spectral range in a single measurement.

There are several difficulties in making precise checks of the CBR spectrum, beyond the problem of detector sensitivity which has been solved. First, the atmosphere is opaque at wavelengths where the CBR is most intense ($\lambda \simeq 2$ mm) so measurements must be carried out at high altitude. Mountain tops, balloons, rockets and satellites have all been employed to this end. Second, the low temperature of the source forces the use of cryogenics to avoid contaminating emission from the instrument itself. Establishing the zero point of an instrument's response with sufficient precision is a crucial and difficult task. Third, the CBR is of such low intensity that one needs to take care to avoid contamination from much hotter foreground sources such as the earth, the sun or dust in the plane of our galaxy. In this context starlight is of no concern but rejection of radiation from the instrument's sidelobes at a level exceeding 10^{-8} is called for because thermal radiation from the earth is so bright.

THE EXPERIMENT

Based on experience gained in earlier attempts to measure the spectrum of the CBR from high altitude[9–11] a new rocket instrument was developed and tested during the 1980's. The instrument is shown very schematically in Figure 1. Radiation from the sky is directed to one side of a symmetric two beam interferometer by means of a hyperbolic horn telescope. The second side of the interferometer is illuminated by an identical telescope terminated, however, with a black body simulator whose temperature can be controlled. Moving mirrors vary the optical path difference in the interferometer, producing interference patterns at each of two detectors at the output

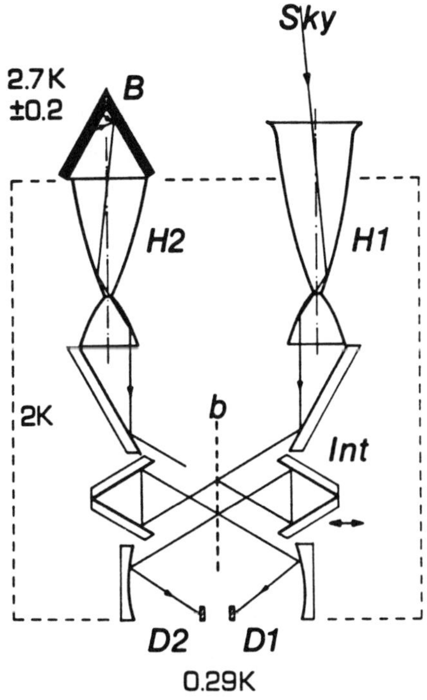

Fig. 1. A diagram showing the principle of the apparatus. H1 and H2 are similar horn type telescopes of 6° field of view. H1 receives radiation from the sky whereas H2 is illuminated by a black-body simulator B. Radiation issuing from the horns enters a two beam interferometer *Int*, with a beam-splitter at *b*, from whence it emerges to be focussed on two bolometric detectors, D1 & D2. As the path difference in the interferometer is changed the signal generated by each detector (interferogram) is proportional to the *difference* in intensity of the sky and the black body. The numbers indicate temperatures of various sections of the spectrometer.

side of the instrument. The amplitude of each interferogram is proportional to the *difference* in intensity between the sky and the black body. If the sky has a Planck spectrum one would obtain null interferograms when the temperature of the internal source matches the sky temperature. Deviations from a null interferogram when the sky and internal source have the same average temperature show immediately that the CBR is distorted.

The absolute sensitivity of the instrument is obtained by periodically changing the temperature by steps of 0.2K through a range of three values near the expected sky temperature. Spectra are obtained from the interferograms by a Fourier analysis. Measuring spectra in flight at different internal radiator temperatures allows calculation of the sky emission spectrum without achieving a null interferogram and without reference to pre- or post-flight calibration data.

During extensive testing of the instrument in the laboratory before flight the aperture of the sky telescope is filled with an external blackbody simulator made of a large hollow cone of Eccosorb.[12] Using the known properties Eccosorb[13] and the geometry of the cone one calculates that the emissivity of this blackbody is $\epsilon \geq 0.999$. This lower limit is calculated disregarding the fact that radiation emitted outside the acceptance angle of the horn telescope is returned to the blackbody, which increases the effective emissivity of the assembly even further. The back surface of the Eccosorb cone is in contact with an independent liquid helium bath which establishes thermal uniformity over the entire blackbody. The temperature of the blackbody can be varied independently of the temperature of the spectrometer by varying the vapour pressure above the liquid helium.

A series of runs was performed in which the external blackbody temperature, inferred from the vapour pressure of the associated helium bath, and the temperature of the internal blackbody radiator, inferred from thermometers on it,[14] were matched to within 1mK at many different temperatures over the range from 1.8K to 4.2K, and interferograms were collected at each temperature. In an ideal instrument these would be null interferograms. In our case they are non-zero but quite small. These offset interferograms are the same in every nulled measurement, independent of source temperature. Since the Planck spectrum changes substantially through this temperature range one concludes that the internal blackbody has an emissivity which also exceeds 0.999 and that the optical efficiency of the instrument is balanced with respect to the two input telescopes to the same level. (In fact one concludes that the product of instrumental imbalance times the internal emissivity, $\alpha(\nu)\ \epsilon(\nu)$ is $1 \pm .001$ at each frequency. Operationally, these two statements are equivalent in terms of determining the sky radiance.) Furthermore, these offset interferograms remained the same in separate experiments even when vibration testing and complete disassembly and reassembly of the instrument occurred between two measurements. We conclude that the shape and amplitude of these offsets depends only on very stable properties of the instrument. The offset measured for each detector is the only part of laboratory calibration data used in analysis of flight data.

The crucial design features of this instrument with respect to minimizing systematic errors are as follows. The instrument is calibrated in flight. It is a differential instrument completely symmetric with respect to the locations of the sky and our calibration source. The calibrator has a very high emissivity and its temperature is well established. Lastly, the instrument operates above the atmosphere with no windows or warm optics. All optical elements are cooled cryogenically.

The instrument was launched from the White Sands Missile Range, New Mexico, on 20 January 1990 at 23:05 MST. Prior to launch the sky telescope port is closed by a vacuum door whose inside surface is 20K. During ascent this door was opened at 150km elevation allowing observation of deep space. The door is reclosed by motor at 100 km altitude on the descent leg, approximately 5 minutes later. The apogee was 250km. In the observing period 166 interferograms were recorded for each detector. During this period the telescope, which has a field of view of 6° located on the payload rotation axis, was pointing in the direction b = 25° l = 184° in galactic coordinates. This region was chosen with reference to sub-millimeter maps of diffuse galactic emission[15] since it is free of strong sources and near to the zenith at the time of flight. To save weight and expense we flew without a guidance system, spinning the payload for stability. The optic axis precessed around a 2° diameter circle during observations.

During the flight the Sun and Moon were both beneath the Earth's horizon, which itself was always more than 95° from the optic axis. Laboratory tests of the telescope beam pattern limit the response at that angle to be less than 10^{-9}.

RESULTS

Interferograms measured late in the flight at three different internal blackbody temperatures are shown in Figure 2. The signals shown are a combination of the interferograms measured by the two detectors. Since one bolometer measures the sky radiance minus the internal blackbody radiance and the second bolometer measures the internal blackbody minus the sky, taking the difference between the two detector voltages enhances the signal and, fortuitously cancels a microphonic signal from our mirror transport mechanism which was damaged during the vibration testing of the payload performed prior to launching. The small offset interferograms obtained from laboratory measurements have been subtracted. In addition to the obvious purpose of removing signals generated within our instrument this subtraction has two desirable effects. It removes the effect of a few mK residual temperature inhomogeneity in the internal blackbody and it refers the calibration of the thermometers on the internal blackbody to the vapour pressure temperature measured on the external calibrator before flight.

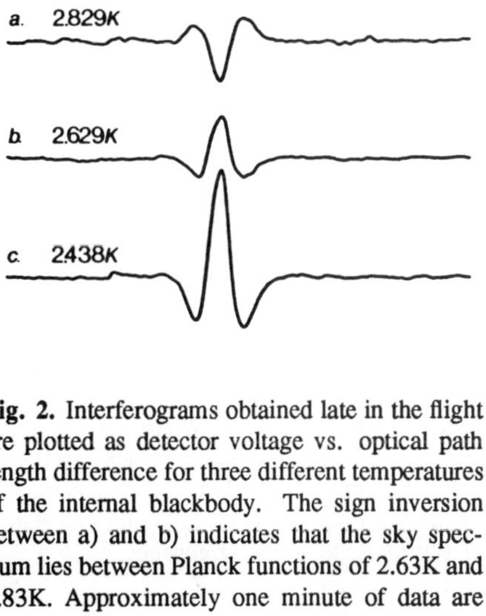

Fig. 2. Interferograms obtained late in the flight are plotted as detector voltage vs. optical path length difference for three different temperatures of the internal blackbody. The sign inversion between a) and b) indicates that the sky spectrum lies between Planck functions of 2.63K and 2.83K. Approximately one minute of data are shown here.

The interferograms are plotted as a voltage against optical path length difference in the interferometer. Each interference pattern shows a peak at zero path length difference, where radiation at all wavelengths interferes constructively, whose amplitude is proportional to the total power in the difference spectrum. It is obvious from interferograms a) and b) that the sky spectrum is very close to a Planck function half way between T = 2.63K and T = 2.83K.

Spectra, obtained from these interferograms by Fourier transform, are shown in Figure 3A. The horizontal scale is the frequency in wave numbers ($\sigma = 1/\lambda$, 10 cm^{-1} corresponds to λ = 1 mm f. ex.) and the vertical scale is in volts per cm^{-1}.

The frequency dependent conversion from detector signal in volts/cm^{-1} to radiance measured in Watts/cm^{-1} cm^2 str depends on detector sensitivity, instrumental Area-solid angle product, filter response and other factors. At any particular frequency this conversion factor can be determined from a plot of response versus radiance of the internal blackbody which is determined from the temperature. As an illustration, the instrumental response at σ = 9.55 cm^{-1} is the slope of the line in Figure 3B where the vertical scale is identical to that for 3A and the horizontal scale is the radiance of the internal blackbody. Notice how linear our response is to radiance in the region of interest.

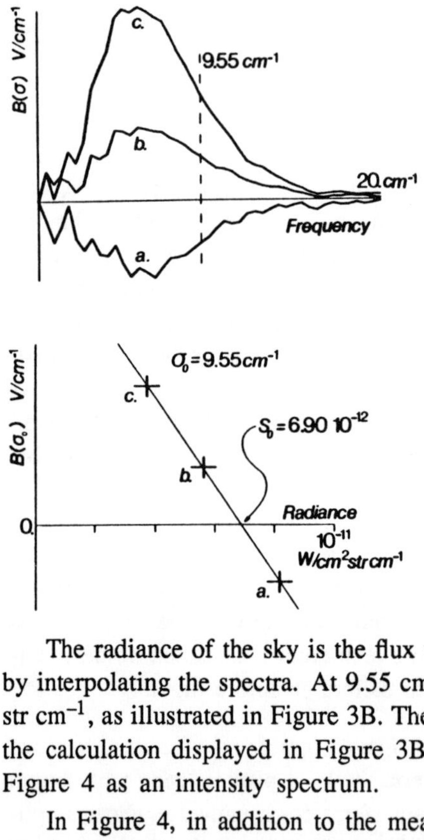

Fig. 3. A. Power spectra obtained from each of the interferograms in Figure 2 by Fourier transform. These are uncalibrated difference spectra comparing the sky to our internal black body. Notice that the sky spectrum (null) lies half way between a) and b) at all frequencies.

B. An illustration of the method for obtaining the instrumental calibration and sky radiance from the spectra in part A at σ = 9.55 cm^{-1}. The values of the power spectra are plotted against internal black body radiance. The slope is the instrumental response, G(σ) and the zero intercept is the sky radiance, S(σ). The particular value shown, 6.9 x 10^{-12} W/cm^2 str cm^{-1}, corresponds to the radiance of a Planck function at 9.55 cm^{-1} when T = 2.735 K. The spectrum in Figure 4 is obtained by repeating the calculation of sky radiance at each frequency.

The radiance of the sky is the flux which would null our instrument, determined by interpolating the spectra. At 9.55 cm^{-1} this radiance is S. = 6.90 $\times$ 10^{-12} W/cm^2 str cm^{-1}, as illustrated in Figure 3B. The spectrum of the sky is obtained by repeating the calculation displayed in Figure 3B at each frequency. The result is shown in Figure 4 as an intensity spectrum.

In Figure 4, in addition to the measured spectrum, a Planck function has been drawn corresponding to a temperature of 2.736 Kelvin. The difference between the data and this Planck function is also graphed. The fit is quite good. In the region $3 \leq \sigma \leq 16$ cm^{-1}, where the signal to noise ratio is high, the rms scatter of the deviation is 9.5 mK, $\frac{1}{3}$% of the mean. With the exception of residual microphonic signals at 3, 16 and 25 cm^{-1} no single point deviates from the Planck function by 1% of peak flux.

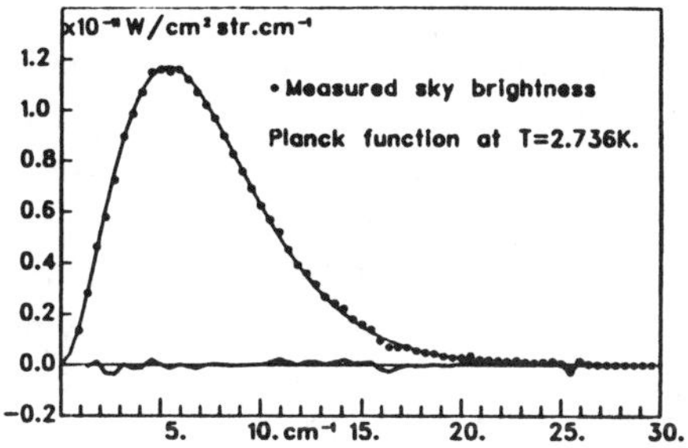

Fig. 4. The intensity of the sky as a function of wavenumber from the data in Figures 2 and 3. The smooth curve is a Planck function which fits the data well. The curve oscillating along the abscissa axis is the difference between the experimental and Planck curves.

The statistical uncertainty of the mean temperature of the measured spectrum is 2 mK, but there are several systematic sources of uncertainty which exceed this: (1) non uniformity of the internal calibrator contributes 5 mk; (2) uncertainties in offset corrections, 3 mK (the uncorrected offsets are of the order of 30 mK); (3) possible non cancellation of microphonics, 2 mK; and (4) uncertainty in instrumental emission due to heating of the sky telescope during lift-off by temporary coupling to the warm (20K) door, 5 mK (this is an upper limit for this contribution obtained by comparing interferograms collected at various telescope temperatures as the horn cooled to its initial temperature during the flight). It is expected that further analysis will reduce the magnitude of uncertainties (1), (3) and (4).

The linear sum of these uncertainties is 17 mK. This very conservative estimate of the uncertainty of the temperature of the CBR is the smallest of any measurement yet reported and provides a reference against which other measurements can be compared in a search for spectral distortions. One must be very careful in handling systematic errors and it would be desirable to repeat this measurement independently to assess their magnitude. In this context it is intriguing that the mean temperature determined by the FIRAS[16] experiment on COBE, albeit with much larger error bars, agrees with our value within 1 mK. However, this agreement at much less than the estimated uncertainties of either experiment must be regarded as a coincidence at the present time.

To aid comparison with other measurements the measured flux is plotted as an effective temperature at each wavelength in Figure 5. This measurement falls slightly below previous single frequency values from balloon borne experiments[17] and some CN measurements[18,19] and particularly below the well publicized Berkeley-Nagoya[8] rocket measurements, which differ from our results by 4, 12 and 17 standard deviations in their respective bands. The latter were interpreted as a large infrared distortion of the CBR; this distortion is not consistent with the present measurement.

The fact that a Planck function fits so closely to the measured spectrum allows one to set close constraints on various distortion parameters. Using the stated temperature errors, and our data only, one finds $y \leq 1 \times 10^{-3}$ at 95% confidence. This places a

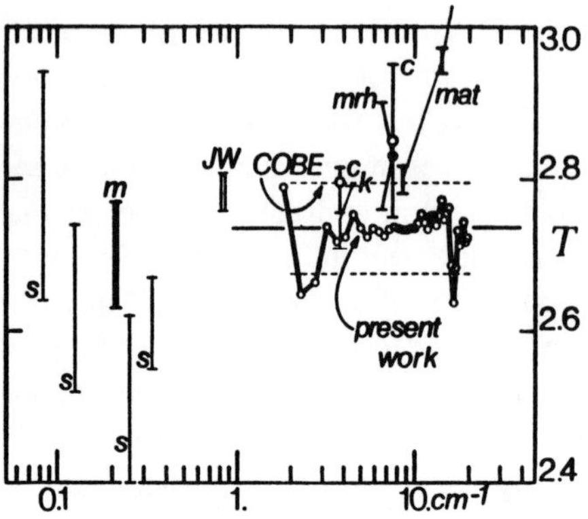

Figure 5. The equivalent black body temperature as a function of frequency. Results inferred from the data in Fig. 4 are shown as circles. The scatter at the band ends was caused by microphonic pickup. Other recent data has been plotted on the same graph: s – Smoot et al.[7]; JW – Johnson and Wilkinson[17]; c – Crane et al.[18]; mrh – Meyer et al.[19]; K – Kaiser and Wright[20]; mat – Matsumoto et al.[8]; m – Mandolesi et al.[21]. The dotted lines are upper and lower limits found by FIRAS on COBE[16].

limit on post recombination energy input to ionized matter by any process although many models of galactic formation can be adjusted to fit this limit. Similarly, a limit of $\mu \leq 8 \times 10^{-3}$ can be set from these results alone assuming μ to be independent of frequency. In addition we can set a limit on the density of meV mass weakly decaying particles with lifetimes $\tau \simeq 10^{10}$ years to be $n < 1/cm^3$. This substantial reduction of this limit provides new information, although particles of this description, which were postulated to explain distortions in the CBR spectrum, are not particularly expected. In addition models which produce a microwave background radiation entirely from thermal emission from dust produced by pregalactic stars[6] are ruled out by these measurements. In a universe with a hot big bang this dust would produce spectral distortions just at the limit of detection of the present results.[6,22]

ACKNOWLEDGMENTS

We thank Vittorio De Cosmo and Wynn Lewis-Bevan for their help.

The instrument was developed under grants from the Natural Sciences and Engineering Research Council. The payload development and launch were coordinated and funded by the Canadian Space Agency. The success of this experiment is a tribute to the high level of expertise in the mechanical and electrical shops of the UBC Physics department where the instrument was built.

REFERENCES

1. R.H. Dicke, P.J.E. Peebles, P.G. Roll and D.T. Wilkinson, Astrophys. J. ***142***, 414–419 (1965).
2. A.Z. Penzias and R.W. Wilson, Astrophys. J. ***142***, 419–421 (1965).
3. H.P. Gush, M. Halpern and E.H. Wishnow, Phys. Rev. Lett. ***65***, 537 (1990).
4. R.A. Sunyaev and Ya. B. Zel'dovich, Ann. Rev. of Astron. and Astrophys. ***18***, 537–560 (1980).

5. L. Danese, C. Burigana, L. Toffolatti,, G. DeZotti and A. Franceschini, "The Microwave Background Radiation, 25 Years Later", Proceedings, L'Aquila, Italy, June 1989.
6. I. Hawkins and E.L. Wright, Astrophys. J. ***324***, 46–59 (1988).
7. G. Smoot in "Particle Astrophysics, Forefront Experimental Issues", edited by E.B. Norman (World Scientific, Singapore, 1989).
8. T. Matsumoto, S. Hayakawa, H. Matsuo, H. Murakami, D. Sato, A.E. Lange and Paul Richards, Astrophys. J. ***329***, 567 (1988).
9. H. Gush, Can. J. Phys. ***52***, 554–561 (1974).
10. H. Gush, Phys. Rev. Lett. ***47***, 745–748 (1981).
11. H. Gush in Gamow Cosmology, edited by F. Melchiorri and R. Ruffini (Italian Physical Society, Bologne, Italy), 260–283 (1986).
12. Manufactured by Emerson & Commings Inc., Gardena, Ca., U.S.A.
13. M. Halpern, H.P. Gush, E. Wishnow and V. De Cosmo, Applied Optics ***25***, 565–570 (1986).
14. V. De Cosmo, H.P. Gush, M. Halpern and A. Leung, Rev. Sci. Inst. ***58***, 441–443 (1987).
15. M. Halpern, R. Benford, S. Meyer, D. Muchlner and R. Weiss, Astrophys. J. ***332***, 596–614 (1988).
16. J.C. Mather, E.S. Cheng, R.E. Eplee, R.B. Isaacman, S.S. Meyer, R.A. Shafer, R. Weiss, E.L. Wright, C.L. Bennett, N.W. Boggess, E. Dwek, S. Gulhis, M.G . Hauser, M. Janssen, T. Kelsall, P.M. Lubin, S.H. Moseley, T.L. Murdock, R.F. Silverberg, G.F. Smoot and D.T. Wilkinson, Astrophys. J. Lett. ***324***, L37–41 (1990).
17. D.G. Johnson and D.T. Wilkinson, Astrophys. J. Lett. ***313***, L1–4 (1987).
18. P. Crane, D.J. Hejyi, M.L. Kutner and N. Mandolesi, Astrophys. J. ***346***, 136–142 (1989).
19. D.M. Meyer, K.C. Roth and I. Hawkins, Astrophys. J. Lett. ***343***, L1–4 (1989).
20. M.E. Kaiser and E.L. Wright, Astrophys. J. Lett. ***356***, L1–4 (1990).
21. N. Mandolesi, P. Calzolari, S. Cortiglioni, G. Morigi, L. Danese and G. de Foitt, Astrophys. J. ***310***, 561–567 (1986).
22. J.R. Bond, B.J. Carr and C.J. Hogan, submitted to Astrophys. J. (1990).

QUESTION (Ofer Lahav): An upper limit of 10^{-3} on the Comptonization parameter y is not a small number in the sense that spectral distortion due to hot I.G.M. could still be hidden in the "perfect" Blackbody spectrum. What are the prospects to get an even lower limit on y?

ANSWER: We are not detector noise limited so if we make progress understanding instrumental emission and acoustic pick up our precision in measuring y will improve. Whether this results in a lower limit or a positive detection one cannot say in advance, of course.

We do not expect to do as well on y as we expect COBE to do. Please see John Mather's answer to the same question.

A DECADE OF LONG-WAVELENGTH CMB MEASUREMENTS

A. Kogut, M. Bensadoun, M. Bersanelli, G. De Amici,
S. Levin, M. Limon, and G.F. Smoot

Lawrence Berkeley Lab and Space Sciences Lab, University of California,
Berkeley, CA 94720

ABSTRACT

We have continued the efforts of an international collaboration begun in 1979 to measure the long-wavelength (Rayleigh-Jeans) spectrum of the cosmic microwave background (CMB). We have repeated measurements at a number of wavelengths over many years, using different equipment and techniques to investigate the possibility of unsuspected systematic errors in the measurements. We conclude that our results show no evidence for significant undiagnosed systematic effects. Our results have improved in precision over ten years and yield significant limits on physical processes in the early universe. The CMB spectrum is consistent with a blackbody over >3 decades of wavelength.

INTRODUCTION

The cosmic microwave background (CMB) carries a record of the interacting radiation and matter fields in the evolving universe. Any transfer of energy between the matter and radiation fields must alter the CMB spectrum from a blackbody distribution. Energy releases at redshifts $z < 5 \times 10^6$ are unable to re-establish a blackbody spectrum; the size and shape of the present distortion depend on the size, redshift, and mechanism of the energy transfer. Possible mechanisms for such energy transfer include dark matter decay, the dissipation of shock and sound waves associated with adiabatic density perturbations, dissipation of gravitational wave energy, phase transitions in the early universe, or energy released through the isotropization of an anisotropic universe.

Within two years of the discovery of the CMB in 1965, a number of long-wavelength measurements confirmed its existence and approximate blackbody spectrum [1]. Nearly all subsequent measurements concentrated on the short-wavelength (Wien) region. Recognizing that the previous measurements of the Rayleigh-Jeans region were not very precise, an international collaboration involving institutions at Berkeley (G.F. Smoot), Bologna (N. Mandolesi), Haverford (R.B. Partridge), Milano (G. Sironi), and Padova (L. Danese and G. De Zotti) conducted a program in the early 1980's to measure the long-wavelength CMB spectrum at wavelengths 12, 6.3, 3.0, 0.9, and 0.33 cm [2]. The Berkeley and Milano groups have continued the effort in recent years, adding measurements at 50, 36, 21, 7.9, and 4.0 cm [3,4,5,6]. When possible, the measurements at each wavelength have been repeated with radical changes in equipment and observing technique, to test for changes in the results arising from unsuspected systematic errors. We review here some results of the long-wavelength measurements and analysis of systematic effects.

EXPERIMENTAL CONCEPTS

We measured the intensity of the CMB at each wavelength using a single-mode microwave radiometer, a receiver whose output was proportional to the power within the receiver passband intercepted by its antenna. Each radiometer was calibrated in units of antenna temperature T_A, proportional to the power P per unit bandwidth B

$$T_A = P/kB \tag{1}$$

where k is Boltzmann's constant. The antenna temperature is related to the thermodynamic temperature T of a blackbody emitter completely filling the antenna aperture by the relation

$$T_A = \frac{xT}{e^x - 1} \tag{2}$$

where x is the dimensionless frequency

$$x = h\nu / kT, \tag{3}$$

h is Planck's constant, and ν is the frequency. Each radiometer measured the temperature of the zenith sky by comparing the output signal as the radiometer alternately viewed the zenith and an absolute cryogenic reference target.

$$T_{A,zenith} - T_{A,load} = G(S_{zenith} - S_{load}) - \Delta T_{Inst} \tag{4}$$

where $T_{A,zenith}$ is the antenna temperature of the zenith, $T_{A,load}$ is the antenna temperature of the absolute reference target, G is the calibration coefficient of the radiometer, $S_{zenith} - S_{load}$ is the measured signal difference between the zenith and the reference target, and ΔT_{Inst} is an instrumental signal generated by possible changes in instrument performance (e.g., changes in internal reflection or absorption coefficients) between observations of the zenith and the load. The antenna temperature of the zenith is the sum of many signals,

$$T_{A,zenith} = T_{A,CMB} + T_{A,Atm} + T_{A,Gal} + T_{A,ground} \tag{5}$$

where the subscripts CMB, Atm, Gal, and Ground refer to the antenna temperatures of the CMB, atmospheric emission, galactic emission, and terrestrial emission in the antenna sidelobes, respectively. Other foreground signals (e.g., from solar or lunar emission or man-made RFI) can also be present. The CMB temperature is thus

$$T_{A,CMB} = G(S_{zenith} - S_{load}) + T_{A,load} - T_{A,Atm} - T_{A,Gal} - T_{A,ground} - \Delta T_{Inst} \tag{6}$$

We neglect atmospheric attenuation in this discussion, although not in the actual analysis. We measured each term in Eq. (6) using the same instrument; where possible, we used more than one technique to measure each signal. We measured $T_{A,Atm}$ by performing tip scans and correlating the signal change with airmass. We estimated $T_{A,Gal}$ by scaling galactic maps made at lower frequencies and cross-checked by measuring the galactic profile predicted by these maps. We measured $T_{A,ground}$ by a beam pattern integration and by using reflectors to block varying amounts of the ground. We measured ΔT_{Inst} by moving the radiometer (as if for a CMB measurement) with targets of various temperatures securely fastened to the antenna aperture.

The experiment was designed to minimize or avoid many potential systematic errors. The cryogenic absolute reference target was oversized to minimize reflections and emission from the walls. We calibrated the radiometers at the antenna aperture, cancelling horn emission and internal reflections to first order in the sky-cold load comparison. We measured foreground signals with the same instrument as the zenith sky measurement, minimizing effects of gain uncertainties. We performed the measurements from remote, dry, high-altitude sites, minimizing the magnitude and variability of atmospheric emission and man-made interference. We measured atmospheric emission at or close to the same time as the zenith sky measurement, minimizing the effects of remaining atmospheric variations.

LIMITS TO SYSTEMATIC EFFECTS

Receiver sensitivity was not a limiting factor in our measurements. The uncertainties in $T_{A,CMB}$ were dominated by limits to systematic effects in Eq. (6). We measured directly all known contributions to the error budget. As a cross-check, and to insure against the possibility of undiagnosed systematic errors affecting the measurements, we repeated each CMB measurement over several years. Each year we changed one or more aspects of the equipment and experimental techniques. The degree to which the calculated values $T_{A,CMB}$ agreed despite radical changes in equipment and technique indicated that systematic errors were well-described by the assigned uncertainties.

a) ABSOLUTE REFERENCE TARGET

Most of the radiometers observed a common reference target (cold load). It consisted of a microwave absorber immersed in liquid helium within a large open-mouth cryostat. An aluminum-coated cylinder ("radiometric wall") surrounded the absorber and prevented radiation from warm parts of the cryostat from reaching the aperture. Two thin polyethylene windows at the aperture prevented condensation of atmospheric gasses within the cold load; a fraction of the helium boiloff gas was heated and circulated between the windows to prevent frost or condensation on the upper window. The antenna temperature of the cold load was given by the boiling temperature of the liquid helium, with minor contributions (<50 mK) from emission and reflection of the windows and radiometric wall.

Since it was a common reference, the cold load was a potential source of error common to all measurements. We measured the emission and reflection properties of all surfaces in the cold load, both individually and as a completed target. Our original cold load was built in 1980 and optimized for operation between 2.5 and 90 GHz [7]. To allow observations at lower frequencies, and to test for unsuspected sources of emission or reflection, we built in 1988 a new cold load optimized for operation between 1 and 10 GHz [8]. The two cold loads are shown in Figure 1. The new cold load resembled the old one, with larger diameter, smoother transitions along the radiometric wall, and a thicker absorber to ensure opacity at long wavelengths. The new cold load used low-pass Teflon-impregnated glass cloth windows to reduce the infrared heat leak to the cryogen, replacing a cumbersome mechanical shutter used in the original cold load. Contributions to $T_{A,load}$ for both cold loads are also shown in Figure 1. Emission and reflection were both small and frequency-dependent.

The longest-wavelength radiometers had horns too large to calibrate at the antenna aperture. The measurements at 0.6 GHz and 0.8 GHz used a coaxial termination immersed in liquid helium [9]. The 2.5 GHz radiometer in 1989 observed both this cold load and the free-space cold load. Similar values for $T_{A,CMB}$ obtained with the free-space cold loads at many frequencies indicated a lack of serious systematic error in our evaluation of $T_{A,load}$; analysis of the 1989 results should provide a further test.

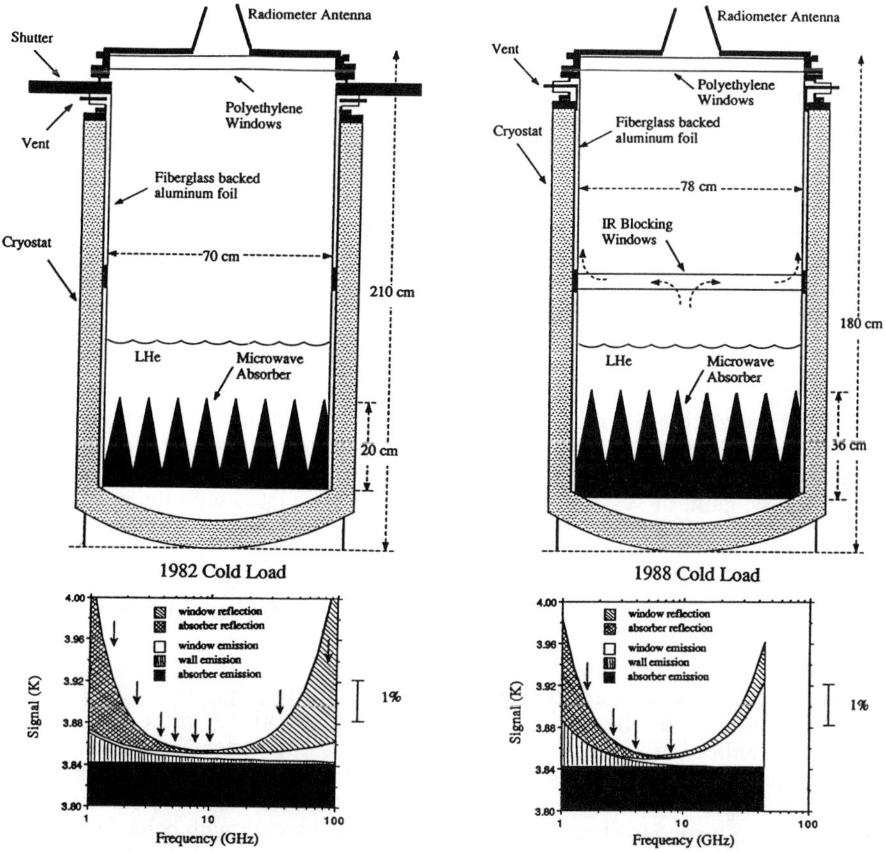

Figure 1. Schematic diagrams and approximate contributions to the antenna temperature of the cold loads used from 1982-1987 and 1988-1989. Arrows indicate frequencies used for CMB measurements. The bars represent 1% of the total cold load emission.

b) SIDELOBE RESPONSE

The Earth is roughly 100 times brighter than the CMB, requiring antennas with good sidelobe rejection. We used corrugated horn antennas, with FWHM typically of order 15 degrees and sidelobe rejection >50 dB more than 60 degrees off beam axis. We reduced $T_{A,ground}$ further by surrounding the antenna with metal reflectors, which diverted to the sky some fraction of the beam which otherwise would have viewed the ground. The reflectors ended in quarter-wave traps or large flared sections to further reduce diffraction over the reflector top edge.

We estimated $T_{A,ground}$ in several ways. By alternately extending and lowering the reflecting shields, we modulated $T_{A,ground}$ and obtained a lower limit [6]. As a cross-check, we numerically integrated the beam pattern (measured with the reflecting shields in place) and horizon profile. We also compared the signal change as the reflecting shields were alternately placed over the antenna and completely removed, changing the effective beam pattern from that of the antenna alone to that of the antenna-reflector combination. The various methods gave good agreement, with typical values for $T_{A,ground} < 30$ mK.

A further test was provided in 1989 by a change of observing site. From 1982 to 1988, we performed the measurements from the White Mountain Research Station at 3800 m elevation in eastern California. In 1989, we performed the measurements from Amundsen-Scott Station at the South Pole. The new site provided an ideal (flat) horizon and lowered the absolute temperature of the ground by ~20%. We were also able to cover large (100 m^2) sections of the surface near some radiometers with reflectors to further reduce ground emission. All tests from the White Mountain site, and preliminary results from the South Pole site, indicated that the ground was not contributing significantly to the sky signal.

c) Instrumental Effects

Both the zenith sky measurement and atmospheric tip scans depended on the radiometer's response remaining independent of orientation. We have tested extensively to limit the effect of small changes in radiometer reflection coefficient, insertion loss, gain, or system temperature that are correlated with the position of the instrument. Instrumental effects may be modelled as

$$\Delta T_{Inst} = \Delta T_{Sys} + \frac{\Delta G}{G}(T_{Sys} + T_{Target}) + \Delta R\,(T_{Broad} - T_{Target}) + \Delta A\,(T_{Inst} - T_{Target}) \qquad (7)$$

where T_{Sys} is the system temperature, T_{Target} is the target temperature, T_{Broad} is the antenna temperature of power broadcast by the radiometer, T_{Inst} is the temperature of emissive elements of the radiometer, and ΔG, ΔR, and ΔA refer to changes in radiometer calibration, reflection coefficient, and absorption coefficient, respectively. We measured ΔT_{Inst} by attaching targets at widely separated temperatures to the antenna aperture and searching for signal changes correlated with position. The observed effects were small and tended to change from year to year, consistent with behaviour expected from small reflections as the instruments were repeated disassembled and rebuilt.

As a precaution against similar instrumental effects in all radiometers, we have used different designs. The measurements at 12, 6.3, 3.0, 0.9, and 0.33 cm wavelength used differential, Dicke-switched radiometers, while the measurements at 21, 7.9, and 4.0 cm used total power instruments. The instruments at 50 and 36 cm did not invert to observe an external cold load. We observed no correlation of ΔT_{Inst} with radiometer design.

d) Atmospheric Scans

All measurements shared the atmosphere as a common foreground. To prevent a single systematic error from contaminating all atmospheric measurements, we have used both reflective mirrors and direct tip scans to allow the beams to view varying airmass. The measurement at 3.0 cm used both methods as a cross-check on their relative accuracy; no significant difference was found [10].

The change of site in 1989 from White Mountain to the South Pole provided an additional cross-check on the atmospheric measurements. With low column density of water vapor (<1 mm precipitable water vapor), atmospheric emission at the Pole was stable over a period of days. By removing atmospheric fluctuations, we were better able to test the contribution of ground emission and instrumental effects on the calculated atmospheric antenna temperature.

e) RFI

Radio-frequency interference (RFI) from communications and video transmission is an increasing problem for cm-wave radiometry. We measured the RF spectrum before and after CMB measurements by replacing the detector diode of the radiometer with a spectrum analyzer. Although this placed a limit on individual sources, it did not rule out intermittent transmissions or multiple sources just below the threshold of detection. The change to a South Pole observing site provided significantly better limits to RFI.

f) GALAXY

The Galaxy was another foreground common to all measurements; it dominated the error budget of the longest-wavelength measurements. We subtracted the galactic signal based on frequency-scaled maps at 408 MHz [11] and compilations of HII regions at 2.7 GHz. Galactic emission off the plane of the Galaxy was dominated by diffuse synchrotron emission; its uncertainty was dominated by uncertainty in the zero-point of the maps and was highly correlated between the radiometers.

As a cross-check on the galactic model, we measured the galactic profile as a function of right ascension (e.g., at constant elevation and atmospheric signal). The results confirmed the model predictions. Further analysis of galactic emission from our longest-wavelength absolute measurements may reduce the galactic uncertainty further. Since the galactic signal was strongly frequency-dependent (<10 mK above 4 cm) it was unable to appreciably affect the shorter-wavelength measurements.

RESULTS

Contributions to Eq. (6) are shown in Table 1 for all years in which we performed measurements. Major changes to potential systematics included:

Year	Change
1983	Rebuilt 90 GHz radiometer to canted-horn design Rebuilt 4.75 GHz radiometer to use reflector reference Rebuilt 2.5 GHz radiometer RF chain
1984	Rebuilt 10 GHz to measure atm with mirror or tipped antenna Rebuilt cold load/antenna interface
1986	Alpe Gera observing site (0.6 GHz)
1987	Rebuilt 1.5 GHz radiometer RF chain Rebuilt 3.8 GHz radiometer RF chain
1988	New cold load Rebuilt 1.5 GHz RF chain
1989	New South Pole observing site Rebuilt 2.5 GHz to use coax cold load, rectangular antenna

Figure 2 shows the CMB thermodynamic temperatures derived from Table 1. The data are consistent with a blackbody with $T_{CMB} = 2.65 \pm 0.04$ K. The fit to a

TABLE 1
CONTRIBUTIONS TO T_{CMB}

Freq [GHz]	λ [cm]	Year	$T_{A,load}$ [mK]	$G(S_{zenith}-S_{load})$ [mK]	ΔT_{Inst} [mK]	$T_{A,Ground}$ [mK]	$T_{A,Atm}$ [mK]	$T_{A,Gal}$ [mK]	T_{CBR} [K]	Site	Ref
0.60	50	1986	5950±300 (A)	-120±120	(B)	5350±700	1170±300	7010±870	3.00±1.20	AG	3
0.82	36	1989	(A, C)	(C)	(C)	(C)	(C)	(C)	(C)	SP	--
1.41	21.3	1986	3800±350	-60±30	0±90	9±8	830±100	800±160	2.11±0.38	WM	4
1.47	20.4	1988	3769± 30	(C)	0±300	30±15	830±100	915±100	(C)	WM	--
		1989	3854± 23	(C)	(C)	(C)	(C)	(C)	(C)	SP	--
2.5	12	1982	4999±212	-1401±100	(D)	12±10	950±50	148±30	2.62±0.25	WM	14
		1983	5518±130	-1557±58	(D)	6±10	950±50	200±30	2.79±0.15	WM	15
		1989	(A, C)	(C)	(C)	(C)	(C)	(C)	(C)	SP	--
3.7	8.1	1986	3735±57	-314±16	-35±20	30±7	870±108	66±30	2.59±0.13	WM	16
3.8	7.9	1987	3742±38	-347±8	-30±20	10±5	898±64	57±10	2.56±0.08	WM	5
		1988	3697±31	-37±1	12±8	7±4	955±55	60±10	2.71±0.07	WM	5
		1989	3762±19	(C)	(C)	6±8	(C)	55±15	(C)	SP	--
4.75	6.3	1982	3716±31	940±80	960±160	30±10	1000±100	40±30	2.73±0.22	WM	17
		1983	3682±10	-45±13	0±20	20±20	997±70	35±25	2.70±0.07	WM	17
7.5	4.0	1988	3623±30	-13±47	-52±34	3±7	1175±78	10±5	2.60±0.07	WM	6
		1989	3685±18	(C)	(C)	22±15	(C)	7±7	(C)	SP	--
9.5	3.2	1982	-	-	-	30±10	1030±70	< 40	-	WM	18
		1983	-	-	-	8±5	1050±70	< 15	-	WM	19

Freq [GHz]	λ [cm]	Year	$T_{A,load}$ [mK]	$G(S_{zenith}-S_{load})$ [mK]	ΔT_{Inst} [mK]	$T_{A,Ground}$ [mK]	$T_{A,Atm}$ [mK]	$T_{A,Gal}$ [mK]	T_{CBR} [K]	Site	Ref
10	3.0	1982	3562±13	61±5	0±40	<3	1190±113	3±3	2.91±0.17	WM	10
		1983	3562±13	43±7	0±30	<3	1200±130	4±2	2.64±0.14	WM	10
		1984	3562±13	-17±8	0±30	<3	1122±120	4±2	2.65±0.21	WM	10
		1986	3568±13	-13±47	6±17	<3	1222±65	8±4	2.56±0.08	WM	10
		1987	3564±13	47±49	0±20	<3	1173±86	6±3	2.62±0.09	WM	20
33	0.9	1982	3068±9	3780±40	-100±50	1±1	4850±140	1±1	2.82±0.21	WM	21
		1983	3068±9	3520±30	-30±30	1±1	4530±90	1±1	2.81±0.14	WM	21
		1984	3068±9	3330±30	-30±30	1±1	4340±90	1±1	2.81±0.14	WM	22
90	0.33	1982	2083±37	11520±60	0±50	1±1	12600±570	<1	$2.58^{+0.68}_{-0.79}$	WM	23
		1983	2083±37	8780±40	10±9	12±15	9870±90	<1	2.57±0.12	WM	23
		1984	2083±37	10220±50	13±5	12±15	11300±130	<1	2.53±0.18	WM	23
		1986	2087±37	14010±50	-8±4	12±15	15020±100	<1	2.68±0.14	WM	24
		1987	2087±37	12760±30	17±8	16±12	13840±35	<1	2.60±0.11	WM	24
		1988	-	-	-43±10	15±15	9360±40	<1	-	WM	--
		1989	-	-	10±21	22±22	8800±20 (E)	<1	-	SP	--

Key to notes:

(A) - coaxial cold load
(B) - the radiometer was not moved during measurements
(C) - analysis of the data not yet completed
(D) - these data are not available separately, but have been included in the corrections to the target temperature
(E) - preliminary result

AG - Alpe Gera, Italy
SP - Amundsen-Scott South Pole Station, Antarctica
WM - White Mountain Research Station, US

blackbody spectrum has χ^2 of 6.3 for 7 degrees of freedom. The data have scatter about the mean consistent with a Gaussian distribution and show no evidence for undiagnosed systematic errors.

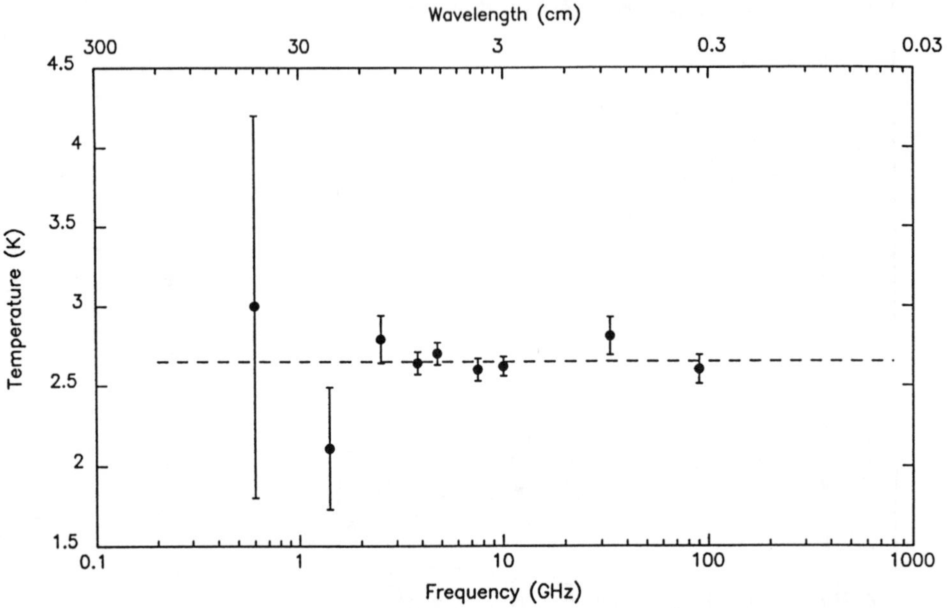

Figure 2. CMB thermodynamic temperature as a function of frequency for our long-wavelength measurements. The dashed line is a fitted blackbody at 2.65 K.

DISCUSSION

Our results achieve maximum sensitivity in the long-wavelength (>3 cm) region of the CMB spectrum. Comparison with measurements at shorter wavelengths allows a wider search for spectral distortions. Figure 3 shows recent precise CMB measurements. The CMB is consistent with a blackbody spectrum across more than 3 decades of wavelength. The weighted average of all CMB measurements yields $T_{CMB} = 2.74 \pm 0.01$ K ($\chi^2 = 37.0$ for 40 DOF).

The long-wavelength results are in marginal agreement with precise results at shorter wavelengths. Although the mean of long-wavelength measurements lies below the short-wavelength mean, a significant fraction of the uncertainties are correlated. No measurement lies more than 2 standard deviations from 2.74 K. Without significantly improved error bars, there is no compelling evidence for distortions in the Rayleigh-Jeans portion of the CMB spectrum.

Figure 3 shows typical distortions produced by a Bose-Einstein spectrum and by free-free emission from an ionized universe. The curves are 1-sigma deviations from a blackbody spectrum. The precise short-wavelength results of Mather et al. [12] and Gush et al. [13] limit the Bose-Einstein chemical potential μ, but do not provide significant limits to free-free emission. The best limits to free-free emission are from our long-wavelength measurements. It is interesting to note that the current data do not rule out the possibility that the universe never passed through a neutral stage.

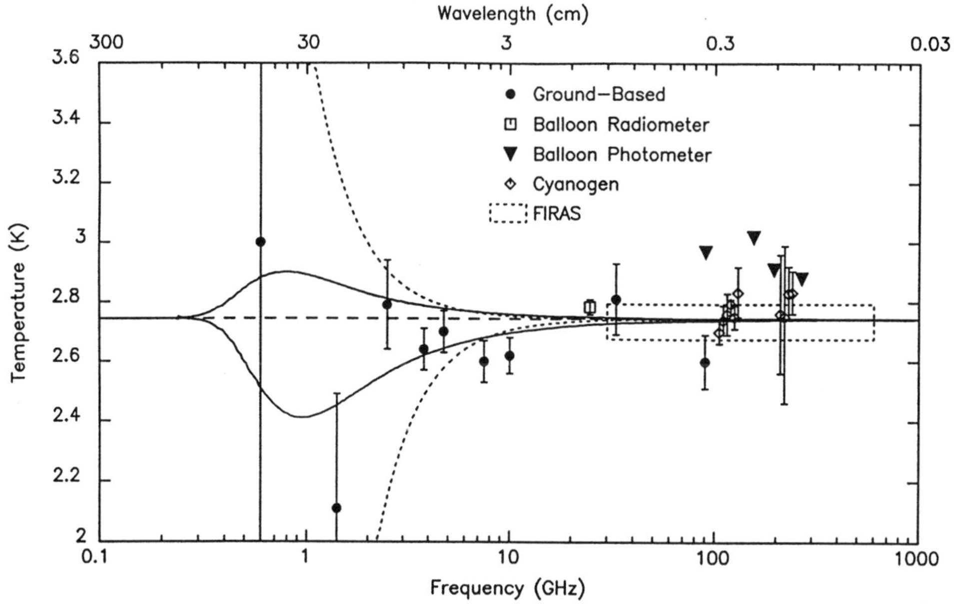

Figure 3. Recent precise CMB measurements. The solid lines are predicted 1σ Bose-Einstein distortions, the dotted lines are free-free emission, and the dashed line a blackbody at 2.74 K.

CONCLUSIONS

We have measured the Rayleigh-Jeans portion of the CMB spectrum over many years, each time changing major portions of the experimental apparatus and techniques to search for unsuspected systematic effects. The resultant data set is consistent with a blackbody CMB spectrum and shows no evidence for significant undiagnosed systematic errors. Comparison with measurements at higher frequencies shows the CMB to be consistent with a blackbody across more than 3 decades of wavelength. Further improvement of our understanding of the CMB spectrum will require roughly a factor of 4 improvement in the uncertainties of measurements in the 3—60 cm range. Achieving this goal will require a new generation of precise measurements, including a better understanding of galactic emission.

We gratefully acknowledge the contributions of our collaborators, including C. Witebsky and S. Friedman, over the course of this experiment. We have been aided by the skilled help of many people, particularly H. Dougherty, J. Aymon, F. Mitschang, and the staffs and crews of the White Mountain Research Station and Amundsen-Scott Station. This work has been supported in part by NSF grants PHY-8015694, AST-8302843, AST-8406187, DPP-8716548, by the Department of Energy under contract DE-AC03-76SFF000098, and by the Italian Consiglio Nazionela delle Ricerche, fellowships 203.2.13 and 203.3.15.

REFERENCES

1. Weiss, R., 1980, *Ann. Rev. Astr. Ap.*, **18**, 489.
2. Smoot, G.F., *et al.*, 1985, *Ap.J. Lett.*, **291**, L23.
3. Sironi, G., *et al.*, 1990, *Ap.J.*, **357**, 301.
4. Levin. S.M., *et al.*, 1988, *Ap.J.*, **334**, 14.
5. De Amici, G., *et al.*, 1990, *Ap.J.*, **359**, 219.
6. Kogut, A., *et al.*, 1990, *Ap.J.*, **355**, 102.
7. Smoot, G.F., *et al.*, 1983, *Phys. Rev. Lett.*, **51**, 1099.
8. Bensadoun, M., *et al.*, 1991, *Rev. Sci. Inst.*, (submitted).
9. Limon, M., Marchioni, C., and Sironi, G., 1989, *J. Phys. E -- Sci. Inst.*, **22**, 963.
10. Kogut, A., *et al.*, 1988, *Ap.J.*, **325**, 1.
11. Haslam, C.G.T, Salter, C.J., Stoffel, H., and Wilson, W.E., 1982, *Astr. Astroph. Suppl. Ser.*, **47**, 1.
12. Mather, J.C., *et al.*, 1990, *Ap.J. Lett.*, **354**, L37.
13. Gush, H., Halpern, M., and Winshowe, A., 1990, *Phys. Rev. Lett.*, **65**, 537.
14. Sironi, G., *et al.*, 1984, *Phys. Rev. D.*, **29**, 2687.
15. Sironi, G., *et al.*, 1986, *Ap.J.*, **311**, 418.
16. De Amici, G., *et al.*, 1988, *Ap.J.*, **329**, 556.
17. Mandolesi, N., *et al.*, 1986, *Ap.J.*, **310**, 561.
18. Partridge, R.B., *et al.*, 1984, *Phys. Rev. D.*, **29**, 2683.
19. Partridge, R.B., *et al.*, 1984, priv comm.
20. Kogut, A., 1989, Ph.D. thesis, University of California at Berkeley.
21. De Amici, G., *et al.*, 1985, *Ap.J.*, **298**, 710.
22. Smoot, G.F., *et al.*, 1985, in *Soc. Ital. Fisica Conf. Proc.*, ed. F.Melchiorri, **1**, 27.
23. Witebsky, C., *et al.*, 1986, *Ap.J.*, **310**, 145.
24. Bersanelli, M., *et al.*, 1989, *Ap.J.*, **339**, 632.

COBE Limits on Explosive Galaxy Formation Scenarios

Janna J. Levin and Katherine Freese
Massachusetts Institute of Technology
77 Massachusetts Avenue, Cambridge, MA 02139
and
David N. Spergel
Princeton University Observatory
Princeton, NJ 08544

ABSTRACT

Hydrodynamic models have been proposed to generate large-scale structure from small-scale initial density perturbations. These models invoke high energy explosions to catalyze the collapse of material on large mass scales. Through inverse Compton scattering, a distribution of explosions in the early universe produces a distortion in the spectrum of the primordial photon background. We calculate the overall spectral distortion and compare our calculations to the recent measurements of the COBE satellite. We find that existing upper limits on the spectral distortion of the background radiation are beginning to challenge this theory of galaxy formation.

INTRODUCTION

Explosions in hydrodynamic models eject a substantial amount of energy. As a result, blastwaves are generated in the ambient medium. We consider the instantaneous injection of energy at initial redshifts, z_i, between 20 and 7. Very early in the blastwave evolution, the volume of the bubble is filled with a tenuous hot gas. It is reasonable to expect that the random thermal motions of the heated gas account for $\approx 2/3$ of the total energy in the blast, E_{in}. As the bubble expands, the interior pressure is converted into work needed to power the expansion. In addition the hot diffuse gas cools through inverse Compton scattering and through thermal bremsstrahlung. Since the explosions occur at large redshifts, bremstrahlung cooling is unimportant and inverse Comptonization of the cosmic background radiation (CBR) dominates the cooling process.

The net consequence of Compton scattering is a redistribution of the photons over frequency. This distorts the shape of the CBR spectrum away from that of a perfect blackbody. The predicted spectral distortion is completely described by the Compton y-parameter (Sunyaev and Zeldovich 1980),

$$y = \int \frac{k_B T_e}{m_e c^2} \sigma_T \, n_e \, d\ell \ ,$$

where $\sigma_T = 6.7 \times 10^{-25}$ cm^2 is the Thomson cross-section, n_e is the electron number density, T_e the electron temperature, m_e the electron mass, and ℓ is the path length. Below we derive a general prescription for the y-distortion due to a distribution of explosions throughout the universe.

PRESCRIPTION

We can average the change in the Compton y-distortion per unit path length over the spherical volume of a single bubble

$$\left\langle \frac{dy}{d\ell} \right\rangle = \frac{\sigma_T}{m_e c^2} \frac{1}{V_b(t)} \int_{V_b(t)} n_e(r,t) k_B T_e(r,t) \mathrm{dV} \ , \tag{1}$$

$V_b(t)$ is the volume of a bubble at cosmic time t. Now consider a universe populated with explosions. We define the filling factor $f(t)$ as the fraction of the volume of the universe occupied by bubbles, $f(t) = \frac{V_b(t) N_b}{V_u(t)}$, where $V_u(t)$ is the volume of the universe at time t and N_b represents the number of bubbles. Define the quantity

$$\frac{d\bar{y}}{d\ell} = \left\langle \frac{dy}{d\ell} \right\rangle \times f(t) \ . \tag{2}$$

$\frac{d\bar{y}}{d\ell}$ is the product of the average contribution to y which the photons acquire as they traverse a bubble, with the fraction of time the photons spend in bubbles. The net distortion of the CBR due to some distribution of explosions in the early universe is then, $y_{tot} = \int_{t_i}^{t_f} \frac{d\bar{y}}{d\ell} \times \frac{d\ell}{dt} dt$, or

$$y_{tot} = \frac{\sigma_T}{m_e c^2} \frac{f_0}{V_{b0}} \int_{t_i}^{t_f} \frac{V_{u0}}{V_u(t)} \left[\int_{V_b(t)} n_e(r,t) k_B T_e(r,t) \mathrm{dV} \right] \frac{d\ell}{dt} dt \ , \tag{3}$$

where t_i is the time at which the blastwave begins to evolve and t_f marks the time at which inverse Compton cooling is no longer efficient. The subscript 0 denotes present day values. Thus an estimate of y_{tot} requires only an estimate of the thermal content, or equivalently of the pressure, in the blast.

The entire problem of determining the pressure, and thus the y-parameter, rests on energy conservation. The conservation of energy equation, averaged over one bubble volume, can be written as

$$\frac{d < P >}{dt} - \gamma \frac{< P >}{< \rho >} \frac{d < \rho >}{dt} = -(\gamma - 1)\Lambda \tag{4}$$

where γ is the ratio of specific heats, $< P >$ is the average pressure, or equivalently $(\gamma - 1)$ times the thermal energy density, and $< \rho >$ is the average density. The cooling function, Λ, is the net rate of energy loss per volume due to Compton scattering. This expression can be integrated in time to obtain

$$< P(t) > \approx P_i \left(\frac{< \rho(t) >}{< \rho_i >} \right)^{-\gamma} \times F(z), \tag{5}$$

where

$$F(z) = \exp \left[-4.9 \times 10^{-21} (\gamma - 1) \mathrm{sec}^{-1} \int_{z_i}^{z} (1 + z')^4 \frac{dt}{dz'} dz' \right] \tag{6}$$

According to equation (3) of the previously outlined prescription, the total y-distortion due to a universe populated with explosions will be

$$y_1 \approx \frac{1}{3}\frac{\sigma_T}{m_e c^2}(\gamma-1)\frac{f_0}{V_{b0}}\int_{z_i}^{z_H}\frac{V_{u0}}{V_u(t)}E_{in}\left[\frac{1+z}{1+z_i}\right]^{3(\gamma-1)} \times F(z)\frac{d\ell}{dz}dz \qquad (7)$$

The limit of integration z_i is the redshift at which the bubble interior is thermalized and the shockwave evolution begins. The upper limit of integration z_H corresponds to one Hubble time. The numerical value of y depends on specific choices for Ω and for γ. In figures 1 and 2, we show the y-parameter as a function of initial redshift z_i generated in an $\Omega_B = 1$ cosmology and an $\Omega_B < 1$ cosmology. (Ω_B is the ratio of the energy density in baryons to the critical density.)

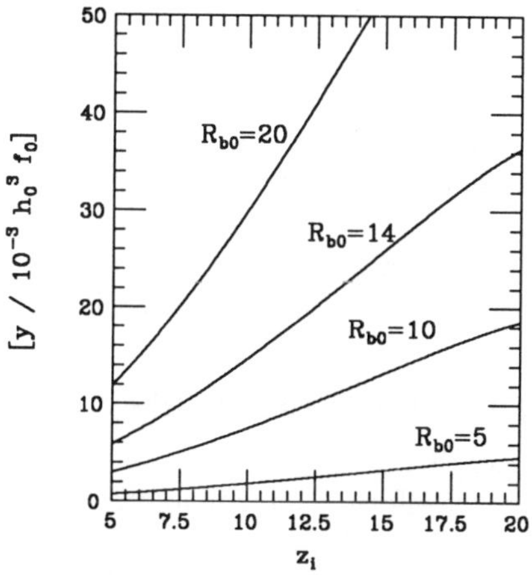

Fig. 1--Ω_B=1
y as a function of z_i and R_{b0}.

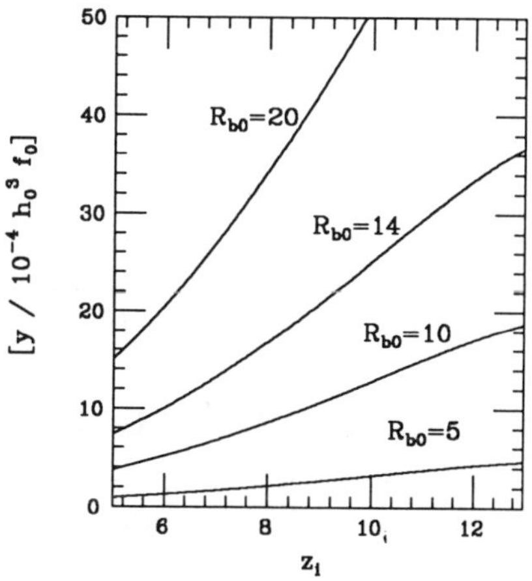

Fig.2--Ω_B=0.1
y as a function of z_i and R_{b0}.

SUMMARY and CONCLUSIONS

Observations of the CBR from COBE reveal a Planckian spectrum at all wavelengths (from 1 cm to 0.5 mm in wavelength)(Mather *et al* 1990). The FIRAS instrument on the satellite actually measures y, the deviation of the spectrum from a Planckian distribution. Currently, Mather *et al* (1990) find $y_{obs} \leq 0.001$.

For brevity, we spare the reader the lengthy results which figures 1 and 2 reflect. Still, to facilitate comparison with the observations, we present an expression which neglects the variation of y with z_i. For a void radius today of R_{b0}, the Compton y-parameter is roughly

$$y \gtrsim 10^{-3}[R_{b0}/(5\ \text{Mpc})]^2 h_0^3 f_0 \Omega_B \ ,$$

where f_0 represents the fraction of the universe occupied by bubbles and h_0 is the Hubble constant in units of 100 km/s/Mpc.

If galaxies are to form within the blastwave shell, then one expects the shell thickness, Δr, to be a few times a galactic radius. The thickness of the shell is a small fraction of the void radius, typically, $\Delta r/R_b \approx 10^{-2}$. Thus an explosive scheme which ultimately generates galaxies must lead to bubbles at least a few Mpc across. As figures 1 and 2 show, if $R_{b0} = 5$ Mpc, then $y \simeq 4 \times 10^{-3}\ \Omega_B$. Thus, the y-distortion due to galactic scale bubbles will escape detection if $\Omega_B \lesssim 0.25$.

Deep sky surveys indicate a bubbly or sponge-like structure on extremely large scales. Galaxies appear clustered on the surfaces of great bubbles which surround voids with average radii on the order of 15 Mpc. The qualitative features of explosive models predict such bubbly structure on very large scales. However, since y grows with R_{b0}^2, hydrodynamic models which attempt to explain structure on very large-scales runs into problems. For instance, if $\Omega_B = 0.1$, $f_0 = h_0 = 1$, and $R_{b0} \sim 15$ Mpc, then $y \sim 4 \times 10^{-3}$. However, a slow Hubble expansion ($h_0 \approx 0.5$) can bring y below the observed value.

In short, we can conclude that if $R_{b0} \lesssim \mathcal{O}(5)$ Mpc and $\Omega_B \lesssim 0.25$, then hydrodynamic models are consistent with the observations. However, explosive scenarios which lead to voids of much larger extent have difficulty satisfying the observational constraints. In particular, one cannot use explosive scenarios to explain the honeycomb structure on large-scales as a universe percolated with bubbles.

ACKNOWLEDGEMENTS and REFERENCES

We want to thank Edmund Bertschinger, Steve Meyer, Jerry Ostriker, Paul Schechter, and Ethan Vishniac for generously offering their insights. K.F. and J.L. acknowledge support from NSF (Presidential Young Investigator Award and grant PHY88-05895), the Sloan Foundation (Grants 26722 and 26623), and NASA (Grant No. NAGW-1320).

Bertschinger, E., 1983, *ApJ*, **268**, 17.
de Lapparent, V., Geller, M.J., and Huchra, J., 1986, *ApJ*, **302**,L1
Hogan, C., 1984,*ApJ*, **284**,L1
Mather, J.C. *et. al.*, 1990, *ApJ*, **354**, L37.
Nath, B.B., and Eichler D., 1989, *ApJ*, **337**, 1.
Ostriker, J.P., and Cowie, L.L., 1981, *ApJ*, **243**, L127.
Ostriker, J.P., and McKee, C.F., 1988, *Rev.Mod.Phs.*, **60**, 1.
Sunyaev, R.A., and Zel'dovich, Y.B., 1980, *Ann. Rev. Astr. Ap.*, **18**, 537.
Vishniac, E., Ostriker, J.P., and Bertschinger, E., 1985, *ApJ*, **291**, 399.

CONSTRAINTS IMPOSED BY THE COBE SPECTRAL MEASUREMENTS

R. A. Daly
Joseph Henry Laboratories
Princeton University
Princeton NJ 08544

ABSTRACT

Upper bounds on the magnitudes of spectral distortions of the microwave background radiation are used to constrain the amplitude of perturbations present prior to recombination. These constraints are further used to constrain the index of the initial power spectrum of density perturbations in the isocurvature and adiabatic models. The current and projected limits imposed by the COBE spectral measurements are discussed.

INTRODUCTION

Processes occurring at redshifts as high as about 6×10^6 can distort the spectrum of the microwave background radiation (MBR). If the spectrum of the background deviates from that of a blackbody at redshifts greater than about 6×10^6 the processes of double compton scattering, bremsstrahlung emission, and compton scattering cause the spectrum to relax to that of a blackbody, with a temperature determined by the total energy of the plasma and radiation field.

One process that will distort the spectrum of the MBR prior to recombination is the damping of pressure waves. These waves will be present in the early universe if fluctuations of the mass density are present, since fluctuations of the mass density imply fluctuations of the energy density of the radiation field. When these perturbations of the pressure of the radiation field enter the horizon, pressure gradients lead to the formation of pressure waves. The dimensionless amplitude of waves with a particular comoving wavelength remains constant as the universe expands, as long as the universe is radiation dominated. The processes of nonlinear dissipation and photon diffusion damp the oscillations of the waves, and may produce a spectral distortion of the MBR depending upon the epoch of the damping. Detailed considerations of the distortions resulting from these damping mechanisms are addressed elsewhere; for a more thorough presentation of the results discussed here see Daly (1991).

The damping of pressure waves during the redshift interval from about 6×10^6 to about 3×10^5 will distort the long wavelength tail of the MBR, producing a chemical potential; here a standard value (Olive *et al.* 1990) of the mean mass density of baryons relative to the critical value of $\Omega_b \simeq 0.017\ h^{-2}$ has been assumed, where Hubble's constant is $H_o = 100\ h$ km s^{-1} Mpc^{-1}. The damping of pressure waves during the redshift interval from about 3×10^5 to about 10^3 will produce a Compton y parameter, which is observable on the Wien side of the spectrum of the MBR.

The magnitude of the distortion resulting from the damping of pressure waves is proportional to the square of the amplitude of the waves that are damped. Waves with a dimensionless amplitude δ_r which are damped during the redshift interval from about 6×10^6 to about 3×10^5 will produce a dimensionless chemical potential of $\mu_* \simeq 0.18\ \delta_r^2$. The current upper bound on μ_* of 10^{-2}

(Danese and De Zotti 1978; Smoot *et al.* 1988; Mather *et al.* 1990; Gush *et al.* 1990) implies that the waves damped over the redshift interval from about 6×10^6 to about 3×10^5 must have dimensionless amplitudes of less than about 0.2 to 0.3. Waves damped over the redshift interval from about 3×10^5 to about 10^3 will produce a Compton y parameter of $y \simeq 0.03\ \delta_r^2$. The current bound on y of 10^{-3} (Mather *et al.* 1990; Gush *et al.* 1990) indicates that $\delta_r \lesssim 0.2$ on scales comparable to the damping scales during this epoch.

On very large scales, such as scales greater than, or on the order of, the horizon size at recombination, the primordial perturbations of the energy density of the radiation field are constrained by MBR anisotropy experiments to be less than about 10^{-3} to 10^{-4}. This means that if the dimensionless amplitude of the primordial perturbations is constant at horizon crossing (or decreasing with increasing comoving wavenumber k) then the damping of waves will produce distortions on the order of (or less than) about 10^{-7}, which is much smaller than one could hope to observe. Hence, it is assumed that the amplitudes of the perturbations of the energy density of the MBR at horizon crossing are increasing with increasing comoving wavenumber k. In this case, a chemical potential will be produced by the damping of waves at a redshift of about 6×10^6, and a Compton y distortion will be produced by the damping of waves at a redshift of about 3×10^5. So, the observationally determined bounds on (or a detection of) the magnitude of μ_* and y translate into bounds on the dimensionless amplitude of perturbations of the energy density of the radiation field with a length scale on the order of the damping scale at each of these respective redshifts.

The damping scale at each of these redshifts depends upon the damping mechanism. Two damping mechanisms are relevant: nonlinear dissipation (weak shock formation) and photon diffusion. Conveniently, the magnitude of the distortion indicates the operative damping mechanism.

Values of the chemical potential greater than about 5×10^{-4} will be produced by nonlinear dissipation, while values less than this will be produced by photon diffusion. Therefore, unless the dimensionless chemical potential is observationally constrained to be less than about 5×10^{-4}, which seems unlikely, the relevant damping mechanism will be nonlinear dissipation. In this case the damping length-scale is proportional to μ_*, and the corresponding mass-scale is proportional to the third power of μ_*. The current upper bound on a chemical potential of 10^{-2} constrains the amplitude of perturbations of the energy density of the radiation field with a mass scale of about $3 \times 10^4\ \Omega_t\ M_\odot$, where Ω_t is the ratio of the mean mass density in the universe at the current epoch to the critical value; here and below a value of $h = 0.75$ and $\Omega_b \simeq 0.03$ are assumed. Note that, for $h = 0.75$, this is the maximum value of Ω_b allowed observationally if there are three effectively massless neutrino species present with equilibrium abundance during primordial nucleosynthesis, as discussed in detail by Daly (1991). As the value of the dimensionless chemical potential decreases, the length scale of the perturbations whose damping could produce the distortion decreases; when this value decreases through 5×10^{-4} the corresponding mass scale remains at a fixed value since the damping scale set by photon diffusion is independent of the value of the chemical potential. This mass scale is about $2.5\ \Omega_t\ M_\odot$. Hence, a chemical potential will be produced by the damping of pressure waves with mass scales between about $3 \times 10^4\ M_\odot$ and about $0.1\ M_\odot$.

Values of the Compton y parameter greater than about 4×10^{-4} will result from nonlinear dissipation, while values less than this will be produced by photon diffusion. The current upper bound on y of 10^{-3} indicates a mass scale for the damping of about $4 \times 10^7\ \Omega_t\ M_\odot$. As the value of the Compton y parameter decreases through about 4×10^{-4} the relevant damping mechanism becomes photon diffusion and the damping scale is then independent of the value of the Compton y parameter; this mass scale is about $2 \times 10^6\ \Omega_t\ M_\odot$. Therefore, a Compton y parameter will be produced by the damping of pressure waves with a mass scale between about $4 \times 10^7\ M_\odot$ and about $10^5\ M_\odot$.

CONSTRAINTS ON THE INITIAL POWER SPECTRUM

As described above, an observational bound on μ_* or y constrains the amplitude of perturbations of the energy density of the radiation field δ_r with a wavelength on the order of the damping scale, λ_d. The amplitude $\delta_r(\lambda_d)$ is related to the amplitude of the perturbations of the mass density on the same scale $\delta_m(\lambda_d)$, in a manner which depends on whether the perturbations are of the isocurvature or adiabatic type. Thus, the upper bound on the magnitude of a spectral distortion implies an upper bound on the amplitude $\delta_m(\lambda_d)$. This bound on $\delta_m(\lambda_d)$ is combined with an estimate of the amplitude of the perturbations of the mass density on some other scale λ_m, $\delta_m(\lambda_m)$, which indicates an upper bound to the index of the initial power spectrum of the mass distribution.

The estimate of $\delta_m(\lambda_m)$ depends upon the cosmological model, since it is obtained by scaling the observed variance of galaxy counts at the current epoch back to early epochs; the observation that $\delta_m(8\ h^{-1}\ \mathrm{Mpc}) \simeq 1$ at the current epoch is used. Thus, the initial value $\delta_{mi}(8\ h^{-1}\ \mathrm{Mpc})$ is approximated by $\delta_{mi}(8\ h^{-1}\ \mathrm{Mpc}) \simeq (1 + z_f)/(1 + z_i)$ where z_i is the redshift at which the perturbations begin to grow and z_f is the redshift at which growth ceases. For cosmological models with zero cosmological constant, $(1 + z_f) \simeq \Omega_t^{-1}$. However, in cosmological models with low Ω_t and a cosmological constant $\Lambda/(3\ H_o^2) \equiv \Lambda' = 1. - \Omega_t$, the growth of structure proceeds almost until the current epoch (Peebles 1984). Hence, in a cosmological model with $\Lambda' = 1. - \Omega_t$, δ_{mi} is decreased by about Ω_t from the value inferred assuming that $\Lambda' = 0$ (when Ω_t is low).

In the isocurvature model the baryons are strongly clumped, and photon production is likely to be significant. Therefore, a chemical potential is unlikely to be produced, while a Compton distortion is unlikely to be affected by photon production (see Daly 1991). In an open universe with only baryons, no cosmological constant, and a value of $h = 0.75$, which implies $\Omega_b \simeq 0.03$, an upper bound on y of 10^{-3} implies an upper bound on n, the index of the initial power spectrum of density perturbations, of $n \lesssim 0.9$. An upper bound on y of 10^{-4} implies an upper bound on n of $n \lesssim 0.25$. The choice $n = 0$ predicts a value of y of $y \simeq 2 \times 10^{-5}$. Note that this assumes that the cosmological constant is zero.

If the cosmological constant is nonzero and has a value at the current epoch of $\Lambda' = 1. - \Omega_b$, then the initial amplitude of the perturbations of the mass density on the scale of $8\ h^{-1}$ Mpc is decreased by about Ω_b from the $\Lambda = 0$ value. Hence, for a given value of n, the predicted value of y decreases by approximately Ω_b^2.

In the adiabatic model the damping of the pressure waves also erases the perturbations of the baryons. Hence, if galaxies are to form by the gravitational growth of initial density perturbations weakly interacting dark matter must be present in the universe, since this will not be affected by the damping of the pressure waves. Observations indicate that Ω_t is low (Peebles 1986). For $\Omega_t = 0.15$, and $h = 0.75$, an upper bound of $y \lesssim 10^{-3}$ implies that $n \lesssim 2.5$, while an upper bound on y of $y \lesssim 10^{-4}$ implies that $n \lesssim 1.9$. A scale invariant spectrum, $n = 1$, predicts $y \simeq 2 \times 10^{-7}$.

Photon production is expected to be insignificant in the adiabatic model, so the constraints from the chemical potential apply. For $\Omega_t = 0.15$ and $h = 0.75$ an upper bound on μ_* of $\mu_* \lesssim 10^{-2}$ implies $n \lesssim 2.1$, and a bound of $\mu_* \lesssim 10^{-3}$ implies $n \lesssim 1.7$. A scale invariant spectrum, $n = 1$, predicts $\mu_* \simeq 10^{-6}$. As in the case for isocurvature perturbations, a value of Λ' of $\Lambda' = 1. - \Omega_t$ would decrease the predicted value of either y or μ_* by the factor Ω_t^2 at a fixed value of n.

The constraints mentioned above are obtained assuming that the galaxy distribution traces the mass distribution in a one to one fashion, that is, that the galaxy distribution is unbiased. If the galaxies are more strongly clustered than the mass, then these constraints are relaxed. For typical values of the bias parameter of about two, the predicted value of n is increased by about 0.1 to 0.2 (see Daly 1991).

The constraints described here also assume that the initial density perturbations are described by a power spectrum which is a power law from the relatively small mass scales on which the damping occurs to the large mass scales of about $10^{15}\ \Omega_t\ M_\odot$ at which the normalization $\delta_m(\lambda_m)$ is determined. The constraints from the Compton y parameter are more reliable than those resulting from μ_* because the mass scales whose damping produce a Compton distortion are much larger than those which produce a chemical potential.

This work was supported in part by the National Science Foundation.

REFERENCES

Daly, R. A. 1991, *Ap. J.*, **371**, in press.
Danese, L., and De Zotti, G. 1978, *Astron. Astrophys.*, **68**, 157.
Gush, H. P., Halpern, M., Wishnow, E. H. 1990, *Phys. Rev. Lett.*, **65**, 537.
Mather, J. C. *et al.* 1990, *Ap. J. (Letters)*, **354**, L37.
Olive, K. A., Schramm, D. N., Steigman, G., and Walker, T. 1990, *Phys. Lett. B*, **236**, 454.
Peebles, P. J. E. 1984, *Ap. J.*, **284**, 439.
Peebles, P. J. E. 1986, *Nature*, **321**, 27.
Smoot, G., Levin, S. M., Witebsky, C., De Amici, G., and Rephaeli, Y. 1988, *Ap. J.*, **331**, 653.

EARLY FIRAS RESULTS AND SUPERCONDUCTING COSMIC STRINGS

M.SIGNORE & N.SANCHEZ

Ecole Normale Supérieure, 24 rue Lhomond, 75005 PARIS, FRANCE
Observatoire de Meudon, 5 pl. Jules Jansen, 92190 MEUDON, FRANCE

ABSTRACT

The preliminary results from FIRAS/COBE (Far Infrared Absolute Spectrometer/Cosmic Background Explorer) and the available anisotropy in the intensity of the Cosmic Microwave Background (CMB)for an angular separation of 1° place new constraints on Superconducting Cosmic Strings (SCS)and their cosmological evolution independently of numerical simulations. The early FIRAS results and the latest PSR 1937 + 21 timing measurements converge to the same upper limits on $G\mu$. The limits continue to descend but the breaking point is not yet reached. We wait for the final FIRAS results.

SCS PARAMETERS

A local Cosmic String (CS) is characterized by the dimensionless parameter (Ref.1):

$G\mu = \left(\frac{\eta}{m_{planck}}\right)^2$, where μ is the tension, η the energy scale of the symetry breaking, $m_{planck} \sim 10^{19}$GeV. For a Grand Unification Theory (GUT) symmetry breaking scale : $\eta \sim 10^{16}$GeV, $\mu \sim 10^{22}$ g/cm, $G\mu \sim 10^{-6}$; for an electroweak scale $\eta \sim 10^2$ GeV, $G\mu \sim 10^{-34}$. During the expansion of the Universe : loops form by self intersection of strings; then closed loops oscillate and radiate gravitational waves until they decay (Ref.2); if a loop is formed at t , its initial length is $L_0 \sim \alpha t$,

its loop formation rate is given by : $\frac{dn}{dt} \sim \beta\, t^{-4}$

and the gravitational radiation rate is such that : $\frac{dEg}{dt} \sim \gamma G \mu^2$, ($\gamma \sim 100$, typically)

CS can carry very large electrical currents and behave like superconducting wires: they are SCS (Ref.3).Oscillating SCS loops emit electromagnetic radiation(Ref.4.5).The electromagnetic radiation rate of SCS is : $\frac{dEem}{dt} \sim f\, \frac{dEg}{dt}$, f may approximately be kept constant during the expansion of the Universe.

GRAVITATIONAL BACKGROUND CREATED BY SCS

Gravitational waves emitted by various loops at different epochs add up into a stochastic gravitational background (Ref.2). If one defines the energy density per logarithmic frequency interval : $\Omega g\,(\omega) = \frac{\omega}{\rho_{crit}} \frac{d\rho g}{d\omega}$

where $\rho_{crit} = 4\,10^{-29}$ g/cm^3 , we can calculate the "gravitational spectrum" created by SCS. There are three relevant frequency regions (Ref.6.7.8):

In region I , loops form, radiate and decay during the radiative era ($t < t_{eq} \sim 10^{11}$s)

$$\Omega_{gI} \sim 1.8\; 10^{-3}\, \alpha^{3/2}\, \beta\, \gamma^{1/2}\, (G\mu)^{1/2}\, (1+f)^{-3/2}\;,\quad \omega >> \omega_1 \qquad (1)$$

In region II, loops form, radiate during the radiative era, then radiate and decay during the matter dominated era ($t > t_{eq}$)

$$\Omega_{gII} \sim 1.1\;\; 10^{-9/2}\, \alpha^{3/2}\, \beta\, \gamma^{-1}\, \omega^{-1/2}\, (1+f)^{-3/2}\,,\quad \omega_1 > \omega > \omega_2 \qquad (2)$$

In region III, loops form, radiate and decay during the matter dominates epoch $t > t_{eq} \sim 10^{11}$s

$$\Omega_{gIII} \sim 36\pi\;\omega t_o\;(\alpha\beta\gamma)\,(G\mu)^2\;(1+f)^{-3/2}\;,\quad \omega < \omega_2 \qquad (3)$$

with $\omega_1 \sim (\gamma G\mu)^{-1} (1+f)^{-1} \ 10^{-15}$ Hz; $\omega_2 \sim (\gamma G\mu)^{-1} (1+f)^{-1} \ 10^{-17}$ Hz
Note that this approximative spectrum is valid as long as $\alpha >> \gamma G\mu$ ie for $\alpha >> 10^{-4}$.

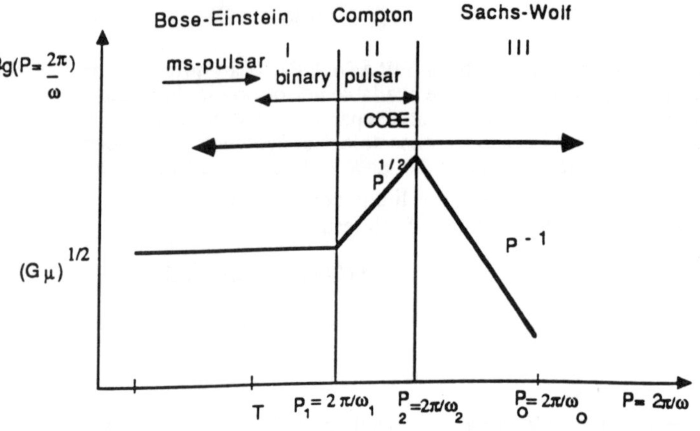

Fig.1 Shematic logarithmic "gravitational spectrum" Ωg (P) as a function of the present period $P = \frac{2\pi}{\omega}$

ON CMB DISTORTIONS & ANISOTROPIES CREATED BY SCS

Electromagnetic waves created by SCS cannot propagate through the cosmic plasma. They can only heat the Universe back to very early epochs. SCS induce in the CBR spectrum, spectral distortions : Bose Einstein and Compton distortions due to their electromagnetic radiation (Ref.4.5.8). Moreover CS and SCS induce angular temperature anisotropies-via Sachs-Wolfe-effect - due to their gravitational radiation (Ref.8).
In the three following sections, we assume no dissipation and an homogeneous heating. We express the distortions and anisotropies created by SCS from $t \sim 10^{-32}$s (SCS formation epoch) till now. We confront them to the latest observations.

CMB ANISOTROPIES DUE TO CS AND SCS

The stochastic and anisotropic gravitational background due to SCS (or CS) induces a photon frequency change equivalent to a temperature anisotropy $\Delta T/T$ characterized by the anisotropy angular distribution function: $C(\phi) = < \frac{\Delta T}{T}(\psi,t) \ , \frac{\Delta T}{T}(\psi',t') >$ where ψ and ψ' are the angular coordinates of the observed fields and $\phi = |\psi - \psi'|$ is the angular separation.
It is related to the observational limits in $\frac{\Delta T}{T}$ through $2 | C(o) - C(\phi) | < [\frac{\Delta T}{T}(\phi)]^2$.
One can show that
$2 | C(o) - C(\phi) | \sim 1.9 \ 10^{-8} \ z^2_s \ \Omega_{gIII}$
Where $\phi \sim 1°$, z_s is the redshift of the last scatterring surface and $\Omega gIII$ is the gravitational density of the region III, eq.(3). If one considers the observational limit (Ref.9)$\Delta T/T$ for $\phi \sim 1°$: $\Delta T/T < 6 \ 10^{-5}$. An upper limit on $G\mu$ in terms of this observational limit is given by (Ref.8)

$$G\mu \leq 2.4\ 10^{-5}\ (1+f)^{3/4}\ (\alpha\beta\gamma)^{-1/2}\ \left(\frac{\Delta T/T}{6\ 10^{-5}}\right) \qquad (4)$$

BOSE-EINSTEIN DISTORTION DUE TO SCS

Heating of the cosmic plasma by SCS during the epoch : $z_{BE} \sim 10^{4.7} < z < z_{ph} \sim 10^{6.4}$ results in a Bose-Einstein distortion. In the Rayleigh-Jeans side of the spectrum , the signature of the distortion is $\frac{\Delta T}{T} = \frac{\mu_o}{r}$ where $r = \frac{h\nu}{kTe}$ and μ_o is the chemical potential which is related to the amount of the energy released between z_p and z_{BE} by SCS through (Ref.8) :

$$\mu_{oISCS} \sim 0.2\, f\,(1+f)^{-3/2}\,\alpha^{3/2}\,\beta\,\gamma^{-1/2}\left(\frac{G\mu}{10^{-6}}\right)^{1/2} \qquad (5)$$

COMPTON DISTORTIONS DUE TO SCS

The heating of the cosmic plasma by SCS during the epoch : $z< z_{BE} \sim 10^{4-7}$ results in Compton distortions.The characteristic signatures of the distortions are $\Delta T/T=2y$ in the Rayleigh-Jeans side and $\Delta T/T=ry$ in the Wien side of the CMB spectrum. y is the comptonization parameter which is related to the amount of the energy released by SCS - for $z < z_{BE}$- through (Ref.8):

$$y_{IISCS} \sim 0.4\ f\,(1+f)^{-3/2}\,\alpha^{3/2}\,\beta\,\gamma^{1/2}\left(\frac{G\mu}{10^{-6}}\right)^{1/2} \qquad (6)$$

One must note that there is a simple relation between the Bose-Einstein distorsion generated in region I and the Compton distortion generated in region II.Eqs.(5 and (6) (Ref.8):

$$y_{IISCS} \sim 2\mu_{oISCS} \qquad (7)$$

IMPLIFICATIONS FROM PRELIMINARY FIRAS RESULTS

Very recently J.Mather et al.(Ref.10) reporting on the very first preliminary results FIRAS results have claimed a pure black body at 2.735 ± 0.06 K with 1% of the peak intensity over the range 1cm down to 500 μm.

Using their conservative 1% error bands, Mather et al. set a 3σ - upper limit on the Comptonization parameter

$y< 10^{-3}$ for $T \sim 2.735 \pm 0.06K$

For a fit to a pure Bose-Einstein spectrum with a chemical potential μ_o independent of frequency they also gave the 3σ- limit : $\mu_o< 10^{-2}$

From the conservative upper limit on the y parameter, and our relation (Ref.7) one can infer

$y_{ISCS} < 10^{-3}$ $\mu_{OIISCS} \leq 0.5\ 10^{-3}$ (Ref.11.12)

An upper limit on G- in terms of the COBE upper limit in y_{II} - is given by

$$G\mu \leq 6.2\ 10^{-12}\ (1+f)^3\ f^{-2}\ (\alpha^{3/2}\beta)^{-2}\ \gamma\left(\frac{y_{II}}{10^{-3}}\right)^2 \qquad (8)$$

Moreover from COBE results and the $\Delta T/T$ (ϕ ~1°) limit,we find for the SCS evolution parameters (Ref.11.12):

$$\alpha^5\left(\frac{\beta}{\gamma}\right)^3 \sim 6.8\ 10^{-14}(1+f)^{9/2}\ f^{-4}\left(\frac{y_{II}}{10^{-3}}\right)^4\left(\frac{\Delta T/T}{6\ 10^{-5}}\right)^{-2} \qquad (9)$$

LATEST RESULTS FROM MS PULSAR TIMINGS

From the latest PSR 1937 + 21 data, one has an upper limit on Ω_{gI} (Ref.13):

$$\Omega_{gI} \leq h^{-2}\, 10^{-7} \text{ for } \omega/2\pi \geq 0.14 \text{ cy / yr} \quad , h= H/100 \text{ Km s}^{-1} \text{ M pc}^{-1} = 0.7$$

the string parameter $G\mu$ in terms of present observational limit on Ω_{gI} is such that :

$$G\mu \leq 1.2 \;\; 10^{-8}\, (1+f)^3\, (\alpha^{3/2}\beta)^{-2}\, \gamma \left(\frac{\Omega_{gI}}{2 \;\; 10^{-7}}\right)^2 \tag{10}$$

while we find for the CS & SCS evolution parameters

$$\alpha^5 \left(\frac{\beta}{\gamma}\right)^3 \sim 0.25 \;\; 10^{-6}\, (1+f)^{9/2} \left(\frac{\Omega_{gI}}{2 \;\; 10^{-7}}\right)^4 \left(\frac{\Delta T/T}{6 \;\; 10^{-5}}\right)^{-2} \tag{11}$$

CONCLUSIONS

- The present orders of magnitude obtained from the present observational limits on $\Delta T/T$ (1°), y, $\Omega g(\omega)$ are :

$$f \sim 2.3 \;\; 10^{-2}, \quad \alpha^5\, (\beta/\gamma)^3 \sim 0.25 \;\; 10^{-6} \;;\; y_{SCS} \leq 10^{-3} \;;\; \mu_{SCSo} \leq 0.5 \;\; 10^{-3} \tag{12}$$

- Preliminary FIRAS results and PSR 1937+21 timing data converge to the same upper limit on $G\mu$ - if $\alpha \gg 10^{-4}$:

$$G\mu \leq 1.2 \;\; 10^{-8} (\alpha^{3/2}\beta)^{-2}\gamma \tag{13}$$

that is to say :

. for generic values (Ref.1.2) ($\alpha \sim 1, \beta \sim 1, \gamma \sim 100$) : $G\mu \leq 1.2 \;\; 10^{-6}$

. for the latest simulations(Ref.14) ($\alpha \sim 2.8\; 10^{-2}, \beta \sim 850$): $G\mu \leq 0.75 \;\; 10^{-7}$

. for "kinky" strings(Ref.15) ($\alpha \sim 10^{-2}$, $\beta \sim 750$) : $G\mu \leq 0.22 \;\; 10^{-7}$

- These limits continue to descend but the breaking point is not yet reached : ie present orders of magnitude of SCS paramaters are such that :

. SCS are GUT strings : $G\mu \sim 10^{-6} - 10^{-8}$,

· SCS may lead to the large scale structures of the Universe (voids and clusters)via explosion scenarios (Ref.4.5),(work in preparation)

. Springs-or loose strings- cannot develop within the "thin" string model and the small current limit,(work in preparation).

- The present "absence of distortion" lies within 1% of the peak intensity , but the FIRAS sensitivity for each spectral element is expected to be 1/1000 of the peak of the 2.735K spectrum. We wait for the final FIRAS results.

REFERENCES

1 .Vilenkin A.1985, Phys.Rep.121, 263
2 . Vachaspati T., Vilenkin A. 1985, Phys.Rev. D 31, 3052
3 . Witten E. 1985, Nucl.Phys.B 249,557
4 . Ostriker J.P., Thompson C., Witten E. 1986 Phys.. lett.B 180, 231
5 . Ostriker J.P. and Thomson C. 1987, Ap.J. 323, L97
6 . Accetta F.S. and Krauss L.M. 1989, Phys. Lett.B 233,93
7 . Sanchez N. and Signore M. 1988, Phys. lett B 214 ,14
. Signore M. and Sanchez N. 1988, Mod. Phys. lett. A4, 799
8 . Sanchez N. and Signore M. 1989, Phys. Lett B 219, 413
9 . Wilkinson D.T. 1986, Phil.Trans. R.S. London A 320, 595
. Partridge R.B., 1988, Rep. Prog. Phys. 51, 647
10. Mather J. et al. 1990, Ap.J. 354, L37
11. Signore M. and Sanchez N. 1990 Mode Phys. A. in press
12. Sanchez N. and Signore M. 1990, Phys. Lett B 241, 332
13. Stinebring D.R. et al. 1990, Phys. Rev. Lett, 65, 285
14. Bennett D. and Bouchet F.R. Princeton University Pre Print PUPT 89-1126
15. Allen B. and Cadwell R.R. 1990, Submitted to Phys. Rev.D

POST-COBE MEASUREMENTS OF THE COSMIC MICROWAVE BACKGROUND SPECTRUM

Suzanne Staggs and David Wilkinson
Physics Department, P.O. Box 708
Princeton University
Princeton, NJ 08544-0708

ABSTRACT

Two recent measurements of the Cosmic Microwave Radiation (CMR) find a spectrum like that of a blackbody with 1% accuracy from $\lambda = 1$ cm to 0.5 mm. Done in space with cold instruments, these experiments closely approximate the ideal conditions of no interfering radiation and thermal equilibrium between instrument and radiation. Future direct measurements of the CMR spectrum will concentrate on longer wavelengths, $\lambda > 1$ cm. We expect that the field will develop along two lines: balloon-based experiments for $\lambda < 10$ cm to avoid atmospheric emission; and ground-based experiments for $\lambda > 20$ cm where Galactic emission becomes the dominant source of error. Experiments of both types have been done but by no means perfected. Improvements will move the measurements closer to the ideal conditions nearly achieved by the space experiments. This paper reviews experiments for $\lambda > 1$ cm and points out some ways that those experiments can be improved to achieve higher accuracy.

INTRODUCTION

In a fitting celebration of the 25th anniversary of the discovery of the Cosmic Microwave Radiation, two measurements of the spectrum reported in January, 1990 have substantially changed the field. Data from the Cosmic Background Explorer (COBE) satellite[1)] and the University of British Columbia rocket[2)] experiments agree with the hot Big Bang model prediction of a blackbody spectrum. Measured sky brightness fits the spectrum of a 2.735 K blackbody to $\pm$ 1% from $\lambda = 1$ cm to 0.5 mm. The goal of verifying the Big Bang model has essentially been achieved. Indeed, there is little need or justification for further work in this wavelength range until new technical developments make even greater accuracy possible. For example, someday one might hope to check the standard model by looking for spectral lines[3)] expected from the decoupling process at z = 1000. This is hard since the lines are weak, $\Delta I/I \sim 10^{-5}$, and wide, $\Delta\lambda/\lambda \sim 10\%$. Of course, if the COBE team should find a distortion, as analysis procedures are improved and more data are used, new experiments in space will be important.

The new results and a sample of other measurements of the CMR temperature are shown in Figure 1. At wavelengths shorter than 0.5 mm, emission from Galactic dust (temperature $\sim$ 20 K) begins to confuse measurements of the CMR spectrum and isotropy. It will be difficult to distinguish cosmic distortions from Galactic ones. At long wavelengths, $\lambda > 20$ cm, a similar situation arises with Galactic radio emission – mostly synchrotron radiation. With radiation temperature increasing as $\lambda^{2.8}$, the Galactic radio brightness swamps the CMR at $\lambda = 75$ cm; $T_{GAL} \sim 20$ K in the coolest regions near the poles. So, at long wavelengths, experimental error due to atmospheric emission is small compared to the error in measuring T_{GAL}. At wavelengths shorter than 10 cm, errors in measuring the atmospheric emission dominate ground-based experiments[4], and balloon-based experiments[5] look more promising. The choice between a ground-based or a balloon-based experiment in the range $\lambda = 10$ cm to 20 cm comes down to details of the technique, keeping in mind that antennas are larger and harder to shield from ground radiation at longer wavelengths.

The next two sections of this paper discuss the advantages and possible improvements in experiments from balloons at centimeter wavelengths and experiments on the ground at decimeter wavelengths.

BALLOON-BASED MEASUREMENTS

Measurements of the CMR temperature from the ground for wavelengths between 3 mm and 20 cm have been limited mainly by atmospheric emission. To reduce errors in T_{ATM} experiments moved from sea level to high altitudes[6] in the late 1960's. In 1980 a large US/Italian collaboration[4] set out to improve ground-based measurements by refining the basic technique and by using simultaneous multi-frequency measurements to improve atmospheric models[7]. Most recently members of this collaboration have reported[8–9] on experiments at the South Pole. Averaging the collaboration's results[4] gives $T_{CMR} = 2.64 \pm .03\ K$; averaging the ground-based results in Peebles' 1971 book (*Physical Cosmology*) gives $T_{CMR} = 2.72 \pm .08\ K$. So, the newer results agree with the older ones and reduce the error by a factor of about two.

To effectively eliminate errors in T_{ATM}, experiments were moved to balloons. Groups at M.I.T.[10], Berkeley[11], and Queen Mary College, London[12] pioneered this work, flying bolometers to make measurements on and shortward of the spectral peak – wavelengths now well-measured by the COBE and UBC work. In 1987, a balloon-based experiment[5] at $\lambda = 1.2$ cm obtained the result $T_{CMR} = 2.783 \pm 0.025$ K, the first measurement with 1% accuracy. The main improvements over ground-based experiments were the elimination of atmospheric emission ($T_{ATM} = 2 \pm 2$ mK at 25 km altitude) and the cooling of most parts of the radiometer front-end to 2.7 K. The first advantage is obvious; the second is much more important than it seems at first. If atmospheric emission is removed, the next largest sources of unwanted radiation are warm metallic surfaces in the instrument – antenna, waveguide, and warm walls at the top of the liquid helium cold load used as a reference. For example, the radiation temperature coming out of a horn antenna pointed at T_{CMR} is

$$T_{OUT} = T_{CMR}(1 - \alpha) + \alpha\, T_o\,, \tag{1}$$

where α ($<< 1$) is the absorption of the antenna and T_o is the physical temperature of the antenna walls. A way to eliminate the effects of wall absorption and emission is to operate all components exposed to T_{CMR} at $T_o = T_{CMR}$. A similar argument holds for eliminating errors due to reflections from the horn, reference load, and receiver.

The balloon experiment is shown in Figure 2. The radiometer front-end was operated near 2.7 K by simply venting the liquid helium bath to the ambient pressure. By using a correlation radiometer only passive components were needed in the dewar and hot emission from the radiometer's mixers was blocked by cold isolators. The cold load which is used to get the radiometer zero point is also operated near 2.7 K; it is shown in the position used to measure its emission.

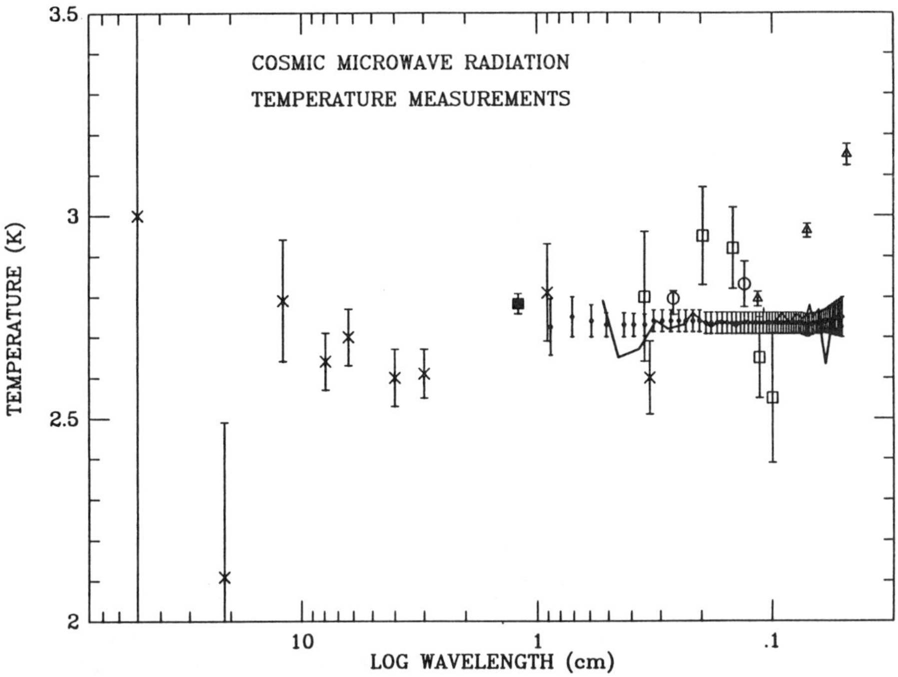

Figure 1. Measurements of the temperature of the CMR. A horizontal straight line in this plot is a Planck spectrum; the data are well fit by a temperature of 2.74 K except for the two points at short wavelength.

× Ground-based measurements discussed in reference 4).

■ Balloon-based heterodyne measurement discussed in reference 5) and below.

□ Balloon-based bolometer measurement, reference 11).

○ Recent results from optical measurements of CN excitation discussed in references 21).

• Preliminary results from the far infrared spectrometer aboard COBE using nine minutes of data, reference 1). Error flags are upper limits to systematic errors affecting the relative intensity, i.e. spectral *shape*. An overall calibration uncertainty could move all points up or down by as much as 60 mK; this range will get smaller as will the error bars on individual points as more data are analyzed.

— Results of the UBC rocket flight, reference 2). Errors are estimated by the scatter in points, shown here connected by line segments. A temperature of $T_{CMR} = 2.736 \pm 0.017$ K is derived from points between $\lambda = 3$ mm and 0.6 mm where the data are unperturbed by instrument microphonics.

△ Results from a Berkeley/Nagoya rocket flight, reference 22). Disagreement with the COBE and UBC results suggests that some unknown local radiation contaminated the data.

To view the sky the cold load is moved aside and the horn is lowered to couple with its throat section. The reference load is heated to calibrate the radiometer (in Volts/Kelvin).

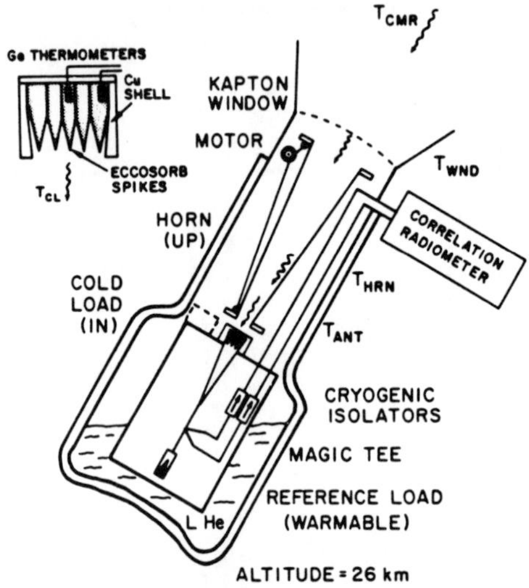

Figure 2. A balloon experiment operated close to 2.7 K. The apparatus was tipped to avoid radiation emitted and scattered from the balloon. The angle was varied to check for ground and balloon radiation in the antenna sidelobes.

Two compromises were made to deal with the environment at balloon altitudes. A window was used to keep air away from cold parts of the instrument and to allow control of the bath pressure and temperature. The upper part of the horn antenna (where absorption is small) was operated at the ambient temperature to avoid driving convection in the dewar neck. These two out-of-equilibrium components contributed most of the error to the experiment by emitting spurious radiation: $T_{WND} = 35 \pm 12$ mK and $T_{HRN} = 50 \pm 12$ mK. It is not hard to imagine ways to reduce or eliminate these effects. The extensive error budget in reference 5) can be used to see which effects come in next. Also, the experiment was done mainly with old parts from isotropy balloon flights. New cryogenic amplifiers offer ways to achieve great simplification.

The ideal of having the apparatus in thermal equilibrium with the measured radiation was nearly achieved in the COBE and UBC experiments. One departure was that bolometric detectors were operated as cold as possible for maximum sensitivity. So, their emission back toward the antenna and cold load is less than T_{CMR}; reflections at the horn of CMR radiation and instrument emission do not compensate.

GROUND-BASED MEASUREMENTS AT VERY LONG WAVELENGTHS

At longer wavelengths, the practicality of mounting experiments on balloon platforms is diminished, both because of the increasingly large dimensions of waveguides and antennas and because of the growing Galactic signal, which must be separated from the CMR. Most schemes for recognizing the Galactic signal require some flexibility in pointing the horn at selected regions of the sky. Historically the solution has been to use ground-based experiments, which must contend with the problems of measurement or calculation of both the atmospheric and Galactic contributions to the antenna temperature. Since size limitations typically result in at least parts of the waveguide and horn being at ambient temperature, emission from the warm walls must also be calculated and/or measured. Similarly, reference cold loads must be joined to the room temperature apparatus, resulting in corrections for wall emission between the detector and the reference temperature cold load.

At wavelengths greater than about 20 cm, Galactic emission is the dominant source of error. Recent data on the CMR temperature at 50 cm come from the Sironi, et al. collaboration[13)], who have used a horn antenna of beamwidth 15° × 16° to measure the sky temperature. Their result is $T_{CMR} = 3.0 \pm 1.2$ K. Much of the error is due to an unusually large ground contribution to the signal arising from the fact that the horn was not shielded from the mountainous horizon at the experimental site, Alpe Gera (where the average elevation is 20°). Another large component of the error results from their method of removing the Galactic signal from the data, which relies on the use of a Galaxy map[14)] at 408 MHz, which has uncertainties in zero point and scale. The group intends to make future measurements at other wavelengths (12 and 36 cm) in order to internally determine the Galactic contribution at 50 cm which should reduce this error significantly.

The first CMR measurements at very long wavelengths were published by Howell and Shakeshaft[15)] in 1967 and although the measurements suffer from a lack of low-noise electronics, some of the fundamental techniques of the experiments seem worth emulating today. In this experiment, measurements were made at 74 cm and 49 cm with scaled horns, which were shielded from the horizon. Drift scans from these horns were used to make an independent determination of the spectral index of the Galactic radiation between the two wavelengths. (The wavelength dependence of the Galactic signal is assumed to be λ^{β}.) The ground, wall, and atmospheric emissions were calculated from models and removed from the data. The final data comprised three numbers: the extraterrestrial sky temperatures T_{SKY} at 74 and 49 cm and the spectral index β. The following coupled equations relate these values to the unknown quantities of interest:

$$T_{SKY}(74) = T_{GAL}(74) + T_{CMR}(74),$$

$$T_{SKY}(49) = T_{GAL}(74)\left(\frac{49}{74}\right)^{\beta} + T_{CMR}(49)\,.$$

Here T_{CMR} and T_{GAL} are the CMR and Galactic contributions as functions of wavelength in cm. For $\beta = 2.8$, the measured quantity is therefore roughly

$$T_{CMR}(74) - \left(\frac{1}{3.1}\right) T_{CMR}(49)\,.$$

Thus, the experiment analyzed in this way measures a weighted average of the CMR temperatures at wavelengths 49 cm and 74 cm and is not sensitive to any spectral deviation between those two wavelengths.

At wavelengths near 20 cm, atmospheric, Galactic and wall radiation are all of comparable magnitude. An early experiment at 20.7 cm is also the work of Howell and Shakeshaft[16]. They used the Hogg model[17] to calculate the atmospheric temperature and assigned a 10% error to the calculation (about 200 mK at sea level). The Galactic signal was obtained by scaling the Pauliny-Toth and Shakeshaft 404 MHz map[18] with an assumed spectral index of 2.7. The experimenters assigned a 200 mK error to the Galactic signal, corresponding to an error of 0.2 in the spectral index. The reference load used in this measurement was a resistive termination mounted in a flask of liquid helium at the end of 75 cm of coaxial cable. Calculation of the coaxial emission led to another 200 mK error, as did calculation of wall emission from the antenna. Errors in the signal detection and in the ground emission calculation were each 100 mK, so that the overall CMR temperature measured was 2.8 ± 0.4 K, if the six errors mentioned above are added in quadrature – a standard practice in this field. (Howell and Shakeshaft quote a more modest 600 mK error because they added the signal and reference load errors linearly.)

A more recent 21.3 cm experiment has been reported by Levin, et al[19]. In this measurement, the atmospheric temperature is determined by scaling with atmospheric models from a direct atmospheric measurement made concurrently at 8.1 cm. The error assigned to T_{ATM} is 10% (or 100 mK at 3800 m above sea level). The Galactic contribution is estimated by scaling the 408 MHz map[14] with a spectral index of $\beta = 2.8 \pm 0.2$. The experimenters fit their data at 1408 MHz to the 408 MHz map in order to fix β to this range of values. The error in β leads to a 200 mK error in the result. The final error arising from emission in the reference cold load and antenna is 300 mK. With careful design and mapping of their antenna, Levin, et al. reduced the error due to ground emission to a negligible 8 mK. Similarly, their detector is sufficiently quiet that the error in detecting the signal is only 30 mK. The final result is $T_{CMR} = 2.11 \pm 0.38$ K.

In summary, the status of 20 cm experiments is that the errors due to the atmosphere, the Galaxy, and the reference load and/or antenna wall emission are the dominant errors and are each of the order of 100 to 200 mK. To improve the atmospheric temperature accuracy, direct measurements of the atmosphere at 20 cm are needed. Dependable measurements would verify atmospheric models and provide data for empirical models for use in real-time monitoring of fluctuations using parameters like the ambient pressure, temperature, and humidity, for example. The main problem with tipping a 20 cm horn to measure the atmosphere directly, of course, is that at different tip angles the antenna views different parts of the Galaxy. Thus, either the Galactic emission must be well-understood or one must assume that the atmosphere is stable to the desired accuracy over periods of several hours, so that a single point in the Galaxy can be tracked in zenith angle. An added complication is the usual one of needing a good understanding of the (necessarily low) sidelobe pickup of the antenna.

Relying on scaling Galactic maps at other frequencies to the frequency of interest is problematic for several reasons. The first is that the absolute zero levels of most Galactic maps are poorly understood. Also, many maps are plagued with striping due to the variation of ground pickup with declination angle. Thirdly, the spectrum of the Galactic synchrotron emission steepens at low frequencies, so that a power law is only locally valid. Finally, Galactic maps are typically made using one of the circularly polarized components of the radiation, while CMR measurements are typically made with linearly polarized horns. Bingham's data[20] indicate that even for 15° diameter spots in the sky, the differences between signals from two orthogonal linear polarizations may be on the order of 100 mK at 1407 MHz.

A promising method of collecting information about the Galactic emission is to obtain frequency information about the sky temperature. Over a suitably small range of frequencies, the signals from the CMR and the atmosphere are approximately flat while the Galactic emission has a frequency signature. By observing the sky temperature at several nearby frequencies, one could hope to use this frequency signature to excise the Galactic signal from the data.

The solution to the problems of wall emission and of obtaining an accurate reference load is, of course, to run the whole experiment in thermal equilibrium. As mentioned above, this option is not practically available at longer wavelengths. However, it is possible to operate the most absorptive portions of the apparatus in thermal equilibrium: some of the waveguide and all of the coax prior to the first stage of amplification, including the coax leading to the reference load.

In an effort to improve the measurements of the CMR spectrum for long wavelengths, we have a 21 cm experiment underway at Princeton. We will take our equipment to the NRAO Green Bank Observatory to reduce problems with interference. Measurements at wavelengths up to 50 cm are planned, using scaled horn antennas to facilitate intercomparison of data from several frequencies to help in removing Galactic emission.

This research was supported by the National Science Foundation and the Department of Education. We thank our colleagues in the 21 cm spectrum experiment – Norm Jarosik, Chris Smith, and Ed Wollack – for many helpful discussions.

REFERENCES

1) Mather, J. L., et al. 1990, *Ap. J. Lett.*, **354**, L37. Mather, J., et al., these proceedings.

2) Gush, H., Halpern, M., and Wishnow, E. 1990, *Phys. Rev. Lett.*, **65**, 537. Halpern, M., et al., these proceedings.

3) Dubrovich, V. K., et al. 1977, *Sov. Astron.*, **54**, 727. Sunyaev, R. A. and Zel'dovich, Ya. B. 1980, *Ann. Rev. Astron. Ap.*, **18**, 537, for a review.

4) Smoot, G., et al. 1988, *Ap. J.*, **331**, 653. Kogut, A., et al. 1990, *Ap. J.*, **355**, 102.

5) Johnson, D. G. and Wilkinson, D. 1987, *Ap. J. Lett.*, **313**, L1.

6) Wilkinson, D. 1967, *Phys. Rev. Lett.*, **19**, 1195. Stokes, R., et al. 1967, *Phys. Rev. Lett.*, **19**, 1199. Boynton, P., et al. 1968, *Phys. Rev. Lett.*, **21**, 462.

7) Danese, L. and Partridge, R. B. 1989, *Ap. J.*, **342**, 604, and references therein.

8) Kogut, A., et al., these proceedings.

9) Sironi, G. 1990, *Second Recontres de Blois, "Physical Cosmology"*.

10) Muehlner, D. and Weiss, R. 1973, *Phys. Rev. Lett.*, **30**, 757.

11) Peterson, J. B., Richards, P. L. and Timusk, T. 1985, *Phys. Rev. Lett.*, **55**, 332.

12) Robson, E. I., et al. 1974, *Nature*, **251**, 591.

13) Sironi, G., et al. 1990, *Ap. J.*, **357**, 301.

14) Haslam, C. G. T., et al. 1982, *Astr. Ap. Suppl.*, **47**, 1.

15) Howell, T. F. and Shakeshaft, J. R. 1967, *Nature*, **216**, 753.
16) Howell, T. F. and Shakeshaft, J. R. 1966, *Nature*, **210**, 1318.
17) Hogg, D. C. 1959, *J. App. Phys.*, **30**, 1417.
18) Pauliny-Toth, I. I. K. and Shakeshaft, J. R. 1962, *Mon. Not. R. astr. Soc.*, **124**, 61.
19) Levin, S. M., et al. 1988, *Ap. J.*, **334**, 14.
20) Bingham, R. G. 1966, *Mon. Not. R. astr. Soc.*, **134**, 327.
21) Crane, P., et al. 1989, *Ap. J.*, **346**, 136. Palazzi, E., et al. 1990, *Ap. J.*, **357**, 14, and references therein.
22) Matsumoto, T., et al. 1988, *Ap. J.*, **329**, 567.

Isotropy of the Cosmic Microwave Background

Preliminary DMR Measurements of the CMB Isotropy

G.F. Smoot[1], C.L. Bennett[2], A. Kogut[3], J. Aymon[1], C. Backus[4], G. De Amici[1], K. Galuk[4], P.D. Jackson[4], P. Keegstra[4], L. Rokke[4], L. Tenorio[1], S. Gulkis[5], M.G. Hauser[2], M. Janssen[5], J.C. Mather[2], R. Weiss[6], D.T. Wilkinson[7], E.L. Wright[8], N.W. Boggess[2], E.S. Cheng[2], T. Kelsall[2], P. Lubin[9], S. Meyer[6], S.H. Moseley[2], T.L. Murdock[10], R.A. Shafer[2], and R.F. Silverberg[2]

[1]Lawrence Berkeley Laboratory and Space Sciences Laboratory, University of California, Berkeley, [2]Laboratory for Astronomy and Solar Physics, NASA, Goddard Space Flight Center, [3]National Research Council, NASA/GSFC, [4]ST Systems Inc., [5]NASA Jet Propulsion Laboratory, Pasadena, [6]Massachusetts Institute of Technology, [7]Princeton University, [8]University of California, Los Angeles, [9]University of California, Santa Barbara, [10]General Research Corporation

ABSTRACT

The COBE Differential Microwave Radiometers (DMR) instrument has produced preliminary full-sky maps at frequencies 31.5, 53, and 90 GHz. The redundant channels and matched beams at three frequencies distinguish the DMR from previous large-scale surveys. Galactic emission is seen unambiguously at all three frequencies. The only large-scale anisotropy detected in the cosmic microwave background is the dipole anisotropy. There is no clear evidence for any other large-angular-scale feature in the maps. Without correcting for any systematic effects, we are able to place limits $\Delta T/T_0 < 3\times10^{-5}$ for the rms quadrupole amplitude, $\Delta T/T_0 < 4\times10^{-5}$ for monochromatic fluctuations, and $\Delta T/T_0 < 4\times10^{-5}$ for Gaussian fluctuations (all limits are 95% C.L. with $T_0 = 2.735$ K). The data limit $\Delta T/T_0 < 10^{-4}$ for any feature larger than 7°. We briefly review the DMR and discuss some implications of these results for cosmology.

INTRODUCTION

The cosmic microwave background (CMB) dominates the photon number in the universe. Fortuitously, most of these photons are available to us with little contamination from local foregrounds. In the standard Hot Big Bang model of cosmology, these photons have freely streamed through the universe since an era in which the universe was a hot, dense plasma. As such, the CMB is the most accessible relic from the era over which the universe has evolved to its present highly-ordered state. The CMB intensity distribution maps the large-scale structure of space-time in the early universe. Despite a quarter-century of effort, no intrinsic anisotropy in the CMB has been detected. Paradoxically, its isotropy convinces us of its cosmological origin and of the ultimate necessity of intrinsic anisotropies.

The Differential Microwave Radiometers (DMR) experiment aboard NASA's Cosmic Background Explorer (COBE) is intended to provide precise maps of the full microwave sky at wavelengths in which competing astrophysical foregrounds (galactic synchrotron, HII, and dust emission) are more than 1000 times fainter than the CMB. The resultant maps provide new information on the large scale structure and evolution of the universe. Major scientific objectives include study of the geometry of space-time (isotropy and symmetry of the Hubble flow, global rotation, and causality) and the distribution of matter and energy in the early universe (fossil progenitors of galaxies, gravity waves, dark matter, and cosmic strings).

DMR INSTRUMENT

The DMR consists of six differential microwave radiometers, two independent radiometers at each of three frequencies: 31.5, 53, and 90 GHz (wavelengths 9.5, 5.7, and 3.3 mm). Each radiometer consists of a heterodyne receiver switched at 100 Hz between two identical corrugated horn antennas. The antennas, designed for low sidelobes and compact size, have a main lobe well described by a Gaussian profile with 7° FWHM (Toral *et al.* 1989) and are pointed 60° apart, 30° to either side of the spacecraft spin axis. A single local oscillator provides a common reference signal to independent mixer-preamp assemblies for the two channels at each frequency. The signals in each channel undergo further amplification, detection, synchronous demodulation, and 0.5 s integration before being digitized and stored in an on-board recorder. Both channels share a common enclosure and thermal regulation system. On-board solid state noise sources provide in-flight calibration by injecting broad-band microwave power into the front end of each radiometer every two hours. Laboratory calibration prior to launch determined the power emitted by each noise source, which provide a transfer standard to orbit. A detailed description of the DMR instrument may be found in Smoot *et al.* (1990).

Each radiometer measures the difference in microwave power between two regions of the sky separated by 60°. The combined motions of spacecraft spin (75 s period), orbit (103 minute period), and orbital precession (~1 degree/day) allow the brightness of each sky position to be compared to all others through a massively redundant set of all possible difference measurements spaced 60° apart.

OBSERVATIONS AND DATA PROCESSING

A Delta rocket launched COBE into a 900 km, near-polar (99° inclination) orbit on November 18, 1989. Within a week all DMR channels were collecting stable data. The data from the on-board tape recorder are telemetered daily to a ground station, where a pre-processor strips the DMR data from the telemetry stream, merges the raw DMR data with spacecraft attitude and orbit information, and checks the quality of the data. We reject data of questionable telemetry quality, data without accompanying attitude information, and spurious data written with the instrument science telemetry disabled. This accounts for less than 1% of the data. We reject spikes and transients in the data, discarding all points that lie more than five times the RMS scatter from the daily mean (< 0.1% of the data). There are no celestial sources other than the Moon that would exceed this limit. Finally, the processor flags the data for the presence in the main beam of various celestial sources (e.g., Jupiter and the Moon) which could contaminate the data. The results presented were obtained by discarding all data with the Moon closer than 25° to an antenna.

We subtract a mean baseline, smoothed over a 4-hour period, from the data and apply a calibration factor derived from the in-flight noise source firings. We convert the calibrated data to solar system barycenter but apply no other corrections for systematic effects. A sparse matrix algorithm varies the temperatures of 6144 independent map pixels to provide a least-squares fit to the data. The pixel size is 2.8°, smaller than the 7° DMR beam; consequently, there is some correlation between neighboring pixels for sources in the sky. Further details of the data processing algorithms may be found in Torres *et al.* (1988).

At this time the random uncertainties are dominated by white noise from the receivers. Figure 1 shows a histogram of calibrated observations of a single pixel pair separated by 60°. The noise has a Gaussian distribution with standard deviation 30 mK, compared to the expected value 28 mK measured before launch. The COBE orbit provides full, but not uniform, sky coverage, with the area surrounding the celestial poles observed several times more frequently than the celestial equator. Rejection of lunar-contaminated data further limits observations near the celestial equator. As a consequence, noise levels are not uniform across the maps. Typical RMS noise per pixel is on the order $\Delta T/T_0 \sim 10^{-4}$, while the instrumental noise per spherical harmonic coefficient (containing many pixels) is $\Delta T/T_0 \sim 3\times10^{-6}$ (68% C.L.).

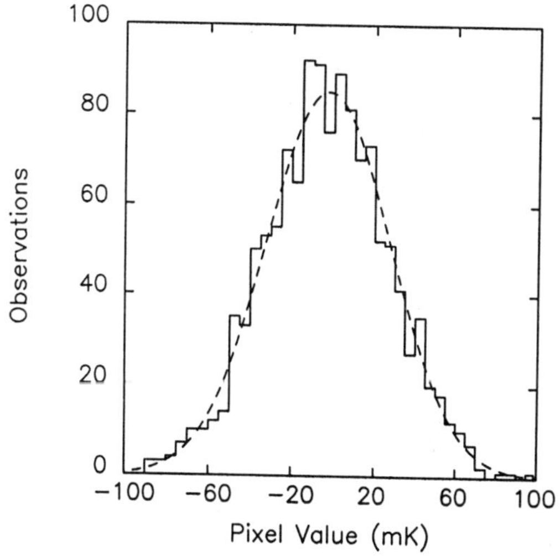

Figure 1. Histogram of differential measurements of two fixed locations on the sky. Each antenna views a single pixel.

The preliminary DMR results presented below are limited not by this statistical uncertainty, but by our current estimates of upper limits on potential systematic effects. These may be classified into three broad categories. The most obvious source of non-cosmological signals is the presence in the sky of foreground microwave sources. These include thermal emission from the COBE spacecraft itself, from the Earth, Moon, and Sun, and from other celestial objects. Non-thermal radio-frequency interference (RFI) must also be considered, both from ground stations and from geosynchronous satellites. Although the DMR instrument is largely shielded from such sources, their residual or intermittent effect must be considered.

A second class of potential systematics is the effect of the changing orbital environment on the instrument. Various instrument components have slightly different performance with changes in temperature, voltage, and local magnetic field, each of which can vary with the COBE orbital position. Longer-term drifts can also affect the data.

Finally, the data reduction process itself may introduce or mask features in the data. The DMR data are differential; the sparse matrix algorithm is subject to concerns of both coverage (closure) and solution stability. Other features of the data reduction process, particularly the calibration and baseline subtraction, are also a source of potential artifacts. All potential sources of systematic error must be identified and their effects measured or limited before maps with reliable uncertainties can be produced.

A variety of techniques exist to identify potential systematics and place limits on their effects. The most general technique is a least-squares fit to the time-dependent effect in the calibrated data. Over a six month period, the orbital inclination and precession combine to decouple orbit-related effects from large-scale structure on the sky. Another technique, suitable for celestial foreground sources, is to produce a sky map in an object-centered coordinate system. The contribution of the source at a given distance from beam center can be read directly from the maps to the noise limit. As a cross-check, and for instances where the time dependence of the signal is not known a priori, we derive upper limits by differencing sky maps produced when the systematic effect is known to be at different strengths. The degree to which the subtracted maps differ from instrumental noise provides a limit to the effect in the DMR sky maps.

The data reduction process itself may introduce artifacts to the sky maps. Examples include the stability of the sparse matrix solution in the presence of noise and signals that do not sum to zero over a closed pixel path (e.g., drifts), cross-talk between nominally independent pixels, and the propagation of signals too small to be fitted to the daily baselines. We use Monte Carlo simulations to limit these effects and conclude that the sparse matrix algorithm is robust and capable of recovering the input sky map with uncertainties consistent with the Gaussian instrument noise per pixel.

TABLE 1

PRELIMINARY 95% C.L. UPPER LIMITS TO POTENTIAL SYSTEMATIC EFFECTS (mK)

Effect	31A	31B	53A	53B	90A	90B
		Signals				
Earth	0.070	0.070	0.040	0.040	0.040	0.040
Moon (>25°)	0.004	0.004	0.004	0.004	0.004	0.004
COBE shield and dewar	0.002	0.002	0.002	0.002	0.002	0.002
Sun	0.001	0.001	0.001	0.001	0.001	0.001
		Environmental Susceptibilities				
Magnetic (dipole)	0.077	0.100	0.260	0.034	0.143	0.059
Magnetic (quadrupole)	0.014	0.017	0.022	0.024	0.079	0.006
Thermal	0.012	0.010	0.012	0.010	0.009	0.007
Voltage	0.010	0.010	0.010	0.010	0.010	0.010
Cross-talk	0.0001	0.0001	0.0001	0.0001	0.0001	0.0001
		Data Reduction and Processing				
Solution Stability	0.0001	0.0001	0.0001	0.0001	0.0001	0.0001
Baseline Drift	0.012	0.033	0.005	0.005	0.009	0.006
Calibration Drift	0.043	0.079	0.031	0.033	0.049	0.039
Absolute Calibration[a]	0.165	0.165	0.165	0.165	0.165	0.165
Antenna Pointing	0.030	0.030	0.030	0.030	0.030	0.030
Total Systematics, Dipole	0.20	0.23	0.32	0.18	0.23	0.19
Total Systematics, $l \geq 2$	0.09	0.12	0.07	0.07	0.11	0.07

[a] Dipole term only

Table 1 summarizes preliminary 95% confidence level upper limits to potential systematic effects in the maps for each of the DMR channels. The results in many cases are limited by sky coverage, signal to noise statistics, and analysis software. We anticipate increasingly tighter limits to potential systematics as coverage, integration time, and software improve. The largest limit is the current 5% uncertainty in the absolute calibration of the instrument. The uncertainty in absolute calibration does not create artifacts in the maps but affects the calculated amplitude of existing features (e.g., the dipole anisotropy). We continue to acquire and analyze calibration information and we anticipate improved calibration in the future. The next largest effect is the modulation of the instrument output in the Earth's magnetic field; the primary effect is on the dipole term of a spherical harmonics expansion. The magnetic effects will be modelled and removed in future analysis. The 53A channel unambiguously shows a magnetic susceptibility; we do not include the 53A channel in the dipole results presented below. The largest potential effects upon the quadrupole and higher-order multipole coefficients are Earth emission and the possibility of undetected calibration drifts. The current 95% C.L. upper limits to combined systematic errors in the DMR maps are $\Delta T/T_0 < 8 \times 10^{-5}$ for the dipole anisotropy and $\Delta T/T_0 < 3 \times 10^{-5}$ for the quadrupole and higher-order terms.

As the DMR gathers redundant sky coverage and as analysis continues, we anticipate refined estimates of, or limits on, these effects. It is important to note, however, that the DMR is free from some of the systematics of previous large-scale sky surveys. The multiple differences generated by various chopping frequencies (spin, orbit, and precession periods) allow separation of instrumental from celestial signals. Two independent full-sky maps, produced with matched beams at each of three frequencies, provide a powerful tool for analysis and removal of possible systematic effects.

RESULTS

Figure 2 shows grey-scale maps of the full microwave sky in galactic coordinates. The independent maps at each frequency enable celestial signals to be distinguished from noise or spurious features: a celestial source will appear at identical amplitude in both channels. The three frequencies allow separation of cosmological signals from galactic foregrounds based on spectral signatures. It is worth noting that the maps cover a total range < 12 mK: the sky at mm wavelengths is featureless to 0.5%.

The most noticeable feature in the maps is the dipole anisotropy. This has a natural interpretation as a Doppler-shifted CMB observed from a frame at motion with respect to the CMB rest frame. Motion with speed $\beta = v/c$ relative to an isotropic blackbody radiation field yields the observed Doppler-shifted temperature distribution

$$T(\theta) = T_0 \, (1 - \beta^2)^{1/2} \, / \, (1 - \beta \cos\theta \,)$$
$$\sim T_0 \, [1 + \beta \cos\theta + (\beta^2/2) \cos 2\theta + O(\beta^3) \,]$$

(Peebles and Wilkinson 1968). The first (monopole) term is the unshifted CMB temperature. The second term is a dipole, varying as the cosine of the angle between the direction of observation and the velocity. The third term is a quadrupole, varying as the cosine of twice the angle with amplitude $\beta/2$ times the dipole amplitude. The temperature distribution for non-Planckian spectra can be derived by noting that $I(\nu)/\nu^3$ is a Lorentz invariant, where $I(\nu)$ is the spectral brightness. The spectrum of all published dipole parameters, including those from the FIRAS experiment (Cheng *et al.* 1990), is consistent with a Doppler-shifted blackbody origin.

The preliminary DMR-determined dipole amplitude 3.3±0.2 mK implies a Solar-system barycenter velocity of 365±18 km s^{-1} toward $(\alpha,\delta) = (11.2^h \pm 0.2^h, -7^\circ \pm 2^\circ)$, or $(l,b) = (265^\circ \pm 2^\circ, 48^\circ \pm 2^\circ)$. The solar system velocity with respect to the local standard of rest is estimated at 20 km s^{-1} toward $(l,b) = (57^\circ, 23^\circ)$ (Kerr and Lynden-Bell 1986),

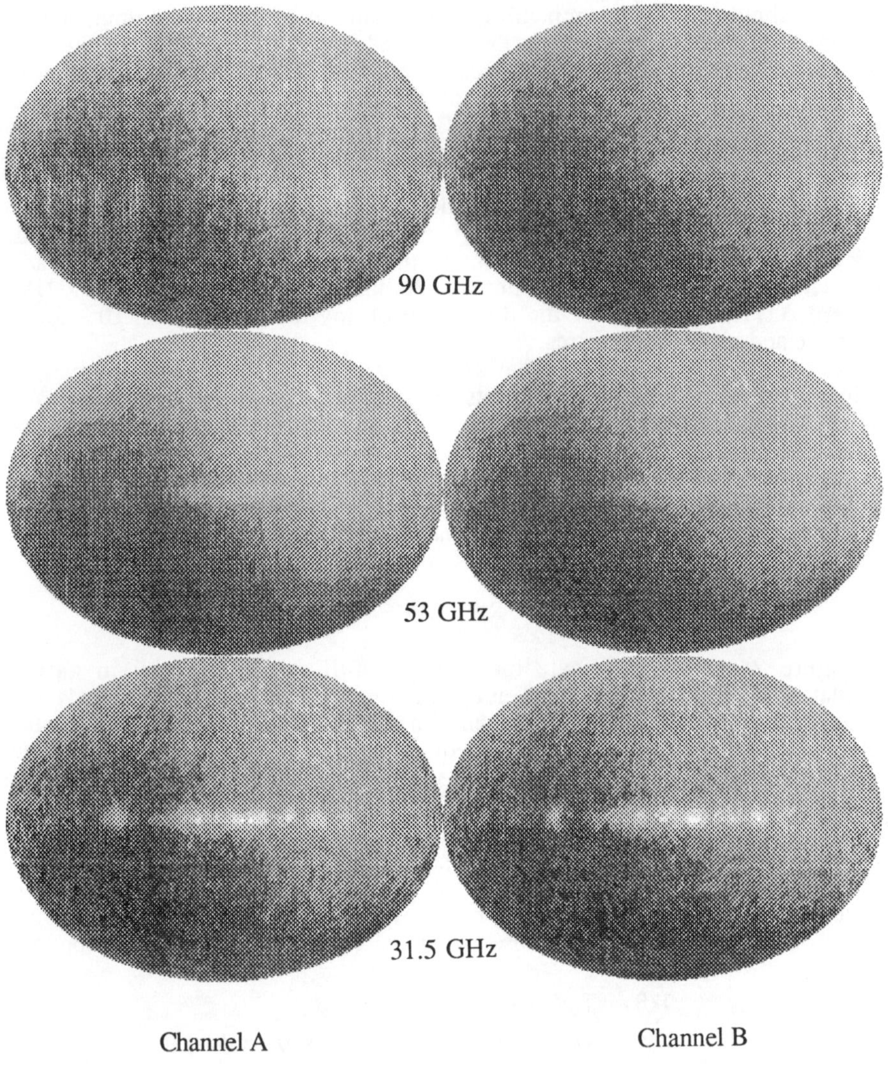

Figure 2. COBE DMR full sky maps of the relative temperature of the sky at frequencies 31.5, 53.0, and 90.0 GHz. The maps are in galactic coordinates and have been corrected to solar system barycenter.

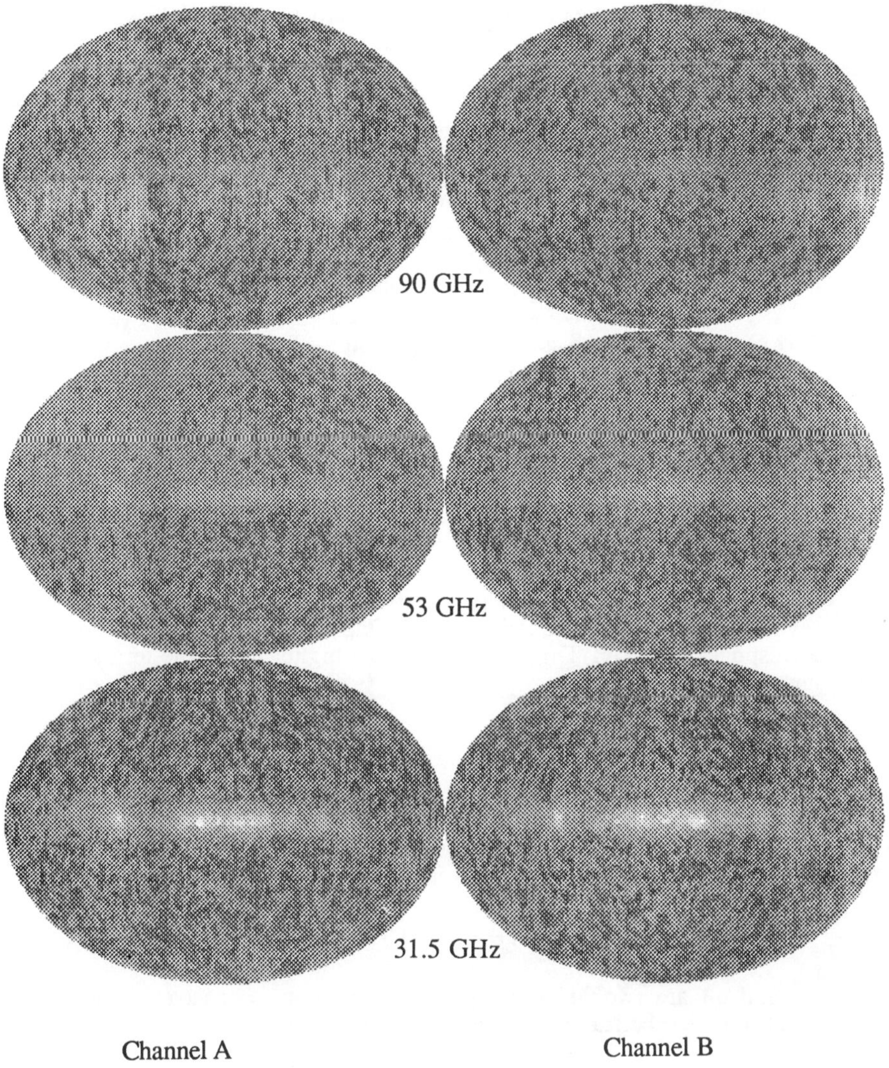

Figure 3. COBE DMR full sky maps of the relative temperature of the sky at frequencies 31.5, 53.0, and 90.0 GHz. A dipole corresponding to a solar system barycenter motion of 365 km/s through a 2.735 K blackbody has been subtracted from the maps in Figure 2.

while galactic rotation moves the the local standard of rest at 220 km s^{-1} toward (90°,0°) (Fich, Blitz, and Stark 1989). The DMR results thus imply a peculiar velocity for the Galaxy of $v_g = 547 \pm 17$ km s^{-1} in the direction (266°±2°, 29°±2°).

Figure 3 shows the DMR maps with this dipole removed from the data. The only large-scale feature remaining is galactic emission, confined to the plane of the galaxy. This emission is present at roughly the level expected before flight and is consistent with emission from electrons (synchrotron and HII) and dust within the galaxy. The ratio of the dipole anisotropy (the largest cosmological feature in the maps) to the Galactic foreground reaches a maximum in the frequency range 60—90 GHz. There is no evidence of any other emission features.

The temperature distribution on the sky, expanded to order l=2, may be written as

$$\begin{aligned} T(\alpha,\delta) = T_0 &+ D_x \cos\delta \cos\alpha + D_y \cos\delta \sin\alpha + D_z \sin\delta \\ &+ Q_1 (3\sin^2\delta - 1)/2 + Q_2 \sin2\delta \cos\alpha + Q_3 \sin2\delta \sin\alpha \\ &+ Q_4 \cos^2\delta \cos2\alpha + Q_5 \cos^2\delta \sin2\alpha \end{aligned}$$

The quadrupole is often quoted in terms of the $\Delta T/T_0$ RMS normalization,

$$Q_{rms}^2 = \frac{4}{15} \left(\frac{3}{4} Q_1^2 + Q_2^2 + Q_3^2 + Q_4^2 + Q_5^2\right)$$

We have made a series of spherical harmonic fits to the data, excluding data within several ranges of galactic latitude. We find a limit of $Q_{rms} < 0.07$ mK (95% C.L.), corresponding to $(\Delta T/T_0)_{rms} < 3 \times 10^{-5}$. To higher multipole terms, the CMB temperature distribution may be written as

$$T(\theta,\phi) = \Sigma_l \Sigma_m k \left[b_{lm}\cos(m\phi) + b_{l-m}\sin(m\phi)\right] \left[\frac{(2l+1)(l-m)!}{4\pi\,(l+m)!}\right]^{1/2} P_{lm}(\cos\theta)$$

where $P_{lm}(\cos\theta)$ are the associated Legendre functions, k takes the values $\sqrt{2}$ (for m=0) or 1 (for $m \neq 0$), and the b_{lm} are real numbers. The CMB correlation function $C(\alpha)$ is the average over all pairs of points in the sky separated by angle α,

$$C(\alpha) = \langle T(\mathbf{n}_1)\, T(\mathbf{n}_2) \rangle$$

where $\mathbf{n}_1$ and $\mathbf{n}_2$ are two unit vectors such that $\mathbf{n}_1 \cdot \mathbf{n}_2 = \cos(\alpha)$. $C(\alpha)$ may be determined from the spherical harmonics expansion.

$$C(\alpha) = \frac{1}{4\pi} \Sigma_l \left[\Sigma_m |b_{lm}|^2\right] P_l(\cos\alpha)$$

where $P_l(\cos\alpha)$ is the Legendre polynomial of order l. Excluding components with l=0 (monopole), l=1 (dipole) and l>20, we find $|C(\alpha)| < 0.01$ mK^2 for the range $15° < \alpha < 165°$, corresponding to $\left\langle \frac{\Delta T(\mathbf{n}_1)}{T_0} \frac{\Delta T(\mathbf{n}_2)}{T_0} \right\rangle < 10^{-9}$.

Monochromatic limits are those which present a pattern covering the full sky and are computed from the spherical harmonic amplitudes in a manner similar to the correlation function,

$$(\Delta T)^2 = \frac{1}{4\pi} \Sigma_m |b_{lm}|^2$$

We find a limit $\Delta T/T_0 < 4\text{x}10^{-5}$ (95% C.L.) for monochromatic effects.

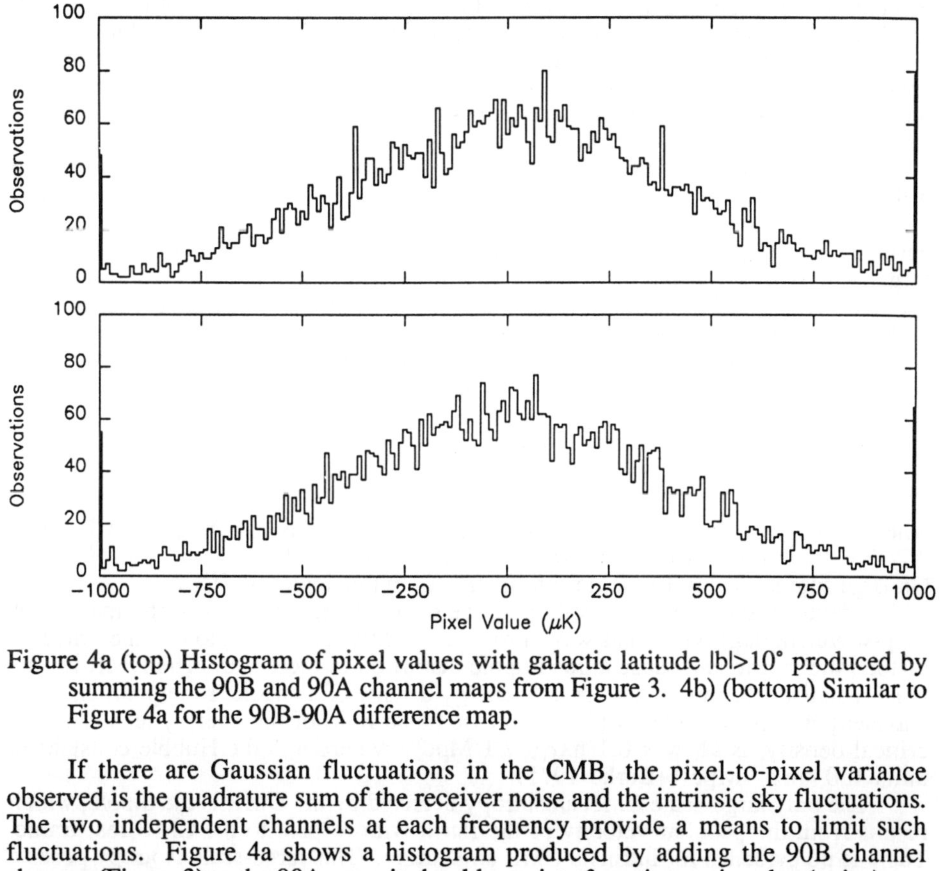

Figure 4a (top) Histogram of pixel values with galactic latitude |b|>10° produced by summing the 90B and 90A channel maps from Figure 3. 4b) (bottom) Similar to Figure 4a for the 90B-90A difference map.

If there are Gaussian fluctuations in the CMB, the pixel-to-pixel variance observed is the quadrature sum of the receiver noise and the intrinsic sky fluctuations. The two independent channels at each frequency provide a means to limit such fluctuations. Figure 4a shows a histogram produced by adding the 90B channel skymap (Figure 3) to the 90A map; it should consist of receiver noise plus (twice) any cosmological signals and whatever systematic effects are the same in both channels. Figure 4b shows a similar histogram produced by subtracting the 90A channel map from the 90B map. Since celestial sources will appear at equal amplitude in both channels, the difference map should consist only of receiver noise and whatever systematic effects are different between the channels. Using the difference maps to estimate the receiver noise per pixel (not constant across the map owing to unequal sky coverage), we compute a limit $\Delta T/T_0 < 4\text{x}10^{-5}$ (95% C.L.) for Gaussian fluctuations.

DISCUSSION

The DMR limits relate directly to the large-scale geometry of the early universe. The data limit any cosmological feature on angular scales larger than 7° to

$\Delta T/T_0 < 10^{-4}$. From our vantage point, the Hubble expansion is uniform to this level. If we assume the Copernican Principle that we do not occupy a special place in the universe, the combination of isotropy and homogeneity implies that the large-scale geometry is described by the Robertson-Walker metric, with perturbations of order $\Delta T/T_0$ or smaller.

The limits to quadrupole and higher-order anisotropies limit global shear and vorticity in the early universe. If the universe were rotating (in violation of Mach's Principle), the resultant metric causes null geodesics to spiral. In a flat universe, the resultant anisotropy is dominated by a quadrupole term (Collins and Hawking 1973); further limits may be established from the correlation function $C(\alpha)$ (Barrow, Juskiewicz, and Sonoda 1985). The DMR results limit the global rotation of the universe to $\omega/H_0 < 10^{-6}$, or less than one ten-thousandth of a turn in the last ten billion years.

Arguments of isotropy and homogeneity constrain perturbations in the metric to order $\Delta T/T_0$. Cosmic strings provide one mechanism for such local perturbations. They are nearly one-dimensional topological defects in the metric predicted by many particle physics gauge theories, and are characterized by large mass per unit length μ (e.g., Vilenkin 1985 and references therein). The tension μ causes transverse oscillations at speeds near that of light, and the gravitational potential causes lensing with bending angle independent of impact parameter. Although the lensing causes no anisotropy in the rest frame of the string (the surface brightness of the CMB is nearly constant), in the rest frame of an observer there is a step discontinuity of magnitude

$$\Delta T/T_0 \sim 8\pi\,\beta\gamma\frac{G\mu}{c^2}$$

where β and γ are the usual relativistic parameters (Stebbins 1988). The DMR limits $\Delta T/T_0 < 10^{-4}$ imply a limit on strings with angular scale > 20° such that $G\mu/c^2 < 10^{-5}$. On large angular scales the metric appears free from topological defects.

Gravity waves provide another source of local perturbations in the metric. A long-wavelength gravitational wave propagating through this region of the universe distorts the metric and cause a quadrupole distortion in the CMB. The DMR results place a statistical limit on the possible energy density of a chaotic superposition long-wavelength gravity waves, which, expressed as a ratio of the energy density to the critical density, is $\Omega_{GW} < 0.1(h\lambda_{GW} / 1\ \mathrm{Mpc})^{-2}$ where h is the Hubble constant in units 100 km s^{-1} Mpc^{-1} (Burke 1975).

The only large-scale CMB anisotropy detected by the DMR is the dipole, which results from the peculiar velocity of the Earth with respect to the CMB rest frame. There is no evidence for intrinsic CMB anisotropy. This large-scale isotropy poses a major problem for cosmology. In standard hot Big Bang models, the last significant interaction between the CMB photons and matter was Thomson scattering, which ceased when the universe cooled sufficiently to form neutral hydrogen atoms. This recombination occurred over a small range (± 4%) of redshift at $z \sim 1300$. A region at the particle horizon at redshift z has a present length $l \sim 2r_h(\Omega z)^{-1/2}$, where r_h is the Hubble radius, and subtends an angle $\theta \sim (\Omega/z)^{1/2} \sim 2°$. The surface of last scattering, observed at microwave wavelengths today, contains over 1000 causally disconnected regions which nevertheless have the same temperature to a part in ten thousand.

Inflationary scenarios provide one solution to the causality problem. In these models, the universe undergoes a spontaneous phase transition $\sim 10^{-32}$ seconds after the Big Bang, causing a period of exponential growth in which the scale size increases by 30 to 40 orders of magnitude. The entire observed universe would then originate from a small pre-inflationary volume in causal contact with itself, eliminating this horizon problem. In the simplest inflationary models, the pre-inflationary matter and radiation fields are diluted to zero along with any pre-existing anisotropies. Standard inflation, however, generates scale-free anisotropies with a Harrison-Zel'dovich spectrum which result in small but detectable CMB anisotropies in the present universe (Gorski 1990, Abbott and Wise 1984). Although current DMR limits are an order of magnitude above the predicted spectrum, we anticipate achieving sufficient sensitivity over the planned mission to test predictions of inflationary models.

A second major problem in cosmology is the growth of structure in the universe. The largest structures in the current universe (walls and voids) are observed to have density fluctuations $\Delta\rho/\rho$ of order unity on scale sizes $\gtrsim$50 Mpc. Structures of this size are at the horizon scale at the surface of last scattering; consequently, the primordial density fluctuations are small and most of the growth is in the linear regime. The assumption of linear growth requires peculiar velocities ~ 0.01c in order to move the matter the required 10^8 light years of co-moving distance in the ~10^{10} years estimated to have elapsed since the surface of last scattering, an order of magnitude greater than the peculiar velocity inferred from dipole anisotropy. To explain the observed structure without violating limits on CMB anisotropy, and to generate the critical density required by inflationary models, many astrophysicists have turned to cosmological models in which most of the matter in the universe (> 90%) is composed of weakly interacting massive particles (WIMPs). The dynamical properties of this "cold dark matter" allow it to clump faster than the baryonic matter, which later falls into the WIMP gravitational potential wells to form the structures observed today. The gravitational potential and motion of the dark particles produce CMB anisotropy whose amplitude depends on the angular scale size. For scale size ~ 10°, most models predict $\Delta T/T_0 \sim 1—3 \times 10^{-5}$, depending on the average density of the universe (Bond and Efstathiou 1984). Although current observations do not provide significant limits to these models, we anticipate that the DMR will provide a stringent test of such models as data collection and analysis continue.

CONCLUSIONS

The DMR instrument is working well and continues to collect data. The current limits reflect "quick-look" data processing before removal of any potential systematic effects. The data show the expected dipole anisotropy, consistent with a Doppler-shifted thermal spectrum. Galactic emission is present at levels close to those expected before launch, and is largely confined to the plane of the Galaxy. There is no evidence for any other large-scale feature in the maps. We find a limit $\Delta T/T_0 < 3\times10^{-5}$ for the rms quadrupole amplitude, $\Delta T/T_0 < 4\times10^{-5}$ for monochromatic fluctuations, and $\Delta T/T_0 < 4\times10^{-5}$ for Gaussian fluctuations. There is no feature with $\Delta T/T_0 > 10^{-4}$ on angular scales > 7°. The results are consistent with a universe described by the Robertson-Walker metric and show no evidence of anisotropic expansion, rotation, or localized defects (strings).

The DMR is expected to continue operating until it is shut off. At the end of two years' operation, we can expect random noise on the order of $\Delta T/T_0 \sim 2\times10^{-5}$ per 7 degree field of view. We have begun the process of identifying and removing systematic effects from the data. More data and analysis will provide improved calibration and improved limits on possible systematic effects in the maps.

The National Aeronautics and Space Administration/Goddard Space Flight Center (NASA/GSFC) is responsible for the design, development, and operation of the Cosmic Background Explorer. GSFC is also responsible for the software development through to the final processing of the data. The COBE program is supported by the Astrophysics Division of NASA's Office of Space Science and Applications. We gratefully acknowledge the many workers whose contributions ensured the success of the COBE mission.

REFERENCES

Abbott, L.F., and Wise, M.B., 1984, *Ap.J. Lett.*, **282**, L47.
Barrow, J.D., Juskiewicz, R., and Sonoda, D.H., 1985, *MNRAS*, **213**, 917.
Bond, J.R. and Efstathiou, G, 1984, *Ap.J. Lett.*, **285**, L45.
Burke, W.L., 1975, *Ap.J.*, **196**, 329.
Cheng, E., *et al.*, 1990, *BAPS*, **35**, 971.
Collins, C.B., and Hawking, S.W., 1973, *MNRAS*, **162**, 307.
Fich, M., Blitz, L., and Stark, A., 1989, *Ap.J.*, **342**, 272.
Gorski, K., 1990, *Ap.J. Lett.* (submitted).
Kerr, F.J., and Lynden-Bell, D., 1986, *MNRAS*, **221**, 1023.
Peebles, P.J.E., and Wilkinson, D.T., 1968, *Phys. Rev.*, **174**, 2168.
Smoot, G.F., *et al.*, 1990, *Ap.J.*, **360**, 685.
Stebbins, A., 1988, *Ap.J.*, **327**, 584.
Toral, M.A., *et al.*, 1989, *IEEE Transactions on Antennas and Propagation*, **37**, 171.
Torres, S., *et al.*, 1989, *Data Analysis in Astronomy*, ed. Di Gesu et al., Plenum Press.
Vilenkin, A., 1985, *Physics Reports*, **121**, 263.

LIMITS ON GAUSSIAN FLUCTUATIONS IN THE COSMIC MICROWAVE BACKGROUND AT 19.2 GHZ

S. P. Boughn
Haverford College, Haverford, PA 19041

E. S. Cheng
NASA Goddard Space Flight Center, Greenbelt, MD 20771

D. A. Cottingham
Center for Particle Astrophysics, University of California, Berkeley, CA 94720

D. J. Fixsen
USRA, NASA Goddard Space Flight Center, Greenbelt, MD 20771

Abstract

The Northern hemisphere data from the 19.2 GHz full sky survey are analyzed to place limits on the magnitude of Gaussian fluctuations in the cosmic microwave background implied by a variety of correlation functions. Included among the models tested are the monochromatic and Gaussian-shaped families, and those with power law spectra for $-2 \leq n \leq 1$. We place an upper bound on the quadrupole anisotropy of $\Delta T/T < 3.2 \times 10^{-5}$ rms, and an upper bound on scale-invariant ($n = 1$) fluctuations of $a_2 < 4.5 \times 10^{-5}$ (95% confidence level). There is significant contamination of these data from Galactic emission, and improvement of our modelling of the Galaxy could yield a significant reduction of these upper bounds.

I. INTRODUCTION

In this paper we present an analysis of the limits that can be set on fluctuations at angular scales greater than 1° by the Northern hemisphere portion of the 19.2 GHz survey [5, 2, 4]. This survey was carried out at an angular resolution of 3° FWHM and a typical sensitivity of 1.5 mK per resolution element. The Northern hemisphere portion covers declinations $-15° < \delta < +75°$.

The instrument used to obtain these data is a balloon-borne single-horn radiometer which is stabilized by chopping between the sky and an internal reference load. The radiometer is a low-noise ruby maser operating at 4.2 K, followed by a conventional heterodyne receiver. The maser is sensitive to a 400 MHz bandwidth centered at 19.2 GHz. The antenna scans in a circle at a constant zenith angle of 45°, rotating at about 1.5 rpm. Pointing information is provided by magnetometers, and is confirmed by observations of the Galactic plane and the moon. A typical observing altitude is 27 km. The Northern hemisphere data were collected in two flights from the National Scientific Balloon Facility in Palestine, Texas, each of which obtained about 8 hours of science data.

We consider only fluctuations which obey Gaussian statistics. A Gaussian field is completely specified by its two-point correlation function. The upper limit that a given experiment can place on the rms fluctuation depends on the assumed correlation function. To preserve this distinction, all of our results are stated as pertaining to a particular correlation function, and not just to an angular scale. We quote the size of fluctuations as $\Delta T/T$ rms (with one noted exception). In Section II we give a brief overview of the method of statistical analysis employed, and the manner in which the data set is prepared. The various correlation functions tested and the upper limits obtained are presented in Section III.

II. ANALYSIS

In this section we briefly cover the method used to determine upper limits on fluctuations; a more detailed presentation can be found in [3]. The method we use to place an upper bound on the magnitude of fluctuations present in the data is the familiar one of selecting a statistic, and seeing where the measured value of this statistic falls on the distribution predicted by a particular model. The rms of fluctuations in the model is adjusted so that the probability that we would have measured a larger value of the statistic is equal to the confidence level (say, 0.95); we then know that if the rms were larger than this value, our observation would be quite unlikely (probability < 0.05) so this value of rms is a reasonable upper bound.

The statistic that we use to compare the observations t_i to a model described by the correlation function $C(\theta)$ is

$$S = \sum_{ij} \frac{t_i t_j}{\sigma_i^2 \sigma_j^2} C'(\theta_{ij}) \tag{1}$$

where σ_i is the statistical error in measurement t_i, $C'(\theta)$ is the correlation function smeared by the antenna pattern of the instrument, and θ_{ij} is the angle between the observations t_i and t_j. We determine the distribution of S, given that the t_i are distributed according to $C(\theta)$, by Monte Carlo analysis.

It is interesting to compare these upper bounds with the upper bounds one would, in some sense, expect from the experiment. One can derive the probability distribution of S under the assumption that there are no fluctuations in sky temperature by the same Monte-Carlo method; the median value of this distribution is a good estimate of the expected value of S in the absence of fluctuations. An upper bound on the rms derived using this value for S as the "measured value" is then a reasonable definition of the "expected upper bound." We refer to this upper bound as the "upper bound sensitivity" of the experiment. A statistician would refer to this definition as the level of fluctuations for which the power of a given confidence level test is 50%. A large discrepancy between the upper bound sensitivity and the actual upper bounds obtained indicates that some correlation is present in the data; if the correlation is due to contamination by systematic effects or foreground sources, the upper bound sensitivity indicates the best that can be achieved by learning how to correct for the contamination, if no actual CMB fluctuation is encountered. We present both the upper bound and upper bound sensitivity for each model tested.

The upper limits quoted in this paper are based on 200 Monte Carlo samples. From examining the scatter between different runs, we conclude that this is sufficient to locate the correct 95% confidence level to within ±5%. Those results which are presented only in graphical form are based on 100 samples, which should be correct to ±10%. In addition, the calibration of the 19.2 GHz survey is accurate to 3%, and the instrumental noise is measured to 5%, giving a total systematic error of ±8%.

The data for the Northern hemisphere are prepared by subtracting an *a priori* model of Galactic emission. This model was constructed by scaling the 408 MHz survey [8] with a spectral index of −2.75. To this was added a compilation of bremmstrahlung sources which had been normalized by fitting to the 25.4 GHz survey [6, 7] and then scaled to 19.2 GHz with a spectral index of −2.1. In those portions of the sky where comparison is possible (*i. e.* in the Galactic plane) the model agrees with our observations to within a factor of two. This poor agreement suggests that there may be significant Galactic emission off the plane which this model does not remove. We then fit and remove the constant and dipole components. The portion within 20° of the Galactic plane is removed, as well as the region near Orion.

III. RESULTS

In this section we present the results of comparing several correlation functions to the data. Since a least-square best fit dipole and offset is subtracted from the data, the correlation functions also have these components removed. Since our coverage of the sky is not uniform, it is not sufficient to simply eliminate the $l = 0$ and 1 components when constructing the Monte Carlo samples; for each sample we determine a least-square best fit dipole and offset and subtract it.

We first compare our data to two families of generic models. Generic correlation functions are conveniently parametrized by an angular scale; we adopt the convention that the angular scale θ_0 is the correlation length. (Note that with this convention the angular scale θ_0 of correlation functions which exclude the $l = 0$ and 1 components cannot exceed the correlation angle of a quadrupole, which is $\sqrt{1/3}$ radians or about 33°.) The first family of generic models, dubbed monochromatic, are defined as having only one non-zero component of the power spectrum C_l; then $C(\theta) = P_l(\cos\theta)$, and $C'(\theta) = H_l^2 P_l(\cos\theta)$. The second generic family is called Gaussian-shaped because the power spectrum has the shape of a Gaussian function; this use of the term "Gaussian" should not be confused with its use in describing the statistics of the fluctuations. These models are defined by

$$C_l = Ne^{-\frac{1}{2}(l/l_0)^2},\ l \geq 2, \tag{2}$$

n	Upper Bound $\Delta T/T \times 10^6$	Sensitivity $\Delta T/T \times 10^6$
+1	45	19
0	53	27
−1	50	32
−2	42	38

Table 1: Upper limits on a_2 for power-law spectrum models.

where N is a constant chosen to normalize $C(0) = 1$.

The results of testing monochromatic and Gaussian-shaped models are shown in Figures 1a and 1b respectively. These figures also show the upper bound sensitivity (as defined in Section II) of our experiment to fluctuations of these forms. Note in particular that the upper limit on the amplitude of a quadrupole ($l = 2$) anisotropy is $\Delta T/T < 2.8 \times 10^{-5}$ rms; the upper bound sensitivity to the quadrupole is $\Delta T/T < 3.2 \times 10^{-5}$ rms. In cases (such as this one) where the upper bound is less than the upper bound sensitivity we prefer to quote the more conservative number. The fact that the limits we have set are in many other cases so far above the sensitivity of the experiment is a clear indication that the map is correlated. It is apparent that we have detected the presence of fluctuations in our data, but the magnitude of Galactic emission in these data and the poor fit of our Galactic model in those regions of the map where comparison is possible preclude identification of these fluctuations as CMB anisotropy. The possibility that the detected anisotropy is due to Galactic emission is supported by the fact that both the upper and lower bounds on all models examined increase significantly when the width of the Galactic plane cut is decreased.

Many theoretical models have as their large angular scale limit a power law spectrum. For these models, the correlation function is given by [1]

$$C_l = (a_2)^2 \frac{\Gamma\left(l + \frac{n-1}{2}\right)\Gamma\left(\frac{9-n}{2}\right)}{\Gamma\left(l + \frac{5-n}{2}\right)\Gamma\left(\frac{3+n}{2}\right)}, \quad n < 3, \quad l \geq 2, \tag{3}$$

and $C(\theta) \approx \theta^{1-n}$. The $n = 1$ case corresponds to scale-invariant fluctuations in two dimensions. For these models we deviate from our practice of quoting an upper limit on the total rms fluctuation, necessitated by the fact that for $n > 0$ these models have infinite rms. Instead, we quote an upper limit on the parameter a_2. Our 95% confidence level upper limits and upper bound sensitivities are presented in Table I.

For most of the models we have presented, the upper bound sensitivity of the experiment is well below the upper bound that we are able to set, suggesting that if we are able to improve the removal of foreground emission, we may be able to significantly lower these upper bounds.

We would like to thank Steve Meyer, Lyman Page, Jeff Peterson, Peter

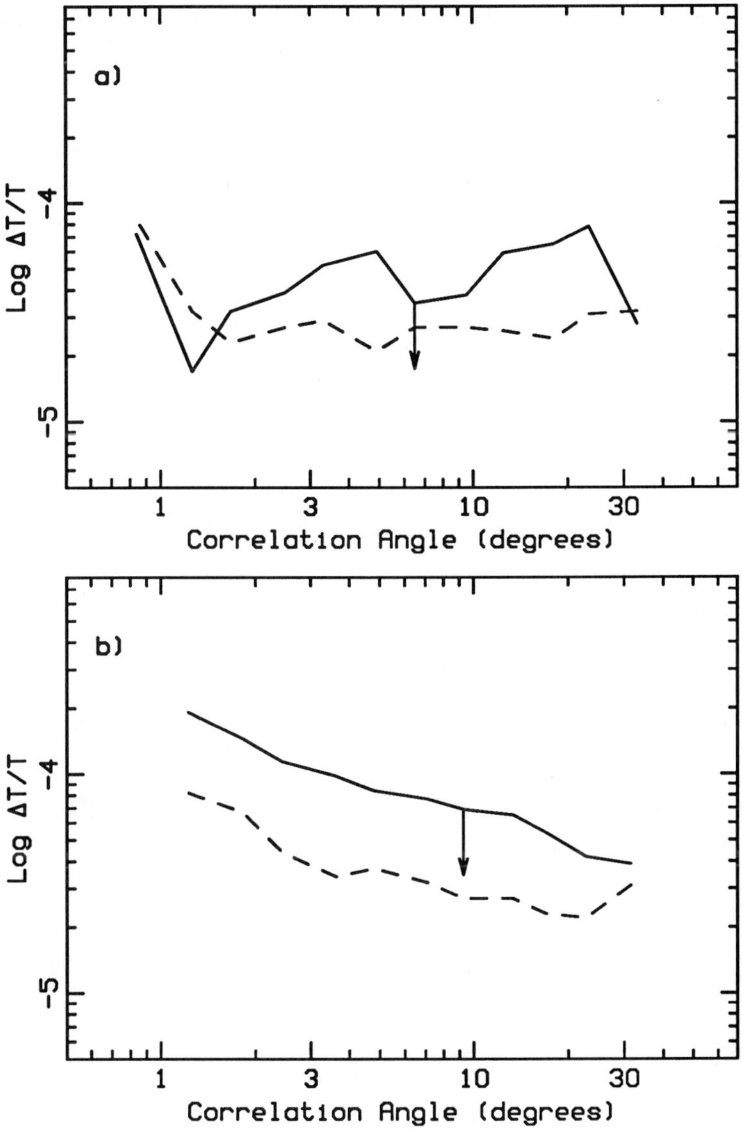

Figure 1: Upper bounds on $\Delta T/T$ rms (95% confidence level). a) Monochromatic models, b) Gaussian-shaped models. In both, the solid line is the 95% confidence level upper bound, and the dashed line is the upper bound sensitivity of this experiment, as defined in the text.

Timbie, Juan Uson, and Nicola Vitorrio for many helpful discussions. This work received a great deal of guidance and support from David Wilkinson. We also benefitted from the discussions at the 1990 Winter Physics Conference at the Aspen Center for Physics. Princeton University was the home of this project at its inception and through the gathering of the Northern hemisphere data. This work was supported by NASA and in part by the NSF.

References

[1] Bond, J. R. and Efstathiou, G. 1987, *M. N. R. A. S.* **226** 655.

[2] Boughn, S. P., Cheng, E. S., Cottingham, D. A., and Fixsen, D. J. 1990, *R. S. I.* **61** 158.

[3] Boughn, S. P., Cheng, E. S., Cottingham, D. A., and Fixsen, D. J. 1990, submitted to *Ap. J. (Letters)*.

[4] Boughn, S. P., Cheng, E. S., Cottingham, D. A., and Fixsen, D. J. 1990, in preparation.

[5] Cottingham, D. A. 1987, Ph. D. thesis, Princeton University.

[6] Fixsen, D. J. 1982, Ph. D. thesis, Princeton University.

[7] Fixsen, D. J., Cheng, E. S., and Wilkinson, D. T. 1983, *Phys. Rev. Lett.* **50** 620.

[8] Haslam, C. G. T., Salter, C. J., Stoffel, H., and Wilson, W. E. 1982, *Astron. and Astrophy. Suppl.* **47** 1.

EARLY RESULTS FROM THE MIT MILLIMETER AND SUB-MILLIMETER BALLOON-BORNE ANISOTROPY MEASUREMENT

Stephan S. Meyer
Edward S. Cheng[1]
Lyman A. Page
Massachusetts Institute of Technology
Cambridge, MA 02139

ABSTRACT

The MIT balloon-borne bolometric search for Cosmic Microwave Background Radiation (CMBR) anisotropies places the the most stringent constraints to date on the fluctuations in the the CMBR. Four maps of half of the Northern Hemisphere at 1.8, 1.1, 0.63 and 0.44 mm wavelength, have a beam size of 3.8° with a 1 sigma sensitivity of less than 0.1 mK (thermodynamic) per FOV in each of the first two channels. Analysis of the sky map at 1.8 mm wavelength using a likelihood ratio test for galactic latitudes $|b| \geq 15°$ yields a 95% confidence level (CL) upper limit on fluctuations of the CMBR at $\Delta T/T \leq 1.6 \times 10^{-5}$ with a statistical power of 92% for Gaussian fluctuations at a correlation angle of 13°. Between 3° and 22°, the upper limit of fluctuations is $\Delta T/T \leq 4.0 \times 10^{-5}$ (95% CL). An anisotropy is detected in the map but it cannot yet be attributed to primordial sources. The ultimate sensitivity for this experiment is $\Delta T/T \sim 7 \times 10^{-6}$ (95% CL) over this angular range for Gaussian fluctuations.

INTRODUCTION

The MIT Cosmic Microwave Background Radiation (CMBR) anisotropy measurement has mapped half of the Northern Hemisphere at 1.8, 1.1, 0.63 and 0.44 mm wavelength, designated Channels 1 through 4, in a flight on October 6, 1989. The radiometer was also flown on October 15, 1988 and May 15, 1990. The results from the 1988 flight are presented in Page, Cheng and Meyer (1990). The instrumentation is is unchanged since 1988 with the exception of the addition of a sensitive vertical gyroscope system to measure the orientation of the optical axis with respect to the zenith, and the addition of an input blocking filter effective at wavelengths shortward of 20 μm. The sensitivity to variations in the CMBR temperature of the sky is 0.63 mK for each one second integration. Only Channel 1 data from the October 6, 1989 flight are discussed.

The flight acquired a total of 10 hours of data with 2.3×10^4 seconds used

[1] NASA Goddard Space Flight Center

to produce a map of the sky covering about half the Northern Hemisphere. The remaining time was devoted to calibration and checking for systematic errors. At an average altitude of 36.3 km, the gondola rotates in azimuth once per 40 seconds with the beam at a zenith angle of 44°. During the flight the zenith position changed from $\alpha = 21.13^{\mathrm{h}}$, $\delta = 34.5°$ to $\alpha = 3.63^{\mathrm{h}}$, $\delta = 34.6°$. In galactic coordinates, the map lies within the region from approximately $\ell = 45°$ to 200° and $b = +60°$ to $-80°$.

DATA REDUCTION

The data stream in each channel consists of four synchronously integrated samples per 4.5 Hz chopper cycle. Each integration is thus 56 ms in length. The bolometric detectors, sensitive to all forms of thermal power, respond to cosmic rays with a positive spike of variable strength depending on the deposited energy. To remove the spikes, the data are filtered after having the sky contribution subtracted. One second of data is removed for each detected event since the detectors and signal electronics require some time to recover. Roughly 7% of the total data set is removed as a result of cosmic ray hits.

The data are divided into six one hour sections by 12 minute calibration sequences. Each hour of the spike-removed data is digitally high-pass filtered with a cutoff frequency corresponding to two gondola rotations. This removes the drift in the offset which is due to both the changing atmospheric pressure (the gondola altitude varies with a period of about 8 minutes or 12 rotations) and the slow cooling of the input optics. The data are binned into 1.3° × 1.3° pixels on the sky. The variance in each sky pixel is minimized by fitting a cubic spline to the time ordered data with one breakpoint for every 8 rotations.

The full map of the sky is made by concatenating these one hour segments. For the data in Channel 1 the best calibration comes from the detection of the CMBR dipole (the dipole is detected in Channels 2 and 3 as well). The high pass filter has removed the information on the DC level of the signal so the data at galactic latitudes greater than 15° are fit to a dipole with fixed amplitude and unknown direction. The value of the dipole amplitude is assumed to be 3.3 mK (Cottingham 1990, Lubin *et al.* 1985). There is one multiplicative and one additive fit parameter for each one hour segment and two parameters for the direction of the dipole. The six hours of data are thus fit with 14 parameters. The dipole direction, after correcting for the earth orbital motion, is $\delta = -1.3° \pm 3.°$ and $\alpha = 10^{\mathrm{h}}.68 \pm 0^{\mathrm{h}}.19$ with the error being dominated by a $\pm 3°$ pointing uncertainty.

After removing data with galactic latitude $|b| \leq 15°$, the map has 3500 pixels each with a temperature and variance. The average error is $\sqrt{\langle \sigma^2 \rangle} = .26$ mK where σ is the standard deviation of the average value in a given pixel determined from the variance of the data samples which fell in the pixel over the one hour intervals. The χ^2 of the fit to the dipole is 1.7 per degree of freedom. The variance in each pixel is 1.2 times what is expected from ground measurements of the radiometer NET indicating no large contribution from residual atmospheric noise.

ANISOTROPY

The map is analyzed for anisotropy using a modified likelihood ratio test. In this initial analysis only Gaussian fluctuation models over a range of correlation angles from 3° to 22° are tested. Following the method and implementation of Cottingham (1987) and Boughn *et al.* (1990), a statistic, S, is formed which is related to the likelihood ratio, λ of Lawrence, Readhead and Myers (1988). The calculation of the formal likelihood ratio is intractable for a data set with several thousand correlated points on the sky. The distribution of S is determined with a Monte Carlo simulation of the experiment with the sky coverage, beam profile, and variance in each pixel taken into consideration. For each correlation angle 300 manifestations of the experiment are created with no CMBR fluctuations and 300 with fluctuations of amplitude η, where η is the total rms fluctuation predicted by a model $\langle T(\Theta_i)T(\Theta_j)\rangle = \eta^2 C(\theta_{ij})$, with the correlation function, C, normalized so that $C(0) = 0$. θ_{ij} is the angle between directions Θ_i and Θ_j. For this analysis the correlation function is Gaussian and the correlation angle is $\theta_0^{-2} = -\left[\frac{1}{C}\frac{d^2C}{d\theta^2}\right]_{\theta=0}$. When η is chosen such that the value of S from the Monte Carlo simulations is higher than the observations 95% of the time, then this value is the 95% confidence upper limit for fluctuations at this correlation angle. The distribution of S, assuming no fluctuations, is used to determine the statistical power of the test, defined as the fraction of this null distribution below S for the data. The values of η are shown in Figure 1 for each tested correlation angle. The power is greater than 90% for each upper limit plotted in Figure 1. Also plotted in Figure 1 is the ultimate sensitivity of the experiment. This is the value of η for which 95% of the simulations exceed the mean S for the null distribution. In other words, if the sky were isotropic to a level well below the ultimate sensitivity, η is the most probable 95% CL upper limit this measurement would produce.

DISCUSSION

The reason for the relatively high statistical power of the upper limits at each correlation angle is that the observed map has more structure than is statistically likely. This can be seen from the reduced χ^2 for the fit to the dipole. A value of $\chi^2/\nu = 1.7$ is statistically unlikely. Also, the 95% sensitivity is more than a factor of two lower than the upper limits. There is a clear detection of some sort of anisotropy. There are several possible sources: galactic emission due to high galactic latitude dust; non-isotropic stationary emission from the atmosphere; systematic errors due either to instrumental effects or the earth in the sidelobes of our beam; and the CMBR. Each of these is discussed below.

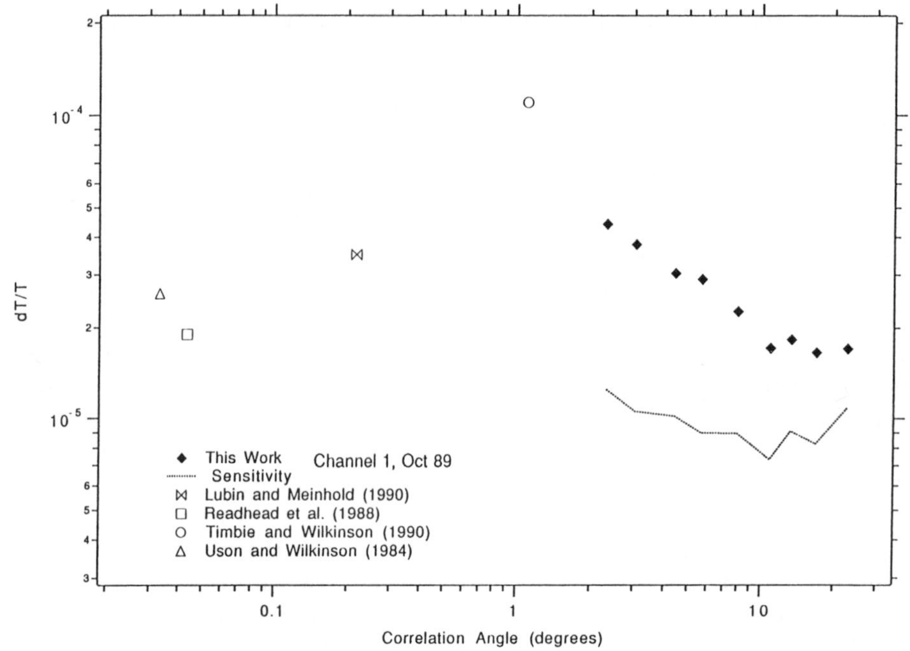

Figure 1. The 95% confidence upper limits for Gaussian fluctuations in the CMBR as a function of correlation angle. The black diamonds are the 95% CL upper limits of this work, the line is the sensitivity of this experiment. The statistical power of the test for each of these points is over 90%. Also plotted are published values for the Gaussian limits of others.

Galactic emission at high galactic latitude is a source which cannot be ruled out with the data at just this wavelength. The IRAS 100 μ map scaled to 1.8 mm using Figure 4 in Page, Cheng and Meyer (1990a), suggests structure at the $\Delta T/T_{\mathrm{CMBR}} \sim 10^{-5}$ level.

Residual atmospheric emission, seen with high signal to noise in Channel 3, could also give rise to apparent anisotropy. The atmosphere must be stationary over the time of the observation of a pixel (5 minutes at the shortest). Any variation over the time scale of 5 minutes will decrease the sensitivity of the experiment. A residual atmospheric "cloud," stationary in position, could produce the observed signal. In the 1988 October flight, such a cloud was observed. In the first hour of the 1989 October flight similar clouds were observed; however, after an hour at the observing altitude their contrast diminished considerably.

The instrument has several characteristics which could result in systematic errors. There is a geomagnetic sensitivity due to the magnetic nature of our cryogenic chopper drive (Page, Cheng, and Meyer 1990c). This effect could produce an apparent dipole at a level of ≤ 0.3 mK. Any dipole is subtracted from the map before analysis for anisotropy and could potentially cause a 10% error in the calibration along with a small error in dipole direction. Significant higher moments are not produced by this mechanism. The instrument is also sensitive to gondola swinging which causes a variable observed airmass along the line of sight and also modulates the cryogenic chopper which is not perfectly balanced. The gyroscopes measure this swinging to 1 arcminute. To detect the variation of the chopper operation we carefully measure the chopper coil EMF during the flight. Using laboratory calibrations we can limit the size of the chopper and airmass effects to < 0.03 mK. Diffracted earthshine could produce anisotropy signals. The total brightness of the earth diffracted over the edge of the ground screens is $T_{\rm ant} \sim 6.0\ \mu$K, and only variations in this could produce detected effects. Radiometric contributions from the balloon and ground screen emission are smaller.

The characteristics of the map production has been tested by dividing the full map into two pieces and determining the upper limits for each piece separately. The results were similar to those for the total map but with the sensitivity averaging 1.5 higher for each piece and the 95% CL upper limit higher by a few percent. In addition, maps using only 5 of the 6 hours of flight data were made and the expected results were obtained. From this it is clear that there is no one position on the sky which is giving rise to the detected signal. The algorithms for the detection of anisotropy from the maps have also been tested by forming a simulated map which has the same σ in each pixel and the same sky coverage as the flight data. The simulated map is produced so that the signal is Gaussian distributed with the correct σ in each pixel. This null map, when fit to zero, has a $\chi^2 = 0.98$, and produces a very low 95% CL upper limit with no statistical power when analyzed in the same manner as the flight data.

Finally, the detected signal or some part of it could be of cosmological origin. A change in the CMBR temperature produces a unique spectral signature in the four channels and can be confirmed by further multispectral analysis. In addition, the 1990 May flight has 30% overlap with the 1989 October flight. This will aid in the separation of atmospheric from CMBR and galactic sources.

CONCLUSIONS

There is a clearly some sort of anisotropy in the data at the level of $\Delta T/T = 1{-}3 \times 10^{-5}$ at scales greater than 4°. However, reasonable extrapolations of dust emission from the IRAS data and previous measurements of the atmosphere suggest that these are possible sources for the signal. Whatever the cause, the upper limit determined from this experiment serves to constrain CMBR anisotropy at new levels. Since these measurements are not limited by instrument sensitivity, further analysis which includes the shorter wavelength data will clarify the origin of the anisotropy.

We wish to thank David Cottingham for many helpful discussions and the use of his routines to analyze the maps for anisotropy. Many people helped in the construction of the instrument. We thank the undergraduate students whose work helped make this flight possible: Michael Chandross, Deborah Dougherty and Emmin Shung. We thank the National Scientific Balloon Facility for a flawless flight. Lyman Page was partially supported by a NASA Graduate Student Fellowship. This research was supported by NASA grant NAGW-1841.

REFERENCES

Boughn, S.P., Cheng, E.S., Cottingham, D.A., Fixsen, D. J., and Wilkinson, D. T. 1990. *Ap. J.*,, submitted.

Cottingham, D. A., 1987 Ph.D. Thesis, Princeton University.

Cottingham, D. A., 1990 (private communication).

Lawrence, C. R., Readhead, A. C. S., and Myers, S. T. 1988 , in The Post Recombination Universe, ed. N. Kaiser and A. N. Lasenby (Dordrecht: Kluwer), p. 173.

Lubin, P., Villela, T., Epstein, G., and Smoot, G. 1985. *Ap. J. (Letters)***298**, L1.

Meinhold, P. R. and Lubin, P. L. 1990. *Ap. J. (Letters)*, submitted.

Page, L. A., Cheng, E. S., and Meyer, S. S. 1990a. *Ap. J. (Letters)***355**, L1.

Page, L. A., Cheng, E. S., and Meyer, S. S. 1990b, in preparation.

Page, L. A., Cheng, E. S., and Meyer, S. S. 1990c, Appl. Optics. submitted.

Readhead, A. C. S., Lawrence, C. R., Myers, S. T., Sargent, W. L. W., Hardebeck, H. E., and Moffet, A. T. 1989. *Ap. J.*,**346**, 556.

Timbie, P. T., and Wilkinson, D. T. 1990. *Ap. J.*,**353**, 140.

Uson, J. M., and Wilkinson, D. T. 1984. *Ap. J.*,**283**, 471.

INFLATION, LARGE SCALE STRUCTURE, AND COBE

Robert K. Schaefer
Bartol Research Inst., Univ. of Delaware
Newark, DE 19716

ABSTRACT

The predictions of the inflationary cosmology are compared with the expected sensitivity of the COBE satellite. It is noted that detection of large angle temperature fluctuations in the usual biased cold dark matter version of inflation requires COBE to reduce systematic errors to the level of the detector sensitivity. However, the difficulty the usual biased cold dark matter scenario has in fitting recent large scale structure observations leads us to consider a variety possible modifications of inflation. These changes generally increase the amplitude of the temperature fluctuations to a more detectable level, although the amount of increase depends the specific modification. Thus we would expect that COBE will easily see temperature fluctuations in a realistic model of large scale structure formation based on inflation.

I. INFLATION

The attractive inflationary model of the early universe makes some strong predictions:

- Space is "flat", implying $\Omega \equiv \rho/\rho_{crit} = 1$. Primordial nucleosynthesis constrains baryonic matter to be less than about 10% of this value, so the universe is dominated by dark matter.
- Quantum fluctuations during inflation produce Gaussian distributed density fluctuations which have an amplitude independent of scale. The fluctuation variance, specified at horizon crossing, is:

$$\left\langle \left(\frac{\delta\rho}{\rho}\right)^2_{t_{HOR}} \right\rangle = \epsilon_H^2 = \text{constant}.$$

  In Fourier space these density fluctuations are characterized by the power spectrum $P_I(k) = k$. Unfortunately the amplitude ϵ_H is not predictable from the particle physics of inflation, so ϵ_H is normalized to the observed structure in the universe.

II. BACKGROUND RADIATION ANISOTROPY FROM INFLATION

Primordial density fluctuations can be detected through their gravitational (Sachs-Wolfe) effect on the cosmic background radiation. These large angle temperature anisotropies can be characterized in two ways:

- *MULTIPOLES* - coefficients of spherical harmonics $Y_{\ell,m}(\hat{e})$. The fractional temperature fluctuations (minus the dipole) can be expanded as:

$$\frac{\delta T}{T}(\hat{\mathbf{e}}) = \sum_{\ell=2}^{\infty} \sum_{m=-\ell}^{+\ell} a_{\ell,m} Y_{\ell,m}(\hat{\mathbf{e}}).$$

and are plotted in Figure 1. We plot the values of the moments scaled by the unknown initial amplitude ϵ_H^2. The ℓth multipole is then defined as $a_\ell^2 = \sum_{m=-\ell}^{+\ell} |a_{\ell,m}|^2$, and can be calculated[1]. The multipoles are stochastic variables with a χ^2 distribution[1,2,3], each with $2\ell + 1$ degrees of freedom. We have calculated the variance of the multipoles and correlation function and plotted standard error bars (square root of the variance) on our graphs.

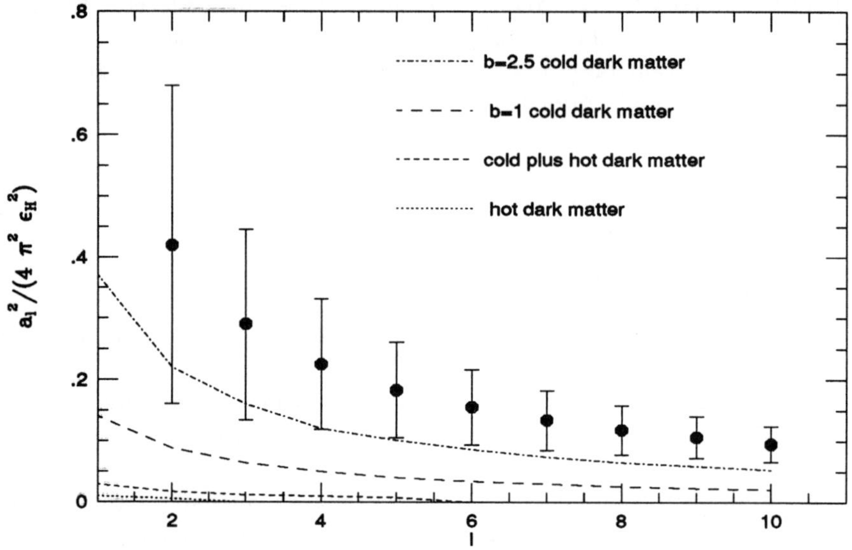

Fig. 1). The first few inflationary multipole moments. The sensitivity of COBE to the various models is represented by the scaled curves.

- *TEMPERATURE ANGULAR CORRELATION FUNCTION*[4],

$$C(\alpha) \equiv \left\langle \frac{\delta T}{T}(\hat{\mathbf{e}}_1)\frac{\delta T}{T}(\hat{\mathbf{e}}_2)\right\rangle = \frac{1}{4\pi}\sum_{\ell} a_\ell^2 P_\ell[cos(\alpha)]$$

where $cos(\alpha) = \hat{\mathbf{e}}_1 \cdot \hat{\mathbf{e}}_2$, and P_ℓ are Legendre polynomials. The finite resolution of the COBE detector with Gaussian width σ_{DMR} can be treated by multiplying[5] the multipole moments by $exp[-(\ell+1/2)^2\sigma_{DMR}^2]$ in the correlation function equation, which has been done in figure 2. Again, we have scaled the correlation function by the undetermined amplitude ϵ_H^2

III. COBE SENSITIVITY

The ultimate sensitivity of the COBE satellite maps is not yet known because the actual number depends on the ability to reduce systematic errors, especially those arising from removal of galactic emissions. Also the COBE instrument has nonuniform sky coverage, giving a directional dependence to the sensitivity. Ignoring these concerns, we can expect a theoretical limit[6] for the COBE sensitivity $|a_{\ell,m}| \geq 1 \times 10^{-5}$. Multipole moments which satisfy $a_\ell^2 \geq 10^{-10}/(2\ell+1)$ can then be detected. In our plots we scale the COBE sensitivity curves instead of the multipoles, so the COBE sensitivity is

$$\frac{a_l^2}{4\pi^2\epsilon_H^2} \geq \frac{1 \times 10^{-10}}{4\pi^2\epsilon_H^2(2\ell+1)}$$

Thus we have drawn a separate line for the COBE sensitivity to each model.

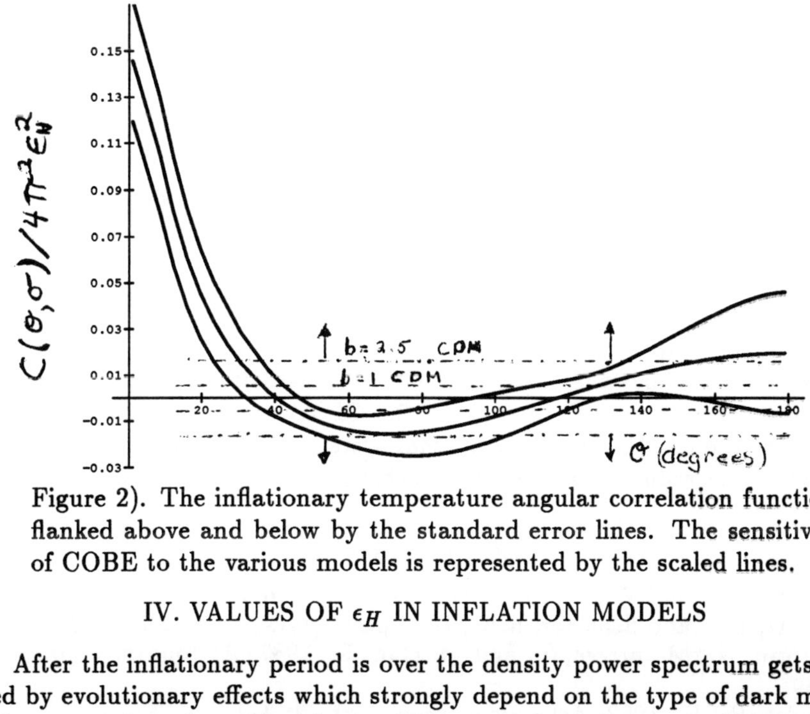

Figure 2). The inflationary temperature angular correlation function, flanked above and below by the standard error lines. The sensitivity of COBE to the various models is represented by the scaled lines.

IV. VALUES OF ϵ_H IN INFLATION MODELS

After the inflationary period is over the density power spectrum gets modified by evolutionary effects which strongly depend on the type of dark matter. This means that the value of ϵ_H will in turn depend on our choice of dark matter. We consider three types of dark matter, cold (CDM), hot (HDM), and a mixture of cold plus hot (CPH) with $3\rho_{cold} = \rho_{hot}$. This latter mix of dark matter is very useful for explaining[7] large scale structure. The values of ϵ_H for these models are shown in Table 1, where we have normalized the models to galactic correlations. Galaxies are assumed to represent $b\sigma$ fluctuations of the density field and "biasing" implies $b \neq 1$. The relative sensitivity of these models is shown in figures 1 and 2. As evidenced by many talks at this workshop, CDM (long considered the "standard model") has difficulty explaining large scale power apparent in observations. In Table I, we show this deficit by comparing the predicted large scale bulk flow velocities with those derived from observations[8]. We also show how the other dark matter models compare.

TABLE I. Normalizations and Bulk Velocities (in km/sec)

Dark Matter Model	$\epsilon_H \times 10^6$	$V(40\ h^{-1}\,\mathrm{Mpc})$	$V(60\ h^{-1}\,\mathrm{Mpc})$
Cold Dark Matter	$4.2/b$	$290/b$	$240/b$
Cold Plus Hot D. M.	$5.4/b$	$390/b$	$320/b$
Hot Dark Matter	$9.2/b$	$690/b$	$550/b$
Observed Velocity		390 ± 70	330 ± 80

Another way to fix the large scale power deficit inflation within the minimum anisotropy producing CDM model is to modify the inflation process itself in a way which breaks the scale invariance of $P_I(k)$. A variety of ways have been proposed to do just this: power law inflation[9,10], double inflation[11], and "non-standard" inflation[12]. These modifications introduce other free parameters which we must fit to structure, although there is no compelling reason from the models why the parameters should have such values in addition to the already needed value of ϵ_H. If we adjust the parameters to explain both the galaxy correlations and the bulk flow velocities, we find that power law inflation and double inflation will give signals at least as large as the CPH model. "Non-standard" inflation can be adjusted to give a COBE signal as close to the CDM model as you like. The real descriminator of this model will be with smaller angle anisotropy measurements.

V. CONCLUSIONS

Inflationary models of structure formation produce temperature fluctuations which are above the theoretical COBE sensitivity limit. However, the so-called "standard model" (biased $b = 2.5$ CDM) produces signals which are so close the the COBE detector limit that systematic errors in the COBE data will have to be reduced to levels approaching this detector limit, not an easy task.

REFERENCES

1. Abbott, L.F. and Wise, M.B., 1984, *Astrophys. Jour. Lett.*, **282**, L47.
2. Abbott, L.F. and Schaefer, R.K., 1986, *Astrophys. Jour.*, **308**, 546.
3. Vittorio, N., Matarrese, S., and Lucchin, F., 1987, *Astrophys. Jour.*, **328**, 69.
4. Peebles, P.J.E., 1982, *Astrophys. Jour. Lett.*, **263**, L1.
5. Scaramella, R. and Vittorio, N., 1988, *Astrophys. Jour. Lett.*, **331**, L53. Scaramella, R. and Vittorio, N., 1990, *Astrophys. Jour.*, **353**, 372.
6. J. Mather, 1982, *Optical Engin.*, **21**, 749.
7. Schaefer, R.K., Shafi, Q., and Stecker, F.W., 1989, *Astrophys. Jour.*, **347**, 575.
8. Bertschinger, E., Dekel, A., Faber, S.M., and Dressler, A., 1990, preprint.
9. Abbott, L.F. and Wise, M.B., 1984b, *Nucl. Phys.*, **B244**, 541.
10. Lucchin, F. and Matarrese, S., 1985, *Phys. Rev. D*, **32**, 1316.
11. Silk, J. and Turner, M.S., 1987, *Phys Rev.*, **D 35**, 419.
12. Salopek, D.S., Bond, J.R., Bardeen, J., 1989, *Phys. Rev.*, **D 40**, 1753. Hodges, H.M., Blumenthal, G.R., Kofman, L.A., Primack, J.R., 1990, *Nuclear Physics*, **B** (in press.)

A MILLIMETER-WAVE ANISOTROPY EXPERIMENT (MAX) TO SEARCH FOR ANISOTROPY IN THE COSMIC BACKGROUND RADIATION ON MEDIUM ANGULAR SCALES

M. L. Fischer, D. C. Alsop, A. C. Clapp, D. A. Cottingham,
A. E. Lange, P. L. Richards, G. Smoot
Department of Physics and Space Sciences Laboratory,
University of California, Berkeley, CA 94720

J. O. Gundersen, T. C. Koch, P. R. Meinhold, P. M. Lubin
Department of Physics, University of California, Santa Barbara, CA 93106

E. S. Cheng
Code 685, NASA Goddard Space Flight Center, Greenbelt, MD 20771

E. Kreysa
Max-Planck-Institut für Radioastronomie, Auf dem Hügel 69,
D-5300 Bonn 1, West Germany

ABSTRACT

We report preliminary results from two balloon flights of a millimeter-wave telescope designed to measure anisotropy in the cosmic microwave background (CMB) on angular scales from 0.3 to 3 degrees. The receiver used in the first flight, in November 1989, was a dichroic, ^{3}He-cooled bolometric photometer with passbands centered at 3, 6, 9, and 12 cm^{-1}. During this flight we measured the spectrum of the brightness of the galactic plane at a galactic longitude of ℓ_{II} = 24 degrees and searched for CMB anisotropy at nine points centered near the north celestial pole (NCP). The noise in the 6 cm^{-1} band integrated down to one sigma errors of 140 μK per point over the 40 minute observation near the NCP. After further analysis these data will set an upper limit on CMB anisotropy of $\Delta T/T < 1\text{-}2 \times 10^{-4}$ (95% CL) for Gaussian correlation functions with a correlation length of 0.5 degree. The second flight, in July 1990, employed improved bolometric detectors. The 3 cm^{-1} band was removed in order to increase the efficiency of the remaining bands. Preliminary analysis of the data gives a factor of 3 improvement in sensitivity over the first flight in the 6 cm^{-1} band. Future plans include a new receiver to provide an additional factor of 5 improvement in sensitivity and a new balloon-borne telescope which is optimized for low-background bolometric detectors at millimeter wavelengths.

INTRODUCTION

Measurements of the anisotropy of the CMB on angular scales from 0.3 to 3 degrees provide information on primeval density perturbations that were just entering the causal horizon at decoupling and place powerful constraints on the cold dark matter scenario of galaxy formation (Vittorio et al., 1989, Bond et al., 1989). Sensitive experiments which probe these angular scales are now being conducted by groups using a variety of instrumentation ground-based and balloon-based platforms (de

Bernardis et al., 1990, Timbie and Wilkinson, 1990, Meyer, 1990, Meinhold and Lubin, 1990). As the sensitivity of the instruments improves, systematic effects and/or confusion with galactic emission will limit the accuracy of the observations. Observations at several wavelengths will be necessary in order to discriminate between anisotropy in the galactic and cosmic backgrounds. At present it is not known which combination of instruments and observing sites will provide the best control of systematic errors at those wavelengths. All promising avenues should be pursued. This paper describes two flights of a balloon-borne millimeter-wave telescope developed to provide a low background environment for bolometric receivers.

INSTRUMENT

The instrument is a pointed, 1-meter off-axis balloon-borne telescope (Meinhold, 1989) which has been modified with a new chopping secondary mirror and a multi-band bolometric receiver (Fischer et al., in preparation). The feed horn accepts radiation from the sky in a 0.5 degree (FWHM) Gaussian beam. All of the mirrors are underfilled to minimize beam spillover at the mirror edges (Fig. 1). The beam is sinusoidally chopped at 5.4 Hz between two points on the sky separated by 1.2 degrees. The gondola azimuth can be wobbled back and forth over a fixed number of positions on the sky, each separated by the effective chopper throw, or scanned continuously. The receiver is a dichroic photometer which simultaneously observes

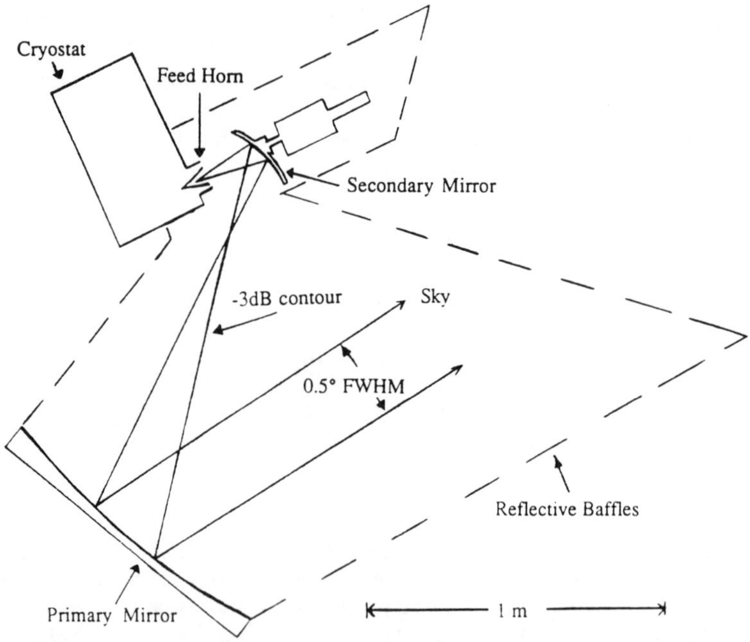

Figure 1. Sideview of telescope. Optical rays indicate nominal -3 dB contours. Dashed lines indicate reflecting baffles which block direct earthshine paths to optical surfaces.

four passbands centered at 3, 6 ,9 , and 12 cm^{-1}. The detectors are ^{3}He-cooled bolometers operating near 0.3 K. The system is calibrated every 30 minutes during the flight by moving a partially reflecting membrane into one lobe of the chopped antenna beam near the prime focus. The membrane reflects a known fraction (between 10^{-3} and 10^{-2}) of an ambient temperature blackbody source into that lobe. The reflectivity of the calibrator membrane is measured in the laboratory to vary from 10^{-3} to 10^{-2} at frequencies from 3 to 12 cm^{-1}.

GALACTIC EMISSION

During the first flight, in Nov. 1989, we measured the integrated emission from the galactic plane in four bands at a galactic longitude of 24 degrees. The telescope scan direction was inclined at an angle of 64 degrees to the galactic plane, measured from the direction of increasing galactic longitude. The beam was chopped along the scan direction, producing the differential measurement of the sky brightness as a function of the scan angle shown in Fig. 2. The data are strongly correlated with the signal expected from interstellar dust at 100 μm, calculated by convolving the 100 μm IRAS map with the effective shape of our chopped antenna beam.

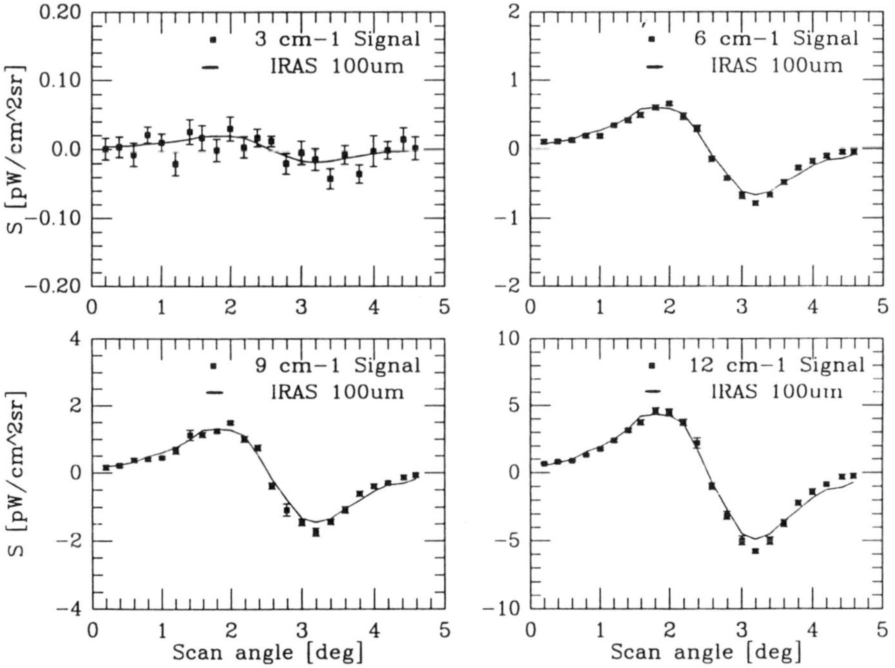

Figure 2. Measured differential signals from the galactic plane emission at $\ell_{II} = 24$ degrees. The solid line shows the signal calculated by convolving the 100 μm IRAS signal with the effective shape of the chopped antenna beam. The amplitude of the calculated signal has been fit to the measured data separately for each passband.

Figure 3 shows the IRAS correlated component of the measured sky brightness as a function of frequency for the 6, 9, and 12 cm^{-1} bands. For comparison, we also show three plausible models for the thermal dust emission, each normalized to the IRAS brightness at 100 cm^{-1}. The models for dust emissivity assume a power law frequency dependence for the emissivity with index n = 0, 1, and 2 and a dust temperature T_{dust} = 20 K. Because the filter bands have a finite bandwidth, the effective frequency of each passband moves toward higher frequency for the steeper dust spectra. This effect is so strong for the 3 cm^{-1} passband that it does not constrain the data. Three points are displayed for each of the other bands, corresponding to the three dust indices n = 0, 1, and 2.

These data are consistent with an emissivity index n between 1 and 2. Given such a steep spectrum, it should be possible to correct the CMB measurements for contamination from dust emission. Galactic dust emission can be identified in the 12 cm^{-1} band and subtracted from the 3 or 6 cm^{-1} bands. It will be important to study regions off of the galactic plane to verify that this spectrum of galactic emission is representative of other galactic latitudes.

CMB ANISOTROPY

During the first flight we observed nine points near the NCP, each separated by 1 degree in scan angle. The telescope was wobbled back and forth in a step-and-integrate mode. The integration time between steps was 1 minute. The total observation time was 40 minutes. Figure 4 shows the binned 6 cm^{-1} data as a function of scan angle. The other bands did not have sufficient sensitivity to provide interesting upper

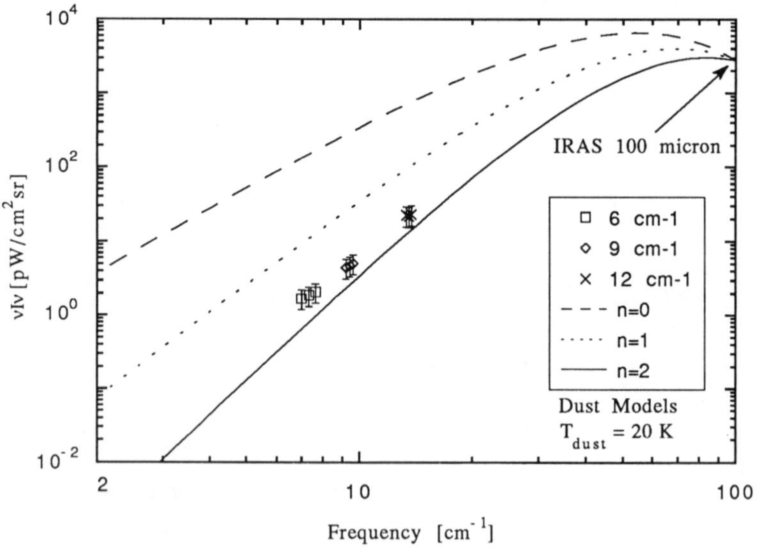

Figure 3. Measured galactic plane brightness as a function of frequency. The curves indicate three thermal dust emission models with emissivity indices n= 0, 1, and 2, and dust temperature T_{dust} = 20 K.

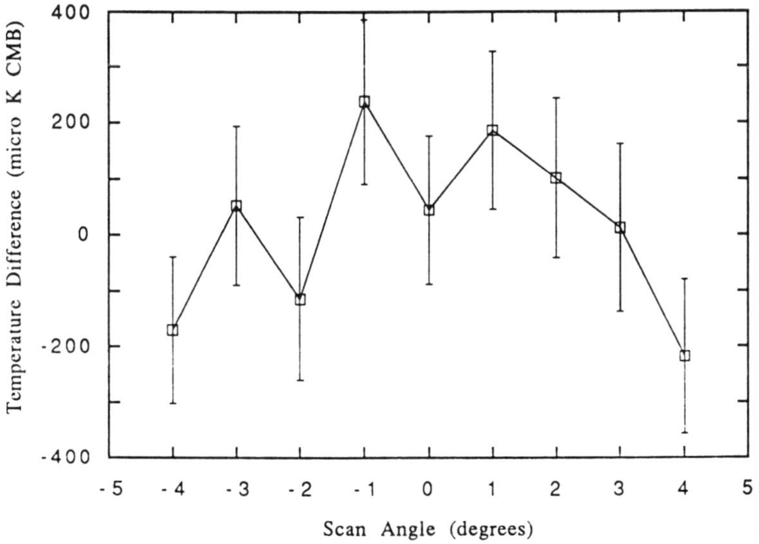

Figure 4. Measured temperature differences in the 6 cm^{-1} band vs. scan angle. These data were measured during the NCP scan of the first flight.

limits and are not shown. A constant offset was subtracted from the binned data. The noise in the 6 cm^{-1} data has a level consistent with detector noise. The noise is Gaussian distributed as judged by a histogram and integrates down as $t^{-1/2}$.

The 1-sigma error bars for each binned point are about 140 μK (2.7K thermodynamic). The χ^2 for these data is 10 for 8 degrees of freedom (9 points minus one for offset subtraction). The probability of obtaining a χ^2 this large or larger is 0.26 which is reasonably consistent with no sky signal. Using a standard likelihood ratio analysis these data will set an upper limit to CMB anisotropy of $\Delta T/T < 1\text{-}2 \times 10^{-4}$ (95% CL) for Gaussian correlation functions with a correlation length of 0.5 degree.

Analysis of the data from the second flight is still in progress. We have determined that the sensitivity of the 6 cm^{-1} band was improved by a factor of 3 over the first flight. During the second flight we obtained 1.3 hours of integration time searching for CMB anisotropy. If the noise integrates down then we may expect to set the limits to $\Delta T/T$ shown in the table below.

Flight Date	Sensitivity Relative to First Flight	Integration Time [Hours]	$\Delta T/T$ [95% UL]
Nov. 89	1	0.7	$1\text{-}2 \times 10^{-4}$
July 90	0.3	1.3	(3×10^{-5})
Spring 91	(0.3)	(6)	(1×10^{-5})
Fall 91	(0.07)	(6)	(2×10^{-6})

(Note that the quantities in parentheses are anticipated but not yet achieved.)

Future plans include a flight of the same receiver in the spring of 1991 with a goal of 6 hours of integration time. We plan to fly a new receiver in the fall of 1991 with an additional factor of 5 improvement in sensitivity. This receiver will use bolometric detectors cooled to 100 mK with an adiabatic demagnetization refrigerator developed for use on SIRTF (Timbie, Bernstein and Richards, 1989). These measurements have the potential to reach the levels of sensitivity shown in the above table.

As the receiver sensitivity continues to increase systematic effects will become more important. A new telescope is being developed which is optimized for balloon-borne bolometric observations at millimeter wavelengths. The issues the new design will directly address are as follows: 1) minimum sidelobe response and low emissivity optics, 2) minimum telescope mass which will allow high altitude flights to minimize atmospheric emission, 3) the choice of observing strategies which maximize our immunity to time varying signals and provide the best diagnostic information on any sources of drift.

We gratefully acknowledge the assistance of the staff of the National Scientific Balloon Facility. M.L.F., D.C.A, and P.R.M. were partially supported by NASA Graduate Student Research Program Fellowships. Work at the University of California was funded by NASA grant NSG-7205 through the Space Sciences Laboratory, and by the Center for Particle Astrophysics, a National Science Foundation Science and Technology Center operated by the University of California under Cooperative Agreement No. AST-8809616. Work at the University of California at Santa Barbara was funded by NASA grant NAGW-1062, and the NSF's Center for Particle Astrophysics.

REFERENCES

Bond, J. R., Efstathiou, G., Lubin, P. M., Meinhold, P. R., 1990, Phys. Rev. Lett., submitted.

de Bernardis, P. et al., 1990, Ap. J. (letters), **360**, L31.

Meinhold, P. R., Lubin, P. M., 1990, Ap. J., submitted.

Meyer, S. S., 1990, private communication.

Timbie, P. T., Wilkinson, D.T., 1990, Ap. J., **353**, 140.

Timbie, P. T., Bernstein, G. B., Richards, P. L., 1990, Cryogenics, **30**, 271

Vittorio, N., Meinhold, P. R., Muciaccia, P. F., Lubin, P. M., Silk, J., 1990, Ap. J., submitted.

CMBR ANISOTROPIES : SMALL SCALE MEASUREMENTS AND THE OVRO EXPERIMENTS

Steven T. Myers
Canadian Institute for Theoretical Astrophysics
60 St. George, Toronto, ON M5S1A1

Anthony C. Readhead, Charles R. Lawrence
California Institute of Technology
Mail Code 105-24, Pasadena, CA 91125

ABSTRACT

This is a short overview of current CMBR anisotropy experiments on small angular scales. Ongoing efforts and future prospects in relation to the testing of cosmological theories are summarized. Details of previously published as well as new observations from the Owens Valley 20 GHz experiment are presented, with limits on the anisotropy on arcminute scales. The new experiment, designated as the RING, is a survey of 96 fields arranged in an interlocked ring-like geometry. The statistical analysis of the RING data results in the upper limits $\delta T/T < 5.0 \times 10^{-5}$ (95%) and $\delta T/T < 6.1 \times 10^{-5}$ (99.87%), for Gaussian fluctuations with a correlation angle of $2\overset{\prime}{.}6$. A significant signal is detected above the noise level, however because of the high probability of residual discrete source contamination these results are quoted as upper limits on intrinsic background fluctuations.

SMALL ANGLE EXPERIMENTS

Since the discovery experiment of Penzias and Wilson (1965), many observations on a variety of angular scales with ever-increasing sensitivity have been made probing for anisotropy in the microwave background radiation. There are two basic angular scales of importance for primary anisotropies : the angle subtended by the horizon scale at recombination ($\sim 2°$) giving the maximum angular scale for causal contact of perturbations, and the angle below which fluctuations are smeared by the non-zero "width" of the last scattering surface ($\sim 10'$). It is customary to categorize observations as small, medium, or large scale based upon these fiducial angles. Experiments such as the balloon (Bernstein *et al.* 1989, de Bernardis 1990, Boughn *et al.* 1990, Meyer, Cheng and Page 1990) and the RELIKT (Klypin *et al.* 1987) and COBE (this conference) space-based experiments probe the medium to large-angle domain, as well as some ground-based observations like the Tenerife experiment (Davies *et al.* 1987) and the interferometric observations of Timbie and Wilkinson (1988). Other experiments have concentrated upon the intermediate angular scales from $10'$ to $2°$, such as the South Pole experiment of Meinhold and Lubin (1990).

Small angular scale experiments entered the limelight with the Uson and

Wilkinson (1984) observations, who reported an upper limit of $\delta T/T < 2.1 \times 10^{-5}$ (95% confidence) for a plane-wave perturbation with wavelength 8′ and $\delta T/T < 2.4 \times 10^{-5}$ (95%) for fluctuations uncorrelated between their switched beams. The measurement was made using the NRAO 140-ft radiotelescope at Green Bank, at a frequency of 19.5 GHz. A "double-differencing" switching scheme (see below) was used with Gaussian beams (1.′5 FWHM) separated by 4.′5. Using a similar system and telescope at the Owens Valley Radio Observatory (OVRO), we have placed an upper limit of $\delta T/T < 1.7 \times 10^{-5}$ (95%) for single uncorrelated beams (Readhead *et al.* 1989, see next section). As a result of these experiments, standard adiabatic baryonic models were ruled out, and the more exotic models using non-baryonic dark matter or isocurvature fluctuation modes with non-standard recombination gained plausibility.

On even smaller angular scales, the Very Large Array (VLA) has been used to probe fluctuations at 5 GHz (Martin and Partridge 1988 and Fomalont *et al.* 1988) and at 15 GHz (Hogan and Partridge 1989). Because of the lower observing frequency, contamination from discrete extragalactic radio sources within the imaged field has plagued the 5 GHz efforts, and after careful subtraction of the contribution of these objects, 95% limits of $\delta T/T < 6 \times 10^{-5}$ on the scale of 1′ and $\delta T/T < 1.2 \times 10^{-4}$ on 18″ have been reached. The 15 GHz limits are somewhat worse due to poorer system performance of the VLA at this frequency. Forthcoming observations at 8.5 GHz using the new low noise receivers should produce limits competitive with the filled-aperture experiments.

For primary anisotropies, the dominant effect at small angles is the width of the last scattering surface. When calculating the perturbations keeping only the first-order terms, contributions from the crests and troughs of fluctuations with wavelengths smaller than this scale will cancel out along the line of sight, reducing the anisotropy measured on angular scales smaller than this characteristic size. As was mentioned above, the relevant angle is $\theta_{LS} \sim 10'$ for standard recombination at $z \sim 1500$. However, in scenarios where structure forms early, such as the isocurvature models, the recombination process can be extended, delayed, or even avoided, and the corresponding smearing angle therefore increases, reaching $\theta_{LS} > 1^\circ$ for a universe that never recombines. This mechanism for decreasing the small-scale fluctuations to avoid violation of the Uson and Wilkinson and OVRO limits will however still produce fluctuations on the intermediate and large angular scales, where they can be constrained by the appropriate observations such as the South Pole experiment (Bond, Efstathiou, Lubin and Meinhold 1990, hereafter BELM).

We would like to depict the sensitivity of experiments to fluctuations of different wavelengths along with the predictions of various theories. One useful representation is to calculate the angular power spectrum C_ℓ of the multipoles $\{\ell = 1, \ldots\}$ for the model in question (see Bond 1988, Bond 1989). Likewise, we can calculate a response function F_ℓ for a given observational setup. The resulting functions for several models and experiments are shown in Figure 1, reproduced from BELM. The solid curves show the filter functions for the OVRO, South Pole, Tenerife, and COBE DMR experments, superposed with the spectra for CDM and Baryon Isocurvature (BISOC) models with standard (SR) and no recombination

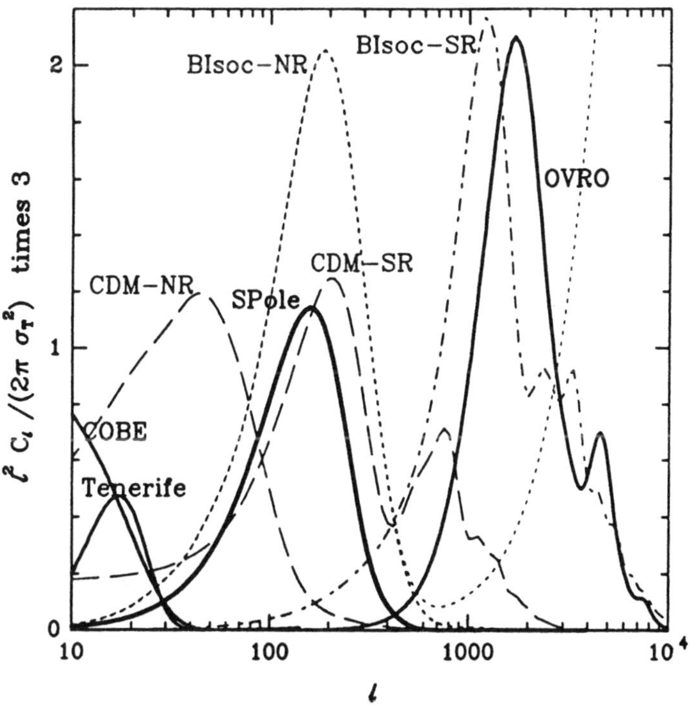

FIGURE 1. Angular power specta C_ℓ (solid curves) and instrumental response functions F_ℓ (broken curves) from Bond, Efstathiou, Lubin and Meinhold (1990).

(NR). Also shown extending to high multipoles from the BISOC-NR model is the contribution from the second-order terms in the perturbations that do not destructively interfere across the last scattering surface (Vishniac 1987). We see that the OVRO experiment is most sensitive to models with enhanced small-scale power, like the BISOC-SR, while the complementary South Pole experiment can catch the BISOC-NR models as well as standard CDM-SR. Note that the large-angle observations of Tenerife and COBE DMR are not as efficient at probing these theories as the medium and small angle experiments.

Small angle observations have particular sensitivity to certain types of secondary distortions such as the Sunyaev-Zeldovich (SZ) effect from keV-temperature gas in galaxy cluster potential wells, and dust emission from protogalactic objects at high redshift. At radio frequencies, the SZ effect causes a reduction in the effective temperature of the microwave background radiation (c.f. Sunyaev and Zeldovich 1981) that can be measured in a background experiment such as OVRO (Birkinshaw, Gull and Hardebeck 1984, Birkinshaw 1987). The cumulative effect of the superposition of clusters in a galaxy-formation scenario may be detectable in these experiments (Schaeffer and Silk 1988, Cole and Kaiser 1989). The dust models of Bond, Carr, and Hogan (1986,1990) are constrained by anisotropy measurements mentioned above and particularly by sub-millimeter and infrared wavelength observations with IRAM (Kreysa and Chini 1988), JCMT (Church *et al.* 1989), as well as results from COBE FIRAS and DIRBE. The proposed SIRTF satellite should have an angular-scale range similar to the OVRO ground-based observations, but at far-infrared wavelengths will be much more sensitive to dust emission. Angular power spectra and instrumental filter functions for these models are given in the contribution by Carr, Bond, and Hogan (these proceedings).

THE OVRO EXPERIMENTS

We now summarize the results of the Caltech observational program for microwave background anisotropies. The 40-meter radiotelescope at the Owens Valley Radio Observatory (OVRO) has been used very effectively to probe anisotropies on arcminute angular scales. At the frequency of 20 GHz, the antenna produces two Gaussian beams of $1\overset{'}{.}8$ full width at half-maximum (FWHM) each separated by $7\overset{'}{.}15$ on the sky, which are differenced in the radiometer every 0.1 seconds. The telescope is moved between "on" and "off" positions also separated by $7\overset{'}{.}15$ and difference formed once again. The effective response pattern of this "double-differencing" is a Gaussian main beam minus the average of two reference beams $7\overset{'}{.}15$ on either side (in the azimuth direction on the sky). The beam-switching is necessary to remove the effect of structure in the atmosphere that also contributes to the measurement, but in the process applies a bandpass filter to the background fluctuation spatial-frequency spectrum.

Two observational programs have been carried out at OVRO using the 20 GHz system. The first is a set of deep measurements of 8 fields near the North Celestial Pole (Readhead *et al.* 1989), the second a survey of 96 fields also near the pole (Myers 1990,Myers,Readhead,and Lawrence 1991). In the 8-field program, designated as NCP, the pointing centers are located at

$$\alpha = 1^{\mathrm{h}}, 3^{\mathrm{h}}, \ldots, 15^{\mathrm{h}} \qquad \delta = 89^{\circ} \quad (1985.0)$$

and thus are separated by nearly $30'$; these are well approximated as independent 3-beam measurements. The 96-field experiment, designated as the RING, consists of 96 "fields" located at

$$\alpha = 0^{\mathrm{h}}, 0^{\mathrm{h}}15^{\mathrm{m}}, \ldots, 23^{\mathrm{h}}45^{\mathrm{m}} \qquad \delta = 88^{\circ}10'42'' \quad (1987.0).$$

The geometry is such that the pointing centers are separated by $7\overset{'}{.}15$ meaning that the main beam for a given field is a reference beam for the adjacent fields. In principle, this means that it is possible to reconstruct the deviations of the signal from the mean for the set of single beams, although unfortunately in practice the presence of noise degrades senstivity to the larger scale fluctuations. Because of the triple-beam convolution, real signal from the background will average to zero over the entire 96 fields while systematic offsets either in the receiver or in the atmosphere will produce a nonzero data mean, while isolated bright sources will show the characteristic $(-0.5, +1.0, -0.5)$ signature.

The NCP observations were made in the period November 1984 to January 1987, with a total of 140 days yielding 400 hours of integration time. The RING observations were carried out beginning in November 1986 and completed in March 1989, with a total of 173 days on the telescope used to obtain 554 hours of integration time. Observations consisted of 2-hour scans of 80-second integrations for the NCP while 13-minute scans of 40-second data points were obtained for the RING; each was centered about the time of transit across the local meridian, when the target was highest in the sky. The 8 fields in the NCP program received an average of 49.7 hours of intergration time each with a 1σ mean error level of 30.9 μK, yielding a sensitivity of 13 mK $s^{1/2}$. In the RING survey, each of the 96 fields received an average of 5.1 hours (after final editing) with a mean error of 113 μK, for a sensitivity of 15 mK $s^{1/2}$. Detailed breakdowns of the various components of the noise budget are given in Readhead *et al.* 1989 and Myers 1990, along with a discussion of the editing and quality control procedures. For the RING, the weighted mean of the 96 field measurements was $\overline{T} = -6 \pm 12$ μK, consistent with zero. No mean level was subtracted nor is there evidence for systematic bias in the data.

The calibrated results of the NCP and RING experiments are presented in Figure 2. In the upper panel, the signal from a confusing discrete radio source is seen in the NCP field at 7^h. The difference in the signal and sensitivity levels between the two experiments is obvious. A Maximum-Likelihood estimate of the RING field mean temperatures reveals an excess *rms* signal of $\hat{\sigma}_{sky} = 133\mu$K above that expected from the noise alone. The largest deviation in the RING is 4.1σ. We know that in a Gaussian error distribution, there is a probability of 12% of finding at least one 3σ or greater deviation in a sample of 96 points, while in the RING dataset we find four. If the largest fluctuations are due to sources smaller than the beam, then they would be expected to exhibit the characteristic signature detailed above and the signal can be enhanced by passing the RING through a "matched filter" (-0.5,+1.0,-0.5). When this is done, we find seven fields with $\pm 3\sigma$ mean levels and three more which can be interpreted as the sidelobes of the others : 4^h15^m,7^h0^m,10^h15^m, 11^h30^m,13^h45^m,17^h0^m (positive), 19^h45^m (negative) and 6^h45^m,11^h45^m, 17^h15^m (negative sidelobes).

We can perform a Bayesian statistical analysis on the RING data in much the same manner as was done for the NCP (Readhead *et al.* 1989), which for 7 fields (7^h excluded) yielded an upper limit of $\sigma_{sky} < 58\mu$K on the *rms* sky signal between 3-beam differences. If we treat the 96 RING fields as independent "measurements" then we can place 99.87% limits on the signal present of $85 < \sigma_{sky} < 195$ μK. We prefer to quote the 99.87% credible limit rather than the traditional 95% limit, as it corresponds to a more robust 3σ threshold. If we remove the seven fields singled out above along with the 14 adjacent fields, then a similar analysis of the 75 remaining points gives only an upper limit at the 99.87% level.

The most likely candidate for the source of the signal is the population of extragalactic radio sources. Estimates of the contribution of radio sources to the *rms* of microwave background experiments has been previously calculated assuming that the confusion is the limiting factor (Danese,De Zotti and Mandolesi 1983,

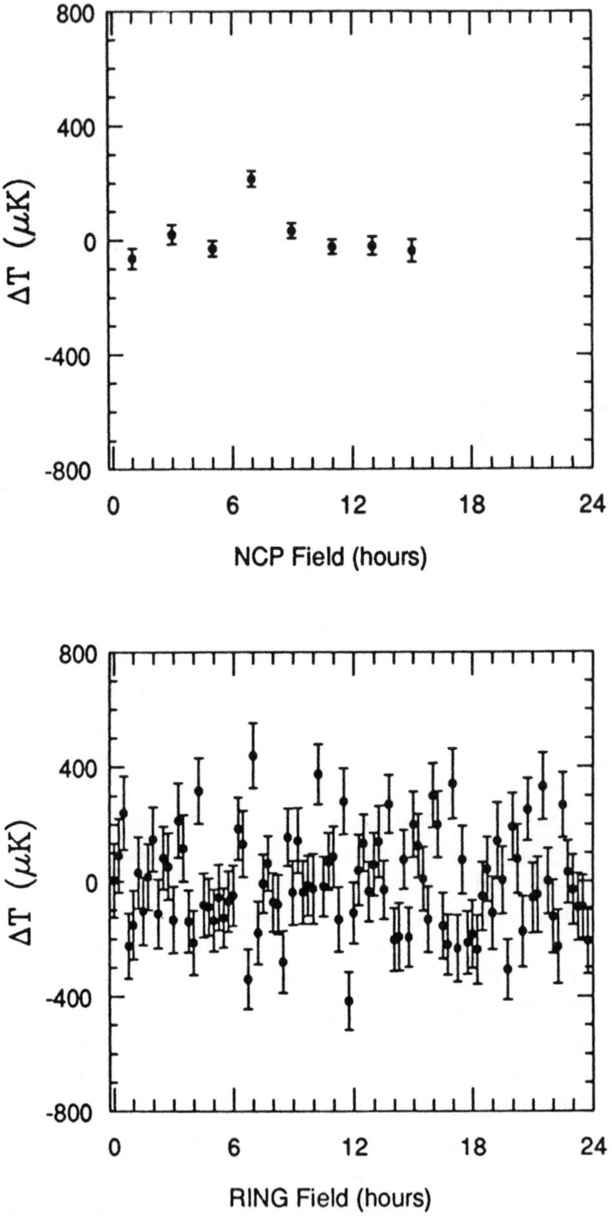

FIGURE 2. Results from Owens Valley experiments. Upper panel — NCP measurements (Readhead *et al.* 1989). Lower panel — uncorrected RING measurements (Myers 1990, Myers, Readhead and Lawrence 1991). Plotted are calibrated field values with 1σ error bars.

Franceschini *et al.* 1989). In a noise-limited experiment such as ours, this will underestimate the contamination level as well as ignore the non-Gaussian nature of the signal. The best estimate of faint source counts at 5 GHz is given by Fomalont *et al.* 1988 from a VLA survey. They estimate the differential number-flux count to be

$$n = 45 S^{-2.1} \quad \mathrm{Jy}^{-1}\ \mathrm{sr}^{-1}$$

in the flux range important to the RING. Taking an average source spectrum $S_\nu \propto \nu^{-0.5}$, the amplitude of the number counts are reduced by a factor of two when extrapolated to 20 GHz. We have performed a Monte-Carlo simulation of RING sample using these counts and found that of the 54% of the trials with deviations between the data limits of ± 2 mJy, the median *rms* was 42μK and the 99%-ile was 80μK.

The direct method of investigation of contamination by radio sources is to observe the fields with a high-gain, high resolution instrument such as the VLA. The entire RING survey area was observed at 1.4 GHz with the VLA in May 1988 with relatively poor sensitivity. We also obtained 5 hours on the VLA in July 1989 to observe 20 of the RING fields at 8.5 GHz in C-array. These maps covered the entire OVRO primary beam and achieved a sensitivity level of 58μJy (1σ) corresponding to a threshold of 78μK (6σ for 1024^2 pixel images) in the OVRO experiment. In the 20 fields observed, 12 were found to contain a total of 18 objects above the cutoff. Of these, only four were found to contain sources both strong and close enough to affect the RING observations, in fields 4^h15^m,11^h30^m, 13^h45^m,and 17^h0^m, all of which showed large signal in the RING. The brightest sources were detected at both 8.5 and 1.4 GHz and one was observed directly at 20 GHz at OVRO. Extrapolation using these spectra to 20 GHz for the sources in 4^h15^m, 11^h30^m, and 17^h0^m produces flux densities sufficent to account for the RING signal (residuals less than 1.3σ). There were three separate sources in the field 13^h45^m, the sum of which accounts for only a small fraction of the observed signal (residual $\sim 2.3\sigma$). Of the seven candidate fields, 7^h0^m, 10^h15^m, and the negative field 19^h45^m contained no significant sources, and remain, with the weaker signal in 13^h45^m, the best candidates for fluctuations.

Using the VLA and OVRO direct measurements of the contaminating radio sources, we have generated a corrected RING dataset. We have subtracted the estimated contributions for the four fields as detailed above, from both the main and adjacent fields affected. No corrections were made to the remaining fields. We have chosen to remove the estimated source contribution in this manner, rather than excluding the field from the RING analysis, in order to keep the RING interlocking intact. A Bayesian analysis of the corrected RING gives $60 < \sigma_{sky} < 170\ \mu$K(99.87%) with a Maximum-Likelihood value of $\hat{\sigma}_{sky} = 111\mu$K. A total of 48μK *rms* was removed from the dataset with these sources. Because of the incompleteness of our VLA observations, the uncertainty in extrapolation to 20 GHz, and the high probablility of variability of the flat-spectrum components of these radio sources, we must assume that there is some level of source confusion remaining in the RING data, and any limits derived from this data are to be considered as *upper limits only.* Supplementary OVRO and VLA observations are planned as a follow-up to this experiment.

We now wish to compare our RING data to the predictions of various cosmological and galaxy formation models, taking into account the spatial correlations induced by the sampling geometry. If we assume the fluctuations are uncorrelated between the main and reference beams, then the single-beam fluctuation limits are $\delta T/T < 1.7 \times 10^{-5}$ (95%) and 3.8×10^{-5} (99.87%) for the 7-fields of the NCP and $\delta T/T < 4.5 \times 10^{-5}$ (95%) and 5.5×10^{-5} (99.87%) for the corrected RING. Another standard approach is to use a parameterized form for the autocorrelation function of a Gaussian random field (see Readhead *et al.* 1989 for example, as applied to the NCP data, and references therein to other applications). Because the RING was intended to probe non-Gaussian fluctuation models, as well as the fact that the sensitivy of the RING to Gaussian perturbations is significantly poorer than that of the NCP, we will note only that if a Gaussian functional form is assumed for the autocorrelation function, then our best limits are for a correlation angle $\phi_C = 2\farcm6$, with upper limits of $\delta T/T < 1.9 \times 10^{-5}$ (95%) and 4.2×10^{-5} (99.87%) for the NCP and $\delta T/T < 5.0 \times 10^{-5}$ (95%) and 6.1×10^{-5} (99.87%) for the corrected RING.

The power-spectrum of the fluctuations produced by realistic models are often reasonably complex (see reviews Bond 1988,Bond 1989), and it is appropriate to compare our data directly to the family of models under investigation. The OVRO NCP observations and the South Pole experiment have been compared to predictions of CDM, HDM, Baryon Isocurvature, and various other models using Monte-Carlo frequentist methods (Vittorio *et al.* 1990) and Bayesian statistics (BELM). We, along with J.R. Bond, have performed similar analyses for the RING data with and without the South Pole and NCP added, and find limits significantly higher than those obtained for the other observations, due to the larger error bars and residual signal. Limits for a variety of models and cosmologies using the NCP and Tenerife are also given in Holtzman (1989) and Sugiyama, Gouda, and Sasaki (1990).

Our original interest in undertaking the RING observations was to place reasonably strong limits on non-Gaussian fluctuations from models such as cosmic strings (Bouchet, Bennett, and Stebbins 1988). As it turns out, we are confronted with two problems : the non-Gaussian nature of the statistics and the large size of the RING dataset. The fact that there is no guaranteed *sufficient* statistic analogous to the mean and two-point correlation function for Gaussian random fields greatly increases the complexity of the statistical calculations needed to characterize the model. This problem is endemic in varying degrees to all non-Gaussian distributions; in the general case all of the N-point distribution functions are required for sufficiency. The second problem, which aggravates the first, is the relatively large number of data-points and the degrees of their interdependence. This problem is not confined to the RING — all-sky or nearly all-sky surveys such as those from balloon experiments and COBE contain so many data points that even the analysis of Gaussian models is tedious. We are currently exploring methods of analysis, particularly involving Monte-Carlo simulations and map generation, to enable constraints to be placed on these theories.

CONCLUSION

We have tried to present a brief summary of the current experiments probing small angular scale microwave background fluctuations along with a discussion of the OVRO observational program, highlighting the new 96-field survey. The signal in this experiment appears to be dominated by contamination from faint radio sources, indicating the limitations of deep surveys at this frequency. Selection of target fields based upon the absence of sources seen in high-frequency VLA images is necessary to avoid this problem, although higher frequency observations should be less affected (barring interference from galactic dust). Several interesting candidate fields for possible fluctuations were found, with follow-up observations scheduled for OVRO. Because of various uncertainties in the subtraction of the VLA-identified radio sources from the RING dataset, we present only upper limits to Gaussian background fluctuations.

The next few years promises to be extremely exciting for microwave background cosmologists. Besides the forthcoming results from COBE, we can expect to hear from the South Pole groups where state-of-the-art bolometers have pushed the instrumental sensivities to the 10^{-5} level. At the Owens Valley observatory, we are commencing operation of a 5-meter dedicated telescope to observe at a frequency of 32 GHz with performance equal or superior to that of the OVRO 20 GHz system; this will probe the angular-scale range of $7'$–$25'$. The smaller aperture and higher frequency will reduce the sensitivity to point-source contamination by more than an order of magnitude. On the other hand, the increasing size of the observational database from these and in particular the COBE and balloon-based surveys will call for an increasingly powerful and sophisticated arsenal of statistical weapons with which to attack the problem of quantification of constraints on theoretical models. The difficulties encountered in the analysis of the RING and the balloon experiments should provide a cautionary indicator of the problems to be expected.

REFERENCES

Bernstein, G.M., Fischer, M.L., Richards, P.L., Petersen, J.B. and Timusk, T. (1989), *Ap. J.*, **337**, L4.

Birkinshaw, M. (1987), in *Observational Cosmology, I.A.U. Symposium 124*, ed. A. Hewitt, G. Burbidge and L.Z. Fang (Dordrecht: Kluwer) pp. 83–86.

Birkinshaw, M., Gull, S.F. and Hardebeck H.E. (1984), *Nature*, **309**, 34.

Bond,J.R. (1988), in*The Early Universe*, ed. W.G. Unruh and G.W. Semenoff (Dordrecht: Reidel) pp. 283–334.

Bond,J.R. (1989), in*Frontiers of Physics—From Colliders to Cosmology*, ed. B. Campbell and F. Khanna (Singapore: World Scientific).

Bond, J.R., Carr, B.J. and Hogan, C.J. (1986), *Ap. J.*, **306**, 428.

Bond, J.R., Carr, B.J. and Hogan, C.J. (1989), *Ap. J.*, in press.

Bond, J.R.,Efstatiou, G.,Lubin, P.M. and Meinhold, P.R. (1990), *Phys. Rev. Lett.*,

in press.
Bouchet, F.R., Bennett, D.P. and Stebbins, A. (1988), *Nature*, **335**, 410.
Boughn, S.P.,Cheng, E.S.,Cottingham, D.A. and Fixsen,D.J. (1990), preprint.
Church, S.E.,Lasenby, A.N. and Hills, R.E. (1990),preprint.
Cole, S. and Kaiser, N. (1989), *M.N.R.A.S.*, **233**, 637.
Danese, L., De Zotti, G., and Mandolesi, N. (1983), *Astron. Astrophys.*, **121**, 114.
Davies, R.D., Lasenby, A.N., Watson, R.A., Daintree, E.J., Hopkins, J., Beckman, J., Sanchez-Almeida, J. and Rebolo, R. (1987), *Nature*, **326**, 462.
de Bernardis, P., Amicone, L., De Luca, A., De Petris, M., Epifani, M., Gervasi, M., Guarini, G., Masi, S., Melchiorri, F., Natale, V., Boscaleri, A., Natali, G. and Pedichini, F. (1990), *Ap. J. Lett.*, **360**, L31.
Fomalont, E.B., Kellerman, K.I., Anderson, M.C., Weistrop, D., Wall, J.V., Windhorst, R.A. and Kristian, J.A. (1988), *Astron. J.*, **96**, 1187.
Franceschini, A., Toffolatti, L., Danese, L. and De Zotti, G. (1989), *Ap. J.*, **344**, 35.
Gull, S.F. (1989), in *Maximum Entropy and Bayesian Methods*, ed. J. Skilling (Dordrecht: Kluwer) pp. 53-71.
Hogan, C.J. and Partridge, R.B. (1989), *Ap. J. Lett.*, **341**, L29.
Holtzman, J.A. (1989), *Ap. J. Suppl.*, **71**, 1.
Klypin, A.A., Sazhin, M.V., Strukov, I.A. and Skulachev, D.P. (1987), *Soviet. Astr. Lett.*, **13**, 104.
Kreysa, E. and Chini, A. (1988), in*Proc. 3rd ESO-CERN Symp. on Astronomy,Cosmology and Fundamental Particles*, ed.Bologna.
Martin, H.M. and Partridge, R.B. (1988), *Ap. J.*, **324**, 794.
Meinhold, P. and Lubin, P. (1990),*Ap. J. Lett.*, in press.
Meyer, S.S.,Cheng, E.S. and Page, L.A. (1990),preprint.
Myers, S.T. (1990), thesis, Caltech.
Myers, S.T., Lawrence, C.R. and Readhead, A.C.S. (1991), in preparation.
Penzias, A.A. and Wilson, R.W. (1965), *Ap. J.*, **142**, 414.
Readhead, A.C.S., Lawrence, C.R., Myers, S.T., Sargent, W.L.W., Hardebeck, H.E. and Moffet, A.T. (1989), *Ap. J.*, **346**, 566.
Schaeffer, R. and Silk, J. (1988), *Ap. J.*, **333**, 509.
Skilling, J. (1989), in *Maximum Entropy and Bayesian Methods*, ed. J. Skilling (Dordrecht: Kluwer).
Sugiyama, N., Gouda, N. and Sasaki, M. (1990), *Ap. J.*,**365**, 432.
Sunyaev, R.A. and Zel'dovich, Y.B. (1981), *Astrophys. Space Sci. Rev.*, **1 (E)**, 1.
Timbie, P.T. and Wilkinson, D.T. (1988), *Rev. Sci. Instrum.*, **59**, 914.
Uson, J.M. and Wilkinson D.T. (1984c), *Nature*, **312**, 427.
Vishniac, E.T. (1987), *Ap. J.*, **322**, 597.
Vittorio, N., Meinhold, P., Muciaccia, P.F., Lubin, P. and Silk, J. (1990),*Ap. J. Lett.*, submitted.

DISCUSSION

Rogier Windhorst: I would like to reply to Jim Peebles question to Steve Myers. I think Myers used about the right slope for the sub-mJy counts. In our ultra-deep VLA surveys, we found that the slope of the sub-mJy counts is remarkably similar, $(dN/dS) \propto S^{-2.1}$, for all frequencies studied (1.4 GHz, Windhorst *et al.* 1985; 5 GHz, Fomalont *et al.* 1988; 8.4 GHz, Partridge *et al.* 1990). This is consistent with a steep spectrum weak sources population dominating the faint counts, with an average $\langle\alpha\rangle$ ($S_\nu \propto \nu^{-\alpha}$) depending on selection frequency ($\alpha \sim 0.6$ at 5 GHz). We might just be lucky that we can extrapolate this to 20 GHz, because the median redshift ... (rest lost in transmission)

FINE-SCALE MICROWAVE FLUCTUATIONS AT $\lambda = 3.6$ cm

E. B. Fomalont, N.R.A.O., Charlottesville, VA, 22903
James Lowenthal, University of Arizona, Tucson, AZ, 85721
(reported by)
R. B. Partridge, Haverford College, Haverford, PA, 19041
and R. A. Windhorst, Arizona State University, Tempe, AZ, 85281

On behalf of my colleagues and myself, I want to report some tentative results from an attempt to improve limits on the amplitude of cosmic background radiation (CBR) fluctuations on subarcminute scales. We used the new 3.6 cm wavelength receivers at the VLA with the instrument in its close-packed D-array. The range of angular scales covered by our work is $8'' \lesssim \theta \lesssim 60''$. We observed two separate fields both known to be free of bright radio sources (6 cm fluxes < 1 mJy). We have good photographic or CCD observations of both. On one field, we have ~ 50 hrs of integration time; on the other ~ 40 hrs. The rms noise in the raw maps, uncorrected for the presence of weak radio sources, ranges from 3-7 μJy for angular scales ranging from 8"-60".

The combination of longer integrating times, lower noise receivers and a selection of fields free of bright sources should enable us to lower published upper limits on $\Delta T/T$ on subarcminute scales substantially. We plan to derive limits on $\Delta T/T$ from the maps once they have been corrected for the presence of weak radio sources as in Martin and Partridge (1988) or Fomalont et al. (1988). We will also look directly at the u-v plane data (after making some necessary corrections for griding and other analysis artifacts). Interferometric data is recorded as fringe visibilities as a function of base line spacing--"u-v data." Maps of the radio sky are conventionally constructed by taking the two-dimensional Fourier transform of the u-v data. Thus the u-v data itself provides directly information about the spatial frequency power spectrum of the sky, and we will present limits on background fluctuations calculated directly in this fashion.

Since we have deep 6 cm and 20 cm maps of these same regions, we will also be able to examine the effect of weak radio sources (which tend to be stronger at longer wavelengths) on our 3.6 cm maps.

Preliminary analysis of the data in one field suggests that we should be able to set limits on $\Delta T/T$ of $\leq 7\text{-}8\times10^{-5}$ at the resolution of the array, 8-10" and $\Delta T/T < 1\text{-}2\times10^{-5}$ at an angular scale of 1 arcminute.

REFERENCES

Martin, H. M. and Partridge, R. B. 1988, *Ap. J.* 324, 794.
Fomalont, E. B., Kellermann, K. I., Anderson, M. C., Weistrop, D., Wall, J. V., Windhorst, R. A., and Kristian, J. A. 1988, *A. J.*, 96, 1187.

Observational bounds on the Cosmic Microwave Background Anisotropy at Small and Intermediate Angular Scale

Nicola Vittorio
Dipartimento di Fisica
Universita' dell'Aquila, Italy

Pio Francesco Muciaccia
Istituto di Astronomia
Universita' di Roma, "La Sapienza"

We re-analyzed the data of three CMB anisotropy experiments, performed on small and intermediate angular scales. The analysis is based on a likelihood ratio test, and consider flat and open CDM models. We discuss: i) the method of data analysis; ii) the statistical significance of the obtained upper limits; iii) the constraints one can pose on the density parameter and the baryonic abundance of these models.

1. INTRODUCTION

The comparison between theoretical predictions of, and observational upper limits on the Cosmic Microwave Background (CMB) anisotropy is a classical test for galaxy formation scenarios. It tells us about the validity of a certain model, and the gain in sensitivity necessary to rule out that model or possibly to detect CMB anisotropy.

The comparison between theoretical predictions and observational bounds has not yet been properly addressed, in spite of the large number of papers devoted to this subject. From one hand, observers have often used *ad hoc* models and assumptions in reducing their data; from the other hand, theoreticians have used observational upper limits *at face value*, taking into account only two, among the possible, parameters which characterize an experiment: the finite antenna resolution (defined as the dispersion σ_B of the Gaussian approximating the angular response of the receiver), and the beam-throat (the characteristic scale θ of a single or a double subtraction experiment). As a result, some models might have been overkilled, without a solid statistical argument.

An unambiguous approach for comparing theory and observation in a statistically significant way is to perform Monte Carlo simulations of the microwave sky expected in a given theoretical scenario, and analyze them as is done in the actual experiment. In this way, one can take into account effects such as data sampling, sky coverage, receiver noise and chopping strategy, that contribute to the final uncertainties in any quantitative upper limit on $\delta T/T$. In principle, as the simulations are based on a given model, the derived upper limits on the CMB anisotropy become model-dependent.

Observations of galaxy clustering (*e.g.* Davis and Peebles, 1983) and superclustering (Bahcall, 1988), of large voids (Geller, Huchra, and de Lapparant 1986) and apparently coherent bulk flows (*e.g.* Gunn 1989) have constrained possible options for the gravitational instability theory that describes how infinitesimal fluctuations in the early universe account for these observations. The flat, biased cold dark matter (CDM) model (*e.g.* Blumenthal et al., 1984) remains an attractive option, despite of its mixed fortune in confronting observations on large scales, because of its theoretical appeal in the context of the inflationary cosmology (*e.g.* Blau and Guth, 1987). This models sets the standard against which rival theories, such as a baryonic dark matter universe with primeval isocurvature fluctuations (Peebles,1987), or the explosive galaxy formation scenario (Ostriker and Cowie, 1981), or cosmic strings (*e.g.* Turok, 1987), are judged.

In CDM cosmologies CMB anisotropies are unambiguously predicted: the primeval fluctuation spectrum is uniquely specified, and erasure of fluctuations due to the fuzziness of the last scattering surface is negligible (in the standard CDM model an early reionization of the intergalactic medium is not expected).

Here we want to report the results of the analysis of a recent anisotropy experiment (Meinhold and Lubin, 1990; Vittorio *et al.* , 1990). We consider both flat CDM models (allowing for values of the baryonic density parameter, Ω_b, larger than the standard few per cent), and open unbiased CDM models (with $\Omega_b << 1$).

We want also to present preliminary results (Muciaccia and Vittorio, 1990) of a re-analysis of the Uson and Wilkinson (1984) and the Readhead *et al.* (1989) experiments and to discuss the statistical significance of the available CMB anisotropy bounds.

2. THEORETICAL BACKGROUND

On small angular scales ($\lesssim 30'$), the CMB anisotropy is induced by intrinsic fluctuations in CMB brightness at decoupling: $\delta T/T \sim 0.3\ \delta\rho/\rho$. Adiabatic fluctuations of the mass-energy density field can be visualized as a true compression of the photon-baryon fluid: where there is an excess of baryons, there is an excess of photons, and the region appears hotter than the background. On intermediate angular scales ($\simeq 1^o$), the CMB anisotropy is determined by Doppler effects. The baryonic material during decoupling starts to fall into gravitational wells, due either to selfgravity (as in pure baryonic universes) or to external X-ino's gravity. As a result, the photons at decoupling are last scattered by moving electrons and we expect, as an order of magnitude, $\delta T/T \sim v/c$. On large angular scales ($> 1^o$), the dominant effect is the gravitational redshift of photons that have to climb the potential well of the perturbation inside which they were last scattered. This effect, discussed by Sachs and Wolfe in 1967, implies $\delta T/T \sim \delta\phi/c^2$, where $\delta\phi = -GM/R(\delta\rho/\rho)$ is the potential fluctuation of a perturbation of mass M and proper size R. In a flat universe the potential fluctuation is time indipendent. In fact, both the density contrast and the size of the perturbation grow linearly with the scale factor. Moreover, if perturbations of the density field constitute a random Gaussian process, characterized by an initially featureless spectrum $[P(k) = Ak^n]$, the contribution per unit logarithmic wavenumber interval to $\delta\phi$ (or to $\delta T/T$) goes as $k^{(n-1)/2}$. Then, if the initial conditions are scale–invariant (*i.e.* $n = 1$) as in CDM, the contribution to the gravitational potential is independent of the perturbation wavenumber.

The detailed calculations of the CMB anisotropy on small and intermediate angular scales rests on the numerical evolution of the CMB brightness trough the matter- radiation decoupling. The calculation scheme was first outlined by Peebles and Yu (1970), and most of the authors follow the same method.

The 2-D pattern of the CMB temperature fluctuations on small and intermediate angular scales is a realization of a 2-D random Gaussian field. It is then completely described by the radiation power spectrum, $P_{rad}(|\mathbf{q}|)$. Here $\mathbf{q}$, is a 2-D wavenumber in the last scattering surface. In the so-called small angle approximation we neglect any curvature of the celestial sphere and approximate the observed patch of the sky as a flat region. The effect of the antenna beam is clearly that of a low band pass filter, as it smooths out all the irregularities on angular scales much smaller than the beam size. The CMB two point autocorrelation function is easily expressed in terms of the radiation power spectrum:

$$\langle C(\theta, \sigma_B) \rangle = C_0 \; \times \; \frac{\int_0^\infty q dq P_{rad}(q) J_0(q r_0 \theta) exp(-q^2 r_0^2 \sigma^2)}{\int_0^\infty q dq P_{rad}(q) exp(-q^2 r_0^2 \sigma^2)} \tag{1}$$

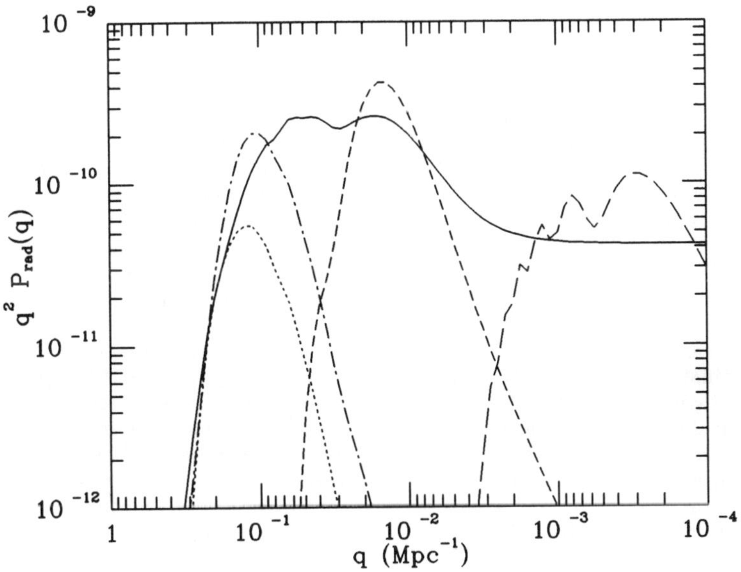

Fig.1: CMB radiation power spectrum (solid line) for the standard flat CDM model ($\Omega_b \lesssim 0.03$ and h=0.5). Effective q-band probed by the Uson and Wilkinson (dotted line), the Readhead et al. (dot-dashed line), the Meinhold and Lubin (dashed line) experiments, and by the COBE satellite (long-dashed line).

TABLE I

Theoretical predictions for CDM cosmologies

Ω_0	Ω_b	h	$b\Delta_{rms}(4'5, 1'27)$	$b\Delta_{rms}(7'.15, 0'.76)$	$b\Delta_{rms}(1^o, 13')$
1.0	0.03	0.5	$7.32 \cdot 10^{-6}$	$1.39 \cdot 10^{-5}$	$2.12 \cdot 10^{-5}$
1.0	0.10	0.5	$9.91 \cdot 10^{-6}$	$1.80 \cdot 10^{-5}$	$2.71 \cdot 10^{-5}$
1.0	0.20	0.5	$1.23 \cdot 10^{-5}$	$2.20 \cdot 10^{-5}$	$3.69 \cdot 10^{-5}$
1.0	0.30	0.5	$1.51 \cdot 10^{-5}$	$2.69 \cdot 10^{-5}$	$4.82 \cdot 10^{-5}$
1.0	0.40	0.5	$1.87 \cdot 10^{-5}$	$3.33 \cdot 10^{-5}$	$6.13 \cdot 10^{-5}$
1.0	0.50	0.5	$2.31 \cdot 10^{-5}$	$4.11 \cdot 10^{-5}$	$7.66 \cdot 10^{-5}$
0.8	0.03	0.5	$1.17 \cdot 10^{-5}$	$2.09 \cdot 10^{-5}$	$2.74 \cdot 10^{-5}$
0.6	0.03	0.5	$2.00 \cdot 10^{-5}$	$3.41 \cdot 10^{-5}$	$3.85 \cdot 10^{-5}$
0.4	0.03	0.5	$3.68 \cdot 10^{-5}$	$6.08 \cdot 10^{-5}$	$6.01 \cdot 10^{-5}$

where: $r_0 = 2c/(\Omega_0 H_0)$, the exponential takes into account the high-"frequency" cut-off due to the antenna beam, J_0 is the Bessel function, $C_0 \equiv \langle C(0, \sigma_B)\rangle$ is the smoothed variance of the CMB temperature fluctuation field, and the symbol $\langle\rangle$ indicates a mean over the ensemble. The *rms* differential CMB anisotropy, $\Delta_{rms}(\theta, \sigma_B)$, can be expressed as $2[C_0 - \langle C(\theta, \sigma_B)\rangle]$ or as $3C_0/2 - 2\langle C(\theta, \sigma_B)\rangle + \langle C(2\theta, \sigma_B)\rangle/2$ for a single or double subtraction experiment, respectively. A differential measurement acts as a high band pass filter: in fact, it is less sensitive to fluctuations on angular scales larger than the beam-throat θ. Theoretical predictions take these two parameters (θ and σ_B) very easily into account. In Fig. 1 we plot the intrinsic 2-D radiation power spectrum for the standard CDM model ($\Omega_0 = 1$, $\Omega_b = 0.03$, h=0.5, n=1) and the effective q-band probed by the various experiments. Note the flat behaviour of $q^2 P_{rad}$ at low wavenumbers, where the Sachs-Wolfe effect dominates, the peak due to the Doppler effect, and the cut-off at high wavenumbers due to the smearing of the last scattering surface. In Table I, we present $\Delta_{rms}(\theta, \sigma_B)$ for a grid of CDM models, normalized to the observed galaxy clustering on small scales: the *rms* mass density fluctuations, averaged over an 8 h^{-1}Mpc sphere, is b^{-1}. The biasing parameter, b, is usually introduced to reconcile the dynamical determinations of the density parameter with the flatness of the CDM models (see, Dekel and Rees, 1987, for a review). The chosen values of θ and σ_B both in Fig.1 and in Table I are appropriate for the Uson and Wilkinson ($\theta = 4'.5$ and $\sigma_B = 1'.27$), the Readhead *et al.* ($\theta = 7'.15$ and $\sigma_B = 0'.76$), and the Meinhold and Lubin ($\theta = 1^o$ and $\sigma_B = 13'$) experiments.

3. DATA ANALYSIS

Given the theoretical predictions of Table I, we want to assess the constraints that an experiment is able to set on these models. The goal of the data analysis is to test two alternative hypotheses. The first, $\mathcal{H}_1$, assumes that CMB temperature fluctuations do exist in the sky. The CMB pattern is determined, for a given model, by the shape of the temperature autocorrelation function, the only free parameter being its amplitude, C_0. The alternative null hypothesis, $\mathcal{H}_0$, assumes instead a completely uniform CMB sky, with no intrinsic fluctuations, *i.e.* $C_0 = 0$. In order to distinguish between these two cases, we use a likelihood ratio analysis, the likelihood ratio being defined as $\lambda \equiv L(C_0)/L(0)$, where

$$L \equiv (2\pi)^{-N/2}|\mathbf{M}|^{-1/2} exp[-\frac{1}{2}\sum_{i=1}^{N}\sum_{j=1}^{N}\Delta_i \mathbf{M}_{ij}^{-1}\Delta_j] \tag{2}$$

is the likelihood function, $\{\Delta_i\}$ is a set of N CMB temperature differences, and $\mathbf{M} \equiv \langle\Delta_i\Delta_j\rangle$ is the covariance matrix of these differences.
Here we briefly discuss the approach followed by Vittorio *et al.* (1990) for analyzing the Meinhold and Lubin experiment. This experiment operated from the South Pole at a frequency of 90 GHz, and at constant declination. As a result, nine CMB temperature difference were obtained on the angular scale of 1 degree. The system consists of a microwave telescope and a very low noise SIS detector, with the main lobe of the telescope response pattern being well approximated by a Gaussian of dispersion $\sigma_B = 13'$. The error per point, σ_i, is approximately 60 μK.
The distribution of the likelihood ratio either under hypothesis $\mathcal{H}_1$ and $\mathcal{H}_0$, is evaluated by Monte Carlo simulations of the sky. Under hypothesis $\mathcal{H}_1$, one can generate, for each of the theoretical models, 10,000 maps ($12^o \times 12^o$), by assigning a value σ_{sky} to the parameter C_0. Each map is sampled as the Meinhold and Lubin experiment sampled the real sky, explicitely taking into account the sinusoidal modulation of that experiment. As a result, nine CMB temperature differences are obtained, to which it is added appropriately distributed Gaussian noise, to simulate the experimental errors, $\{\sigma_i\}$. In addition, a linear component from each simulated data set is subtracted in order to be more consistent with the analysis of the real data (Meinhold and Lubin, 1990). In Fig. 2 we show 9 simulated data set for a flat unbias CDM model together with the real data set ($\delta T/T$ is given in μK). Under hypothesis $\mathcal{H}_0$, we proceed likewise, but impose $C_0 = 0$.

The standard approach
We want to have a low probability, α, of rejecting hypothesis $\mathcal{H}_1$ when $\mathcal{H}_1$ is true (the so-called type I error). So, we tune σ_{sky} in the simulations in order to have only a fraction $\alpha = 5\%$ of the cosmic observers that measure a likelihood ratio $\lambda \leq \lambda_{obs}$ (λ_{obs} is the likelihood ratio obtained using the actual data). Then, *a posteriori*, we evaluate the power of the test, namely the fraction β of cosmic observers that, under hypothesis $\mathcal{H}_0$, would also measure $\lambda \leq \lambda_{obs}$. If β is small, we have a large probability $(1-\beta)$ of accepting $\mathcal{H}_1$ when hypothesis $\mathcal{H}_0$ is true (the so-called type II error).
The result of this analysis sets the upper limit on C_0 and, then, on the differential (single subtracted) temperature fluctuation: $\Delta_{rms}(1^o, 13') < 4.7 \cdot 10^{-5}$, with a confidence level $1-\alpha = 95\%$, and a power of the test $\beta = 56\%$. This upper bound is actually model *independent*, at least for the class of models we considered here: this is due to the fact that the experiment probe a very narrow region of the power spectrum (see Fig.1), and is then quite insensitive to changes in the power spectrum shape, due to variation of the cosmological parameters. The derived upper limit allows, for example, to set a lower bound to the density parameter, as in open universes the efficiency of density fluctuation growth is reduced: we found $\Omega_0 \gtrsim 0.6$, for open unbiased CDM models with $\Omega_b \lesssim 0.03$ and $h = 0.5$. It must

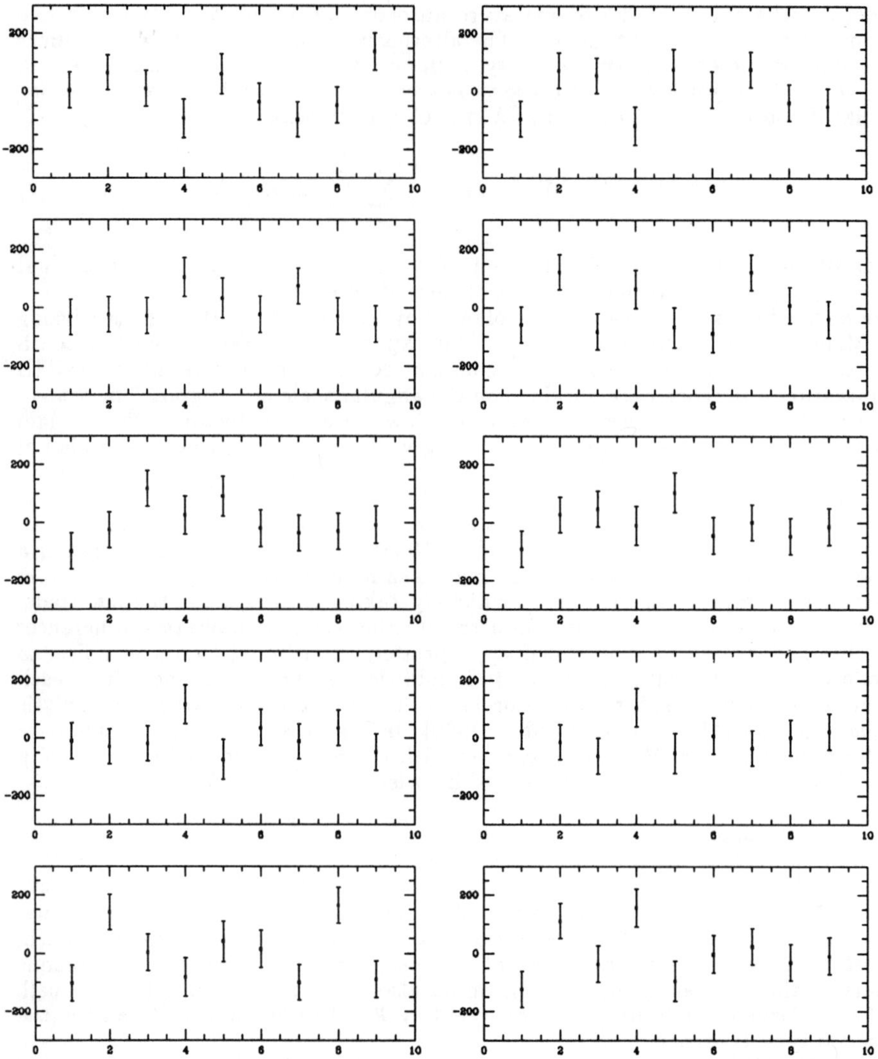

Fig.2: Simulated CMB (single subtracted) temperature differences for the Meinhold and Lubin experiment. Here we consider a flat, unbiased CDM model with $\Omega_b =$ 0.03 *and h=0.5. The data set in the third row and second column is the real one. The angular distance between data points is* 1°.

be kept in mind, however, that this lower limit on Ω_0 must be quoted *together with* its own power of the test, *i.e.* $\beta = 56\%$. Flat unbiased CDM models with a baryonic fraction of 20%, the maximum allowed by standard nucleosynthesis (*e.g.* Boesgard and Steigman, 1985), are marginally consistent with the Meinhold and Lubin experiment. The standard $\Omega_0 = 1$, $\Omega_b = 0.03$, h=0.5 unbiased model predicts an anisotropy which is only a factor of ~ 2.2 below the observational upper limit. A biasing factor $b \simeq 1.5$, suggested by the large scale observations, implies that at the present level of sensitivity this experiment would have probably missed a detection by a factor of few.

A more conservative approach

As theoretical predictions and observational bounds are getting so close, we should be conservative in judging the viability of a given model, without running the risk of overkilling some scenarios.

We propose to act as follows. Let us *define* as upper limit on $C_0^{1/2}$ the value σ_{sky} such that (*c.f.* Fig.3):

$$P(\lambda < \lambda_{oss}|\mathcal{H}_1) = \bar{\alpha}; \quad P(\lambda < \lambda_{obs}|\mathcal{H}_0) = \bar{\beta} \tag{3}$$

or, equivalently,

$$P(\lambda < \lambda_*|\mathcal{H}_1) \equiv \alpha_* = 5\%; \quad P(\lambda < \lambda_*|\mathcal{H}_0) \equiv \beta_* = 95\% \tag{4}$$

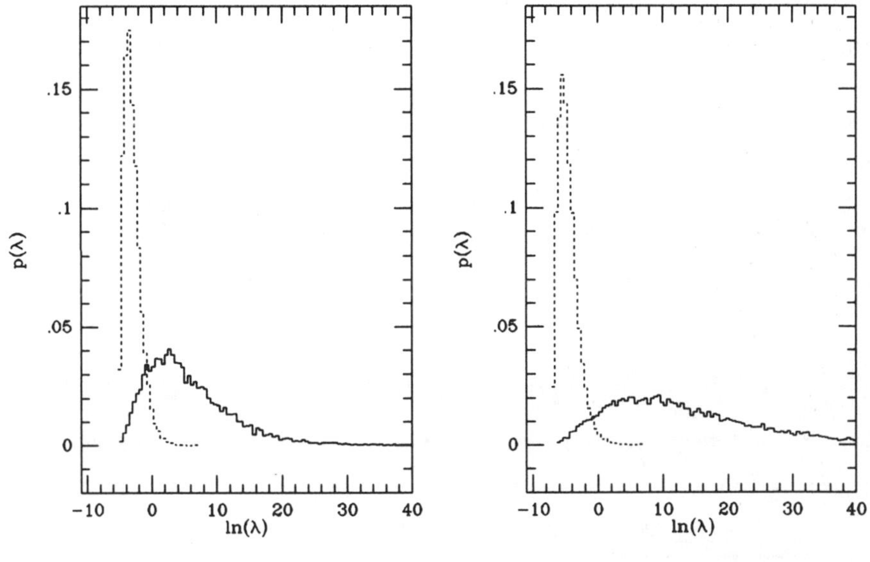

Fig.3: Distribution of likelihood ratios under hypothesis $\mathcal{H}_1$ (solid line) and $\mathcal{H}_0$ (dotted line). The upper limit on $C_0^{1/2}$ has been evaluated in the left panel by fixing the size of the test [$P(\lambda < \lambda_{oss}|\mathcal{H}_1) = 5\%$] and reading the power of the test a posteriori, while in the right panel both the size and the power of the test have been fixed a priori to 5% and 95%, respectively.

One can always find values for $\bar{\alpha}$, $\bar{\beta}$, and λ_* that verify the previous conditions. It is clear from Eq.(4) that if $\lambda > \lambda_*$ we accept hypothesis $\mathcal{H}_1$ (against $\mathcal{H}_0$) with a very large power of the test ($\beta_* = 95\%$). Viceversa, if $\lambda < \lambda_*$, we accept hypothesis $\mathcal{H}_0$ (against $\mathcal{H}_1$) with again a large power, now given by $1-\alpha_* = 95\%$. If $C_0^{1/2} > \sigma_{sky}$, we would most probably ($> 95\%$) have a detection: σ_{sky} is then an upper limit in the common sense of the word. In Eq.(4) we have in practise fixed the power β_* of the test *a priori*, rather than *a posteriori* as described above: this procedure offers several advantages. First, different experimental results can be compared selfconsistently, given that a likelihood ratio analysis is performed on the data assuming the same model and the same *a priori* chosen power of the test. Second, one can avoid problems with low χ^2 data sets, that occur 50% of the times under hypothesis $\mathcal{H}_0$. But the most relevant adavantage in our opinion is to be able to accept or reject a model with both a high confidence level, 95%, and a large *a priori* chosen power, 95%.

If Eq.(3) and Eq.(4) have to be verified at once, for the same value σ_{sky} of $C_0^{1/2}$, then $\bar{\alpha} << 5\%$ and $\bar{\beta} << 95\%$. In other words, if we follow the standard approach and we find that $\bar{\beta}$ is small, we have to increase correspondingly the confidence level $1-\bar{\alpha}$ of the upper limit, in order to obtain a meaningful [according to Eq.(4)] upper bound on the CMB anisotropy.

On the basis of these considerations, we performed a re-analysis of the Uson and Wilkinson, Readhead *et al.* and Meinhold and Lubin experiments, fixing the power of the test *a priori*, and allowing for values of $\beta_* = 50\%$ (comparable to the value obtained for the Meinhold and Lubin experiment), 75% (comparable to the value obtained for the Readhead *et al.* experiment) and 95% respectively, in order to show how sensitive the results of this analysis are to the power of the test.

4. DISCUSSION AND CONCLUSIONS

The preliminary results of our analysis are given in Table II as upper limits on Δ_{rms}. These upper limits at the 95% confidence level are given as a function of the power of the test. They are model *independent*, as already mentioned, at least for the CDM models we have considered, because different experiments probe a very narrow region of the power spectrum.

TABLE II

Observational bounds on Δ_{rms} in CDM cosmologies

	published	$\beta_* = 50\%$	$\beta_* = 75\%$	$\beta_* = 95\%$
Uson and Wilkinson	$3.0 \cdot 10^{-5}$	$6.1 \cdot 10^{-5}$	$7.7 \cdot 10^{-5}$	$1.0 \cdot 10^{-4}$
Readhead *et al.*	$2.1 \cdot 10^{-5}$	$1.7 \cdot 10^{-5}$	$2.2 \cdot 10^{-5}$	$3.0 \cdot 10^{-5}$
Meinhold and Lubin	$4.7 \cdot 10^{-5}$	$4.4 \cdot 10^{-5}$	$5.4 \cdot 10^{-5}$	$6.8 \cdot 10^{-5}$

The conclusion that can be drawn from our analysis is that the more stringent upper limit on CDM cosmologies from CMB anisotropy bounds on small and intermediate angular scales is provided by the Readhead *et al.* experiment: Δ_{rms} $(7', 0'.76) < 3 \cdot 10^{-5}$, at the 95% confidence level and 95% power of the test. This upper limit is, however, still too loose for contraining flat CDM models, especially if one allows for a biasing parameter $b \sim 1.5$. Still, we are able to exclude flat CDM universes with $\Omega_b \gtrsim 0.35$ if $b = 1$, and $\Omega_b \gtrsim 0.5$ if $b \sim 1.5$.
As already mentioned, an upper limit on Δ_{rms} implies a lower limit on the density parameter Ω_0. The Readhead *et al.* experiment is able to set a quite stringent lower bound for the density parameter in CDM cosmologies: $\Omega_0 \gtrsim 0.6$, at the 95% confidence level and 95% power. This is much stronger than the constraint derived either from the Uson and Wilkinson or from the Meinhold and Lubin experiment, because of its large power of the test. Note also that the published Uson and Wilkinson upper limit was used to set a lower limit to the density parameter in CDM dominated models (*e.g.* Vittorio and Silk, 1984; Bond and Efstathiou, 1984). The lower bound derived there ($\Omega_0 \gtrsim 0.4$ for h=0.5) is much less reliable than the one presented here, because the Uson and Wilkinson published upper limit has a very low power of the test, $\lesssim 20\%$.
In conclusions, we have proposed a way of analyzing CMB anisotropy data, in order to extract statistically robust upper limits. In order to accept or reject theoretical models, is not enough to compare theoretical *rms* predictions for the CMB anisotropy directly with the observational bounds. It is important to analyze a set of experimental CMB temperature differences, taking into account specific models for the formation and evolution of the large scale structure of the universe, as the CDM models considered here. In the frequentist approach we have used, it is important to derive upper limits with both high confidence level and high power of the test. Only by *a priori* fixing the power of the test (*e.g.* 95%), we are able to derive a real, statistically robust upper limit, without overkilling a specific model. This carefulness in comparing theory and observations seems to be extremely important in view of the fact that the sensitivity of the experiments is likely to go down in the near future, while is very difficult to imagine how to lower the flat biased CDM model predictions.

REFERENCES

Bahcall, N., 1988: Ann. Rev. Astron.Astrophys,26,631
Blau, S.K., and Guth, A.H., 1987: in "300 Years of Gravitation",eds. S.W.Hawking and W.Israel,(Cambridge Un.Press, Cambridge).
Blumenthal, G., Faber, S., Primack, J., and Rees, M. 1984: Nature, 301,584
Boesgard, A.M. and Steigman, G. 1985: Ann.Rev.Astron.Astrophys.,23,319
Bond, J. and Efstathiou, G. 1984: Ap.J.Lett.,285,L44
Davis, M. and Peebles, P.J.E. 1983: Ap.J.,267,465.
Dekel, A., and Rees, M. J. 1987: Nature,326,455
Geller, M.J., Huchra, J.P. and de Lapparent, V. 1986: Ap.J. Lett.,308,L1
Gunn, J.E., 1989: In 'The Extra-Galactic Distance Scale', A.S.P. Conference series, eds C. Pritchet and S. van den Bergh.
Meinhold and Lubin, 1990: Ap.J.Lett., in press
Muciaccia, P.F., and Vittorio, N. 1990: in preparation
Ostriker, J.P. and Cowie, L. 1981: Ap.J.Lett.,243,L127
Peebles, P.J.E. and Yu 1970, Ap.J., 162, 815

Peebles, P.J.E. 1987: Nature,327,210
Readhead, A.C.S., Lawrence, C., R., Meyers, S.,T., Sargent, W.,L.,W., Hardebeck, H.E., Moffet, A.T., 1989: Ap.J.,346,566
Sachs, R.K., and Wolfe, M.A., 1967, Ap.J., 147, 73.
Turok,N. 1987: in 'Nearly Normal Galaxies', ed. S. M. Faber, Springer-Verlag
Uson, J. and Wilkinson, D.T. 1984: Ap.J.,277,L1
Vittorio, N. and Silk, J. 1984: Ap.J. Lett.,285,L39
Vittorio, N., Meinhold, P., Muciaccia, P.F., Lubin, P., and Silk, J. 1990: Ap.J. Lett.,in press

IN AN OPTICALLY THICK UNIVERSE, THE COSMIC MICROWAVE BACKGROUND CAN APPEAR SMOOTHER THAN THE UNIVERSE ACTUALLY WAS

Dennis W. Marks
Valdosta State College, Valdosta, GA 31698

ABSTRACT

The relativistically invariant equation of radiative transfer with arbitrary velocity flows, temperature and density gradients, and gravitational fields can be expanded in projected, symmetric, trace-free tensors. In the optically-thick limit, the monopole moment of the radiation field is Planckian plus a small term that depends on the proper velocity dotted with the temperature gradient (if any) divided by the absorption (but not the scattering) cross-section. The dipole moment contains a term proportional to the temperature gradient and another term proportional to the proper acceleration, both divided by the sum of the absorption cross-section and the scattering cross-section. The quadrupole term is proportional to the rate of strain (if any), divided by the sum of the absorption cross-section and nine-tenths the scattering cross-section. All higher-order moments vanish identically. Thus the cosmic microwave background should not show higher-order multipole moments, in agreement with COBE observations. Moreover, the multipole expansion shows terms proportional to the temperature gradient, but not to the density gradient. The absence of temperature fluctuations in the CMB does not imply the absence of density fluctuations in the early universe. The smoothness of the cosmic microwave background may be a weaker constraint on theories of galaxy formation than currently thought.

INTRODUCTION

The high degree of isotropy of the cosmic microwave background is one of the most basic astronomical observations, but one of the most puzzling. These observations present theoreticians with at least two problems: the problem of the universe's initial conditions and the problem of the formation of galaxies, which exhibit a much higher density variation. These problems may, however, not be as severe as they first appear, since any opaque gas will appear nearly isotropic if viewed from within. The observed near-isotropy may be a consequence of the scattering of photons after emission, rather than a consequence of isotropy of the sources of the photons.

RELATIVISTICALLY INVARIANT RADIATIVE TRANSFER

At space-time position $x^\alpha = (ct,x,y,z)$, let $F(p_\beta,x^\alpha)$ be the photon distribution function. In terms of the observed intensity,

$$2F = \frac{c^2}{h\nu^3} I_\nu \tag{1}$$

where the factor *2* arises from summing over polarization states. Intensity is not a Lorentz invariant, however, so we will work with the photon distribution function, which is a Lorentz scalar.

A photon has zero mass ($p_\alpha p^\alpha = 0$). Viewed by an observer moving with four-velocity v^α ($v_\alpha v^\alpha = -c^2$), the photon has energy $h\nu = -p_\alpha v^\alpha$ and direction vector $n^\alpha = cp^\alpha/h\nu - v^\alpha/c$. The direction vector is a space-like unit vector ($n_\alpha n^\alpha = 1$), orthogonal to the four-velocity ($n_\alpha n^\alpha = 0$). The photon distribution function can be expressed as $F(\nu, n^\beta, x^\alpha)$.

In thermodynamic equilibrium, F is the Planck function,

$$F_o = [\exp(-p_\alpha v^\alpha/kT) - 1]^{-1}. \tag{2}$$

Out of thermodynamic equilibrium, the departures of F from the Planck function can be expanded in spherical harmonics of n^α:

$$\begin{aligned} F(\nu,n^\beta,x^\alpha) - F_o(\nu,x^\alpha) &= A(\nu,x^\alpha) + B_\beta(\nu,x^\alpha)n^\beta + \frac{1}{2}C_{\beta\gamma}(\nu,x^\alpha)n^{\beta\gamma} \\ &+ \frac{1}{6}D_{\beta\gamma\delta}(\nu,x^\alpha)n^{\beta\gamma\delta} + \ldots \end{aligned} \tag{3}$$

Here $n^{\alpha\beta} = n^\alpha n^\beta - h^{\alpha\beta}/3$ and $n^{\alpha\beta\tau} = n^\alpha n^\beta n^\tau - (n^\alpha h^{\beta\tau} + n^\beta h^{\tau\alpha} + n^\tau h^{\alpha\beta})/5$, where $h_\alpha^{\ \beta} = \delta_\alpha^{\ \beta} + v_\alpha v^\beta/c^2$ is the projection tensor onto the space-like hypersurface orthogonal to v^α ($v^\alpha h_\alpha^{\ \beta} = 0$). The tensors A, B_β, $C_{\beta\tau}$, $D_{\beta\tau\delta}$, . . . and 1, n^β, $n^{\beta\tau}$, $n^{\beta\tau\delta}$, . . . are of the form Thorne (1981) calls PSTF- projected, symmetric, trace-free.

The equation of radiative transfer in a moving fluid can be written as

$$p^\alpha \frac{dF}{dx^\alpha} = \frac{-h\nu}{c^2}\left[\frac{A}{\tau_0(\nu)} + \frac{B_\beta n^\beta}{\tau_1(\nu)} + \frac{1}{2}\frac{C_{\beta\gamma}n^{\beta\gamma}}{\tau_2(\nu)} + \frac{1}{6}\frac{D_{\beta\gamma\delta}n^{\beta\gamma\delta}}{\tau_3(\nu)} + \ldots\right] \tag{4}$$

where the right-hand side has been constructed by Straumann (1976) as the most general possible scalar function linear in the departures from thermodynamic equilibrium. In this form the equation of transfer describes the effects of scatterers and absorbers not on individual photons, but on the moments of the photon distribution. The dependences of A, B_β, $C_{\beta\tau}$, $D_{\beta\tau\delta}$, *etc.*, on frequency and position can be found by taking moments of the equation of transfer:

$$\oint p^\alpha \frac{dF}{dx^\alpha} d\Omega = -4\pi \frac{h\nu}{c^2}\frac{A}{\tau_0} \tag{5a}$$

$$\oint n_\beta p^\alpha \frac{dF}{dx^\alpha} d\Omega = -\frac{4\pi}{3}\frac{h\nu}{c^2}\frac{B_\beta}{\tau_1} \tag{5b}$$

$$\oint n_{\beta\gamma} p^\alpha \frac{dF}{dx^\alpha} d\Omega = -\frac{4\pi}{15}\frac{h\nu}{c^2}\frac{C_{\beta\gamma}}{\tau_2} \tag{5c}$$

$$\oint n_{\beta\gamma\delta} p^\alpha \frac{dF}{dx^\alpha} d\Omega = -\frac{4\pi}{105}\frac{h\nu}{c^2}\frac{D_{\beta\gamma\delta}}{\tau_3} \tag{5d}$$

Close to thermodynamic equilibrium, the left-hand sides of these equations can be evaluated using the equilibrium form of F, namely the Planck function. The resulting integrations yield

$$A(\nu,x^\alpha) = c\tau_0\, F_0'\frac{h\nu}{kT}\left[\frac{\nu^\beta}{cT}\frac{\partial T}{\partial x^\beta} + \frac{1}{3}\frac{\nu^\beta{}_{;\beta}}{c}\right] \tag{6a}$$

$$B_\beta(\nu,x^\alpha) = c\tau_1\, F_0'\frac{h\nu}{kT}\left[\frac{\nu^\beta}{cT}\frac{\partial T}{\partial x^\beta} + \frac{\nu^\gamma \nu_{\beta;\gamma}}{c^2}\right] \tag{6b}$$

$$C_{\beta\gamma}(\nu,x^\alpha) = c\tau_2\, F_0'\frac{h\nu}{kT}\left[h_\beta^\delta\frac{\nu_{\gamma;\delta}}{c} + h_\gamma^\delta\frac{\nu_{\beta;\delta}}{c} - \frac{2}{3}h_{\beta\gamma}\frac{\nu^\delta{}_{;\delta}}{c}\right] \tag{6c}$$

$$D_{\beta\gamma\delta}(\nu,x^\alpha) = 0 \tag{6d}$$

where $F_q{}'$ is the derivative of the Planck function with respect to its argument $h\nu/kT$ and semicolon denotes the covariant derivative.

The monopole moment of the cosmic microwave background should be Planckian plus a term proportional to the divergence of the fluid flow and a term that depends on the proper velocity of the observer dotted with the temperature gradient (if any). The dipole moment contains a term proportional to the temperature gradient and a term proportional to the proper acceleration. The quadrupole term is proportional to the rate of strain (if any). The monopole, dipole, and quadrupole moments may be (but are not required to be) non-zero; all moments higher than quadrupole are identically zero because F_o is independent of n^β.

The dipole moment is usually given in terms of the observer's velocity relative to the surface of last scattering. The velocity term does not show up in the optically thick case, however, because the observer does not see a fixed set of scatterers, but whichever scatterers were one mean-free-path away one mean-free-time ago. The dipole moment also contains terms proportional to the temperature gradient (if any) and to the four-acceleration (if any). In any case, the dipole moment Lorentz transforms like a vector and hence shows aberration. The difference in the dipole moments observed six months apart shows clearly in the COBE data (Smoot 1990).

In addition, the mean free times are not be the same for the monopole, dipole, and quadrupole moments. For a combination of atomic absorption and Thomson scattering

$$1/c\tau_0(\nu) = \kappa(\nu) \tag{7a}$$

$$1/c\tau_1(\nu) = \kappa(\nu) + n_e\sigma_{Th} \tag{7b}$$

$$1/c\tau_2(\nu) = \kappa(\nu) + 9n_e\sigma_{Th}/10 \tag{7c}$$

where σ_{Th} is the Thomson scattering cross section (Masaki 1971). If departures from a pure Planckian spectrum are detected in the cosmic microwave background, these departures need not be the same for the monopole, dipole, and quadrupole moments. In particular, the monopole departure depends on the absorption cross-section, but not on the scattering cross-section: scatterers cannot change a monopole distribution. Scatterers do have a different effect on the dipole and quadrupole moments, which could be measured if the quadrupole were not otherwise zero.

CMB IMPLICATIONS

In this paper we have considered the view that the observed smoothness of the cosmic microwave background is the result of the opacity of the material of the universe rather than a manifestation of the distribution of the material in the early universe. From inside an optically thick gas, one should not be able to see multipole moments higher than quadrupole. Higher order moments are simply smeared out by the "Compton fog" (Hogan 1980).

The proof in the preceding section entails an expansion of the photon distribution function in multipole moments about a local equilibrium distribution; that is, the temperature in the Planck function may depend on position. Weinberg (1972) argued that parts of the observable universe were not causally connected, so there would not have been the opportunity for the photon distribution function to have come to an equilibrium distribution; thus parts of the microwave background could have very different temperatures. Others have argued that a universal temperature was imposed at the Planck time or that the observable universe expanded by inflation from a part that was in equilibrium. All these arguments concern a global thermodynamic equilibrium (one temperature for the entire universe); this paper assumes only local thermodynamic equilibrium. Even with temperature inhomogeneities allowed, no higher-order moments should be observable in the cosmic microwave background of an optically thick universe.

Moreover, to first order in the mean free times, the multipole expansion contains terms proportional to the temperature gradient, but not to the density gradient. The absence of temperature fluctuations in the observed cosmic microwave background simply does not imply the absence of density fluctuations in the early universe. The early universe could have had larger density fluctuations than the observed temperature fluctuations. The regions of higher density would have a proportionately curtailed mean free path, and so the observed cosmic microwave background would not show as much contrast as the actual density fluctuations (if any) in the early universe. Even if the early universe were slightly inhomogeneous, the cosmic microwave background should appear less inhomogeneous than the universe actually was. (Isothermal fluctuations in the density, for example, would produce no temperature fluctuations in the cosmic microwave background.) If density fluctuations in the early universe were actually greater than the observed temperature fluctuations, then the smoothness of the cosmic microwave background may be only a weak constraint on theories of galaxy formation.

The author would like to thank Roger H. Lamprey and Bradley Brown for assistance with computer graphics and to acknowledge support received from the Center for Faculty Development and Instructional Improvement of Valdosta State College.

REFERENCES

Hogan, C. J. 1980, M.N.R.A.S., 192, 891.
Masaki, I. 1971, Publ. Astr. Soc. Japan, 23, 425.
Raine, D. J. 1981, The Isotropic Universe (Bristol: Adam Higler)
Smoot, G. 1990, this conference.
Straumann, N. 1976, Helvetica Physica Acta, 49, 269.
Thorne, K. S. 1981, M.N.R.A.S., 194, 439.
Weinberg, S. 1972, Gravitation and Cosmology (New York: Wiley)

STATE REDUCTION AND COSMIC ANISOTROPY

C. H. Woo
Department of Physics
University of Maryland
College Park, MD 20742

ABSTRACT

Between the quantum domain and the classical world of facts there is an intermediate domain for which a "classical" probabilistic description is appropriate. As a result of decoherence, such a description obeys the classical additivity of probabilities. In contrast to the usual interpretation in classical probability theory, however, the "classical" probabilistic description which emerges from an underlying quantum theory is meant to be applicable to a single system, such as the universe.

It is argued that small inhomogeneities and anisotropies in the very early universe should be described probabilistically. If, furthermore, the transition between this probabilistic description and a later deterministic description happens as late as after the decoupling of photons, then the smoothness of the microwave background receives a different interpretation: it is not because the fluctuations at decoupling were extremely small, but because the contributions from different fluctuations add to smooth things out.

DECOHERENCE AND ROBUSTNESS

Between the quantal and the classical there is an intermediate domain for which a "classical" probabilistic description is appropriate. We will first explain what is meant by this intermediate domain using the Copenhagen language, and then re-phrase it in terms of "decoherence via coarse-graining".

The Copenhagen school emphasized the need to bring a measurement to a close by an irreversible act of amplification that results in a communicable record [1]. The words "irreversible", "amplification", and "communicable" each requires a brief comment. The underlying dynamics is presumably reversible, but the coarse-grain variables of interest can gain from or lose information to the **unobserved** variables; thus irreversibility refers to the effective behavior of a subset of variables. The term "amplification" should include the establishment of the absence of a signal which, were it present, would be amplified. The reference to "communicable" may seem to be excessively anthropocentric. But if one takes a slightly broader view of communication, any lasting record which can be decoded by intelligent beings later is communicable; thus the requirement does not have to refer to human beings, nor to our decoding capability at any moment. Communicability serves to introduce some objectivity, as well as a near certainty in a basically probabilistic framework through the existence of corroboration among a great many communicators.

Bohr's emphasis on what brings a measurement to a close, and on the need to treat the apparatus and the system being measured as a unit, reflects his distrust of any notion of "being" or "reality" that lacks any empirical or operational means of verification. Developsments in quantum measurement theory since the early days show Bohr's caution to be prudent--attempts to attribute reality to a wider set of entities led to severe conceptual difficulties time and again. Since, however, most of these difficulties are due to quantum interference, the question arises: suppose the relative phase between two alternative outcomes is almost certainly non-retrievable, and yet there is no communicable record to pin down which alternative was realized in our world, what is the proper mode of description then? If one accepts Bohr's conservative attitude towards "reality", then the absence of any means of verifying a unique outcome implies that the probabilistic description of quantum mechanics is still appropriate, but the lack of interference means that the additivity of classical probabilities holds. The interpretation of such "classical" probabilities, however, requires care, particularly in the way they are applied to a single system, as we shall explain below.

Some recent work focuses attention on "decoherent histories", to which a quantum theory assigns probabilities [2,3,4]. In many circumstances coarse-graining produces decoherence:the reduced density matrix describing the coarse-grained variables readily becomes nearly diagonal in a basis determined by the initial state and the dynamics [5]. Like the many-world interpretation, this formalism is meant to be applicable to a single system such as the universe; thus classical phenomena should emerge automatically from the theory. The robustness of facts in the classical world must then emerge also.

Consider, e.g., the coarse variable which describes the pointer position of a meter. The pointer behaves according to classical deterministic laws most of the time, but not always. When the meter measures the outcome of a quantum event, the predictability of its behavior is interrupted. And yet the statement "before the event the pointer was at x and afterwards it was at y" should be a fact. What makes it a fact is that the pointer is correlated with many other coarse variables which also behave according to classical deterministic laws most of the time. Thus there must be a classically verifiable network of facts to provide the context for the unpredictable outcome of a quantum event to be recognized. Its connection to this body of robust facts is what justifies the total neglect henceforth of other outcomes(thus "closing" this particular branching process).

It is clear that the robustness of a classical branch involves more than decoherence. After coarse-graining, the master equations satisfied by the reduced density matrices usually contain damping and fluctuation terms, which describe the residual influence of the traced over "environment variables" on the coarse variables. Although in some cases the environment can be effective in enforcing decoherence while introducing negligible damping and fluctuations, in general the problem is quite complex [7]. But even when the residual influence of the environment is small, the interactions among the coarse variables themselves still play an important role in determining if quasi-stable structures form to facilitate record keeping, if chaos develops and many coarse variables become sensitive to even small noises, and in particular if quantum fluctuations become easily amplified in the setting. In general, a decohered branch is not the same as a robust classical branch;then the diagonal elements of the reduced density matrix do not represent separate possible classical worlds--as emphasized by Bell [6], the "and" has not yet become "or".

EXCLUSIVE AND NONEXCLUSIVE MEASUREMENT RESULTS

Although the existence of an intermediate domain for which an "and"-type probabilistic description holds even for a single system seems to us quite plausible, our argument suffers from the lack of precision which plagues this kind of discussion. One way the issue may become clearer is via additional empirical evidence. We therefore ask: what would be the empirical consequences of the existence of such intermediate domains? We would need measurement results which are correlated with the possible outcomes at an earlier time, but which do not force the answer to be only consistent with a unique outcome. For example, suppose there were two possible outcomes A and B. An ideal exclusive measurement result would be such that it is consistent only with A alone or B alone; whereas by a nonexclusive result we mean some numerical data which can be consistent with A alone or B alone, but also with a probability p for A and (1-p) for B, depending on what the data are. The design of such "poor measurements" presents a novel challenge: although it is not hard to conceive of an arrangement in which the different outcomes have little chance of interfering, the hard part is to make sure that at the same time the event with the attendant measurement setup leaves no exclusive records behind. The feasibility of such laboratory tests will be discussed elsewhere.

Because of the paucity of stable structures in the early universe, some events at an early time may belong in the intermediate domain. To be specific, consider a density fluctuation 10^{-12} second after the big bang. On the one hand, the fluctuations can be small but already macroscopic, and different fluctuation configurations can decohere. On the other hand, the criteria for the selection of one specific outcome via its tight connection to a body of facts are not met--at least so it seems. For example, such a fluctuation may involve a local excess of matter over the average value by $\delta m \simeq 10^{-2}$ gm in a volume of 1 cc. Model studies of decoherence [8,9] suggest that decoherence for the coarse variables occurs rapidly, with a ratio of decoherence time to other dynamical times of the order $\hbar^2/((\Delta x)^2 kT\delta m)$. It has been suggested that such an estimate may have fairly general validity [9]. If that is true, one such fluctuation configuration would have decohered from another at 10^{-12} sec. Since, however, such an inhomogeneity corresponds to $\delta\rho/\rho \simeq 10^{-17}$, it would be amazing if there are enough fingerprints in the macro-world to allow the identification by classical physics one specific small inhomogeneity distribution at 10^{-12} sec. as being the unique predecessor.

Thus if the early fluctuations are small, it is necessary to describe the state as a probabilistic distribution of different density fluctuations. As lumps grow, however, surely a specific large scale inhomogeneity distribution becomes classically traceable later. When does the transition from a probabilistic description to a deterministic one take place? We do not know for sure. It is of some interest to consider, however, the radical hypothesis that this juncture may have occurred as late as after the decoupling of photons. Then the microwave background provides just the kind of nonexclusive result which yields only imperfect information about the past. A smooth microwave background would be consistent with extremely small density fluctuations at decoupling, but it can also be consistent with a symmetric distribution of not so tiny fluctuations (perhaps as large as $\delta\rho/\rho \simeq 10^{-3}$ at decoupling). The question one must ask is: Other than a smooth microwave background, what are the consequences of associating the early universe with a probabilistic distribution of density fluctuations? For instance, how is the microwave temperature affected? Unfortunately, to answer such questions requires knowing how one decoherent component in the distribution affects the evolution of another, and this is a difficult task. If one could ignore such mutual influence (which is not justified), and if one assumes a symmetric distribution of small fluctuations, one finds the deviation from a single temperature spectrum $n(E,T)$ to be larger at higher photon energies E, with a spectrum $\simeq n(E,T)[1+O((\delta\rho/\rho)^2(E/kT)^2)]$. It should be emphasized, however, that even when the initial condition is assumed to be homogeneous and isotropic, the summation over density fluctuations does not necessarily have symmetric weights, because the probability of observing photons is conditional on the present world being what it is, which is inhomogeneous. We have only checked in a toy model that the later asymmetry in the matter sector need not destroy the symmetry of the decoupled quanta. Nowadays computer simulations exist of how small specks accrete into larger lumps [10]. It would be interesting if the experts would try to run a backward simulation. That is, after picking a desired configuration at some post-decoupling time, one can try to find all the prior configurations that would evolve into this specific one, and then see how symmetric or asymmetric the distribution of prior configurations was at the decoupling time.

I wish to thank the following people, but also make clear that the usual caveat that "discussion does not imply approval" is especially appropriate in this case. For their criticisms I thank D. Brill, J.B.Hartle, B.L.Hu, T.Jacobson,C.Misner, E.J.Squires,W.H.Zurek.

Footnotes

[1] N. Bohr, Atomic Physics and Human Knowlege, Wiley, New York, 1958.
[2] R.J.Griffiths, J. Stat. Phys. 36, 219, 1971.
[3] R. Omnes, J. Stat. Phys. 53, 893, 1988.
[4] J.B.Hartle and M.Gell-Mann, Complexity,Entropy and the Physics of Information, ed. W.H.Zurek, Addison Wesley, Redwood City, 1990.
[5] E.Joos and D.H.Zeh, Z. Phys., B59, 223, 1985.
[6] J.S.Bell, Phys. World, Aug. 1990, p. 33.
[7] B.L.Hu, Univ. of Md. PP #91-096, 1990.
[8] A.O.Caldeira and A.J.Leggett, Phys. Rev. A31, 1059, 1985.
[9] W.H.Zurek in Frontiers of Nonequilibrium Statistical Physics, ed. G.T.Moore and M.O.Scully, Plenum, New York, 1986.
[10]See, e.g., M.Davis et.al., Astrophys. J. 327, 507, 1988.

The IR Background: Early Star and Galaxy Formation

THE DIFFUSE INFRARED BACKGROUND: COBE AND OTHER OBSERVATIONS

M. G. Hauser, T. Kelsall, S. H. Moseley, Jr.,
R. F. Silverberg
Laboratory for Astronomy and Solar Physics
Goddard Space Flight Center
Greenbelt, MD 20771

T. Murdock (General Research Corporation)
G. Toller, W. Spiesman, J. Weiland
(General Sciences Corporation)

ABSTRACT

The Diffuse Infrared Background Experiment (DIRBE) on the Cosmic Background Explorer (COBE) satellite is designed to conduct a sensitive search for an isotropic cosmic infrared background radiation over the spectral range from 1 to 300 micrometers. The cumulative emissions of pregalactic, protogalactic, and evolving galactic systems are expected to be recorded in this background. The DIRBE instrument, a 10 spectral band absolute photometer with an 0.7 deg field of view, maps the full sky with high redundancy at solar elongation angles ranging from 64 to 124 degrees to facilitate separation of interplanetary, Galactic, and extragalactic sources of emission. Initial sky maps show the expected character of the foreground emissions, with relative minima at wavelengths of 3.4 micrometers and longward of 100 micrometers. Extensive modelling of the foregrounds, just beginning, will be required to isolate the extragalactic component. In this paper, we summarize the status of diffuse infrared background observations from the DIRBE, and compare preliminary results with those of recent rocket and satellite instruments.

INTRODUCTION

The search for cosmic infrared background radiation (CIBR) is a relatively new field of observational cosmology. Measurement of this distinct radiative background, expected to arise from the cumulative emissions of pregalactic, protogalactic, and galactic systems, would provide new insight into the cosmic 'dark ages' following the decoupling of matter from the cosmic microwave background radiation. Observationally, there have been no corroborated detections of the CIBR, though possible evidence for an isotropic infrared background in data from recent rocket experiments has been reported (Matsumoto et al. 1988a; Lange et al. 1990). The Diffuse Infrared Background Experiment (DIRBE) on the COBE spacecraft is the first satellite instrument designed specifically to carry out a systematic search for the CIBR. In this paper we shall report the status of the DIRBE investigation, which is still in its early phase, and will compare preliminary DIRBE measurements of the infrared sky brightness with those of the recent investigations from instruments on sounding rockets and the IRAS satellite. Current conservative upper limits on the CIBR will be compared with representative theoretical predictions to illustrate the status and challenges of measuring this important cosmic fossil.

COSMOLOGICAL MOTIVATION FOR CIBR STUDIES

Cosmological motivations for searching for an extragalactic infrared background have been discussed in the literature for several decades (early papers include Partridge and Peebles 1967; Low and Tucker 1968; Peebles 1969; Harwit 1970; Kaufman 1976). Both the cosmic redshift and reprocessing of short wavelength radiation to longer wavelengths by dust act to shift the short wavelength emissions of cosmic sources toward or into the infrared. Hence,the wide spectral range from 1 to 1000 micrometers is expected to contain much of the energy released since the formation of luminous objects, and could potentially contain a total radiant energy density comparable to that of the cosmic microwave background radiation (CMBR).

The CIBR has received relatively little attention in the theoretical literature compared to that devoted to the CMBR (Negroponte 1986), which has a central significance to Big Bang cosmology and quite distinctive and definite predictions as to its character. However, advances in infrared instrumentation, and especially the introduction of cryogenically cooled infrared instruments on space missions, has stimulated increasing attention to prediction of the character of the CIBR (Fabbri and Melchiorri 1979; Bond, Carr, and Hogan 1986; McDowell 1986; Fabbri et al. 1987; Fabbri 1988; Bond, Carr, and Hogan 1990). Though it is outside the scope of this paper to review these predictions (some of which are discussed in other papers at this Workshop), it is increasingly clear that measurement of the spectral intensity and anisotropy of the CIBR would provide important new insights into intriguing issues such as the amount of matter undergoing luminous episodes in the pregalactic Universe, the nature and evolution of such luminosity sources, the nature and distribution of cosmic dust, and the density and luminosity evolution of infrared-bright galaxies.

MEASURING THE ABSOLUTE BRIGHTNESS OF THE INFRARED SKY

Observing the CIBR is a formidable task. Bright foregrounds from the atmosphere of the Earth, from interplanetary dust scattering of sunlight and emission of absorbed sunlight, and from stellar and interstellar emissions of our own Galaxy dominate the diffuse sky brightness in the infrared. Even when measurements are made from space with cryogenically cooled instruments, the local astrophysical foregrounds strongly constrain our ability to measure and discriminate an extragalactic infrared background. Furthermore, since the absolute brightness of the CIBR is of paramount interest for cosmology, such measurements must be done relative to a well-established absolute flux reference, with instruments which strongly exclude or permit discrimination of all stray sources of radiation or offset signals which could mimic a cosmic signal.

Table 1 lists recent experiments capable of making absolute sky brightness measurements in the infrared (for a compilation including some earlier measurements, see Negroponte 1986). Instruments or detector channels designed specifically to measure that part of the spectrum dominated by the CMBR have been excluded. Murdock and Price (1985) flew an absolute radiometer with strong stray-light rejection on a sounding rocket in 1980

Table 1. ABSOLUTE INFRARED BACKGROUND MEASUREMENTS

REFERENCE	WAVELENGTH (μm)	BEAM SIZE (DEG)	SKY COVERAGE	SOLAR ELONGATION* (DEG)	DATES
Murdock & Price 1985	2 - 30	0.1x0.25	40°-80°rad cone b=-23, l=155	($\lambda-\lambda_{\odot}$= 22-180)	8/18/80; 7/31/81
Matsumoto et al. 1988a	0.7 - 5	4	32° rad cone b=55, l=150	($\lambda-\lambda_{\odot}$= 113-180)	1/13/84
Lange et al. 1990	100 - 300	7.6	15° rad cone b=35, l=203	140-170	2/22/87
IRAS (ZOHF, vers. 3)	12 - 100	0.5x0.5	All	80-100; 60-120	1/26/83-11/22/83
COBE/DIRBE	1 - 300 1 - 3.4 pol.	0.7x0.7	All	64-124	11/18/89-9/21/90
COBE/DIRBE	1 - 4.5 1 - 3.4 pol.	0.7x0.7	All	64-124	9/21/90-present
COBE/FIRAS (high freq.)	100 - 500	7	All	94	11/18/89-9/21/90

*In this paper, elongation is the angle between the instrument line-of-sight and the observer-Sun line. The geocentric ecliptic longitude is denoted by λ.

and 1981. Their primary objective was measuring scattering and emission from interplanetary dust, and no attempt was made to extract an extragalactic component. Matsumoto et al. (1988a) flew a near-infrared experiment on a rocket in 1984. They have reported possible evidence for an isotropic residual near 2 micrometers, perhaps in a line feature, for which they cannot account in their models of emission from the interplanetary medium and the Galaxy. This group has flown a modified instrument early in 1990 to investigate further this result (Matsumoto, private communication). Lange et al. (1990) report a detailed analysis of the data in the high-frequency channels of the rocket experiment flown by Matsumoto et al. (1988b) in 1987. After correlating with galactic and interplanetary signals, upper limits and a possibly significant isotropic residual are reported. The IRAS sky survey instrument, though not specifically designed for absolute background measurements, had an approximate zero point calibration, and was, within the limits of long-term stability, capable of good relative total sky brightness measurements. Uncertainties in the IRAS absolute calibration have impeded efforts to extract an estimate of the CIBR (Rowan- Robinson 1986). Table 1 includes two entries for the DIRBE instrument on COBE: one for the period of operation with liquid helium in the dewar, when measurements were made over the 1 - 300 μm spectral range, and one for the period following helium depletion, when the 1 - 4.5 μm sensors continued to operate. The DIRBE experiment is discussed in some detail below. Finally, Table 1 lists the high frequency channel (100 to 500 μm) of the COBE FIRAS experiment. Though results are not yet available from these data, the FIRAS, with its all-sky coverage, excellent stray light rejection, absolute calibration, and high sensitivity, promises to be an important instrument for CIBR studies. Quantitative comparison of the measurements from the experiments listed in Table 1 and a summary of current CIBR limits are discussed below.

THE DIFFUSE INFRARED BACKGROUND EXPERIMENT

The Diffuse Infrared Background Experiment (DIRBE) on the COBE mission is the first space experiment designed primarily to measure the CIBR. The aim of the

DIRBE is to conduct a definitive search for an isotropic CIBR, within the constraints imposed by the local astrophysical foregrounds. The experimental approach is to obtain absolute brightness maps of the full sky in 10 photometric bands (J[1.2μm], K[2.3μm], L[3.4μm], and M[4.9μm]; the four IRAS bands at 12, 25, 60, and 100 μm; and 120-200 and 200-300 μm bands). In order to facilitate discrimination of the bright foreground contribution from interplanetary dust, linear polarization is also measured in the J, K, and L bands, and all celestial directions are observed hundreds of times at all accessible solar elongation angles (depending upon ecliptic latitude) in the range 64 to 124 degrees. The instrument is designed to achieve a sensitivity for each field of view of λI_λ = 10^{-13} W cm^{-2} sr^{-1} (1 σ, 1 year). This level is well below estimated CIBR contributions (e.g., Bond, Carr, Hogan 1986; see Fig. 3a). For further general information about the COBE mission, see the descriptions by Mather (1982) and Gulkis et al. (1990). Early scientific results from the mission have been summarized by Mather et al. (1990).

The DIRBE instrument is an absolute radiometer, utilizing an off-axis folded Gregorian telescope with a 19-cm diameter primary mirror. The optical configuration (Magner 1987) is carefully designed for strong rejection of stray light from the Sun, Earth limb, Moon or other off-axis celestial radiation, or parts of the COBE payload (Evans, 1983). Stray light rejection features include both a secondary field stop and a Lyot stop, super-polished primary and secondary mirrors, a reflective forebaffle, extensive internal baffling, and a complete light-tight enclosure of the instrument within the COBE dewar. Additional protection is provided by the Sun and Earth shade surrounding the COBE dewar, which prevents direct illumination of the dewar aperture by these strong local sources. The DIRBE instrument, which is maintained at a temperature below 2 K within the dewar, measures absolute brightness by chopping between the sky signal and a zero-flux internal reference at 32 Hz using a tuning fork chopper. Instrumental offsets are measured by closing a cold shutter located at the prime focus. All spectral bands view the same instantaneous field-of-view, 0.7 deg x 0.7 deg, oriented at 30 degrees from the axis about which the COBE spacecraft spins at 0.8 rpm. The rotation modulates the solar elongation angle of the DIRBE line-of-sight by 60 degrees during each rotation, and allows the DIRBE to sample fully 50% of the celestial sphere each day. Four highly reproducible internal radiative reference sources can be used to stimulate all detectors when the shutter is closed to monitor the stability and linearity of the instrument response. The redundant sky sampling and frequent response checks provide precise photometric closure over the sky for the duration of the mission. Calibration of the photometric scale is obtained from observations of isolated bright celestial sources. Careful measurements of the beam shape in pre-flight system testing and during the mission using scans across bright point sources allow conversion of point-source calibrations to surface brightness calibrations.

Mapping of the sky with the DIRBE instrument began on November 21, 1989, though checkout and tuning of spacecraft and instrument operations occupied much of the time until December 10. Routine surveying of the sky continued until depletion of the liquid helium on September 21, 1990. The interior of the dewar has subsequently warmed slowly to a present temperature of about 40 K. Though the detectors at wavelengths longer than 5 μm provide no useful data following helium depletion, the instrument mechanisms continue to operate normally and the J, K, L, and M band detectors (all of which are InSb photovoltaic detectors) continue to provide usable data at reduced sensitivity. The present plan is to continue to operate the DIRBE through a second year if the quality of data warrants it. This will allow completion of sky mapping over one full orbit around the Sun to aid modeling of the interplanetary dust foreground, and will enable a search for temporal variation in this foreground during the second year.

The data obtained during the helium-temperature phase of the mission, based upon quick-look examination, are of excellent photometric quality, showing good sensitivity, stability, linearity, and stray-light immunity. Few artifacts are apparent other than those induced by energetic particles in the South Atlantic Anomaly and variations in instrument temperature. Both of these effects will be removed in final data processing. Strong rejection of off-axis radiation sources is confirmed by the absence of response to the Moon (which saturates the response in all detectors when in the field of view) until it comes within about 3 degrees of the field of view. The sensitivity per field of view (shown in Figure 3b), based upon noise measured with the shutter closed and response determined from measurements of known celestial sources, is better than the goal stated above except at 150 and 250 μm. The sensitivities are worse than the design goal by factors of 10 and 4 respectively at these wavelengths. Nevertheless, averaging over many high galactic latitude fields-of-view will allow very sensitive measurements at these wavelengths. The nuclear radiation environment in orbit caused very little response change (<1%) in all detectors except the Ge:Ga photoconductors used at 60 and 100 μm. Thermal and radiative annealing procedures applied to the Ge:Ga detectors following passages through the South Atlantic Anomaly will allow response correction to about 1% at these wavelengths. It is expected that fully reduced DIRBE sky maps will have photometric consistency over the sky better than 2% at each wavelength, nearest-neighbor band-to-band (color) brightness accuracy of 3% or better, and absolute intensity scale accuracy better than 20%.

Qualitatively, the initial DIRBE sky maps show the expected character of the infrared sky. For example, at 1.2 μm stellar emission from the galactic plane and from isolated high latitude stars is prominent. Zodiacal scattered light from interplanetary dust is also prominent. At fixed ecliptic latitude the zodiacal light decreases strongly with increasing solar elongation angle, and at fixed elongation angle it decreases with increasing ecliptic latitude. These two components continue to dominate out to 3.4 μm, though both become fainter as wavelength increases. A composite of the 1.2, 2.3, and 3.4 μm images was presented by Mather et al. (1990). Because extinction at these wavelengths is far less than in visible light, the disk and bulge stellar populations of the Milky Way are dramatically apparent in this image. At 12 and 25 μm, emission from the interplanetary dust dominates the sky brightness, again strongly dependent upon ecliptic latitude and elongation angle. At wavelengths of 60 μm and longer, emission from

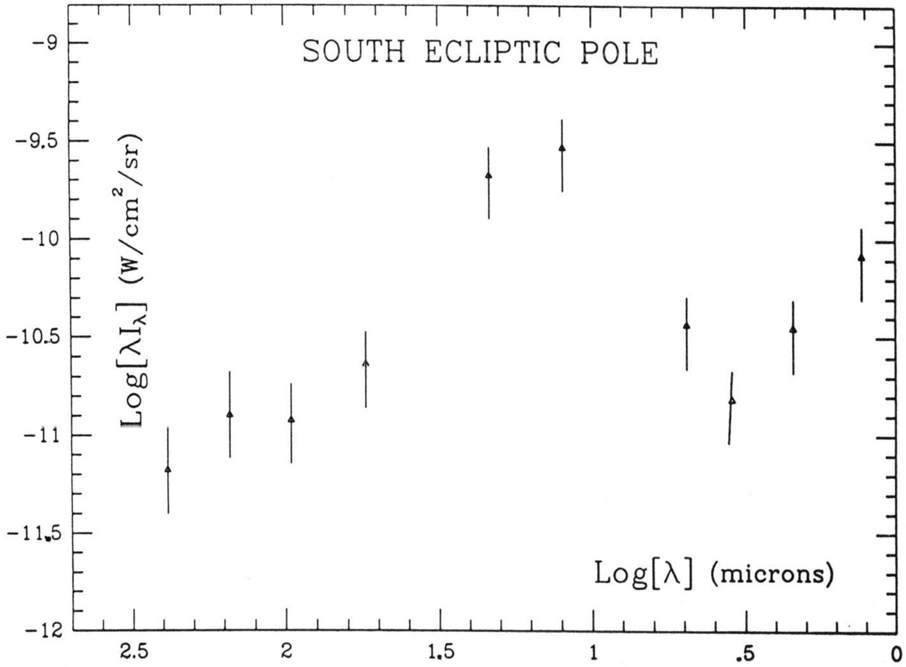

Figure 1. DIRBE measurements of the sky brightness at the South Ecliptic Pole.

---

the interstellar medium dominates the galactic brightness, and the interplanetary dust emission becomes progressively less apparent. The patchy infrared cirrus noted in IRAS data (Low et al. 1984) is evident at all wavelengths longer than 25 μm. The DIRBE data will clearly be a valuable new resource for studies of the interplanetary medium and Galaxy as well as the search for the CIBR.

In searching for the extragalactic infrared background, the most favorable conditions are directions and wavelengths of least foreground brightness. In general, because of the strong interplanetary dust foreground and the relatively modest gradient of that foreground over the sky, the infrared sky is faintest at high ecliptic latitude. A preliminary DIRBE spectrum of the sky brightness toward the south ecliptic pole is presented in Figure 1. This figure shows the strong foreground from starlight and scattered sunlight at the shortest wavelengths, a relative minimum at 3.4 μm, emission dominated by interplanetary dust peaking around 12 μm, and generally falling brightness from there out to submillimeter wavelengths. The 3.4 μm "window", between interplanetary dust scattering and emission, and the submillimeter "window", between cirrus emission and the CMBR, present the best opportunities for the CIBR search. The status of this search is discussed further below.

COMPARISON OF INFRARED SKY BRIGHTNESS MEASUREMENTS

There have been a number of recent measurements of the absolute infrared sky brightness (Table 1). Before proceeding with a discussion of their implications, we first wish to compare DIRBE data with those from other experiments. The comparison of brightnesses measured in a given celestial direction is somewhat confused by the presence of a varying foreground arising from the interplanetary medium, which depends upon the position of the Earth at the time of observation, as well as upon any intrinsic variability in the interplanetary dust cloud. We have attempted to minimize differences attributable to foreground changes by selecting DIRBE data obtained on the same day of year (day number), and hence solar elongation angle, as that of the comparison experiment. We emphasize that these comparisons are made between as-observed sky brightnesses, not components of emission inferred from the measurements.

The near-infrared data of Matsumoto et al. (1988a) toward the north galactic pole are compared with DIRBE data in Figure 2a. The absolute brightnesses are consistent within the indicated error bars at 2.2 and 4.9 μm. However, at 1.2 μm the DIRBE measurement is a factor of 1.5 above the rocket data, and at 3.4 μm the DIRBE result is a factor of 1.6 less. We can with some confidence conclude that the shape of the energy distribution over this spectral range is quite different in the two experiments, with the DIRBE finding a much steeper slope over the 1.2 to 3.4 μm range. This conclusion is strengthened by the fact that the indicated error bars for the DIRBE data are entirely dominated by the systematic absolute calibration uncertainty. For

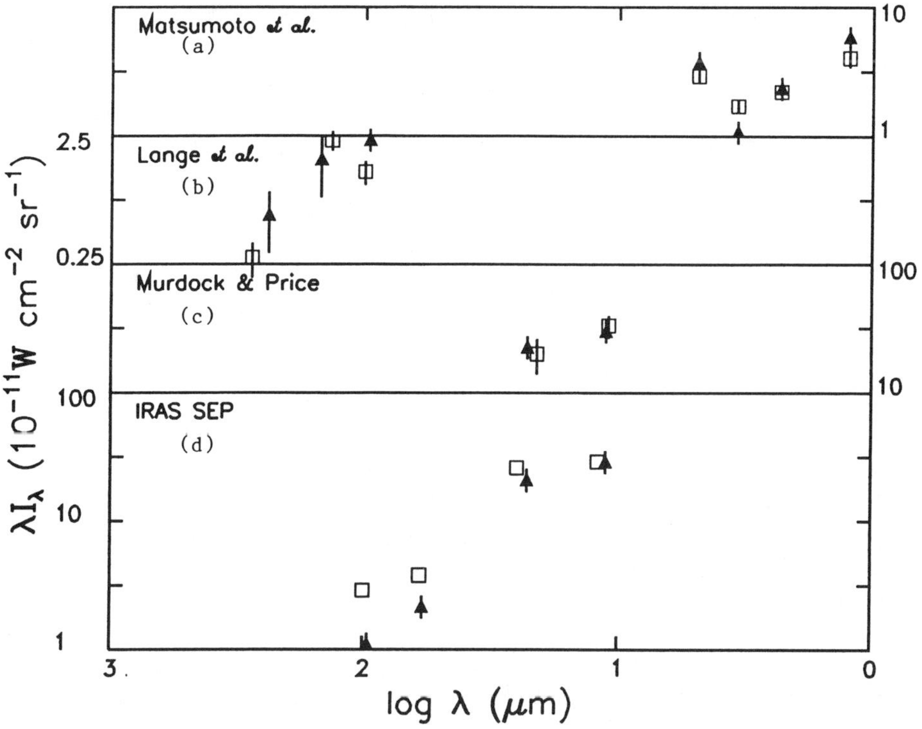

Figure 2. Comparison of absolute sky brightness measurements made with the DIRBE with those of previous experiments. In all cases, the DIRBE data are shown as filled triangles, with error bars including estimated systematic calibration uncertainties. Table 1 gives observational parameters of the experiments compared with the DIRBE results. Note that the scale on the ordinate is logarithmic in each panel, with labels alternately on the left and right. (a) Matsumoto et al. (1988a) rocket instrument observations near the north galactic pole. (b) Lange et al. (1990) presentation of the high frequency rocket data of Matsumoto et al. (1988b) near b = 20 deg, l = 198 deg. (c) Murdock and Price (1985) rocket observations at the north ecliptic pole. (d) IRAS data from Vers. 3 of the Zodiacal Observation History File observed toward the south ecliptic pole on 19 Feb 83.

these four wavelengths, however, the DIRBE celestial calibration was derived from observations of the same star (Sirius), yielding band-to-band calibration uncertainties much smaller than the error bars shown in Fig. 2a. Since a steeper 1.2 to 3.4 μm energy distribution is more consistent with the expected contributions from starlight and zodiacal light than the data of Matsumoto et al. 1988a (e.g., see Fig. 10 of their paper), these DIRBE data raise some doubt about their inference of an extragalactic component near 2.2 μm.

The far-infrared data of Lange et al. (1990) are compared with DIRBE data in Fig. 2b. The paper of Lange et al. presents an analysis of data from the high frequency channels of the measurements described by Matsumoto et al. (1988b), including the variation of sky brightness over the small circle scanned during their flight. Since their data were obtained at a solar elongation angle larger than reached by DIRBE, we have chosen to compare with the measured point at their smallest elongation (140 degrees), which is near b = 20 deg, l = 198 deg. We have averaged the DIRBE data over the area of their 7.6-deg beam in this direction. The largest elongation reached by the DIRBE observations in this direction is 124 degrees. Since the zodiacal emission is not bright at long wavelengths, and varies little with elongation at large angles, no attempt has been made to correct for the slight difference in observing conditions. The DIRBE preliminary calibration is particularly uncertain at 150 and 250 μm; within these large uncertainties, the two experiments are reasonably consistent at these wavelengths. At 100 μm the DIRBE result is a factor of 1.8 brighter, a difference larger than the estimated 20% uncertainty in DIRBE absolute calibration.

The north ecliptic polar brightnesses at 10 and 20 μm measured by Murdock and Price (1985) are compared with DIRBE data in Fig. 2c. These measurements agree within 10%, well within the estimated uncertainties.

IRAS measurements toward the south ecliptic pole are compared with DIRBE data in Fig. 2d. The two experiments are seen to agree reasonably well at 12 and 25 μm, but the DIRBE results are substantially fainter at 60 and 100 μm, reaching a factor of 2.6 at 100 μm. We have made a more detailed comparison with IRAS data at several points on the sky. By choosing points of different brightness, and assuming the DIRBE photometry to be correct, we can determine both zero point and responsivity 'errors' for the IRAS data. We find evidence for zero point errors, evidently not constant in time, of a few MJy/sr at all IRAS wavelengths. These are most significant (as a fraction of total brightness) at the longest wavelengths. We also find that the adopted IRAS dc responsivity at 60 and 100 μm was evidently too low, making reported sky brightnesses too high. The responsivity problem can be at least qualitatively understood in terms of the detector characteristics and the data reduction procedures used for IRAS image-format data (see the IRAS Explanatory Supplement (1988), pp. IV-9 and IV-10 for a discussion of the dependence of detector responsivity on frequency and incident power). The combined effect of the zero point and responsivity discrepancies is particularly large at 100 μm toward faint sky regions, as illustrated in Fig. 2d. Since IRAS detector responsivity depended on source angular scale (IRAS was a scan-modulated instrument) and source brightness, no simple conversion between the total brightness data from the two experiments is universally applicable. It should be noted that this discussion applies only to surface brightness measurements; point source flux calibration is not subject to similar problems.

CURRENT LIMITS ON THE CIBR

To meet the cosmological objective of measuring the CIBR, the foreground light from interplanetary and galactic sources must be discriminated from the total observed infrared sky brightness. This task requires extensive careful correlation studies and modelling, which in the case of the DIRBE investigation is just beginning. A conservative upper limit on extragalactic light is the total observed brightness in a relatively dark direction. The sky brightness at the south ecliptic pole presented in Fig. 1 is a fair representation of the best current limits from the DIRBE. As noted above, the faintest foregrounds occur at 3.4 μm, in the minimum between interplanetary dust scattering of sunlight and re-emission of absorbed sunlight by the same dust, and

Table 2. COSMIC INFRARED BACKGROUND LIMITS

REFERENCE	WAVELENGTH (μm)	λI_λ (10^{-11} W cm^{-2} sr^{-1})
Matsumoto et al. 1988a[1]	1.27	<0.6
	2.16	1.0±.37
	3.8	1.1±.34
	5.0	<0.5
Lange et al. 1990[1]	100	<0.32
	135	0.55±0.16
	275	<0.29
DIRBE (SEP)[2]	1.2	8.3±3.3
	2.3	3.5±1.4
	3.4	1.5±0.6
	4.9	3.7±1.5
	12	29 ±12
	22	21 ±8
	55	2.3±1
	96	1.2±0.5
	151	1.3±0.7
	241	0.7±0.4

NOTES:
1. values or upper limits for an isotropic component of the total observed sky brightness.
2. value of the total observed sky brightness in a dark direction, with no separation of components.

longward of 100 μm, where interstellar dust emission begins to decrease. Through careful modelling, we hope to be able to discriminate isotropic residuals at a level as small as 1 percent of the foregrounds. These near-infrared and submillimeter 'windows' will allow the most sensitive search for, or limits upon, the elusive cosmic infrared background.

The DIRBE data at the south ecliptic pole are listed in Table 2 along with CIBR data from the Matsumoto et al. (1988a) and Lange et al. (1990) papers. These investigators have attempted to discriminate the various

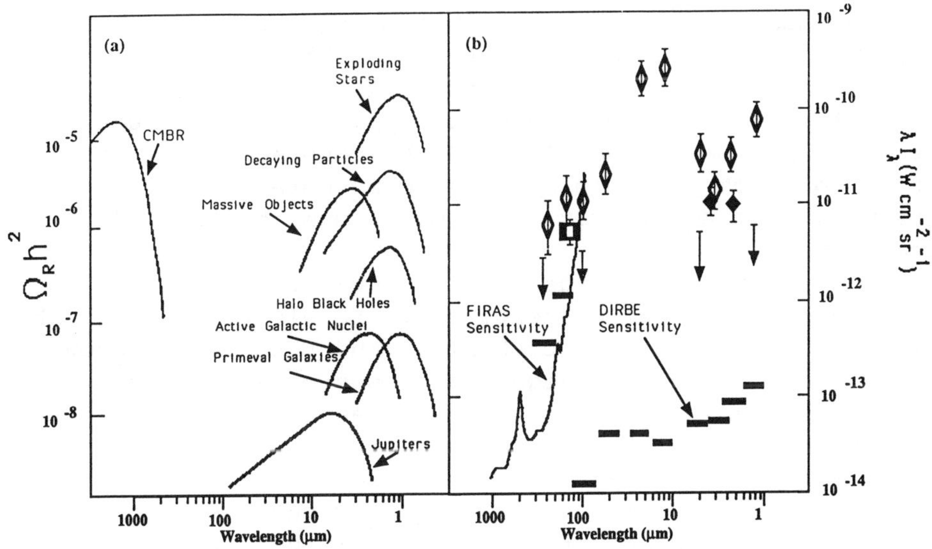

Figure 3. (a) Representative theoretical estimates for backgrounds arising from pre- and proto-galactic objects in a dust-free universe (Carr 1988; Bond, Carr, and Hogan 1986). Ω_R is the energy density in radiation per logarithmic frequency interval, in units of the critical density for closure, $\rho_C = 2 \times 10^{-29}\ h^2\ g\ cm^{-3}$, where h is the Hubble constant in units of 100 km/sec per Mpc. (b) Current observational limits on the isotropic extragalactic infrared background radiation. The scale on the ordinate corresponds to that in panel (a). The data are from Table 2: the south ecliptic pole sky brightness measured by the DIRBE (open diamonds); estimated near-infrared isotropic residuals from Matsumoto et al. (1988a) (filled diamonds and upper limits); estimated far-infrared isotropic residuals from Lange et al. (1990) (open square and upper limits). The estimated minimum detectable brightness (1 σ) per instrument field of view by the COBE experiments in a nominal 1 year mission is also shown: a solid horizontal bar for each DIRBE band; the solid curve for the FIRAS high frequency channel (per spectral resolution element).

components of emission contributing to their measurements, and have reported possible isotropic residuals they cannot account for in their solar system or galactic emission models, or upper limits on such residuals. Because of limited sky coverage and little time to check possible systematic measurement errors, results from the rocket experiments must be treated cautiously.

In order to put the observational data of Table 2 into a cosmological context, they are plotted in Fig. 3b. The sensitivity of the COBE DIRBE and FIRAS measurements is also illustrated. Figure 3b shows clearly that foreground emission, rather than measurement sensitivity, is the major challenge in searching for the CIBR. These data are to be compared with the theoretical curves of Figure 3a, which show estimated contributions to the CIBR from pregalactic and protogalactic sources in a dust-free universe. The present conservative observational limits are beginning to constrain some of the theoretical models, though in a dusty universe energy from these sources can be redistributed farther into the infrared than shown in Fig. 3a. If the foreground components of emission can confidently be identified, the current COBE measurements will seriously constrain (or identify) the CIBR across the infrared spectrum. However, the spectral decade from about 6 to 60 μm will have relatively weak limits until measurements are made from outside the interplanetary dust cloud.

CONCLUSION

The cosmic infrared background radiation promises to enhance our understanding of cosmic evolution since decoupling. The high quality and extensive new measurements of the absolute infrared sky brightness obtained with the DIRBE and FIRAS experiments on the COBE mission promise to allow a definitive search for this elusive background, limited primarily by the difficulty of distinguishing it from bright astrophysical foregrounds. Careful photometric reduction and modelling of the COBE data will yield important new cosmological and astrophysical insights.

ACKNOWLEDGMENTS

The authors gratefully acknowledge the contributions to this report by their colleagues on the COBE Science Working Group and the many other participants in the COBE Project, especially H. Freudenreich, J. A. Skard, and A. Panitz. We thank A. Lange for permission to quote unpublished results, and useful discussions of his results. The COBE mission is funded by NASA's Astrophysics Division. Mission management, design, development, operation, and data reduction are the responsibility of the Goddard Space Flight Center.

REFERENCES

Bond, J. R., Carr, B. J., and Hogan, C. J. 1986, Ap. J., 306, 428.
Bond, J. R., Carr, B. J., and Hogan, C. J. 1990, "Cosmic Backgrounds from Primeval Dust", Ap. J. (to be publ.).
Carr, B. J. 1988, Comets to Cosmology, A. Lawrence (ed.), Proc. of the Third International IRAS Conference, London, July 6-10, 1987, pp. 265-278, Springer Verlag.
Evans, D. C. 1983, SPIE Proc., 384, Generation, Measurement, and Control of Stray Radiation III, R. P. Breault (ed.), pp. 82-89.
Fabbri, R. 1988, Ap. J., 334, 6.
Fabbri, R. and Melchiorri, F. 1979, Astr. and Ap., 78, 376.
Fabbri, R., Andreani, P., Melchiorri, F., and Nisini, B. 1987, Ap.J. 315, 12.
Gulkis, S., Lubin, P. M., Meyer, S. S., and Silverberg, R. F. 1990, Sci. Am., 262, 132.
Harwit, M. 1970, Rivista del Nuovo Cimento Vol. II, 253.
IRAS Explanatory Supplement 1988, Infrared Astronomical Satellite (IRAS) Catalogs and Atlases, Vol. 1, C. A. Beichman, G. Neugebauer, H. J. Habing, P. E. Clegg, and T. J. Chester (eds.), NASA RP-1190.
Kaufman, M. 1976, Ap. Sp. Sci., 40, 369.
Lange, A. E., Richards, P. L., Hayakawa, S., Matsumoto, T., Matsuo, H., Murakami, H., and Sato, S. 1990, "The 100 to 300 micron Diffuse Background: Spectrum of the Infrared Cirrus and New Limits on the Extragalactic Background", (private comm.).
Low, F. J. and Tucker, W. H. 1968, Phys. Rev. Lett., 21, 1538.
Low, F. J., et al. 1984, Ap. J. (Lett.), 278, L19.
Magner, T. J. 1987, Opt. Eng., 26, 264.
Mather, J. C. 1982, Opt. Eng., 21, 769.
Mather, J. C., et al. 1990, Observatories in Earth Orbit and Beyond, Proc. of IAU Colloquium 123, Y. Kondo (ed.), Kluwer Acad. Publ., Dordrecht.
Matsumoto, T., Akiba, M., and Murakami, H. 1988a, Ap. J., 332, 575.
Matsumoto, T., Hayakawa, S., Matsuo, H., Murakami, H., Sato, S., Lange, A. E., and Richards, P. L. 1988b, Ap. J., 329, 567.

McDowell, J. C. 1986, MNRAS, 223, 763.
Murdock, T. L., and Price, S. D. 1985, Astr. J., 90, 375.
Negroponte, J. 1986, MNRAS, 222, 19.
Partridge, R. B. and Peebles, P. J. E. 1967, Ap. J., 148, 377.
Peebles, P. J. E. and Partridge, R. B. 1967, Ap. J., 148, 713.
Peebles, P. J. E. 1969, Phil. Trans. Royal Soc. London, A, 264, 279.
Rowan-Robinson, M. 1986, MNRAS, 219, 737.

AUDIENCE QUESTIONS

B. Partridge: Have fluctuations in sky brightness yet been analyzed in the DIRBE data?

Answer: This has not yet been done, though we shall attempt to do so. Identifying fluctuations of cosmological significance will be observationally difficult due to the patchiness of the foreground arising from the Galaxy.

COBE ATTITUDE REFINEMENT

E. L. Wright (UCLA, Los Angeles, CA 90024-1562), V. K. Kumar,
I. Freedman, A. Rae, S. Basappa, R. Venkataraman
(STX, Lanham, MD 20706), F.S. Patt (GSC, Laurel, MD 20707)

ABSTRACT

Methods used to determine and validate the fine-aspect attitude (which contributes to the instrumental anisotropy limit) are presented for the Cosmic Background Explorer satellite. The focus is on the use of the Diffuse Infrared Background Experiment (DIRBE) star observations to refine the attitude and the methods used in the gyroscope calibration and attitude validation.

INTRODUCTION

The anisotropy experiment on COBE, the Differential Microwave Radiometer (DMR), will have enough sensitivity after 2 years of integration to detect large-scale anisotropies as small as $\Delta T/T \approx O(10^{-6})$. Since the magnitude of the dipole anisotropy is $\Delta T/T = 10^{-3}$, the attitude solution needs to be good enough to avoid generating false anisotropies at the part per million level. Since individual DMR data samples have a very low signal-to-noise ratio, it is only essential that the systematic errors of the attitude solution be less than 1 milliradian.

CURRENT STATUS

Considerable progress has been made in understanding the attitude determination process, and several significant systematic errors have been found and removed from the solution. The preliminary coarse attitude solution presented at the 175th AAS meeting had errors of 15' RMS, which were primarily systematic: there was a pitch error with a period of one cycle per orbit generated by the large time derivative of the Earth scanners' outputs, and there was a sawtooth error in yaw caused by miscalibration of the spin axis gyro. The period of this sawtooth is an input parameter, the "smoothing interval," used by the attitude determination process. Long smoothing intervals allow a large number of DIRBE star hits to be averaged in the Fine Aspect solution but amplify the sawtooth error. The spin axis gyro scale factor calibration is now known to depend on the temperature of the spacecraft electronics, leading to orbital and seasonal effects, and rapid changes during the eclipse season.

The current quality of the DIRBE Fine attitude solution is shown by two scatter plots of star hit residuals for the time period 12/25/89, 0913-1421 UT. The first plot, using the coarse attitude, was presented at the January

1990 AAS meeting. The square box on each plot shows the size of the 0.7 degree square DIRBE beam. The second plot shows the accuracy achieved without the gyro temperature correction, which limits the smoothing interval to 30 minutes. In this case the interquartile range of the along-scan errors is less than 0.04 degree, so the "RMS" error is about 0.03 degree. The cross-scan error should cover the range -0.35 to +0.35 degree, and in fact 97% of the star hits fall within this range.

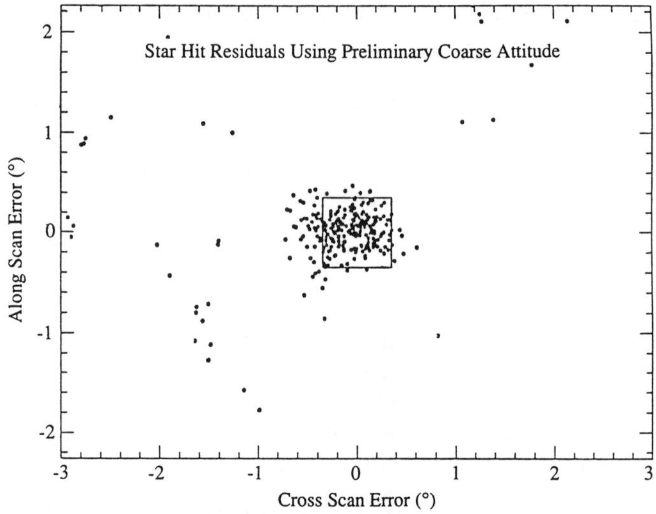

Figure 1a. The star hit residuals for the time range 0913 to 1421 UT on 12/25/89. Residuals are resolved into along-scan errors and cross-scan errors, which should show a uniform distribution over -0.35 deg. to +0.35 deg. The box in the middle shows the DIRBE beam. Stars far from the box are misidentified peaks. The systematic shift of the points to the left of the box shows a slight change in the alignment of the DIRBE beam relative to the spin axis. The attitude solution shown in this plot used the Sun sensor, Earth sensor, and gyro information only. The DIRBE Fine Aspect solution will use the star hit information to reduce the residuals.

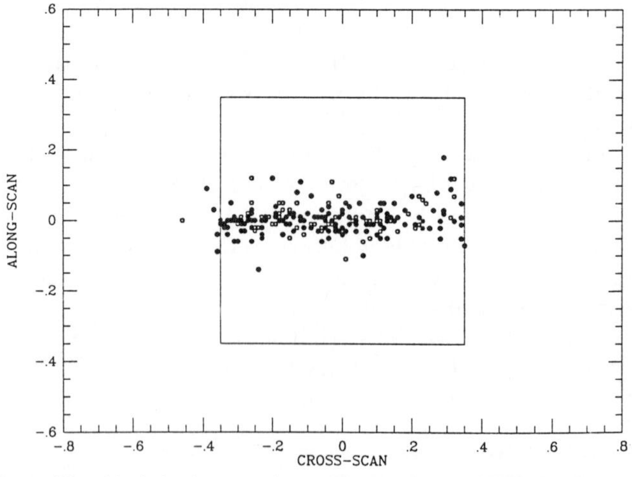

Figure 1b. The star hit residuals for the same data as Fig. 1a using the DIRBE Fine Aspect without the gyro temperature correction.

In order to extend the smoothing interval and bridge over gaps in DIRBE coverage caused by the South Atlantic Anomaly, the temperature correction to the gyro scale factor is now being implemented. Three charts below show the time history of the along-scan error for all the star hits during 24 hours starting 19 September 1990 at 09:30 UT. The top two charts show data analyzed with 30-minute and 50-minute smoothing intervals when the gyro temperature correction is used. The interquartile range is 0.035 degree, and the errors are not highly correlated.

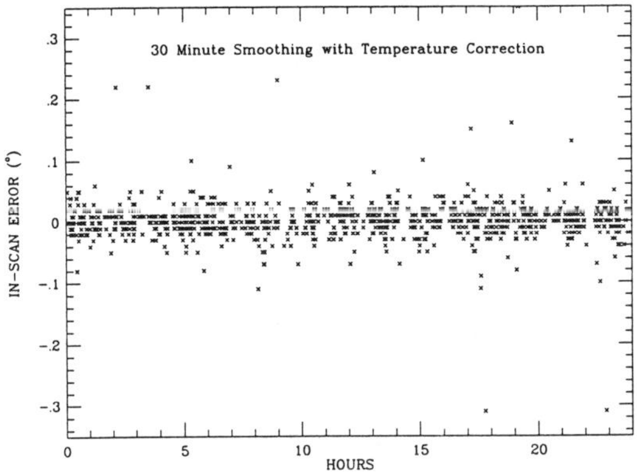

Figure 2a. The time history of the along scan error for all of the star hits during 24 hours starting 19 September 1990 at 09:30 UT. The data is analyzed with the gyro temperature correction and a 30-minute smoothing interval.

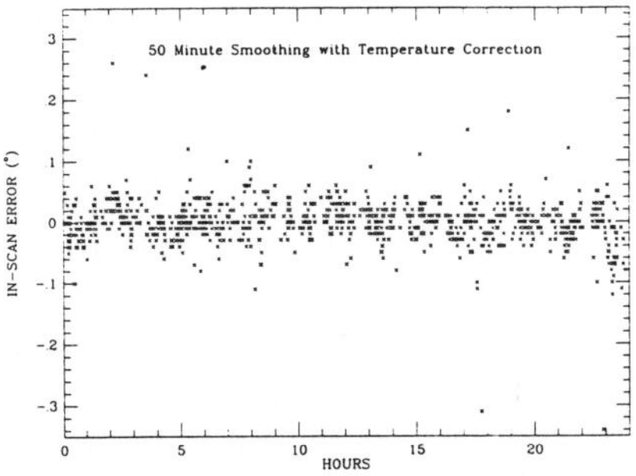

Figure 2b. The same data as Fig. 2a analyzed with the gyro temperature correction and a 50-minute smoothing interval.

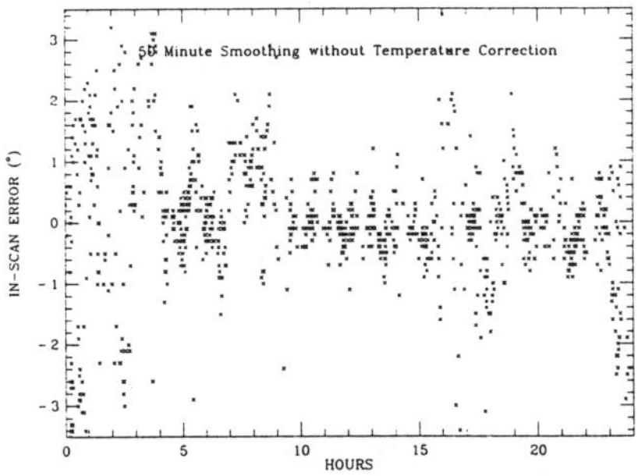

Figure 2c. The same data as Fig. 2a analyzed WITHOUT the gyro temperature correction and a 50-minute smoothing interval.

The bottom chart shows the effect of a 50-minute smoothing interval without the temperature correction. The interquartile range is now 0.09 degree, and the errors are clearly systematic. But even these large errors should be acceptable to the DMR as long as the smoothing intervals are not phase locked to the orbit, the time of day, or the season.

An overview of the attitude accuracy is given in the chart below. The interquartile range of the along-scan star hit residuals is plotted versus day of the year. The points show the coarse solution, which does not use DIRBE star hits. The open circles are DIRBE Fine Aspect solutions but still show the effect of errors in the spin axis gyro scale factor calibration.

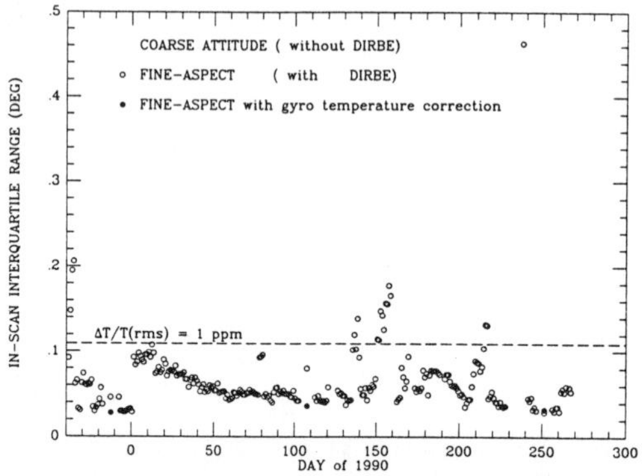

Figure 3a. The interquartile range of the along-scan star hit residuals is shown versus day of the year. The points show the coarse solution, which does not use DIRBE star hits. The open circles are the DIRBE Fine Aspect solution but still show the effects of errors in the spin axis gyro scale factor correction. The closed circles indicate the results of recent tests of the temperature-corrected fine-aspect solutions.

With the temperature correction to the gyro scale factor implemented, we expect that the next recomputation of the attitude solution will achieve accuracies on individual star hits that are 2-3 times better than the level required for 1 part per million anisotropy measurements. The closed circles indicate the results of recent tests of the temperature-corrected fine-aspect solutions. The systematic errors will then be smaller still. Finally, we note that the short wavelength DIRBE detectors used to detect stars for the Fine Aspect solution are still working after the depletion of the liquid helium.

ACKNOWLEDGEMENTS

The National Aeronautics and Space Administration/Goddard Space Flight Center (NASA/GSFC) is responsible for the design, development, and operation of the Cosmic Background Explorer (COBE) under the guidance of the COBE Science Working Group. GSFC is also responsible for the development of the analysis software and for the production of the mission data sets.

Special thanks are due to the scientists and engineers who designed and built the DIRBE and to the staff in the attitude control system branch at GSFC.

RECONCILING ZIP AND IRAS OBSERVATIONS OF THE ZODIACAL BACKGROUND

L. J Rickard
Center for Advanced Space Sensing, Naval Research Laboratory, Washington, DC 20375-5000

S. W. Stemwedel
Applied Research Corporation, Landover, MD 20785

S. D. Price
Air Force Geophysical Laboratory, Hanscom AFB, MA 01731

INTRODUCTION

The rocket-borne observations of the Zodiacal Infrared Project (ZIP; Murdoch and Price 1985, *Astr. J.*, **90**, 375) provide near-IR ($2\mu m < \lambda < 30\mu m$) measurements of the zodiacal dust cloud over a broad range of solar elongation angle ($22° < \epsilon < 180°$), at moderate spectral resolution. This broad elongation coverage has not been matched by the subsequent extensive surveys of zodiacal emission made by the Infrared Astronomy Satellite (IRAS) and the Cosmic Background Explorer (COBE). However, a long-standing obstacle to the use of these data in extending models of the zodiacal background has been apparent discrepancies (at common elongations) in such basic geometrical characteristics as the brightness ratio of the ecliptic poles to the ecliptic plane.

As part of the SDIO Celestial Backgrounds Program, we have performed an in-depth reanalysis of the ZIP database.

CROSS CALIBRATION

Although it is possible to use ZIP measurements of IRAS-identified point sources for cross calibration, the structure of the ZIP focal plane and the way it was scanned (circles concentric about δ And) make this approach impractical. We hypothesized that common observations of the diffuse component of the Galactic background could be used instead. So, we converted the 55 individual detector databases for the two ZIP flights to all-sky images. At low elongations, the Galactic plane crossings are overwhelmed by the zodiacal foreground, but usable Galactic plane detections are seen at large elongations, over Galactic longitudes $35° < \ell < 97°$. Figure 1 shows sections of scans from both flights, superimposed on a contour map of the Galactic plane as observed by IRAS at $12\mu m$, but convolved and rebinned to the same resolution as the ZIP data.

To reduce the noise, we averaged the data in strips of constant Galactic latitude, producing one-dimensional scans. These data were then fit with a

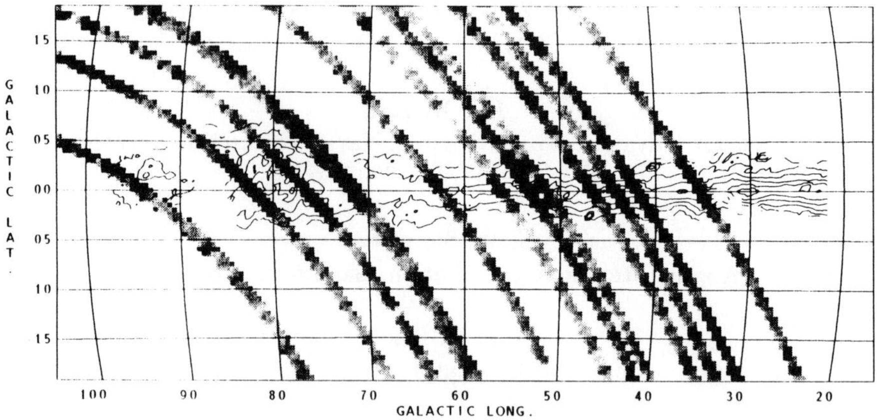

Figure 1. ZIP scans crossing the IRAS Galactic plane.

Gaussian model (to represent the Galactic plane) and a linear baseline (to represent the zodiacal foreground emission). The fitted parameters were used to test for internal consistency. Correlations of results obtained by pairs of detectors under the same filter, and of results obtained by different flights for those scans that cross at nearly identical Galactic longitudes, indicated that some of the ZIP detector pairs were discrepant. Accordingly, we limited our further analysis to the three detector pairs that showed the least confusion in their Galactic plane results — detectors 9 and 10 ($\lambda_{eff} \approx 12\mu$m), 15 and 16 ($\lambda_{eff} \approx 13\mu$m), 19 and 20 ($\lambda_{eff} \approx 17\mu$m).

For comparison, we reduced IRAS HCON 1 Galactic plane data to an angular resolution and map format consistent with the ZIP data. We subtracted linear baseplanes fit to the regions more than 10° away from the plane, and then fit Gaussians to oblique slices along the positions of the ZIP scans. Figure 2 shows that the fitted amplitudes for the IRAS data and the ZIP detector 10 data differ essentially by a scaling factor ~ 2.

ECLIPTIC LATITUDE ANALYSIS

We extracted from the ZIP Payload database those data points that fell within 4° of a selected set of elongation angles. These constant elongation datasets were then matched to hypothetical variations with ecliptic latitude, by means of a nonlinear least-squares fitting algorithm. The Lorentzian model

$$I(\beta) = Constant + Scale \times \frac{(\Gamma/2)}{(\Gamma/2)^2 + (\beta - \beta_\circ)^2}$$

was used by Clark *et al.* (in preparation) to characterize IRAS data, on the assertion that the physical processes controlling the dust distribution are likely to yield such a structure. It also has the virtue of using few parameters, and

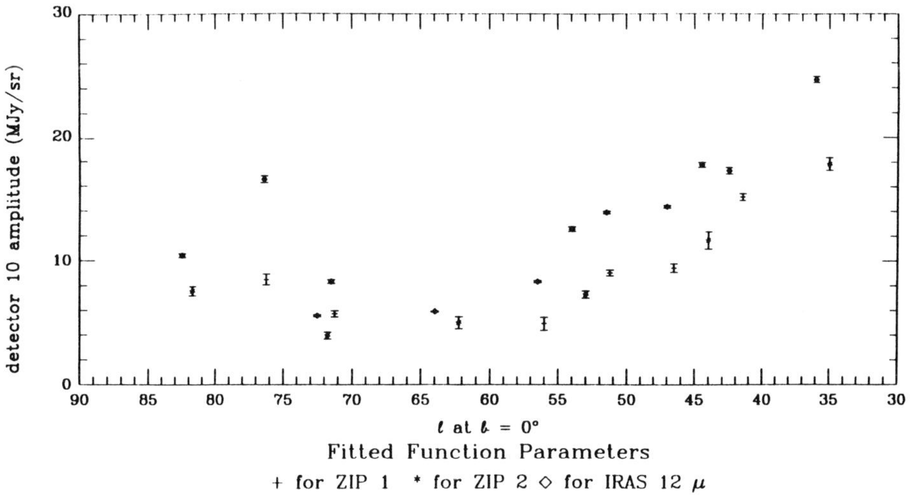

Figure 2. ZIP and IRAS amplitudes as a function of Galactic longitude.

being well behaved during the course of the nonlinear fitting. The "revised Hauser function"

$$I(\beta) = I_{SYM} + k\{\beta_C\,|csc(\beta - \beta_\circ)|[1 - exp(-x - \alpha x^2)] - 1\}$$

with $x = (\beta - \beta_\circ)/\beta_C$ has been used by Vrtilek (in preparation) to model the IRAS data. This function makes a smooth transition from a $csc(\beta)$ law near the poles to a Gaussian in the plane. However, it uses more free parameters, to fit the north and south sides of the ecliptic plane independently.

Many of the constant elongation strips show clear deviations in which the data appear to be "tilted" with respect to the best-fit function, no matter which is chosen. We suspected that this represented some time-dependent variation, and so we rebinned the payload data into separate rolls, ordered by time. We then refit the standard functions with two modifications. First, we included a linear drift term to model the extra time variation. Second, in order to cope with the broad range of elongation angles covered by individual rolls, we incorporated an extra exponential scale factor based on the model elongation variation fit to the Vrtilek sample. The resulting fits were quite successful. Further analysis demonstrated that the drift term results from an error in the correction for foreground atmospheric emission.

IMPLICATIONS

The fitted parameters for the individual roll data now provide a revised description of the zodiacal IR background according to ZIP. The extrapolated fluxes at the north ecliptic pole are in good agreement with the IRAS values

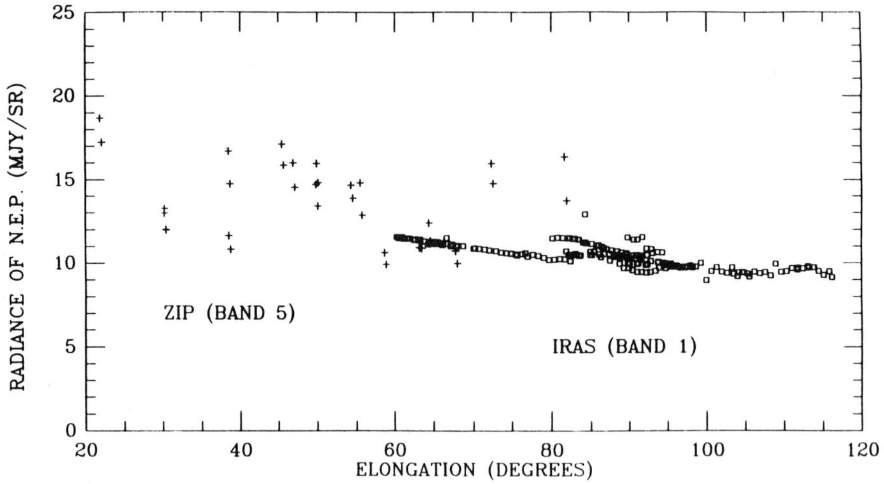

Figure 3. The brightness of the North Ecliptic Pole, from Lorentzian fits.

(figure 3). The apparent pole-to-plane contrast is reduced, but not eliminated, by removing the drift term from the latitude variation.

The database of fitted parameters also shows how the ZIP results extend the IRAS measurements. For example, figure 4 shows the characteristic latitude width of the zodiacal cloud inferred from the Lorentzian fits. The ZIP results form a smooth extrapolation of the IRAS results in toward the sun.

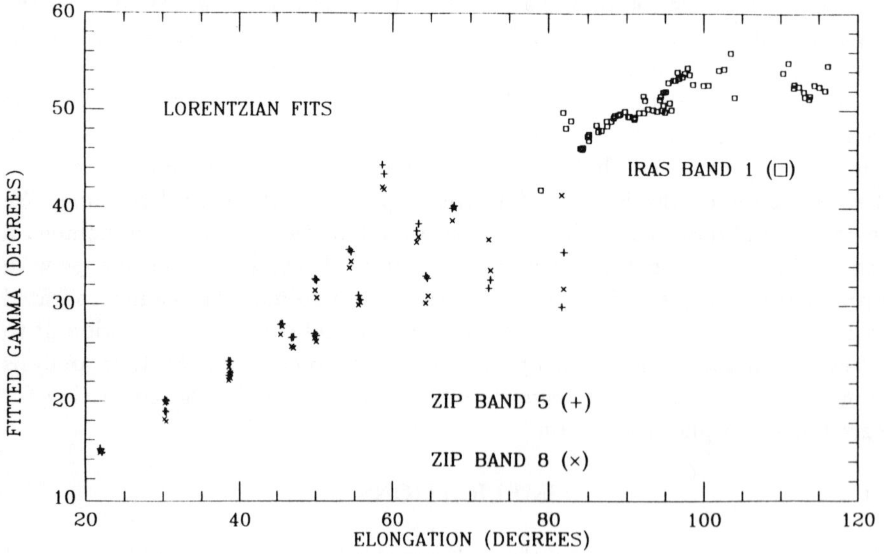

Figure 4. The fitted latitude width of the zodiacal emission

COSMOLOGICAL IMPLICATIONS OF COBE

B.J.Carr
Astronomy Unit, Queen Mary & Westfield College, London E1 4NS

J.R.Bond
CITA, University of Toronto, Ontario M5S 1A1

C.J.Hogan
Astronomy Department, University of Washington, Seattle, WA 98195

ABSTRACT

We examine the constraints placed by FIRAS and DIRBE on radiation production in the early Universe. IRAS galaxies must produce a submillimetre background at some level and, if evolutionary effects are large, one can already place an interesting upper limit on the epoch of galaxy formation. A variety of pregalactic sources could have produced radiation, in particular, the VMOs whose black hole remnants are sometimes invoked to explain the dark mass in galactic halos. In some circumstances such pregalactic radiation would have been reprocessed into the far-IR or submillimetre bands by cosmological dust. This problem has been studied in detail by Bond, Carr & Hogan (1990), with particular emphasis on the anisotropies expected, and this paper summarizes the main results of that work. The VMO-plus-dust scenario is still consistent with FIRAS data but it may soon be excluded, in which case the radiation would need to be in the near-IR.

INTRODUCTION

The FIRAS experiment shows[1] that the cosmic background radiation (CBR) is well fit by a perfect black-body spectrum of temperature 2.735±0.06 K over the entire waveband 500μ to 5000μ. Besides supporting the hot Big Bang model for the origin of the background, this also places important constraints on the variety of processes in the early Universe which would be expected to produce deviations from a black-body spectrum. These constraints derive from three important upper limits: the chemical potential μ is below 10^{-2}; the Compton parameter y is below 10^{-3}; and any excess flux $\Delta I(\lambda)$ must be less than 1% of the CBR peak flux $I(\lambda_{peak})$ over the relevant waveband. These limits each have implications for the early Universe.

* $\mu<10^{-2}$. This places constraints[2] on any energy released in the redshift range $10^6>z>10^4$. In this period the Compton interactions of energetic electrons with the CBR photons produce a Bose-Einstein spectrum with a chemical potential μ which is measure of the energy imput relative to the CBR density. However, there is no time for Bremmstrahlung processes to produce the extra photons necessary to produce a black-body spectrum except at very low frequencies. Only for $z>10^6$ would any heat input be completely thermalized. The limit on μ places interesting constraints, for example, on superconducting cosmic strings[3,4], evaporating primordial black holes[5] , and the dissipation of primordial density perturbations[6,7].

* $y<10^{-3}$. This places constraints on energy released in the redshift range $z<10^4$. In this period complete relaxation to a Bose-Einstein distribution is not possible but Compton effects still remove photons from the Rayleigh-Jeans region of the CBR and redeposit them in the Wien region. This process is important both before and after decoupling, the y parameter being a measure of the integral of the electron pressure over redshift[8]. The limit on y may already exclude the explosion[9] and mock-gravity[10] scenarios for large-scale structure formation, and it constrains dark matter scenarios involving decaying elementary particles[11,12]. It also restricts the thermal history and present density and temperature of any hot intergalactic medium[13], implying in particular that a smooth IGM could only provide a tiny fraction of the observed X-ray background[14,15].

* $\Delta I(\lambda)<10^{-2}I(\lambda_{peak})$. This places a direct constraint on any photons produced in the period after decoupling. Of course, the first two effects also imply an excess $\Delta I(\lambda)$ over some range of λ but there the effect is indirect in that the distortion is only produced by the interactions of energetic electrons with the CBR photons. The most important source of photons in the FIRAS band is probably dust[16-25] and this will be the main focus of attention in this paper.

The DIRBE and DMR experiments also give interesting cosmological information. In particular, DIRBE places upper limits on the density of any cosmic background radiation in near-IR to far-IR bands[26] and this implies further constraints on astrophysical processes in the period between decoupling and galaxy formation. The DMR measurements at 9.5mm, 5.7mm and 3.3mm, besides confirming the dipole anisotropy due to our own peculiar motion, also constrain[27] any cosmological temperature anisotropies $\Delta T/T$ on scales above 7° to be less than 10^{-4}. Together with anisotropy constraints from other ground-based experiments, this has a number of important cosmological implications. In particular, it places constraints on the form of the density fluctuations at decoupling[28], various features of large-scale structure[29], the spectrum of primordial gravity waves[30], and the existence of cosmic strings[31]. It also constrains models for spectral distortions produced by dust[18,25], an issue we address in more detail later.

BACKGROUND LIGHT LIMITS

We begin by summarizing the observational upper limits on the background light density obtained from COBE and other experiments. These are conveniently expressed in critical density units by defining a quantity

$$\Omega_R(\lambda) \equiv \frac{4\pi\nu I(\nu)}{c^3\rho_{crit}} = 7\times10^{-7}h^{-2}\left[\frac{\lambda}{100\mu}\right]^{-1}\left[\frac{I(\nu)}{MJy/sr}\right] \quad (1)$$

where h is the Hubble parameter in units of 100 km/s/Mpc. For comparison, the CBR peaks at $\lambda_{peak}=1400\mu$ with a density $\Omega_R(\lambda_{peak})=1.8\times10^{-5}h^{-2}$. The limits are summarized in Figure (1).

FIGURE (1). This compares the observational constraints on the background radiation density from DIRBE, FIRAS, IRAS and the Nagoya-Berkeley experiment (NB) with the backgrounds expected from an evolving population of IRAS galaxies and a population of VMOs whose black hole remnants provide galactic halos.

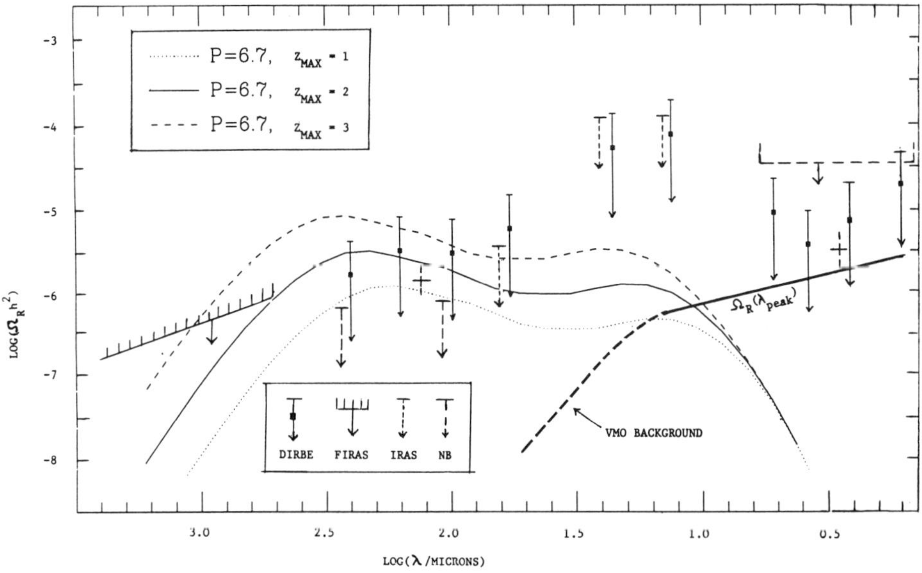

The current FIRAS results[1] imply that the CBR is so well fit by a black body spectrum that any excess background must have an intensity less than 1% of the peak CBR intensity over the range 500-5000μ. This implies a constraint

$$\Omega_R(\lambda) < 2\times10^{-7}h^{-2}(\lambda/\lambda_{peak})^{-1} \qquad (2)$$

although the limit could eventually be strengthened by a factor of 10. The DIRBE results at the south ecliptic pole[26] give upper limits in the J, K, L, M, 12μ, 25μ, 60μ, 100μ, 120-200μ and 200-300μ bands. However, the limits indicated in Figure (1) are very conservative since they do not include any subtraction for the foreground backgrounds from interstellar and interplanetary dust.

Stronger but less firm constraints in the far-IR band come from the most recent analysis[32] of the Nagoya-Berkeley data[33] (assuming the measurements in this waveband are unaffected by the problem which led to a spurious measurement at 700μ). After careful modelling of the interstellar and interplanetary dust contributions, they get $\Omega_R(275\mu)<7\times10^{-7}h^{-2}$, $\Omega_R(135\mu)=(1.4\pm0.4)\times10^{-6}h^{-2}$ and $\Omega_R(100\mu)<8\times10^{-7}h^{-2}$. The 135$\mu$ result may contain an isotropic component, although it would not necessarily be extragalactic. Some of the IRAS limits are also shown in Figure (1). At one stage it was claimed[34] that IRAS data indicated a 100μ background with $\Omega_R(100\mu)=3\times10^{-6}h^{-2}$. However, this is inconsistent with both the Nagoya-Berkeley and DIRBE results, the latter indicating that the IRAS offset is wrong by a factor of 3 at 100μ. Both the DIRBE and IRAS limits are of order $10^{-4}h^{-2}$ at 12μ and 25μ. They are very weak there because the interplanetary dust emission is so large. Indeed it may never be possible to get good cosmological light limits in this waveband.

Recently Rowan-Robinson et al.[35] have tried to use the IRAS and DIRBE limits in conjunction with detailed dust models to place constraints on the far-IR background. They attribute 46% of the isotropic 100μ intensity to zodiacal light and 54% to interstellar emission. They then extrapolate these proportions to the other wavebands by assuming that the zodiacal emission has a black-body spectrum with temperature 255K and that the high latitude interstellar emission is as described by Rowan-Robinson et al.[36]. This accounts for most of the observed DIRBE background intensity at 60μ, 100μ, 150μ and 250μ, and they infer a constraint $\Omega_R < 10^{-6}(\lambda/100\mu)^{-1}h^{-2}$ at all four wavelengths. However, it should be stressed that this conclusion is very model-dependent (so it is not included in Figure 1) and much more work will be required before this limit can be made more secure.

In the near-IR, a 1984 Japanese rocket experiment[37] gives a limit $\Omega_R(1\text{-}5\mu) < 3\times10^{-5}h^{-2}$, apart from the possible detection of a "line" at 2.2μ with $\Omega_R(2.2\mu) = 3\times10^{-6}h^{-2}$. It is claimed that this background can have a bandwidth of at most 20% on the basis of the limits either side of 2.2μ. However, it must be stressed that the neighbouring limits come from a different instrument, so the line interpretation may not be secure. In any case, the Japanese claim is controversial because of the problem of subtracting starlight and zodiacal emission.

THE CONTRIBUTION FROM IRAS GALAXIES

The most prominent sources of far-IR radiation at the present epoch are IRAS galaxies and so it is important to ask what constraints the COBE results place on models for galaxy evolution[38]. Rowan-Robinson et al.[35] have argued that even the conservative DIRBE results place interesting restrictions on the epoch of galaxy formation. This follows from the claim[39] that the IRAS redshift surveys indicate strong evolution in the 60μ luminosity function for galaxies at $z < 0.3$. It may be fitted either by luminosity evolution of the form

$$L(t) = L(t_o)\exp[2Q(1-(1+z)^{-3/2})/3] \qquad (3)$$

with an exponent $Q = 3.2 \pm 1.0$ or by density evolution of the form

$$\Phi(t) = \Phi(t_o)(1+z)^P \qquad (4)$$

with $P = 6.7 \pm 2.3$. Luminosity evolution might arise as a result of the reduction in gas supply within galaxies, while number evolution might be associated with galaxy interactions. Saunders et al.[39] also claim that evolution of these forms is compatible with the 60μ source counts, although the counts cannot distinquish between the two models or indeed some combination of density and luminosity evolution.

In order to compare with the COBE data, one needs a model for the spectrum of the IRAS galaxies. Rowan-Robinson et al. use a two-component model[40] in which one has a cool interstellar dust component (equivalent to the cirrus emission from the disk of our own galaxy) and a warm starburst component, associated with bursts of star formation shrouded by dust. They assume both components evolve in the same way. In the luminosity evolution case, the predicted background is an order of magnitude below the DIRBE limits. However, in the number density evolution case, it exceeds the DIRBE limits unless there is a redshift cut-off at $z<2$ and this would require a late epoch of galaxy formation. This is illustrated in Figure (1).

It must be stressed that the claim for direct evidence of evolution only pertains for $z<0.3$, so there is no need to extrapolate the evolution to the sort of redshift indicated above. All one can strictly infer is that density evolution cannot be described by a single parameter P all the way back to $z=2$. Nevertheless, the calculation does give an indication of how DIRBE data may be used to constrain models of galaxy evolution. This idea is explored further in other papers in this session. Note that if one had no evolution at all, the predicted background from IRAS galaxies would be about two orders of magnitude below the DIRBE constraints.

A purely theoretical attempt to predict the background from IRAS galaxies - without any recourse to claims for empirical evolution - has been made by Djorgovski & Weir[41]. They fit the spectrum of a typical starburst galaxy by a simple dust model and then constrain the luminosity history of the starburst by the requirement that one does not overproduce metals or dark remnants. Although they were originally trying to fit the Nagoya-Berkeley submillimetre excess, the same argument can applied in conjunction with the COBE data.

PREGALACTIC SOURCES OF RADIATION

As emphasized in Bond, Carr & Hogan[18] [BCH1], there are many possible sources of radiation in the era between recombination and galaxy formation. The total density Ω_R associated with various scenarios is indicated below:

SOURCE	$\Omega_R/10^{-6}$
CMB	$26h^{-2}$
Pregalactic stars	$4\left[\frac{\Omega_*}{0.01}\right]\left[\frac{1+z_*}{10}\right]^{-1}\left[\frac{\epsilon}{0.004}\right]$
Black hole accretion	$\left[\frac{\Omega_{B,acc}}{10^{-4}}\right]\left[\frac{1+z_*}{10}\right]^{-1}\left[\frac{\epsilon}{0.1}\right]$
Decaying particles	$5B_X\Omega_{X,i}\left[\frac{1+z_d}{10^5}\right]^{-1}$
Bound structures	$0.01\left[\frac{\Omega_c}{0.1}\right]\left[\frac{V}{300km/s}\right]^2\left[\frac{1+z_c}{5}\right]^{-1}$

In the "pregalactic stars" scenario, the stars are assumed to have a density parameter Ω_* (normalized to the density associated with visible material) and to burn at a redshift z_* (normalized to a value somewhat larger than that associated with galaxy formation). They are also assumed to produce radiation with an efficiency ϵ (normalized to the value associated with very massive stars). As discussed later, the sort of massive stars whose remnants have been invoked[42] to provide galactic halos would have $\Omega_*=0.1$ but z_* might then as large as 100, so Ω_R need not be much larger. In the "black hole accretion" scenario, the holes are supposed to accrete a gas density $\Omega_{B,acc}$ at a redshift z_* and to produce radiation from accreted material with an efficiency ϵ (normalized to the optimal value of 0.1). Note that $\Omega_{B,acc}$ could be much less than the density of the black holes themselves. However, there are some circumstances in which one could expect the holes to at least double their mass through accretion. This might apply, for example, to the sort of massive black holes supposed to lurk in galactic nuclei. The normalization of 10^{-4} for $\Omega_{B,acc}$ corresponds to a scenario in which one has a $10^8 M_\odot$ black hole at the centre of every galaxy. In the "decaying particles" scenario, the particles are assumed to have an initial density $\Omega_{X,i}$ and to decay at a redshift z_d, with a fraction B_X of their rest mass going into photons. The background produced in the "bound structures" scenario is associated with the energy released via cooling in the development of gravitationally bound objects from density fluctuations. In this case, the radiation density depends on the density of the cooling gas Ω_c, the redshift of cooling z_c, and the velocity dispersion V in the final bound structure. The parameters are normalized to the values appropriate for galaxies. One would also have a larger X-ray contribution from clusters of galaxies but this would arise too late to be reprocessed by dust.

One sees that, in all cases except the last, the background density is expected to be of order 10% of the CBR density. While some of the sources are rather exotic, the first one should certainly exist at some level. Most of the backgrounds would currently be in the near-IR if the radiation was affected only by redshift effects. If they were reprocessed by dust, they would be shifted into the far-IR or submillimetre bands but the energy densities indicated above would still apply.

Let us now consider the effect of dust on this pregalactic radiation. In BCH1 we adopted a rather simplistic analysis, in which the dust cross-section for photons of wavelength λ is assumed to be

$$\sigma_d = \frac{\pi r_d^2}{1+(\lambda/r_d)^\alpha} \tag{5}$$

where r_d is the grain radius. Thus one has a geometric cross-section for $\lambda<<r_d$ but σ_d falls as $\lambda^{-\alpha}$ for $\lambda>>r_d$. This implies that a smoothly distributed population of grains with density parameter Ω_d would absorb photons with $\lambda<r_d$ for

$$1+z_d > 10\left[\frac{\Omega_d}{10^{-5}}\right]^{-2/3}\left[\frac{r_d}{0.1\mu}\right]^{2/3}\left[\frac{\rho_{id}}{3g/cm^3}\right]^{2/3} h^{-2/3}\Omega^{1/3} \tag{6}$$

where Ω is the total density parameter and ρ_{id} is the internal grain density. Here we have normalized Ω_d to the sort of value appropriate for galaxies; the pregalactic value is necessarily less than this unless there is subsequent dust destruction. Even if there is no pregalactic dust, the dust within galaxies themselves could still reprocess any pregalactic radiation providing eqn (6) with $\Omega_d=10^{-5}$ is satisfied and providing galaxies cover the sky. The latter condition requires that the redshift of galaxy formation z_G satisfy

$$1+z_G > 11\left[\frac{R_G}{10\mathrm{kpc}}\right]^{2/3}\left[\frac{\Omega_{GB}}{0.01}\right]^{-2/3}\Omega^{1/3}h^{-1/3} \qquad (7)$$

where R_G is a typical galactic radius and Ω_{GB} is the density parameter associated with the baryons in galaxies. We have normalized R_G and Ω_{GB} to the sort of values appropriate for galaxies like our own, although dwarf galaxies might have a somewhat larger covering factor. In this context, note that the term "pregalactic" is rather ambiguous since the pregalactic clouds whose stellar fragments are supposed to be generating any pregalactic dust might themselves be described as "dwarf galaxies". However, we here regard the term "galaxy" as referring to objects like the Milky Way.

If either pregalactic or galactic dust does reprocess the pregalactic radiation, then the simple analysis of BCH1 implies that the dust radiation should currently peak at a wavelength

$$\lambda_{peak} = 700\mu\left[\frac{\Omega_R h^2}{10^{-6}}\right]^{-1/5}\left[\frac{r_d}{0.1\mu}\right]^{1/5}\left[\frac{1+z_d}{10}\right]^{1/5} \quad (\alpha=1) \qquad (8)$$

where we have normalized the radiation density Ω_R to the sort of value anticipated in Table (1). The crucial point is that the wavelength is very insensitive to the various parameters appearing in eqn (8) because the exponents are so small. (The exponents would be different for $\alpha=2$ but still small.) The fact the Nagoya-Berkeley experiment[23] appeared to find a submillimetre excess peaking at almost exactly the wavelength predicted was, of course, very exciting and prompted us to embark on a much more detailed attempt to fit the data. The Nagoya-Berkeley excess has now been disproved but the need for a more careful analysis remains since there must be reprocessed pregalactic light at some level.

In the more sophisticated analysis of Bond et al.[25] (BCH2), we characterize the dust by a function $A_d(\lambda)$, related to the grain radius r_d and grain absorption cross-section σ_d by

$$A_d(\lambda) = \left[\frac{\sigma_d}{\pi r_d^2}\right]\left[\frac{3\lambda}{8\pi r_d}\right] \quad , \qquad (9)$$

and allow for the fact that the exponent α in eqn (5) could itself be wavelength-dependent. A_d is the imaginary part of the trace of the polarizability tensor of the grain; it is a useful parameter because it is constant over the wavelength domain in which $\alpha=1$. The grain model is then specified by three parameters rather than two: an "effective" grain radius r_s for the source radiation ($r_s=r_d$ for a geometrical cross-section), the value of A_d at 100μ (denoted as A_{d100}), and the value of α in the far-IR. We also introduce a more realistic model for the source luminosity history $L_s(z)$, allowing for both "burst" and "continuous" models. In the latter, $L_s(z)$ is assumed to evolve as $(1+z)^P$ between a turn-on redshift z_1 and a turn-off redshift z_o. We assume that the dust density $\Omega_d(z)$ is fixed below some redshift z_d.

With these assumptions, the dust temperature evolves according to

$$\frac{T_d(z)}{T_c(z)} = \left[1+ \left[\frac{10}{1+z}\right]^{\alpha} \left[\frac{r_s A_{d100}}{100\mu}\right]^{-1} \left[\frac{\Omega_{RS}(z)}{\Omega_c(z)}\right] \right]^{1/(4+\alpha)} \tag{10}$$

where Ω_{RS} and Ω_c are the source radiation and CBR density at redshift z. For $r_s=r_d$ and constant α, this would be equivalent to eqn (8) but the parameter Ω_{RS} must be interpreted carefully. If the UV opacity up to the epoch corresponding to redshift z is small, $\tau_{uv}(z)<1$, then Ω_{RS} is just associated with the integral of L_s over time until then. (This was essentially the case considered in BCH1). However, if $\tau_{uv}(z)>>1$ and P<4, then the energy density at any point comes only from the region within one UV optical depth of that point. In this case, Ω_{RS} is roughly the radiation density generated over a cosmological time at redshift z divided by $\tau_{uv}(z)$. In particular, this includes the case in which the sources are completely shrouded in local dust clouds (cf. IRAS galaxies). If $\tau_{uv}>>1$ and P>4, then the emission from the turn-on redshift z_1 dominates and one effectively has a "burst" with $\Omega_{RS}(z)$ being the radiation density at z_1 divided by $\tau_{uv}(z_1)$ and multiplied by $\exp[\tau_{uv}(z)-\tau_{uv}(z_1)]$.

In BCH2 we consider 14 different models (corresponding to different values of α, Ω_d, Ω_R, z_1, z_o and P) and the resulting spectra are shown in Figure (2). Also shown are the current and ultimate FIRAS limits, the 275μ Nagoya-Berkeley limit, and the limit of Gush et al. (GHW). Although the spectra were obtained numerically, they can also be understood analytically. In an analytic treatment of the continuous models, one can characterize the situation by a single dust temperature T_d and the (wavelength-dependent) optical depth $\Delta\tau$ across the redshift region or "shell" where most of the emission occurs. In the burst case with $\tau_{uv}<1$, the dust temperature hardly changes and eqn (8) is a good approximation. In the burst case with $\tau_{uv}>>1$, the radiation is absorbed over a narrow redshift range and $\Delta\tau$ corresponds to the distance over which the dust cools by 1/e.

GALACTIC HALOS AS MASSIVE STELLAR REMNANTS

The most recent cosmological nucleosynthesis arguments[43] indicate that the baryonic density parameter must lie in the range 0.04 to 0.06 for a Hubble parameter of 50, whereas the density associated with visible galaxies is only 0.01. This indicates that a large fraction of the baryons must be dark. It is possible that the dark baryons are in a hot intergalactic medium[13], although the Gunn-Peterson test and COBE constraints on the Compton distortion of the CBR require that its temperature be in the range 10^4 to 10^6K. More likely, the baryons have been processed into dark remnants through a first generation of "Population III" stars. In this case, their density could at least suffice to explain the dark matter in galactic halos, although one would still need non-baryonic dark matter to explain the critical density required

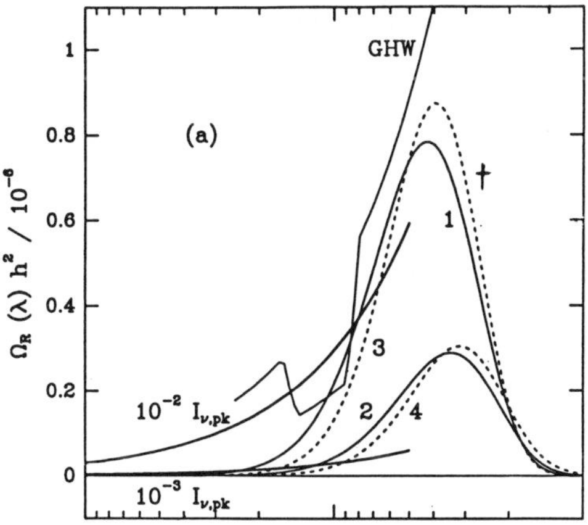

FIGURE (2a). All these models have the same energy release ($\Omega_R h^2=10^{-6}$) but the quantity and character of dust varies. $\alpha=1$ in 1 and 2; $\alpha=2$ in 3 and 4; $\Omega_d=10^{-5}$ in 1 and 3; $\Omega_d=10^{-6}$ in 3 and 4. All the models have $z_1=9$ and $z_0=0$ and the energy is released in a burst.

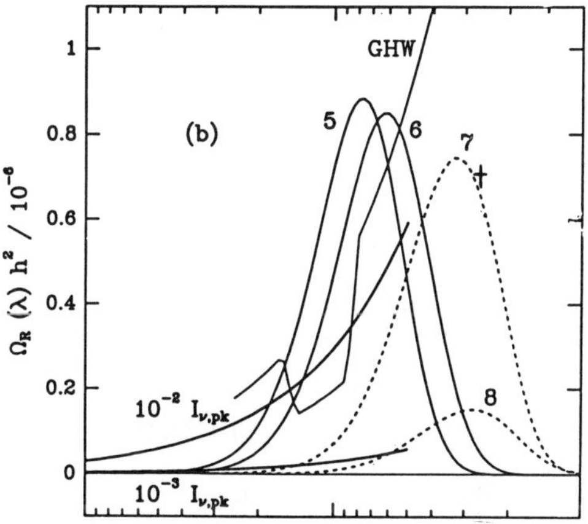

FIGURE (2b). In these models the dust properties are fixed ($\alpha=1.5$) but the amount and characteristic redshifts are varied. $\Omega_d=10^{-5}$ in 5 and 7; $\Omega_d=10^{-6}$ in 6 and 8; $z_1=50$ and $z_0=25$ in 5 and 6; $z_1=5$ and $z_0=0$ in 7 and 8. All the models have $\Omega_R h^2=10^{-6}$. The energy is released in a burst in 7 and 8.

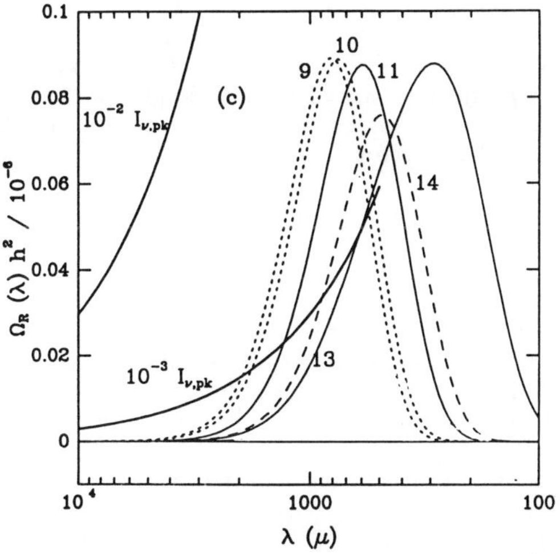

FIGURE (2c). All these models have $\alpha=1.5$ and $\Omega_R h^2=10^{-7}$ but the other dust characteristics vary. $\Omega_d=10^{-5}$ in 9, 11 and 14; $\Omega_d=10^{-6}$ in 10 and 13; $z_1=50$ and $z_0=25$ in 9 and 10; $z_1=9$ and $z_0=0$ in 11 and 13; $z_1=5$ and $z_0=0$ in 14. The energy is released in a burst in 11 and 14.

by the inflationary picture. A variety of arguments[44] show that the Population III objects must be either the black hole remnants of "Very Massive Objects" (VMOs) larger than 100 $M_\odot$ which collapse during their oxygen-burning phase or "brown dwarfs" which are too small to burn hydrogen at all[45].

Carr et al.[42] have argued for the first option because of its more exciting observational consequences. The most striking signature of the VMO scenario would be the background radiation generated during the stars' hydrogen burning phase. (There would also be far-IR radiation generated by a population of brown dwarfs, but it would would be presently undetectable either as a background[46] or as emission from individual galaxies[47].) This can be predicted rather precisely since all VMOs have a surface temperature T_S of about 10^5K and generate radiation with efficiency $\epsilon \approx 0.004$. (Both quantities are almost independent of the VMO mass[48].) We assume that the VMOs have a density parameter $\Omega_* = 0.02h^{-2} = 0.1h_{50}{}^{-2}$, corresponding to the sort of density required to explain the dark mass in galactic halos, and that they produce black-body radiation with temperature T_S. If the radiation is affected only by cosmological redshift, it should then have a peak wavelength and density

$$\lambda_{peak} = 4\left[\frac{1+z_*}{100}\right]\mu, \qquad \Omega_R(\lambda_{peak}) = 10^{-6}h^{-2}\left[\frac{1+z_*}{100}\right]^{-1}\left[\frac{\Omega_* h_{50}{}^2}{0.1}\right] \qquad (12)$$

at the present epoch, where z_* is the redshift at which the VMOs burn their nuclear fuel. We can place an upper limit on z_* by noting that the main-sequence time of a VMO is $t_{MS} = 2\times10^6$y (independent of mass), so that z_* cannot exceed the redshift when the age of the Universe is t_{MS}. This implies $z_* < 240h^{-2/3}$ and hence

$$\lambda_{peak} < 15\mu, \qquad \Omega_R > 5\times10^{-7}h^{-2} \qquad (13)$$

We can place a lower limit on z_* from UV-IR background light limits. As discussed by McDowell[19], these imply a constraint on the density of VMOs burning at any redshift z_*. If one requires $\Omega_* = 0.1h_{50}{}^{-2}$, this places a lower limit on z_*, partly because one needs the radiation density to have been redshifted away sufficiently and partly because one needs the wavelength to have been redshifted into the near-IR band, where the background light limits are weaker. In the absence of neutral hydrogen absorption, one requires $z_* > 30$ and this implies

$$\lambda_{peak} > 1\mu, \qquad \Omega_R < 3\times10^{-6}h^{-2} \qquad (14)$$

Limits (13) and (14) imply that the peak of the VMO background must lie somewhere on the heavy line in Figure (1). This means that the spectrum must lie to the right of the heavy broken line.

The situation would be somewhat different if the radiation from the VMOs were absorbed by neutral hydrogen, either in the background Universe or in the clouds where the VMOs form[49]. In this case, the spectrum would be cut off shortward of $9[(1+z_*)/100])\mu$, with most of the radiation appearing in redshifted recombination lines. The UV-IR limits would then allow z_* to be as low as 4, so the limit on Ω_R would be a factor of 6 higher than indicated in eqn (14).

Let us first ask whether the direct light from VMOs could explain the background reported at 2.2μ by Matsumoto et al.[37] If this is a line, as claimed, then it would be natural to identify it with the redshifted 1200A Lyman-α line. In this case, the redshift can be specified uniquely as $z_*=16$ and, if most of the radiation goes into the line, the predicted density is $\Omega_R=6\times10^{-6}h^{-2}$. To compare this with the observed background ($\Omega_R=3\times10^{-6}h^{-2}$), one must multiply by the ratio of the bandwidth of the actual line to that of the observations. The observed background is probably too small unless the line is exceedingly narrow but that is unlikely because it would require that the VMOs form over a very narrow range of redshifts. In fact, we have seen that the claim that one needs a line may be unjustified. If one identifies λ_{peak} in eqn (12) rather than the redshifted Lyman-α wavelength with 2.2μ, one would need $z_*=50$ and hence $\Omega_R=2\times10^{-6}h^{-2}$. This fits the data somewhat better. If the 2.2μ background is not confirmed, Figure (1) shows that one could still hide the VMO light at around 10μ since the background limits on Ω_R are very weak there.

One might also wonder whether the VMOs could be protogalactic rather than pregalactic. However, in this case z_* would be less than 10; eqn (12) then implies $\lambda_{peak}<0.4\mu$ and $\Omega_R>10^{-5}h^{-2}$, which is definitely incompatible with observation. One would therefore need to suppose that the VMO light was reprocessed by dust in the parent galaxy. This happens in IRAS galaxies, of course, but the emission from IRAS galaxies peaks at around 100μ and the limits in Figure (1) show that there is no waveband longward of that where $\Omega_R>10^{-5}h^{-2}$ is permitted. It therefore seems that a population of VMOs numerous enough to produce galactic halos would need to be pregalactic.

We now consider the possibility that the light from VMOs is reprocessed not by dust within the parent clouds but by cosmological dust. We have seen the such dust could either be pregalactic in origin or confined to galaxies themselves if galaxies cover the sky. In this case, eqn (8) implies that the spectrum should peak at a wavelength

$$\lambda_{peak} = 400\mu \left[\frac{1+z_*}{100}\right]^{1/5} \left[\frac{r_d}{0.01\mu}\right]^{1/5} \left[\frac{1+z_d}{10}\right]^{1/5} \qquad (\alpha=1) \qquad (15)$$

where we have used eqn (12) to express Ω_R in terms of z_* since this should still specify the radiation density. Since one needs $z_d>10$ if galaxies are to cover the sky and also $z_*>10$, this implies $\lambda_{peak}>250(r_d/0.01\mu)^{1/5}\mu$ for $\alpha=1$. The peak wavelength would be somewhat smaller for $\alpha=2$. The question is whether such a background could still be compatible with the FIRAS results. Since the FIRAS limit, given by eqn (2), extends all the way down to 500μ, one clearly needs r_d to be as small as possible and α as large as possible. However, r_d could hardly be less than 0.01μ and α never goes above 2, so the situation is clearly marginal. To consider the question more precisely, one can apply the BCH2 analysis to the VMO scenario by merely noting that each value of Ω_R corresponds to a a value of z_*. Comparison with the far-IR and COBE constraints in Figure (2) shows that the VMO-plus-dust scenario is still viable in all models except 1, 5 and 6. For example, if $z_*=100$, $z_1=9$ and $z_o=0$, then one is below the constraints for $\alpha=2$ but above them for $\alpha=1$. However, the anticipated strengthening of the COBE constraint could in principle eliminate all of the models.

If the COBE results do eventually eliminate the VMO-plus-dust scenario, it should be stressed that one would not necessarily expect dust reprocessing anyway. There may no pregalactic dust at all and protogalactic dust would only suffice if the conditions specified by eqns (6) and (7) were satisfied. Although galaxies might be expected to cover the sky at redshifts exceeding 4 in the CDM picture[50], the analysis of Fall et al.[51] indicates that the dust-to-gas ratio in primordial galaxies may only be in the range 1/20 to 1/4 that of the Milky Way for $2<z<3$, which makes condition (6) difficult to satisfy. Therefore, if COBE does eventually eliminate the far-IR backround predicted, the most straightforward implication would be that eqns (7) and (8) are not satisfied. In this case, the VMO light would have to be in the near-IR (as discussed earlier).

ANISOTROPIES

One of the important implications of the above discussion is that the mere detection of a far-IR or submillimetre background will not identify its astrophysical source, since the expected spectrum depends only weakly on the epoch and mode of radiation production. However, crucial information about the source could come from the detection of any anisotropies in the background. Indeed, since the peak of dust background is so close to that of the CBR, interesting constraints on dust models can already be inferred from CBR anisotropy limits.

In BCH1 we argued that the dust emission can be regarded as coming from a "shell" of thickness $\sim 1h^{-1}$Gpc at a distance $6h^{-1}$Gpc. We then calculated the anisotropies on the assumption that the dust clumps in the same way as galaxies, with a statistical fluctuation in the number of galaxies within each beam. If the galaxy correlation function[52] has the form $\xi_g(x)=(x/x_0)^{-\gamma}$ with $\gamma\simeq1.8$ for $x<x_0$, then one just has Poisson fluctuations for beam sizes σ exceeding the angular scale θ_0 associated with the correlation scale x_0 but the fluctuations are enhanced for $\sigma<\theta_0$. One can infer that the rms intensity fluctuations have the form[18]

$$\left\langle\left[\frac{\Delta I}{I}\right]^2\right\rangle^{1/2} = \begin{cases} \left[\frac{x_0}{ct}\right]^{1/2}\left[\frac{\sigma}{\theta_0}\right]^{-1} & (\sigma>\theta_0) \\ \left[\frac{x_0}{ct}\right]^{1/2}\left[\frac{\sigma}{\theta_0}\right]^{(1-\gamma)/2} & (\sigma<\theta_0) \end{cases} \qquad (16)$$

Here ct is the horizon size and x_0 is the correlation length at the epoch of dominant emission. This would be about $5h^{-1}(1+z)^{-1}$Mpc if there were no evolution of the clustering pattern, corresponding to $\theta_0\simeq1'$, so eqn (16) gives fluctuations of a few percent on arcmin scales. The distinctive feature of the dust anisotropies is that they are wavelength-dependent and much larger than the primary anisotropies at small angular scales (where the latter are expected to be erased).

In BCH2, we confirm that the concept of an emission shell is a good one but we have refined the anisotropy estimates by allowing for fluctuations in the temperature of the emission shell as well its optical depth. (The temperature fluctuations are just associated with spatial fluctuations in the luminosity of the sources.) We have also extended the analysis to the high optical depth case, so as to include, for example, the situation with IRAS galaxies where the dust and sources are highly clumped. Our analysis includes both the linear regime, appropriate when the density fluctuations are small, and the "shot noise" regime, appropriate after galaxies have formed.

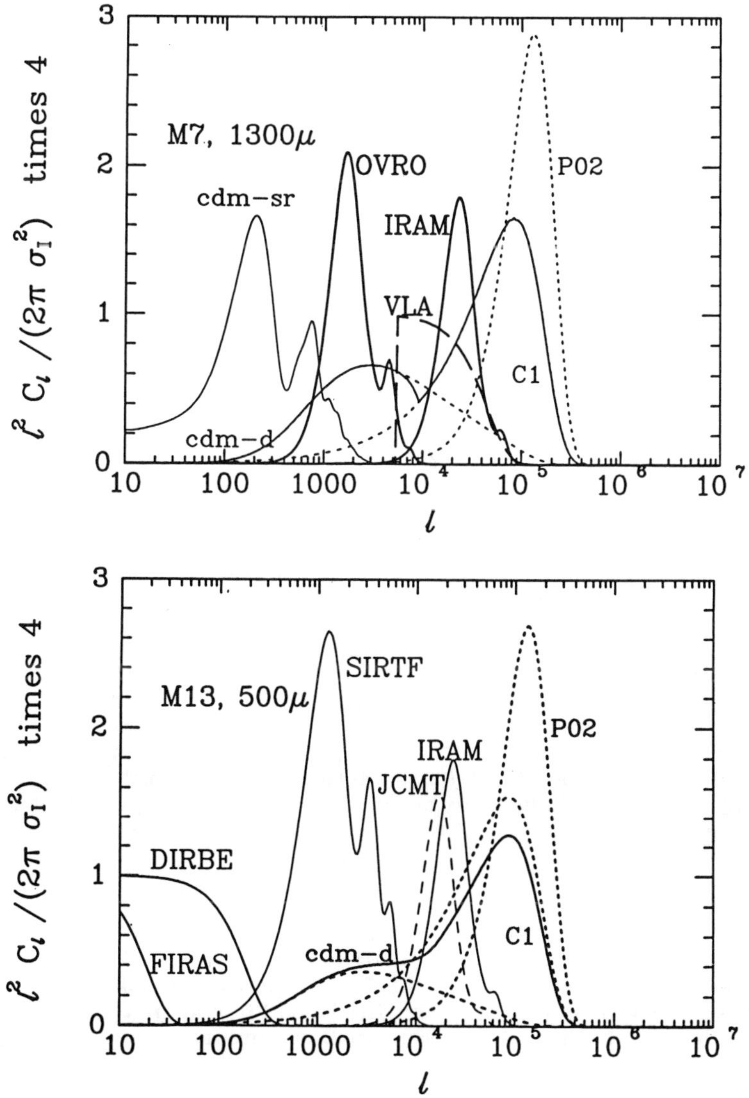

FIGURE (3). This shows the ability of various experiments to probe the anisotropies predicted in several dust scenarios discussed in the text.

Some of the results of our analysis are summarized in Figure (3), where we compare the predicted anisotropies with the current limits from the IRAM experiment of Kreysa & Chini[53], the OVRO experiment of Readhead et al.[54], the JCMT experiment of Church et al.[55], the VLA experiment of Hogan & Partridge[56], and the FIRAS, DIRBE and SIRTF experiments. Figure (3) represents the ability of each experiment to probe the submillimetre anisotropy as a filter in multipole space. (Note that the ℓ-pole probes an angular scale of $3438/\ell$ arcmin.) When one multiplies this by the angular power spectrum $C_l(\lambda)$ of the background and integrates over $d\ln\ell$, one obtains the rms anisotropy for the experiment. The overall normalization is the rms energy density fluctuation $\sigma_I = \langle(\Delta I/I)^2\rangle^{1/2}$ and this is straightforwardly related to the rms temperature fluctuation $(\Delta T/T)_{rms}$.

The figure shows the fluctuations associated with model 7 at 1300μ and model 13 at 500μ. In the first case, we also show the primary fluctuations associated with the CDM scenario (cdm-sr indicates the "standard recombination" case). "PO2" is the Poisson fluctuation model in which the galaxies are placed randomly; "C1" is the $\xi \sim r^{-1.8}$ continuous clustering model, with the amplitude appropriate for observed galaxies; "cdm-d" assumes the usual CDM linear fluctuation spectrum with biasing and linear growth. One can also consider a hybrid of the last two cases in which one has both Poisson and continuous clustering contributions. The figure confirms that the dust fluctuations peak on a smaller scale ($\ell=10^3$ to 10^6) than the primordial density fluctuations. The optimal filter scale would be around 1" but the resolution achievable with submillimetre telescopes is generally above 10". One can also infer from the figure which experiments place the most interesting constraints on each dust scenario.

As a specific example, consider a CDM-like scenario with galaxies and $\alpha=1.5$ dust forming at $z_G=5$ and the burst of radiation having energy $\Omega_R h^2=10^{-6}$ (cf. model 7). The Kreysa & Chini experiment[53] at 1300μ uses a beam with $\theta_{fwhm}=11''$ and $\theta_{throw}=30''$ and gives a constraint $(\Delta T/T)_{rms}<2.6\times10^{-4}$. This implies that the number density of the sources must satisfy $n_G^*>0.06(h^{-1}Mpc)^{-3}$ and their correlation length must be $x_0<4h^{-1}Mpc$. If one took the background at 1300μ to have the maximum density compatible with the COBE constraints, then one would need $n_G^*>1(h^{-1}Mpc)^{-3}$ and $r_0<0.7h^{-1}Mpc$. Future anisotropy experiments will be able to strengthen these limit on n_G^* and r_0, thus providing a very promising probe of the formation of protogalaxies at high redshift.

REFERENCES

1. Mather, J.C. et al., 1990, Ap.J.Lett., 354, L37.
2. Danese, L. & De Zotti, G., 1980, Astr.Astrophys., 84, 364.
3. Ostriker, J.P. & Thompson, C., 1987, Ap.J.Lett., 323, L97.
4. Signore, M. & Sanchez, N., 1990, Int.J.Mod.Phys.A, in press.
5. MacGibbon, J. & Carr, B.J., 1990, Ap.J., in press.
6. Daly, R.A., 1990, Ap.J., in press.
7. Barrow, J.D. & Coles, P., 1990, Mon.Not.R.Astr.Soc., in press.
8. Zeldovich, Ya.B. & Sunyaev, R.A., 1969, Astrophys.Space Sci., 4, 301.
9. Yoshioka, S. & Ikeuchi, S., 1987, Ap.J.Lett., 323, L7.
10. Hogan, C.J., 1989, Ap.J., 340, 1.

11. Field, G.B. & Walker, T.P., 1989, Phys.Rev.Lett., 63, 117.
12. Fukugita, M. & Kawasaki, M., 1990, Ap.J., 353, 384.
13. Bartlett, J.G. & Stebbins, A., 1990, Ap.J.Lett., in press.
14. Lahav, O., Loeb, A. & McKee, C.F., 1990, Ap.J.Let., 349, L9.
15. Taylor, G.B. & Wright, E.L., 1989, Ap.J., 339, 619.
16. Rowan-Robinson, M., Negroponte, J. & Silk, J.,1979, Nature, 281, 635.
17. de Bernardis, P., Masi, S., Malagoli, A. & Melchiorri, F., 1985, Ap.J., 288, 29.
18. Bond, J.R., Carr, B.J. & Hogan, C.J., 1986, Ap.J., 306, 428 [BCH1].
19. McDowell, J.C., 1986, Mon.Not.R.Astr.Soc., 223, 763.
20. Negroponte, J., 1986, Mon.Not.R.Astr.Soc., 222, 19.
21. Hayakawa, S. et al., 1987, Pub.Astr.Soc.Japan, 39, 941.
22. Draine, B.T. & Shapiro, P., 1989, Ap.J.Lett., 344, L45.
23. Naselsky, P.D. & Novikov, I.D., 1989, Space Res. Inst. Preprint 1424.
24. Adams, F.C., Freese, K., Lenon, J. & McDowell, J.C.,1989,Ap.J.,344, 24.
25. Bond, J.R., Carr, B.J. & Hogan, C.J., 1990, Ap.J., in press [BCH2].
26. Mather, J.C. et al., 1990, Ap.J., in press; Hauser, M.G., this volume.
27. Smoot, G.F. et al., 1990, COBE preprint 90-04; also this volume.
28. Bond, J.R. & Efstathiou, G., 1984, Ap.J.Lett., 285, L45.
29. Bertschinger, E., Gorski, K.M. & Dekel, A., 1990, Nature, 345, 507.
30. Burke, W.L., 1975, Ap.J., 196, 329.
31. Stebbins, A., 1988, Ap.J., 327, 584.
32. Lange, A.E. et al., 1990, Ap.J., in press.
33. Matsumoto, T. et al., 1988, Ap.J., 329, 567.
34. Rowan-Robinson, M., 1986., Mon.Not.R.Astr.Soc., 219, 737.
35. Rowan-Robinson, M., Oliver, S. & Saunders, W., 1990, preprint.
36. Rowan-Robinson, M., Hughes, J., Vedi, K. & Walker, D.W., 1990, Mon.Not.R.Astr.Soc., 246, 273.
37. Matsumoto, T., Akiba, M. & Murakami, H., 1988, Ap.J., 332, 575.
38. Rowan-Robinson, M. & Carr, B.J., 1988, in "The Post-Recombination Universe", ed. Kaiser, N. & Lasenby, A. (Dordrecht, Reidel).
39. Saunders, W. et al., 1990, Mon.Not.R.Astr.Soc., 242, 318.
40. Rowan-Robinson, M. & Crawford, J.,1989,Mon.Not.R.Astr.Soc., 238, 52.
41. Djorgovski, S. & Weir, N., 1990, Ap.J., 351, 343.
42. Carr, B.J., Bond, J.R. & Arnett, W.D., 1984, Ap.J., 277, 445.
43. Olive, K.A., Schramm, D.N., Steigman, G. & Walker, T., 1990, Phys.Lett.B., 236, 454.
44. Carr, B.J., 1990, Comm.Astrophys., 14, 257.
45. Ashman, K.M. & Carr, B.J., 1988, Mon.Not.R.Astr.Soc., 234, 219.
46. Karimabadi, H. & Blitz, L., 1984, Ap.J., 283, 169.
47. Adams, F.C. & Walker, T.P., 1990, Ap.J., 359, 57.
48. Bond, J.R., Arnett, W.D. & Carr, B.J., 1984, Ap. J., 280, 825.
49. Carr, B.J., McDowell, J.C. & Sato, H., 1983, Nature, 306, 666.
50. Fall, S.M., Pei, Y.C. & McMahon, R.G., 1989, Ap.J.Lett., 341, L5.
51. Najita, J., Silk, J. & Wachter, K.W., 1990, Ap.J., 348, 383.
52. Peebles, P.J.E., 1980, "The Large-Scale Structure of the Universe", Princeton University Press.
53. Kreysa, E. & Chini, A., 1989, Particle Astrophysics Workshop, Berkeley, Dec.1988.
54. Readhead, A.C.S. et al., 1989, Ap.J., 346, 566.
55. Church, S.E., Lasenby, A.N. & Hills, R.E., 1990, preprint.
56. Hogan, C.J. & Partridge, R.B., 1989, Ap.J.Lett., 341, L29.

BACKGROUND RADIATION FROM GALAXIES AND COBE

Boqi Wang
Astronomy Dept, University of California, Berkeley, CA94720

ABSTRACT

We assess the general possibility of detecting the integrated background radiation from galaxies at far-infrared wavelengths. We model the evolution of interstellar dust in disk galaxies, and obtain analytical solutions for the dust abundance as a function of the galaxy age. The total emission from dust at far-infrared wavelengths in a galaxy is calculated. Consistent with IRAS observations, we find that the integrated far-infrared background from galaxies peaks at around 100-130 μm, with total radiation density 0.5 to 6% that of the cosmic microwave background. We calculate deviations due to galaxies at submillimeter wavelengths from the blackbody spectrum of the cosmic microwave background; in terms of the temperature variation, we find $\Delta T_{CMB}/T_{CMB} = (0.5-2)\times 10^{-2}$ at 0.4 mm. The autocorrelation of fluctuations in the far-infrared background due to the galaxy clustering is also calculated. The effects of the integrated radiation from galaxies are expected to be detected by COBE.

I. INTRODUCTION

Observations by the Infrared Astronomy Satellite (IRAS) proved that galaxies are powerful radiation emitters at far-infrared wavelengths ($\sim 10-100\mu$m). Some galaxies have a luminosity in the far-infrared comparable to that of quasars. The local luminosity density of galaxies at far-infrared wavelengths obtained from the IRAS survey is about one third of the optical luminosity density. If the far-infrared luminosity of galaxies at early times is comparable to the present values, the integrated radiation from galaxies may constitute a distinct component of the extragalactic background radiation. With the unprecedented sensitivity of the various experiments on Cosmic Background Explorer (COBE), we expect that the positive detection of the far-infrared background (FIRB) from galaxies may be possible. We here present the main results of our calculations of the FIRB from normal galaxies. More detailed description of our work can be found elsewhere (Wang 1990a, 1990b).

II EVOLUTION OF FIR LUMINOSITY IN DISK GALAXIES

The far-infrared luminosity of galaxies is dominated by the emission from dust. Observations and theoretical studies of interstellar dust indicate that dust is efficiently destroyed in the interstellar medium; in our Galaxy the time scale for destroying dust by supernova shocks is estimated to be shorter than the time scale for injection of dust from stellar outflows (McKee 1989 and references therein). Therefore a continuous supply of dust besides the usual stellar source is required to explain the amount of dust we observe today in the interstellar medium. We assume that dust can form rapidly enough in molecular clouds (Herbig 1970, Field 1974) to replenish amount of dust destroyed by supernovae. We solve the equations for evolution of gas, heavy elements and dust abundance in disk galaxies. To be consistent with the chemical evolution of our Galaxy, either prompt initial enrichment of heavy elements or accretion of matter from outside of the disk is assumed. In the former case, the disk of galaxies is supposed to start with a non-vanishing initial metallicity, whereas in the latter case gas is constantly accreted onto the disk. We consider (1) a constant star formation rate, and (2) a star formation rate that decreases exponentially with time. The time scale for the exponential star formation rate is determined by the amount of gas observed today. We take the e-folding time $\tau = (0.33 - 0.44) \times t_0$, corresponding to the mass fraction of gas in the disk equal to 5 and 10% respectively, where t_0 is the present age of galaxies. In both cases, analytical solutions for the metallicity and dust abundance are obtained.

Observations of galaxies at submillimeter and millimeter wavelengths have not been without uncertainties (Chini *et al.* 1986, Eales *et al.* 1989), but Eales *et al.* (1989) found no evidence for a distinct population of cold dust, and that the spectra of galaxies from far-infrared to submillimeter wavelengths can be fit by the thermal emission of dust with a single temperature $30 - 50$K, contrary to that predicted by the two-component model (Helou 1986). Comparisons between the Hα emission line and IRAS 40-120 μm luminosities for spiral galaxies also suggest that massive stars are responsible for both the Hα and far-infrared emission (Devereux and Young 1990). We assume here that warm dust with a temperature $\sim$ 30 K is responsible for most of the far-infrared radiation from $\sim 20\mu$m to submillimeter wavelengths, consistent with the mean 60/100 μm flux ratio of galaxies observed by IRAS. We adopt a dust emissivity law, $\propto \nu^n$, with $n = 1$ or $n = 2$. We assume that the total power emitted by each grain of dust is proportional to the number of late-type stars, and hence, to the star formation rate. Given the amount of dust and the dust temperature, we then can compute the far-infrared radiation from a galaxy.

III. SPECTRUM OF THE INTEGRATED BACKGROUND

We integrate the radiation from galaxies at different epochs, taking into account the redshift effects. We model the rate of galaxy formation per volume as a Gaussian in redshift z, with the maximum rate occurring at z_f and a Gaussian width σ_z. To cover possible range of galaxy formation epochs predicted by different theories, we adopt the following parameters: $(1) z_f = 2, \sigma_z = 1$, $(2) z_f = 5, \sigma_z = 2.5$ and $(3) z_f = 20, \sigma_z = 10$. We take the Hubble constant $H_0 = 50$ km/s/Mpc and density parameter $\Omega = 1$. The luminosity density is assumed equal to the local luminosity density at 60 μm, $(2.1 \pm 0.2) \times 10^7$ $L_\odot$ Mpc^{-3} based on IRAS observations (Saunders *et al.* 1990).

In figure 1, we present the integrated FIRB spectrum λI_λ as a function of wavelength, λ, for $z_f = 5$ and $\sigma_z = 2.5$. For comparison, the spectrum of the cosmic microwave background (CMB) with a blackbody temperature $T_{CMB} = 2.74$K is also shown. Different curves are results from different models for the chemical evolution and the star formation rate. Because the luminosity density is normalized to the present observed level at 60μm, models with an exponential star formation rate imply stronger star-forming activities at early times, thus more radiation, than models with a constant star formation rate. Similarly, earlier galaxy formation implies stronger FIRB. The total radiation density of the FIRB is from 0.5 to 6% that of the CMB. At wavelengths longward of 400μm, the CMB dominates, but the background from galaxies produces a distinct component, distorting the blackbody spectrum of the CMB. The temperature variation of the CMB due to the FIRB from galaxies, $\Delta T_{CMB}/T_{CMB}$, is found to be $(0.5-2) \times 10^{-2}$ at 0.4 mm. The predicted FIRB should be detected by the COBE's Diffuse Infrared Background Experiment (DIRBE) at $\sim 100\mu$m; the FIRB has an intensity $\lambda I_\lambda \simeq (0.4 - 4) \times 10^{-5}$ ergs/cm^2/s/sr, compared with the designed sensitivity of DIRBE of $\sim 1 \times 10^{-6}$ ergs/cm^2/s/sr (Hauser 1990). The deviation of the blackbody temperature due to the FIRB of galaxies should be observed by COBE's Far Infrared Absolute Spectrophotometer (FIRAS) at submillimeter wavelengths, which is expected to reach below 0.1% of the peak intensity of the CMB (Mather 1990). We have also calculated the anisotropy of the FIRB resulting from the clustering of galaxies, but due to the limited space here, we refer to the paper by Wang (1990b).

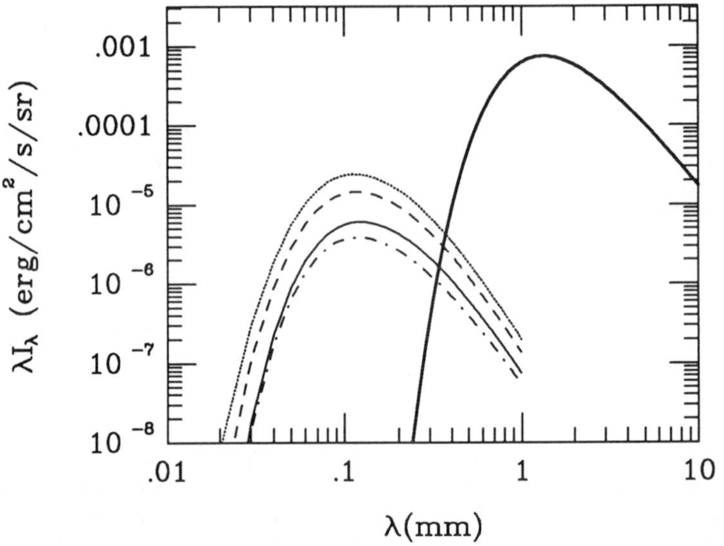

Figure 1. *The intensity of the integrated background radiation from galaxies at various wavelengths. The parameters of galaxy formation are taken to be* $z_f = 5$ *and* $\sigma_z = 2.5$. *Different curves correspond to values predicted by different models of chemical evolution: the prompt initial enrichment models with a constant star formation rate (solid curve) and a star formation rate that decreases exponentially with time (dotted curve:* $\tau = 0.33t_0$; *dashed curve:* $\tau = 0.43t_0$*), and the accretion models with a constant star formation rate (dash-dotted curve), respectively. For comparison, the intensity of the cosmic microwave background is also plotted (thick line to the right). Notice that the designed sensitivity of FIRAS and DIRBE is* $\lambda I_\lambda \sim 1 \times 10^{-6}$ *erg/cm*^{2}*/s/sr.*

IV. DISCUSSION

The precise spectrum and the exact magnitude of the FIRB are uncertain due to our incomplete knowledge of dust in external galaxies, and the evolution and formation of galaxies. However, the fact that the luminosity density of IRAS galaxies at $\sim 10-100\mu$m, j_{IRAS}, is about one third of the total optical luminosity density (Saunders *et al.* 1990) demonstrates the importance of the far-infrared radiation from galaxies, independent of models. If the galaxy luminosity function at far-infrared wavelengths does not change significantly in a Hubble time, the total radiation density contributed by galaxies is approximately $j_{IRAS}t_0 \simeq 2 \times 10^{-15}$ erg/cm^3, roughly 0.5% of the CMB density. This is in agreement with the lower bound of our estimate of the FIRB if essentially no evolution of the infrared luminosity is present (the star formation rate is constant with time). The evolution of galaxies could produce a FIRB that is an order of magnitude larger.

The implications of the FIRB are twofold. First, although the contribution by discrete radio sources may cause temperature fluctuations of the CMB longward of its peak wavelength ($\sim$ 1 mm) (Franceschini *et al.* 1989), emission from dust in galaxies can dominate the fluctuations of the CMB at submillimeter wavelengths. Since observational limits both on the deviation from the blackbody spectrum and on the anisotropy of the CMB have become increasingly small, one would naturally expect that at some level the fluctuations by extragalactic sources must be encountered, possibly even obscuring the intrinsic anisotropy from primeval density fluctuations that lead to galaxy formation. In particular, we expect that provided Galactic contamination can be removed, FIRAS and DIRBE on COBE should be able to detect the FIRB from galaxies at far-infrared and submillimeter wavelengths in the near future. Precise spectral information by COBE in principle can help us distinguish the FIRB from other sources that may cause fluctuations in the CMB, such as the Compton scattering due to hot electrons, because the spectrum of FIRB is uniquely specified by the dust emission properties. Secondly, as evident from their contribution to the background radiation, observations of galaxies at far-infrared and submillimeter wavelengths can provide another window in studying the formation and evolution of galaxies. One of the results derived from our calculations is that earlier galaxy formation would produce a stronger FIRB. The prompt initial enrichment of heavy elements in disk galaxies also predicts a larger amplitude for the FIRB than the accretion models. Consequently, the detailed spectral and anisotropy data of future observations of the FIRB would certainly constrain and refine models of galaxy formation, improving our understanding of chemical and dust evolution in galaxies.

I thank G. B. Field for his valuable comments on the present work. This work is supported in part by the NASA grant NAGW931.

REFERENCES

Chini, R., Kreysa, E., Krugel, E., and Mezger, P. G. 1986, Astr. Ap., **166**, L8.

Devereux, N. A. and Young, J. S., 1990, *Ap. J. (Letters)*, **350**, L25.

Eales, S. A., Wynn-Williams, C. G., and Duncan, W. D. 1989, *Ap. J.*, **339**, 859.

Field, G. B. 1974, *Ap. J.*, **187**, 453.

Franceschini, A., Toffolatti, L., Danese, L., and De Zotti, G. 1989, *Ap. J.*, **344**, 35.

Hauser, M. G., 1990, this volume.

Helou, G. 1986, *Ap. J. (Letters)*, **311**, L33.

Herbig, G. H. 1970, *Mem. Soc. Roy. Sci. Liege*, Ser. 5, **19**, 13.

Mather, J. C. 1990, this volume.

McKee, C. F. 1989, in *IAU Symposium 135, Interstellar Dust*, eds. Allamandola, L. J., and Tielens, A. G. G. M., (Dordrecht:Kluwer), p.431.

Saunders, W, Rowan-Robinson, M., Lawrence, A., Efstathiou, G., Kaiser, N., Ellis, R. S., and Frenk, C. S. 1990, *M.N.R.A.S.*, **242**, 318.

Wang, B. 1990a, submitted to *Ap. J.*.

Wang, B. 1990b, submitted to *Ap. J.*.

The submillimetre background expected from normal galaxies

Steve Eales
Department of Astronomy, University of Toronto, Toronto, Ontario M5S 1A7, Canada

Yesterday there was alot of debate about whether or not COBE (Mather 1982) will be able to rule out different cosmological models (Vittorio, this volume). This morning Hauser (this volume) has discussed some of the exotic kinds of object that COBE may discover. I now want to talk about something that COBE will definitely be able to do. COBE's measurements of the background emission from normal galaxies will place important limits on the strength and type of the cosmological evolution experienced by normal galaxies. As I will show in this talk, the mean intensity of, the spectral shape of, and the size of the fluctuations in this background, all depend on the form and size of the evolution. COBE's measurements of this background will definitely be able to put limits on the evolution, because even if there is no evolution, COBE will be sensitive enough to detect the background.

To calculate the submillimetre background, one needs to know three things: (a) the cosmological correction that has to be made because the emission from a galaxy is not independent of frequency (the K-correction); (b) the current space density of galaxies as a function of luminosity; (c) the evolution function—that is, how the space density depends on redshift. I'll now discuss these in turn.

To calculate the K-correction, one needs to know something about the spectral energy distributions of galaxies in the far-infrared and submillimetre wavebands. Fig. 1 shows far-infrared and submillimetre fluxes for all the normal galaxies that have reliable submillimetre flux densities (all those of which I am aware). So you can get some idea of the range of spectral energy distributions of the galaxies, I've normalised all the fluxes so that each galaxy has a flux of 100 Jy at 100 μm. As there are so few galaxies with measured submillimetre fluxes, to calculate the K-correction I assumed that all galaxies have the same spectral energy distribution, the one shown by the solid line in the figure. This consists of the sum of the thermal emission from dust at 25 K and the thermal emission from dust at 75 K, with there being 1923 times as much cold dust as hot dust, and assuming an emissivity function $\propto \nu^2$. I adjusted the temperatures and the proportion of cold and hot dust so that the curve passes, as nearly as I can make it, through the median flux at each wavelength. I don't claim any significance for this model; it's merely the simplest model I could think of that provides a reasonable fit to the data.

To calculate the background emission at a frequency ν, one ideally wants to know the local space density as a function of L_ν, in which L_ν is the luminosity at that same frequency. Although this is not known at submillimetre frequencies, it is known at far-infrared frequencies, because of all the work done by the IRAS satellite. If it's known how the space density depends on luminosity at one frequency, it's possible to estimate how the space density should depend on luminosity at some other frequency by using a common spectral energy distribution for all galaxies, an assumption that I anyway had to make to calculate a K-correction. In practise, I used the IRAS 60 μm luminosity function (the space density as a function of the luminosity at 60 μm—Lawrence et al. 1986) and calculated the space density at other wavelengths by assuming that all galaxies have the spectral energy distribution shown by the solid line in Fig. 1.

These first two terms, the K-correction and the space density, can be reasonably accurately estimated; the third, the evolution, is not well known at all—indeed, the interest of the calculation lies in seeing how the background depends on the form and strength of the evolution. Nevertheless, there are already upper limits to the strength of the evolution. In principle, an upper limit comes from the belief that metals and the far-infrared and submillimetre emission from galaxies have the same ultimate source: massive stars. Thus the present-day metal abundance of galaxies is one constraint on the strength of the evolution (cf. Cowie 1988). In practise, however, it is a rather soft constraint, because of uncertainty about how much of the far-infrared emission is coming from massive stars and how much is coming from low-mass stars (Soifer, Houck, and Neugebauer 1987; Eales, Wynn-Williams, and Duncan 1989; Young, 1990). A more practical constraint arises because of the strong correlation between the radio emission and far-infrared emission from spiral galaxies (Devereux and Eales 1989). If a model for the evolution in the far-infrared/submillimetre wavebands has evolution that is too strong, more faint radio sources are predicted than are actually observed. I have tried 4 models for the evolution that, I think, cover the range of possible forms and strengths for the evolution and do not exceed the limits produced by the counts of faint radio sources.

A: No evolution; the space density is independent of redshift.

B: The space density of galaxies is independent of redshift, but the galaxies' luminosities depend on cosmic time in the following way: $L_\nu = L_o e^{-3.0(t-\tau_u)/\tau_u)}$, in which L_o is a galaxy's current luminosity and τ_u is the age of the universe. Apart from the assumption that the space density is constant, this model implicitly assumes that all galaxies evolve in the same way and that the rate of change of luminosity is independent of cosmic time. I chose the value of 3.0 to give the strongest evolution that is still consistent with the radio source counts.

C: Same as A except that there is assumed to be no dust in the universe beyond a redshift of 1, and hence no far-infrared/submillimetre emission from galaxies. Fall, Pei, and McMahon (1989) have found evidence that the dust-to-gas ratio is much less at $z \simeq 3$ than it is today. I invented this model to see how big an effect this possible lack of dust at high redshifts might have on the submillimetre background.

D: Same as B except that there is assumed to be no dust in the universe beyond a redshift of 1. The reason for this model is the same as that for C.

Fig. 2 shows the background predicted for the four models. There are a number of interesting points to notice. Firstly, even if there is no evolution, COBE will be sensitive enough to detect the background. Secondly, the background does not depend very strongly on the strength of the evolution; there is only a factor of 2 increase in brightness between the no-evolution model and the model with maximum evolution. Thirdly, the signature of a redshift cutoff is a more peaked spectral shape. One might argue that these models do not cover the full range of possible forms and strengths for the evolution. In particular, model B, which assumes luminosity evolution and a constant space density, is directly in conflict with galaxy formation models based on cold dark matter, in which the space density is not constant, because galaxies form by the merging of smaller galaxies (e.g. White 1989). If, however, the assumption that the space density is constant is wrong, the effect on the background is unlikely to be huge for the following reason. CDM galaxy formation models predict that most of the merging occurs at $1 < z < 2$ (White 1989), and so the space density should be changing most rapidly at these high redshifts. The background, though, because of the shape of the K-correction, is relatively insensitive to differences in the assumed evolution at $z > 1$. This can be seen by comparing models B and D. One has the strongest possible luminosity evolution at $z > 1$; the other has no emission from galaxies at z > 1 at all; the difference between the predicted backgrounds is not large.

Fig. 3 shows the predicted level of the fluctuations for the different models. The level of the fluctuations is simply the root of the variance of the background, scaled to the 0.7° DIRBE beam (Mather 1982). As I have assumed no clustering, the figure really shows predicted lower limits for the level of the fluctuations.

Cowie, L.L. 1988, *The Post-Recombination Universe*, eds N. Kaiser and A.N. Lasenby.
Devereux, N.A. & Eales, S.A. 1989, *Astrophys. J.*, 340, 708.
Eales, S.A., Wynn-Williams, C.G. & Duncan, W.D. 1989, *Astrophys. J.*, 339, 859.
Fall, S.M., Pei, Y.C. & McMahon, R.G. 1989, *Astrophys. J. (Letters)*, 341, L5.
Lawrence, A., Walker, D., Rowan-Robinson, M., Leech, K.J. & Penston, M.V. 1986, *Mon. Not. R. astr. Soc.*, 219, 687.
Mather, J.C. 1982, *Opt. Eng.*, 21, 769.
Soifer, B.T., Houck, J.R. & Neugebauer, G. 1987, *Ann. Rev. Astr. Ap.*, 25, 187.
Stark, A.A., Davidson, J.A., Platt, S., Harper, D.A., Pernic, R., Loewenstein, R., Engargiola, G. & Casey, S. 1989, *Astrophys. J.*, 337, 650.
White, S.D.M. 1990, *The Interstellar Medium in Galaxies*, eds H.A. Thronson & J.M. Shull, p371.
Young, J.S. 1990, *The Interstellar Medium in Galaxies*, eds H.A. Thronson & J.M. Shull, p67.

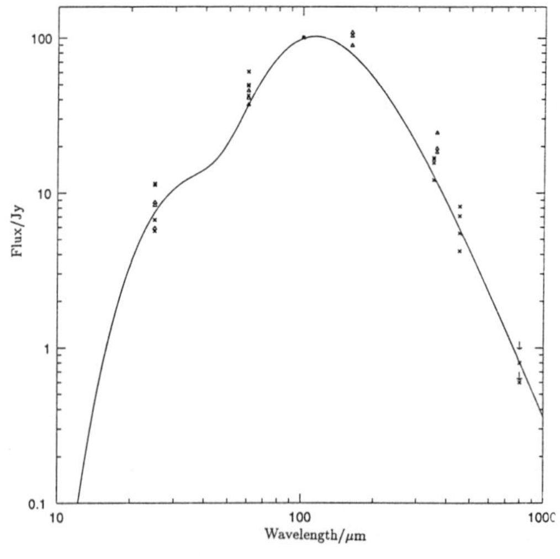

Fig. 1: Far-infrared and submillimetre fluxes for all the normal galaxies that have reliable submillimetre fluxes. The triangles show data from Stark et al. (1989); the crosses show data from Eales, Wynn-Williams, and Duncan (1989). All the fluxes have normalised, so that each galaxy has a flux of 100 Jy at 100 μm. The solid line is described in the text.

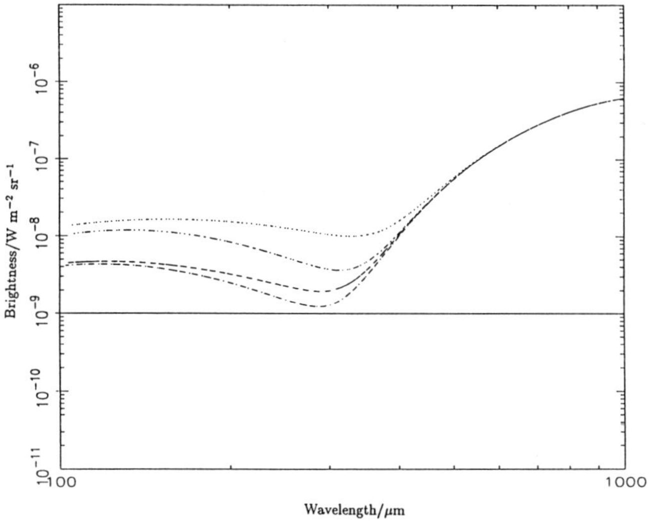

Fig. 2: The brightnesses of the backgrounds predicted for the different models. In each case, the line shows the sum of the background from the galaxies and the background from the 2.7-K background. The dotted line shows the background predicted by model B; the line with dashes interspersed with sequences of three dots is for model D; the dashed line is for model A; and the dot-dash line is for model C. The horizontal solid line shows the sensitivity limit that COBE will achieve (Mather 1982).

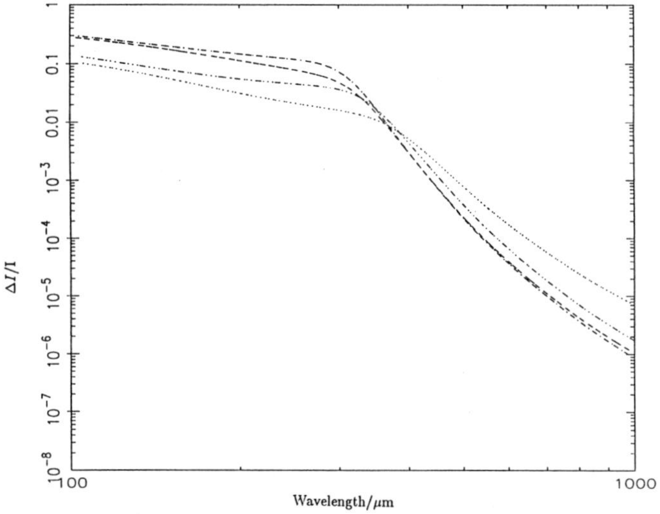

Fig. 3: The level of fluctuations predicted by the models. This is simply $\Delta I/\mathrm{I}$, where $(\Delta I)^2$ is the variance of the background in a 0.7° beam (FWHM) and I is the flux from the background, from both the galaxies and the 2.7-K background, in that beam. 0.7° was chosen because this is the beam size of DIRBE (Mather 1982). The code for the models is the same as in Fig. 2.

THE PREGALACTIC FORMATION OF GLOBULAR CLUSTERS

Edward I. Rosenblatt
The Center for Astrophysics and Space Sciences
University of California, San Diego.

The baryonic Jeans mass after decoupling in the early universe was 10^5 $M_\odot$ (Peebles and Dicke 1968). For a flat or monotonically declining density fluctuation spectrum, the first structures to collapse by gravitation would have had this minimum mass. Peebles and Dicke suggested that such structures can be identified with present-day globular clusters.

One of the major objections to this idea, however, is the observation of a radial metallicity gradient for the galactic population of globular clusters. If globular clusters formed well before the collapse of the proto-galaxy, metallicity should not be correlated with radial position. A solution to this metallicity gradient problem is provided by Zinn (1985) who found convincing evidence that galactic globular clusters, based on dynamical and spatial grounds, can be divided into two distinct components: a disk and a halo population. If one examines each component separately, there is little or no evidence for a metallicity gradient. The apparent gradient for the cluster system as a whole is instead due to the changing ratio of the number of disk to halo globular clusters versus radial distance.

Another potential problem with the pregalactic formation of globular clusters is their non-zero metallicity abundances and the small spread in metallicity among the stars within a particular cluster. If truly primordial, globular clusters must be self-enriched by a process that produces the observed homogeneity. Much research has recently addressed this issue. For example, Murray and Lin (1990) have modeled the fragmentation of proto-cluster clouds and found that global fragmentation leads to self-enrichment and homogeneity. Another possibility involves a variable IMF such that the first stars to form within a proto-cluster were predominantly massive, thus enriching and homogenizing the gas (Smith and McClure 1986; Yoshii and Saio 1986).

In light of a gradient free population of globular clusters, Rosenblatt, Faber, and Blumenthal (1988, hereafter RFB) re-examined the pregalactic formation of globular clusters in a universe dominated by cold dark matter (CDM). Their results, which are valid assuming any monotonically declining density fluctuation spectrum such as CDM, are summarized here. However, it must be emphasized that the discussion of pregalactic formation of globular clusters pertains only to the halo population. It is likely that the disk population formed by some other mechanism such as a two-phase thermal instability (Fall and Rees 1985), or by cloud-cloud collisions involving shock-induced cooling (Gunn 1980). Moreover, observations of globular cluster-like observations in other galaxies such as the LMC (*e.g.,* Freeman 1980) have provided evidence for the relatively recent formation of clusters. Such clusters may have formed, for example, from galaxy-galaxy interactions. It is thus probable that globular clusters in general formed from several distinct processes. However, the predictions resulting from a primordial formation scenario of *halo* clusters closely agree with several independent observations as will be shown below.

The basic premise of the pregalactic scenario is that 10^{5-6} $M_\odot$ density fluctuations, which lie above some threshold $n\sigma$, survive protogalaxy collapse to form a halo population of globular clusters. This threshold is defined globally so that fluctuations at a given σ level are correlated to the same specific density at all points in the universe. Thus, clusters would be exceedingly rare in intergalactic space but relatively common in high-σ parent fluctuations (Bardeen *et al.* 1986). From an examination of the mass fraction of globular clusters in local group and Virgo cluster galaxies (See Fig. 1 of RFB for details), it is found that n = 2.8. Thus, fluctuations of n $\geq 2.8\sigma$ are assumed to survive and ultimately form the halo population of globular clusters in galaxies.

Following the method of Blumenthal *et al.* (1984), the physical characteristics of galactic globular clusters were derived. Figure 1 shows cluster baryonic density versus internal virial temperature (see Fig. 2 of RFB for more details). The solid nearly horizontal curves represent collapse loci for dissipationless collapse of 1 σ, 2 σ, and 3 σ perturbations from a CDM fluctuation spectrum (dashed lines are perturbations embedded in 10^{12} $M_\odot$ parent fluctuations). The diagonal lines show the total mass of a perturbation including baryonic and dark matter. The line leading downward from the 2.8 σ and $M_{tot} = 10^6$ $M_\odot$ intersection is a schematic dissipation track which indicates that collapse factors of order 10 are required to explain present epoch globular structure. Thus, if globular clusters collapsed by roughly this factor, the formation of globular clusters from primordial density flucations satisfactorily accounts for their observed radii and densities.

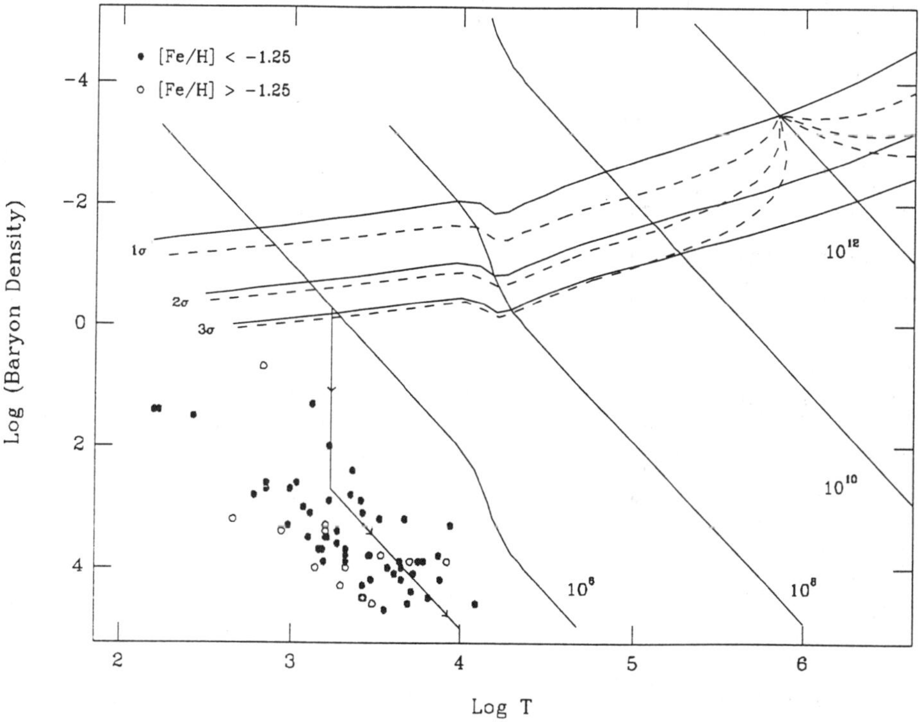

Fig. 1 Baryonic density (n_b) vs virial temperature (T) for theoretical collapsed objects and Galactic globular clusters. The dots represent Galactic halo globular clusters; the open circles are disk globulars shown for comparison. The precepts used to define n_b and T are identical to those in Fig. 3 of Blumenthal *et al.* (1984). Also see RFB Fig. 2 for more details.

A comparison can also be made between the radial distribution of galactic halo globular clusters and two models that represent dark matter distribution (see Fig. 2). If globulars are completely pregalactic, they would have collapsed dissipationlessly during galaxy formation along with the dark matter. Any difference between the two distributions today must therefore reflect a difference in the initial distributions prior to collapse. However, in certain scenarios such a diffference is expected on theoretical grounds. As noted above, with a flat spectrum as in CDM, high-σ peaks are expected to be statistically less common in the outer parts of protogalaxies (Bardeen *et al.* 1986). Depending on exactly how protoclusters survive, the final distribution of primordial clusters could be more centrally concentrated than the dark matter distribution.

Relative to the isothermal sphere, the halo infall model is more centrally concentrated (*i.e.*, has a steeper slope), but both models fit the inner part of the cluster distribution adequately within the errors. However, beyond R $\approx$ 10 kpc the observed clusters fell well below both models, indicating a significant paucity of clusters relative to dark matter in the outer parts of the Milky Way. A radius of 15–20 kpc is a reasonable minimum radius to assume for the dark matter halo, and by that point the divergence amounts to about a factor of 4 as predicted.

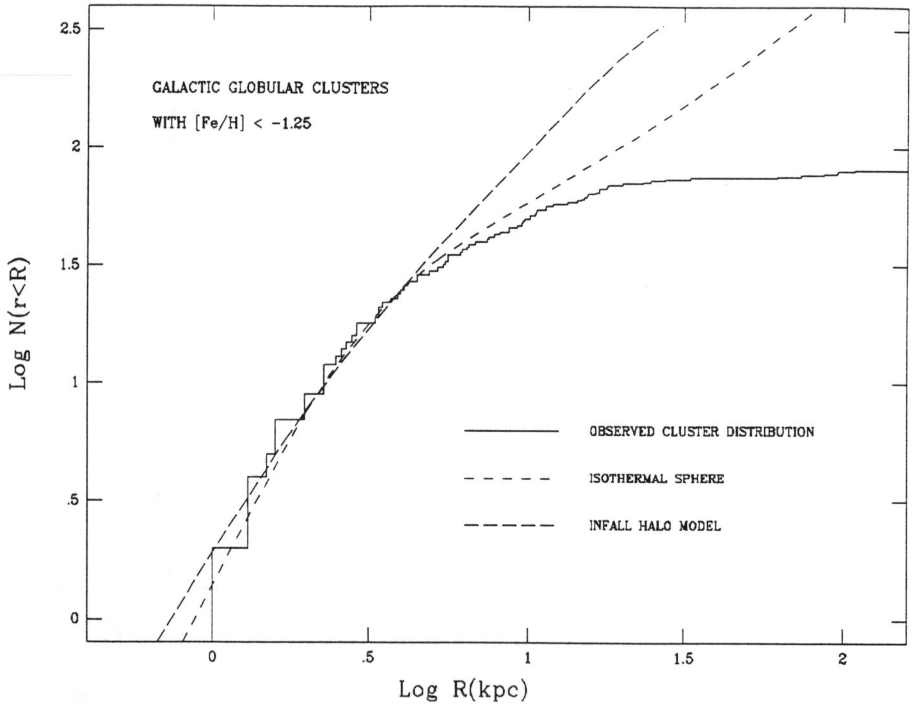

Fig. 2 The cumulative radial distribution of Zinn's halo clusters in the Galaxy compared to two models for the dark matter. Zinn's halo sample has been trimmed at [Fe/H] $\equiv -1.25$ to eliminate a few possible disk clusters. The two DM models are the standard isothermal sphere (*short dashes*) and a model showing the effect of baryonic infall on the DM distribution (Blumenthal *et al.* 1986). Relative to either DM distribution, there is a distinct paucity of observed globular clusters in the outer part of the Galaxy.

A final comparison can be made with the observed luminosity function of globulars in the Local Group and Virgo cluster. This function is roughly Gaussian in magnitude, with $<M_V> = -7.3$ and $\sigma = 1.2$ mag (Harris and Racine 1979). For $M_V/L = 1.6$ (Illingworth 1976), this corresponds to a logarithmic mean baryonic mass of $\sim 10^5$ $M_\odot$ and a total initial mass of $\sim 10^6$ $M_\odot$ before any tidal stripping.

The existence of the lower mass cutoff at 10^{5-6} $M_\odot$ can naturally be explained as the Jeans mass (Peebles and Dicke 1968), or alternatively as the threshold mass for efficient baryon dissipation and cooling (Blumenthal *et al.* 1984). The mass cutoff on the upper end, however, requires a separate explanation. To the extent that hierarchical clustering in this mass range is basically a self-similar process, it alone cannot preferentially select any particular mass value. However, since the number of globulars left behind today depends critically on survival, it seems to be possible to couple this idea with hierarchical clustering in order to provide a natural decline at higher masses. The cluster mass fractions (Fig. 1 of RFB) suggest that roughly 0.25% of all low-mass perturbations survive as globulars. In purely self-similar clustering, this percentage would remain constant at each level of the clustering hierarchy. Since at each level of the hierarchy the number of objects of a given mass, $dN(M)/d \log M$, scales as M^{-1}, we expect the number of surviving objects to scale likewise. Recalling that $M \propto L_V$ for constant M/L, and that $\Phi(M_V)/dM_V$, we have

$$\Phi(M_V) = \text{const } \frac{dN(M)}{dlogM} \propto M^{-1} \propto L_V^{-1} \qquad (1)$$

and also therefore

$$N(M) \propto L_V^{-1} \qquad (2)$$

These expressions should be valid on all mass scales much less than a galaxy mass, $M \ll 10^{12}$ $M_\odot$, where the clustering is nearly self-similar. Note that in deriving equations (1) and (2), the usual multiplicity function was not used, as the basic hypothesis here is that clusters of different masses are the densest survivors from successive stages of the hierarchy, which is a diachronic concept. The multiplicity function is a snapshot of the mass distribution of fluctuations that exist at a given epoch and is a synchronic concept.

Relations (1) and (2) were compared to observed luminosity functions in Fig. 5 of RFB. The relations fit the data fairly well within two magnitudes above the peak, but predict too many clusters brighter than this level. For a total Local Group cluster population of $\sim$450 globulars, relation (1) predicts 11 objects between $M_V = -9.8$ and -11.3, compared to an observed population of only five. The Poisson probability of seeing so few objects is 0.04. One plausible explanation for the deficit of brightest globulars is the preferential effect of dynamical friction on higher mass objects, causing them to spiral into the central region of galaxies (Tremaine 1976; Quinn and Goodman 1986). Using equation (8) of Tremaine (1976), we estimate that the spiraling-in time from a distance of 10 kpc in M31 is less than a Hubble time for clusters above 2×10^6 $M_\odot$. Allowing for the fact that perhaps one-half or two-thirds of this mass might today be the inner remains of a dark halo, we find that this mass corresponds to $M_V \approx -9$ mag, which is about where the M^{-1} functions begin to depart from the data.

The major results are summarized as follows. An attempt was made to compare theory with the observed properties of globular clusters. It is assumed that all fluctuations above a fixed threshold σ defined globally over the whole universe survive as globulars. This means that the absolute value of the density threshold decreases somewhat with cluster mass but is independent of local environment. For a density threshold level of roughly 2.8 σ, adequate fits are achieved to cluster radii and densities, the mass fraction of globulars versus Hubble type, the radial density profiles of globulars within galaxies, and the globular luminosity function.

References

Bardeen, J.M., Bond, J.R., Kaiser, N., and Szalay, A.S. 1986, *Ap. J.*, **304**, 15.

Blumenthal, G.R., Faber, S.M., Primack, J.R., and Rees, M.J. 1984, *Nature*, **311**, 517 (BFPR).

Fall, S.M., and Rees, M.J. 1985, *Ap. J.*, **298**, 18.

Freeman, K.C. 1980, in *IAU Symp. 85, Star Clusters*, ed. J.D. Hesser (Dordrecht: Reidel), p. 317.

Gunn, J.E. 1980, in *Globular Clusters*, ed. D. Hanes and J.D. Hesser (Cambridge: Cambridge University Press), p. 301.

Harris, W.E., and Racine, R. 1979, *Ann. Rev. Astr. Ap.*, **17**, 241.

Illingworth, G. 1976, *Ap. J.*, **204**, 73.

Murray, S.D., and Lin, D.N.C. 1990, *Ap. J.*, **363**, 50.

Peebles, P.J.E., and Dicke, R.H. 1968, *Ap. J.*, **154**, 891.

Rosenblatt, E.I., Faber, S.M., and Blumenthal, G.R. 1988, *Ap. J.*, **330**, 191 (RFB).

Smith, G.H., and McClure, R.D. 1986, *Ap. J.*, **316**, 206.

Yoshii, Y., and Saio, H. 1986, *Ap. J.*, **301**, 587.

Tremaine, S.D. 1976, *Ap. J.*, **203**, 345.

Quinn, P.J., and Goodman, J. 1986, *Ap. J.*, **309**, 472.

Zinn, R.W. 1985, *Ap. J.*, **293**, 424.

POST-DECOUPLING EVOLUTION OF PRIMORDIAL CLOUDS: THE ROLE OF RAYLEIGH SCATTERING

Leonid M. Ozernoy
IGPP, Los Alamos National Laboratory, Los Alamos, NM 87545

ABSTRACT
Soon after decoupling, the opacity of contracting primordial clouds is determined mainly by Rayleigh scattering which, even without accounting for cooling due to molecules and metals, would lead to a Jeans mass less than 0.1 $M_\odot$.

The cosmological redshift range between decouplng ($z \approx 1000$) and the copious formation of molecules ($z \lesssim 100$) define an epoch, our current incomplete knowledge of which is inversly proportional to its potential importance for the structure formation. A process that has been omitted so far for this epoch is Rayleigh scattering by neutral hydrogen. This process becomes essential for a contracting primeval cloud after its density exceeds $n \approx 10^8$ cm^{-3} when the cooling due to *f-b* (free-bound) emission dominates over Compton cooling off microwave background radiation. Lyman alpha photons released in the course of *f-b* emission are absorbed again within the cloud until they eventually split up, due to two-photon emission, into lower energy quanta. As the density of the cloud increases, the opacity due to Rayleigh scattering increases until the diffusion time of photons becomes larger than the free-fall time. At this moment, the Jeans mass becomes

$$M_J = \frac{\pi^4}{3}\left(\frac{kT}{m_{\rm H}}\right)^{5/2}\frac{\sigma_R}{cG^2 m_{\rm H}}, \tag{1}$$

where $\sigma_R = \sigma_T[\nu_{\rm eff}^2/(\nu_{\rm eff}^2 - \nu_0^2)]^2 \approx 4 \cdot 10^{-3}\sigma_T$; σ_R and σ_T are the Rayleigh and Thomson scattering cross sections, correspondingly; $\nu_{\rm eff}$ is an effective frequency of *f-b* emission (close to the Balmer continuum edge); and ν_0 is the resonant frequency corresponding to the Lyman continuum edge.

Evaluation of M_J requires calculation of thermal balance in the cloud. Energy gains include shock heating due to (i) collisions of fragments in the turbulent collapsing cloud and (ii) collisions of the clouds themselves. However, a stringent upper limit to Eq. (1) immediately follows if one takes into account that the Compton cooling implies a conservative constraint $T \leq 10^4$ K, which results in $M_J \lesssim 0.1$ $M_\odot$.

Although M_J does not depend explicitly on n, "locking" of photons due to an inequality $t_{\rm diff} \geq t_{\rm free-fall}$ sets in at densities

$$n \geq 0.13(c^6 m_{\rm H}/\sigma_R^6 G^3 M_J^4)^{1/5} \approx 2 \cdot 10^{17}(M_J/0.1\ M_\odot)^{-4/5}\ \rm{cm}^{-3}. \tag{2}$$

The contracting atomic gas is converted to molecular form at much smaller densities, so that a proper account of Rayleigh scattering should be done numerically. However, this cannot alter the conclusion that the first stars in the Universe might form before the production of metals and be of very low mass.

A full account of this work will be published elsewhere in a paper with Dr. V.F. Mukhanov.

THE SHAPE OF THE INVISIBLE HALO: N-BODY SIMULATIONS ON PARALLEL SUPERCOMPUTERS

M. S. Warren, W. H. Zurek
Theoretical Astrophysics, T-6, Mail Stop B288
Los Alamos National Laboratory
Los Alamos, New Mexico 87545
P. J. Quinn
Mount Stromlo and Siding Springs Observatories
The Australian National University
P.B. Weston Creek, A.C.T. 2611, Australia
J. K. Salmon
Caltech Concurrent Supercomputing Facility
California Institute of Technology, 206-49
Pasadena, California 91126

ABSTRACT

We study the shapes of halos and the relationship to their angular momentum content by means of N-body ($N \sim 10^6$) simulations. Results indicate that in relaxed halos with no apparent substructure:

i.) The shape and orientation of the isodensity contours tends to persist throughout the virialised portion of the halo.

ii.) Most ($\approx 70\%$) of the halos are prolate.

iii.) The approximate direction of the angular momentum vector tends to persist throughout the halo.

iv.) For spherical shells centered on the core of the halo the magnitude of the specific angular momentum is approximately proportional to their radius.

v.) The shortest axis of the ellipsoid which approximates the shape of the halo tends to align with the rotation axis of the halo. This tendency is strongest in the fastest rotating halos.

INTRODUCTION

We model the formation of collapsed structures in an Einstein-deSitter ($\Omega = 1$) universe. The initial power spectrum of density perturbations was:

$$P(k) = \begin{cases} A k_b^{-3}(\frac{k}{k_b})^{-2}, & \text{if } k > k_b \\ A k_b^{-3}(\frac{k}{k_b}), & \text{if } k < k_b \; . \end{cases}$$

The wavelength at which the crossover occurs is $2\pi/k_b \approx 1.5$ Mpc. This spectrum was chosen to investigate the inheritance of features on sub-galactic scales. The initial conditions were produced by Fourier transforming the initial complex amplitudes (chosen from a Gaussian random distribution) and using the result to perturb a 128^3 mesh of particles according to the Zel'dovich growing mode formalism.[1,2] The computational volume was a sphere of comoving radius 5 Mpc cut out from this cube. The initial conditions correspond to a redshift $z = 15$, and the simulation was terminated after expansion by a factor of 20. Vacuum boundary conditions were used, with a buffer region of 1.5 Mpc at the outer edge which was excluded from analysis. The power spectrum was normalized such that $\delta M/M$ was initially 0.1 at a scale of 1 Mpc. The equations of motion were solved in physical coordinates, with a time variable $p = t^{0.75}$. The Newtonian force law was smoothed at small scales by a Plummer kernel with a softening length (in physical coordinates) of 5 kpc. The total mass of the system was $3.5 \times 10^{13} M_{\odot}$ and 1097921 particles were used. All results are presented in units which assume a Hubble constant of 50 km sec^{-1} Mpc^{-1}.

Two aspects of the simulation depend greatly on the number of particles. The initial conditions are at best accurate down to a scale of the mean interparticle spacing. Also, analysis of other parameters of the halos is affected by $1/N^{1/2}$ noise. Because the final halos in this simulation contain $\sim 10^4$ particles, their parameters may be determined fairly reliably.

THE PARALLEL TREECODE

A modified version of the Barnes-Hut treecode[3] was used to simulate the dissipationless collapse of objects from the initial conditions. The basic method creates an oct-tree subdivision of space to group particles. The force is calculated by descending the hierarchical tree. This results in an execution time which scales with the number of particles as $N \log N$. The performance and accuracy of hierarchical tree methods in the cosmological context has been investigated by Bouchet and Hernquist.[4] In the original Barnes-Hut algorithm the size of the cell, s, is compared to the distance from the particle to the center of mass of the cell, d. If $s/d \leq \theta$ where θ is an adjustable tolerance parameter, the particles within the cell are treated as a single psuedo-particle. The results presented here use a treecode which differs in its definition of θ. The new parameter, θ_{fd}, uses the distance from the particle to the nearest point on the surface of the cell (face-distance) in place of d. This corrects a problem in the Barnes-Hut algorithm which occurs when a dense object enters the corner of a cell which has a significant mass concentration in the opposite corner.[5] We have used $\theta_{fd} = 1.6$, (roughly equivalent to a Barnes-Hut $\theta = 0.75$) with quadrupole cell interactions.

Our parallel code runs on several message passing, distributed memory parallel computers (hypercubes). This type of machine typically has between 32 and 1024 processors, each with speed and memory comparable to a typical workstation. Computations presented here were done on the 64 processor Caltech/JPL MarkIII hypercube. Each node of the machine contains two Motorola 68000 series microprocessors, along with a 64 bit Weitek floating point unit capable of nearly 3 million single precision floating point calculations per second. The lowest level of the force calculation routine was coded in Weitek assembly language to achieve this performance. Each processing unit has 4 million bytes of local memory, giving a total of 256 Mbytes of memory. The 900 timesteps in the simulation, consisting of 35×10^{12} floating point operations, were completed in about 60 hours of CPU time. The time needed per timestep for the calculation ranged from 190 sec. at $z = 15$, to 280 sec. at $z = -0.3$, demonstrating that the computing time is relatively insensitive to the amount of clustering in the system, in contrast to particle-particle particle-mesh methods.

The efficiency of a parallel program may be established by comparing it to an extrapolated version of a sequential computation. This quantifies the amount of processing speed lost to parallel overhead. The parallel overhead may be further broken down into load imbalance and communication overhead. The time spent sending messages between processors in the code is negligible, and the overall efficiency is close to 85% with 64 processors. This means the program runs only 15% slower than it would on a hypothetical sequential computer with computing power $P_{64} = 64 \times P_1$, where P_1 is the power of a single processor node. It is fortunate that the nature of the treecode algorithm, which is inefficient on vector supercomputers, fits very well within the scheme of "large-grained" parallelism.

In order to allow each processor to take part in the overall computation, a technique termed *domain decomposition* is used. Space is divided up into distinct subvolumes, with each processor responsible for updating the positions of each particle in its region. The straightforward approach of dividing space into a cubical mesh is not feasible, as some regions of space will have many more particles, making some processors do all the work, while others sit idle. The method actually used divides space into irregular cells which gives each processor an equal amount of work, so that the load imbalance inefficiency is minimized.[5]

The parallel treecode is still being developed, and there are many improvements yet to be made. The most pressing is the integration of smoothed particle hydrodynamics into the equations of motion to study how dissipation affects the formation of structure. Additional modification of the subdivision criterion will certainly improve accuracy, and a better integrator would also be welcome. The addition of two-body regularization and a locally variable timestep would also allow applications to globular clusters. There is also the inevitable improvement in parallel supercomputers. Parallel machines are now abreast of the vector supercomputers, and their performance is improving more rapidly.

The practice of quoting energy conservation as a test of accuracy in a cosmological simulation is ambiguous. For $\Omega = 1$ the total energy is very near zero, so comparing the error to this value is of little use. If the error is compared to the potential energy, it must be specified at a given time, relative to either the physical potential energy, or the cosmic potential energy. For this simulation, if the energy is defined in physical coordinates, the ratio of the error at the end of the simulation, divided by the potential energy at the end, was slightly less than 1%.

CHARACTERISTICS OF HALOS

Halos were chosen with a two part algorithm. First all regions in a 3D mesh with an overdensity of 2000 or more were identified. Then these regions were collected into groups that were separated by at least 300 kpc from a neighboring group. Particles within 300 kpc of the center of mass of the group were then collected as a single halo. The exact center of the halo was determined by the mean position of the 10 most tightly bound particles. Visual inspection of the 155 halos selected in this manner identified a few objects where substructure confused the algorithm, and these were eliminated from further analysis.

Using the methods above provided 140 halos to study, ranging in mass from slightly less than $2 \times 10^{10} M_\odot$ to $34 \times 10^{10} M_\odot$. A part of the simulation volume with about 20 halos is shown in Fig. 1a. The halo indicated by the small box is examined in further detail in Figs. 2 and 4. The volume represented in the upper left panel of Fig. 2 is 7×10^5 times smaller than that in Fig. 1a, yet it still contains almost 2000 particles.

(a.) Density profiles and rotation curves

Previous numerical results have indicated that the mass profiles of galactic halos are related to the power law exponent of the initial density fluctuations, and that the dark halos of galaxies carry a strong memory of their initial conditions.[9,10] One goal of this simulation was to look for these effects when using different power laws on small and large scales. The mass profiles of the halos may be established by plotting a rotation curve, defined by $v(r) = \sqrt{GM(r)/r}$, where $M(r)$ is the mass within radius r of the center of the halo. These are shown in Fig. 1b. It is

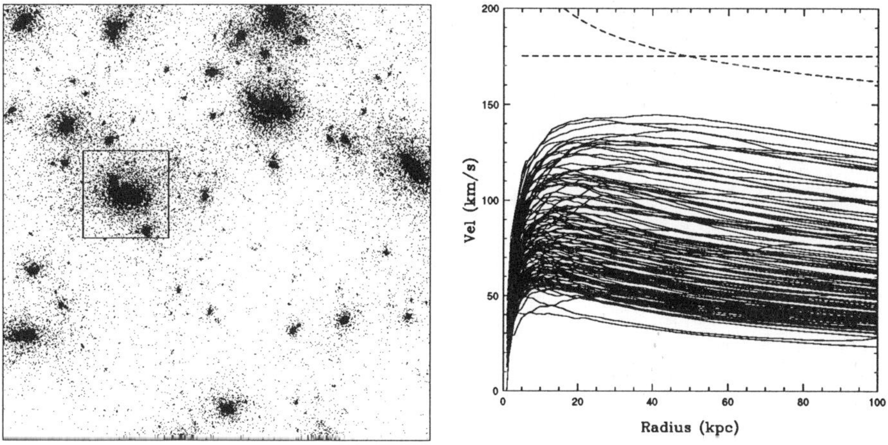

Figure 1. a.) A projected subsection volume 3 × 3 Mpc in area and 10 Mpc deep, which is ≈ 1/10 of the entire volume. Only one of every two particles are plotted. The small box indicates a halo, the time evolution of which is illustrated in Fig. 4. Its structure is analysed in Fig. 2.

b.) Rotation curves for the 140 most massive halos. The slope predicted by the secondary infall paradigm[6,7,8] for a pure Zel'dovich ($P(k) \sim k$) power law is shown by the downwardly sloping dotted line (normalized to 100 km sec^{-1} at 50 kpc, and offset vertically by 75 km sec^{-1}). The flat slope expected from $P(k) \sim k^{-2}$ is also indicated.[7,8,9]

immediately evident that the behavior of all the halos follows the same basic pattern. The slope predicted by the secondary infall paradigm[6,7,8] for a pure $P(k) \sim k$ power law is shown by the downwardly sloping dotted line in Fig. 1b. The slope of every relaxed halo is consistent with this slope at radii larger than around 50 kpc. Each halo formed from a region which was, on average, 1 Mpc across. The power spectrum for scales smaller than 1.5 Mpc was $P(k) \sim k^{-2}$, so it is interesting that the behavior of the slope for a large part of the halo seems to ignore the power on scales smaller than the initial extent of the particles in the halo. There also appears to be a trend for more massive halos to have a larger core, with a more extended flat region. This effect may be related to the merger history of the halo. It is hard to tell if there is a signature of the broken power law in the rotation curves. Further simulations should provide the comparisons necessary to resolve any systematic effects. It is not even completely clear if the soft cores, where the rotation curves are rising, are a real physical result, or the result of the softened potential or relaxation effects.

(b.) The shape of the halo

In order to determine the shapes and orientations of halos, a convergent method was used to find an ellipsoidal isodensity contour constrained to have a given enclosed volume, $V = (4\pi/3)abc$ where a, b, c are the minor, intermediate, and major axes of the ellipsoid, respectively. The method is equivalent to finding

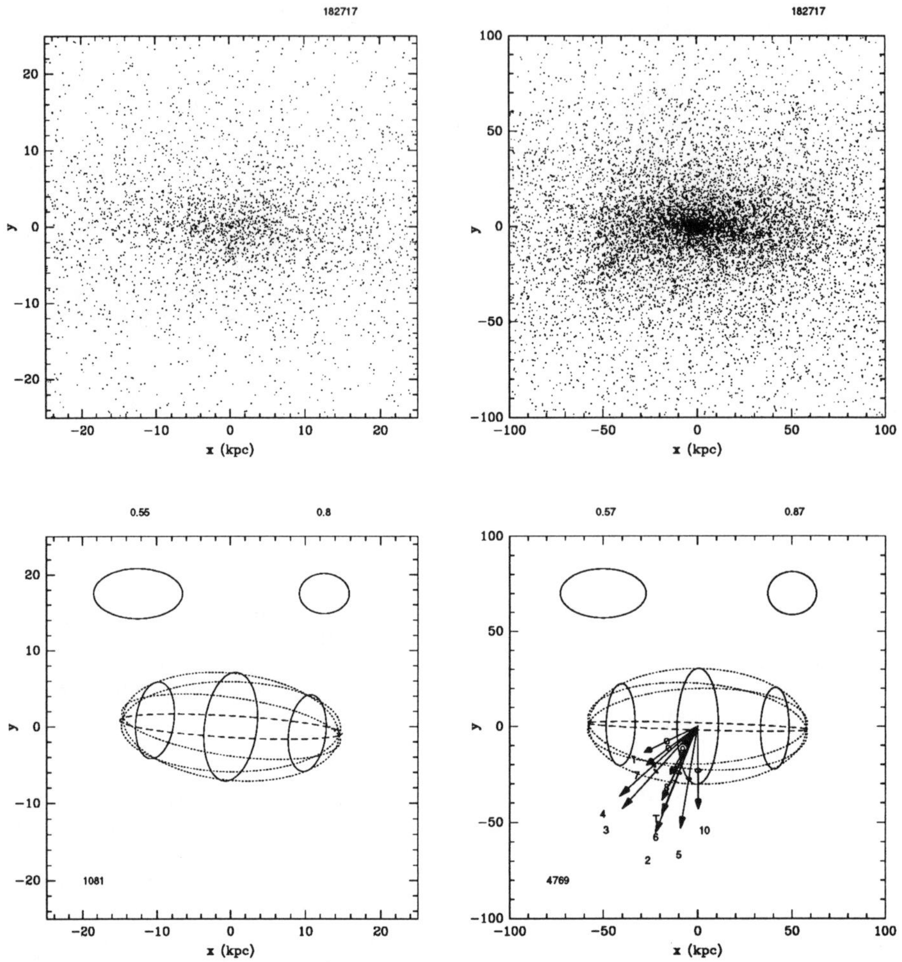

Figure 2. The top two panels show particles in a selected halo at two resolutions, a cube 50 kpc on a side, and a cube 200 kpc on a side. The lower panels show ellipsoids which are a best fit to an isodensity contour of the particle distribution. The ellipses indicate the projected view from along the minor and major axis of the ellipsoid. The arrows in the lower right panel indicate the direction of the angular momentum vector for particles in ten concentric spherical shells, with thickness 10 kpc. Note that the ellipsoids have the same general shape and orientation, and that the angular momentum vectors tend to point along the minor axis of the ellipsoid.

an ellipsoid which maximizes the number of particles contained within this fixed volume, centered about the previously defined center of the halo. The ellipsoids studied had volumes of spheres with radii of 10, 20, and 40 kpc. Plots were made for each halo to assess the alignment of the isodensity contours. In a large majority of the cases these contours maintain their basic shape and alignment. One representative example is shown in Fig. 2. The contours at 10 and 40 kpc are shown, along with the particle distribution at two resolutions.

If the ratios of the axes are plotted against each other (see Fig. 3a) about 2/3 of the halos lie above the line $e_1 = e_2$ ($e_1 = b/c$, $e_2 = a/b$). Purely prolate objects have $e_2 = 1$, and purely oblate have $e_1 = 1$. This result may be affected by halos which have not completely relaxed, as an object with two density centers would be identified as a prolate halo.

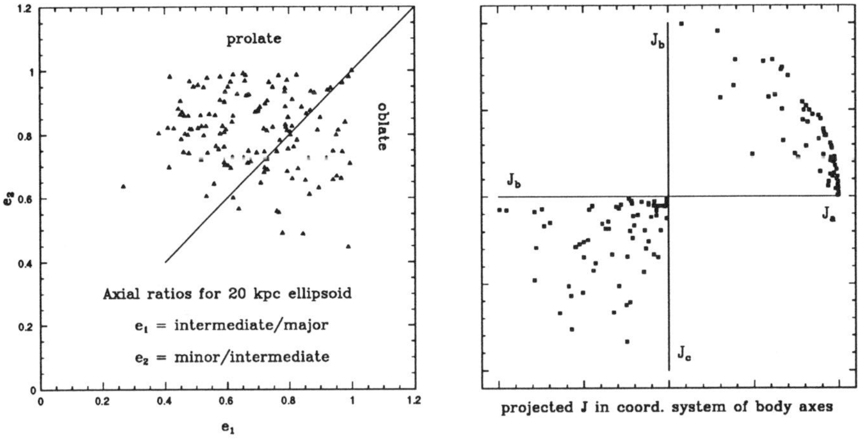

Figure 3. **a.)** The axial ratios of the 140 halos in the sample. Approximately 2/3 of the halos lie on the prolate side of $e_1 = e_2$.

b.) To demonstrate alignment of the angular momentum vector $\vec{J}$ with the minor axis $\vec{a}$ we have plotted $J_a = \hat{J} \cdot \hat{a}$ vs. $J_b = \hat{J} \cdot \hat{b}$ in the upper right corner ($\hat{J} = \vec{J}/|J|$, $\hat{a} = \vec{a}/|a|$, etc.). For a large fraction of the halos $\hat{J} \cdot \hat{a} \simeq 1$, that is, $\vec{J}$ and $\vec{a}$ are aligned. By contrast, in the plot of $J_b = \hat{J} \cdot \hat{b}$ vs. $J_c = \hat{J} \cdot \hat{c}$ in the lower left, most of the halos lie near the origin. That is, their angular momentum vector tends to be normal to the plane defined by the intermediate and major axes of the ellipsoid. Only halos with spin parameter $\lambda > 0.04$ were used in this plot.

(c.) Rotation and shape

The alignment of the angular momentum in the halos is striking. The vectors are correlated among themselves in a given halo, and also tend to align with the minor axis of the halo. As shown in Fig. 3b, the vectors tend to point along the minor axis, and virtually none point along the major axis. The symbols in Fig. 3b represent halos with a spin parameter larger than .04. Approximately 75% of these halos have their total angular momentum vector pointing within 30° of the minor axis. The effect is less pronounced for the the halos with $\lambda \leq 0.04$, but is still significant. Little correlation was seen between the degree of alignment and the eccentricity of the halo.

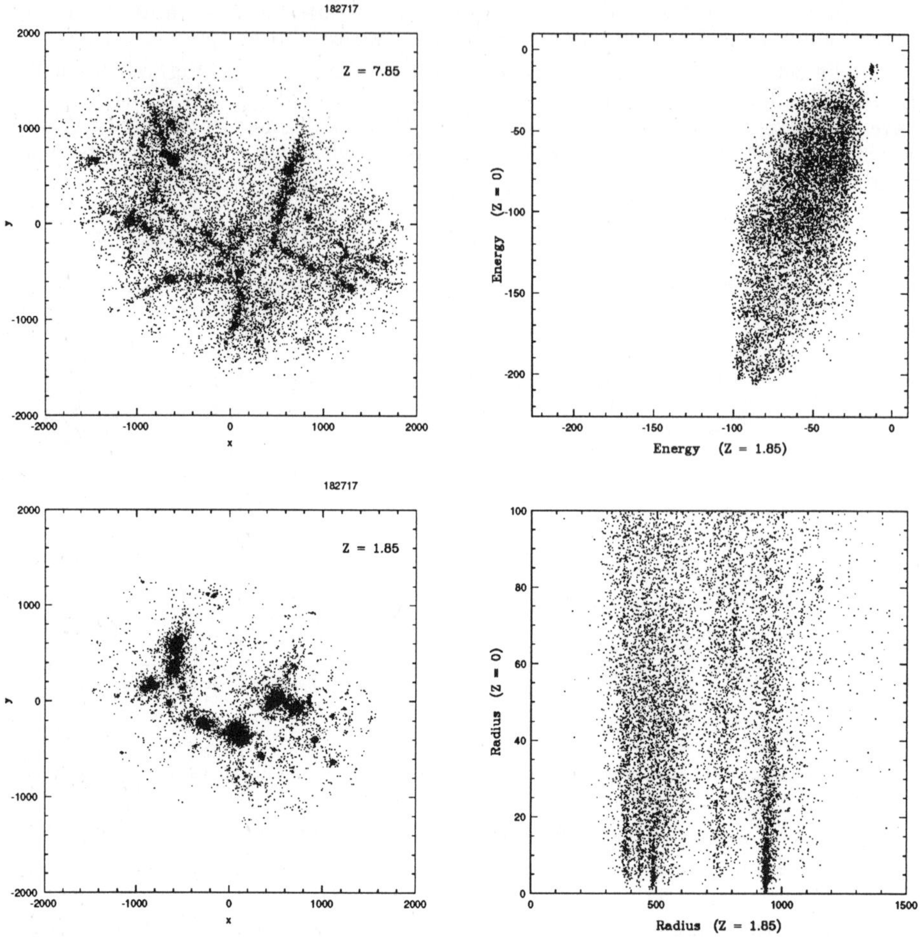

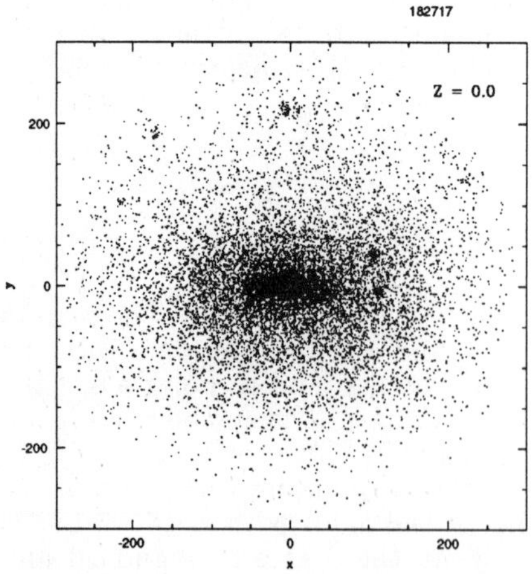

Figure 4. On the left are three snapshots, in comoving coordinates, of particles which end up in the selected final halo. On the upper right, a scatter plot of energy at $z = 1.85$ vs. energy at $z = 0$ suggests a correlation between the binding energies of particles at different epochs. The plot immediately above shows the distance from the center of mass at $z = 1.85$ vs. the final radius. In disagreement with the assumptions of the secondary infall paradigm, no correlation is seen between the initial and final locations of particles.

DISCUSSION

The evidence presented here supports the notion that halo formation is an organized process. The processes of tidal stripping and dynamical friction give rise to orderly deposition of mass, angular momentum and energy. The secondary infall paradigm[6,7,8] models collapse as a self-similar process, in the course of which shells are deposited radially onto an already virialised core. Its predictions are in accord with the density profiles found in the numerical experiment. However, its primary assumption (layer by layer infall onto the central peak) is in violent disagreement with the formation process observed in the simulations. Fig. 4 supports the view that the collapse process appears to be primarily due to the merging of already collapsed objects.[11,12] Quinn and Zurek[13] have addressed this issue, and have conjectured that the binding energy of the particles – and not their distance from the peak – is the important parameter. This also implies particles in the cores of progenitor clumps tend to end up in the core of the final halo.[11] Regularities we have observed in the numerical experiments may lead to analytic models which relate density profiles, angular momenta and other parameters of the collapsed halo to the initial cosmological or pre-merger conditions.

We would like to thank James Kuyper for writing the assembly code for the force calculation subroutine and the Caltech Concurrent Supercomputer Facility for providing time on the MarkIII hypercube. One of us (MSW) was supported in part with a pre-doctoral fellowship by a grant from IGPP.

References

1. Ya. B. Zel'dovich, *Astr. Ap.*, **5**, 84–89, 1970.
2. G. Efstathiou, M. Davis, C. S. Frenk, and S. D. M. White, *Ap. J. Suppl.*, **57**, 241–260, 1985.
3. J. Barnes and P. Hut, *Nature*, **324**, 446–449, 1986.
4. F. R. Bouchet and L. Hernquist, *Ap. J. Suppl.*, **68**, 521–538, 1988.
5. J. K. Salmon, *Parallel Hierarchical N-body Simulations*, Ph. D. Thesis, California Institute of Technology, 1991.
6. J. E. Gunn and J. R. Gott, *Ap. J.*, **176**, 1, 1972.
7. Y. Hoffman and J. Shaham, *Ap. J.*, **297**, 16–22, 1985.
8. J. A. Fillmore and P. Goldreich, *Ap. J.*, **281**, 1–8, 1984.
9. P. J. Quinn, J. K. Salmon, and W. H. Zurek, *Nature*, **322**, 329–335, 1986.
10. G. Efstathiou, C. S. Frenk, S. D. M. White, and M. Davis, *M.N.R.A.S*, **235**, 715–748, 1988.
11. W. H. Zurek, P. J. Quinn, and J. K. Salmon, *Ap. J.*, **330**, 519–534, 1988.
12. P. J. Quinn, W. H. Zurek, J. K. Salmon, and M. S. Warren, "The Formation of Halos via Mergers: the organized and organizing dynamics of mergers", in *Proceedings of 1989 Heidelberg Conference on Dynamics and Interactions of Galaxies*, ed. A. Toomre and R. Wielen, Springer-Verlag, 1990.
13. P. J. Quinn and W. H. Zurek, *Ap. J.*, **331**, 1–18, 1988.

Large Scale Structure from Galaxy Distributions

MAPS AND MOTIONS OF GALAXIES IN THE LOCAL UNIVERSE

Alan Dressler, The Observatories of the Carnegie Institution of Washington, 813 Santa Barbara St., Pasadena, CA 91101

ABSTRACT

The view of large-scale structure in the universe has evolved considerably over the last decade as the prevalence of large superclusters and voids has become more apparent. Highlights of the observational steps that led to this new picture are presented, including efforts to map the distribution of galaxies and, through their peculiar motions, the distribution of dark matter on large scales. Comparison of these two maps for the local volume show rough agreement, though it is still not clear if luminous galaxies trace the distribution of dark matter with any precision.

1. A NEW VIEW OF THE UNIVERSE

Describing the distribution of galaxies and dark matter in the local universe has become a principal goal of observational cosmologists in recent years. This surge in activity is largely hinged on the hope that large scale structure is a reflection of Big Bang physics, a legacy of particles and fields that characterize an energy regime beyond the reach of any earthbound experiment.

The linear relation between recession velocity and distance discovered by Hubble in 1929, based primarily on redshifts from V. Slipher, was a conceptual breakthrough in our understanding of the birth and evolution of the universe, but it also forged a powerful tool for making three-dimensional maps of galaxy distributions. Ironically, it was the consistency of the Hubble relation in various directions of space, established in the 1930's through the 1970's, that helped engrain the notion of a homogeneous, isotropic universe, a paradigm that discouraged the enormous effort that would be required to show otherwise. Reinforced by the discovery of the isotropic distribution of radio sources, the uniform Hubble flow entrained the picture of a quite smooth distribution of galaxies punctuated by rich clusters and groups. Sandage's claim that deviations from uniform Hubble flow were less than 100 km s^{-1}, based on the absence of blueshifts for nearby galaxies, reinforced this notion that the distribution of matter was remarkably smooth.

Thus it was that 30 years after the discovery of the importance of the redshift, this fundamental quantity had been measured for only about 500 galaxies, after 50 years, only about 2000. Until the 1970's, obtaining redshifts was tedious work that gobbled-up large amounts of scarce telescope time; combined with the expectation that redshift maps of a homogeneous, isotropic universe would be rather uninteresting, it is perhaps not surprising that few researchers had been willing to devote much time to the effort. But the proliferation of 2-meter to 4-meter telescopes in the 1970's, added to vast improvements in spectrograph/detector efficiencies, changed the nature of the effort to one where thousands of redshifts could be garnered in a few years with a telescope of modest aperture.

The first concerted efforts to map systematically a large volume of space, the Harvard-Smithsonian Center for Astrophysics (CfA) survey (Huchra *et al.* 1983) doubled the number of available redshifts, but it did little to dispel the prevailing notion of a relatively homogeneous universe. The theoretical underpinning came from Peebles (1980), who developed the model that gravitational instability had produced a hierarchical clustering of galaxies from a very homogeneous distribution left by the hot Big Bang. By the 1970's, measurements of the smoothness of the cosmic background radiation (CBR) were beginning to constrain the size of the initial fluctuations, though there was little clue as to their origin. Peebles took them as gaussian and randomly phased, with a "white noise" spectrum - all "initial conditions" put in by hand. As with the N-body simulations of cold and hot dark matter that Peebles' work would sire, the fluctuation amplitude was the principle adjustable parameter with which to reproduce clustering at the present epoch. The "figures of merit" to test the ability of the models to match observations like the CfA survey were the power-law slope and strength of the galaxy-galaxy (two-point) correlation function.

This image of an universe rapidly approaching homogeneity beyond the correlation length of $\sim 5h^{-1}$ Mpc was largely dissolved by the publication of the second CfA survey - the "slice of the universe" - by de Lapparent, Geller, and Huchra, in 1986. Its dense sampling of a large volume of space offered a new portrait of large-scale structure. With its $\sim 50h^{-1}$ Mpc voids and thin, arc-like chains of galaxies, the "slice" raised questions not only about the initial fluctuation spectrum, but even the viability of the bottom-up model of structure formation through gravitational instability. But, as with the fall of most paradigms, there is a strong tendency to credit the decisive blow and forget important, perhaps even compelling, evidence that had previously indicated the prevalence of large-scale inhomogeneity.

Fritz Zwicky and Gerard de Vaucouleurs were on the right track in the 1950's, Zwicky through his prodigious efforts at mapping the distribution over the northern sky and association of all galaxies into superclusters, de Vaucouleurs through his persistent promotion of the idea that our Galaxy sits on the outskirts of a coherent, flattened overdense region - the Local Supercluster. These ideas contradicted the picture of large-scale homogeneity, but they failed to win general acceptance because large numbers of redshifts were not available to defend against the counterclaim that superclustering was a result of projection effects, analogous to stellar constellations. As late as 1980, when Jim Peebles produced a stunning visualization of the Lick counts that showed a motley structure on large-scales, researchers (including Peebles) rejected the notion that these features were real.

The turning of the tide can be located at studies like that Tifft and Gregory (1976), who used new redshifts to argue that the Coma cluster was embedded in a halo of galaxies that extended well beyond what had traditionally been considered to be its boundary. They concluded: "When grouped by redshift and position on the sky, virtually all the galaxies are shown to belong to groups or clusters. Truly isolated galaxies are nonexistent or very rare." Tifft and Gregory also were the first to call attention to large voids, noting "The greater part of the foreground [of the Coma cluster] is completely devoid of galaxies." A somewhat deeper sampling by Gregory and Thompson (1978) went further, establishing a link from the Coma cluster all the way to the to the next closest rich cluster, Abell 1367 - a distance of $\sim 50h^{-1}$ Mpc. (This structure is part of what has been become known as the "Great Wall".)

The evidence for very large-scale inhomogeneities was markedly strengthened by the discovery of the Great Void in Boötes by Kirshner, Oemler, Schechter, and Shectman (1981). What made this a particularly significant discovery was that it was serendipitous - the survey was aimed at a random part of the sky rather than at one of the filamentary, high-density regions in the sky maps, like the Coma-A1367 or Perseus-Pisces Superclusters. (Since attention had been drawn to these regions by their appearance, the discovery of large-scale structure could be discounted as a selection effect.) The scale of the Great Void is approximately $100h^{-1}$ Mpc, though its shape is controversial and it is now known that it is not completely empty, but rather a region whose galaxy density is somewhere between 4 and 10 times below average.

Geller and Huchra (1989) have added two adjacent slices to the original CfA slice of the universe; this increase in volume has strengthened the case that this survey, too, is now sampling a typical region. There is considerable continuity across all three regions; in particular, they combine to show a long, filamentary structure named the "Great Wall" which has dimensions of roughly $200h^{-1}$ Mpc x $50h^{-1}$ Mpc, and a typical thickness of $10h^{-1}$ Mpc or less! It is not just the total size, but the thinness of such a structure, that challenges models based on the idea that gravitational instability is the dominant mechanism in the formation of large-scale structure. However, it must also be born in mind that these are maps in *redshift space*. Peculiar motions of order 500 km/sec, which appear to be fairly common, can falsely accentuate the thinness of structures like the Great Wall. Viewed perpendicular to the line-of-sight, a thicker structure, its Hubble expansion retarded by the gravity of its greater-than-average density, would appear significantly squeezed.

Margaret Geller consistently makes the point that all surveys have found a structure as large as they *could* find, *i.e.*, a coherent feature that stretches across the entire surveyed volume. To the extent that this is true, it has pressed the need for more extensive surveys, which, by the necessity of limited observational resources, have not been filled surveys like the CfA slice. Pencil beam surveys, which push a basically one-dimensional probe out to large redshifts, are one way to cover huge scales. The results of Broadhurst *et al.* (1990) have been particularly provocative. Not only has this survey appeared to confirm the prevalence of structures on a scale size of $100h^{-1}$ Mpc, the size of the Great Void and the Great Attractor (discussed below), but it has provided evidence for periodicity in the structure. For the first beam in particular, which traced a line from the north to south Galactic caps, both amplitude and phase coherence suggest some sort of an honeycomb-like structure with a modest variation in cell size around $100h^{-1}$ Mpc. Subsequent pencil beams by the group have shown greater variation, and the team has been cautious is claiming too much about the periodicity of the phenomena. However, they believe that the result of a "preferred scale" in large-scale structure is robust.

On the other hand, the early results of a huge new "fan" survey by Kirshner, Oemler, Schechter, and Shectman give little indication of any such cellular structure. When completed, their "unfilled" survey will contain some 10,000 redshifts (typical $z = 0.1$) in two areas. Already it would seem, however, that the distribution of galaxies in these enormous areas, each several times bigger than the CfA slices, is approaching homogeneity, *i.e.*, a structure as big as the entire survey is not readily apparent. Although this might have been forecast from the two-point correlation function and the isotropy of radio counts, the possibility that a fairly homogeneous map may be on the way can understandably be welcomed with a sigh of relief.

2. THE LOCAL VOLUME

The previous discussion carries with it a major assumption that is, as yet, relatively untested: that galaxies are accurate tracers of the dark matter. It is the distribution of this dark component, which contains at least an order-of-magnitude more mass than found in the stars and gas of galaxies, that is predicted N-body simulations based on weakly interacting particles.

The volume defined within 60 h^{-1} Mpc is particularly important in this regard because, even though it contains only a small amount of space compared to the surveys described above, for this volume it is possible to make a map of the <u>mass</u> distribution as well as the galaxy distribution. This is accomplished by measuring the deviations from uniform Hubble flow that are produced by inhomogeneities in the matter distribution, including both luminous and dark. Thus, in this smaller volume we can test the N-body models directly and, by comparing the distribution to fair galaxy samples, determine to what extent and how to use the distribution of galaxies alone to indicate the mass distribution, out to distances that are too great for accurate measurements of peculiar motions.

It is reassuring that this "local volume" includes the salient features seen on the larger scales - several large superclusters and voids. The Local Supercluster, with its remarkably flat distribution of galaxies roughly $25h^{-1}$ across, defines what de Vaucouleurs called the *Supergalactic Plane*. If extended around the sky, the Supergalactic plane intercepts two regions of high galaxy density: the Perseus-Pisces supercluster at a recession ($V \approx 5000$ km s^{-1} and Hydra-Centaurus supercluster ($V \approx 3000$ km s^{-1}) are prominent concentrations on nearly opposite sides of the sky. In addition, there is a sizeable void in front of the Perseus-Pisces supercluster. These structures show up well in optical surveys, but a good deal of attention in recent years has been focussed on the galaxy catalogs produced from the *IRAS* satellite. Advantages afforded by this sample are its uniformity and that the far-IR fluxes of these observations are much less susceptible to dust absorption in our Galaxy; principle disadvantages are that the far-IR flux is not particularly well correlated with galaxy mass or luminosity, and early morphological types (E, S0, Sa), the dominant components of high-density regions, are poorly represented.

Nevertheless, the distribution of *IRAS*-selected galaxies shows the same large-scale structure (Strauss and Davis 1988).

It is worth pointing out that redshift surveys over large sky areas containing the Perseus-Pisces (Giovanelli, Haynes, and Chincarini 1986) and Hydra-Centaurus, Pavo-Indus superclusters (the Great Attractor region, Dressler 1988, 1990) show the same sharp peaks in redshift space as has been seen in the pencil beam surveys by Broadhurst *et al.* All in all, it looks like we live in a fairly representative volume of space.

3. MEASURING PECULIAR MOTIONS

Not only can we measure the space positions of galaxies, we can also measure their *motions*. Although the major component of galaxy motion is due to the expansion of the universe, an overdensity in the matter distribution will impart additional *peculiar velocities*, deviations from pure Hubble flow, in its direction. Thus, galaxies will stream towards overdense regions and away from underdense regions. Averaged over large scales these motions are still small enough that the perturbations can be modeled with linear perturbation theory, which simplifies the recovery of the underlying mass distribution.

Rubin *et al.* (1976) made the first concerted effort to measure departures from uniformity in the Hubble flow by measuring the motions of hundreds of galaxies in various directions, galaxies chosen to be roughly equidistant from the Milky Way. They interpreted the asymmetry they measured in the velocity field as due to a motion of our own Galaxy of $V_{pec} \sim 500$ km s^{-1}, an unheard of speed according to conventional wisdom. All doubt that the Milky Way was, in fact, in rapid motion (although not in the direction that Rubin *et al.* had found), was soon removed when an unmistakable anisotropy was detected in the temperature of the cosmic microwave background radiation (Smooth, Gorenstein, and Muller 1977, Cheng *et al.* 1979). The detection of this dipole anisotropy, interpreted as the Doppler shift caused by a 600 km s^{-1} peculiar motion of our Galaxy (Smooth and Lubin 1979, Fixen, Cheng, and Wilkensen 1983), raised new questions. Over what range are other galaxies sharing this motion? To what are they streaming?

It is quite simple to measure galaxy motions, of course, at least the component toward or away from us. Spectroscopic observations of galaxies and stars in our own Galaxy are compared to obtain a redshift, assumed to be a measurement of the Doppler shift due to motion of the galaxy. Most of this measured velocity is due to the *expansion* of the universe; the peculiar motion is a small part that can be found only after subtracting the cosmic contribution. This step requires a knowledge of the distance to the galaxy, since, in an Hubble flow unperturbed by a inhomogeneous mass distribution, velocities are a linear function of distance.

Methods for finding distances amount to trying to predict the size or luminosity of a galaxy based on the measurement of some property that can be measured independent of distance. Tully and Fisher (1977) showed that for spiral galaxies the total luminosity is well correlated with galaxy rotation speed. The relation has been tested and calibrated using galaxies in clusters, for which the common distance of all objects means that relative *apparent* brightness indicates true relative luminosity. Tests on some dozens of clusters shows that rotation speed correlates with galaxy luminosity with an rms scatter of about 40%, which translates into a distance error of about 20% per galaxy. Such poor accuracy means that for a galaxy whose distance would correspond to a recessional velocity of 5000 km s^{-1} in a unperturbed Hubble flow, the subtraction of cosmic contribution will result in an uncertainty in the peculiar motion of order 1000 km s^{-1}. With present techniques, then, we must measure the motions of many galaxies in a particular region of space and combine the results to obtain the needed accuracy of a few hundred kilometers per second.

4. MAPS OF PECULIAR VELOCITIES

In 1982 the Tully-Fisher relation was applied to a sample of spiral galaxies in a fundamental study of the peculiar motions of galaxies in the Local Supercluster. Aaronson *et al.* (1982) measured to distances to several hundred spiral galaxies and subtracted the expansion field component from the observed recession velocity, leaving a field of peculiar motions. Their pioneering study showed a clear pattern of "infall" of galaxies toward the region of highest galaxy density.[1] This confirmed that the higher local density of galaxies was indeed outlining a region of higher *mass* density, and that the effect of gravity over the denser region could be seen. These first measurements of large-scale gravity confirmed that there is at least 10 times as dark matter as can be seen in the visible galaxies.

Davis and Peebles (1983) used this and other measurements when they emphasized the importance of comparing the Milky Way's motion of ~600 km s^{-1} with respect to the CBR to *local* measurements of peculiar motion. More than a dozen studies had measured the peculiar motion toward Virgo at that time; they had found values ranging from 0 - 500 km s^{-1}, to compare with the ~400 km s^{-1} component of Local Group motion with respect to the distant CBR. Davis and Peebles favored the high values, but some of the best determinations of peculiar motion, like the Aaronson *et al.* work, indicated that only 50-75% of the CBR-referenced motion toward Virgo could be attributed to motion that originated within the Local Supercluster (Dressler 1984). Furthermore, this single component is less than half of the total motion indicated by the CBR dipole. These measurements suggested that major sources of gravitational acceleration lay beyond the Local Supercluster, an important clue that mass inhomogeneities were on a very large

---

[1]Actually, the Hubble expansion velocity is larger than the infall, so it is more correct to say that expansion of the Local Supercluster is *retarded* relative to the average expansion rate for the universe. Except for the immediate vicinity of the Virgo and Ursa Major clusters, galaxies are still receding from each other. The Aaronson *et al.* study modeled the infall as a spherical pattern with a fall in peculiar velocity proportion to 1/r, but Burstein (1990) has emphasized that the data are better matched by a shear flow with velocity directly proportional to distance from a "spine" running between the Ursa Major and Virgo clusters.

scale.

A map of peculiar velocities for a larger volume of space became one of the goals of the *Seven Samurai*, the collective name given to a group composed of Burstein, Davies, Faber, Lynden-Bell, Terlevich, Wegner, and myself, who were studying the intrinsic properties of elliptical galaxies. The Seven Samurai measured the velocity dispersions of some 400 elliptical galaxies selected by uniform criteria over the entire sky (Davies *et al.* 1986, Faber *et al.* 1989) Like rotation speed, velocity dispersion a is measured spectroscopically, with the width of absorption lines indicating the characteristic stellar speed. For each galaxy we also made *photometric* observations (Burstein *et al.* 1986), recording the total light intensity from the object and its the run of intensity with radius. It was discovered that in a three-space of velocity dispersion, total brightness, and average (surface) brightness, the distribution of elliptical galaxies is nearly planar (Dressler *et al.* 1987a, Djorgovski and Davis 1987). This allowed the construction of new distance indicator which, like the Tully-Fisher relation, used a distance independent parameter, velocity dispersion, to predict a linear combination of parameters that are distance-dependent. The accuracy of this method for predicting distances was found to be about the same as the method for spirals, about 20%.

When the method was applied to the sample of elliptical galaxies, the result was surprising. A map of the peculiar velocities exhibited a large-scale flow pattern over a region nearly 5 times the size of the Local Supercluster (Dressler *et al.* 1987b). The Samurai confirmed that galaxies within the Local Supercluster streaming along with the Local Group, and that there was no obvious end to the flow, as would be expected if the source of the gravitational pull had been reached. This result was described simply as "bulk flow", though later we became convinced that the data would allow extraction of terms higher than the dipole. In particular, it seemed quite significant that galaxies in the direction of the flow out ahead of the Milky Way were moving more rapidly, typically at 1000 km s^{-1}. This included the rich association of Centaurus clusters, previously considered a likely source of the flow, now found to be racing ahead to some more distant gravitational center. Conversely, galaxies "at our back" were observed to be moving in the same direction but more slowly than the Milky Way and its companions.

This gradient was easily interpreted in terms of a quadrupole contribution to the velocity field, the expected effect of the pull of a distant mass concentration. This quadrupole term had previously been discovered in the velocity field of the Local Supercluster (Lilje, Yahil, and Jones 1986). This evidence encouraged us to model the flow pattern expected for a gravitational source just outside the well-sampled volume. We found that a distributed, spherical mass in the direction Galactic longitude = 307°, Galactic latitude = 9°, whose center lay at a distance of 4350 km s^{-1} from our Galaxy, would induce a peculiar velocity field that matched the observations in surprising detail (Lynden-Bell *et al.* 1987). Normalizing the model to the observed peculiar velocities implied a mass of $\sim$3-5 x 10^{16} solar masses - equivalent to a huge supercluster. The introduction of this "Great Attractor (GA) had greatly reduced the chi-square with the addition of a handful of free parameters.

An all-sky galaxy map (Fig. 8 of Lynden-Bell *et al.*) of the galaxy distribution in this part of the sky, produced by Ofer Lahav from three galaxy catalogs, showed that the GA region, though partly obscured by dust from the plane of our own galaxy, is the "most luminous" part of the sky. Apparently a substantial overdensity of galaxies accompanies the mass overdensity revealed by the measurements of peculiar motions. At least in a rudimentary sense, galaxies were found to trace the hidden mass.

The GA model offered a serious challenge for the cold-dark-matter (CDM) paradigm. N-body simulations by Bertschinger and Juskiewicz (1988) showed that such a massive and extensive concentration was extremely improbable, especially in the prevalent model where galaxies were assumed to cluster more strongly than the dark matter, (*biasing* factor $b = 2.5$). They pointed out, however, that the GA *model* was not well constrained by the data then available, and a number of less extreme mass distributions would also fit the data. Further tests of the GA model could clearly provide a strong constraint on CDM.

New measurements of peculiar motions in the GA region have become available. Large-amplitude peculiar motions were found in the GA region by Aaronson *et al.* (1989) for cluster spirals, and by Lucey and Carter (1988) for cluster ellipticals. Although these independent observations were generally supportive, they also uncovered anomalous motions that suggest that, not surprisingly, the GA model is an oversimplification.

Sandra Faber and I spent three more observing seasons at the Carnegie Institutions's Las Campanas Observatory and obtained data for 134 elliptical galaxies in the GA region and 117 spiral galaxies, both in and out of clusters (Dressler and Faber 1990a,b; Burstein, Faber, and Dressler 1990) These new data not only represented a manifold increase in the number of galaxies in the region with peculiar velocities, but also probed further into and beyond the putative center of the GA. This is crucial because the unambiguous signature of a region in gravitational contraction is a change in sign of peculiar velocities as the center of mass is crossed. The new data show just such an effect. Compared to the straight line that indicates a uniform Hubble expansion, galaxies on the near side of 4000 km s^{-1} moving too fast for their distance, *i.e.*, they have a net positive peculiar motion. Beyond this the peculiar motions drop rapidly, crossing zero at a distance of approximately 5000 km s^{-1}, and they turn negative beyond. Although the most distant points are subject to greater noise and increasing systematic effects, it would appear that galaxies on the far side of the Great Attractor are seen "falling back". A completely independent data set for spirals by Mathewson and collaborators (private communication) shows the same pattern. The full S-wave pattern, so named because the data describe a full oscillation around the Hubble line, is the expected signature for a region in which the Hubble expansion has been slowed. Hubble diagrams done in exactly the same way for galaxies in other directions in the sky show no such evidence for a large gravitational distortion.

5. DO GALAXIES TRACE THE MASS?

Though the GA model is undoubtedly an oversimplification, the data do seem to bear out the notion that this one large overdensity dominates the velocity field over a region of space as large as $100h^{-1}$ Mpc across. This is still a small volume of the universe, much less than 1%, but it does represent a significant reach to the largest structures yet discovered. Though of similar scale to the Great Wall and other prominent superclusters, the Great Attractor is unique in that it is the first large-scale feature mapped in the *mass* distribution, the more fundamental quantity of large-scale structure.

It is important to emphasize that, although it was clear from 1986 that the Great Attractor might amount to nothing more than the dark matter associated with several superclusters of galaxies in this region of the sky, without an extensive and complete galaxy map this could not be concluded without prejudice. Thus, the worth of the GA model was to describe the mass distribution *independently* of the galaxy distribution, as a step toward making an unbiased assessment of whether galaxies actually trace the distribution of dark matter. Recently, the British press carried a report of the "death of the Great Attractor", citing a paper by Rowan-Robinson *et al.* (1990) in which they conclude that luminous galaxies in the region, if embedded in a dark matter distribution at the closure density, are sufficient to explain the peculiar motions of galaxies. This work, though an important step in comparing the maps of galaxies and mass, misrepresented the concept of the Great Attractor as due to dark, mysterious matter unassociated with galaxies, and thus failed to emphasize the important point that, based on their survey, light appears to trace mass over the volume of space studied.

Because present techniques limit the accuracy of distance measurements to ~20% per galaxy, it is difficult to determine peculiar velocities in the more distant Great Wall in order to see if a similar underlying *mass* structure is present. No doubt there are many other "great attractors" that are just beyond the reach of present techniques. To compare the distribution of light and mass, then, we must focus our attention on the smaller volume within 6000 km s^{-1}.

Comparisons have been made to galaxy samples done with catalogs using visible and far-infrared light, the latter from observations made with the *IRAS* satellite. Redshifts for several thousand galaxies detected by *IRAS* have been used to construct galaxy maps, with the assumption that this subset is representative of *all* galaxies within the volume (Strauss and Davis 1988, Rowan-Robinson 1989) The results vary somewhat between two different *IRAS* samples, and also for the optically-selected sample, but a basic feature of all surveys is that there are two main galaxy concentrations in the volume, one associated with the Great Attractor and another identified as the Perseus-Pisces supercluster. Estimates on the relative strength of these concentrations range from near parity to a 2:1 ratio in favor of the Great Attractor.

The issue of whether light traces mass in the volume may rest on whether the Perseus-Pisces supercluster is as rich in galaxies as the Great Attractor region. There is a fairly striking dissimilarity between the map of the mass distribution calculated from the velocity field using the POTENT method by Bertschinger and Dekel (1988) compared to one of the IRAS samples. The galaxy map implies that we are caught midway between two comparable overdensities, yet the peculiar velocity field shows no such bimodality - a single mass feature appears to dominate. However, the "one-in-six" sparse IRAS sample done by the Rowan-Robinson group shows the effect less strongly (Perseus-Pisces is less prominent), thereby muddling the issue. Furthermore, data for peculiar motions in the Perseus-Pisces region remain sparse, particularly in the densest concentrations in the direction of the Perseus constellation.

It is crucial to realize that studies using galaxy surveys alone to find the sources of the 600 km s^{-1} Local Group motion with respect to the CBR suffer two serious drawbacks. First, they *assume* that light traces mass; second, by fitting only the dipole term, they are unable to locate the distance of these sources. Shot noise and small-number statistics for distant galaxies in the *IRAS* samples result in very poor knowledge of the scale over which the dipole originates. On the other hand, using the full map of peculiar velocities has the great advantage that gradients in the velocity field locate the sources exactly through a local measurement. It is through the comparison of the full velocity field and the local galaxy counts that one can meaningfully address the questions of whether light traces mass and the value of Ω for clustered matter.

Recently, Willick (1990) has presented preliminary evidence that galaxies on the Pisces side of the supercluster exhibit only a minor infall pattern - less than half as strong as in the GA - and that the entire Pisces region is moving in the general direction of the Local Supercluster/Great Attractor at some 400 km s^{-1}. If this result holds up, and is bolstered by measurements now being processed for galaxies in the Perseus region, it will exacerbate the difference in the distributions of light and mass. Furthermore, since the pull of the Great Attractor, even when combined with the Local Supercluster, should be too weak to pull Perseus-Pisces as well, such a result may imply that a significant part of the motion may arise from a yet more distant, even greater attractor (Scaramella *et al.* 1989, Raychadbury 1989).

6. FUTURE RESEARCH DIRECTIONS

Within the next five years it will be possible to assemble a truly representative galaxy survey and to extend measurements of peculiar motions to thoroughly test the degree to which galaxies trace the mass distribution, and the degree to which the mass distribution is dominated by large concentrations like the Great Attractor. If it turns out that galaxies are good tracers of the underlying mass distribution, we can proceed with confidence to study large-scale structure solely from the more-easily obtained distribution of galaxies. However, the early evidence may offer some reason to believe that galaxies trace the mass poorly, and in a way that may not be easily predictable. Such an important claim requires much better proof, but if true, reading and interpreting large-scale structure of the universe solely from galaxy maps will be significantly complicated or even blocked. On the other hand, discovering that galaxies don't follow precisely the distribution of dark matter would be a powerful clue to the question of how galaxies formed and the seeding of structure.

Observations of many types and from many sources continue to indicate that vast organizational patterns, at least 100 Mpc across, are a common feature of the universe. On the other hand, observations the CBR radiation continue to paint a picture of a remarkably smooth young universe. For example, recent measurements of temperature fluctuations on scales of 0.1 - 1.0 degrees set upper limits for $\Delta T/T < 3 \times 10^{-5}$ (Partridge 1989). Yet model-independent, robust calculations (Bertschinger, Gorski, and Dekel 1990) show that the fluctuation that grew into the Great Attractor would have generated a temperature fluctuation $\Delta T/T \geq 1 \times 10^{-5}$: photons lose this much energy as they climb out of the gravitational potential well of the proto-Great Attractor. Failure to have yet detected the fluctuations that grew into galaxies, clusters, and superclusters still leaves acceptable models, but the "squeeze" is on (Readhead *et al.* 1989, Meinhold and Lubin 1990). The extent to which these two types of observations are in conflict represents the greatest challenge to those who model the growth of structure in the early universe.

REFERENCES

Aaronson, M., Huchra, J, Mould, J, Schechter, P. L., and Tully, R. B. 1982, *Ap.J.*, **258**, 64.

Aaronson, A. *et al.* 1989, *Ap.J.* **338**, 654.

Bertschinger, E., and A. Dekel 1989, *Ap.J.Letters*, **336**, L5.

Bertschinger, E. Gorski, K. M., and Dekel, A. 1990, *Nature* **345**, 507.

Bertschinger, E., and Juskiewicz, R. 1988, *Ap.J.Letters* **334**, L59.

Broadhurst, T. J., Ellis, R. S., Koo, D. C., and Szalay, A. S. 1990, *Letters to Nature*, **343**, 726.

Burstein, D. 1990, *Rept. Prog. Phys.* **53**, No. 4.

Burstein, D., Davies, R. L., Dressler, A., Faber, S. M., Stone, R. P. S., Lynden-Bell, D., Terlevich, R. J., and Wegner, G. 1987, *Ap.J.*, **64**, 601.

Burstein, D., Faber, S. M., and Dressler, A. 1990, *Ap.J.*, **354**, 18.

Cheng, E. S., Saulson, P. R., Wilkenson, D. T., and Corey, B. E. 1979, *Ap.J.Letters* **232**, L139.

Davies, R. L., Burstein, D., Dressler, A., Faber, S. M., Lynden-Bell, D., Terlevich, R. J., and Wegner, G. 1987, *Ap.J.Suppl.*, **64**, 581.

Davis, M., and Peebles, P. J. E. 1983, in *Annual Reviews of Astronomy and Astrophysics*, **21**, 109.

de Lapparent, V., M. J. Geller, and J. P. Huchra 1986, *Ap.J.Letters*, **302**, L1.

Djorgovski, S., and Davis, M. 1987, *Ap.J.*, **313**, 59.

Dressler, A. 1984, *Ap.J.*, **281**, 512.

Dressler, A. 1988, *Ap.J.*, **329**, 519.

Dressler, A. 1990, *Ap.J.Suppl.*, in press.

Dressler, A., Lynden-Bell, D., Burstein, D., Faber, S. M., Terlevich, R. J., and Wegner, G. 1987a, *Ap.J.*, **313**, 42.

Dressler, A., Faber, S. M., Burstein, D., Davies, R. L, Lynden-Bell, D., Terlevich, R. J., and Wegner, G. 1987b, *Ap.J.Letters*, **313**, L37.

Dressler, A., and Faber, S. M. 1990a, *Ap.J.* **354**, 13.

Dressler, A., and Faber, S. M. 1990b, *Ap.J.Letters*, **354**, L54.

Faber, S. M., Wegner, G., Burstein, D., Davies, R. L, Dressler, A., Lynden-Bell, D., and Terlevich, R. J. 1989, *Ap.J.Suppl.*, **69**, 763.

Fixsen, D. J., Cheng, E. S., and Wilkenson, D. T. 1983, *Phys.Rev.Letters* **50**, 620.

Geller, M. J., and Huchra, J. P. 1989, *Science*, **246**, 897.

Giovanelli, R., M. P. Haynes, and G. L. Chincarini 1986, *Ap.J.*, **300**, 77.

Gregory, S. A., and Thompson, L. A. 1978, *Ap.J. 222*, 784.

Kirshner, R. P., Oemler, A., Schechter, P. L., and Shectman, S. A. 1981, *Ap.J.Letters*, **248**, L57.

Huchra, J. P., Davis, M., Latham, D., and Tonry, J. 1983, *Ap.J.Suppl.* **52**, 89.

Lilje, P., Yahil, A., and Jones, B. J. T. 1986, *Ap.J.* **307**, 91.

Lucey, J. R., and Carter, D. 1988, *M.N.R.A.S* **235**, 1177.

Lynden-Bell, D., Faber, S. M., Burstein, D., Davies, R. L., Dressler, A., Terlevich, R. J., and Wegner, G 1988, *Ap.J.* **326**, 19.

Meinhold, P., and Lubin, P. 1990, *Ap.J.Letters*, in press.

Partridge, R. B. 1989, in *Astronomy, Cosmology, and Fundamental Physics*, (Kluwer, Dordrecht), 105.

Peebles, P. J. E. 1980, *The Large Scale Structure of the Universe*, (Princeton: Princeton University Press).

Raychadbury, S. 1989, *Letters to Nature*, **342**, 251.

Readhead, A. C. S., Lawrence, C. R., Meyers, S. T., Sargent, W. L. W., Hardebeck, H. E., and Moffet, T. 1989, *Ap.J.* **346**, 566.

Rowan-Robinson, M. 1989, in *Large-Scale Structure and Motions in the Universe*, (Kluwer, Dordrecht), 19.

Rowan-Robinson, M. *et al.* 1990, *M.N.R.A.S.* **247**, 1.

Rowan-Robinson, M. in {\it Large-Scale Structure and Motions in the Universe}, 19-39, (Kluwer, Dordrecht, 1989)}

Rubin, V. C., Ford, W. K. Jr., Thonnard, N., Roberts, M. S., and Graham, J. A. 1976, **81**, 687.

Scaramella, R., Baiesi-Pillastrini, G., Chincarini, G., Vettolani, G., and Zamorani, G. 1989, *Nature* **338**, 562.

Smoot, G. F., Gorenstein, M. V., and Muller, R. A. 1977, *Phys.Rev.Letters*, **39**, 898.

Smoot, G., and Lubin, P. M. 1979, *Ap.J.Letters* **234**, L83.

Strauss, M. A., and Davis, M. 1988, in *Large Scale Motions in the Universe: A Vatican Study Week*, (Princeton University Press, Princeton (1989), 255.

Tifft, W. G., and Gregory, S. A. 1976, *Ap.J.* **205**, 696.

Tully, R. B., and Fisher, R. 1977, *Astr.Ap.*, **54**, 661.

Willick., J. A. 1990, *Ap.J.Letters* **351**, L5.

REDSHIFT SURVEYS

Margaret J. Geller and J.P. Huchra

Harvard-Smithsonian Center for Astrophysics

I. INTRODUCTION

Between 1976 and 1990, the number of galaxies with measured redshifts has risen steeply from 2,700 to more than 40,000. A field once data poor is becoming data rich. The new data are changing our perception of large-scale structure in the universe.

The largest project nearing completion — the Center for Astrophysics (CfA) redshift survey (Section II)— indicates that bright galaxies are distributed on thin surfaces — two-dimensional structures— which surround (or nearly surround) vast voids. Recent completion of a southern hemisphere survey (da Costa *et al.* 1988) gives some support to this picture. The narrow peaks in the deep surveys by Broadhurst *et al.* (1990) may be the intersection of the probes with structures similar to the "Great Wall" seen in the CFA survey (Section III) (Geller and Huchra 1989) and to the sheet in the Perseus-Pisces region. The message of all surveys is now clear: large structures are a *common* feature of all surveys big enough to contain them.

We take $H_o = 100h$ km $s^{-1}Mpc^{-1}$ with h = 1 unless otherwise specified.

II. THE LARGE-SCALE GALAXY DISTRIBUTION

Over the last few years each new approach to mapping out the distribution of individual galaxies has uncovered unexpectedly large structures. Surveys like the one of the Boötes (KOSS 1987) region are an efficient method of finding large voids. Surveys like the CfA redshift survey extension (Geller and Huchra 1989) which are complete over a region of large angular scale are less efficient for identifying individual large structures, but they are necessary for quantitative characterization of the distribution of galaxies over a range of scales.

The goal of the Center for Astrophysics redshift survey extension is to measure redshifts for all galaxies in a merge of the Zwicky *et al.* (1961-1968) and Nilson (1973) catalogs which have $m_{B(0)} \leq 15.5$ and $| b_{II} | \geq 40$ °. There will be about 15,500 galaxies in the complete survey; more than 11,500 of these already have measured redshifts. The survey is now very nearly complete in 5 slices of the northern Galactic Hemisphere (a "slice" is a 6°-wide strip at constant declination) as well as 5 slice in the southern Galactic hemisphere. The northern strips cover the right ascension range $8^h \leq \alpha \leq 17^h$ and the declination ranges $20.5° \leq \delta < 44.5°$ and $8.5° \leq \delta < 14.5°$ (see Geller and Huchra 1989).

In the southern Galactic hemisphere a 30° -wide region (5 "slices") covering $6° \leq \delta > 36°$ with $0^h \leq \alpha \leq 4^h$ and $20^h \leq \alpha \leq 24^h$ is complete. More than 75% of the redshifts were measured with the 1.5-meter telescope and the MMT at Mt. Hopkins. The mean external error in the redshift measurements is ~35 km s^{-1} .

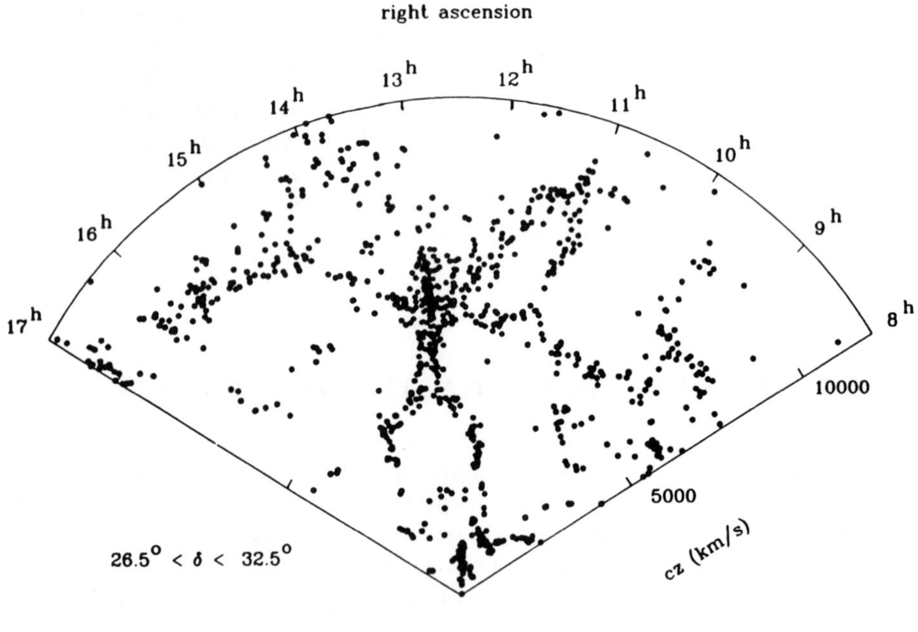

Figure 1: Observed velocity versus right ascension for the survey strip

centered at $\delta = 29.5°$. The strip is 6° in declination. Only the galaxies with velocities $\leq$ 15,000 km s^{-1} are shown.

Figure 1 is a plot of the observed velocity versus right ascension for the first strip centered at 29.5° (de Lapparent, Geller, and Huchra 1986): the strip is 6° wide in declination. The plot includes only the 1067 galaxies with redshifts $\leq$ 15,000 km s^{-1}. A galaxy of characteristic luminosity, M^*, is at roughly 9,000 km s^{-1} in this survey. Nearly every galaxy in this slice lies in an extended thin structure. The boundaries of the low density regions are remarkably sharp. The only pronounced velocity finger in this slice is the Coma cluster at $\sim 13^h$; however, many groups of galaxies are embedded in the structures (Ramella, Geller, and Huchra 1989).

Several of the voids are surrounded by thin structures in which the intergalaxy separation is small compared with the radius of the empty region. The density of the survey is clearly important for definition of the structure. The largest low density region in this survey is located between 13^h20^m and 17^h with $4000 \lesssim cz \lesssim 9000$ km s^{-1} . The diameter of the void is $\sim$ 5000 km s^{-1} or 50 Mpc in the absence of large-scale flows. This void is comparable with the void in Boötes but is much more densely surveyed. The galaxy number density contrast for this void is $\delta\rho/\rho \simeq -0.8$, again comparable with the estimates for the Boötes void. Recent infra-red Tully-Fisher distances (Bothun *et al.* 1991) indicate that the diameter of the void is the same in physical space as in redshift space.

The first slice alone demonstrates that the thin structures in the distribution of galaxies are cuts through two-dimensional sheets, *not* one-dimensional filaments. A total of four complete slices covering the declination ranges $26.5° \leq \delta < 44.5°$ and $8.5° \leq \delta < 14.5°$ support the *qualitative* picture suggested by the first slice. Figure 2 shows a 3-D representation of these four slices. The most pronounced structure in this slice — the band of galaxies running across the entire right ascension range at velocities between 7500 km s^{-1} and 10,000 km s^{-1} — is the "Great Wall". About half of the galaxies in the survey are in this feature.

The apparent extension of the "Great Wall" in both right ascension and declination is limited only by the extent of the survey; we do not yet know the full extent of the structure. The equivalent spatial extent in these two dimensions is $\sim 60h^{-1}$ Mpc $\times$ 170 h^{-1} Mpc. The typical thickness of these sheets (approximately along the redshift direction) is $\lesssim 5h^{-1}$ Mpc FWHM. The density contrast between this wall and the mean for the survey is $\Delta\rho/\rho \simeq 5$. If we take $\Omega \simeq 0.2$, the mass in this observed portion of the Great Wall is $\sim 2 \times 10^{16} M_{\odot}$.

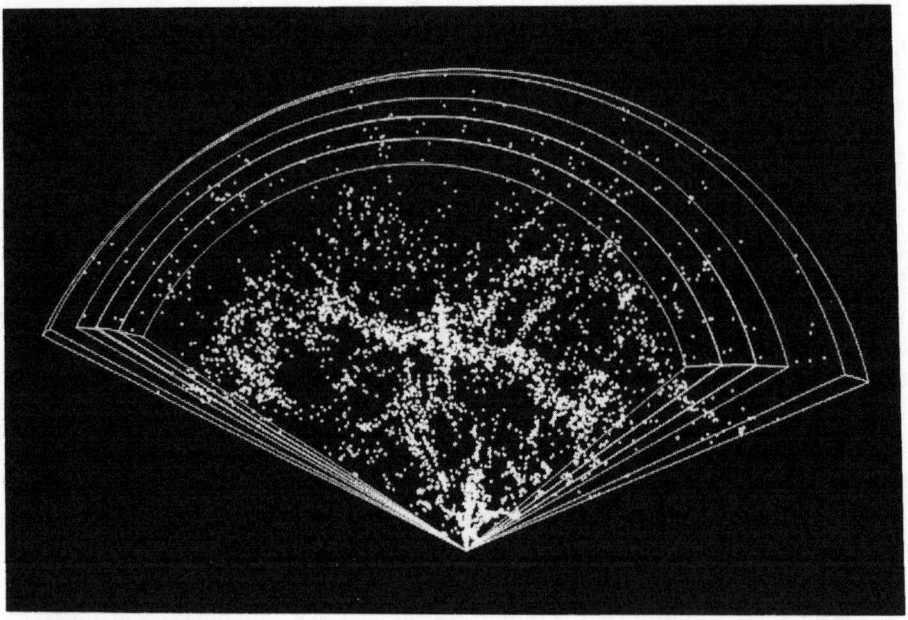

Figure 2: Projection of a three-dimensional display of four complete slices of the redshift survey. These slices cross the northe Galactic pole. The large feature running nearly horizontally throughout the survey is the "Great Wall".

The CfA survey is also complete in a portion of the region covered by the Arecibo surveys. Individual structures appear to be smaller in this region than in the North; however, the general character is similar. The similarity of the structure revealed by these surveys underscores the insensitivity of large-scale structure to the properties of individual galaxies. Figure 3 shows the 30°-wide "slice" which spans the declination range $6° < \delta \leq 36°$ and the right ascension range $20^h \leq \alpha \leq 4^h$. Comparison of these plots with Figures 2a—c in Haynes and Giovanelli (1986) shows that the structure in the 21-cm surveys are the same as those in the magnitude limited optical survey. The sampling density in the optical survey is greater than in the 21-cm survey by Giovanelli and Haynes, and the effective depth of the optical survey to $m_{B(0)} \leq 15.5$ is somewhat greater than the depth of the 21-cm surveys.

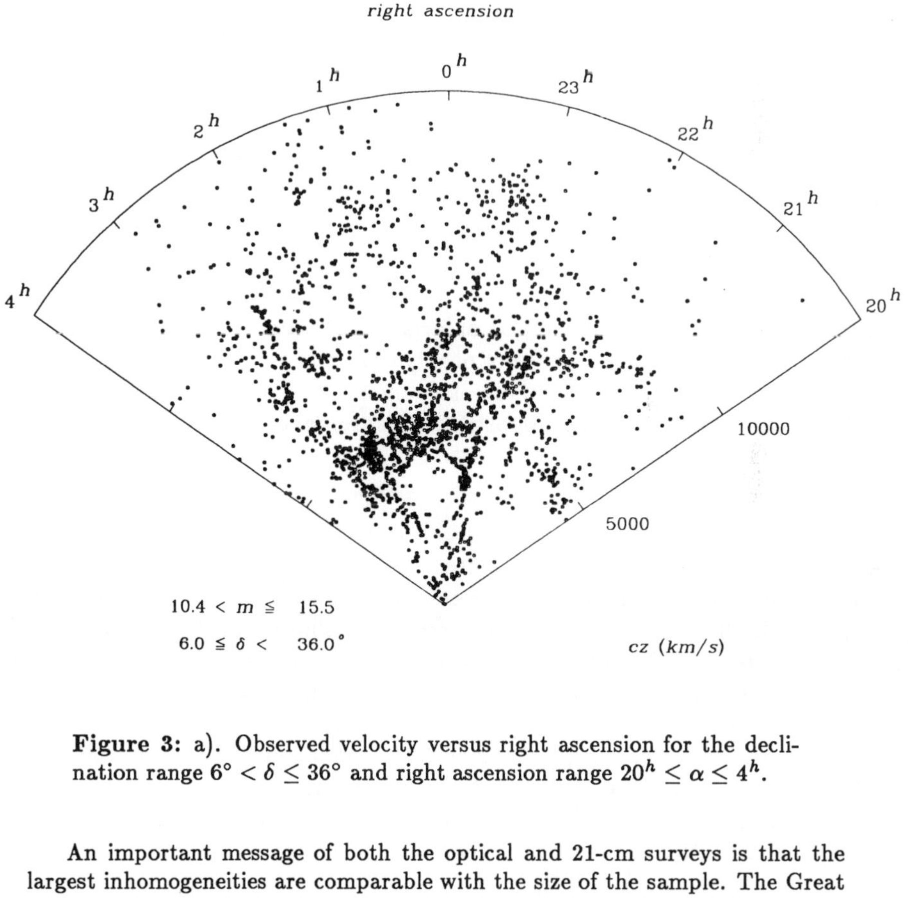

Figure 3: a). Observed velocity versus right ascension for the declination range $6° < \delta \leq 36°$ and right ascension range $20^h \leq \alpha \leq 4^h$.

An important message of both the optical and 21-cm surveys is that the largest inhomogeneities are comparable with the size of the sample. The Great Wall is a clear demonstration of this point. None of the existing redshift surveys are thus large enough to be "fair".

The samples of the redshift survey in Figures 2 and 4 show that the scale of density fluctuations is comparable with the extent of these survey regions. Comparison of the redshift distributions for the three adjacent slices in the North and the 5 slices in the South provides a direct quantitative demonstration that these samples are not large enought to be "fair" (Vogeley, Geller and Huchra 1990). Each of these samples cover $\sim$ 0.6 sr and includes $\sim$ 2500 galaxies. Vogeley *et al.* apply a two-tailed K-S test to the two velocity distributions: the probability that they are drawn from the same underlying distribution is 10^{-24}. N-body simulations must exhibit similar variations in the velocity distribution of comparably large independent samples.

The largest inhomogeneities we detect are the largest we *could* detect within the limits set by the extent of the survey — we have few direct limits on larger structures in the distribution of light-emitting matter. Several deeper redshift surveys designed to address this issue are under way.

Deeper reliable galaxy catalogs are an obvious prerequisite for deeper redshift surveys. Maddox *et al.* (1990) discuss the first results of one such project. They derived a uniform catalog of $\gtrsim$ 2 million galaxies brighter than $b_J = 20.5$ from machine scans of 185 UK Schmidt plates. The angular correlation function derived from these data implies more large-scale clustering than predicted by the standard models for the formation of large-scale structure. The authors suggest that the large-scale features they observe are more distant "Great Walls". An independent machine-based survey of the Southern sky (Heydon-Dumbleton *et al.* 1990) will provide another important measurement of large-scale clustering in deep samples.

Deeper redshift surveys (Broadhurst *et al.* 1989 (BEKS) provide more direct indications of "Great Walls" at larger redshift. The BEKS survey contains a deep probe toward the North Galactic pole (Koo, Kron, and Szalay 1987) and one toward the South Galactic pole (Broadhurst, Ellis, and Shanks 1988) combines with shallower data in both directions. The effective mean depth of these surveys is $z \simeq 0.2$. BEKS observe appproximately equally spaced peaks in the galaxy distribution (with a spacing of $\sim$120 h^{-1} Mpc). They interpret these peaks as "Great Walls". Quantitative comparison of their results with the CfA redshift survey observations confirms that their suggestion is reasonable (de Lapparent, Geller, and Huchra 1991; Ramella, Geller, and Huchra 1991). All of these surveys confirm that large-scale inhomogeneities are a common feature of the galaxy distribution.

III. STATISTICS OF THE CfA REDSHIFT SURVEY

Lack of adequate statistical characterizations of the data makes comparison of different surveys difficult. Although the situation is improving rapidly, the limited extent of surveys compromises measures on scales much larger than

$10h^{-1}$ Mpc. Measuring high order properties of the distribution is an important challenge.

Most studies of galaxy clustering apply correlation function techniques (Peebles 1980 and references therein). However, calculation of the N-point correlation functions is only practical for small N. Statistics which depend upon the higher order moments of the galaxy distribution include measures of the topology of the distribution (Gott *et al.* 1989; Ryden *et al.* 1989), percolation statistics (Shandarin and Zeldovich), the void probability function (Fry *et al.* 1989; Ostriker and Strassler 1989; Vogeley *et al.* 1991), and the filling factor (de Lapparent, Geller, and Huchra 1991).

de Lapparent, Geller and Huchra (1991) show that galaxies fill $\lesssim 25 \pm 5\%$ of the volume of the CfA survey; the remaining $\gtrsim 75 \pm 5\%$ of the volume is devoid of bright galaxies ($M_{B(0)} \leq -18.0 + 5\log h$ at 5000 km s^{-1} ; $M_{B(0)} < -19.5 + 5\log h$ at 10,000 km s^{-1}). In other words, the voids occupy ~75% of the volume of the survey; the sheets fill ~25%. By examining galaxy counts in cells de Lapparent, Geller, and Huchra (1991) and Ramella, Geller, and Huchra (1991) derive an average surface density for structures like the "Great Wall". The *average* surface number density is ~ 0.3 galaxy h^2Mpc^{-2} with $M_{B(0)} \leq -18.2 + 5\log h$.

The deep probes of BEKS have an effective spatial extent of ~$5h^{-1}$ Mpc at $z \simeq 0.2$, comparable with the correlation length for individual galaxies. We can use the "Great Wall" in the CfA survey to study the way the BEKS probes would "detect" similar structures at redshifts ~ 0.2. About 50% of the time, these narrow probes detect the wall at a signal-to-noise $N_{struc}/N_{unif} \gtrsim 3-4$ where n_{struc} is the number of galaxies seen in the structure and N_{unif} is the number expected in the redshift interval (Δv=1500 km s^{-1})for a random distribution. Nearly all of the peaks in this great wall simulation contain samll groups of galaxies (Ramella, Geller, and Huchra 1991). Thus one might expect that the surface density of walls calculated from detection in the deep probes would be an overestimate and so might the size of voids.

IV. FOR THE FUTURE

Analyses of galaxy redshift surveys raise a number of important issues. The samples are probably not large enough to be representative; the largest well-defined structures are comparable with the extent of the surveys. It now appears that both shallow extended surveys and narrow deep probes uncover thin sheets like the "Great Wall".

A general description of large-scale structure also depends upon having surveys which are large enough to be representative. Deeper surveys extending over large angular scale are important for enabling the identification of structures larger than those which have already been observed; the largest structures provide one of the tightest constraints on models. At least two surveys are underway to meet this goal: one in the North (Geller *et al.* 1991) and one in the

South (KOSS 1991). Both of these surveys reach to a limiting apparent magnitude $m_{B(0)} = 17.5$ and span ~100° across the sky. More ambitious projects are also plannes by several groups.

Further progress in mapping the large-scale structure of the universe requires reliable photometric catalogs. Even at the depth of the Zwicky catalog, there is no uniform survey of the sky. Systematic variations in the magnitudes from one region to another in a single catalog (not to mention variations from one catalog to another) almost surely compromise detailed analyses of the properties of the structures. The advent of large format CCD's has made fundamentally important digital surveys possible. An intermediate depth ($m \leq 20^{th}$ magnitude), multicolor survey should be the next major step in the study of large scale structure.

This work is supported in part by NASA Grant NAGW-201, by the Smithsonian Scholarly Srudies Program and by the Digital Equipment Corporation.

REFERENCES

Bothun,G.D., Beers,T.C., Mould,J., and Huchra,J.P. 1986, *Ap. J.* **308**, 510.

Bothun, G.D., Geller, M.J., Huchra, J.P., Kurtz, M.J., and Schild, R. 1991, in preparation.

Broadhurst, T.J., Ellis, R.S., Koo, D.C., and Szalay, A.S. 1990, *Nature*, submitted.

da Costa, L.N., Pellegrini, P.S., Sargent, W.L.W., Tonry, J., Davis, M., Meiksin, A., and Latham, D.W. 1988, *Ap.J.*, in press.

Davis, M. Efstathiou, G., Frenk, C. and White, S.D.M. 1985, *Ap.J.*, **292**, 371.

Fry, J.N., Giovanelli, R., Haynes, M.P., Melott, A.L., and Scherrer, R.J. 1989, *Ap.J.*, **340**, 11.

Geller, M.J. and Huchra, J.P. 1989 *Science*, **246**, 897.

Geller, M.J. *et al.* 1991, in preparation

Giovanelli, R., Haynes, M.P., and Chincarini, G. 1986, *Ap.J.*, **300**, 77.

Gott, J.R. III *et al.* 1989, *Ap.J.*, **340**, 625.

Haynes, M.P. and Giovanelli, R. 1986, *Ap. J. (Letters)*, **306**, L55.

Heydon-Dumbleton, N.H., Collins, C.A., and MacGillivray, H.T. 1990, *M.N.R.A.S.*, **238**, 379.

Huchra, J.P., Geller, M.J., de Lapparent, V., and Corwin, H. 1990, *Ap.J. Suppl.*, in press.

Kirshner, R.P., Oemler, A. Jr., Schechter, P.L., and Shectman, S.A. 1987, *Ap.J.*, **314**, 493.

Kirshner, R.P., Oemler, A. Jr., Schechter, P.L., and Shectman, S.A., in preparation.

Koo, D. Kron, R. and Szalay, A. 1986 in *13th Texas Symposium on Relativistic Astrophysics*, M. Ulmer, ed, (World Scientific: Singapore).

de Lapparent, V., Geller, M.J., and Huchra, J.P. 1986, *Ap.J. (Letters)*, **202**, L1.

de Lapparent, V., Geller, M.J., and Huchra, J.P. 1991, *Ap.J.*, in press.

Maddox, S.J., Efstathiou, G., Sutherland, W.J., and Lovejay, J., *M.N.R.A.S.*, **242**, 43p .

Nilson, P. 1973, *Uppsala General Catalogue of Galaxies, Uppsala Astr. Obs. Ann*, **6**.

Ostriker, J.P. and Strassler, M.J. 1989, *Ap.J.*, **338**, 579.

Ramella, M., Geller, M.J., and Huchra, J.P. 1989, *Ap.J.*, **344** 57.

Ryden, B.S. *et al.* 1989, *Ap.J.*, **340**, 647.

Shandarin, S.F., and Zeldovich, Ya. B. 1989, *Rev. Mod. Phys.*, **60**, 1.

Vogeley, M. J., geller, M.J., and Huchra, J.P. 1991, in preparation.

Zwicky, F.,Herzog, W., Wild, P, Karpowicz, M. and Kowal, C. 1961-1968, *Catalog of Galaxies and of Clusters of Galaxies*, (Pasadena: California Institute of Technology)

LOCAL LARGE-SCALE STRUCTURE: DYNAMICS AND STATISTICS

Nick Kaiser
CIAR Cosmology Program
CITA, University of Toronto

Summary

The acceleration of the Local Group inferred from IRAS galaxies agrees well in direction with the Local Group motion relative to the microwave background and, if IRAS galaxies trace the mass, this indicates that the universe has a density close to the critical value. A deep (0.6 Jy) redshift survey of IRAS galaxies showns that the predicted motion of other galaxies also correlates quite well with those observed, giving a largely independent confirmation of a high value for Ω, in contrast with earlier dynamical studies using the Local Supercluster. While the dynamical state of the superclusters revealed by IRAS is in generally good agreement with the inflationary prediction, statistical measures of the large-scale clustering (both counts in cells and power spectrum analysis) show that there is more large scale structure than predicted in the standard cold dark matter model.

1 Introduction

The theory of inflation predicts (and 'naturalness' dictates) that the universe should have density very close to the critical value, and that the spectrum of primordial fluctuations should have the scale-invariant, 'Harrison-Zel'dovich' form with fractional rms fluctuations in density $\propto \lambda^{-2}$ (or equivalently that the power spectrum should be a power law with $P(k) \propto k$). If we add the extra assumptions that the fluctuations should be Gaussian (i.e. that the phases of the Fourier components be uncorrelated) and that the density of the universe is dominated by some particle whose random thermal velocity was always smaller than the Hubble expansion across any scale of interest, we arrive at the 'standard' cold dark matter model—a hypothesis of enormous predictive power. The purpose of this paper is to describe various tests of these predictions using observations of superclustering in the local universe. The questions to be addressed are first the dynamical issue: Can Ω really be unity? And second the statistical question: Does $P(k)$ have the shape predicted by standard CDM?

The standard CDM model is an example of a *hierarchical* theory; i.e. one in which non-linear structures form first on small scales and these evolve by a complicated process of merging to make larger structures as time goes on. On galactic scales one needs to grapple with both the problems of non-linear gravitational evolution of the dark-matter and also the influence of non-gravitational physics on the baryonic component—it is strongly suspected (see e.g. Blumenthal *et al.* 1984) that galaxies are largely the result of collisional cooling which will allow the baryons to segregate and settle into the centres of the confining dark haloes of an appropriate size and binding energy, but the details of this are for from being well understood. On supercluster scales the dynamical situation is much simpler—the largest structures we can see are still in the linear regime or close to it—but we still need to worry about the possibility of a bias between the distribution of the *mass*, which is what the theory predicts, and the distribution of luminous tracers (e.g. galaxies), which is what we can actually observe.

There are a number of mechanisms which can plausibly segregate mass and light. I have already mentioned collisional cooling which can separate baryons from dark matter to make galaxies. Thus, when we find that in the luminous parts of galaxies a dynamical mass-to-light ratio of $\sim 3-10$ in solar units (which, if this mass-to-light ratio were representative of the universe as a whole, would imply a density of less than one percent of the critical value) we do not consider this damning evidence against the inflationary prediction, just that we need to cast our net wider to measure the total matter content. On somewhat larger scales (from galaxy haloes through groups to clusters of galaxies) we are still dealing with systems with dynamical times much shorter than the age of the universe, and we can again apply the virial theorem to estimate mass-to-light ratios and, multiplying by the luminosity density of the universe inferred from redshift surveys, obtain estimates of Ω (again assuming that the mass-to-light ratios of these systems are representative of the universal value). The results of this exercise show an enormous increase in M/L with scale, and in rich clusters one finds mass-to-light ratios of a few hundred (for $h = 1$), corresponding to $\Omega \simeq 0.2$.

Thus, while these studies provide strong evidence for copious amounts of dark matter greaty exceeding the mass of baryons we see in stars, if Ω is really unity, it must be that the mass-to-light ratios in the luminous parts of rich clusters are still a factor 5 or so smaller than the value for the universe a whole. There are at least two two plausible mechanisms by which such a bias might arise. In the first, galaxies are initially (at their formation at a redshift of a few) distributed just like the dark matter on scales larger than individual galaxy haloes, but each galaxy has a strongly segregated baryonic core-dark matter halo structure. Numerical simulations (Barnes, 1983, 1984; Evrard, 1987, West and Richstone, 1988) show that if several core-halo structures merge to form a group, the resulting group inherits, to a large degree, the segregation of its subunits—the high phase-space density material in the inner parts of the subunits finding itself after merging in the inner parts of the newly formed group—and could explain why, when we measure the mass-to-light ratio in the luminous parts of a group, we get a value much smaller than that of the entire system, or of the universe as a whole. As several groups come together to form a cluster, the process will be repeated, and the simulations suggest that this kind of *inherited segregation* could plausibly be strong enough to account for the observed M/L's, even if the universe has $\Omega = 1$.

Another possibility way to segregate mass and light in clusters etc, is if galaxy formation was more efficient than average in protoclusters. The idea here is that while the amplitude of density fluctuations is decreasing with increasing scale, it does so rather slowly (the CDM spectrum being rather 'red' on the scales between galaxies and clusters), and there may be significant 'cross-talk' between cluster scales (going non-linear today) and galaxy scales (which went non-linear at $z \simeq 3$ or so). One clue to how this might come about comes from the clustering properties of high peaks in Gaussian random fields. It is easy to show that if one takes such a field and clips off the regions above some high threshold (density constrast $> \nu$ in units of the rms σ), then these high peaks will be more strongly clustered than the underlying field. For large-scales where the correlations are weak one finds that the two-point function is enhanced by a factor $b^2 = (\nu/\sigma)^2$. This process can easily be visualised (e.g. BBKS, 1986) if one thinks of a field with small scale fluctuations superposed on larger scale, 'background' fluctuations $\Delta_b(\mathbf{r})$. One finds that for a high threshold, the space density of peaks is modulated roughly as $n_{\rm pk} \propto \exp(\nu\Delta_b/\sigma)$, and the long wavelength limit mentioned above just comes from taking a first order expansion in Δ_b, since the correlation function goes as the square of the density fluctuation.

The process can be applied with some justification to rich clusters today—the threshold being the overdensity required for collapse by the present—and nicely explains why clusters today should be more strongly clustered than the mass, and it is very appealing to imagine that the formation of galaxies at $z \simeq 2-3$ is analogous to cluster formation today. However, for galaxies, there is no obvious physical cause for the threshold that should be invoked. Perhaps a better way to think about biasing of galaxy formation is to think of how a 'background' perturbation Δ_b would affect an individual galactic-scale condensation. For a positive Δ_b, the fluctuation will turn around and collapse sooner, and have a deeper potential well. If one hypothesises that galaxies generate a luminosity determined mainly by the depth of the confining potential well, then this can introduce a bias. Calculations based on Press-Schechter theory (Cole and Kaiser, 1989) suggest that this may produce a significant bias on the scale of clusters of galaxies (the Press-Schechter calculations are well described by an exponential modulation $n \propto \exp(\beta\Delta_b)$ at given L, and if one assumes that the luminosity is exactly proportional to the 4th power of the halo velocity dispersion, one finds $\beta \simeq 0.4$, giving perhaps a factor 2 suppression of M/L in clusters from this process). These calculations are only meant as an illustrative model for what might really be happening. While assuming $L \propto v_{\rm halo}^4$ has some empirical appeal, it is by no means mandatory. It may be possible to believe that galaxies are not biased in this way (Peacock, 1990), or that the effect is somewhat stronger than this crude model would suggest. It seems quite likely that this 'statistical bias' may augment the 'inherited segregation' described above, and so the situation is likely to be quite complicated. One important distinction between these two types of biasing mechanisms is that if we go to very large-scales, there should be no inherited segregation, but the statistical bias should give a constant bias:

$\delta n/n = b\delta\rho/\rho$, with $b = 1 + \beta$.

In the absense of a complete theory for galaxy formation the situation is necessarily somewhat confused. Stimulated by the 'high-peak' model, simulations (e.g. White *et al.*, 1987) were made where the galaxies were laid down as expected for such peaks, with the peak threshold set to a rather high value and giving a strongly biased distribution $b \simeq 2.5$. While these simulations gave encouraging results, is was subsequently realised that there is some indication that galaxies may not be as strongly biased as this on large-scales, and that one should at least consider a much milder statistical bias (as expected from the simple model described above), perhaps augmented on scales of clusters and below by the inherited segregation effect. Even if one were to back the specific $L \propto v_{\mathbf{halo}}^4$ model, or some variant thereof, the predicted large-scale bias factor b is very dependent of the form of the spectrum, and since, as we shall see, there is now some indication that the CDM spectrum does not work very well on large scales, there is no very strong reason to believe that it works in detail on scales between galaxies and clusters. If so, one should probably consider b to be a constant best determined by observation. The expectation that, for sufficiently large-scales, one would expect $(\delta n/n)/(\delta\rho/\rho)$ to tend to a constant is a common feature of all the schemes described above (though one would most naturally expect b to be a function of the luminosity or type of galaxy). Of course, is may be that there is some other effect which can more radically decouple mass and light even on the rather large-scales we will consider here. Again, it is probably best to test for this empirically rather than simply speculate.

2 Weighing the Universe with Supercluster Dynamics

As mentioned above, for dense systems ($\rho \gg \overline{\rho}$, $t_{\mathbf{dyn}} \ll H^{-1}$) we can estimate $L \sim lz^2H^{-2}$, and $M \sim \sigma^2\theta/H$, from observations of the apparent luminosity l, the redshift z, the angular size θ and the internal velocity dispersion σ, and form M/L. Multiplying by the ratio of the known (or at least estimated) luminosity density of the universe to the critical density gives an estimate of Ω which one can think of a a partial accounting of the mass in the universe. When we go to supercluster scales we are dealing with systems which are still expanding ($\rho - \overline{\rho} <\sim \overline{\rho}$, $t_{\mathbf{dyn}} \sim H^{-1}$) and one must proceed differently. For linear perturbations (and there is good reason to think that most of our motion is produced by fluctuations in the linear regime), the velocity of a particle (or galaxy) should be proportional to the acceleration times an effective time over which the acceleration has been operating: $\mathbf{v}_{\mathbf{pec}} = \mathbf{g}t$. The acceleration depends on Ω and on the bias factor b, and the effective age also depends on Ω, and the result is

$$\mathbf{v}_{\mathbf{pec}} = \frac{\Omega^{0.6}}{b} H \int d^3r \, \frac{\Delta n}{n} \frac{\hat{\mathbf{r}}}{r^2}. \qquad 1$$

An equivalent statement is that

$$\nabla \cdot \mathbf{v}_{\rm pec} \propto \frac{\Omega^{0.6}}{b}\frac{\Delta n}{n}, \qquad 2$$

which makes is more explicit that all one is really testing is the continuity equation.

An immediate problem is that one cannot determine Ω and b separately from observations in the linear regime. It might be thought that by going into the weakly non-linear regime that one might be able to determine the two separately. This would be true if, for instance, one were prepared to assume that the bias factor was constant even when the density perturbations start to become non-linear, but this is highly unlikely to be the case; certainly in the models discussed in the Introduction this would not be the case, the statistical contribution to the bias (defined as $(\delta n/n)/(\delta\rho/\rho)$) varying roughly exponentially with the *initial* density contrast, and in these models the departures from constant bias set in at about the same point that any departures from dynamical linearity set in. Given the imperfections in our understanding of dynamical non-linearity and the imperfections in the velocity data any hope of determining Ω directly from this type of measurements is a forlorn one, and one must live with the ambiguity. In the past, it was popular to assume that $b = 1$—that galaxies fairly trace the mass on very large-scales—and the ratio of the velocity to the apparent acceleration was then interpreted as an estimate of $\Omega^{0.6}$, often expressed as a mass-to-light ratio of the universe. Nowadays, it is more common to assume that $\Omega = 1$, and express the result as a value for b. No matter how the result is expressed, the test is still the same. The important question is: Does the apparent acceleration correlate with the observed motion? and, if so, is the value of b (or Ω) plausible? If not, one would have to reject the class of theory described in the Introduction. If it is, then one has some justification in assuming that fluctuations in the light reflect fluctuations in the mass (though in a biased manner) and proceeding to compare the spectrum of light fluctuations with the spectrum of mass fluctuations predicted in the CDM model.

To measure b (or Ω) one needs to measure some appropriate combination of velocities and galaxy density contrast. There are a range of possibilities. One natural division is between those tests which exploit the velocity of the Local Group inferred from the dipole anisotropy of the microwave background and those which use velocities of other galaxies (usually relative to the frame of the Local Group) as determined from Tully-Fisher or $D_n - \sigma$ techniques. In the former class one must of course estimate the acceleration of the local group and one can distinguish two sub-classes of this type of test: those which use only angular information (and exploit the property that since both gravity and luminosity obey inverse square laws, the dipole moment of the extragalactic sky should be proportional to the acceleration), and those which use redshift information. The latter allow one, in principle, to check whether the acceleration appears to have converged within the survey volume, whereas with the former one must simply assume that the survey is deep enough to encompass the source of our motion (though one has some indication of whether this is true by looking at the discrepancy in angle between the dipole and the actual motion). In the second class of test (using the motions of other galaxies), there are also two sub-classes which can be usefully distinguished: One can either start from peculiar velocities, calculate $\nabla \mathbf{v}$, and compare with the distribution of galaxies $\delta n/n$ or work in the other direction: measure $\delta n/n$, calculate $\mathbf{g}(\mathbf{r})$ and then compare with the observed motion. The main advantage of the former is that the relation between $\nabla \mathbf{v}$ and $\delta n/n$ is a local one; perturbations outside the observed region do not contribute to $\nabla \mathbf{v}$, whereas with the latter one needs a deep survey with nearly full sky coverage. Probably the biggest disadvantage with the former is that one starts with the less reliable data—there is always the worry that the small offsets in the Tully-Fisher or $D_n - \sigma$ relations may be influenced by environment—and any errors (systematic or otherwise) will propagate through the analysis. I do not think that there is any 'best' solution; the different appoaches can be used in a complementary manner, and it is best to apply all of these, and using as many diverse techniques as possible.

Historically, this kind of dynamical test was applied to the Local Supercluster. The Local Supercluster is certainly a prominent density enhancement, and dominates the view of the world obtained from surveys such as the RSA catalogue (Yahil *et al.*, 1980). It is also very clear that the direction to the Local Supercluster does not agree very well with the Local Group motion, so one is really forced to resort to tests using the motion of other galaxies with concomitant uncertainties. The situation as of the early eighties was reviewed by Davis and Peebles (1983). Various groups had worked on this problem using a variety of data sets. The estimated density contrast (within our radius) was taken variously from the RSA and CfA surveys; the concensus value being a density contrast of 2-3. For the most part, the perturbation was modelled as an isolated spherical perturbation with parameters describing the radial profile etc. At the time of the review by Davis and Peebles, there was quite a range estimates of the slowing of the Hubble expansion by the Local Supercluster, ranging from about 150 km/s up to 300 km/s or more, and this gave a correspondingly wide range of estimates of Ω (but all below about 0.3). Since then, the estimates have tended to reduce the estimated infall velocity, and, using the same estimate for the density contrast one would infer $\Omega \simeq 0.1$ or even lower. At face value, this is a big problem for a biased $\Omega = 1$ model (at least with what I would consider a plausible—i.e mild—bias parameter) since one would surely have expected to find a value of Ω larger than that obtained from virial analysis of clusters whereas, if anything, the indication is for a smaller Ω. As I will now describe, there has since been a significant increase in the size and quality of data avalable, and there is now a broad concensus that the best estimate of Ω is a lot larger, and, if IRAS galaxies trace the mass at least, then the current best estimate of Ω is very close to unity. I do not want to go into a detailed *post-mortem* of what might have gone wrong with the old Virgo infall calculations save to mention that i) there was always some considerable doubt as to whether the adopted density contrast was correct; ii) it was always dubious to assume a single isolated spherical perturbation when the direction to the LSC was so discrepant from the motion of the Local Group, and it is now amply apparent that the LSC is really a rather minor perturbation in the whole picture, and iii) the state of the art has now progressed far beyond the need for assumptions of spherical infall, and all of the applications I will mention below make no restrictive assumptions of this kind.

Tests using the Local Group motion: The big observational development here has been the advent of the IRAS point source catalogue and the galaxy catalogues which can be extracted from it by a suitable colour selection. The remarkable result is that the flux weighted dipole moment (e.g. Harmon *et al.*, 1987) agrees in direction with the motion of the Local Group to a precion of about 10 degrees. The most natural expanation of this is that in this catalogue we are seeing deep enough and a large enough solid angle to encompass the source of our acceleration, and that this survey might give a useful estimate of b for IRAS galaxies. Given an estimate of the physical depth of the survey from a redshift survey of a limited patch of sky (Lawrence *et al.*, 1986) it is possible to obtain an estimate of b, and Kaiser and Lahav (1989) found $b \simeq 1.6$. The indication from IRAS galaxies then is that the large-scale dynamics may be compatible with a high value for Ω. The only important assumption about the geometry of the source of our acceleration is that it is not so distant as to be beyond the scope of the IRAS survey and that there is not a very large component of acceleration hiding in the plane of the galaxy. To the extent that the two vectors agree quite well there is empirical indication that the effect of such unseen superclusters is quite small. It seems from this result that if one wanted to believe that Ω is really much smaller than unity, then one would have to say that a good deal of our acceleration comes from much greater distance, and happens to align in direction coincidentally with the acceleration from the stuff that is seen by IRAS. As I will shortly describe, one can make tests of this idea to some extent, and the results so far seem not to support this.

A further observational development made possible by IRAS are the all sky redshift surveys. Strauss and collaborators (e.g. Strauss and Davis, 1988) have made a complete survey to a flux limit of ca. 2 Jy, and the QDOT survey (Lawrence, *et al.*, 1991) is a 1-in-6 random sample to 0.6 Jy. The former survey has a higher space density nearby, but the cross-over is at ca. 70 Mpc/h, beyond which the QDOT survey has the higher sampling rate. Both of these surveys give a 3-dimensional view of the structures around us (though of course in redshift space) and can be used to see if the acceleration does converge. Both of these surveys show that a major source of our acceleration is the supercluster surrounding the clusters of Hydra and Centaurus (coindicing roughly with the Great Attractor concentration inferred from peculiar motions). In the 2 Jy survey, the Perseus-Pisces supercluster appears to be amost as prominent, but in the QDOT survey it is much less so, for reasons which are not well understood (by me at least). Both of these surveys have been used to explore the convergence of our acceleration with distance. The 2 Jy survey seems to show a reasonable convergence (to a direction 20-30 degrees away from the Local Group motions) within about 5000 km/s. In the QDOT survey there is a hint that there may be some additional acceleration (out to perhaps 8000 km/s), but the problem is that in both surveys the shot noise diverges rapidly at large distances. Both studies gave estimates of b (or Ω) very close to unity in rough agreement with the results from the flux-weighted dipole. It is not at all clear that the precision of this test is really any better than that from the flux dipole alone, despite the extra redshift information. What these 3-D studies do, however, is give confidence that the results are not severely affected by uncertainty in the acceleration from very nearby. This is reassuring since the uncertainty in the flux-weighted dipole is dominated by the brightest galaxies.

Tests using motions of other galaxies: As mentioned above, there are two possible routes here: one can either estimate $\nabla \mathbf{v}$ from the velocity data and compare with $\delta n/n$ or estimate $\mathbf{g}$ from $\delta n/n$ and compare with $\mathbf{v}$ (or the line of sight component thereof). The former route has been pioneered by Bertschinger and Dekel (BD), and their results are described by Dekel in these proceedings. One problem with this method is that the data are inhomogeneous in spatial coverage, and the noise becomes large at large distance. Bertschinger and Dekel have tried to circumvent this by smoothing the velocity field derived, but it turns out that a rather large smoothing radius is required. Albert Stebbins and I (Stebbins and Kaiser, 1991; see also Kaiser, 1990) have tried an alternative approach; what we did was to make a regularised fit to a 'non-parametric' model for the velocity field (i.e. one with an effectively infinite number of degrees of freedom). Our model is a set of Fourier modes for the density field with velocity derived from that assuming linear theory. Thus, just as in the BD technique, the assumption that the velocity field is curl-free is built in from scratch. What we did was to calculate the most probable velocity field given the data (we used the compliation of Burstein (1990)) and an assumed power spectrum for the density field. This is a Bayesian approach—the power spectrum defines our prior probability distribution for velocity field configurations—and the technique has similarites to maximum entropy modelling: e.g. the solution is the 'smoothest' map compatible with the data and therefore tends to relax to zero in the absence of data. Our initial results were very encouraging indeed; several prominent (and real!) features could be seen in the reconstructed density map, and, as in our approach the resolution is automatically adjusted to extract the most information at each point, the Local Supercluster for instance appears quite prominently. In order to explore the range of allowed results we vary both the slope and amplitude of the assumed spectrum and this gives a reasonable indication of which results can be believed. We find that the Great Attractor, for instance, appears quite robust. What we have subsequently discovered is that there is a bias in this kind of technique as we initially implemented it which makes the interpretation rather difficult. The problem is that if one starts with a map of the peculiar velocity as a function of *estimated* distance, then simply the effect of distance estimate errors will mimic the effect of gravity. What happens is that if one has a supercluster of galaxies, these tend all to lie at a similar redshift. In estimated distance space, the errors (on the order of 20 %) will scatter the galaxies front and back, and so one will see an apparent infall pattern (which is, of course, spurious) just due to the effect alone. The overall effect is then a complicated combination of thus Malmquist-type effect and real infall. We have tried an alternative solution working from a map of peculiar velocity in *redshift* space (in linear theory, at least, this is justifiable); here the effect of distance errors is to look like random thermal velocities rather than gravitational infall, and we have found that the resulting maps are unfortunately quite sensitive to what technique is used. We consider that this type of technique is certainly promising, and with better data and a fuller understanding of the type of bias described above this may yield a reliable estimate of the bias factor. The results obtained so far with both techniques indicate a rather mild bias (high Ω) which is in accord with other indications.

The IRAS QDOT survey allows one to exploit the second approach (Kaiser, *et al*, 1991). What we have done here is to bin the galaxies on a grid with cells whose size increases linearly with distance from us and, after allowing for the radial selection function, incompleteness in the plane of the galaxy etc., construct a map of the density contrast in redshift space. We then correct this for peculiar velocities in an iterative manner using linear theory to make a map of the density field in real space. From the resulting 3-D density contrast map we can calculate the acceleration for each of the 1000 or so galaxies in the Burstein compilation (we actually binned these in the same cells used to make the density map) and thereby compare the predicted and observed motions. In making this test it is best to work in the Local Group frame. This is simply because there is a divergent uncertainty in the acceleration due to distant matter, but if we predict peculiar velocities relative to the local group any spurious acceleration from distant regions tends to cancel out. The result is shown in figure 1.

At first sight this may not look like a very wonderful correlation, but perhaps the best way to read this figure is to divide the points into 2 according to predicted velocity. It is then quite apparent that the mean velocity of the the galaxies which IRAS predicts to be moving away from us is indeed positive and very significantly (at about the 4-sigma level) larger than the velocity of the other set of galaxies. The uncertainty in the estimate of b obtained here (the slope of the correlation) is similar to that obtained using the Local Group motion, and the estimate here independently confirms the result obtained there that (if IRAS galaxies trace the mass) the best estimate of the density parameter is close to unity.

While there is on the whole reasonable agreement between the predicted and observed motions (and this should be regarded both as support for the theory of gravitational instability and a testimony of the ability of observers to measure distances accurately), it is of interest to look at the residuals. There

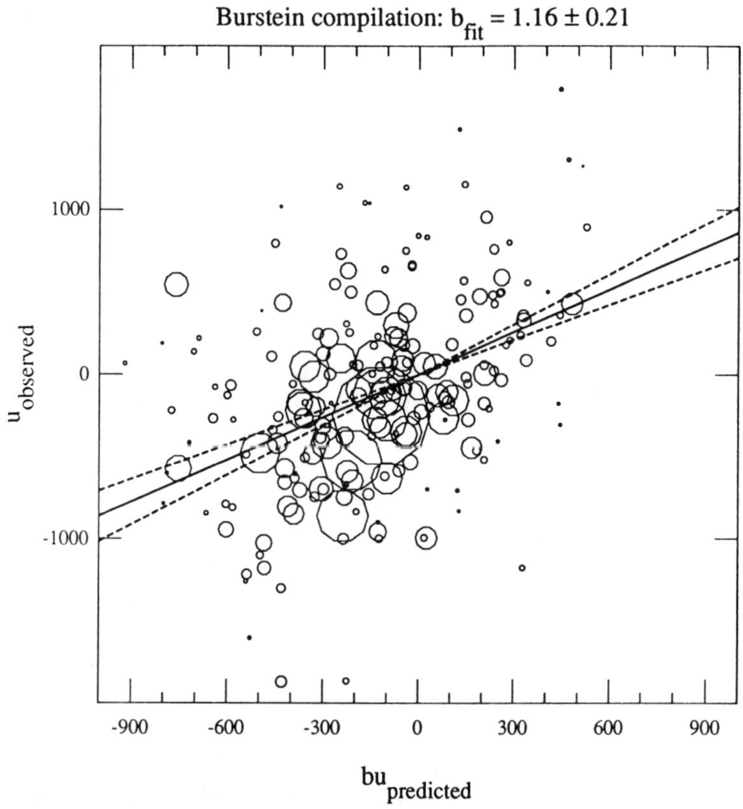

Figure 1. Scatter plot of observed peculiar velocities against those predicted from the spatial distribution of IRAS galaxies assuming gravitational instability. The galaxies with velocities have been binned into cells, and the area of the symbol is proportional to the statistical weight. With no peculiar velocity the χ^2 is about 1500 and is reduced to ca. 800 by the IRAS predictions. This is still bigger than the number of cells (ca. 200), so there is some excess scatter here over and above that from measurement errors. The solid line is the best linear fit, and the dashed lines correspond to 1-sigma errors in b.

is some excess scatter in the plot in figure 1 over and above that expected from velocity measurement errors alone, and this is at the level of about 300 km/s in the rms. One interestingly large residual is right around the Hydra-Centaurus region where (according to IRAS) the acceleration relative to the Local Group is negative whereas the galaxies are actually moving quite rapidly away from us. The fact that these galaxies are (in the MBR frame) moving faster than the Local Group seems to be a robust feature of the data (it is for this reason that the Great Attractor model puts the centre of the attractor just beyond the major visible concentration of galaxies). This is worrying. Possible explanations are i) there is something awry with the distance estimates there, ii) there is stuff hiding in the plane which IRAS doen't see, iii) there is a more distant attractor having a big influence there (the Shapley concentration perhaps). If the latter explanation is correct then this probably means that our estimate of Ω is too large, but more data is needed to clarify the situation.

3 Power Spectrum from the IRAS QDOT Survey

The QDOT survey has been studied via counts in cells analysis (Efstathiou *et al.*, 1990) and shows that there is a problem with the CDM model in that there seems to be more power on large scales than is predicted in that theory. The survey can also be used to give an estimate of the power spectrum of galaxy clustering. What we have done is to bin the galaxies on a regular grid in redshift space. We then calculate the expected number of galaxies (in the absence of clustering but allowing for incompleteness) on a similar grid, and then take the Fourier transform of the difference, multiplied by a radial weight function. If we multiply this by its complex conjugate the result is a convolution of the true power spectrum with a 'point spread function' which is just the power spectrum of the weighted selection function. For the QDOT survey this psf is a compact ball in k-space, so we obtain a straightforward estimate of $P(k)$ this way. If one makes the additional assumption that the large-scale Fourier components have random phases then one can also calculate the covariance matrix which tells us how the neighbouring estimates of $P(k)$ are correlated. The weight function is chosen to give equal weight per volume nearby (where the counting statistics due to the discreteness of galaxies is negligible) and equal weight per galaxy at large distances where the counting statistic fluctuations dominate over the real density fluctuations. The actual function used is a smooth function which extrapolates between these two limiting regimes. The resulting power spectrum is shown on figure 2, and is compared to the power spectrum expected from CDM.

This shows graphically the problem already apparent from the counts in cells analysis that there seems to be more power on large scales than is predicted.

For the results shown the 'break' in the weight function was set to be 100 h^{-1} Mpc, and so significant weight comes from that radius and somewhat beyond, including the giant superclusters Hercules, Pisces-Cetus etc. If the break is set to a smaller value so as to downweight these superclusters then the result is much less problematic for CDM. This is similar to what we found from the counts in cells analysis; the nearby region shows a little more power than predicted by CDM, but not drastically so. This means that there is a slight loophole here for CDM if the distant superclusters—where we see only intrinsically bright galaxies—are enhanced by statistical bias. What are needed to close this loophole are good peculiar velocities around e.g. the Hercules supercluster to see if the dynamical imprint of this structure (if indeed it is a real mass enhancement) is there.

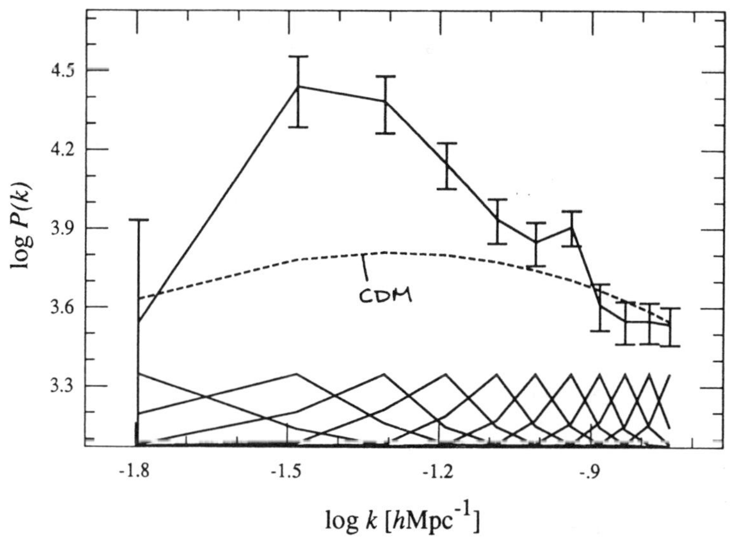

Figure 2. Power spectrum estimated from IRAS QDOT redshift survey. The curves along the bottom show the covariance between the power estimates, showing that neighbouring estimates are somewhat correlated, but at larger separations the estimates are essentially uncorrelated. The estimated power drops beyond $k = 30h/\mathrm{Mpc}$ (ca. 180 Mpc/h wavelength), but this may be an artefact caused by internally estimating the selection function. The dashed curve is for CDM with $\sigma_8 \simeq 1$ normalisation.

4 Summary

In §2 I have shown that a wide variety of tests and techniques now give a consistent picture that i) the acceleration predicted by IRAS agrees quite well with the motions both of the Local Group and of other galaxies, and ii) if IRAS galaxies trace the mass then $\Omega \simeq 0.8 \pm 0.3$. This is quite a change from earlier studies using the LSC and assuming spherical symmetry which gave a much lower estimate of Ω. The estimate of Ω obtained by comparing predicted and observed motions of other galaxies relative to the Local Group gives a result which is independent from but which confirms earlier estimates obtained by comparing with the Local Group motion. Problems and questions for the future are: What is happening in the Great Attractor region? Does the rather large discrepancy found there indicate mass hiding in the plane? a more distant attractor? a problem for $D_n - \sigma$? Other questions are: What about ordinary galaxies—are they more strongly biased than IRAS galaxies? Does the bias factor depend on luminosity? What will be found if these dynamical tests are pshed to even larger scales?

Concerning the statistics of large scale structure: The success of the dynamical tests give some confidence that the fluctuations in galaxy counts do reflect underlying mass fluctuations and that the IRAS QDOT counts in cells analysis and the more recent power spectrum analysis so indeed suggest a serious problem for the standard CDM model. The main *caveat* here is that the result relies strongly on the influence of the Hercules supercluster and other superclusters at large distance (> 100 Mpc/h), and it would be nice to have direct confirmation from distortion of the Hubble flow that these are 'real' density fluctuations.

References

Blumenthal, G.R., Faber, S.M., Primack, J.R. & Rees, M.J., 1984. Nature, **311**, 517.

Barnes, J., 1983. MNRAS, **203**, 223

Barnes, J., 1984. MNRAS, 208, 873

Evrard, A.E., (1987) Ap.J., 316,36

West, M.J., and Richstone, D.O., 1988. Ap.J., **335**, 532

Cole, S., and Kaiser, N., 1989. MNRAS, **237**, 1127.

Peacock, J.A., 1990, MNRAS, **243**, 517

White, S.D.M., Frenk, C.S., Davis, M. & Efstathiou, G., 1987. Ap.J., **313**, 505.

Yahil, A., Sandage, A. & Tammann, G.A., 1980. Ap.J., **242**, 448.

Davis, M. & Peebles, P.J.E. 1983. *Annu. Rev. Astron. Astrophys.*, **21**, 109.

Harmon, R.T., Lahav, O. & Meurs, E.J.A., 1987. MNRAS, **228**, 5p.

Kaiser, N. and Lahav, O., 1989. MNRAS., **237**, 129.

Lawrence, A., Walker, D., Rowan-Robinson, M., Leech, K. J. & Penston, M. V., 1986. MNRAS, **219**, 687.

Strauss, M.A. & Davis, M. 1988. In: *Large Scale Structure of the Universe, IAU Symp. 130*, eds. J. Audouze & A. Szalay, Reidel, Dordrecht,

A. Lawrence, M. Rowan-Robinson, J. Crawford, I. Parry, X-Y Xia, R.S. Ellis, C.S. Frenk, W. Saunders, G. Efstathiou, N. Kaiser. 1991, In preparation

Stebbins, A., and Kaiser, N., 1991. In preparation

Kaiser, N., 1990. Contemporary Physics, **31**, 149

Efstathiou, G., Kaiser, N., Saunders, W., Lawrence, A., Rowan-Robinson, M., Ellis, R.S. and Frenk, C.S., 1990, MNRAS, **247**, 10p.

ANALYSIS OF THE LARGE SCALE STRUCTURE WITH DEEP PENCIL BEAM SURVEYS

A.S. Szalay[1,2], R.S. Ellis[3], D.C. Koo[4] and T.J. Broadhurst[5]
1. Department of Physics and Astronomy, The Johns Hopkins University, Baltimore MD 21218
2. Department of Physics, Eötvös University, Budapest, H-1088
3. Department of Physics, University of Durham, DH1 3LE
4. Lick Observatory, University of California, Santa Cruz, CA 95064
5. Mathematics Department, Queen Mary & Westfield College, London E1 4NS

ABSTRACT

Recent observations of the large scale structure of the universe suggest inhomogeneities on scales larger than previously thought. A deep redshift survey with a 'pencil beam' geometry of galaxies at the Galactic Poles indicates strong clustering, with a provocative regularity on a scale exceeding 100 h^{-1} Mpc. We examine in detail the role of such deep 'pencil beam' surveys in studies of the large scale structure using the 1-D power spectrum of the distribution to illustrate the statistical significance of the feature claimed by Broadhurst *et al* (1990). We demonstrate how deep probes complement sparse-sampled wide angle local surveys providing new information on large scales. To test the results found from our original survey, we examine future sampling strategies that would delineate the topology more clearly and briefly discuss the implications of such results in theories accounting for the origin of structure. We quantify the sensitivity of pencil beam surveys to the topology of large-scale structures and compare them to sparse sampled wide-angle local surveys. In particular, we also present a more detailed statistical analysis for the Galactic Pole data set, and show that it remains significant at the 2×10^{-4} level.

1. INTRODUCTION

Many of the most fundamental cosmological questions relate to the formation of the large scale structure; these connect the smallest and largest physical scales known. Those processes responsible for small quantum fluctuations in the early universe occur at energies and length scales inaccessible to modern accelerators. Recent results concerning the large scales over which galaxies may be organized suggest a new tool for examining processes in the early Universe.

To model the structure seen, cosmologists consider a variety of initial conditions and follow subsequent evolution under the assumption of gravitational instability, with combined analytic and numerical techniques, making certain extra assumptions. The predictions are compared with the distribution of galaxies and fluctuation limits for the cosmic background radiation (CBR).

The most popular galaxy formation theory is the biased Cold Dark Matter (CDM) model (Blumenthal *et al* 1984), in which most of the mass density is in the form of non-interacting dark particles. Together with a scale invariant Zeldovich–Harrison spectrum for the initial density fluctuations, the theory satisfies many observational constraints on small (< 100 h^{-1} Mpc) scales (White *et al* 1987). However, to explain the absence of CBR fluctuations, its proponents invoked the concept of 'biasing' whereby galaxies only form at high peaks of the mass fluctuations. Much stronger correlations in the distribution of visible galaxies are predicted than for the underlying mass (Kaiser 1984, Bardeen *et al* 1986).

CDM predicts the universe should be relatively homogeneous on scales above 30-40 h^{-1} Mpc and it is thus important to test this observationally. Surprisingly little accurate data exists. The apparent strong clustering for rich clusters (Bahcall and Soneira 1983) is in marginal conflict with theory. Large scale motions of galaxy samples offer an alternative test since they are related to density fluctuations. Recent results indicate coherent streaming within 80 h^{-1} Mpc scales (Rubin *et al* 1976, Burstein *et al* 1986, Lynden-Bell *et al* 1988) but these analyses rely on environment-independent distance indicators whose robustness is largely untested.

Recent galaxy redshift surveys suggest significant deviations from uniformity on large scales. Kirshner *et al* (1983) found a 60 h^{-1} Mpc sphere with a large underdensity in the galaxy distribution. deLapparent, Geller and Huchra (1986) delineated the 'Great Wall' in their CfA surveys - a structure connecting several known Abell clusters over a spatial extent of 100 by 50 h^{-1} Mpc. Chincarini *et al* (1983) and Giovanelli *et al* (1986) had earlier shown the Perseus-Pisces supercluster extends also to large distances in the transverse directions.

Are such 'walls' representative features of the universe? N-body simulations are able to generate apparently similar structures from CDM initial conditions (Park 1990), but few quantitative tests have been done so far. Statistical tests are readily constructed when the topology is understood, but that is evidently not yet the case.

Our task in this paper is to examine the ways in which evidence for large scale power might be assembled. In §2 we begin by contrasting two rather different observing strategies – the sparse-sampled local survey and the deep pencil beam survey demonstrating the unique features of each. In §3 we examine the latter strategy quantitatively, then in §4 we discuss the statistical consequences of the one-dimensional nature of the pencil beams. In §5 we present a more detailed statistical analysis of the recent survey of Broadhurst *et al* (1990, BEKS), then in §6 we discuss ongoing research and the extension of our strategy to answer further questions.

2. STRATEGIES FOR MAPPING LARGE SCALE STRUCTURE

First we address the question of the best strategy for delineating the large scale distribution of galaxies, given that its precise topology remains unclear. With the exception of dedicated telescopes, a typical redshift survey can only measure a few thousand redshifts in a few years' observation. Indeed, there is probably no single strategy optimal in all situations since the choice is strongly influenced by what presumptions hold about the statistical properties and the topology of the large scale galaxy distribution. There is a fundamental difference between the strategies and goals of the sparse-sampled wide angle surveys, and the deep pencil beam surveys. Their correlation functions measure different statistical aspects of the same distribution. In this section, aimed towards studies of the large scale structure, we will concentrate our attention on these two strategies, demonstrating that for a full picture both are required.

In general, answers depend on what questions were asked. When discussing large scale structure, one should first specify what do these words mean. For some, large scale structure is equal to the large scale behaviour of the galaxy two point correlation function. Others like to draw maps of the galaxy distribution, and call that the large scale structure. These data sets contain different information, of course, and therein lies the fundamental difficulty in presenting a coherent view of the same universe. We will elaborate on these differences below, and focus on finding large coherent structures: our use of the words Large Scale Structure...

There are two competing sources of statistical noise in redshift surveys. The 'sampling noise' is determined by the number of redshifts obtained, and the 'clustering noise' is from the observed small scale correlation function of galaxies; the latter depends on the shape and size of the survey volume. For a given geometry, there is an optimum number of galaxies to observe, at which point the 'sampling noise' equals the 'clustering noise'. Beyond this critical number, the signal to noise for large scale correlations is not significantly improved. Such an argument has been used, e.g. to determine the optimum 'sparse sampling' rate for wide angle or all-sky surveys (Kaiser 1986).

If fluctuations in the universe are strictly Gaussian, their full statistical description is contained in the two-point correlation function. In this case the 'sparse sampled' survey is indeed the best way to obtain this information. If the universe, however, contains some very sharp large scale features, like the 'Great Wall', a sparsely sampled survey may fail to identify those and it may thus not be possible to test whether the sample is strictly 'fair', i.e. representative.

Another strategy that can be adopted uses a smaller solid angle, sampling fainter galaxies in the same amount of telescope time. There is some freedom in how the shape of the volume is chosen. Restricting the survey to a thin slice increases the dynamic range of relative distances compared to a wide angle, symmetric cone. Choosing a pencil beam enables us to go even deeper, at the expense of becoming much narrower. Because of this, the CfA slice delineated well the transverse spatial coherence of the 'Great Wall' (de Lapparent, Geller and Huchra 1986).

The slices or pencil beams cannot be arbitrarily thin, due to the well-known small scale clustering of galaxies

$$\xi(r) = \left(\frac{r}{r_o}\right)^{-1.8} \tag{1}$$

where $r_o = 5\ h^{-1}$ Mpc is the correlation length (Peebles 1980). Any linear dimension chosen to be less than r_o will ensure noise from small scales will dominate any signal from large scale structure (see §3). Thus there is a practical limit on how thin such slices should be; the CfA slice is fairly optimal.

The presence of structures like the 'Great Wall' also means, that either higher order correlations are present, or the phases of the fluctuations are correlated: clearly different from a homogenous isotropic Gaussian random field. We can see this using a simple model, due to Peebles (1989). Consider an extreme distribution with all galaxies on sheets, spaced at separations λ, to form a three dimensional cubic lattice. The 3-D correlation function can be calculated analytically

$$\xi(r) = \frac{1}{6}\left[\frac{\lambda}{r} + 2\frac{\lambda}{r}\text{lnt}\frac{r}{\lambda} - 2\right]. \tag{2}$$

The correlation function obviously reveals a feature at $r = \lambda$, but with a fairly small amplitude, $\xi(\lambda) \approx 0.16$, since in different directions the sheets are seen under different inclination angles, smoothing out the correlations. However, if we observe the same structure with pencil beam surveys, every beam would still go through a sequence of sheets, yielding very strong features in each 1-D correlation function. An infinite number of pencil beams would provide a much weaker signal, in a similar but not equivalent manner to the 3-D correlations. The key difference is that sharp walls are non-Gaussian structures whose phases in the Fourier decomposition are strongly correlated. A 3-D correlation function – equivalent to the power spectrum i.e. no phase information – will give a poor characterization of such a distribution.

From these it is obvious, that sharp sheets are well detected in pencil beams, and the one dimensional correlation functions can have a very large amplitude at their typical separation, while the three dimensional correlations – not to mention the projected angular correlations – would be quite small. This, of course does not guarantee that the detected features are all walls.

If 'Great Walls' are numerous, i.e. if they are typical of the large scale structure, from the surface density of galaxies seen, we can estimate what would constitute a 'fair sample'. Suppose a fraction f of the bright galaxies are on walls with a surface density equal to that observed, $\mu = 0.3$ galaxies Mpc^{-2} (deLapparent *et al* 1991). The characteristic 'cell' size can then be estimated from the corresponding average volume density of bright galaxies, $n = 0.02$ galaxies Mpc^{-3} (deLapparent *et al* 1991). We estimate the fraction f from the CfA survey itself, the wall contains 906 galaxies out of 2536 (Ramella *et al* 1991) thus $f = 906/2536 \approx 0.36$. Given a topology, we can determine the mean surface-to-volume ratio. This number is remarkably robust, thus for simplicity we can consider a spherical topology and count only half the surface area, since walls separate two volumes. With these numbers the typical size of the voids would be

$$\lambda = 2R = \frac{3\mu}{nf} = 126\ h^{-1}\,\mathrm{Mpc} \tag{3}$$

This number is an order of magnitude estimator, of course, but shows that if the 'Great Wall' is a typical structure of the universe, it cannot be replicated much more often than λ. In this case, a strategy penetrating depths several times larger than this scale is required. For a fixed number of redshifts, ignoring practical considerations of multiple object techniques for the moment, it is then advantageous to narrow the slice further into a pencil beam of a small solid angle. The beam size constraint can be understood intuitively as follows: if the radius is small ($< r_o$), one cannot be certain whether the detected overdensities in the galaxy distribution indicate large coherent structures, or small scale gravitational clustering with a typical size $\simeq r_o$. Larger radii overcome this problem, but then the depth of the survey is sacrificed since the sampling rate cannot be decreased significantly.

An optimum geometry can be defined using exactly the same considerations as developed by Kaiser (1986) for the wide-angle sparse-sampled case viz. balancing 'sampling noise' and 'clustering noise' under certain topological assumptions. For example, if the large scale structure is dominated by walls with a well defined surface density μ, separated by large voids of diameter λ, how can we best determine if such structures are common? Evidently, the sparse-sampled local surveys are *far from optimum* in this case. We will consider a 1D correlation function along a pencil beam, and show, that in the presence of walls, it can be very effective in detecting them.

3. OPTIMUM CHOICE OF THE PENCILBEAM GEOMETRY

To quantify these arguments in a simple analytical model we approximate such a pencil beam survey as a cylinder of length D, cross-sectional radius R, and calculate the clustering noise assuming a correlation function measured in bins of length L. The number of independent cells contributing to a given bin in the large scale correlation function is then $M = D/L$. If we assume that there is no large-scale clustering other than that of Eq.(1) truncated on scales over 30 h^{-1} Mpc, then the relative variance of the one-dimensional correlation function (the pair count N_p) along the line of sight can be written as

$$\epsilon_\xi^2 = \frac{\langle \Delta N_p^2 \rangle}{\langle N_p \rangle^2} = \frac{4}{M}\left(\frac{1}{n_o} + \xi_o\right) + \frac{1}{M}\left(\frac{1}{n_o} + \xi_o\right)^2 \approx 4\left(\frac{1}{N_T} + \frac{\xi_o}{M}\right), \tag{4}$$

where n_o is the mean number of galaxies in a cell, and $N_T = Mn_o$ is the total number of galaxies observed. For larger radii the first term is the important one, but for small R the quadratic term dominates. This can be considered as a 1-D equivalent of Kaiser's (1986) ε_ξ. Here ξ_o is the mean of the small scale correlation function averaged over a cell of radius R and length L, well approximated by an expression depending only on the geometry, obtained from numerical integration:

$$\xi_o = 2.24\ r_o^{1.8} L^{-0.75} R^{-1.05}. \tag{5}$$

The optimum number of galaxies to be observed is given by balancing the two criteria: $N_{crit} \approx M/\xi_o$, since oversampling will not reduce ϵ_ξ considerably further. In practice, if there are only a few pencil beams, it may be advisable to oversample slightly, by a factor of three or so, to make evident, that the data are not affected by sampling, but for a large scale program in the future this may not be necessary.

Now we address the choice of the best geometry, which is further complicated by constraints imposed by the field of view of the telescope, and other considerations, like the detection of individual walls. Substituting the appropriate numerical values for the number density of galaxies and the geometry relevant for the pencil beams, and assuming a sampling rate f, we find that ϵ_ξ scales inversely with D,

$$\epsilon_\xi^2 = \frac{1}{M} F(R,f) \propto D^{-1} F(R,f) \tag{6}$$

where $F(R,f)$ is the dimensionless noise per cell in the survey: the more independent cells there are, the smaller the total statistical noise becomes. For example, if the large scale topology is dominated by walls separated by $\lambda \approx 120\ h^{-1}$ Mpc, then the number of walls found by a pencil beam is $N_w = D/\lambda$. If the walls are clustered on scales $\ll D$, the signal to noise in the detection of a nonzero clustering amplitude is increasing with the depth of the survey in the usual manner

$$\Delta_\xi \propto N_w^{1/2} \propto D^{1/2} \tag{7}$$

Such a detection can be compared with the null-hypothesis of no large scale correlations quantitatively, e.g. in order to reject the null hypothesis at some significance level, or to compare with the results of N-body simulations. Any pencil beam redshift survey should then be optimized for achieving this using the smallest possible amount of telescope time. The parameter to maximize is the ratio $\eta = \Delta_\xi/\epsilon_\xi$.

$$\eta = \frac{\Delta_\xi}{\epsilon_\xi} \propto D\ F(R,f)^{-1/2} \tag{8}$$

Assuming Euclidean geometry for simplicity at this stage, the number of galaxies and the total observing time depend on R and D as

$$\begin{aligned} N_T &\propto fR^2 D, \\ T_o &\propto fN_T D^2 \propto fR^2 D^3. \end{aligned} \tag{9}$$

We can express the effective depth that can be reached, given T_o and R, scaling as

$$D \propto T_o^{1/3} R^{-2/3} f^{-1/3} \tag{10}$$

Substituting this into Eq.(8) yields:

$$\eta \propto T_o^{1/3} f^{-1/3} R^{-2/3} F^{-1/2} \tag{11}$$

For detection at a given significance level, we can alternatively ask what amount of observing time is required for different radii. Expressing the observing time T_o as a function of R, at a fixed value of η

$$T_o \propto \eta^3 f R^2 F^{3/2} \tag{12}$$

Results of a more complete numerical analysis, taking realistic values, and assuming a sampling rate close the complete $f = 1$ are presented in Figure 1. This curve has a well defined minimum at $R \approx r_o$, confirming our intuitive estimate very well. In a real situation, i.e. that our spectra are sky limited, and taking cosmological effects into account would further strengthen the point, that one should not go *arbitrarily* deep and narrow.

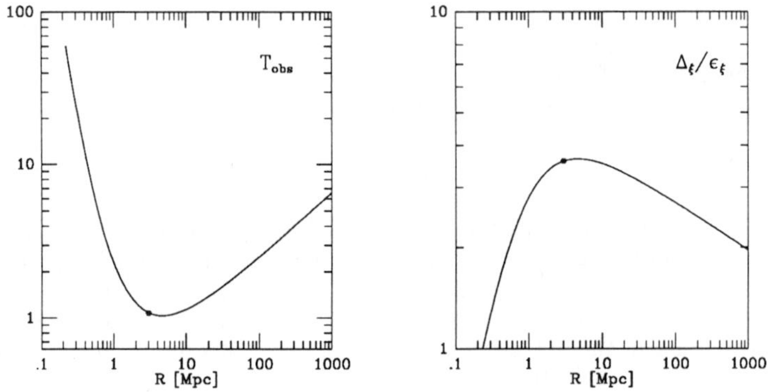

Figure 1. The dependence of the observing time T_o and the efficiency η on the radius of the cylindrical pencil beam, when detecting an existing signal in the large scale correlation function at a given significance. The small dot indicates the strategy of the current project, close to the optimal.

We have to consider another question: how frequently does one detect the individual walls. This problem has been recently addressed by deLapparent *et al* (1991), and Ramella *et al* (1991), and discussed by Geller (1990). They consider the 'Great Wall' data set, and place the wall at different redshifts. Since the wall does not have a uniform surface density, rather it is quite 'patchy', they infer that the BEKS pencil beams would typically detect more than 5 galaxies in 50 % of the walls in the redshift range of 0.1 - 0.4. This number is very sensitive on the diameter of the pencil beam, for a 3 h^{-1} Mpc diameter the probability is 20 %, while for 10 h^{-1} Mpc it is 90 %. They also note – entirely correctly – that there may be a bias in the surface densities from the pencil beams, since 'misses' are not counted, but detections are from the higher than average surface areas.

We can raise one additional point here: in these simulations the walls were assumed to be almost perpendicular to the line-of-sight, as the 'Great Wall' itself. If we consider a random inclination angle, the surface area increases as $1/\cos i$. The mean of this quantity is divergent, we have to truncate at the aspect ratio of the wall: $\ln(100/5) \approx 3$. Thus the typical diameter is increased by about $\sqrt{3} = 1.73$, from 6 to about 10 h^{-1} Mpc, making it more likely that the pencil beams would detect almost every 'Great Wall'-like structure.

Above a somewhat arbitrary threshold of 5 galaxies has been set. This can be further quantified: in order to detect a wall, the number of galaxies detected is compared to the small scale variance of the galaxy counts at that redshift. Consider cells of length $\Delta L \approx 5\ h^{-1}$ Mpc along the survey, equal to the approximate width of the walls (deLapparent *et al* 1986), and close to the resolution of our redshifts (300 km/s^{-1}). Those cells containing a wall should have a typical galaxy count N. The expected count n_o and its variance σ_N for the same cell are

$$\begin{aligned} N &= fR^2\pi\mu/\cos i \\ n_o &= fnR^2\pi\Delta L \\ \sigma_N^2 &= n_o^2\left(\frac{1}{n_o} + \xi_o\right) \approx nR^2\pi\Delta L\xi_o \end{aligned} \tag{13}$$

for small ΔL. If we set the detection threshold to $N > n_o + \alpha\sigma_N$, in the first order the detection is independent of f, assuming that the clustering noise dominates. For a realistic survey, with $\alpha = 3$ and our canonical numbers we get $f > 0.1$, and $N > 2.75$. The previously used $N = 5$ corresponds to a marginally better detection at the 3.3σ level. One can see now the origin of this quantity. If $N \approx 1$, there is little use for this argument, since we will never detect fractional galaxies, this is another reason to consider $N = 5$ as a very reasonable detection threshold. So generally, $f \ll 0.1$ seems to exclude the detection of the individual walls, leaving only little freedom in choosing our sampling by statistical considerations, unless we are ready to give up the individual detections.

To consider this problem in its completeness can be quite complicated, involving higher order statistics. If we wanted to set the optimal threshold for separating walls from small scale clumps, we should study the distribution P_N, the probability of finding N galaxies in a given cell. The null hypothesis of the small scale gravitational clustering can yield such a distribution, and this should be compared to the similar one derived from the cell counts within the 'Great Wall', as discussed in Ramella *et al* (1991). Then one should chose the threshold for the optimal separation of the two. However, the P_N distribution contains contributions from all N-point correlation functions.

4. STATISTICS OF ONE-DIMENSIONAL PROJECTIONS

As it is well known in statistics, one-dimensional subsets or projections of three-dimensional random fields have rather different statistical properties from the original one. If we have a density fluctuation field $\delta(\mathbf{x})$, and an observational selection function $W(\mathbf{x})$, we measure $\Delta(\mathbf{x}) = \delta(\mathbf{x})W(\mathbf{x})$, their product, excluding sampling noise, etc. If $\delta(\mathbf{x})$ is a random field with a known power spectrum $P_3(\mathbf{k})$, the power spectrum of the observed field Δ can be written using the convolution theorem:

$$\langle|\Delta(\mathbf{k}')|^2\rangle = \langle|\delta * w|^2\rangle = \frac{1}{(2\pi)^3}\int d^3\mathbf{k}\ P_3(k)|w(\mathbf{k}' - \mathbf{k})|^2 \tag{14}$$

where $w(\mathbf{k})$ is the window function, the Fourier transform of the selection function $W(\mathbf{x})$. In obtaining the above relation, we have only used the translational invariance of the ensemble average, requiring that

$$\langle \delta(\mathbf{k}')\delta^*(\mathbf{k})\rangle = (2\pi)^3 \delta^{(3)}(\mathbf{k}' - \mathbf{k})\ P_3(\mathbf{k}) \tag{15}$$

where $\delta^{(3)}$ is the three dimensional Dirac-delta. For simplicity let us consider an extreme window, an infinitely thin pencil along the z-axis, extending to infinity. The window function in this case is $w(k_1, k_2, k_3) = 2\pi\delta^{(1)}(k_3)$. It is easy to show, that in this case the ensemble averaged one-dimensional power spectrum $P_1(k)$ becomes

$$P_1(k_3) = \frac{1}{(2\pi)^2} \int dk_1 dk_2\ P_3(k_1, k_2, k_3) \tag{16}$$

The integral describes a projection: in a given point k_3 along the survey axis in Fourier-space we project all power in the (k_1, k_2) plane, perpendicular to the axis, intersecting it at k_3, into a single value $P_1(k_3)$, a lot of power is aliased from small to large scales. Now we can consider various options for the three-dimensional power spectrum, from the most general case to the homogenous isotropic process. Until now, our treatment is fully general. If $\delta(\mathbf{x})$ is not a homogenous random process, like the cubic lattice discussed in §2, generally $P_3(\mathbf{k})$ can have sharp spikes, that can show up in $P_1(k_3)$ as well. On the other hand if $\delta(\mathbf{x})$ is a homogenous and isotropic random process, the power spectrum $P_3(\mathbf{k})$ will depend on $k = |\mathbf{k}|$ only, and $P_1(k)$ can be written as

$$P_1(k) = \frac{1}{2\pi} \int_k^\infty d\ell\, \ell\, P_3(\ell) = P_1(0) - \frac{1}{2\pi} \int_0^k d\ell\, \ell\, P_3(\ell). \tag{17}$$

In this latter case there is an other implication: the one dimensional power spectrum originating from a three dimensional homogenous isotropic random process (not necessarily Gaussian) is always monotonically decreasing. As the simplest example, plane waves with their wave vectors perpendicular to the line of sight will contribute to fluctuations in the DC level of the 1D power.

As one can see in the form of Eq.(17) if there is a large scale (low frequency) cutoff in P_3 at the wave number k_0, then P_1 is constant from 0 to k_0 : it is a 'white noise', with the level $P_1(0)$. The noise level can be physically interpreted as the three dimensional variance $\sigma_3^2 = \langle |\delta(\mathbf{x})|^2 \rangle$ multiplied with the expectation value $\langle 1/\ell \rangle / \pi$, the inverse wave number where the maximum contribution to P_1 comes from

$$P_1(0) = \frac{1}{2\pi} \int_0^\infty d\ell\, \ell\, P_3(\ell) = \frac{1}{\pi} \langle \frac{1}{\ell} \rangle \left[\frac{1}{2\pi^2} \int_0^\infty d\ell\, \ell^2\ P_3(\ell) \right] \tag{18}$$

The values of $\Delta(k)$ at different k are not independent, they correlate due to the convolution with the window function:

$$\langle \Delta(\mathbf{k}'')\Delta^*(\mathbf{k}')\rangle = \frac{1}{(2\pi)^3} \int d^3\mathbf{k}\ P_3(k) w(\mathbf{k}'' - \mathbf{k}) w^*(\mathbf{k}' - \mathbf{k}) \tag{19}$$

For realistic survey geometries we have to calculate the appropriate window function, and with a basic model for $P_3(k)$, like CDM or a minimal truncated power law correlation function we can predict the 1D power spectrum. If we approximate the survey as a cylinder of fixed radius R and effective depth D, there are two separable window functions, a radial and a transverse one. For more complicated geometries generally the transverse window function depends on the z coordinate, for a fixed solid angle the z dependence is relatively simple. There is a natural coherence scale in k-space due to the large D, roughly $k_* = 1/D$. The projection in k-space takes place then in disks of this thickness, and 3D power is projected only from within a transverse window of conjugate radius of approximately $2\pi/R$. For a cylindrical survey the transverse window function is independent of the z coordinate

$$w(\ell R) = 2\frac{J_1(\ell R)}{\ell R}, \quad (20)$$

where ℓ is the magnitude of the wave vector in the plane of the projection. The physical importance of these two length scales can be easily understood, the depth D will enter through the number of independent cells along the line of sight, while the transverse scale R determines the level of the small scale clustering noise entering the problem. Interestingly enough, if the survey geometry consists of multiple pencilbeams, sparsely sampling a larger surface area with a radius B, the important scale will be this larger B, rather than the small R (see §6 for more). Let us consider two such transverse windows (both two dimensional), the first describing the individual pencil beams with radius R, $W_R(\mathbf{x})$, and the wider one with the scale B, $W_B(\mathbf{x})$, describing the pattern of the sparsely distributed pencils. The total selection window becomes the convolution,

$$W(\mathbf{x}) = \int d^2\mathbf{x}'\, W_B(\mathbf{x}')\, W_R(\mathbf{x}-\mathbf{x}'). \quad (21)$$

Due to the convolution theorem, the window functions in Fourier space will be the products, thus $w(\mathbf{k}) = w_B(\mathbf{k})\, w_R(\mathbf{k})$. The total width of $w(\mathbf{k})$ will determine how much small scale (large k) power will be included in the integrals. The window function narrower in k-space will dominate, thus the relevant scale is the wider radius B. If W_B consists of N points over the area size B, $\langle|w_B(\mathbf{k})|^2\rangle$ will have a DC component of $1/N$, thus the correlation noise due to the width of a single pencil beam will be attenuated by this amount.

5. APPLICATION TO THE BEKS SURVEY

The advent of efficient multi-object spectrographs on large optical telescopes has considerably accelerated the progress towards completing redshift surveys of faint galaxies. The instrumentation used in such surveys typically sample 10-50 galaxies within fields of 10-40 arcmin diameter and thus 'pencil beam' geometries are produced which are characteristically different from strategies used to map the galaxy distribution locally. With 40' fields on 4m telescopes, the diameter of the survey at the median redshift of about $z \approx 0.3$ is 6 h^{-1} Mpc; such geometries are close to optimal for detecting wall-like topologies on scales comparable to those revealed in the CfA surveys (§2).

The deep surveys (Koo, Kron and Szalay 1986, Broadhurst, Ellis and Shanks 1988) were originally motivated by the need to determine the redshift distribution of magnitude-limited samples as a means of discriminating between various models explaining the excess numbers of faint counts (see Ellis 1991 for a review). However, since such surveys reach considerably deeper than earlier work, the redshift distributions have also been used to quantify the large scale distribution of galaxies (BEKS 1990).

BEKS provided evidence from a combined sample of galaxies in the North and South Galactic poles for structures on scales >100 h^{-1} Mpc with a provocative regularity. The BEKS survey consisted of two deep surveys spanning 2000 h^{-1} Mpc – the deepest so far – and two previous brighter surveys by others (Kirshner *et al* 1983, Peterson *et al* 1986), which together cover a volume well approximated by a cylinder of a constant comoving radius (Figure 2). The two Northern fields lie within the CfA slice, and the 'Great Wall' is readily detected. Surprisingly, however, at large radial distances most galaxies lie in a few discrete 'spikes' separated typically by 130 h^{-1} Mpc. The spread of spike separations is, in fact, surprisingly small $\simeq$23%. This is revealed using the 1-D pair counts (Figure 3) and the 1D power spectrum (Figure 4) which has a very sharp peak at the wave-number corresponding to 130 h^{-1} Mpc.

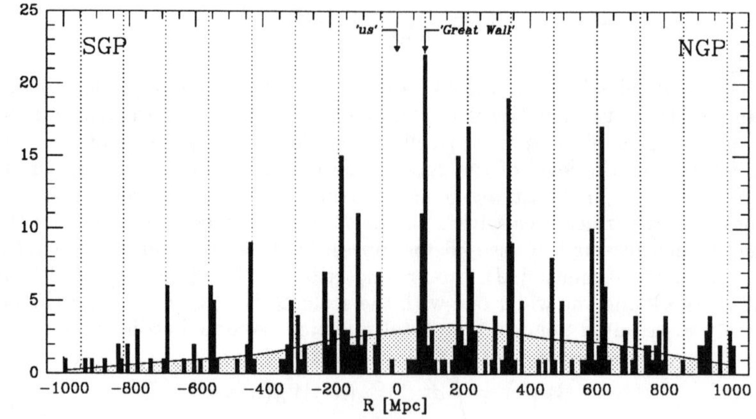

Figure 2. The combined redshift distribution of the four data sets, consisting of a deep and narrow (30 arc mins), and a bright and wide, survey at each of the Galactic Poles (South to the left, North to the right). The full selection function (smoothed from the data) is shown as a shaded curve. We used $\Omega = 1$ and $h = 1$ to convert redshifts to comoving distance. Both our position and that of the 'Great Wall' is indicated with small arrows.

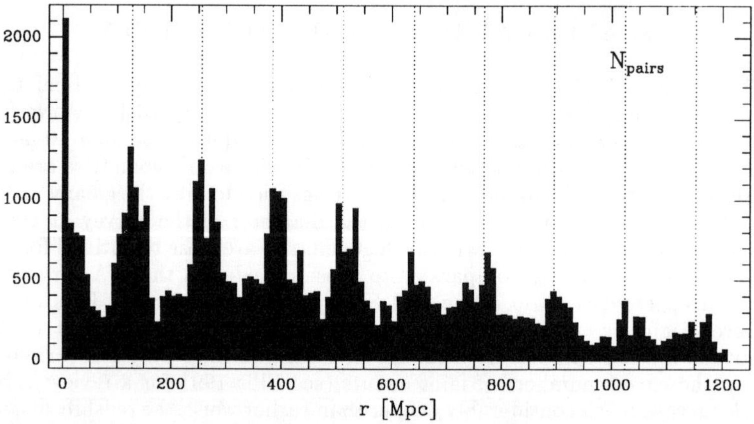

Figure 3. One-dimensional pair count correlation for all the redshift data available in the NGP-SGP axis. The dashed line indicates the multiples of 128 h^{-1} Mpc. The relative excess of the pair counts over the smooth background is at least 1.5 at the first peak. Note the anticorrelation between 50 and 100 h^{-1} Mpc.

We can approximate our radial selection function roughly with a Gaussian, this multiplies the true galaxy distribution. In Fourier space the true power spectrum of the galaxy distribution is convolved with the square of the window function, also a Gaussian with an approximate conjugate width of $k_* = 0.001$ Mpc^{-1}. This is the natural line width of our power spectrum, the coherence length in k-space. For this reason we chose this as our units for the wave numbers. The width of the spectral feature at $k = 7.7$ is close to 1, hardly larger than the natural line width due to the selection window, implying that the intrinsic coherence length in r space is similar to the range of our observations!

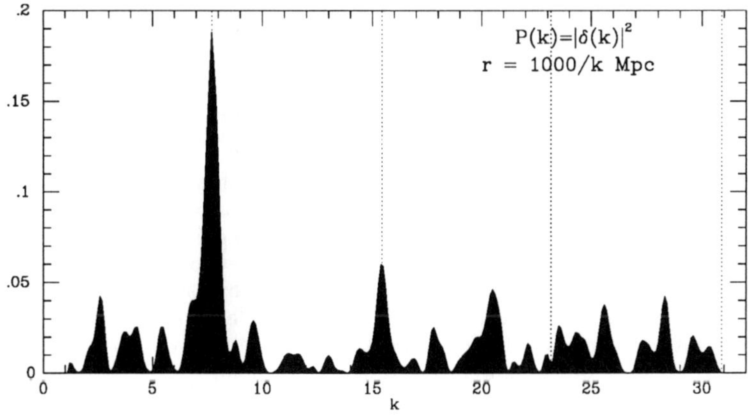

Figure 4. One-dimensional power spectrum of the BEKS sample. The dashed line corresponds to a separation of 128 h^{-1} Mpc, the best fit to the first two harmonics. The one-dimensional power spectrum is a non-trivial projection from three dimensions. The observed peak has a formal probability of 2.2×10^{-4}. The physical scale corresponding to the wave number k is $r(k) = 1000/k \ h^{-1}$ Mpc.

The first question we have to ask, how likely is this just due to chance, is it an upward fluctuation, arising from a simple underlying null hypothesis? When estimating this probability, we have to be aware, that this is an *a posteriori* statistics, i.e. we have first chosen the highest peak in the observed power spectrum, and asked statistical questions later. The probability of finding such peaks in the power spectrum by chance is further complicated by the fact that the small scale correlations increase the variance of the number of galaxies. Thus before we proceed any further we specify our minimal null hypothesis:

- galaxies have the well-known correlation function $\xi(r) = (r/5 \ h^{-1}\,\mathrm{Mpc})^{-1.8}$
- they are uncorrelated on larger scales ($> 30 \ h^{-1}$ Mpc)

As we have seen above, this null hypothesis implies a white noise power spectrum on scales above the truncation of the small scale $\xi(r)$. We have to be careful to distinguish between ensemble averages and single realizations. While ensemble averages will be monotonically decreasing, or remain white noise, the individual realizations can have peaks in various places. Also, if our survey is not deep enough to be statistically fair, maybe the hypothesis of a homogenous isotropic process is not valid (e.g. we sample only the few nearest cells).

We denote our Fourier transform of the galaxy distribution as f_k, and the single realization of the 1D power spectrum from the data as A_k, calculated by treating each galaxy as a Dirac delta at its observed position r_n, neglecting redshift distortions,

$$f_k = \frac{1}{N} \sum_{1}^{N} e^{2\pi i k r_n} \qquad (22)$$

$$A_k = |f_k|^2.$$

Under our null hypothesis, the power spectrum amplitudes have an exponential distribution, since the real and complex parts of the Fourier transform are distributed as a Gaussian. Even if parts of the signal in r-space are very non-Gaussian, the power spectrum amplitudes satisfy this property quite well. In this case the probability distribution is described by the noise level A_0:

$$dP(A_k) = \exp(-A_k/A_0)\frac{dA_k}{A_0} \tag{23}$$

All models for the small scale clustering or a more detailed influence of the survey geometry enter through this single number. One can have external and internal estimators of this quantity. All those estimators using a certain two point function are external, in the sense that little is known about the fine details of the two-point correlation function on scales much beyond 15 h^{-1} Mpc, we only rely upon our assumptions.

The amplitude for this white noise is easily calculated analytically from our null hypothesis with the truncated $\xi(r)$, if we assume a simple cylindrical geometry. The noise level is the sum of the sampling and clustering noise:

$$A_0 = \frac{1}{N_T} + \frac{\xi_o}{M}, \tag{24}$$

where N_T is the total number of galaxies in the survey, ξ_o is the mean value of the small scale correlation function over a cell of length 30 h^{-1} Mpc along the pencil beam, and M is the number of independent cells along the line of sight, as discussed in §2. For the BEKS data $N_T = 400$, and $M/\xi_o = 80$, depending only on the geometry. The expected level of white noise is thus:

$$A_0 = \frac{1}{400} + \frac{1}{80} = \frac{1}{67} = 0.015 \tag{25}$$

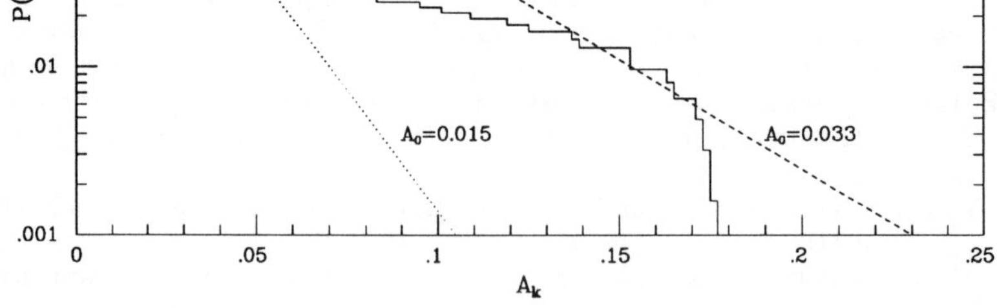

Figure 5. The cumulative distribution of the power spectrum amplitudes A_k from Fig.4. The shoulder is the effect of the large peak, while the small amplitude part shows the exponential behaviour. The lines show various levels of the white noise, 0.015 is the best fit to the low amplitude part.

There is a way to obtain this number independently of any hypotheses, using an internal estimator of the same quantity. We can calculate the cumulative distribution of the actual – numerically determined – power spectrum amplitudes, and fit an exponential to this distribution. The slope of the exponential at small amplitudes gives us the required noise level, which turns out to be

$$A_0 = 0.015, \tag{26}$$

in excellent agreement with the estimate in Eq. (26). Of course, our analytical model is somewhat oversimplified, a proper treatment of the window shapes coming from the radial and angular selection criteria yields a 25 % higher number. This small difference may arise from the assumption that $\xi(r)$ extends out to 30 h^{-1} Mpc, whereas empirical values of J_3, the integral of ξ, imply convergence by 15 h^{-1} Mpc; this would somewhat overestimate the strength of the small scale correlations. Also, a similar variance in the intrinsic estimator is expected from higher (4th) moments of the galaxy distribution.

Such estimates for the general noise level have to be confronted with the observed BEKS peak amplitude $P_1(7.7) = 0.177$. Since under our null hypothesis the amplitude of the power spectrum has an exponential distribution, we can model our a posteriori observation as determining the power spectrum in N_k independent bins, then picking the one with the largest amplitude. One may wonder, whether the formalism developed by Bardeen *et al* (1986) on peaks of random Gaussain fields should be used. However, since we pick the largest peak in the Fourier spectrum, it is also the maximum value of the observed spectrum. The cumulative distribution of the maximum of N samples drawn from an exponential distribution is well known (Kendall and Stuart 1977):

$$G(x|N) = 1 - \exp(-Ne^{-x}) \approx Ne^{-x}, \tag{27}$$

where $x = A_k/A_0 = 11.8$ is the relative height of the peak over the noise. In our case we set the truncation scale at 30 h^{-1} Mpc, corresponding to $k \approx 30$ in our units of the natural line width. This is equal to the number of independent k-bins over the range where the white noise behaviour is expected to hold. Our final probability of finding such a peak in the power spectrum becomes

$$G(11.8|30) = 2.2 \times 10^{-4}. \tag{28}$$

suggesting a statistically significant clustering of galaxies on very large scales. The estimate is not sensitive to the truncation of $\xi(r)$ at 30 h^{-1} Mpc. On the other hand, it is very sensitive on the noise level A_0 : if A_0 turned out to be a factor of 2 larger, 0.030, the corresponding probability would grow to 0.1. This very strong dependence is due to the appearence of A_0 in the double exponential of G. It is easy to see, how a slight mistake or a somewhat incorrect assumption in calculating the external estimator may lead to a very different probability than the one quoted here. This is the reason, why we strongly advocate the use of the intrinsically determined noise level only, and discard the use of the external estimations.

5. DISCUSSION

As BEKS stressed in their paper, further data is urgently needed to address many questions posed by their striking result.

- How sharp is the detected 130 h^{-1} Mpc scale?
- Does the galaxy distribution show similar structures in other directions?
- Do the observed structures have a large transverse extent as would be expected if the large scale topology is cellular?

In order to get a better handle on the errors, and to determine which part of the data contributes most of the signal, we have successively excluded individual spikes in the redshift histogram, as well as whole catalogs, which are combined into our survey. After these eliminations the characteristic scale still shows up in every part of the data set, although the statistical significance decreases according to the expectations.

We have to be aware that the derived radial distances arise directly from the raw redshifts, no velocity corrections have been used. On the other hand, on the relevant scales ($> 100\ h^{-1}$ Mpc), only very substantial velocities can have any appreciable effect. Since the survey volume is still small enough to contain only about 0.2 Abell cluster per cone, it is rather unlikely that objects with cluster-like velocity dispersions are present. We know of one Zwicky cluster in one of the fields, and of course our pencil beam at the NGP passes near Coma.

When we convert from redshifts to comoving distance, the conversion is sensitive to the curvature of the universe. By taking various values of Ω one can shift the high redshift tail of the distribution relative to the near part. This is reflected in the pair correlations. Tests indicate, that to first order the sharpness of the peak in the Fourier spectrum is rather insensitive to our choice of Ω, although it may split up into a doublet, containing the same amount of power for very low values of Ω.

We have observations under way in several other directions. Preliminary analyses of these directions indicate, that those other fields by themselves are at least as clustered on $>100\ h^{-1}$ Mpc scales as the SA57 (NGP) field. Adding the SGP, the NGP and the two bright surveys effectively tripled the dynamical range of the survey, which resulted in a similar improvement in the signal-to-noise ratio of the correlation function. We are currently taking data in the complementary directions to these new surveys, expecting a similar trend in the S/N ratio.

In order to assess the transverse extent of the spikes detected in the data we can use the angular positions within the spikes. This already indicates that some of the spikes extend almost uniformly across the field. As discussed in §4, it is rather advantageous to use a sparsely distributed pattern over a 25-100 square degree area of 6-10 well sampled pencil beams of 0.3 square degrees each. With this we gain on several grounds, without much sacrifice:

- the effective radius grows to several degrees, lowering the correlation noise
- the wall detection is enhanced by the use of multiple beams
- it is well suited for the multiobject spectrographs
- it improves the detection of walls compared to the same number of galaxies from a uniformly sparse survey

Using the Nessie fiber spectrograph at the Kitt Peak 4m Mayall telescope we are in the process of obtaining redshifts (up to 70 each) in several small (40 arc min) pencil beams within 5 degrees of our well sampled areas. These transverse separations correspond to distances of about 30 h^{-1} Mpc. The cross-correlation of these sparse surveys with the well sampled central ones enables us to derive statistically meaningful information about the transverse extent of the walls, even if the individual surveys are undersampled. Already 160 redshifts were obtained around the SA57 field, in agreement with our expectations. More than 400 new redshifts were obtained in the SGP region recently, confirming the previously found Southern spikes over a 10×10 degree area rather dramatically.

We expect that within a year we will have at least four independent directions with over 200 galaxies each, around some of those we will have sparsely sampled with a few adjacent pencil beams, and altogether have a database of over 2000 faint galaxies total. This should be large enough to answer quite a few of the still outstanding questions.

BIBLIOGRAPHY

Bahcall, N. and Soneira, R. 1983. *Ap. J.*, **270**, 20.

Bardeen,J., Bond, J.R., Kaiser, N. and Szalay,A.S. 1986. *Ap. J.*, **304**, 15.

Blumenthal,G.R., Faber, S.M., Primack, J.R. and Rees,M.J. 1984. *Nature*, **311**, 517.

Broadhurst,T.J., Ellis, R.S., and Shanks, T. 1988. *M.N.R.A.S.*, **235**, 827.

Broadhurst,T.J., Ellis, R.S., Koo, D.C. and Szalay, A.S. 1990. *Nature*, **343**, 726.

Burstein, D., Davies, R.L., Dressler, A., Faber, S.M., Lynden-Bell, D., Terlevich, R.J. and Wegner, G., 1986. In *Galaxy Distances and Deviations from the Hubble Expansion*, ed. B. Madore and B. Tully, (Boston:Reidel), p.123

Chincarini, G.L., Giovanelli, R., Haynes, M.P. 1983. *Astron. Astrophys.*, **121**, 5.

deLapparent, V., Geller, M.J. and Huchra, J.P. 1986. *Ap. J.*, **302**, L1.

deLapparent, V., Geller, M.J. and Huchra, J.P., 1991. In preparation.

Geller, M.J., 1990. This volume.

Giovanelli, R., Haynes, M.P. and Chincarini, G.L. 1986. *Ap. J.*, **300**, 77.

Kaiser, N. 1984. *Ap. J.*, **273**, L17.

Kaiser, N. 1986. *M.N.R.A.S.*, **219**, 785.

Kendall, M. and Stuart, A., 1977. *The Advanced Theory of Statistics*, (Macmillan Publishing: New York).

Kirshner, R.P., Oemler, A., Schecter, P.L. and Schectman, S.A. 1983. *Astron. J.*, **88**, 1285.

Koo, D.C., Kron, R.G. and Szalay, A.S., 1987. *Proc. of the 13th Texas Symposium on Rel. Astrophys.*, ed. M.P.Ulmer, (World Scientific) p. 227.

Koo, D.C., Kron, R.G., Munn, J. and Szalay, A.S., 1990. In preparation.

Lynden-Bell, D., Faber, S.M., Burstein, D., Davies, R.L., Dressler, A., Terlevich, R.J. and Wegner, G. 1988. *Ap. J.*, **326**, 19.

Park, C. 1990. *M.N.R.A.S.*, **242**, 59p.

Peebles, P.J.E. 1980. *The Large Scale Structure of the Universe*, (Princeton University Press).

Peebles, P.J.E., 1989. In *Fractals in Physics*, eds. A. Aharony and J. Feder (North Holland).

Ramella, M., Geller, M.J. and Huchra, J., 1991. In preparation.

Rubin, V. Thonnard, N., Ford, W.K. and Roberts, M. 1976. *Astron. J.*, **81**, 719.

Szalay, A.S., Koo, D.C., Ellis, R.S. and Broadhurst, T.J., 1990. In preparation.

Voronoi, G. 1908. *Reine Angew. Math.*, **134**, 198.

White, S.D.M., Frenk, C.S., Davis, M. and Efstathiou, G. 1987. *Ap. J.*, **313**, 505.

THE ORIGIN OF LARGE-SCALE PRIODICITY

Neta A. Bahcall
Princeton University Observatory
Princeton, NJ 08544-1001

ABSTRACT

The apparent ~ 130 h^{-1} Mpc periodic distribution of galaxies recently observed in pencil-beam surveys is shown to originate in known large-scale superclusters. The superclusters, covering the high-latitude sky to $z \leq 0.08$, show the same peaks that appear in the pencil-beam survey, indicating that the galaxy peaks are drawn from these superclusters. Predictions for future pencil-beam surveys in different directions are given. The results suggest a network structure of large (~ 100 h^{-1} Mpc) supercluster surfaces surrounding low density regions.

INTRODUCTION

Recent observations of the redshift distribution of galaxies in narrow (~40') pencil-beam surveys to $z \lesssim 0.3$ (Broadhurst *et al.* 1990; hereafter BEKS) reveal a highly clumped and apparently periodic distribution of galaxies. The distribution features peaks of galaxy counts with an apparently regular separation of 128 h^{-1} Mpc, with few galaxies between the peaks. What is the origin of this clumpy, periodic distribution of galaxies ? What does it imply for the nature of the large-scale structure? We investigate these questions observationally, by comparing the specific galaxy distribution with the distribution of known clusters and superclusters in a wide-angle survey. Statistical analyses of the BEKS galaxy distribution have recently been discussed by several authors; these include the periodicities expected in bubble and sheet models (Kurki-Suonio *et al.* 1990), in CDM model simulations (Park and Gott 1990), in a Voronoi Tessellation model (Ikeuchi and Turner 1990, van de Weygaert 1990), and in power spectrum analysis (Kaiser and Peacock 1990). In the present work, (Bahcall 1990), we identify from observations the specific known structures in which the galaxy peaks originate; we then discuss the implications of these identifications for the nature of the large-scale structure.

ANALYSIS AND RESULTS

The supercluster catalog of Bahcall and Soneira (1984; hereafter BS) was used to compare the superclusters and the narrow-beam galaxy distribution. This catalog is the largest complete sample of superclusters available. The 3-D catalog contains all superclusters to $z \leq 0.08$, covering the entire high-latitude sky ($|b| \geq 30^{o}$) for $\delta \geq -27^{o}$. Superclusters are defined as groups of rich clusters of galaxies above a given spatial density enhancement of clusters. A map of the BS superclusters is presented in Fig. 1. The mean space density of the superclusters (Fig. 1) is ~ 10^{-6} Mpc^{-3}. This corresponds to a mean separation of superclusters of ~ 100 h^{-1} Mpc.

The redshift distribution of the superclusters is presented in a redshift cone diagram in Fig. 2 (for the slice $\delta = 0^{o} - 40^{o}$). The supercluster contours in Fig. 2 represent the observed extent, in redshift and projection, of the densest part of the superclusters, i.e., the most highly clumped group of rich clusters (at a spatial overdensity of clusters of $f \geq 20$). Any more extended tails of superclusters are not shown, but are expected to be present in some cases to scales of $\gtrsim$ 100 h^{-1} Mpc (Chincarini *et al.* 1981, Bahcall and Soneira 1982, Oort 1983).

The redshift distribution of the superclusters is plotted as a histogram (shaded area) in Fig. 3 (with total count normalized to the number of clusters per supercluster). This distribution is superimposed on the narrow-beam galaxy distribution of BEKS (as presented in their preprint for the combined beam surveys, corrected for the observed selection function). It is apparent from Fig. 3 that the supercluster distribution and the BEKS galaxy distribution are essentially identical for $z \lesssim 0.1$. It indicates that the galaxy clumps observed in the narrow-beam survey originate from these superclusters as the beam crosses the superclusters surface. The main contributors to the supercluster clumps are indicated in Fig. 3. The first northern clump originates from the Coma-Hercules superclusters (known also as the Great-Wall, Geller & Huchra 1989; see Fig. 2); the second northern clump is mostly due to the large Corona Borealis supercluster (BS12). The first southern clump is due to the Perseus supercluster (mostly below the limits of the BS catalog, but BS19 is its extension; e.g. Giovanelli *et al.* 1986); the second, rather broad southern clump, is due to a few superclusters in this region (BS 1-3, 20).

The narrow-beam survey of BEKS is directed towards the North and South Galactic Poles; the BS superclusters are distributed widely over the sky, from near the galactic poles to $\gtrsim$ 150 h^{-1} Mpc away in projection (Fig. 1). The position of the superclusters closest to the poles (i.e., nearest to the BEKS survey), to within ~ 50 h^{-1} Mpc, are indicated by arrows in Fig. 3. Since some of the superclusters that contribute to the BEKS clumps are at separations $\gtrsim$ 50 - 100 h^{-1} Mpc from the poles (e.g., Perseus, BS2, 20, as well as Hercules tail), these structures must have large surfaces extending to scales $\gtrsim$ 50 or 100 h^{-1} Mpc, in order to intersect the polar beams. A similar conclusion regarding the extent and shape of superclusters was reached by Bahcall and Soneira (1982) based on studying superclusters around the Bootes void (Kirshner *et al.* 1981). This earlier analysis indicated that the high density supercluster regions are embedded in larger surfaces, $\gtrsim$ 100 h^{-1} Mpc in size, and that these large pancake-like structures surround large voids (i.e., underdense regions).

The observed (BEKS) number of clumps and their typical mean separation are consistent with the mean number density of superclusters and their estimated average extent. The BEKS mean separation of ~ 130 h^{-1} Mpc corresponds to an average supercluster cross-section of $\sim \langle (90\ h^{-1}\ \mathrm{Mpc})^2 \rangle$; this is comparable to the observed extent of supercluster surfaces (eg., $\sim 10^2 \times 10^2 \times 20\ h^{-1}$ Mpc (BS, Chinacarini *et al.* 1981, Bahcall and Soneira 1982, Oort 1983, Giovanelli *et al.* 1986, Postman *et al.* 1986)). The estimated average galaxy overdensity in these superclusters is also comparable with that observed for the average BEKS clump. The filling factor of the extended superclusters is ~ 0.2.

The BS superclusters exhibit weak positive correlations on scales ~ 100 - 150 h^{-1} Mpc (Bahcall and Burgett 1986). This implies that the superclusters, and thus their related galaxy clumps, are not randomly distributed, but are located in some weakly correlated network of superclusters and voids, with typical mean separation of ~ 100 - 150 h^{-1} Mpc.

Based on the above picture of a large-scale supercluster network, what would we expect future narrow-beam surveys in different directions to reveal? Figs. 1 - 3 indicate that surveys in different directions should continue to yield a highly clumped distribution of galaxies, with typical mean separation of ~ 100 - 150 h^{-1} Mpc. The positions of the clumps, however, will change somewhat with direction. Moderate changes in direction (≲ tens Mpc) may yield only small changes in clump positions, since the same large-scale structure may still be in view. If the structure is tilted to the line-of-sight, the clump positions should indicate this gradual redshift shift with position (as is indeed indicated in the BEKS nearby surveys). In the rarer occasions of a beam crossing an edge-on supercluster, an appropriately broad redshift clump would be observed. Large changes of beam directions (≳ 150 h^{-1} Mpc) should yield more significant variation in peak positions within the typically expected range of ~ 100 - 150 h^{-1} Mpc. Higher observational resolution in the redshift direction in the surveys will also yield a detectable shift in the peak position (e.g., note that some of the broader BEKS clumps are composites of more than one supercluster). The apparent periodicity in the galaxy distribution will be greatly reduced when pencil-beam surveys in different directions are combined. However, separate beams in various directions are expected to continue to show a clumped, quasi-periodic distribution of galaxies, with typical mean separation of ~ 100 - 150 h^{-1} Mpc.

REFERENCES

Bahcall, N.A., *Ap.J. Letters* (submitted)
Bahcall, N.A., and Burgett, W.S., 1986, *Ap.J. Letters*, **300**, L35.
Bahcall, N.A., and Soneira, R.M., 1982, *Ap.J. Letters*, **258**, L17.
Bahcall, N.A., and Soneira, R.M., 1984, *Ap.J.*, **277**, 27 (BS84).
Broadhurst, T.J., Ellis, R.S., Koo, D.C., and Szalay, A.S., 1990, *Nature*, **343**, 726 (BEKS).
Chincarini, G., Rood, H.J., and Thompson, L.A., 1981, *Ap.J. Letters*, **249**, L47.
Geller, M.J., and Huchra, J.P., 1989, *Science*, **246**, 897.
Giovanelli, R., Haynes, M.P., and Chincarini, G., 1986, *Ap.J.*, **300**, 77.
Ikeuchi, S., and Turner, E.L., 1990, submitted to *MNRAS*.
Kaiser, N., and Peacock, J.A., 1990, preprint.
Kirshner, R.P., Oemler, A. Jr., Schechter, P.L., and Shectman, S.A., 1981, *Ap.J. Letters*, **248**, L57.
Kurki-Suonio, H., Mathews, G.J., and Fuller, G.M., 1990, *Ap.J. Letters*, **356**, L5.
Oort, J., 1983, *Ann. Rev. Astron. Astrophys.* **21**, 373.
Park, C., and Gott, J.R., 1990, submitted to *MNRAS*.
Postman, M., Geller, M.J., and Huchra, J.P., 1986, *A.J.* **92**, 1238.
van de Weygaert, R., 1990, submitted to MNRAS.

FIGURE CAPTIONS

Figure 1: Projected map of the northern (top) and southern (bottom) clusters and superclusters from the BS catalog (R≥0 case). Outer contour is $|b| = 30^o$. The galactic poles, and thus the directions of the BEKS survey, are at the centers of these polar maps. The supercluster density contours represent the 3-D density enhancement of the superclusters, from f = 20 to f = 400. The numbers refer to the supercluster number. (The Coma supercluster, BS10, appears at f = 10 (or f = 20 for the R≥1 sample), and is marked by a dotted contour).

Figure 2: Redshift cone diagram of the superclusters in the $\delta = 0^o$ - 40° slice. The inner and outer circles (z = 0.02 and 0.08, respectively), and the two diagonal lines (marking the low-latitude zone), represent the completeness boundaries of the catalog. (The known Perseus supercluster (Giovanelli *et al.* 1986), which is below the catalog limit (i.e., at $z < 0.02$ and $|b| < 30^o$), is marked by a dotted contour; its extension - B19 - is at the catalog boundary. We also indicate by a dotted contour supercluster BS1, which is near the SGP (i.e., BEKS survey), but outside the $\delta = 0^o - 40^o$ southern slice). The plotted extent of the superclusters correspond only to their high-density regions (i.e., $f \geq 20$, cf. Fig. 1), and does not show their expected extensions (see text).

Figure 3: Histogram of the redshift distribution of the Fig. 2 superclusters (shaded area), for $z \simeq -0.1$ to 0.1, superimposed on the BEKS galaxy distribution (corrected for selection, BEKS preprint). The location of the superclusters are marked with higher resolution by the top rectangular symbols, and are identified by name. The arrows indicate the superclusters closest ($\lesssim 50\ h^{-1}$ Mpc) to the BEKS survey (poles).

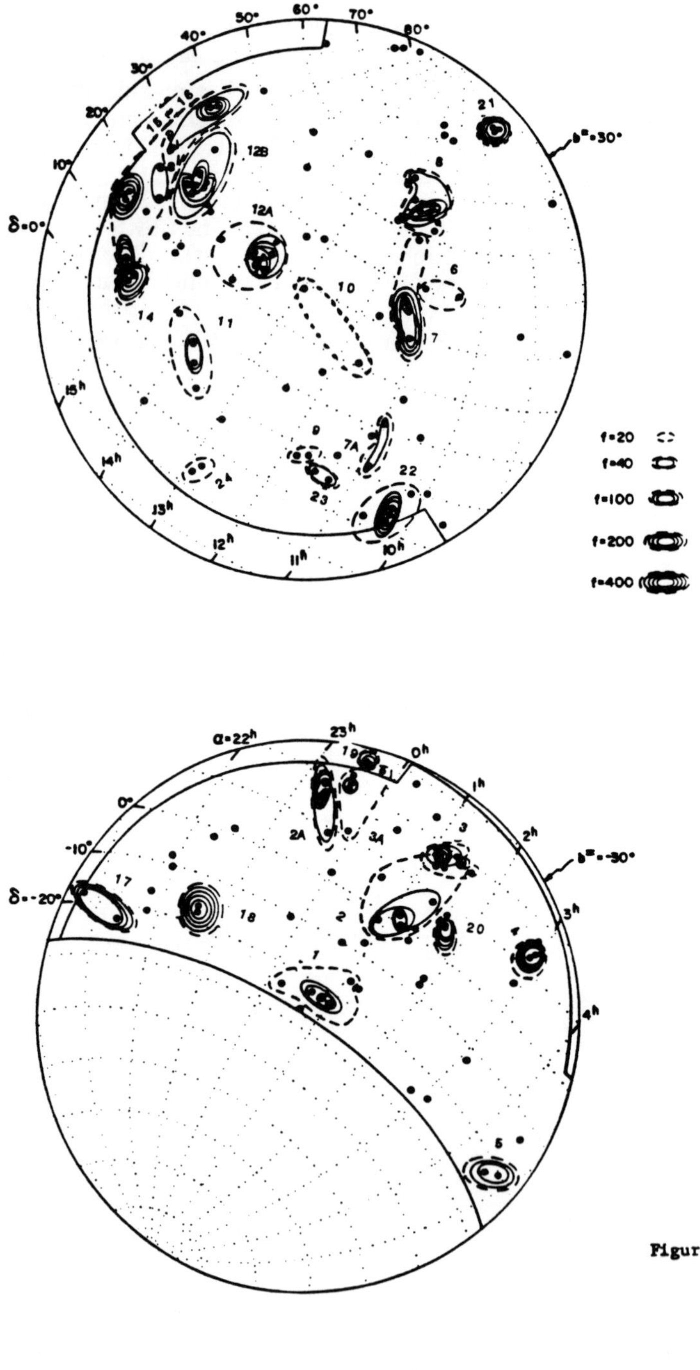
50°
60°
70°
40°
80°
30°
20°
10°
δ=0°
21
b^II=30°
12B
12A
8
6
10
14
11
7
15^h
14^h
13^h
12^h
11^h
10^h
9
7A
22
23
f=20
f=40
f=100
f=200
f=400
α=22^h
23^h
0^h
1^h
2^h
3^h
4^h
0°
-10°
δ=-20°
b^II=-30°
17
18
2A
3A
3
2
20
1
5

Figure 1

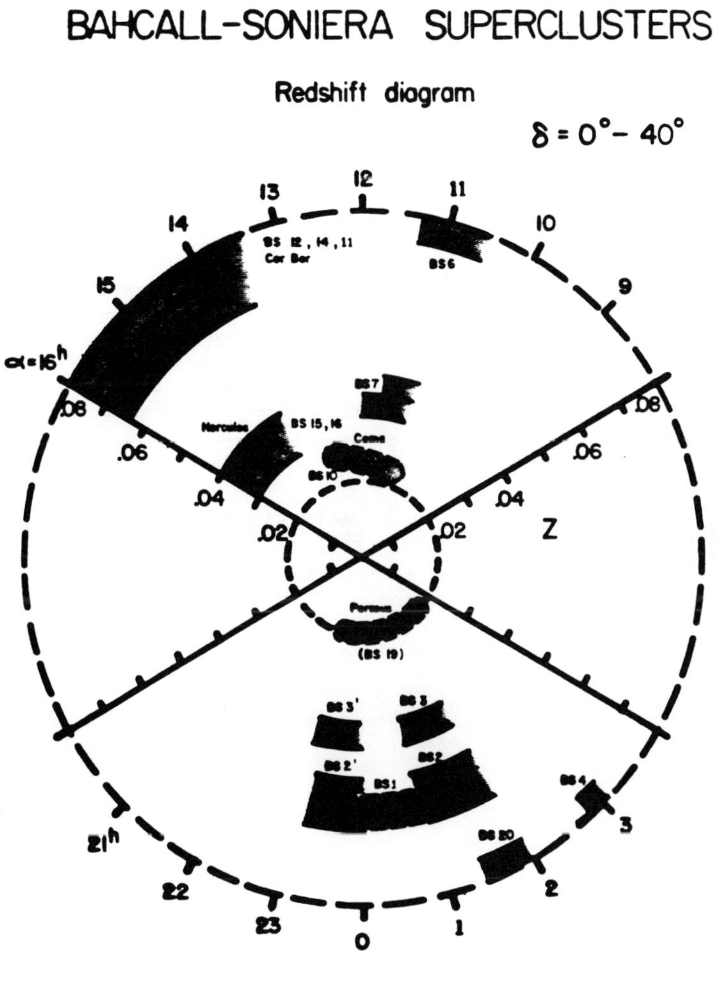
BAHCALL-SONIERA SUPERCLUSTERS
Redshift diagram
δ = 0°– 40°
12
13
11
14
10
15
9
α=16h
.08
.06
.04
.02
.02
.04
.06
.08
Z
BS6
BS7
BS 15, 16
21h
22
23
0
1
2
3

Figure 2

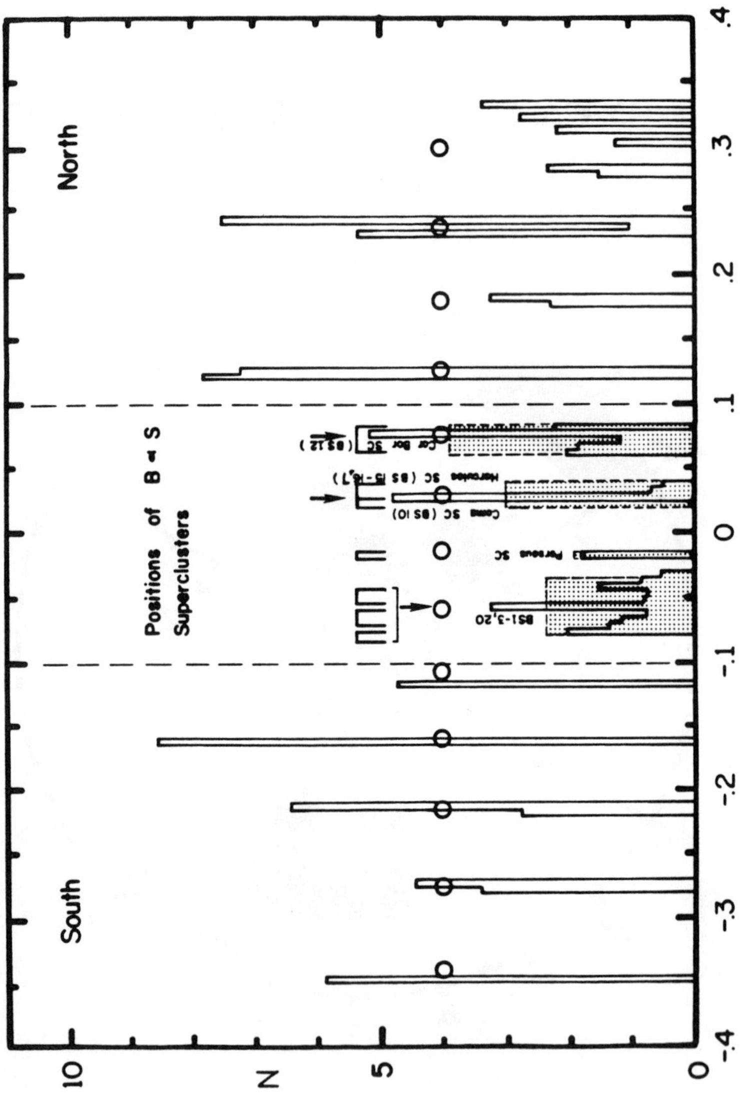

Figure 3

ORIGIN OF REGULARITIES IN THE REDSHIFT DISTRIBUTIONS OF DISTANT GALAXIES AND QUASARS

Leonid M. Ozernoy
IGPP, Los Alamos National Laboratory, Los Alamos, NM 87545

ABSTRACT

A model in which inhomogeneities originated during inflation and transformed eventually into density peaks with maximum wavelength equal to the horizon size at the decoupling epoch ($z_{\rm dec} \approx 10^3$) would give at the present for very large scale structure a periodic distribution: $\ln(1+z) \approx 0.04n + 0.04$, $n = 0, 1, 2, \ldots$. This formula in which the true phase is replaced by its characteristic value fits the data on redshift distribution of distant galaxies (Broadhurst *et al.* 1990) fairly well.

Some relevant implications are discussed in the paper, such as (i) difference between the constants of periodicity $C \equiv \Delta\ln(1+z)$: C_G for distant galaxies and C_Q for quasars; and (ii) a recent finding that quasars close on the sky to low redshifted galaxies show a higher concentration towards peaks in the quasar redshift distribution than do quasars in general (Karlsson 1990).

I. INTRODUCTION

Recent work (Broadhurst *et al.* 1990) has claimed to reveal a remarkable regularity in redshift distribution of distant galaxies in the direction of the Galactic poles, implying a period of about 130 h^{-1} Mpc in their spatial structure on a scale of $\sim$ 1000 Mpc ($h = H_o/100$ km s^{-1} Mpc^{-1}). Although it has been met by the astronomical community with some scepticism (*e. g.* Davis 1990), a rigorous statistical analysis of the deviation from periodicity indicates that the data more closely resemble a regular cellular pattern than a distribution produced by random voids or sheets (Kurki-Suonio *et al.* 1990). One should remember that evidence for a similar periodicity in redshift distribution of quasars even on a much larger scale has been known for about 20 years (Karlsson 1971) and since that time has been strengthened enormously (for recent reviews and a list of earlier references, see Arp *et al.* 1988, and Barnothy and Barnothy 1989).

Several models have been proposed to explain the periodicity in redshift distributions of quasars and distant galaxies. They are as follows: (1) a torus topology of the Universe (Fang 1990); (2) an apparent structure resulting from an oscillating Hubble constant (Moritawa 1990); (3) a real structure formed after decoupling from the perturbations whose origin goes as far back as the inflationary epoch (Ozernoy 1990); and (4) a regular but aperiodic structure associated with an explosive scenario of galaxy formation (Coles 1990). Zero point offsets in the periodicity are found to be different in different directions (Arp *et al.* 1988); this can be used to rule out models 1 and 2.

Below, the main features of model 3 are described and some of its numerous implications are discussed.

II. THE MODEL

The universe is supposed to have an inflationary period so that the adiabatic density perturbations produced during inflation are given by

$$\frac{\delta\rho}{\rho} \approx \frac{H^2}{\dot{\psi}_{\rm h}}, \tag{1}$$

where ψ is some "inflaton" field and $\dot{\psi}_{\rm h}$ is evaluated when the present horizon was stretched beyond H^{-1} during the inflationary epoch (Steinhardt 1990).

By the end of decoupling ($z = z_{\rm dec} \approx 10^3$) the sound waves are transformed into density perturbations. This is accompanied by a jump in the amplitude of the density waves compared with that in the sound waves before recombination, and this effect is maximal on a characteristic length scale that coincides with the horizon size at the moment $t_{\rm dec}$ (Peebles and Yu 1970). In other words, at this moment the *final* power spectrum of some density wave $\delta\rho/\rho$ is peaked on the largest characteristic scale, $ct_{\rm dec}$. A key point is that the subsequent expansion reveals other parts of the same protostructure that has been produced during inflation. Therefore, the galaxies formed eventually as the smallest part of that structure should have a spatially-periodic distribution on very large scales with period $\approx ct_{\rm dec}$.

The coordinate of the peak of $\delta\rho/\rho$ on the largest scale as measured at the present time is given by

$$\int_{r_{\rm dec}}^{0} k(t)dl = 2\pi n + \varphi, \tag{2}$$

where $k(t)$ is the minimum wave number ($= 2\pi/ct$); dl is a distance along a null geodesic; n is an integer (≥ 0); and φ is a phase. For light, $dl = c\,dt$ and it is straightforward to show that the maxima of density inhomogeneities are located at redshifts implicitly given by

$$I(z,\Omega) \equiv \int_0^z \frac{dz}{(1+z)(1+\Omega z)^{1/2}} = \alpha n + \beta. \tag{3}$$

Here α and β stand for the constants (in a given direction) entering Eq. (2):

$$\alpha \equiv \frac{H_{\rm o}\lambda_{\rm dec}(1+z_{\rm dec})}{c} \approx \frac{4}{3}\frac{\Omega^{1/2}}{(1+z_{\rm dec})^{1/2}}, \tag{4}$$

$$\beta \equiv \frac{\varphi}{2\pi}\alpha. \tag{5}$$

The integral in the l.h.s. of Eq. (3) can be easily evaluated to give

$$I(z,\Omega>1) = \frac{2}{(\Omega-1)^{1/2}}\arctan\left\{\frac{[(\Omega z+1)^{1/2}-1](\Omega-1)^{1/2}}{(\Omega z+1)^{1/2}+(\Omega-1)^{1/2}}\right\}, \tag{6a}$$

$$I(z,\Omega=1) = 2\left\{1-\frac{1}{(1+z)^{1/2}}\right\}, \tag{6b}$$

$$I(z,\Omega<1) = \frac{1}{(1-\Omega)^{1/2}}\ln\left\{\frac{[(1+\Omega z)^{1/2}-(1-\Omega)^{1/2}][1+(1-\Omega)^{1/2}]}{[(1+\Omega z)^{1/2}+(1-\Omega)^{1/2}][1-(1-\Omega)^{1/2}]}\right\}$$
$$\longrightarrow \ln(1+z) \;\text{ at } \Omega \ll 1. \tag{6c}$$

Caustics in the z-distributon of the scales entering the non-linear regime correspond to a real very large scale structure that should be two-dimensional and will tend to form perpendicularly to the line-of-sight (McGill 1990).

As the apparent periodicity in redshift distributions for both distant galaxies and quasars indicates $\Delta \ln(1+z) = \text{const}$, the asymptotics in Eq. (6c) seems to advocate the case of low Ω's. If we put $\Omega \approx 0.2$ to 0.3 (following the current data of observational cosmology) into Eq. (4), then $\alpha \approx 0.04$ and whenever $\varphi \sim 2\pi$ Eq. (5) yields $\beta \sim \alpha \approx 0.04$. This set of α and β results in a z-distribution of distant objects given by

$$\ln(1+z) \approx 0.04n + 0.04 \qquad \text{with } n = 0, 1, 2 \ldots , \tag{7}$$

which fits the data of Broadhurst *et al.* (1990) fairly well.

Evidently, phase β in Eq. (3) is strongly direction-dependent. Period α is expected to be direction-dependent too, if the line of sight is not perpendicular to the stratified layers of very large scale structure.

Equation (2) presented above coincides formally with a similar equation given by Fang *et al.* (1982). However, their model does not incorporate inflation and therefore is unable to produce a periodicity in redshift distributions from a random initial spectrum of inhomogeneities.

III. SOME IMPLICATIONS

a) Karlsson Effect

Recently Karlsson (1990), using large samples of the objects, has shown with a high significance that quasars close on the sky to low redshift galaxies show a higher concentration towards the peaks in the quasar redshift distribution than do quasars in general. As selection effects probably cannot explain this phenomenon, Karlsson has concluded that it might indicate a real difference between quasars close to galaxies and other quasars, thereby reviving the old idea of "local" and cosmological quasars. However, this conclusion can be avoided if two features of Karlsson's samples are taken into account:

(i) in each of his two samples A and B *most* quasars lie in the same direction in the sky: at $\alpha \sim 0^h$ in Sample A and at $\alpha \sim 12^h$ in sample B;

(ii) quasars selected by their apparent proximity to galaxies are closer on the sky to each other than are quasars in a random sample.

A combination of these two mutually enhancing effects seems to be the key to explaining the origin of Karlsson effect. The role of the nearby galaxies seems to be solely in providing *a smaller range of phase shifts* for quasars entering the sample. The theoretical foundation of the proposed interpretation follows from the model described in Section II: one of the main corollaries is that the phase of this periodicity should be direction-dependent. (The phase shift given by Eq. (7) is a characteristic value only.) As a result, quasars lying close to one another on the sky would demonstrate the periodicity effect more strongly than quasars in general.

In order to examine this interpretation, sample A and/or B need to be compared with some semi-random sample selected in such a way that it would include quasars taken randomly say from a narrow range $0^h \pm \text{few}^h$ or $12^h \pm \text{few}^h$

of directions. If the proposed interpretation is correct, the difference between these samples would be much smaller than has been found by Karlsson using his random sample.

b) Difference Between The Periods For Galaxies And Quasars

The constant of periodicity found in Section II, $C \equiv \Delta \ln (1+z)$, is close to the value $C_G = 0.043$ which characterizes the z-distribution of distant *galaxies* (Broadhurst *et al.* 1990), but it turns out to be about five times less than the value $C_Q = 0.205$ found by many authors for the z-distribution of *quasars* (*e.g.* Arp *et al.* 1988). Generally, this is not a surprise *if* (i) the total lifetime of a quasar, τ_Q, is smaller than that of its host galaxy, τ_G; *and* (ii) the quasar phenomenon is recurrent. (Notice that if the latter occurs only once during the lifetime of the host galaxy, this only affects the number of quasars seen within each peak, but not the value of C_Q). *If* the quasar phenomenon is recurrent, coherence in 'dying' and 'resurrection' might explain the difference in C_Q and C_G given by $C_Q/C_G = \tau_G/\tau_Q$. If $\tau_Q \sim 2 \cdot 10^9$ yr and $\tau_G \approx 10^{10}$ yr then $C_Q/C_G \sim 5$ as observed.

Another reason for the difference between C_Q and C_G might be selection effects related to the rather small numbers of quasars in the bins of $\Delta \ln (1+z)$. It is remarkable, in this respect, that Chu and Zhu (1989), while studying the periodicity in the redshift distribution of the Ly α forest, were able to uncover a fine structure within the "period" 0.2, since the accuracy of the measurements of the absorption-line redshifts was sufficiently high. Specifically, they found that the ratio of the emission and absorption line redshifts, $(1+z_{\rm em})/(1+z_{\rm ab})$, has a periodic distribution with the period 0.05. Further analysis of richer samples of the Ly α forest as well as the accumulation of data on fainter quasars would make it possible to study the structure of the quasar z-periodicity in finer detail.

IV. DISCUSSION AND CONCLUSIONS

Inflation, in addition to what has already been suggested, performs one more spectacular job: it creates a necessary condition for a periodicity found in the z-distribution of distant galaxies and quasars. Regardless of any specific model for the origin of the perturbation spectrum, inflation prepares coherent perturbations which at the moment of decoupling (redshift $z_{\rm dec} \approx 10^3$) are transformed into density waves whose coherence extends over many horizon scales. Whereas the very phenomenon of spatial periodicity is due to inflation, the value of the period results from the interaction of matter and radiation during the process of decoupling and it is given by the horizon size at recombination, *i. e.* $ct_{\rm dec}$. [1] Its value corresponds at the present moment to a period in the redshift distribu-

[1] It is worth mentioning that the mere presence of the largest scale, $ct_{\rm dec}$, in the final perturbation spectrum, without its coherence over many horizon scales, would result in a cut-off in the correlation function of the largest structures on scales $ct_{\rm dec}(1+z_{\rm dec})$, but not in the spatial periodicity of these structures. The discovery of similar periodicities in other directions, although with somewhat different periods and phases, would provide strong support for the model.

tion of distant objects equal to $63(\Omega h^2)^{-1/2}$ Mpc $= 140h^{-1}$ to $120h^{-1}$ Mpc at $\Omega = 0.2$ to 0.3, which turns out to be consistent with the observed periodicity in the spatial distribution of distant galaxies in the direction of the Galactic poles (Broadhurst *et al.* 1990). Even the phase characterizing the distance to the nearest wall of galaxies (the "Great Wall") appears to be correct, at least as a first approximation, if one puts the value of the phase equal to that of the period. The only free parameter required for such a satisfactory description is the value of the cosmological density parameter Ω, taken to be, in accordance with current data, between 0.2 and 0.3.

Is such a low Ω compatible with the inflation paradigm? As is known, 'chaotic inflation' (Linde 1990) is consistent with $\Omega < 1$, but requires some specially chosen (and probably unrealistic) initial conditions. A recent inflationary scenario, 'extended inflation' (La *et al.* 1989), seems to be more appropriate for low Ω's. This version of inflation accomodates a range (although rather narrow) of initial conditions which give $\Omega \lesssim 0.5$ (Steinhardt 1990). It remains to be seen whether this scenario could give in a consistent way the spectrum of perturbations appropriate to form both galaxies and the very large scale structure.

A potential trouble with the present model (relevant to almost all scenarios for the formation of the universe's structure) is related to temperature inhomogeneities in the microwave background radiation. One expects $\Delta T/T \sim 10^{-4}\Omega^{-1}$ on angular scales $\Theta \sim 1^\circ h^{-1}$ which correspond to the horizon size just before the decoupling epoch. At low Ω's, such a value of $\Delta T/T$ could easily be excluded by COBE observations.

I am thankful to Charles Keller, Director of IGPP, whose hospitality and invariable support I have enjoyed during my stay at LANL.

REFERENCES

Arp, H., Bi, H.G., Chu, Y., and Zhu, X. 1988, MPA Preprint No. 427.
Barnothy, J., and Barnothy, M. 1989, *Comments Astrophys.*, **13**, 179.
Broadhurst, T.J., Ellis, R.S., Koo, D.C., and Szalay, A.S. 1990, *Nature*, **343**, 726.
Chu, Y., and Zhu, X. 1989, *Astr. Ap.*, **222**, 1.
Coles, P. 1990, *Nature*, **346**, 446.
Davis, M. 1990, *Nature*, **343**, 699.
Fang, L.-Z. 1990, Preprint, FERMILAB-Pub-90/26-A.
Fang, L.-Z., Chu, Y.-Q., Liu, Y., and Cao, Ch. 1982, *Astr. Ap.*, **106**, 287.
Karlsson, K.G. 1990, Preprint.
Kurki-Suonio, H., Mathews, G.J., and Fuller, G.M. 1990, Preprint.
La, D., Steinhardt, P.J., and Bertschinger, E.W. 1989, *Phys. Lett.*, **B 231**, 231.
Linde, A.D. 1990, *Inflation and Quantum Cosmology*, Academic Press: San Diego.
Moritawa, M. 1990, UBC Preprint.
McGill, C. 1990, *M.N.R.A.S.*, **242**, 544.
Ozernoy, L.M. 1990, *Bull. AAS*, **22**, 875.
Peebles, P.J.E., and Yu, J.T. 1970, *Ap. J.*, **162**, 815.
Steinhardt, P.J. 1990, *Nature*, **345**, 47.

Cosmology with Oscillating Hubble Parameter

Masahiro MORIKAWA *

Department of Physics, University of British Columbia
Vancouver, B. C. Canada V6T 2A6

Summary

We propose a cosmological model in which resonant energy exchange between gravity and matter is realized. The Hubble parameter becomes oscillatory and causes an apparent density fluctuation without conflicting with the isotropy of the microwave background radiation. We study a non-conformally coupled scalar field with a very small mass about $10^{-31} eV$. *Several new observational consequences are derived: a large scale coherent periodicity in the redshift distribution of galaxies with the periodicity about* $100 Mpc$, *a QSO redshift clustering, a changing Hubble parameter on a scale* $z \approx 0.01$, *a dark matter density that increases with distance, a luminosity distance-redshift relation, and a changing Gravitational constant.*

We study a new type of cosmology in which there is a sufficient energy resonance between gravity and matter. In order to produce the resonance, we introduce a non-conformally coupled scalar field $\phi(t)$, which is supposed to dominate the energy density of the Universe at least at the late stage of the evolution. The mass m of $\phi(t)$ should be small and the coupling parameter ξ to the scalar curvature should be large for a sufficient amount of resonance to take place all the time of cosmic evolution. Typical values we consider will be $m \approx 10^{-31} eV$ and $\xi \approx 10$. If there is no curvature coupling, then the strength of the resonance at present is determined by the ratio of the characteristic scale of the gravity (the present value of the Hubble parameter H_0) and that of matter m, and the resonance is quite small. On the other hand, if we consider the curvature coupling $\xi\phi^2 R$, we can expect sufficient resonance and enough oscillation in the Hubble parameter. This is because the Hubble parameter, for $mt \gg 1$, behaves

$$H(t) \approx 6\xi g\phi\dot{\phi} + \sqrt{g(\dot{\phi}^2 + m^2\phi^2)}, \tag{1}$$

where $g \equiv 4\pi G/3$, and the scalar field is approximately given by the Bessel function: $\phi(t) = \phi_0 \cdot (mt)^{-1/2} J_\nu(mt)$, where $\nu^2 = -(4/3)\xi + (1/4)$ and $\phi_0 = 1/\sqrt{6G}$. The first term on the right hand side of Eq.(1) gives oscillatory part and the second term slowly varying part of the Hubble parameter.

An oscillating Hubble parameter causes apparent density fluctuations without affecting the isotropy of the microwave background radiation. The present model is simple but it yields various new observational consequences, on which we are now going to argue.

* Bitnet address: USERHIRO@UBCMTSG

(i) Periodicity in the redshift distribution of galaxies: Even if the matter distribution is uniform in the constant cosmic time slices, there is an apparent inhomogeneity in the propagation of a light ray emitted by a distant galaxy and arriving at the present observer. This will cause an apparent oscillation in the number count with respect to the redshift. Since this oscillation is uniform and isotropic, it does not disturb the isotropy of the cosmic background radiation. Within a fixed solid angle $d\Omega$, the number of galaxies which have redshift between z and $z + dz$, and absolute luminosity between L and $L + dL$ is just the density $n(L, z)dL$ times the volume element $a^3\chi^2 d\chi d\Omega$, where χ is a comoving distance. If we consider spatially flat Universe, the number count becomes

$$\frac{dN}{d\Omega dLdz} = n(L, t_0)a_0^2\chi^2 H^{-1}, \tag{2}$$

where the suffix zero means the value at present. Therefore, the oscillation in H directly induces that in dN/dz as a function of z. We expect an apparent galaxy concentration on equally separated spherical shells with a common center.

Let us check the observational side. The recent deep narrow-cone survey of the large-scale redshift distribution of galaxies at the Galactic poles revealed a dramatically periodic distribution in the redshift [1]. The survey was performed at both Galactic poles within a narrow angle 0.3°. A single periodicity ($128h^{-1}Mpc$) in the distribution on a large scale (2000 h^{-1} Mpc) in both directions has been found. If we identify this period as the inverse of the mass of ϕ in our model, we can fix the mass: $m \approx 10^{-31}eV$. We give a typical numerical calculation of the number count as a function of redshift in Fig. 1. The analytic result using the approximate Bessel form for ϕ gives almost the same result. The parameter ξ determines the strength of the oscillation in H. We identified the present time so that it reproduce the observed fact: there appear 8 peaks in number count within a redshift region 0-0.5.

If the oscillation of the Universe is the very origin of the periodicity, then the profile of the number count must be the same in any direction of survey. Therefore it is imperative to observe the galaxy distribution in other directions. Actually in the observation [1], the zero point of the redshift is slightly shifted about $\Delta z = 0.011$ to the south, if we identify the nearest peak of the galaxy distribution as the proper zero point. This shift may be caused by the peculiar motion of the Earth with respect to the cosmic background radiation. However, this peculiar velocity $300km/s$ to the direction $l = 260°$, $b = 45°$ [2] only induces a shift in redshift $\Delta z = 0.8 \times 10^{-3}$ to the south (one order less than needed). The reason of this discrepancy between [1] and our explanation may come from inadequate fit of the data in [1]. They fitted the data with an equal spacing in the comoving coordinate ($128Mpc$) throughout the whole range of the observation. However, if our model is correct, then the periodicity is expected in the cosmic time and moreover, there is no reason that the separation of the most inside two peaks is the same as the rest; the present phase of the oscillation is naturally expected to be between a peak and a valley. Therefore, we propose to fit the north data and south data independently with the same separation in cosmic time disregarding the mismatch at the zero point of redshift. The mismatch itself reveals the present phase of the Hubble oscillation.

(ii) Clustering in the redshift distribution of QSO and Lyman-α forest: Clustering in the number count is also observed for the QSO distribution [3] and for the Lyman-α forest [4]. They even report the periodicity in the variable $\ln(1+z)$ with a period 0.2. Our result does not predict this phenomenological rule. However, it predicts similar interval between peaks in the variable $\ln(1+z)$ (See Fig.2.).

(iii) Increasing Hubble parameter with distance: If the Hubble parameter is oscillating, and we are at the phase of decreasing H, then any observation of H results in an increasing Hubble parameter in z. The scale is of order $z \approx 0.01$. (See Fig.3.) Actually in the observational determination of H_0, there is a general tendency that H_0 increases with distance [5]. Ordinarily, this tendency is interpreted as the result of Malmquist bias. However, the indeterminacy in H_0 between the range $50 - 100 km/sec/Mpc$ for long time strongly suggests that H actually varies on the scale $z \approx 0.01$. The solid line in Fig. 3 shows the prediction of the actual Hubble parameter which is defined to be the ratio of the redshift and the luminosity distance.

(iv) Increasing dark matter density with distance: The dark matter distribution has a similar tendency as H in the same scale [6]. This is also naturally explained from our point of view: since our scalar field contributes as dark matter, its density increases with distance.

(v) Luminosity distance-redshift relation: Though violent oscillation in the Hubble parameter is produced, its effect on most of the observations is quite mild. If we draw luminosity distance-redshift relation, the oscillation is negligibly small. (See Fig. 4.) On this diagram, the value of the deceleration parameter q_0 is irrelevant. The indeterminacy of the parameter q_0 for long time indicates that q_0 is actually not a relevant parameter of the present Universe.

(vi) Variation of the Gravitational constant G: Since we are considering non-conformal coupling, the effective gravitational constant changes in time depending on the dynamics of the scalar filed:

$$\frac{\dot{G}}{G} = 12 g \xi \phi \dot{\phi}. \tag{3}$$

In our model, the value is between $-0.5\ 10^{-10} h/year$ and $+0.5\ 10^{-10} h/year$ depending on the present phase of the oscillation. (See Fig. 5.) If we are in a $H-$decreasing phase, then $\dot{G}/G$ becomes negative, while in a $H-$increasing phase, it is positive! The Viking radar measurement [7] gives the present limit of the value of the change of G: $|\dot{G}/G| < 3 - 0.6 \times 10^{-11}/year$. If we would know the signature of $\dot{G}/G$, we can identify the present phase of the oscillation.

I am grateful to Redouane Fakir, Salman Habib, Raymond Laflamme, and Bill Unruh for valuable discussions. I would like to thank The Japan Society for the Promotion of Science for financial support.

References

[1] T. J. Broadhurst, R. S. Ellis, D. C. Koo, and A. S. Szalay, Nature **343** (1990) 726. and the references therein.
[2] D. J. Fixsen, E. S. Cheng, and D. T. Wilkinson, Phys. Rev. Lett. **50** (1983) 620.
[3] L. Z. Fang, Y. Q. Chu, Y. Liu, Ch Cao, Astron. Astrophys. **106** (1982) 287.
[4] Y. Chu and X. Zhu, Astron. Astrophys. **222** (1989) 1.
[5] S. van den Bergh, Nature **299** (1982) 297.
[6] V. Trimble, Ann. Rev. Astron. Astrophys. **25** (1987) 425.
[7] C. M. Will, in *Three Hundred Years of Gravitation*, edited by S. W. hawking and W. Israel, Cambridge Univ. Press (1987).

Figure Captions

Fig.1 A numerical calculation of the Number count as a function of redshift. We have chosen $\xi = 10$, and initial conditions $\sqrt{g}\phi = 0.1$, $\dot{\phi} = 0$, $(H/m = 0.16)$. The selection function is not taken into account. The vertical axes is in a free scale. The Hamiltonian constraint is satisfied within 2% error. Shaded bars are the North Pole observation [1].

Fig.2 Same as Fig.1 but in a global scale.

Fig.3 A numerical calculation of the Hubble parameter as a function of redshift. The broken line shows the bare Hubble parameter in our oscillating Universe. The dotted line shows the bare Hubble parameter in the standard cosmology $(\Omega = 1)$. The solid line shows our prediction of the actual Hubble parameter which is defined to be the ratio of the redshift and the luminosity distance.

Fig.4 A numerical calculation of the luminosity distance-redshift relation.

Fig.5 The change of the gravitational constant $(\dot{G}/G)$ as a function of the cosmic time.

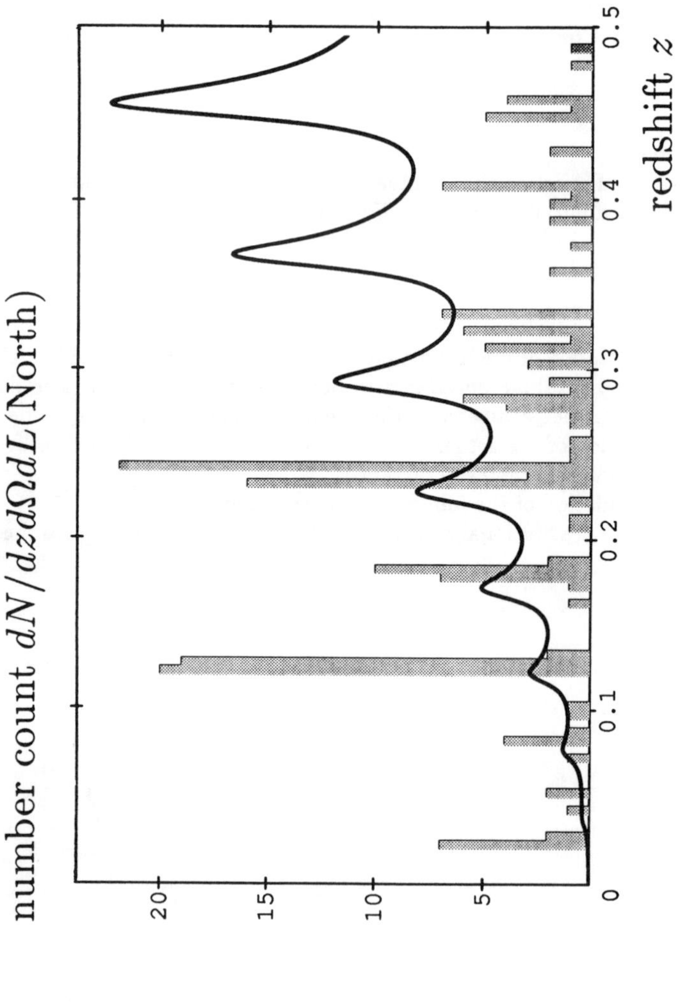
number count $dN/dzd\Omega dL$(North)
20
15
10
5
0
0.1
0.2
0.3
0.4
0.5
redshift z

Fig.1

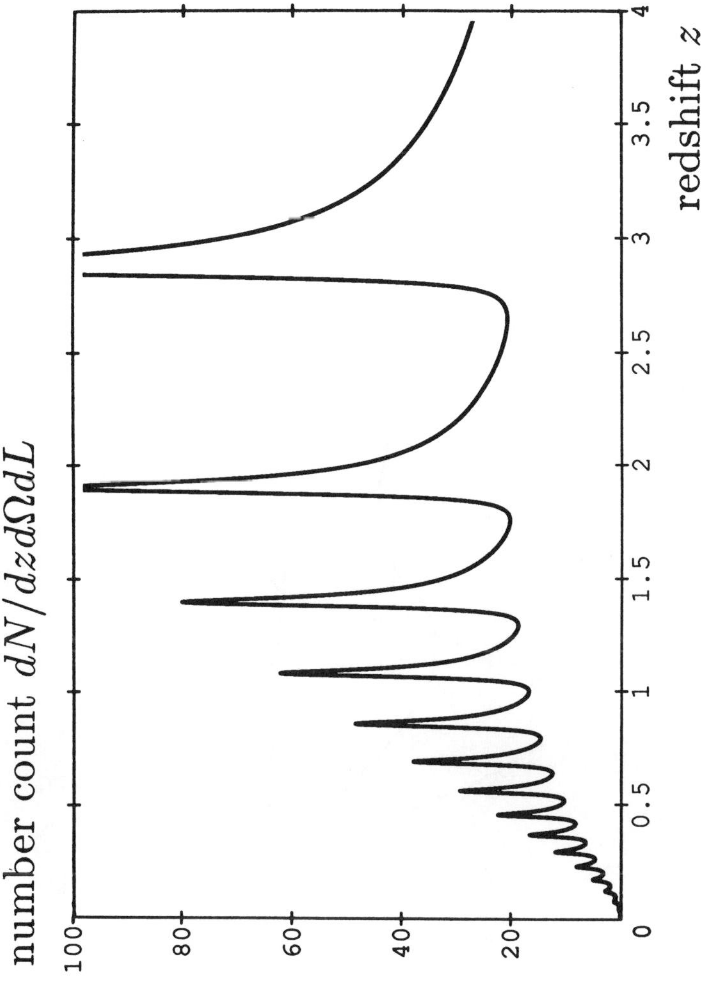
number count $dN/dzd\Omega dL$
100
80
60
40
20
0
0.5
1
1.5
2
2.5
3
3.5
4
redshift z

Fig.2

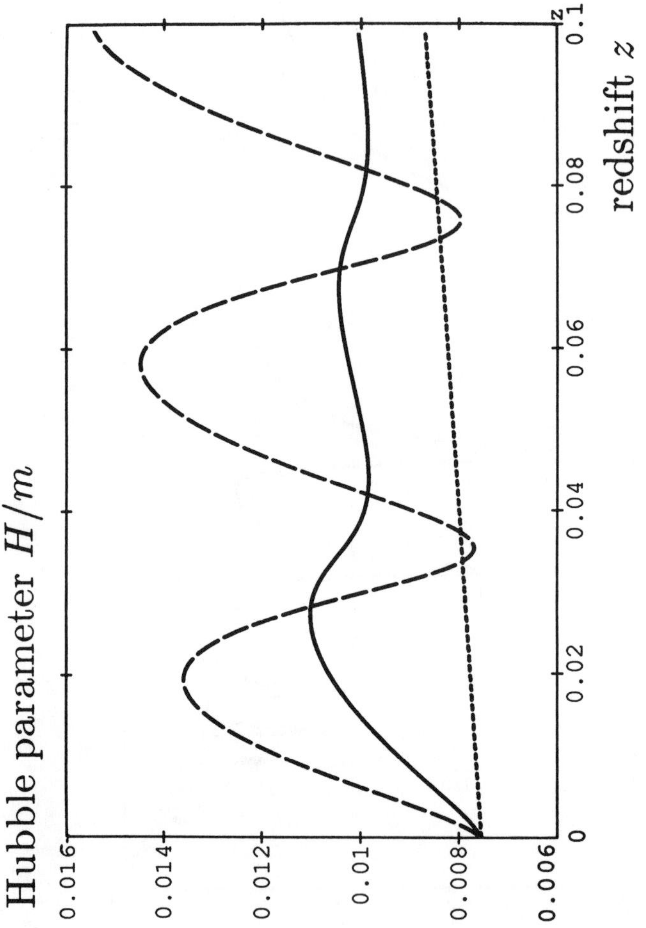
Hubble parameter H/m
0.016
0.014
0.012
0.01
0.008
0.006
0
0.02
0.04
0.06
0.08
0.1
z
redshift z

Fig.3

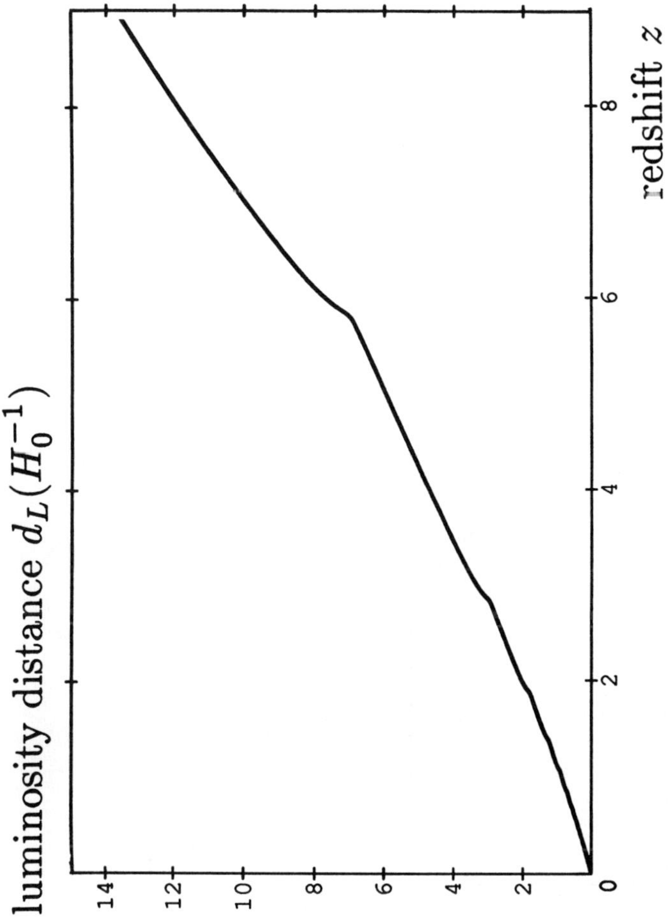

Fig. 4

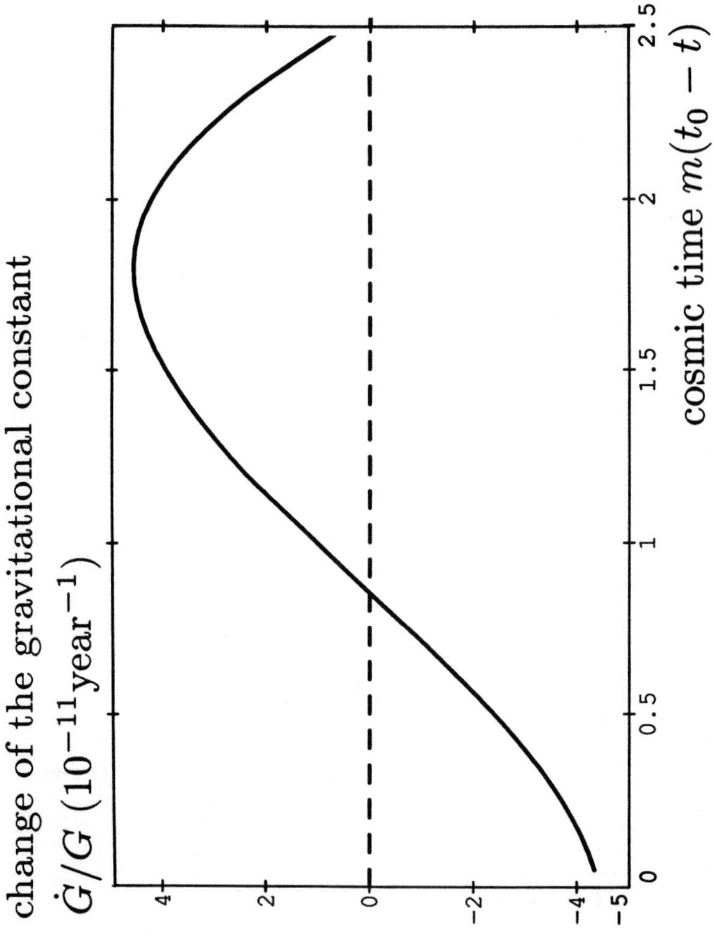
change of the gravitational constant
$\dot{G}/G$ (10^{-11}year^{-1})
4
2
0
-2
-4
-5
0
0.5
1
1.5
2
2.5
cosmic time $m(t_0 - t)$

Fig.5

N-BODY SIMULATIONS: FROM GALAXIES TO COSMOLOGY

Edmund Bertschinger
Department of Physics, MIT, Cambridge, MA 02139

ABSTRACT

Gravitational N-body simulations are increasingly used to complement analytical and observational studies in cosmology. This article reviews some of the key results obtained by and the numerical methods used for N-body simulations in both galaxy dynamics and cosmology. New results are presented from large particle-mesh and particle-particle/particle-mesh simulations.

INTRODUCTION

Gravitational N-body simulations are a widely used theoretical tool in stellar and galactic dynamics and cosmology. Given the dominance of the gravitational force on large scales, the complexity of nonlinear dynamics, and the inability of astrophysicists to perform real experiments, N-body simulations have an important role to play in helping us to understand the formation and evolution of galactic and extragalactic systems. Simulations are neither a sterile computer game, incapable of providing real physical insight, nor are they an oracle that we can consult to tell us which among several is the correct theory of everything. Instead, when used in conjunction with analytical models and analyzed carefully in comparison with observational data, N-body simulations provide a powerful tool for bridging the gap between astronomical observations and theory.

The last few years have seen a rapid growth in the N-body industry, fueled by technological improvements in both computer hardware and software. On the software side, several new numerical methods have been introduced, yielding significant increases in the efficiency of computing gravitational forces in a many-body system. Some of these advances will be described in this article. On the hardware side, supercomputers have become readily accessible for most serious users. Many universities have acquired supercomputers. The National Science Foundation's Supercomputing Centers especially have made a significant difference for U.S. researchers.

Gravitational N-body simulations have been used for many purposes. However, one can make a rough division into four areas of application: the dynamical evolution of star clusters; galaxy dynamics; galaxy formation and clustering; and large-scale structure. Perhaps the most significant contributions of N-body simulations may have been in the area of stellar dynamics, while the most recent widespread interest in their application concerns large-scale structure. It is no coincidence that these areas are the ones having presently the most effective observational input. However, this review will focus instead on the intermediate-scale problems of galaxy dynamics and galaxy formation and clustering. It is these areas where simulations are most needed to substitute for experimental data.

Purely gravitational N-body simulations of galaxies may be a dying breed. Proper study of galaxy formation and evolution clearly requires gas dynamics because shock heating, radiative cooling, star formation and stellar energy input are essential in the formation of the luminous galaxies that we see. However, three-dimensional gas dynamical simulations of galaxy formation and evolution are still in their infancy. Moreover, if most of the mass in the universe is non-baryonic dark matter, it should be an excellent approximation, at least for the dark matter beyond several tens of kpc from galaxy centers, to neglect the gas dynamical effects on the baryonic component. Even with this justification, one still has the very difficult problem of how to identify luminous galaxies in a simulation that includes only dark matter. This problem deserves serious examination rather than being ignored or bypassed by an ad hoc recipe for "biased" galaxy formation. For this reason, I consider it premature to draw strong conclusions from many cosmological simulations of large-scale structure, although other researchers may disagree.

This article will focus on numerical techniques and some of the key questions of galaxy dynamics and clustering being addressed with gravitational N-body simulations. Recent reviews have been given by White (1986, 1988), Sellwood (1987) and Greengard (1990). For a description of the methods, the monograph by Hockney and Eastwood (1981) is a classic, as is the article of Efstathiou *et al.* (1985). Although this article will not discuss cosmological or other gas dynamical simulations, the interested reader may consult the recent papers of Evrard (1988, 1990), Evrard and Davis (1988), Hernquist (1989), Hernquist and Katz (1989), Olson and Kwan (1990*a,b*), Ryu *et al.* (1990), Cen *et al.* (1990), and Cen (1990). The present article also avoids discussion of the adhesion approximation (Kofman and Shandarin 1990).

N-BODY SIMULATIONS OF GALAXIES: QUESTIONS AND METHODS

Gravitational N-body simulations have been very helpful in clarifying many important issues concerning the structure, stability, formation and interaction of galaxies. Several of these successes are mentioned below.

Are stellar disks gravitationally stable? Using a large, two-dimensional particle-mesh simulation, Hohl (1971) found that a disk that was locally unstable according to the Toomre Q criterion was violently bar unstable. Ostriker and Peebles (1973) confirmed this result using a direct $N = 300$ simulation, they gave a global stability condition ($T/|W| \lesssim 0.14$, where T and W are the total rotational kinetic and gravitational energies), and they argued cogently that a hot disk or, preferentially, a dark halo, was needed to stabilize our own galaxy. Their insight was so great that I suspect the N-body simulation was superfluous.

How is spiral structure maintained in disks? If dark halos or hot disk components prevent the global bar instability, how is the local spiral instability nurtured? Without continual encouragement, spiral structure is found to be short-lived in N-body experiments, as bars form or the disk heats itself up. Sellwood and Carlberg (1984), Carlberg and Sellwood (1985) and Carlberg and Freedman (1985) realized, and their grid-based N-body simulations confirmed, that a small amount of gaseous dissipation or infall could both keep the disk cool enough for a swing-amplifier to produce attractive spiral structure and yield a stellar age-velocity dispersion relation in agreement with observations in the solar neighborhood.

How quickly are small satellite companion galaxies eaten as a result of dynamical friction and orbital decay in the halo of a large parent galaxy? Does the Chandrasekhar formula, which gives the drag on a point mass moving through a uniform medium, neglecting the self-gravity of the gravitational wake, apply for one galaxy moving in the halo of another? Early N-body simulations

(White 1978) appeared to confirm the Chandrasekhar formula. Semi-restricted N-body simulations, in which the two-body forces between stars in the parent halo are ignored, were in agreement (Lin and Tremaine 1983). The effects of a fully self-gravitating halo remained unclear for several years. Finally, a combination of analytic work and full N-body simulations using the hierarchical tree algorithm (Hernquist and Weinberg 1989) now shows convincingly that the actual orbital decay rate is two to three times slower than that given by the Chandrasekhar formula when a fully self-consistent parent halo is used.

What happens when galaxies of comparable mass collide? The first gravitational N-body simulation of a tidal encounter between galaxies was Holmberg's remarkably creative study using an analog optical computer: each galaxy was represented by 37 light bulbs, with a photocell and galvanometer measuring the inverse square law (Holmberg 1941). In more modern times, the classic work of Toomre and Toomre (1972) used a restricted three-body method (and a digital computer) to show that bound, prograde encounters induce striking tidal tails and bridges like those seen in many odd galaxy pairs. Beginning with Toomre (1977), van Albada and van Gorkom (1977), and White (1978, 1979), many workers have performed N-body simulations to test Toomre's hypothesis that mergers of disks and/or spheroids produce ellipsoidal remnants indistinguishable from elliptical galaxies (e.g., Barnes 1988; Balcells and Quinn 1990 and references therein). The most exciting results have come from tree code simulations using $\gtrsim 10^4$ particles per galaxy. Barnes (1989) showed that mergers rapidly destroy a compact group of galaxies, while Hernquist (1989), including gas dynamics, showed that tidal encounters and mergers can send gas spiraling into a nucleus, triggering starbursts and nuclear activity.

Some galaxies are surrounded by faint shells or ripples — the elliptical NGC 3923 being the most spectacular known case. How did these shells form? Using restricted N-body experiments, Quinn (1984), Dupraz and Combes (1986), Hernquist and Quinn (1988, 1989), and Heisler and White (1990) showed that the merger of a cold dwarf satellite on a nearly radial orbit with a larger elliptical results in the formation of alternating shells like those observed. Barnes (1989) also found such shells in his fully self-consistent merger remnant.

Now we consider the numerical methods used for N-body simulations of galaxy dynamics. The mathematical problem is to solve the equations of motion for N self-gravitating particles:

$$\frac{d^2\vec{r}_i}{dt^2} = \sum_{j\neq i} \frac{-Gm(\vec{r}_i - \vec{r}_j)}{\left(|\vec{r}_i - \vec{r}_j|^2 + \epsilon^2\right)^{3/2}} \,, \quad i,j = 1 \text{ to } N \,. \tag{1}$$

To prevent the formation of excessively tight binaries, the force is generally softened for separations $r \lesssim \epsilon$ using the Plummer law given here. Galaxy dynamics simulations have four important desiderata: (1) accurate forces (e.g., a two-body Kepler potential over the range of separations $100\,\text{pc} \lesssim r \lesssim 200\,\text{kpc}$); (2) accurate time integration; (3) large N; and (4) isolated boundary conditions. Of these, (2) and (4) are the easiest to achieve.

Several algorithms have been devised for computing forces. The most obvious is direct summation. Considerable artistry is needed to reduce the operations count to significantly less than N^2. Aarseth (1972, 1985) is the acknowledged master. For fewer than several thousand particles, Aarseth's algorithm is the method of choice when accurate forces are needed, but it becomes extremely costly for $N \gg 10^4$. New parallel architectures may revive direct summation (Hillis and Barnes 1987; Makino and Hut 1988).

The hierarchical tree algorithm (Appel 1985; Barnes and Hut 1986) represents a major advance in the computation of the sum in equation (1). In this method, distant clumps of particles are reduced to single pseudo-particles with monopole and quadrupole moments, while nearby and isolated particles are treated as single particles. In the Barnes-Hut algorithm, space is recursively divided into a hierarchy of cubic cells, each cell containing one or more particles. If a cell of size s and distance d satisfies $s/d < \theta$, the particles in this cell are treated as one pseudo-particle located at the center of mass of the cell. This algorithm has several advantages (Hernquist 1987, 1988; Barnes and Hut 1989; Jernigan and Porter 1989). First, it is fast, requiring $O(N \log N)$ operations to compute all forces on N particles. Second, the force errors can be bounded and made small by choosing $\theta < 1$ and including quadrupole moments (higher-order moments not being cost-effective). Finally, the algorithm is fully spatially adaptive and does not slow down as clustering increases. The tree code has already proved its value in several examples described above, and its use is bound to spread. However, it does have disadvantages. First is the fact that it is a memory hog: from $20N$ to $30N$ words of storage are required to compute forces for N particles (Hernquist 1987). For most applications and machines the algorithm is cpu-bound rather than being memory bound, so this is generally not a major disadvantage in practice. Second, the softening length cannot be made much larger than the mean interparticle spacing, with the result that two-body relaxation effects may be nonnegligible. As Hernquist and Barnes (1990) show, however, all N-body algorithms are collisional to some extent, and the tree code is no worse than other algorithms with the same softening length.

Another hierarchical algorithm, promising accurate computation of all forces in only $O(N)$ operations, is the fast multipole method (Greengard and Rokhlin 1987; Greengard 1988, 1990). This method is mathematically much more complicated than the Barnes-Hut algorithm, which may explain why it has yet to be implemented by astrophysicists. However, since the number of particles being simulated is increasing rather rapidly with time, before long the promise of the speed gained with the fast multipole method must overwhelm the pain of implementation.

Harmonics expansion methods approximate the gravitational potential using a truncated expansion in a set of orthonormal basis functions appropriate to the geometry of the problem under consideration. For studies of spheroidal galaxies, spherical harmonics may be used for the angular dependence of the potential. It is possible to implement the expansion method with no spatial grid (Villumsen 1982), with a radial grid only (McGlynn 1984), or with both radial and angular grids (van Albada and van Gorkom 1977; van Albada 1982). The main advantage of the harmonics expansion methods is speed — only $O(N_h N)$ operations are required to compute forces on N particles using N_h expansion coefficients. For highly symmetric systems, it is possible to use $N_h \ll N$. However, the method is ill-suited to most problems because of the geometric restrictions and because the force errors are difficult to quantify and may be large. For most problems a tree code would be a better choice.

In Cartesian coordinates, the appropriate orthonormal basis functions are plane waves, leading to a Fourier series expansion. N-body methods based on a cubic lattice and discrete Fourier transforms have been in widespread use for more than two decades (Hohl and Hockney 1969; Miller 1970; Hockney and Eastwood 1981). In these methods, commonly called particle-mesh (PM) by astrophysicists, the forces are computed in five steps: (1) first, the density is interpolated from the particles to a grid; (2) then, the density is Fourier transformed using the Fast Fourier Transform (FFT) algorithm; (3) the Fourier transform of the potential (or force) is computed by multiplication with a suitable Green's function (Fourier convolution); (4) the potential (or force) is returned to the spatial domain using the inverse FFT; and (5) the force is interpolated to the particles from the potential (or force) on the grid. The main advantage of PM methods is speed. For N particles and N_g grid points, steps (1) and (5) require $O(N)$ operations, step (3) requires $O(N_g)$ operations, and steps (2) and (4) require $O(N_g \log N_g)$ operations. The main disadvantage is that the force resolution is limited by the grid. Unless the method is carefully implemented, the forces may approximate the inverse square law poorly for separations $r \lesssim 6 - 8\Delta x$ for a grid spacing Δx (Bouchet, Adams and Pellat 1985; Bouchet and Kandrup 1985); the interparticle force always goes to zero for $r \lesssim \Delta x$. Also, obtaining isolated boundary conditions requires a factor of 8 waste of storage in three dimensions (Hohl and Hockney 1969). For high-resolution studies of galaxies, the tree code is superior.

COSMOLOGICAL N-BODY SIMULATIONS

N-body simulations of isolated or interacting galaxies have proved very instructive, but they beg the question of how the galaxies formed. In the prevailing gravitational instability scenario, galaxy formation and clustering occurred as a result of nonlinear amplification of small-amplitude density fluctuations generated at high redshift and imprinted on an otherwise locally homogeneous and isotropic, uniformly expanding cosmological model. Cosmological N-body simulations start at large redshift with small fluctuations specified according to some theory. The fluctuations are described by the density power spectrum $P(k)$, defined at a fixed time by

$$\langle\widehat{\delta}(\vec{k}_1)\widehat{\delta}^\star(\vec{k}_2)\rangle = P(k_1)\,\delta_D^3(\vec{k}_1 - \vec{k}_2)\;, \quad \frac{\delta\rho}{\rho}(\vec{x}) = \int d^3k\,\widehat{\delta}(\vec{k})e^{i\vec{k}\cdot\vec{x}}\;, \tag{2}$$

where a subscript D is placed on the Dirac delta function to distinguish it from the density perturbation. The contribution to the variance of $\delta\rho/\rho$ from waves in the wavenumber volume element d^3k is $P(k)d^3k$. For a Gaussian random field, $P(k)$ provides a complete statistical description. Cosmological N-body simulations require specification of the initial power spectrum (and higher density moments, for a non-Gaussian random field) and the cosmological parameters Ω_0 and H_0. Such simulations have been used to address many important questions, several of which are recounted here.

Can hierarchical clustering account for the power-law form of the observed galaxy two-point correlation function? Motivated by the hierarchical clustering of galaxies deduced by Peebles and his associates (see Peebles 1980 for a review), several groups have performed simulations starting from the simplest, scale-free initial conditions: $\Omega_0 = 1$ and white noise, $P(k) \propto k^n$, $n = 0$. Aarseth, Gott and Turner (1979), using direct summation with up to 4000 particles, concluded that this model does not match observations as well as an open model with $\Omega_0 = 0.1$ and $n = -1$. Efstathiou and Eastwood (1981), introducing the particle-particle/particle-mesh (P^3M) algorithm to astrophysics with an impressive (for that time) $N = 20000$, showed that the nonlinear correlation function obtained from the $\Omega_0 = 1$, $n = 0$ model is steeper than the observed $\xi(r) \propto r^{-1.8}$ (see also Miller 1983; Dekel and Aarseth 1984; Peebles 1985). The conclusion from these and other simulations is that, contrary to the suggestions of earlier analytic work, non-Poisson fluctuations are needed to reproduce the observed galaxy clustering.

What kind of dark matter and fluctuations yield good agreement with observations? Cold or hot dark matter? Baryons only? Adiabatic or isocurvature fluctuations? None of the above? After nearly ten years of debate (e.g., Klypin and Shandarin 1983; Centrella and Melott 1983; White, Frenk and Davis 1983; Melott *et al.* 1983; Davis *et al.* 1985; White *et al.* 1987*a*,*b*; Centrella *et al.* 1988; Frenk *et al.* 1988, 1990; Park 1990; Villumsen, Scherrer and Bertschinger 1991), we still do not know the answer to this question, although some progress has been made. Hot dark matter models with adiabatic fluctuations suffer from late galaxy formation, excessive clustering amplitude, and excessive microwave background anisotropy. Cold dark matter models, with a scale-invariant spectrum of primeval fluctuations generated during inflation, are in critical condition due to their apparent lack of sufficient large-scale structure (Saunders *et al.* 1991 report the latest damage). A nonzero cosmological constant would improve the fit (Efstathiou, Sutherland and Maddox 1990; Turner 1991) but would represent a serious modification of the underlying cosmological model (Bertschinger 1990).

Are galaxies "biased" tracers of the mass? Kaiser (1984) showed that, if galaxy formation takes place preferentially at high peaks of the primordial fluctuation field, galaxies are more clustered than the mass. Using $N = 32^3$ P^3M simulations, Davis *et al.* (1985 and later papers by these authors) showed that the $\Omega_0 = 1$ cold dark matter model could, with high peak bias, account for many features of the galaxy distribution on small and intermediate scales. Moderate to strong clustering bias has been seen in many N-body simulations (Evrard 1987; West and Richstone 1988; Carlberg and Couchman 1989) and in the real galaxy distribution (e.g., White, Tully and Davis 1988). Besides having different spatial clustering amplitudes, galaxies and the overall mass distribution also have different small-scale velocity dispersions, an effect that Carlberg, Couchman and Thomas (1990) dubbed cosmological velocity bias. Although the full significance of velocity bias has yet to be understood, it may significantly modify conclusions about the viability of cosmogonical models by boosting the primordial fluctuation amplitude above previous estimates, increasing the amount of power on large scales.

Cosmological N-body simulations have some practical differences with galaxy dynamics simulations. The desiderata for cosmological simulations are similar — accurate forces, accurate time integration, and large N — except that periodic boundary conditions are preferable over isolated boundary conditions in order to simulate an infinite universe using a finite computer. A large fundamental volume is best to minimize unphysical effects arising from periodic replicas. To factor out the background Hubble expansion, cosmological simulations are generally performed in comoving coordinates $\vec{x} = \vec{r}/a(t)$, where $a(t)$ is the expansion scale factor and $\dot{a}/a = H$ is the Hubble parameter. In comoving coordinates, the equations of motion are

$$\frac{d^2\vec{x}_i}{dt^2} + 2H\frac{d\vec{x}_i}{dt} = -\vec{\nabla}_i\phi \,, \quad \nabla^2\phi(\vec{x},t) = 4\pi G\bar{\rho}(t)\frac{\delta\rho}{\rho} = \frac{3}{2}\,\Omega H^2\frac{\delta\rho}{\rho} \,. \tag{3}$$

Appropriate initial conditions are given by the linear growing mode of density perturbations, which is equivalent to small displacements from a uniform grid of comoving Lagrangian positions $\vec{q}$ using the Zel'dovich (1970) approximation:

$$\vec{x}(\vec{q},t) \approx \vec{q} + D(t)\vec{\psi}(\vec{q}) \; , \quad \vec{\nabla}\cdot\vec{\psi} = -\frac{\delta\rho}{\rho} \; , \quad \vec{\nabla}\times\vec{\psi} = 0 \; . \tag{4}$$

Formulae for the linear growing mode growth factor $D(t)$ may be found in, e.g., Peebles (1980). The linear displacement field $\vec{\psi}(\vec{q})$ is found most easily by solving the linearized continuity equation (the second of equations 4) using Fourier transform methods. The initial density fluctuation field $\delta\rho/\rho$ is readily generated as, e.g., a realization of a Gaussian random field with specified power spectrum.

Several algorithms for computing forces are used in cosmological simulations. The first is direct summation, used in the simulations of Aarseth and colleagues cited above. Quinn, Salmon and Zurek (1986) used direct summation in a small (10 Mpc diameter) volume, with mass resolution decreasing with distance from the center, to study the formation of individual galaxy halos at high resolution. Given its superior efficiency, the tree code is replacing direct summation in cosmological simulations. Bouchet and Hernquist (1988) and Suginohara *et al.* (1991) discuss the application of a tree code to cosmology, including fully periodic boundary conditions. Suto and Suginohara (1990) present results from a cosmological tree code simulation with an impressive $N = 128^3$ particles. Even more impressive is the extremely clustered simulation of Warren *et al.* (1991), which followed about 10^6 particles in the same volume studied earlier by Quinn *et al.* (1986).

Particle-mesh methods have been popular for cosmological simulations because of their speed and the naturally periodic boundary conditions. They have the advantage of offering high mass resolution with large N, but suffer from relatively poor force resolution. Better resolution is possible in two space dimensions, and Melott and Shandarin (1989, 1990) and Alimi *et al.* (1990) have used large PM simulations to study the physics of gravitational clustering in two dimensions.

The force resolution limitation of the PM algorithm can be removed by supplementing the forces with a direct sum over pairs separated by $\lesssim 3\Delta x$, resulting in P^{3}M (Hockney and Eastwood 1981). This method has proven its value in the simulations of Davis, Efstathiou, Frenk and White. The only disadvantage of P^{3}M, in my opinion, is the fact that it readily becomes dominated by the direct summation, grinding to a halt for a very highly clustered mass distribution. A logical solution, which no one seems to have yet implemented, would be to replace the local direct summation with the hierarchical tree algorithm. Although the speed of the resulting hybrid may not be significantly greater than a pure tree code, the storage requirements will be less because one does not need large, distant cells in the tree.

Seeking simultaneously better mass and force resolution, Villumsen (1989) introduced a hierarchical particle-mesh algorithm that embeds subgrids of lower-mass particles into a large, periodic PM volume. Forces in the subvolumes are computed using isolated boundary conditions for the fine particles in the subvolume, supplemented by lower-resolution forces from the more massive particles exterior to the subvolume. This ingenious method allows very high resolution in principle, but it remains to be seen how accurately structures are followed after being subdivided across several levels of such a hierarchy.

Couchman (1990), with his mesh-refined P^{3}M, introduces subgrids for a simpler purpose: to speed up the direct summation of nearby neighbors in

P^3M by shifting some of the burden to FFTs on a high-resolution PM subgrid. Couchman's algorithm is only superficially similar to Villumsen's. His subgrids are used only for the force computation, and they do not modify the force law or the particle motions relative to the standard P^3M algorithm. Simple in purpose, the algorithm is powerful in practice, quite possibly making P^3M faster than a tree code for highly clustered distributions. I am a great enthusiast of Couchman's algorithm; in the next section, you will see why.

LARGE COSMOLOGICAL N-BODY SIMULATIONS

Larger and faster computers have made possible a dramatic increase in the size of N-body simulations. Park (1990) performed PM simulations of large-scale structure with more than 4 million particles; Park and Gott (1990) increased N to more than 11 million. Gelb and Bertschinger (1989 and 1991, in preparation) have simulated $256^3 = 16.8$ million particles in a PM code with $384^3 = 56.6$ million grid points and, within one of their highly-clustered large simulations, they followed more than 1 million particles with high force resolution using P^3M. This section presents some of the methods and results of Gelb and myself (see also Gelb 1991).

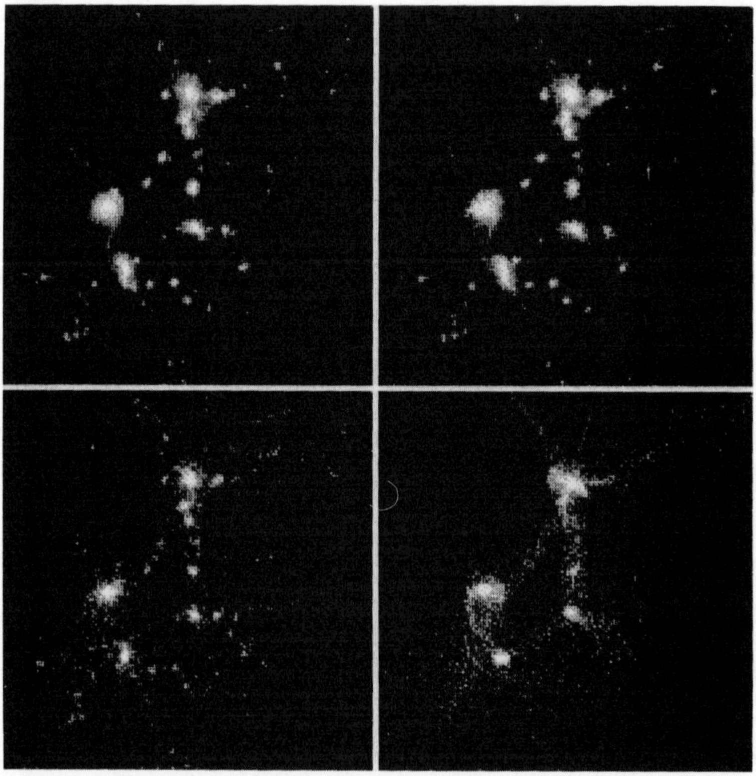

Fig. 1. Comparison of logarithmic mass density images (as positive images to give the best contrast at low density) from a $25.6^2 \times 6.4$ Mpc^3 subvolume of 4 cold dark matter N-body simulations in a 51.2 Mpc cube with $\Omega_0 = 1$ and $H_0 = 50\,\mathrm{km\,s^{-1}\,Mpc^{-1}}$, evolved to linear amplitude $\sigma_8 = 0.7$. The 4 panels show results from simulations starting from equivalent initial conditions but having different mass and force resolution. Top left: $N = 256^3$, 384^3 mesh force. Top right: $N = 128^3$, 256^3 mesh force. Bottom left: $N = 64^3$, P^{3}M forces. Bottom right: $N = 64^3$, 64^3 mesh force. The graininess is due to shot noise for finite N.

Most of our simulations use the PM method with cloud-in-cell (trilinear) interpolation and an optimal Green's function to reduce the anisotropic effect of the grid ("quiet" PM, in the parlance of Hockney and Eastwood 1981). We compute the force directly using FFTs to improve the resolution over the standard central-differencing of the potential; even so, our force is half the inverse square law value at 1.4 grid spacings. This softening is undesirable because it results in wasted mass resolution.

Figure 1 illustrates the effect of varying force and mass resolution. It should make clear that accurate results demand high dynamic range for both mass and force resolution, i.e., large particle number N and a force softening distance significantly less than the mean interparticle spacing.

We have implemented P^{3}M to improve the force resolution in our simulations. For highly clustered mass distributions, the direct summation portion of the standard P^{3}M algorithm (Efstathiou *et al.* 1985) becomes a computational bottleneck. With a partial implementation of Couchman's new algorithm (Couchman 1990) — with only 1 level of subgrid and no direct summation below this — we gained a dramatic improvement in speed. In a simulation of scale-invariant clustering with $\Omega_0 = 1$, $n = -2$ initial spectral slope, $N = 256^3$ particles, and a 384^3 PM grid, we followed more than one million of the particles (in 5% of the volume) with P^{3}M with force resolution given by 1024 Plummer softening lengths across the cube. We began the simulation using the direct summation algorithm, which ground to a halt with the cpu time per timestep increasing as a^4. Changing horses midstream, we switched to Couchman's algorithm, with an immediate 4-fold speedup. Even more happily, this algorithm slowed little as clustering increased, enabling us to expand by another factor of 2.2, for a total expansion factor of more than 30. With 1200 timesteps, we conserved energy to better than 0.25%. At the end of the simulation, the standard P^{3}M direct summation algorithm would have been a factor of 100 slower per timestep. Even so, we still spent 70% of the simulation cpu time computing short-range forces for only 6.5% of the particles. Figure 2 shows the clustering in a small region followed with high force resolution.

Besides being fast, Couchman's P^{3}M algorithm also allows an accurate fit to an arbitrary force law. Figure 3 shows the forces and relative errors for randomly placed pairs using our latest and most accurate implementation of this algorithm, and, for comparison, the standard cloud-in-cell PM algorithm (with central differencing of the potential and no optimization of the Green's function). All simulators are encouraged to display the accuracy of their method in a plot like this (or some variant of it, for the tree code).

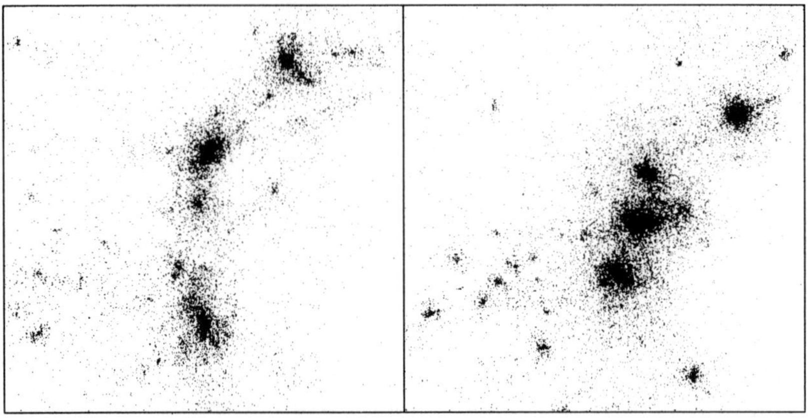

Fig. 2. Particle plots from a $(1/6)^2 \times (1/20)$ subvolume of the $n = -2$ scale-free hybrid PM/P^{3}M simulation. Every fifth particle only is plotted. Left: $a = 0.6a_f$. Right: $a = a_f$.

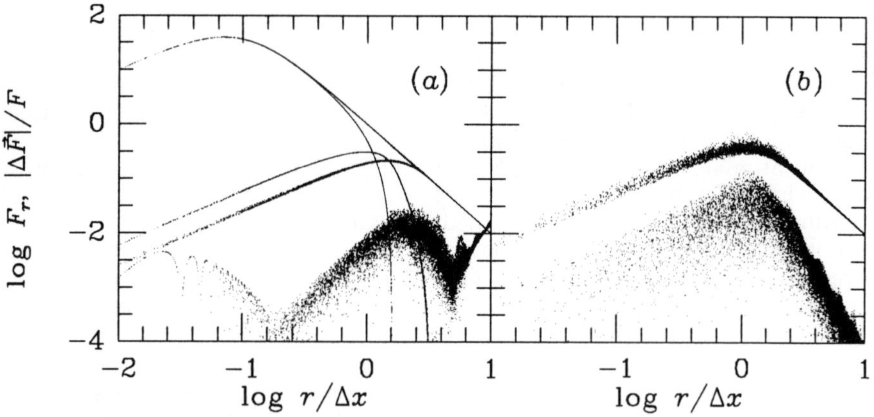

Fig. 3. The gravitational force (radial component) and the relative error for twenty thousand randomly placed pairs, using (*a*) a high-accuracy implementation of Couchman's P^{3}M algorithm and (*b*) the standard cloud-in-cell PM algorithm. (One massive test particle is placed at random in the cube and surrounded by 200 randomly located particles that probe the force field; this procedure is then repeated 99 times.) In (*a*), the force is decomposed into contributions from the coarse PM grid, the PM subgrid, and, for the smallest separations, direct summation. The apparent increase in errors at large separation is due to the periodic boundary conditions of the cube of side $64\Delta x$, modifying the force law from the inverse square law assumed in the error calculation. In (*b*), only the tangential errors are shown.

Our scale-free simulation offers a test of the the scaling behavior expected in an $\Omega_0 = 1$ cosmology with scale-free initial fluctuations. From the stable clustering argument of Davis and Peebles (1977), we expect the nonlinear power spectrum to be self-similar, $P(k,a)/P_{nl}(a) = \tilde{P}[k/k_{nl}(a)]$. The nonlinear wavenumber (defined by extrapolating linear theory until the relative density fluctuations exceed unity on this scale) decreases with expansion factor a as $k_{nl}(a) \propto a^{-\alpha}$, $\alpha = 2/(3+n)$; $P_{nl}(a)$ is defined by extrapolating linear theory to $k = k_{nl}$. For $k \ll k_{nl}$, the linear theory prediction $P \propto a^2 k^n$ is expected to be valid, while, for $k \gg k_{nl}$, the scaling model predicts $P/P_{nl} \propto (k/k_{nl})^{\gamma-3}$, $\gamma = 3(3+n)/(5+n)$.

Figure 4 shows the evolution of the nonlinear scaled power spectrum for our large $n = -2$ simulation and for a smaller (PM with 128^3 particles and a 256^3 grid) white noise ($n = 0$) simulation. Aside from departures at the smallest and largest wavenumbers, due to the neglect of waves longer than the cube and to force softening, respectively, the scaling predictions are confirmed very well. Previous workers have also seen evidence for scaling behavior (e.g., Efstathiou *et al.* 1988), but our large simulations show the best evidence to date for the asymptotic behavior of the power spectrum predicted by Davis and Peebles (1977).

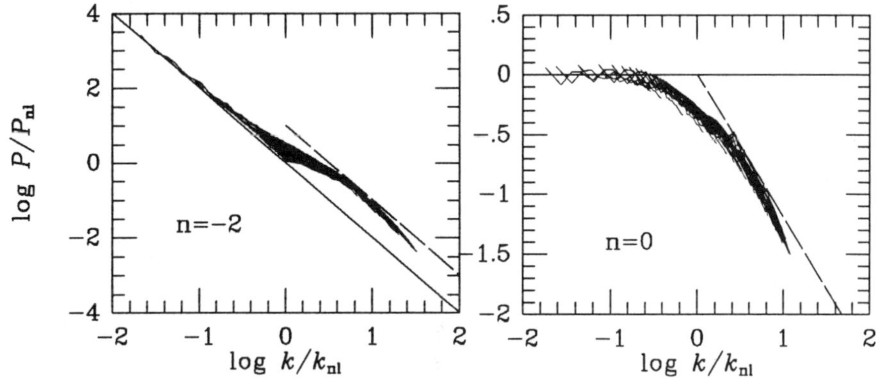

Fig. 4. Power spectra for 2 different simulations starting from scale-free initial conditions $P(k) \propto k^n$. The data curves show the individual power spectra at different times, excluding the longest wave and wavelengths shorter than 4 times the mean interparticle spacing. The curves have been shifted with $k_{nl}(a)$ and $P_{nl}(a)$ given by the scaling argument. The solid straight line shows the linear power spectrum and the dashed line shows the limiting nonlinear slope expected at high wavenumbers (the offset is arbitrary).

These results give confidence that the computer simulations are correct. Moreover, similarity solutions provide useful approximations over broad ranges of length scales for the non-scale-free behavior expected in physically realistic models such as cold dark matter. Even more important, from the departures from scaling noticed above we can deduce the limitations imposed by finite mass and force resolution and a finite box size. These limitations allow us to estimate the domain of validity of numerical results obtained for other models.

In particular, our results imply that a cube of side 51.2 Mpc is too small even for accurate computation of the galaxy two-point correlation function at late times (low bias factor) in the cold dark matter model. A cube twice as large (with the particle mass increased by a factor of eight) would both be large enough and still leave us with adequate mass resolution. Given the current rate of progress, it will not be long before high force resolution simulations — with tree codes or P^3M — are being performed with more than 10 million particles.

ACKNOWLEDGEMENTS

This work was supported by NSF grant AST90-01762. Supercomputer time was provided by the Cornell National Supercomputer Facility and the MIT Supercomputer Facility. The large simulations were performed in collaboration with James M. Gelb; I am grateful to him for his assistance in preparing the figures.

REFERENCES

Aarseth, S. J. 1972, in *Gravitational N-Body Problem,* IAU Colloq. 10, ed. M. Lecar, (Dordrecht: Reidel), p. 373.
Aarseth, S. J. 1985, in *Multiple Time Scales,* ed. J. U. Brackbill and B. I. Cohen (Orlando: Academic), p. 377.
Aarseth, S. J., Gott, J. R., and Turner, E. L. 1979, *Ap. J.*, **228**, 664.
Alimi, J.-M., Bouchet, F. R., Pellat, R., Synget, J.-F., and Moutarde, F. 1990, *Ap. J.*, **354**, 3.
Appel, A. W. 1985, *Siam. J. Sci. Stat. Comput.*, **6**, 85.
Balcells, M., and Quinn, P. J. 1990, *Ap. J.*, **361**, 381.
Barnes, J. E. 1988, *Ap. J.*, **331**, 699.
Barnes, J. E. 1989, *Nature*, **338**, 123.
Barnes, J., and Hut, P. 1986, *Nature*, **324**, 446.
Barnes, J. E., and Hut, P. 1989, *Ap. J. Suppl.*, **70**, 389.
Bertschinger, E. 1990, *Nature*, **348**, 675.
Bouchet, F. R., Adams, J.-C., and Pellat, R. 1985, *Astr. Ap.*, **144**, 413.
Bouchet, F. R., and Kandrup, H. E. 1985, *Ap. J.*, **299**, 1.
Bouchet, F. R., and Hernquist, L. 1988, *Ap. J. Suppl.*, **68**, 521.
Carlberg, R. G., and Couchman, H. M. P. 1989, *Ap. J.*, **340**, 47.
Carlberg, R. G., Couchman, H. M. P., and Thomas, P. 1990, *Ap. J. (Letters)*, **352**, L29.
Carlberg, R. G., and Freedman, W. L. 1985, *Ap. J.*, **298**, 486.
Carlberg, R. G., and Sellwood, J. A. 1985, *Ap. J.*, **292**, 79.
Cen, R. Y. 1990, preprint.
Cen, R. Y., Jameson, A., Liu, F., and Ostriker, J. P. 1990, *Ap. J. (Letters)*, **362**, L41.
Centrella, J. M., Gallagher, J. S., Melott, A. L., and Bushouse, H. A. 1988, *Ap. J.*, **333**, 24.
Centrella, J., and Melott, A. L. 1983, *Nature*, **305**, 196.
Couchman, H. M. P. 1990, preprint.
Davis, M., Efstathiou, G., Frenk, C. S., and White, S. D. M. 1985, *Ap. J.*, **292**, 371.
Davis, M., and Peebles, P. J. E. 1977, *Ap. J. Suppl.*, **34**, 425.
Dekel, A., and Aarseth, S. J. 1984, *Ap. J.*, **283**, 1.

Dupraz, C., and Combes, F. 1986, *Astr. Ap.*, **166**, 53.

Efstathiou, G., Davis, M., Frenk, C. S., and White, S. D. M. 1985, *Ap. J. Suppl.*, **57**, 241.

Efstathiou, G., and Eastwood, J. W. 1981, *M.N.R.A.S.*, **194**, 503.

Efstathiou, G., Frenk, C. S., White, S. D. M., and Davis, M. 1988, *M.N.R.A.S.*, **235**, 715.

Efstathiou, G., Sutherland, W. J., and Maddox, S. J. 1990, *Nature*, **348**, 705.

Evrard, A. E. 1987, *Ap. J.*, **316**, 36.

Evrard, A. E. 1988, *M.N.R.A.S.*, **235**, 911.

Evrard, A. E. 1990, *Ap.J.*, **363**, 349.

Evrard, A. E., and Davis, M. 1988, *Nature*, **333**, 335.

Frenk, C. S., White, S. D. M., Davis, M., and Efstathiou, G. 1988, *Ap. J.*, **327**, 507.

Frenk, C. S., White, S. D. M., Efstathiou, G., and Davis M. 1990, *Ap. J.*, **351**, 10.

Gelb, J. M. 1991, these proceedings.

Gelb, J. M., and Bertschinger, E. 1989, *Bull. Amer. Astr. Soc.*, **21**, 1172.

Greengard, L. 1988, *The Rapid Evaluation of Potential Fields in Particle Systems* (Cambridge: MIT Press).

Greengard, L. 1990, *Computers in Phys.*, **4**, 142.

Greengard, L., and Rokhlin, V. 1987, *J. Comput. Phys.*, **73**, 325.

Heisler, J., and White, S. D. M. 1990, *M.N.R.A.S.*, **243**, 199.

Hernquist, L. 1987, *Ap. J. Suppl.*, **64**, 715.

Hernquist, L. 1988, *Computer Phys. Comm.*, **48**, 107.

Hernquist, L. 1989, *Nature*, **340**, 687.

Hernquist, L., and Barnes, J. E. 1990, *Ap. J.*, **349**, 562.

Hernquist, L., and Katz, N. 1989, *Ap. J. Suppl.*, **70**, 419.

Hernquist, L., and Quinn, P. J. 1988, *Ap. J.*, **331**, 682.

Hernquist, L., and Quinn, P. J. 1989, *Ap. J.*, **342**, 1.

Hernquist, L., and Weinberg, M. D. 1989, *M.N.R.A.S.*, **238**, 407.

Hillis, W. D., and Barnes, J. 1987, *Nature*, **326**, 27.

Hockney, R. W., and Eastwood, J. W. 1981, *Computer Simulation Using Particles* (New York: McGraw-Hill).

Hohl, F. 1971, *Ap. J.*, **168**, 343.

Hohl, F., and Hockney, R. W. 1969, *J. Comput. Phys.*, **4**, 306.

Holmberg, E. 1941, *Ap. J.*, **94**, 385.

Jernigan, J. G., and Porter, D. H. 1989, *Ap. J. Suppl.*, **71**, 871.

Kaiser, N. 1984, *Ap. J. (Letters)*, **284**, L9.

Klypin, A. A., and Shandarin, S. F. 1983, *M.N.R.A.S.*, **204**, 891.

Kofman, L., and Shandarin, S. F. 1990, *M.N.R.A.S.*, **242**, 200.

Lin, D. N. C., and Tremaine, S. 1983, *Ap. J.*, **264**, 364.

Makino, J., and Hut, P. 1988, *Ap. J. Suppl.*, **68**, 833.

McGlynn, T. A. 1984, *Ap. J.*, **281**, 13.

Melott, A. L., Einasto, J., Saar, E., Suisalu, I., Klypin, A. A., and Shandarin, S. F. 1983, *Phys. Rev. Letters*, **51**, 935.

Melott, A. L., and Shandarin, S. 1989, *Ap. J.*, **343**, 26.

Melott, A. L., and Shandarin, S. 1990, *Nature*, **346**, 633.

Miller, R. H. 1970, *J. Comput. Phys.*, **6**, 449.

Miller, R. H. 1983, *Ap. J.*, **270**, 390.

Olson, K. M., and Kwan, J. 1990*a*, *Ap. J.*, **349**, 480.

Olson, K. M., and Kwan, J. 1990*b*, *Ap. J.*, **361**, 426.

Ostriker, J. P., and Peebles, P. J. E. 1973, *Ap. J.*, **186**, 467.

Park, C. B. 1990, *M.N.R.A.S.*, **242**, 59p.
Park, C. B., and Gott, J. R. 1990, preprint.
Peebles, P. J. E. 1985, *Ap. J.*, **297**, 350.
Peebles, P. J. E. 1980, *The Large-Scale Structure of the Universe* (Princeton: Princeton University Press).
Quinn, P. J. 1984, *Ap. J.*, **279**, 596.
Quinn, P. J., Salmon, J. K., and Zurek, W. H. 1986, *Nature*, **322**, 329.
Ryu, D., Vishniac, E. T., and Chiang, W. H. 1990, *Ap. J.*, **354**, 389.
Saunders, W., Frenk, C., Rowan-Robinson, M., Efstathiou, G., Lawrence, A., Kaiser, N., Ellis, R., Crawford, J., Xia, X.-Y., and Parry, I. 1991, *Nature*, **349**, 32.
Sellwood, J. A. 1987, *Ann. Rev. Astr. Ap.*, **25**, 151.
Sellwood, J. A., and Carlberg, R. G. 1984, *Ap. J.*, **282**, 61.
Suginohara, T., Suto, Y., Bouchet, F. R., and Hernquist, L. 1991, *Ap. J. Suppl.*, **76**, in press.
Suto, Y., and Suginohara, T. 1990, preprint.
Toomre, A., and Toomre, J. 1972, *Ap. J.*, **178**, 623.
Toomre, A. 1977, in *The Evolution of Galaxies and Stellar Populations,* ed. B. M. Tinsley and R. B. Larson (New Haven: Yale University Observatory), p. 401.
Turner, M. S. 1991, in *Proceedings of the IUPAP Conference on Primordial Nucleosynthesis and the Early Evolution of the Universe,* ed. K. Sato (Dordrecht: Kluwer), in press.
van Albada, T. S. 1982, *M.N.R.A.S.*, **201**, 939.
van Albada, T. S., and van Gorkom, J. H. 1977, *Astr. Ap.*, **54**, 121.
Villumsen, J. V. 1982, *M.N.R.A.S.*, **199**, 493.
Villumsen, J. V. 1989, *Ap. J. Suppl.*, **71**, 407.
Villumsen, J. V., Scherrer, R. J., and Bertschinger, E. 1991, *Ap. J.*, **365**, in press.
Warren, M. S., Zurek, W. H., Quinn, P. J., and Salmon, J. K. 1991, these proceedings.
West, M. J., and Richstone, D. O. 1988, *Ap. J.*, **335**, 532.
White, S. D. M. 1978, *M.N.R.A.S.*, **184**, 185.
White, S. D. M. 1979, *M.N.R.A.S.*, **189**, 831.
White, S. D. M. 1986, in *Inner Space/Outer Space,* ed. E. W. Kolb, M. S. Turner, D. Lindley, K. Olive, and D. Seckel (Chicago: The University of Chicago Press), p. 228.
White, S. D. M. 1988, in *The Early Universe,* ed. W. G. Unrugh and G. W. Semenoff (Dordrecht: Reidel), p. 239.
White, S. D. M., Frenk, C. S., and Davis, M. 1983, *Ap. J. (Letters)*, **274**, L1.
White, S. D. M., Davis, M., Efstathiou, G., and Frenk, C. S. 1987*a*, *Nature*, **330**, 451.
White, S. D. M., Frenk, C. S., Davis, M., and Efstathiou, G. 1987*b*, *Ap. J.*, **313**, 505.
White, S. D. M., Tully, R. B., and Davis, M. 1988, *Ap. J. (Letters)*, **333**, L45.
Zel'dovich, Ya. B. 1970, *Astr. Ap.*, **5**, 84.

Audience Questions

Windhorst: Since the sky is spectroscopically confusion-limited at V>26 mag, we can never measure x(r) for you at cosmologically interesting distances. What are the prospects for you to compute w(q) for us as a function of apparent magnitude, under reasonable assumptions of the OLF(z), N(z) and spectral evolution models? That way we could directly compare our results.
Bertschinger: Your suggestion is an important one. The main difficulty, of course, is knowing what to call a galaxy in the simulations and how much luminosity to assign to each one. Clearly, a phenomenological treatment will be required. White and Frenk (preprint) have considered many models in their analytical treatment of the galaxy formation problem. A more realistic treatment will require combining these phenomenological models directly with the N-body simulations. Given that the simulations may now have adequate dynamic range for following the assembly and clustering of protogalaxies at high redshift, I am optimistic that a useful comparison can be made with your deep angular correlation function.

Melott: Your CDM simulations appear to be a filamentary net, while Margaret Geller emphasized in the previous session that those observations imply surfaces. This could be purely a matter of visual presentation. I would emphasize that it is possible to measure the topology objectively using the algorithm developed by Gott et al. The code is available by electronic mail, and will soon be published in Physics Reports. The galaxy density and survey volume of the new CfA survey appear to be enough for a determination of fairly high significance.

Rosenblatt: Is the galaxy resulting from the merger of the 5 or 6 individual galaxies [in the Barnes movie] initially in a close group a disk or an elliptical type, and what are some salient implications?
Bertschinger: Your question is answered by Barnes (1989). The merger remnant in his simulation looks very much like an ordinary bright elliptical. Barnes concluded that mergers in compact groups may produce a significant fraction of all ellipticals. Rapid merging seems to be the fate of galaxies with extensive dark matter halos, but there is no observational evidence that the merger rates predicted assuming all galaxies have dark halos are excessive. Whether or not a specific theory, such as cold dark matter, predicts the correct abundance of merger remnants remains to be seen, although it is widely believed that cold dark matter leads to excessive merging of halos. Until merger simulations include gas dynamics, so that the baryons can dissipate into tightly bound cores, we cannot be certain whether there is a problem.

Bland-Hawthorn: When you talk of the topology seen in your very impressive simulation, how can you be sure that you have adequate contrast to compare with observations? Could the perceived topology change with more particles?
Bertschinger: Indeed, the topology can change with the number of particles. We can check for such effects by varying the number of particles in the simulations. As the number of particles tracing a given density field is decreased, Poisson fluctuations from discrete sampling fragment continuous structures. This effect is more pronounced at low density. The density contours shown in the three-dimensional rotation of our large cold dark matter simulation were at a density of only about three times the mean, and the density was binned at the mean interparticle spacing, so that shot noise is not entirely negligible. However, with more smoothing or higher density contrast our topology is robust. On the relatively small scales that we were simulating (a few to 50 Mpc), the evolved cold dark matter distribution clearly has a lumpy topology.

Galaxy Correlation Functions in Redshift Space from Large N-body Simulations

James M. Gelb
Massachusetts Institute of Technology
77 Massachusetts Avenue, Cambridge, MA 02139

ABSTRACT

We computed galaxy-galaxy correlation functions in redshift space with data generated from a set of large Cold Dark Matter N-body (Ω=1) simulations. Galaxy halos were identified as local density maxima in the dark matter distribution and artificial magnitude limited galaxy catalogs were made. We simulated the observing process by choosing several galaxies as observing sites at 3 epochs from 3 simulations, and then by measuring the redshift, angular direction on the sky, and apparent magnitude of the galaxies. We then computed the 2-point correlation function in redshift space, $\xi(r_p, \pi)$, as was done for the CfA (m_B=14.5) redshift survey by Davis and Peebles (1983). The fluctuations from simulation–to–simulation in our small computational box (51.2 Mpc on a side) are too large to accurately distinguish the 3 epochs studied (b=2, 1.4, and 1) when computed as Davis and Peebles did. Furthermore, the simulated contours of constant $\xi(r_p, \pi)$ provide a poor match to the data—they are excessively elongated along the π axis for small projected separations. However, when we computed the velocity dispersion by direct pair counts, the best match is for a bias factor b=1.4 (σ_8=0.7). We need to perform simulations in larger volumes and we need to compute velocity dispersions from $\xi(r_p, \pi)$ in order to draw stronger conclusions.

ANALYSIS

We have studied the gravitational clustering of galaxies in redshift space for simulated Cold Dark Matter universes. The simulations used employ 128^3 particles in a box 51.2 Mpc on a side with $\Omega = 1$ and a Hubble constant of 50 km s^{-1} Mpc^{-1} and use the particle–mesh technique with a 256^3 grid. Distances are reported in Mpc, not h^{-1} Mpc. The effective softening distance of our force law is $\sim$ 280 kpc. The particle mass is $\sim$ 4.4x10^9 M$_\odot$. The box was chosen to be large enough to study clustering on scales up to $\sim$10-20 Mpc but small enough to resolve galaxies with many particles. In a box of this size, however, we found that the the correlation amplitude of $\xi(r)$ measured by direct pair counts was underestimated. This has been confirmed by running a simulation with equivalent initial conditions in a box 102.4 Mpc on a side. A larger box is required to pick up long wavelengths in the CDM spectrum and to avoid saturating on scales comparable to the size of the box. In the 51.2 Mpc box, the limited force resolution affects our results on scales $\lesssim$ 0.5 Mpc and the limited size of our box affects our results on scales $\gtrsim$ 10 Mpc. In order to sample the effective depth of the CfA redshift survey, we periodically replicated our 51.2 Mpc box out to a radius of 200 Mpc. This enabled us to to probe distant pairs where the redshifts were not dominated by peculiar velocities. Unfortunately, however, this replication process is one source of large simulation–to–simulation scatter since the "details" of any one simulations are repeated many times. Three realizations of the initial conditions were simulated.

We identified galaxy halos as local maxima in the density field, smoothed on a 100 kpc scale, at various epochs: b=2, 1.4, and 1. (The bias parameter is $b \equiv a^{-1}$; where the amplitude of the linear power spectrum has been chosen so that the expansion factor a=1 gives the $8h^{-1}$ Mpc spherical tophat linear normalization.) We then defined the circular velocity of each halo to be $\sqrt{GM/R}$, where M is the mass of the bound particles within 300 kpc of the peak and R is 300 kpc. Our method of identifying halos is more realistic than the *ad hoc* method of tagging the initial density field. We found that the peaks in the initial density field were not well correlated with the peaks in the evolved matter distribution.

We illuminated our halos by using a blue Tully-Fisher relationship, $M_{B_T} = -7.3\ log_{10}(2V_{circ}) - 1.13$ (Pierce and Tully 1988). Observing sites were selected on a few galaxies with circular velocities $\sim$220 km s^{-1}, peculiar velocity with respect to the CMB frame $\sim$400 km s^{-1}, and with locations outside of large clusters. We then shifted our box, using periodic boundary conditions, so that the box was centered on the observer. The velocities were corrected to the CMB frame. The hypothetical observer then measures, for each galaxy, the angular position on the sky, m_{B_T}, and the redshift. We included only galaxies brighter than $m_{B_T} = 14.5$ and we eliminated the galaxies beyond 200 Mpc as well as the intrinsically faint galaxies ($M_{B_T} > -20$). We selected these parameters to be as consistent as possible with the analysis of the CfA survey by Davis and Peebles.

The simulated magnitude limited catalog was then cross-correlated with a random catalog in order to determine the 2-point correlation function in redshift space. In order to mimic the effect of a magnitude limit on the random catalog, galaxies were randomly placed in shells about the observer. The number of galaxies placed in each shell was determined by $4\pi r^2\ \Delta r\ n\ f$, where n is the number density of all galaxies and f is the fraction that can be seen at a distance r. The galaxies in the random catalog were assumed to be at rest. This method is slightly different than the method used by Davis and Peebles. We used a luminosity function generated from the simulations themselves. Since gravitational N-body simulations without gas dynamics tend to make too many large halos, our catalogs contain a few excessively bright galaxies. As a consequence, our Poisson errors were manageably small on scales $\gtrsim 1$ Mpc. We used equal weighting for all galaxies.

We then selected pairs of objects in the catalogs. For each pair, with associated redshifts cz_1 and cz_2, we computed the perpendicular projected distance $r_p \equiv 1/H_0\ (cz_1 + cz_2)\tan(\theta_{12}/2)$ and the parallel velocity difference $\pi \equiv (cz_1 - cz_2)$. The angle θ_{12} is the angular separation of the pair of galaxies on the sky. The pair was rejected if $\theta_{12} > 50^0$.

The pairs were then binned in r_p and π. The following definition was used to compute $\xi(r_p, \pi)$,

$$\xi(r_p, \pi) \equiv \frac{DD(r_p, \pi)}{DR(r_p, \pi)} - 1.$$

The DR counts are for pairs when one galaxy was selected from the simulated data catalog (D) and the other galaxy was selected from the random catalog (R). The DD counts are for pairs when both galaxies were selected from the simulated data catalog.

Fig. 1 shows an artificial data catalog generated from one of the simulations at b=1.4. The image is an $x - y$ slice in physical space and is 20 Mpc thick, centered on the observer. There is an apparent magnitude limit, m_{B_T}=14.5, and faint galaxies, $M_{B_T} > -20$, have been eliminated. Fig. 2 shows the same galaxies in redshift space. The periodic stacks are not shown. These types of catalogs were used in the analysis and were cross correlated with random catalogs. For each epoch we chose 3 simulations, 2 observing sites, and 2 random catalogs for a total of 12 observations per epoch. We found that scatter in our results were most sensitive to the simulation–to–simulation differences and much less so to the observer–to–observer locations or the choice of random catalog.

RESULTS AND CONCLUSIONS

Figs. 3 and 4 depict $\log_{10}\xi(r_p,\pi)$ versus π for $0.8 < r_p < 1.6$ Mpc and $1.6 < r_p < 3.2$ Mpc respectively. The solid lines are for b=2, the dotted lines are for b=1.4, and the dashed lines are for b=1. A dot is placed at the point where Davis and Peebles found $\log_{10}\xi = 0$ in their best model of the CfA survey. The results from 3 simulations are shown. The scatter from various observing sites and random catalogs is much less so they are not shown. We have studied other bins of r_p and concluded that the scatter over the range ~ 1 Mpc to ~ 10 Mpc is too large to clearly pick a best value of b.

We have computed the pairwise velocity dispersion, $\sigma(r)$, directly from our simulated galaxy catalogs in physical space for galaxies above various absolute magnitude thresholds. Fig. 5 shows the results for galaxies whose circular velocities exceed 100 $\mathrm{km\,s^{-1}}$ ($M_{B_T} < -18$) for b=2, 1.4, and 1. For scales $r \gtrsim 5$ Mpc the results show little scatter from simulation–to–simulation and are not very sensitive to the circular velocity cut-off. If the modeling of Davis and Peebles is a fair estimate of σ, then our best, preliminary estimate of b is 1.4. Davis and Peebles find σ=400 $\mathrm{km\,s^{-1}}$ at r_p=6.3 Mpc for H_0=50 $\mathrm{km\,s^{-1}\,Mpc^{-1}}$. The center–of–mass velocity of our halos yield smaller values of $\sigma(r)$ than the mass (all particles) does. E. Bertschinger and the author are studying the source of this velocity bias (see Carlberg *et al.*, 1990) as well as issues of dynamic range in length and mass.

Finally, Fig. 6 shows contours of constant $\xi(r_p,\pi)$=4, 2, and 1 (from the center outward). The circles are shapes of contours for perfect Hubble flow. The contours are shown for $b = 2$ (solid lines), $b = 1.4$ (dotted lines), and $b = 1$ (dashed lines). The results were averaged over the 12 observations per epoch to yield cleaner contours. The contours shows excessive elongation along the π axis for small scales of r_p. The level of distortion, however, for the ξ=1 contour is consistent with the results of Davis and Peebles.

In order to get better statistics and to make a more fair comparison with the Davis and Peebles analysis of the CfA survey we need to run simulations in larger boxes and to compute $\sigma(r_p)$ from $\xi(r_p,\pi)$. To date we have been motivated by the importance of dynamic range in mass and the ability to identify halos without the ad hoc procedure of tagging the initial density field.

This work was supported by the National Science Foundation under grant AST90-01762. Ed Bertschinger has provided many helpful suggestions regarding the simulations and analysis. We also would like to thank Paul Schechter for many useful discussions.

This research was conducted using the Cornell National Supercomputer Facility, a resource of the Center for Theory and Simulation in Science and Engineering at Cornell University, which receives major funding from the National Science Foundation and IBM Corporation, with additional support from New York State and members of its Corporate Research Institute. We appreciate the assistance of Paul Schwarz on using the IBM 3090.

REFERENCES

Davis, M. and Peebles, P.J.E., 1983, *Ap.J.*, **267**, 465.

Carlberg, R.G., Couchman, H.M.P., and Thomas, P.A., 1990, *Ap.J. (Letters)*, **352**, L29.

Pierce, M.J. and Tully, B., 1988, *Ap.J.*, **330**, 579.

FIG. 1, b=1.4, 20 Mpc THICK

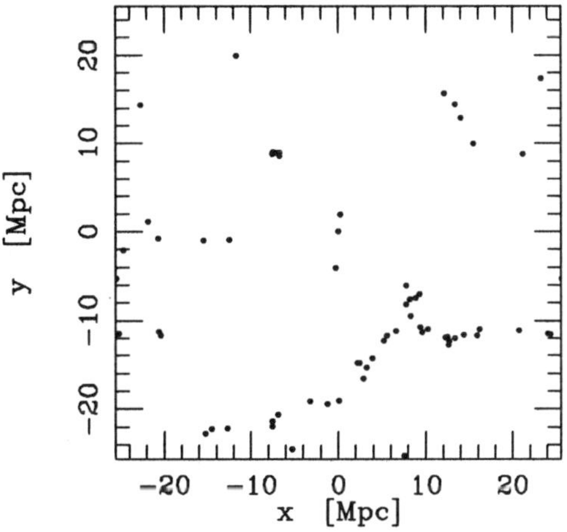

FIG. 4, 1.6<r_p<3.2 Mpc

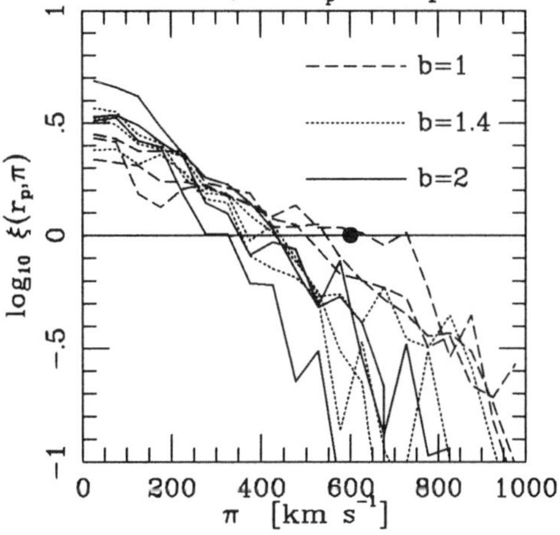

FIG. 2, b=1.4, REDSHIFT SPACE

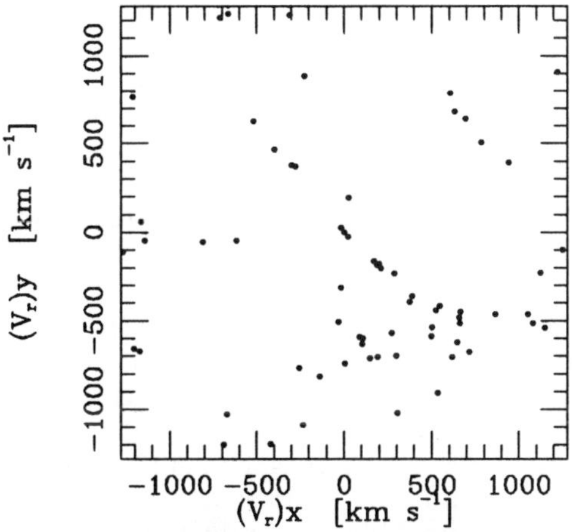

FIG. 5, DIRECT, V_{circ}>100 km s^{-1}

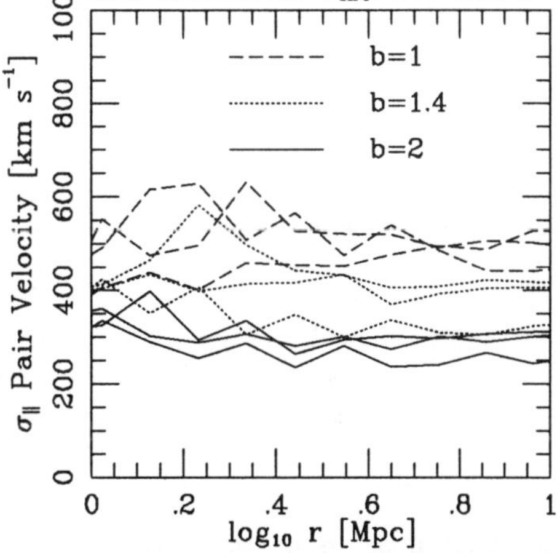

FIG. 3, 0.8<r_p<1.6 Mpc

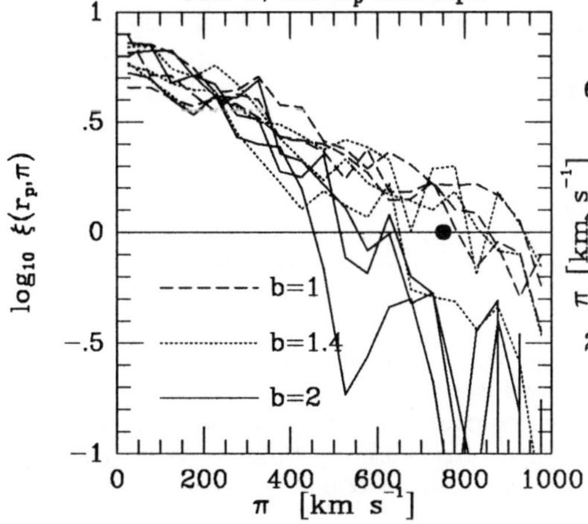

FIG. 6, CONTOURS OF $\xi(r_p,\pi)$

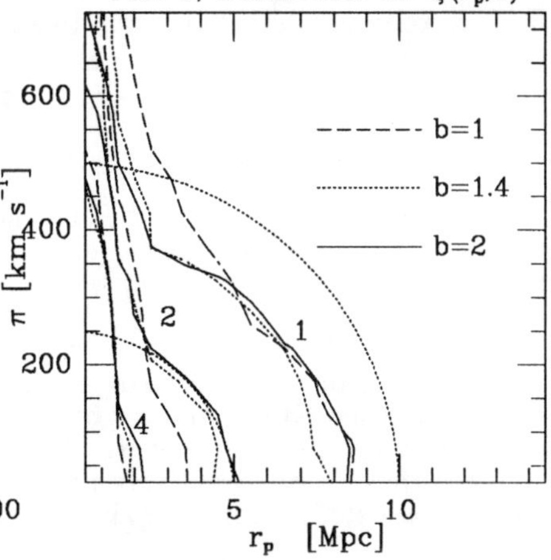

GALAXY CORRELATION FUNCTION DOWN TO V=26 ON 0.5° SCALES

Rogier A. Windhorst and Lyman W. Neuschaefer
Department of Physics and Astronomy
Arizona State University, Tempe, AZ 85187-1504

ABSTRACT

We extend the galaxy 2-point correlation function $w(\theta)$ down to V~26 mag on scales $\theta \lesssim 0.5°$. For $14 \lesssim g \lesssim 26$ mag, $w(\theta)$ roughly maintains a constant slope ($1-\gamma \simeq -0.6$ to -0.8), which has apparently undergone little change over the *entire* redshift range sampled. For $20 \lesssim g \lesssim 24.5$ mag, $w(\theta)$'s amplitude continues to decrease with fainter magnitudes. Beyond $g \gtrsim 25$ mag, our catalogs become gradually incomplete, and $w(\theta)$'s amplitude starts to *rise* again while maintaining roughly the *same* slope. We give several possible explanations for this effect.

1) INTRODUCTION

The galaxy correlation functions, $\xi(r)$ and $w(\theta)$, are important tools to quantify large-scale clustering and its possible evolution with cosmic epoch. The large width of the optical luminosity function combined with the spectroscopic confusion limit at $V \gtrsim 26$ imply that $\xi(r)$ might not be measurable out to cosmological redshifts, even with the next generation telescopes. We can, however, measure $w(\theta)$ down to the faintest possible flux levels, and interpret this with models for the redshift distribution and spectral evolution of faint field galaxies.

The most comprehensive $w(\theta)$ measurements have been done by Groth and Peebles (1977), Maddox et al. (1990), Stevenson et al. (1985), Koo and Szalay (1984), and references therein. At bright magnitudes, $w(\theta)$ shows a break at $\theta \simeq 2$–$3°$ ($\Leftrightarrow$ 7–10h^{-1} Mpc), beyond which its slope steepens. For $\theta < 0.5°$, these surveys find $w(\theta) \propto \theta^{1-\gamma}$ with $\gamma \sim 1.8$, perhaps flattening to $\gamma \sim 1.6$ at $B_J \gtrsim 21$.

Deep redshift surveys (e.g., Colless et al. 1990) show that the median redshift at g~21 is ~0.3, while at g~23 mag z_{med} is estimated to be $\gtrsim 0.4$–0.6 [using $J^+ = B_J = g + 0.39 + 0.37(g-r)$]. At fainter fluxes, the field galaxy redshift distribution is unknown. Combining broad-band magnitudes and colors with spectral evolution models (Koo 1985, 1990; Bruzual 1990) can possibly constrain the sampled redshift range, suggesting that z_{med} may be as large as ~1.25 at $g \simeq 25$. The lack of field galaxies *not* detected in the U-band by Guhathakurta et al. (1990) suggests that $z_{med} \lesssim 3$ at their completeness limit ($g \lesssim 27$). With a 95% completeness limit at $g \simeq 24.5$, our wide area survey may thus reach beyond redshifts of unity.

2) PHOTOMETRIC AND ASTROMETRIC REDUCTIONS

We used the Palomar 200" Four-shooter CCD-array to image two deep radio-optical survey regions – each ~ $(0.43°)^2$ – with nearly contiguous mosaics in Gunn gri down to g~26 mag. Here, we present the first results for Gunn g in one of these regions. Details on the complete survey, data reduction and analysis are given by Neuschaefer (1991). The mosaics were treated as following:

- In addition to the standard processing, the night sky fringe pattern was removed from the Gunn i images by median filtering. In Gunn g and r, large scale additive gradients remained which were removed by fitting a low-order bicubic spline-surface to an iteratively filtered grid of modal sky values. With the sky flat to *few* $\times 10^{-4}$, complete galaxy catalogs could be constructed.

• After the usual two-stage astrometry, random and systematic errors are ~0.4". The (RA,DEC)'s of individual catalogs were transformed onto a master Cartesian grid (l, m). Figure 1 shows all objects with g$\gtrsim$19.75 mag on this grid.
• Global photometric zeropoints were derived from Gunn standards, by renormalizing the sky-mode in each chip, and from the chip-to-chip overlap. After these calibrations, the zeropoints in all chips were consistent within 0.05 mag.

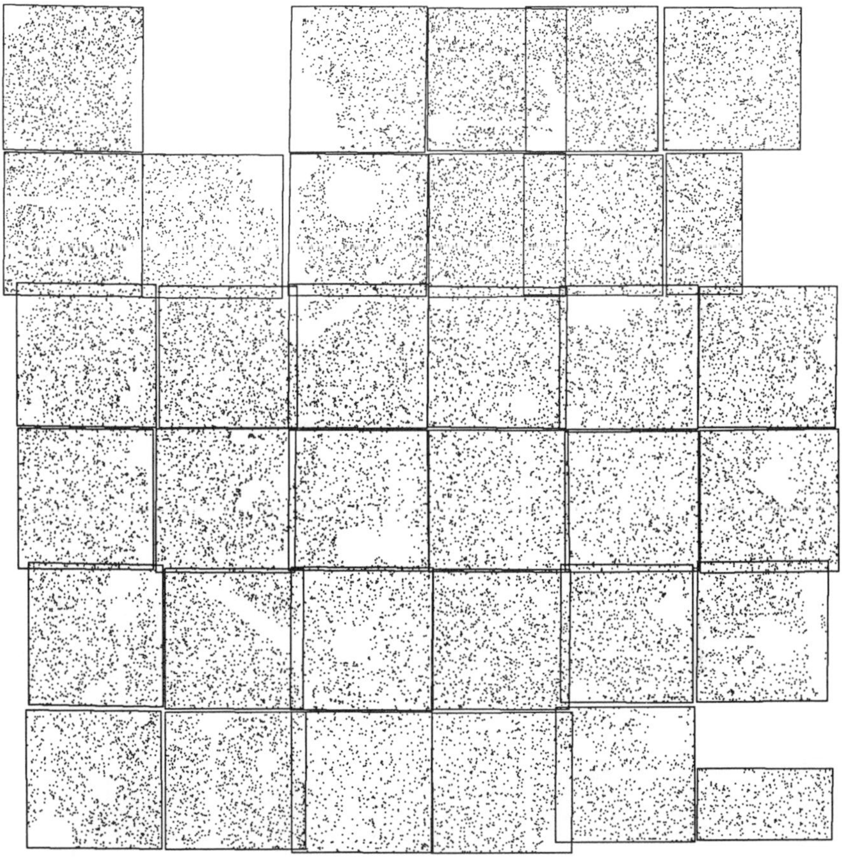

Figure 1: Map of all non-stellar objects in our Four-shooter mosaic with Gunn g$\gtrsim$19.75 mag. Blank areas were masked out as described in the text.

3) COMPILATION AND FILTERING OF CATALOGS

Detection and classification of faint objects was done on individual CCD frames with FOCAS (Valdes 1982). This process comprises:
a) Object detection: To include most real objects in the first pass, a signal threshold of $2.5\sigma_{sky}$ was used, and an area threshold of 6 contiguous pixels.
b) Object classification: Each object profile was fit by a two-parameter template, consisting of a 'broadened' and a stellar PSF-component. Classification was based on the weak, monotonically decreasing relation between object-size and g-magnitude. The resulting starcounts are consistent with models for our Galaxy.

c) Catalog filtering: In each CCD catalog we removed effects due to: 1) overlapping frames; 2) obscuration caused by the guide arm and bright stars blocking background objects; 3) small unexposed borders between chips; and 4) spurious detections caused by bright stars' diffraction spikes, saturation streaks, and other defects. Figure 1 illustrates the resulting areas (slanted bars; circles; horizontal/vertical bars) masked out in the catalogs and computation of $w(\theta)$. The two Four-shooter mosaics yielded ~60,000 faint objects down to g,r~26.

4) MEASUREMENT OF $w(\theta)$

$w(\theta)$ was obtained as the observed count of pair separations normalized by that of a random distribution covering an area of the *same* shape and solid angle as our Four-shooter mosaic (Fig. 1). We tested our code by measuring $w(\theta)$ for a 3-dim CDM simulation field with known $w(\theta)$ provided by Dr. E. Bertschinger.

Our galaxy counts in Gunn g are consistent with Tyson's (1988) deeper B_J counts, from which we estimate that 95% completeness occurs for galaxies of *average* surface brightness at Gunn g=24.5, r=24.5, and i=23.5 mag, respectively. To correct for incompleteness, as well as slight frame-to-frame seeing and focus variations, we weighted every galaxy detection with the ratio of Tyson's galaxy count to the N(m) *observed* in each particular chip. For $g \lesssim 24.5$ mag, these weights are all unity. For $24.5 \lesssim g \lesssim 25.25$, they slowly increase to a factor of ~2. For $g \gtrsim 25.25$, weights become so large that they make the computation of $w(\theta)$ unstable, and are set to 1.0. The data are presented in Figure 2.

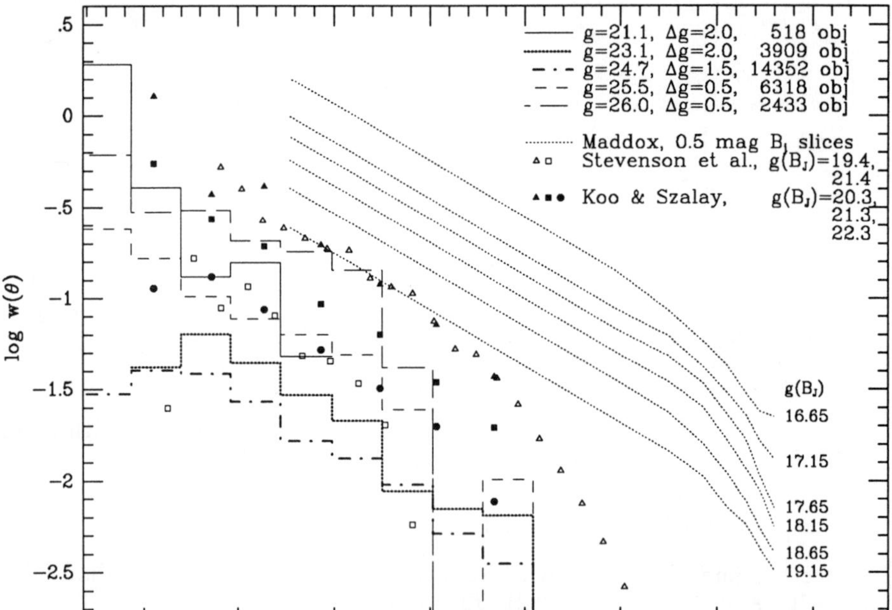

Figure 2: The two-point correlation function $w(\theta)$ for all galaxies in Figure 1 compared to photographic surveys. Histograms indicate data from the current work. Bin centers were weighted with N(m). Δg is the bin width.

For each magnitude range plotted in Figure 2, Table 1 lists the median g-magnitude, the number of detected galaxies N_{gal}, and the θ-range and number of bins N_{bin} used to determine the best fit slope and amplitude of $w(\theta)$. At the smallest θ, some points deviate due to problems in splitting faint objects, and due to residual problems with star-galaxy separation at bright levels. At the largest θ, differential incompleteness (due to seeing or sensitivity variations on scales twice the chip size) may not have been entirely corrected for. Also, for g~23.5–24.5 mag $w(\theta)$ reaches an extremely small amplitude ($\lesssim 0.3\%$ at a *few* arcmin scales – which is smaller than in any of the other surveys). As a consequence, $w(\theta)$ is more sensitive to random and systematic errors at g~24 mag, and goes occasionally negative at the 10^{-3} level on scales of 1–2× the chip size. We solve this by: 1) using many CRAY runs (24 in Fig. 2) to average out any systematics from the random number generators used; and 2) by least-squares fitting the following expression (Koo and Szalay 1984):

$$w(\theta) = \frac{n_{gg}(\theta)/B_k - n_{rg}(\theta)}{n_{rr}(\theta)}$$

Here $n_{rg}(\theta)$ is the number of galaxies paired with random points, and corrects for the effects of frame borders (Hewett, 1982), and B_k is the background factor. A unique solution for the correlation amplitude A_ω, slope $1-\gamma$, and B_k involves several iterations. First, the slope is fitted with B_k fixed to unity. Next, we minimize χ^2 by allowing B_k and A_ω to vary, while the input slope $1-\gamma$ is fixed at its previously determined value. In each next iteration the slope was not allowed to change by more than 1σ since it might otherwise diverge. The process converged after two iterations. Our best fit results are given in Table 1. The number of usable bins in the fit depends somewhat on the value of B_k. The standard error in $1-\gamma$ was computed following Barrow et al. (1984), and becomes large for the extreme magnitude bins that have the least statistics.

Table 1. – *Best fit parameters of w(θ) as function of Gunn g-magnitude*

g_{med} (mag)	N_{gal}	N_{bin}	Used θ-range $log\theta_1(')$	$log\theta_2(')$	$1-\gamma$ (rms)	log A_ω ($\theta = 1°$)	B_k	χ^2
21.1	00518	4	−0.93	−0.14	−1.0 (0.5)	−3.21 (0.26)	1.000	0.80
23.1	03909	8	−0.93	+0.91	−0.70 (0.14)	−2.90 (0.08)	0.984	1.05
24.7	14352	9	−1.20	+0.91	−0.59 (0.13)	−2.95 (0.07)	0.993	1.10
25.5	06318	9	−1.20	+0.91	−0.63 (0.14)	−2.45 (0.07)	0.981	3.54
26.0	02433	7	−1.20	+0.39	−0.57 (0.35)	−1.87 (0.12)	1.000	16.3
[27.0	00405	7	−1.20	+0.39	−0.5 (0.6)	−1.15 (0.15)	1.000	12.9]

5) RESULTS AND DISCUSSION

Figure 2 compares our $w(\theta)$ measurement to earlier work (see section 1). For g=22–24.5 mag, where we have significant statistics and our sample is complete, $w(\theta)$ has least squares slopes $1-\gamma \simeq -0.6$ to -0.7 at the angular scales where we can measure it. For g=24.5-26.5 mag, our catalogs become gradually incomplete, and the data suggest a marginal flattening of $w(\theta)$'s slope (to $\sim$–0.5) at fainter magnitudes *when B_k is kept fixed.* However, when B_k is allowed to vary, the resulting slope remains steady at $1-\gamma \simeq -0.6$ to -0.7. Given the bootstrap errors, the slope of $w(\theta)$ has thus not changed significantly in our magnitude range ($19 \lesssim g \lesssim 26.5$ mag). Given that similar values for $1-\gamma$ ($\simeq -0.6$ to -0.8) are observed in the brighter surveys, we conclude that – within the errors – the slope of $w(\theta)$ has apparently undergone little change over the *entire* redshift range sampled for $14 \lesssim g \lesssim 26.5$ mag. Our discussion in Section 1 suggested that the corresponding redshifts range from $z \simeq 0.03$ to probably of order 1.5–2.5.

Figure 2 shows that the correlation *amplitude* steadily decreases towards fainter fluxes in the range $14 \lesssim g \lesssim 24.5$ mag, where it reaches a minimum. Beyond $g \gtrsim 24.5$ mag, the amplitude of $w(\theta)$ starts to *rise* again towards fainter fluxes (at given θ), and *continues* to do so out to $g \sim 26.5$ mag while maintaining the *same* slope $1-\gamma \simeq -0.6$. Since its amplitude here again exceeds 3–10%, this drastic rise cannot be entirely due to the instrumental systematics mentioned in Section 4. We consider the following other possible explanations, in order of expected importance:

a) Any seeing limited survey only selects the highest surface brightness galaxies at the faintest magnitudes, which is only amplified by the $\propto (1+z)^{-4}$ surface brightness dimming. At z=0, high surface brightness galaxies are known to be more strongly clustered than low surface brightness ones (Dressler 1980). This would boost $w(\theta)$'s amplitude if this clustering property still existed at fainter fluxes. Hence, even though at $g \sim 26$ mag we only detect a *few* % of all existing galaxies, it is at least remarkable that the *detected* galaxies – a good fraction of which show up on overlapping frames of similar quality – yield the same slope for $w(\theta)$ as the brighter magnitudes, where our catalogs are complete.

b) For $g \gtrsim 25$ mag, an increasing fraction of faint objects could be sub-galactic clumps that still have to merge into single galaxies (e.g., Koo 1990). Assuming that the fraction of such clumps increases towards higher redshifts, they could also boost the amplitude of $w(\theta)$ at fainter fluxes.

c) For $g \gtrsim 24.5$ mag, z_{med} may start to decrease again with fainter fluxes *if* the galaxy counts were dominated by a local population of dwarf galaxies. Tyson (1988) and Koo (1990) argued against the existence of such a population.

d) The rise in $w(\theta)$'s amplitude for $g \gtrsim 24.5$ mag could merely indicate that $z_{med} \sim 1.25$ at $g \sim 25$ *and* that $\Omega \simeq 1$, since the θ–z relation would then reach a minimum at z=1.25. The linear scale associated with our field size would then primarily depend on H_0 , and less on the precise value of z_{med} . When all afore mentioned instrumental systematics and selection effects are understood and corrected for, a solid determination of the rise in $w(\theta)$'s amplitude towards the faintest flux levels could provide a strong constraint to the value of Ω.

Since the spectral evolution of field galaxies is unknown at g=24–26 mag, we have not attempted to rescale $w(\theta)$ to the locally measured ones with the relativistic version of Limber's projection equation. For brighter surveys, this can be done, since reasonable models exist for the spectral evolution, and the resulting median redshift as function of apparent magnitude. Instead, we suggest that a new approach be followed in the future to interpret samples of the current depth. We believe it would be better to generate realistic models for the evolution of large scale structure through N-body or CDM-simulations, and convolve these with a plausible range of models for the epoch-dependent optical luminosity function – or equivalently the spectral evolution of field galaxies – which should include dependency on galaxy type, M/L, cosmology, etc. The resulting artificial galaxy fields should then be treated in *exactly the same way* as our data, i.e., be convolved with the known observational selection function (seeing, chip-to-chip and frame-to-frame sensitivity, areal coverage, etc.), and processed through FOCAS. We can then compute their $w(\theta)$ as for the real data, and see which models best compares to the observed $w(\theta)$. This method has two advantages: 1) convolving a family of models is easier than deconvolving data to constrain models; and 2) almost all observational selection effects present in the data will also be present in the convolved models, so that a comparison is justified.

ACKNOWLEDGEMENTS

This research was supported by grant AST-8821016 of the National Science Foundation, a Fellowship of the Arizona State University Graduate College, and the Alfred P. Sloan Foundation. We thank Dave Burstein for his advice.

REFERENCES

Barrow, J. D., Bhavsar, S. P., and Sonoda, D. H. 1984, MNRAS, 210, 19P
Bruzual A., G., 1990, in Evolution of the Universe of Galaxies, ASP Conference Series, Vol.10, ed. R. G. Kron (Provo, UT, BookCrafters, Inc.), p. 185
Colless, M. M. et al. 1990, MNRAS, in press
Dressler, A. 1980, ApJ, 236, 351
Groth, E. J., and Peebles, P. J. E. 1977, ApJ, 217, 385
Guhathakurta, P., Tyson, J. A., and Majewski, S. R. 1990, ApJL, 357, L9
Hewett P. C. 1982, MNRAS, 201, 867
Koo, D. C., and Szalay, A. S. 1984, ApJ, 282, 390
Koo, D. C. 1985, AJ, 90, 418
Koo, D. C. 1990, in Evolution of the Universe of Galaxies, op. cit., p. 268
Maddox, S. J. et al. 1990, MNRAS, 242, 43P
Neuschaefer, L. W. 1991 Ph.D. Thesis, Arizona State University
Stevenson, P. R. F. et al. 1985, MNRAS, 213, 953
Tyson, J. A. 1988, AJ, 96, 1
Valdes, F. 1982, SPIE, 331, 465

QUESTIONS

Bhavsar: The errors you quote on the slope of the correlation function are formal errors. You should do something like a bootstrap to get the intrinsic error. (Note that the points that determine the slope are not independent data points).

Windhorst: You are absolutely right. My short talk did not permit me to address this, but the results of the bootstrap you suggested are discussed in the text. We also Monte Carlo tested systematic errors as function of magnitude and θ, as induced by the different chip sensitivities, differential seeing, random number generators, etc. These constrain the range of slopes to $1-\gamma = -0.6$ to -0.8.

White: Did you cover the area in different bands?

Windhorst: Yes, we did cover our two mosaics about equally deep in Gunn gri, and intend to use this information to constrain the sampled redshift range as a function of apparent magnitude and color. To roughly separate the effects of redshift and galaxy type, we will also need coverage in the U-band, which can now be done efficiently with a UV-flooded TEK2048 on a large telescope.

THE NATURE OF FAINT BLUE GALAXIES

P. Guhathakurta
Institute for Advanced Study, Princeton, NJ 08540

ABSTRACT

Deep CCD images in U (3600 Å), B_J (4500 Å), and R (6500 Å) are used to define and analyse a population of faint, blue galaxies that approach the confusion limit at $B_J = 28$. The average $B_J - R$ and $U - B_J$ colours of these galaxies become bluer as they get fainter and are only slightly redder than those of a flat spectrum source at $B_J \sim 26$. The slope of the number-versus-apparent magnitude relation, measured at three different wavelengths, is indicative of relatively recent evolution in these systems. Indirect methods are used to place broad constraints on the redshift distribution of faint blue galaxies. The galaxies appear to be weakly clustered on the sky, and this suggests that they are progenitors of systems that are intrinsically faint at the present epoch. Further, field galaxy redshift surveys indicate that brighter counterparts of these galaxies ($B_J \lesssim 23$) are at low redshifts ($z \lesssim 0.7$) and are fainter than L^* galaxies. The redshift surveys suggest a scenario where the faint tail of the luminosity function is evolving more rapidly than the bright end. We have applied this type of evolution to the present-day galaxy luminosity function and are able to find a reasonably good match to the B_J galaxy counts derived from our CCD data and the observed trends in the $B_J - R$ and $U - B_J$ colours. The uncertainties in the nature of evolution of faint blue galaxies, however, continue to compromise their usefulness as probes of the curvature of spacetime.

Ultra-deep CCD images of three high-latitude fields were taken using the CTIO 4-m telescope, with cumulative integration times per field of about 10, 2, and 3 hours in the U, B_J, and R filters, respectively. The images through different filters were registered and co-added to enable reliable detection of galaxies as faint as $B_J = 28$. There are $\sim 3 \times 10^5$ predominantly blue, faint galaxies per square degree to these limits (~ 2000 in our sample in an area of 27 arcmin^2) distributed more or less uniformly over the sky (Tyson 1988). Brighter counterparts of these objects were first discovered in photographic surveys over a decade ago (*cf.* Kron 1978). Several authors have suggested that the faint, blue galaxies may represent the bulk of galaxies at an early stage of their evolution history.

The effect of spacetime geometry on the field galaxy number-apparent magnitude relation is overshadowed by the effects of galaxy evolution. While the R band counts are marginally consistent with that expected from no-evolution models, those at shorter wavelengths (B_J and U) are significantly steeper. If galaxies are brighter in the past because of increased star formation rates, the UV radiation associated with hot, young stars is being redshifted into the U and B_J bands. The relatively shallow count slopes in R and I would then suggest that this evolution is quite recent ($z \lesssim 3$). The steepening of the count slope with decreasing wavelength is equivalent to the statement that the galaxies become *bluer* as they get fainter. This is obvious from the $U - B_J$ versus $B_J - R$ colour-colour diagram in Fig 1 (below), where the cluster of points moves down and to the left as one goes from a relatively bright sub-sample ($20 < B_J < 23.5$ in Fig 1a) to a fainter set of galaxies ($25 \lesssim B_J \lesssim 26$ in Fig 1b). At the faint end, the mean galaxy colours are close to that of a flat spectrum, F_ν = constant (solid lines in Fig 1).

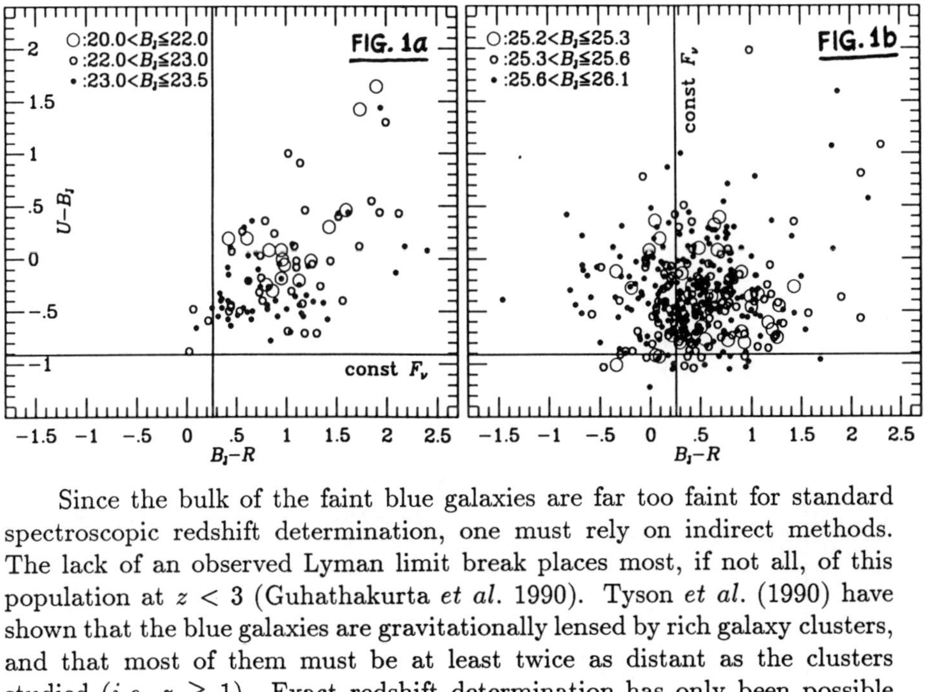

Since the bulk of the faint blue galaxies are far too faint for standard spectroscopic redshift determination, one must rely on indirect methods. The lack of an observed Lyman limit break places most, if not all, of this population at $z < 3$ (Guhathakurta *et al.* 1990). Tyson *et al.* (1990) have shown that the blue galaxies are gravitationally lensed by rich galaxy clusters, and that most of them must be at least twice as distant as the clusters studied (*i.e.* $z \gtrsim 1$). Exact redshift determination has only been possible for brighter counterparts ($B_J < 23$) of the faint blue galaxies (Broadhurst *et al.* 1988; Colless *et al.* 1989), and their average redshift is about 0.5. A naive extrapolation (based on no-evolution, for example) from these relatively bright galaxies to the faint blue galaxies ($B_J \sim 27$) would place the latter at a mean $z \sim 1$, close to the rough lower limit imposed by gravitational lensing arguments.

Broadhurst *et al.* emphasize that the exact *shape* of the z distribution at $B_J < 21.5$ is as though galaxies have not evolved at all, even though the number of galaxies and count slope at these magnitudes is greater than no-evolution predictions. This implies that the faint end of the luminosity function (LF) evolves more rapidly with look-back time (t_z) than the bright end. We have examined such a model parametrizing it as a Schechter function with a steepening faint end slope: $\alpha(z) = -1.25 - k_\alpha t_z/t_\infty$ (see Fig 2 below).

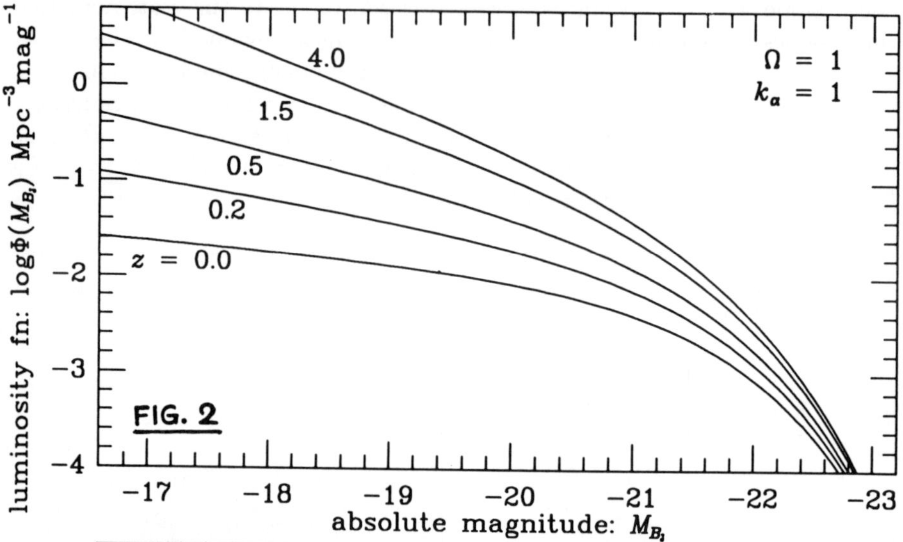

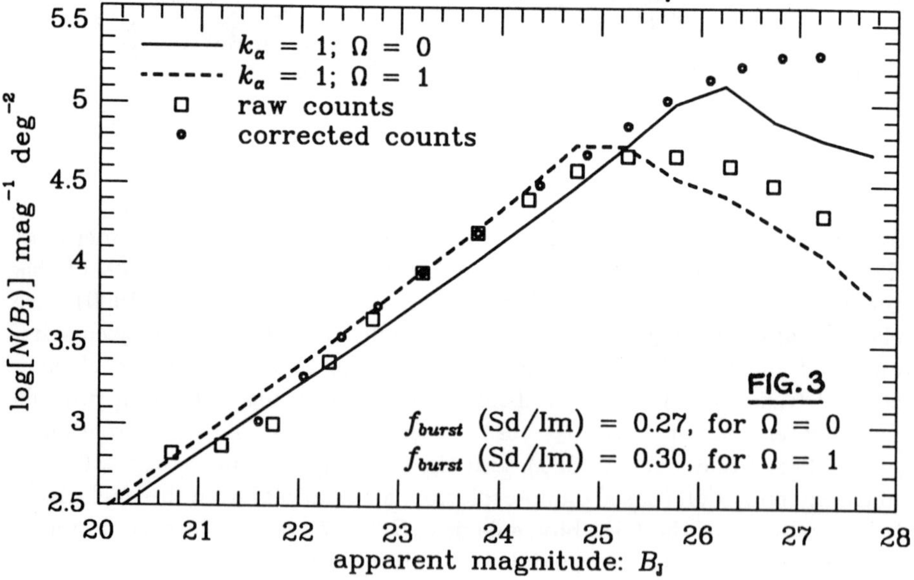

It should be noted that only the overall evolution of the LF is constrained by the count and redshift data, *not* the details of how individual galaxies evolve. The observed stellar content of galaxies has been used to constrain galaxy lifetimes, under the assumption that the LF evolves as in Fig 2 while conserving galaxy number. The late-type spirals (Sd/Im) are assumed to undergo intermittent starbursts, so that only a fraction ($f_{burst} \sim 0.3$) are visible at a given instant. Present-day galaxy spectral shapes are convolved with our filter response functions in order to compute K-corrections. This model has been used to compute B_J galaxy counts and $B_J - R$ and $U - B_J$ color distributions (see Guhathakurta 1990 for details). Our model counts for two different cosmologies are compared to Tyson's (1988) B_J counts in Fig 3 (see previous page).

Even though Fig 3 suggests that an $\Omega = 1$ universe predicts too few galaxies, that conclusion is subject to the validity of the assumption of galaxy number conservation, *i.e.* the assumption that the LF at any z solely contains (brighter) counterparts of objects that the present-day LF already accounts for. A preliminary study shows the galaxy-galaxy correlation function for faint blue galaxies to be very weak, so much so that they must be progenitors of objects that are more weakly correlated than L^* galaxies. This raises the possibility that the majority of faint blue galaxies are brighter examples of objects that we do not even account for in the present-day LF. If this is indeed the case, faint blue galaxies are not telling us much about the evolution of "normal" galaxies. In short, the uncertainties in the evolution history of faint blue galaxies continues to make them unsuitable tracers of q_0.

I am greatly indebted to Tony Tyson and Jim Gunn for making this project possible. I would also like to thank Matthew Colless, George Efstathiou, and Richard Ellis for several useful discussions.

REFERENCES

Broadhurst, T. J., Ellis, R. S., and Shanks, T. 1988, *M.N.R.A.S.*, **235**, 827.

Colless, M. M., Ellis, R. S., and Taylor, K. 1989, in *The Epoch of Galaxy Formation*, eds. C. S. Frenk *et al.* (Dordrecht: Kluwer), p. 359.

Guhathakurta, P. 1990, *Ap. J.*, submitted.

Guhathakurta, P., Tyson, J. A., and Majewski, S. R. 1990, *Ap. J.* (*Letters*), **357**, L9.

Kron, R. G. 1978, PhD thesis, Univ. of California, Berkeley.

Tyson, J. A., Valdes, F., and Wenk, R. A. 1990, *Ap. J.* (*Letters*), **349**, L1.

Tyson, J. A. 1988, *A. J.*, **96**, 1.

Large Scale Structure and the Intergalactic Medium

THE DIFFUSE X-RAY BACKGROUND AND THE INTERGALACTIC MEDIUM

Abraham Loeb
The Institute for Advanced Study
Princeton, New Jersey 08540

ABSTRACT

The possible origin of the x-ray background (XRB) from a hot intergalactic medium (IGM), as opposed to discrete sources, is critically analysed. The Comptonization of the microwave background radiation by the IGM is discussed for an arbitrary electron distribution function, using a multiple–scattering approach for relativistic Comptonization . The recent spectral results from COBE can be used to rule-out the existence of a homogeneous hot IGM as the dominant source of the XRB. The alternative possibility of Bremsstrahlung emission by a clumped IGM is considered.

I. INTRODUCTION

The origin of the x-ray background (XRB) is still unknown although its discovery[1] preceded that of the microwave background. The x-ray spectrum as observed by the HEAO-1 mission can be well fitted by a thermal Bremsstrahlung shape before any discrete sources are subtracted from it[2], namely $dI/dE_0 \approx 5.6(E_0/[3\mathrm{keV}])^{-0.29}\exp(-E_0/[40\mathrm{keV}])$ $\mathrm{keV/keV}\cdot\mathrm{cm}^2\cdot\mathrm{sec}\cdot\mathrm{sr}$, for photon energies $3 \leq E_0 \leq 100$ keV. Below 3 keV there are indications from the *Einstein* IPC data[3] that the spectrum becomes steeper. Most of the properties of the discrete source populations (such as active galactic nuclei) are best determined in the soft x–ray band ($\lesssim 10$ keV), and are not well-known at the energies and flux levels of interest for the XRB. This observational situation serves as a fertile ground for two competing scenarios for the origin of the x–ray background:

a. Contribution from discrete sources, such as: Active Galactic Nuclei (AGN)[4],[5],[6], supernovae or massive x-ray binaries[7].

b. Thermal Bremsstrahlung emission by a hot (50–300 keV) intergalactic medium (IGM)[8],[9],[10],[11],[12].

The most popular candidate for models of class (a) are AGN. The traditional $E^{0.7}$ power-law fit for AGN spectra requires spectral evolution of the sources[4], since it is too steep at low energies and too flat at hard x-rays to account for the entire background. However, various papers have proposed recently to use different AGN spectra, that are still compatible with existing observations, and when properly summed over redshift (without spectral evolution) result in the x-ray background. On phenomenological grounds, Schwartz and Tucker[5] used Log–gaussian or thermal Bremsstrahlung plus power-law spectra with only luminosity evolution for their sources, and obtained good spectral fit to the HEAO-1 data points. Nevertheless, their prediction for a soft x-ray excess below 3 keV was not found in the *Einstein* IPC data[3]. On theoretical grounds, various authors[6] suggested that if an original power-law spectrum of an obscured source is predominantly observed through reflection from a cold Thomson-thick material (e.g, in an accretion disk), it should generically result in a break near 30 keV. The net output spectrum will then be compatible with the observed shape of the XRB. In this context, a redshifted iron fluorescence line should also be detectable below ~ 3 keV. There are indications from GINGA for the above features in several low-redshift Seyfert galaxies[13]. Nevertheless, considering the entire class (a) interpretation, it is still the case that no known class of sources was as yet observed to have exactly the same spectrum as the XRB.

The following discussion will concentrate on interpretation (b) that involves the intergalactic medium. This interpretation is motivated by the resemblance of the observed spectrum to thermal Bremsstrahlung. The simplest model of class (b) involves a homogeneous IGM that was heated over a short period of time and later cooled[9–12] by adiabatic expansion and Compton scattering on the cosmic background radiation (CBR). This version can be ruled-out by the recent spectral results from COBE (see Sec. IIa). The hot homogeneous IGM would have necessarily Compton-scattered the CBR and produced strong deviations from its observed blackbody shape[12]. More advanced models consider a clumped IGM. In principle, such a medium can satisfy the above constraint, since the Bremsstrahlung emissivity is proportional to $\langle n_e^2 \rangle$, while the Comptonization parameter along the line of sight[14], $y \equiv \int n_e (T_e/m_e) \sigma_T dt$ is proportional to $\langle n_e \rangle$. Here, n_e, T_e are the electron density and temperature ($k_B = c = 1$) , t is the Cosmological time, and $\sigma_T = 0.67 \times 10^{-24} \mathrm{cm}^2$ is the Thomson cross-section. Nevertheless, a two-phase IGM in pressure equilibrium can also be ruled-out by COBE (Sec. II*b*); and one is left with the possibility of transient clumps having a density contrast $\gtrsim 20$, that are produced and heated continuously. These necessary ingredients of the IGM model can be satisfied during the formation of large scale structure in the universe (see Sec. II*c*). It is possible to construct a model that will be compatible with COBE and provide a significant fraction of the XRB (especially, in its softer part). The model serves as an alternative to descrete sources, and is mostly constrained by the high degree of isotropy of the XRB and the CBR.

From the observed fluctuations in the 1–3 keV band, Hamilton and Helfand[15] deduced that if most of the XRB originates from point-like sources then their surface density should exceed 5000/deg^2. This result does not limit strongly the clumped IGM hypothesis, since the emission below 3 keV may be dominated by discrete sources or numerous clumps that are formed on small scales. Above 3 keV the isotropy limits are much less stringent. The HEAO-1 observations provide the isotropy constraint[16] of $\Delta I/I \lesssim 2\%$ on $5° \times 5°$ (i.e., $\gtrsim 100$ sources/deg^2) for 3-10 keV. The microwave background isotropy also constrains the clumped IGM model, as discussed in Sec. II*c*.

Finally, it should be mentioned that there is evidence[17] for the existence of a diffuse background at higher, namely gamma–ray energies (0.1–100 MeV). The observational data is much less accurate than that of the XRB (e.g., concerning the isotropy) because of the low sensitivity of the detection process and the high instrumental background; but there is strong evidence for an *MeV bump* in the spectrum. Many theoretical ideas have been proposed for the origin of the γ-ray background, among which are AGN emission[17] (although others claim for only $\sim 10\%$ AGN contribution[18]); the redshifted $p\bar{p}$ or π^0 annihilation in a baryon-symmetric universe[19], cosmic-ray interactions[17] and other processes. In the following sections we shall limit our discussion to the XRB ($\lesssim 100$ keV). The conclusions of this work will be summarized in Sec. III.

II. X-RAY EMISSION BY THE INTERGALACTIC MEDIUM

II*a*. HOMOGENEOUS IGM.

Variants of the hot homogeneous IGM model were presented in many papers, published before COBE was launched[8–12]. Most of these studies assumed that the electrons were heated once; recent studies, which include electron–electron bremsstrahlung, found an IGM density parameter (assuming homogeneous gas of baryons) of $\Omega_b \equiv (8\pi G/3H_0^2)\rho_{\rm IGM} \approx 0.3 \pm 0.1$, where the Hubble constant is $H_0 = 100\ h$ km s^{-1} Mpc^{-1}. This Ω_b is higher than the baryonic density parameter deduced from the standard Big Bang nucleosynthesis[20]. Although the homogeneous IGM model for the XRB has various non-attractive characteristics, it has definite predictions that can be tested by observations. Its detailed study allows one to identify the necessary changes in the hot IGM properties that will allow it to be compatible with existing data (while providing the XRB).

The thermal evolution of the IGM electrons can be computed for an arbitrary distribution function, $f_e(p)$, using the cosmological Boltzman equation[12]; that takes into account electron-electron (e-e) collisions; and electron cooling by the adiabatic cosmological expansion and due to scattering off the CBR photons. It is important to follow the evolution of $f_e(p)$ since its high energy tail has strong effects on the CBR; and since the Coulomb relaxation rate for e-e collisions is comparable to the sum of the Hubble expansion rate and the CBR Comptonization rate at $z \sim 4$; as shown in Fig. 1*a* (note that electron-ion thermalization is slower by the electron to ion mass ratio; thus the ions may have a completely different temperature in this model). Figure 1*b* shows an example[12] for the evolution of an initial top hat electron distribution function, $f_e(p)$; assuming heating at $z_H = 5$. Given the thermal history of the electrons, one can calculate the resulting Bremsstrahlung spectrum by integrating their emissivity over redshift. The integrated spectrum can be fitted to match the HEAO-1 observations; the magnitude of the x-ray flux is controlled by Ω_b while its "flatness" is determined by the maximal value of $T_e(z)/(1+z)$.

A direct constraint on the above model can be obtained from the Compton–scattering effects of the hot IGM electrons on the CBR. Since the electrons may have semi-relativistic energies and arbitrary $f_e(p)$, the Kompaneets equation or its Sunyaev–Zeldovich solutions cannot be used for evaluating these effects. Instead, one can calculate accurately the Comptonization effects of hot (relativistic) electrons on soft photon spectra, by using a multiple scattering approach[10,11,[21]]. This formalism describes analytically, in a compact form, the distortion of the soft photon spectrum using the Fourier transform of the single-scattering probability. In the non-relativistic case, the validity of the known analytical results derived from Kompaneets equation can be extended to arbitrary electron distribution functions and initial photon spectra. In the non-relativistic limit, the distorted spectrum is characterized by the single parameter, $y \equiv \frac{1}{3}\int\langle\gamma\beta^2\rangle\sigma_T dt$; while in the relativistic case, the average shift in the photon frequency is $\langle\Delta\nu/\nu\rangle = (e^{4y_{\rm tot}} - 1)$; where, $y_{\rm tot} \equiv \frac{1}{3}\int\langle\gamma^2\beta^2\rangle\sigma_T dt$, β is the electron velocity $[c = 1]$; $\gamma = (1-\beta^2)^{-1/2}) = p/m_e\beta$ and the average here is over $f_e(p)$. The temperature shift in the Rayleigh-Jeans part of a black-body spectrum is given by, $\Delta T_{\rm RJ}/T_{\rm RJ} = \int\langle\{1-\ln([1+\beta]/[1-\beta])/2\gamma^2\beta\}\rangle\sigma_T dt$. For a detailed description of this formalism, and its analytical and numerical results, see Ref. 21.

Based on the relativistic Comptonization formalism and the thermal history of the IGM electrons, one can calculate self-consistently the predicted effects of the hot IGM on the CBR spectrum. Typical results from this calculation[12] are presented in Figs. 1*c*, 1*d*. Figure 1*c* shows the fitted XRB spectrum, and Fig. 1*d* presents the resulting CBR distortion, for heating at $z_H \sim 5$ to an initial electron temperature of $T_H = 5 \times 10^9$ K (as in Figs. 1*a*, *b*). The spectral distortion differs appreciably from the non-relativistic (Kompaneets) calculation. Typically, the distortion below 500μm is larger by more than an order of magnitude than allowed by the COBE limit[[22]] $y < 10^{-3}$. Following a large sample of runs with a variety of electron distribution functions and thermal histories, one can conclude that a minimal clumpiness of $C \equiv (\langle n_e^2\rangle)^{1/2}/\langle n_e\rangle \gtrsim 20$, is required for obtaining effectively $y < 10^{-3}$ with an instantaneously heated IGM. This minimal clumpiness corresponds to the existence of highly nonlinear structures (clumps) in the IGM, that may be either transient or in a steady-state.

Fig. 1*a*: The evolution of the electron temperature with redshift (solid line). The IGM electrons are heated instantaneously at $z_H = 5$ to $T_H = 5 \times 10^9$ K, with an initially narrow top–hat distribution function. The dotted line shows $T_e \propto (1+z)^2$ for a non-relativistic matter in adiabatic expansion. The area below the dashed line indicates where thermalization by e-e Coulomb collisions occurs.

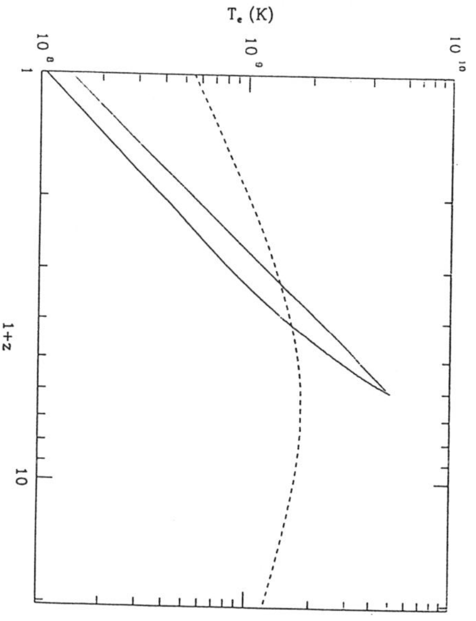

Fig. 1a

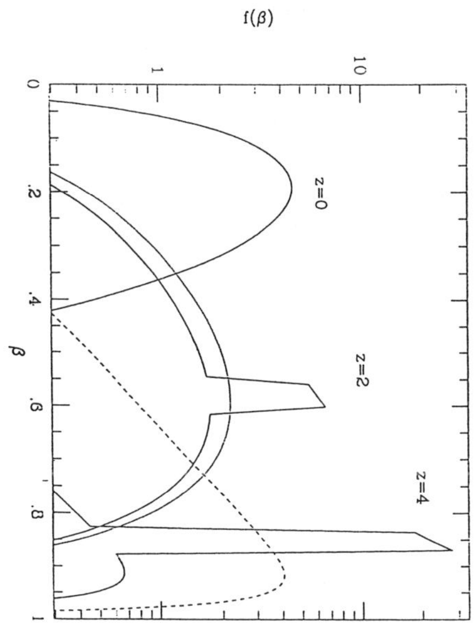

Fig. 1b

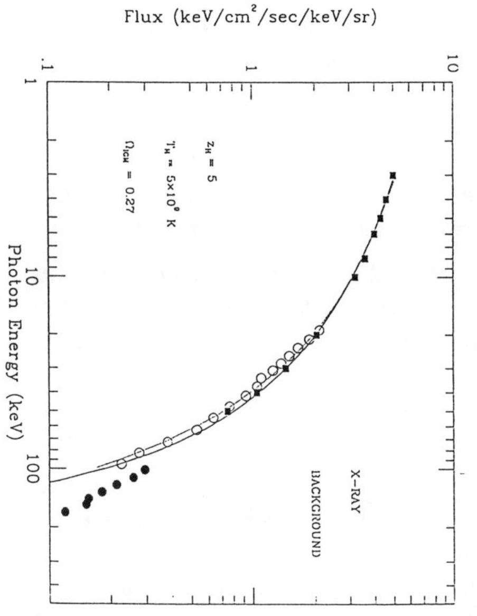

Fig. 1c

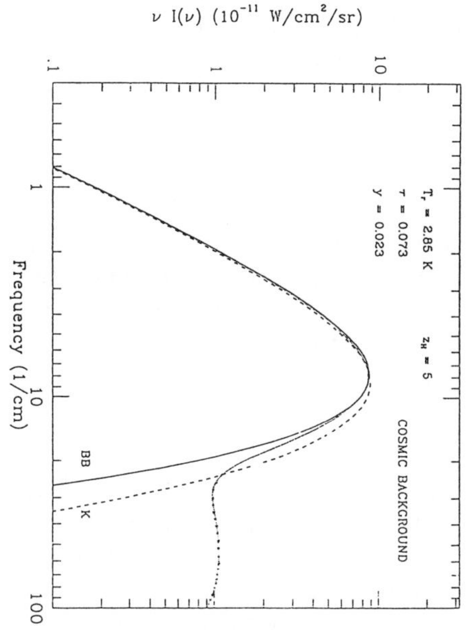

Fig. 1d

Fig. 1b: Snapshots of the evolution of the electron distribution function $f_e(\beta)$ at $z = 4, 2$ and 0. The 'spikes' at $z = 4$ and 2 are the remnants of the initial top-hat in $f_e(p) = \gamma^{-3} f_e(\beta)$ at $z_H = 5$. The dashed and dotted lines show pure Maxwellians at the same temperature as the distribution functions at $z = 4$ and 2. At $z = 0$ the evolved distribution function becomes almost Maxwellian.

Fig. 1c: The relativistic bremsstrahlung from an initial narrow top–hat distribution fitted to the XRB measurements (solid line). The XRB data from the HEAO-1 A2 experiment are marked by filled squares, those from the HEAO-1 A4 Low and Medium Energy Detectors are marked by open and filled circles, respectively. Boldt's fit to the data is the dotted line[2].

Fig. 1d: The predicted cosmic background spectrum (dotted line) given the relativistic Bremsstrahlung model for the XRB from Fig. 1c. The solid line is a pure blackbody (BB) spectrum at 2.85 K and the dashed line is the approximate Kompaneets (K) solution, which depends only on $y \approx 0.02$.

IIb. TWO–PHASE IGM IN A STEADY STATE

If the IGM is heated only once, its pressure will equilibrate within a sound travel time over scales of $\sim 20h^{-1}$Mpc $(T_e/40\text{keV})^{1/2}$. The expansion time is typically much shorter than the radiative cooling time. Thus, the existence of dense clumps must be supported either by an external pressure (two-phase IGM) or by a gravitational field. The first model[9, 23] involves necessarily a larger integrated electron pressure along the line-of-sight than the homogeneous IGM model; because the clumped phase that emits the XRB fills only part of the intergalactic volume and should be in pressure equilibrium with the rarefied hotter gas that surrounds it. Therefore, this model results in a larger value of y than that of a homogeneous IGM, and can also be ruled-out by COBE. The second possibility considers virial equilibrium of the clumped IGM in deep potential wells, and can be ruled-out from the CBR and XRB isotropies and the observed fraction of gravitationally lensed quasars[24].

Based on the above considerations, the IGM must have transient clumps that are continuously heated if it accounts for the XRB. A natural arena for such properties would be the formation of large-scale structure in the universe.

IIc. CLUMPED IGM EMISSION DURING THE FORMATION OF LARGE–SCALE STRUCTURE

In the context of large-scale structure formation, the clumpiness of the IGM and the length scales over which nonlinear clumping occurs, both increase as functions of cosmological time. Therefore, the IGM temperature increases rather than decreases in time[25]. Nonlinearities on scales $\sim 20h^{-1}$ Mpc typically correspond to kinetic energies of $\sim$ 40 keV per baryon; close to the characteristic temperature of the x-ray background. For simplicity, one can parametrize the redshift dependences of the electron temperature and clumpiness as power laws[26],

$\langle T_e \rangle = \langle T_{e0} \rangle/(1+z)^{\alpha}$ and $C^2 \equiv (\langle n_e^2 \rangle)^{1/2}/\langle n_e \rangle = (1+\sigma_m^2/[1+z]^{\delta})$; where, the averages here are over the baryonic mass density and typically[25] $\alpha, \delta \sim (1-3)$ while T_{e0}, σ_m^2 depend on the cosmological power spectrum (CDM, HDM, etc.). The numerical results from a 3-D code that simulates the IGM hydrodynamics together with dark matter dynamics[25], indicates that the highest temperature regions have generically the highest baryon density. Thus, the hot dense clumps that dominate the x-ray emissivity are consistent with the requirements outlined in Sec. 1*a, b*. Generically, the IGM is being continuously heated by breaking of large scale shocks into hot transient clumps. Figure 2 shows an example for the emitted x-ray spectrum[26] assuming, $\langle T_{e0} \rangle = 60$ keV, $\sigma_m^2 = 300$, $\Omega_b h^{3/2} = 0.02$ and $\alpha = \delta = 2$.

The Comptonization in this scenario is non-relativistic ($T_e \lesssim 100$ keV) and is purely characterized by $y \approx \langle n_{e0} \rangle \sigma_T H_0 \left(\langle T_{e0} \rangle / m_e [\alpha - 3/2] \right)$, (for $3/2 < \alpha$ and a fully ionized IGM). With $\langle T_{e0} \rangle = 60$ keV, $\alpha = 2$, and $\Omega_b h = 0.02$, one gets, $y \approx 2.8 \times 10^{-4}$; i.e., the model can be consistent with COBE.

There is another effect that should further decrease the above value of y. The large–scale shocks are expected to heat mostly the ions, since these carry most of the inertia. The transfer of energy from the ions to the electrons may take more than a Hubble time. The hydrodynamics of the two-temperature IGM plasma can be analysed by adding the energy conservation equation of the electrons to the three conservation equations of the ions[27] and solving for the electron temperature in addition to the ion temperature, density and velocity profiles. The flow velocities and densities of the electrons and ions are equal because of the the quasi-neutrality of the plasma. The timescale for electron-ion thermalization by Coulomb collisions in an unclumped IGM, τ_{ei}, is typically much larger than the Hubble time, $H_0 \tau_{ei} \approx 150 \times (0.02/\Omega_b h)(T_e/10\mathrm{keV})^{3/2}$; (note that turbulence or magnetic fields only couple the momentum of the fluids). This situation may yield a variety of interesting effects: it allows high electron temperatures only at very high–densities ($C \gtrsim 10^2, n_e \gtrsim 10^{-3}\Omega_b$ cm^{-3}), and therefore decreases the integrated CBR Comptonization per x–ray flux. The ions act as a "thermal battery" that is being discharged only at the high-density peaks. In addition, *the voids may be filled with hot ions and cold electrons;* having low radiative losses and, effectively, containing baryonic "dark matter".

Observationally, the x-ray fluctuations are constrained by HEAO-1 to be[28], $\Delta I/I \lesssim 2\%$ on $5^\circ \times 5^\circ$ beams for 3–10 keV. The central limit theorem then implies that the number of clumps within the beam, N_{cl}, exceeds 2500. For an effective cosmological emission volume ($z \lesssim 1$) covered by the beam $\sim V_b$, this implies an average distance between neighbouring clumps of $\ell \sim (N_{cl}/V_b)^{-1/3}$. When gas is swept into the clumps and structure becomes nonlinear, the typical temperature obtained is, $T_e \sim m_p(H_0\ell)^2 \sim m_p H_0^2 (\Delta I/I)^{4/3} V_b^{2/3}$; where m_p is the proton mass. For $\Delta I/I \sim 2\%$ and $V_b \sim 0.07$ Gpc^3 one has $T_e \sim 100h^2$ keV. If the hard x–ray background is dominated by emission from a clumped IGM, strong fluctuations ($> 2\%$) in the x-ray sky are expected to exist at high photon energies (40–100 keV), since these photons were mostly emitted from the rare high–temperature peaks. As a result of its Comptonization by the IGM electrons, the microwave background may also have detectable anisotropies of order, $\Delta T_{\rm RJ}/T_{\rm RJ} \approx 2y/\sqrt{N_{cl}} = 10^{-5} \times (y/2.5 \times 10^{-4})/\sqrt{(N_{cl}/2500)}$.

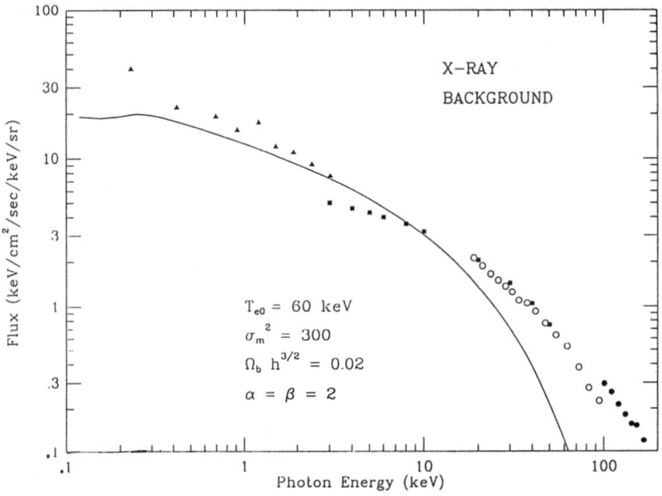

Fig. 2: An example for the x–ray emission spectrum of a clumped hot IGM (solid line) with $\langle T_{e0} \rangle = 60$ keV, $\Omega_b h^{3/2} = 0.02$, $\sigma_m^2 = 300$, and $\alpha = \delta = 2$.

III. CONCLUSIONS

The spectral results from COBE preclude a thermal Bremsstrahlung origin for the XRB from a homogeneous IGM or a clumped IGM that was heated only once. The simplest alternative would be discrete sources. However, no class of objects was observed yet at the appropriate x-ray band to have the needed spectrum. The needed class of objects should be highly populated, as required by the constraints on the XRB isotropy.

In fact, even before COBE, there were many theoretical difficulties with the model of a uniform IGM that was heated instantaneously at a high redshift (~ 5): the high Ω_b required was in conflict with the standard nucleosynthesis bounds; galaxy formation should have been completed before the heating epoch (since the galactic virial temperature is $\ll T_e$); Lyman-α clouds may not survive such a hot IGM; and the energy sources for the heating were not specified. The clumped IGM model; in which the continuous heating source is gravitational instabilities on large-scales, overcomes most of these difficulties. It stands, therefore, as an alternative to discrete sources for the origin of the XRB. The strongest constraint on this model arises from the isotropies of the XRB and the CBR. Its primary advantage is that it specifies the energy source for the x-ray emission, and can therefore be confronted self-consistently with observations. At present, it also merits more detailed theoretical investigation.

The near future promises new data from many observational projects[29] such as: GINGA, ROSAT, BBXRT and GRO. If realised as planned, these projects will probably follow COBE and decrease the number of relevant models per x-ray photon.

ACKNOWLEDGMENTS

This work benefited appreciably from papers and fruitful discussions I had with my collaborators, Ofer Lahav, Chris F. McKee and Jerry Ostriker. The research was supported in part by a W. M. Keck Foundation Fellowship.

REFERENCES

1. R. Giacconi, H. Gursky, F. Paolini & B. Rossi, *Phys. Rev. Lett.* **9**, 439 (1962).
2. E. Boldt, *Physics Reports* **146**(4), 215 (1987).
3. X. Wu, T. Hamilton, D. J. Helfand & Q. Wang, preprint, *Columbia Astrophysics Laboratory Contribution Number 400* (1990).
4. D. Leiter & E. Boldt, *Astroph. J.*, **260**, 1 (1982); E. Boldt & D. Leiter, *Astroph. J.*, **322**, L1 (1987); K. Morisawa & F. Takahara *Publ. Astr. Soc. Jap.* **41**, 873 (1989); A. A. Zdziarski *Mon. Not. R. astr. Soc.* **233**, 739 (1988).
5. D. A. Schwartz & W. H. Tucker, *Astroph. J.*, **332**, 157 (1988); D. A. Schwartz, Y. Qian & W. H. Tucker in *Proc. of the 23rd ESLAB Symp. on Two-Topics in X-Ray Astronomy, ESA SP-296* (Bologna, Italy, 1989).
6. A. C. Fabian, I. M. George, S. Miyoshi & M. Rees, *Mon. Not. R. astr. Soc.* **242**, 14p (1990); K. Morisawa, M. Matsuoka, F. Takahara & L. Piro in *Proc. of the 23rd ESLAB Symp. on Two-Topics in X-Ray Astronomy, ESA SP-296* (Bologna, Italy, 1989) and *Astr. Astroph.* **236**, 299 (1990).
7. See e.g., J. Bookbinder, L. L. Cowie, J. H. Krolik, J. P. Ostriker & M. Rees, *Astroph. J.* **237**, 647(1980); R. Daly, Physics preprint, Princeton University (1990); and references therein.
8. R. Cowsic & E. J. Kobetic, *Astroph. J.*, **177**, 585 (1972); G. B. Field, *Ann. Rev. Astr. Ap.*, **10**, 227 (1972); G. B. Field & S. C. Perrenod, *Astroph. J.*, **215**, 717 (1977); C.F. McKee, *Physica Scripta*, **21**, 738 (1980); X. Barcons, *Astroph. J.*, **313**, 547 (1987).
9. P. W. Guilbert & A. C. Fabian, *Mon. Not. R. astr. Soc.*, **220**, 439 (1986).
10. E. L. Wright, *Astroph. J.*, **232**, 348 (1979).
11. G. B. Taylor, & E. L. Wright, *Astroph. J.* **339**, 619 (1989).

12. O. Lahav, A. Loeb & C. F. McKee, *Astroph. J. Lett.* **349**, L9 (1990).

13. K. A. Pounds, K. Nandra, G. C. Stewart, I. M. George & A. C. Fabian *Nature* **344**, 132 (1990); L. Piro, M. Yamauchi & M. Matsuoka, *Astroph. J. Lett.* **360**, L35 (1990).

14. R. Sunyaev & Ya. B. Zel'dovich, *Ann. Rev. Astr. Ap*, **18**, 537 (1980).

15. T. T. Hamilton & D. J. Helfand, *Astroph. J.*, **318**, 93 (1987).

16. R. A. Shafer & A. C. Fabian, in *Early Evolution of the Universe and its Present Structure, IAU Symp. No. 104* eds. G. O. Abell and G. Chincarini (Dordrecht, 1983), pp. 333–343.

17. See, e.g., P. V. R. Murthy & A. W. Wolfendale, *Gamma-ray Astronomy* (Cambridge University Press, 1986); or W. R. Webber, J. A. Lockwood & G. A. Simpson, In *Proc. of the 17th International Cosmic Ray Conference, Paris* (1981).

18. Y. T. Gao, D. B. Cline & F. W. Stecker, *Astroph. J. Lett.*, **357**, L1, (1990).

19. See e.g., Y.T. Gao, F. W. Stecker, M. Gleiser & D. Cline, to be published in *Astroph. J. Lett.* (1990) and references therein.

20. J. Yang, M. S. Turner, G. Steigman, D. N. Schramm & K. A. Olive, *Astroph. J.* **281**, 493 (1984).

21. A. Loeb, C. F. McKee & O. Lahav, *Institute For Advanced Study preprint IASSNS-AST 90/22*, submitted to *Astroph. J.* (1990).

22. J. Mather *et al.*, *Astroph. J. Lett.* **354**, L37 (1990).

23. X. Barcons & A. C. Fabian, *Mon. Not. R. astr. Soc.* **230**, 189 (1988).

24. R. D. Rogers & G. B. Field, *Harvard CfA preprint No. 3096*, to appear in *Astroph. J.* (1991).

25. R. Y. Cen, A. Jameson, F. Liu, & J. P. Ostriker, *Astroph. J. Lett.* **362**, L41 (1990).

26. A. Loeb & J. P. Ostriker, *Institute For Advanced Study preprint IASSNS-AST 90/31*, submitted to *Nature* (1990).

27. Ya. B. Zel'dovich & Yu. P. Raiser, *Physics of Shock Waves and High Temperature Hydrodynamic Phenomena* (Academic Press, New-York, 1966), Vol. I, p. 421; Vol. II. pp. 515–522.

28. R. A. Shafer & A. C. Fabian, in *Early Evolution of the Universe and its Present Structure, IAU Symp. No. 104* eds. G. O. Abell and G. Chincarini (Dordrecht, 1983), pp. 333–343.

29. R. F. Mushotzki, "Signposts to the Future", in *Proc. of the 23rd ESLAB Symp. on Two-Topics in X-Ray Astronomy, ESA SP-296* (Bologna, Italy, 1989).

SPECTRAL-LUMINOSITY EVOLUTION OF ACTIVE GALACTIC NUCLEI AND THE COSMIC X-RAY BACKGROUND

Darryl Leiter, Science Division, FSTC, Charlottesville, VA. 22901
Elihu Boldt, Goddard Space flight Center, Greenbelt, MD. 20771

ABSTRACT

Black hole accretion disk dynamo processes are generally regarded as the central power source for AGN (Rees 1984). If the precursor active galaxies (PAG) for such AGN are formed at redshift $z \gtrsim 4$ and contain initial central seed black holes $\sim 10^6$ solar masses, then the Eddington limited X-ray radiation emitted during their lifetime will undergo the phenomenon of "spectral-luminosity evolution". When accretion disks are first formed at the onset of galaxy formation the accretion rate occurs at high values of luminosity/size compactness parameter $L/R > 10^{30}$ erg/cm-sec. Such high values of L/R generate dynamic constraints (Cavaliere & Morrison 1980; Guilbert, Fabian, & Rees 1983), which suppress nonthermal black hole accretion disk dynamo processes in favor of thermal processes. This causes the spectrum of X-radiation emitted by early AGN to be predominantly thermal. A superposition of such thermal, comptonized PAG sources can account for the residual cosmic X-ray background (CXB) (Boldt & Leiter 1986, 1987, 1990) and can act as a source of X-ray heating of the intergalactic medium (IGM) for $z \gtrsim 4$, (Collin-Souffrin 1990).

I. PRECURSOR ACTIVE GALAXIES (PAG) AND THE CXB

PAG initially undergo Eddington limited accretion. Since the hard X-radiation from thermal processes in the PAG accretion disk comes from a region on the order of 10 gravitational radii, L/R can be written in terms of L/L_{Edd} as

$$L/R = (L/L_{Edd})\, 10^{32} \text{ erg/cm-sec} \tag{1}$$

where

$$L_{Edd} = 4\pi G M m_p c/\sigma_T \tag{2}$$

is the Eddington luminosity, σ_T is the Thomson scattering cross section, m_p is the proton mass, M is the mass of the central black hole, G is the gravitational constant and c is the speed of light.

Since $L/R > 10^{30}$ erg/cm-sec the PAG emit a flat comptonized spectrum with exponential roll-off energy and spectral index on the order of $E \gtrsim 160$ KeV and $\alpha_x \gtrsim 0$ respectively (Zdziarski 1988). In this context the observed redshifted PAG exponential roll-off energy and roll-off spectral index are consistent with the observational constraints associated with the residual CXB (Leiter & Boldt 1982) given by

$$23 \text{ KeV} \lesssim E/(1+z) \lesssim 35 \text{ KeV}, \qquad 0 \lesssim \alpha_x \lesssim 0.2 \quad . \tag{3}$$

For Eddington limited accretion the initial PAG black hole mass M_i increases exponentially to M_{PAG}. This is given by

$$M_{PAG} = M_i \exp(A_{PAG}\, \Delta t_{PAG}) \tag{4}$$

$$A_{PAG} = 30\ (\ (1 - \epsilon)/\epsilon\ )\ . \tag{5}$$

In the above Δt_{PAG} is the Eddington limited lifetime of the PAG phase in units of $(2/3H_o) = 1.3 \cdot 10^{10}$ years, corresponding to H_o=50 km/(sec-Mpc), and $\epsilon \lesssim 0.3$ is the accretion disk mass-energy conversion efficiency (Thorne 1974).

If z_{MAX} is the maximum redshift associated with the onset of the PAG and z_{MIN} is the minimum redshift for this population, corresponding to the associated onset of daughter AGN such as Seyfert galaxies, then the relationship between light-travel lookback times $t_{MAX,MIN}$ in $(2/3H_o)$ units and redshifts $z_{MAX,\ MIN}$ is given by

$$t_{MAX\ ,\ MIN} = [\ 1 - (1+z_{MAX\ ,\ MIN})^{-3/2}\ ]\ . \tag{6}$$

In this context the total PAG lifetime is given by the relation

$$\Delta t_{PAG} = (\ t_{MAX} - t_{MIN}\ )\ . \tag{7}$$

For $z_{MIN} \leq z \leq z_{MAX}$, the surface density of PAG sources (required to be $\gtrsim$ 5000 sources/deg ; Hamilton, Helfand 1987) is given by

$$\sigma_{PAG} = (\ \sigma(t_{MAX}) - \sigma(t_{MIN})\ )\ \text{sources/deg}^2 \tag{8}$$

$$\sigma(t) = (1.75 \cdot 10^4 \cdot (N_o(PAG)/10^{-4}\ \text{Mpc}^{-3}))\ \ [\ t+3(1-t)^{2/3}-3(1-t)^{1/3}\ ] \tag{9}$$

where N_o(PAG) Mpc^{-3} is the co-moving density of PAG. The ensemble averaged proper bolometric PAG luminosity, emitted in the form of an X-ray spectrum during the Eddington limited PAG lifetime, is

$$L_{PAG}(t) = 13 \cdot 10^{45}\ (M_i/10^8\ M_o) \cdot \exp(\ A_{PAG}(t_{MAX}-t)\ )\quad \text{erg/sec.} \tag{10}$$

where M_i is in units of solar mass (M_o). The total PAG energy flux (I_{PAG}) in X-rays contributed to the RCXB is

$$I_{PAG} = (2/3H_o) \cdot (c/4\pi) \cdot N_o(PAG) \cdot \int_{t_{MIN}}^{t_{MAX}} dt(\ L_{PAG}(t)/(1+z(t)\ ) \tag{11}$$

where HEAO-1 observations require $I_{PAG} \lesssim 2.2\ 10^{-7}$ erg/cm^2-sec-sr.

II. SPECTRAL-LUMINOSITY EVOLUTION OF PAG INTO SEYFERT GALAXIES

At the end of the PAG lifetime the Eddington ratio falls to the value $L/L_{Edd} \lesssim 0.01$ (i.e. $L/R < 10^{30}$ erg/cm-sec) and nonthermal black hole accretion disk dynamo processes (Rees, Begelman, Blandford, & Phinney 1982) become important, generating a broad band of non-thermal radiation including X-rays and gamma rays. Under these conditions the PAG spectrally evolve into Seyfert galaxies. For these Seyfert galaxies the compactness parameter (Luminosity/Size) is connected to the Seyfert Eddington ratio $\lambda_{SEY} = L_{SEY}/L_{Edd}$ by the relationship

$$(L/R) = ((\lambda_{SEY})/(R/r_G)) \cdot 10^{33} \text{ erg/cm-sec} ; \quad r_G = GM/c^2 \text{ cm} . \tag{14}$$

Spectral-luminosity evolution of the thermal PAG state into the nonthermal Seyfert state begins when the PAG compactness parameter falls below $L/R < 10^{30}$ erg/cm-sec. Since the hot optically thin inner region of a PAG accretion disk has a maximum intensity of X-ray emission at radius $R \cong 10\ r_G$, equation (14) implies that spectral evolution of PAG into Seyfert galaxies will occur when the value of the Seyfert supply-limited Eddington ratio is $\lambda_{SEY} \lesssim 0.01$. This criterion dynamically defines the the beginning of the Seyfert epoch. If we assume that PAG-Seyfert spectral evolution occurs in a supply limited regime, the supply limited luminosity of the Seyfert is given by

$$L_{SEY} = 13 \cdot 10^{45}\ (M_{PAG}/10^8\ M_\odot)\ \lambda_{SEY} \quad \text{erg/sec.} \tag{15}$$

Padovani, Burg & Edelson (1990) suggest that the low value observed for the average Seyfert central black hole mass $\frown 2\ 10^7$ Mo favors scenarios in which AGN activity occurs in recurrent bursts over time intervals $\frown 10^8$ years. We adopt this point of view and assume that supply-limited Seyfert AGN activity cycles "on-off" in a quasi-periodic manner related to internal galactic accretion instabilities (Shlosman, Begelman & Frank 1990). If the number of "on-off" cycles of the Seyfert galaxy from light travel look-back time t_{MIN} to the present epoch is given by "n", and the "on time" and "off-time" per Seyfert cycle are given respectively by Δt_{SEY-ON}, $\Delta t_{SEY-OFF}$ then

$$t_{MIN} = n \cdot (\Delta t_{SEY-ON} + \Delta t_{SEY-OFF}) \ . \tag{16}$$

Because Seyfert activity cycles on and off stochastically, the observed co-moving density of Seyfert galaxies in the "on" state is equal to the product of the co-moving density of PAG times the probability of a Seyfert being "on" during t_{MIN}, viz:

$$N_o(SEY) = N_o(PAG) \cdot ((n\ \Delta t_{SEY-ON}) \ /\ t_{MIN}). \tag{17}$$

Note that (17) implies that $N_o(PAG) > N_o(SEY)$. In this context the mass of the Seyfert central black hole in the present epoch is given by

$$M_{SEY} = M_{PAG}\ (1 + A_{SEY} \cdot n \cdot \Delta t_{SEY-ON}) \tag{18}$$

where

$$A_{SEY} = 30\ \lambda_{SEY}\ ((1 - \varepsilon) \ /\ \varepsilon) \ . \tag{19}$$

We can use the Newton-Raphson variation method to solve equations (1)-(19). Assuming a $\gtrsim$ 10% foreground contribution to the CXB from Seyferts and the inputs:

$N_o(SEY) \cong (2 - 4)\ 10^{-4}\ \text{Mpc}^{-3}$, $M_{SEY} \cong 2 \cdot 10^7\ M_o$,

$\lambda_{SEY} \cong (1 - 2)\ 10^{-2}$, $\Delta t_{SEY-ON,OFF} \gtrsim 10^8$ years, we find that for $6 \geq z \geq 4$ the PAG have a comoving density of $N_o(PAG) \cong 10^{-3}$ Mpc^{-3}. These PAG start at $z \cong 6$ with initial black hole mass $M_1 \cong 10^6$ M_o and during their Eddington limited lifetime $\Delta t_{PAG} \cong 5 \cdot 10^8$ years emit an average luminosity in X-radiation $\langle L_{PAG} \rangle \cong 10^{45}$ erg/sec. It can be shown (Collin-Souffrin 1990, Boldt & Leiter 1990) that in addition to acting as the source of the residual CXB, the PAG can act as a source of heating and ionization for the IGM for $z \geq 4$.

REFERENCES

Boldt, E., 1987, "Cosmic X-ray Background", Physics Reports, 146,215

Boldt, E., Leiter, D, 1990, "in preparation"; 1987, Ap. J., 322, 689; 1986, Struc. & Evol. of AGN (Reidel 1986), 383; 1984, Ap. J., 276, 427.

Cavaliere, A., Morrison, P., 1980, Ap. J., 238, L63.

Collin-Souffrin, S., 1990, Ast. & Ap. in press 1990.

Guilbert, P.W., Fabian, A.C., Rees, M.J., 1983, M.N.R.A.S., 205, 593.

Hamilton, T., Helfand, D., 1987, Ap. J. 318, 93.

Leiter, D., Boldt, E., 1982, Ap. J., 260, 1; 1989, "Proc. GRO Science Workshop, 4-14".

Padovani, P., Burg, R., Edelson, R.A. 1990, Ap. J., 353, 438.

Rees, M.J., 1984, Ann. Rev. Astr. & Ap., 22, 471; Rees, M.J., Begelman, M.C., Blandford, R.D., & Phinney, E.S., 1982, NATURE, 295, 17.

Shlosman, I., Begelman, M.C., Frank, J., 1990, NATURE, 345, 679.

Thorne, K, 1974, Ap. J., 191,507.

Zdziarski, A.A., 1988, M.N.R.A.S., 233, 739.

Question: Andrew S. Wilson

I believe that the X-ray spectra of AGN may well match that of the X-ray background even at zero redshift. Recent work (Setti, Woltjer, 1989, Astr. Ap. 224, L21) has identified a large population of active nuclei hidden behind tori of great optical depth on the order of 10^{23} to 10^{24} H-atoms/cm^2. The effect on the X-ray spectrum is a photoelectric cutoff at relatively high energies ~ 3 to 10 KeV. The effect of inclusion of such sources is to flatten the integrated spectrum of AGN which then matches the X-ray background spectrum very well.

Answer: Darryl Leiter

If the majority of AGN sources of the X-ray background are located at low redshift these sources can not play a significant role in heating and ionizing the IGM at high redshift. A key advantage to the PAG spectral-luminosity evolution model is the fact that most of the PAG sources are located at high redshift $z \geq 4$. Hence in addition to being the source of the residual X-ray background the PAG can act as a source of heating and ionization of the IGM for $z \geq 4$ (S. Collin Souffrin, 1990, Astr. Ap. in press).

EVIDENCE FOR EVOLUTION OF THE LUMINOSITY FUNCTION OF CLUSTERS OF GALAXIES

A.C. Edge[1,2], G.C. Stewart[3], A.C. Fabian[1] and K.A. Arnaud[4]

[1] Institute of Astronomy, Madingley Road, Cambridge CB3 0HA
[2] Max-Planck Institut für Extraterrestriche Physik, Garching, FRG
[3] X-ray Astronomy Group, University of Leicester, Leicester LE1 7RH
[4] Goddard Space Flight Center, Greenbelt, Maryland, USA

ABSTRACT

An all-sky, X-ray flux-limited sample of clusters of galaxies shows evidence for a significant deficit in the number density of high luminosity clusters in the redshift range z ~0.1–0.2, compared with that of nearby clusters. This indicates that the X-ray luminous clusters are undergoing strong evolution. The strength of the effect is consistent with hierarchical merging models. The implications of such strong evolution for clusters are discussed.

INTRODUCTION

Clusters of galaxies are the largest gravitationally-bound structures known and therefore provide important opportunities to investigate the nature and evolution of structure on large scales. Previous work on clusters has been limited by uncertainties in the optical identification and selection of clusters (Lucey 1983; Sutherland 1988). The most reliable method of selection of clusters uses their X-ray emission which is unaffected by the projection effects that hamper optical cluster searches. All-sky X-ray surveys have identified samples of clusters (*e.g.* Piccinotti *et al.* 1982; Kowalski *et al.* 1984). These samples suffer from confusion problems and are dominated by clusters at redshifts $z < 0.1$. At larger redshifts, previous *Einstein* data have shown that cluster properties are similar to those seen locally and that the luminosity function is not radically different at z =0.2–0.4 (Henry *et al.* 1982). We have recently produced a sample consisting of the 53 X-ray brightest clusters (2–10 keV) in the sky (Lahav *et al.* 1989; Edge *et al.* 1990). This shows clear evidence for evolution of the X-ray luminosity function of clusters (Edge *et al.* 1990) and is confirmed by the independent *Einstein* Medium Sensitivity Survey (EMSS) sample (Gioia *et al.* 1990). The sense of the evolution is such that the most luminous clusters are more numerous at the present time. Below is an overview of the main result of the all-sky sample and the implications of rapid X-ray evolution for clusters.

RESULTS

We can make a simple estimate of the completeness of our sample by comparing the observed source counts with those expected for a Euclidean distribution. Fig. 1 shows the source counts for high luminosity ($L_x > 8 \times 10^{44}$ erg s^{-1}) and low luminosity clusters for both the all-sky sample. A large difference between the two luminosity sub-samples is apparent. The deficit at low fluxes, seen in the whole high galactic latitude sample, is due to a lack of high luminosity clusters, which are all at redshifts of greater than 0.1. Taking the low luminosity clusters as a 'control' sample, we infer that the selection of clusters is around 90 per cent complete for all clusters and there is a highly significant deficit of high luminosity clusters, as seen in the EMSS sample. The difference between the two sub-samples is clear at a level well above the flux limit used for the analysis, *i.e.* at the Piccinotti *et al.* (1982) flux limit of 3×10^{-11} erg cm^{-2} s^{-1} and at 100 per cent completeness.

Having established the apparent incompleteness in the sample is due to a real deficiency of high luminosity clusters at higher redshifts, the luminosity function was calculated for the high galactic-latitude sample in two ways; taking a flux-limited sample and taking a volume limited sample at a redshift of 0.1 (corresponding to a luminosity of $>8\times10^{44}$ erg s^{-1} at a flux limit of 1.7×10^{-11} erg cm^{-2} s^{-1}). Determining the luminosity function from the $|b| > 20^\circ$ sample, we derive the results shown in Fig. 2. The derived luminosity function is consistent with previous work (Piccinotti *et al.* 1982; Kowalski *et al.* 1984). The difference in space density between the flux- and volume-limited samples for the high luminosity clusters is a factor of three. This shows the strength of the evolutionary effect.

DISCUSSION

Having established that clusters are evolving rapidly, we are faced with the question of why was it not detected it optically? There are several reasons for this such as the difficulty in creating complete samples, but the primary one is that the X-ray data provides a much higher contrast for the richest clusters. Where an optical estimate for the cluster mass is linearly proportional to the number of galaxies, the X-ray estimate is proportional to the mass to the power of three, *i.e.* a factor of two in mass gives eight in X-ray luminosity. Therefore changes in mass at large redshifts, which would be difficult to discern over the fluctuations in the background galaxy count in the optical, are much more easily seen in X-ray data.

The existence of strong X-ray evolution has important implications when considering the well-established correlations between X-ray and optical cluster properties (Mushotzky 1984, Edge & Stewart 1991). For instance the very strong correlation between X-ray luminosity and the fraction of spiral galaxies in the cluster indicates that the galaxy population follows the present state of the ICM. Therefore as a cluster increases in X-ray luminosity through a merger, the number of spiral galaxies in the cluster decreases. So the data points to a scenario of changing galaxy type. Also the strong relation between the absolute magnitude of a brightest cluster member (BCM) and the X-ray properties (Edge 1990) implies that as clusters evolve, their BCMs evolve in tandem. One simple explanation is that, as clusters evolve through mergers with other clusters, the BCMs of each cluster merge soon after their parent clusters. This would result in the BCM luminosity being related to the cluster mass (as is seen) and a population of recently merged clusters where the BCMs were not yet merged (binary clusters such as Coma) or merging (dumbbell galaxies).

CONCLUSIONS

Higher redshift (z >0.1), high luminosity ($L_x > 8 \times 10^{44}\,\mathrm{erg\,s^{-1}}$) clusters are under-represented in X-ray samples, indicating strong evolution at recent epochs. The evolution can be accounted for by hierarchical merging, which predicts a gradual flattening of the luminosity function with time. The number of known X-ray clusters will increase dramatically after analysis of the *ROSAT* survey and *ROSAT* pointed observations (Trümper 1984), which will allow any evolution to be traced out to redshifts as large as 1.0.

ACKNOWLEDGEMENTS

ACE would like to thank the conference organisers and the MPE for their hospitality in München. ACE and ACF thank the Royal Society for financial support. GCS acknowledges receipt of an SERC advanced research fellowship.

REFERENCES

Edge, A.C. 1990. *Mon. Not. R. astr. Soc.*, submitted.

Edge, A.C., Stewart, G.C., Fabian, A.C. & Arnaud, K.A., 1990. *Mon. Not. R. astr. Soc.*, **245**, 559.

Edge, A.C. & Stewart, G.C., 1991. *Mon. Not. R. astr. Soc.*, submitted.

Gioia, I.M., Henry, J.P., Maccacaro, T., Morris, S.L., Stocke, J.T., & Wolter, A., 1990. *Astrophys. J. (Letters)*, **356**, L35.

Henry, J.P., Soltan, A., Briel, U. & Gunn, J.E., 1982. *Astrophys. J.*, **262**, 1.

Kowalski, M.P., Ulmer, M.P., Cruddace, R.G. & Wood, K.S., 1984. *Astrophys. J. Suppl.*, **56**, 403.

Lahav, O., Edge, A.C., Fabian, A.C. & Putney, A., 1989. *Mon. Not. R. astr. Soc.*, **238**, 881.

Lucey, J.R., 1983. *Mon. Not. R. astr. Soc.*, **204**, 33.
Mushotzky, R.F., 1984. *Physica Scripta.* Vol.**T7**, 157.
Piccinotti, G., Mushotzky, R.F., Boldt, E.A., Holt, S.S., Marshall, F.E., Serlemitsos, P.J. & Shafer, R.A., 1982. *Astrophys. J.*, **253**, 485.
Sutherland, W., 1988. *Mon. Not. R. astr. Soc.*, **234**, 159.
Trümper, J., 1984. *Physica Scripta.* Vol.**T7**, 209.

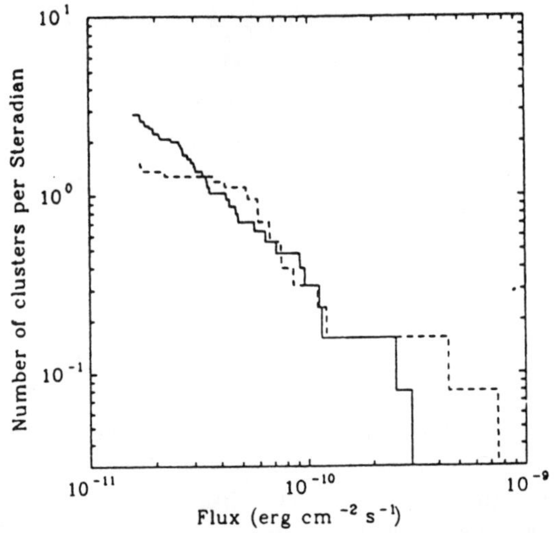

Figure 1: The $\log N - \log S$ distribution of low and high luminosity sub-samples (solid and dashed lines respectively) for the all-sky sample

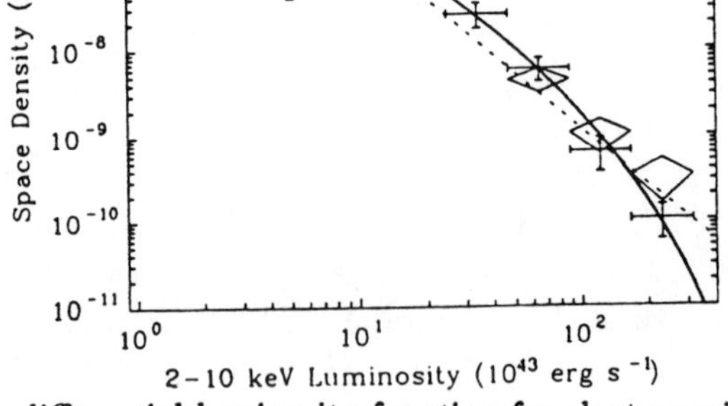

Figure 2: The differential luminosity function for clusters with $|b| > 20°$. The diamonds are the results for the volume-limited sample within $z = 0.1$.

THE EVOLVING INTERGALACTIC MEDIUM: THE UNCOLLAPSED BARYON FRACTION IN A COLD DARK MATTER UNIVERSE

Paul R. Shapiro and Mark L. Giroux
University of Texas at Austin
Austin, TX 78712
and
Arif Babul
Institute of Astronomy, Univ. of Cambridge
Madingley Road, Cambridge CB3 0HA

ABSTRACT

A self-consistent treatment of the thermal and ionization history of the intergalactic medium (IGM) must take into account the growth of structure in the universe, since the mean density of the IGM corresponds primarily to the time-varying *uncollapsed* fraction of the baryon-electron component of the matter. The collapsed fraction, in turn, can have a "feedback" effect on this uncollapsed fraction by releasing ionizing radiation and thermal energy and by contributing to the opacity of the universe. We have calculated the time varying density of the IGM by coupling our detailed, numerical calculations of the thermal and ionization balance and radiative transfer in a uniform IGM of H and He, including the mean effect of an evolving distribution of gas clumps embedded in a smoothly distributed ambient gas, to the linearized equations for the growth of density fluctuations in both the gaseous and dark components in a Cold Dark Matter (CDM) universe, identifying the IGM density with the uncollapsed baryon fraction. We find that, even if the IGM is never reheated, a significant fraction of the baryons remain uncollapsed at redshifts $z \sim 4$. If, instead, the collapsed fraction releases enough ionizing radiation or thermal energy to reionize the IGM by $z > 4$ as required by the Gunn-Peterson (GP) constraint, the uncollapsed fraction at $z \sim 4$ is even higher. The known quasar distribution is insufficient to supply the ionizing radiation necessary to satisfy the GP constraint in this case, and, if stars are responsible, instead, a substantial metallicity must have been produced by $z \sim 4$. The alternative of energy release by supernovae leads to a similar conclusion.

CALCULATION

The comoving mean density of the IGM in units of the critical density is $\Omega_{IGM}(z) = \Omega_b[1 - f_c(z)]$, where Ω_b is the total baryon mass fraction in the universe and $f_c(z)$ is the time-varying, collapsed baryon fraction. We have calculated $\Omega_{IGM}(z)$ by coupling our detailed IGM code[1-3] to the linearized equations for the growth of density fluctuations in both the gaseous and dark matter components in a CDM universe[4,5]. We consider a standard CDM model with $\Omega_{tot} = 1$, $\Omega_b = 0.1$, and $h = 0.5$, with a power spectrum normalized so that the rms density fluctuation evaluated using a "top-hat" filter is $\sigma_0 =$

$1/b$ at $R = 8h^{-1}$ Mpc at present, where b is the bias parameter, and $b = 2.6$ is the so-called "standard bias".[6] A density fluctuation can be said to have collapsed out of the background when its amplitude $\frac{\delta\rho}{\rho}$ predicted by the linear growth rate reaches 1.69, the well known factor which describes the completion of top hat collapse. Accordingly, we take $f_c(z)$ to be the probability that $\frac{\delta\rho}{\rho}$ is greater than or equal to 1.69 at a random point in the Gaussian-filtered baryon density field, filtered on mass scales greater than 10 $M_\odot$.

The growth of baryon density fluctuations is damped relative to that of the CDM fluctuations by the effect of gas pressure, which depends upon the temperature of the background IGM. To calculate this, we have solved the rate equations for ionization and recombination, together with the equations of energy conservation, including the effects of cosmological expansion, radiative cooling, Compton scattering of the CMB, and the diffuse flux emitted by the gas, and radiative transfer. We have included the contribution to the continuum opacity of the universe due to H and He in the observed quasar absorption line clouds (QALC's). We have considered a variety of sources of photoionization, including quasars and primeval galaxies, as well as the possibility that hydrodynamical processes deposit thermal energy in the IGM.

RESULTS AND CONCLUSIONS

i) No Reheating Case. Figs. 1 and 2 show the gas temperature T and collapsed baryon fraction f_c versus redshift when no reheating occurs. Even in this limiting case, the *uncollapsed* baryon fraction at $z \sim 4$ is 0.45 (0.2). [Our notation "# (#)" means "for case $b = 2.6$ ($b = 1$)."]

ii) Sudden Reheating Case. Figs. 3 and 4 illustrate the effect of sudden reheating of the IGM at epoch $z = z_{reheat}$ to some constant temperature T_{reheat}. For $T_{reheat} = 10^4$ K (e.g. a photoionized IGM), the uncollapsed baryon fraction at $z \sim 4$ is 0.8 (0.4) for $z_{reheat} = 15$ and 0.65 (0.25) for $z_{reheat} = 5$. For $T_{reheat} = 10^6$ K (e.g. a collisionally ionized IGM), the uncollapsed baryon fraction at $z \sim 4$ is 0.98 (0.75) for $z_{reheat} = 15$ and 0.65 (0.25) for $z_{reheat} = 5$. The estimates of the uncollapsed baryon fraction for both this and the self consistent reheating case which follows are conservative in that we do not allow collapsed baryons to escape back into the IGM upon reheating.

iii) Self-Consistent Reheating Case. Figs. 5 and 6 show results when the IGM is self-consistently reheated by the time-varying injection of either ionizing photons or bulk thermal energy by the collapsed baryon fraction. We assume a space-averaged emissivity per comoving volume of either ionizing radiation, j_{ph} ($photons\ cm^{-3}\ s^{-1}$), or thermal energy, j_{th} ($ryd\ cm^{-3}\ s^{-1}$), given by j_{ph} (or j_{th}) $= (3/2)H_o n_H^o f_c(z)\ \tilde{\zeta}_{ph}$ (or $\tilde{\zeta}_{th}$), where $\tilde{\zeta}_{ph}$ and $\tilde{\zeta}_{th}$ are the dimensionless constant emission rates in ionizing photons and thermal energy (in Rydbergs), respectively, per collapsed baryon per Hubble time, and $n_H^o = 8.04 \times 10^{-6}\Omega_b h^2$. The constants ζ_{ph} and ζ_{th} are the marginal values required to ensure that the H Ly α GP optical depth $\tau_{GP} \leq 0.1$ at $z = 4.1$[7]

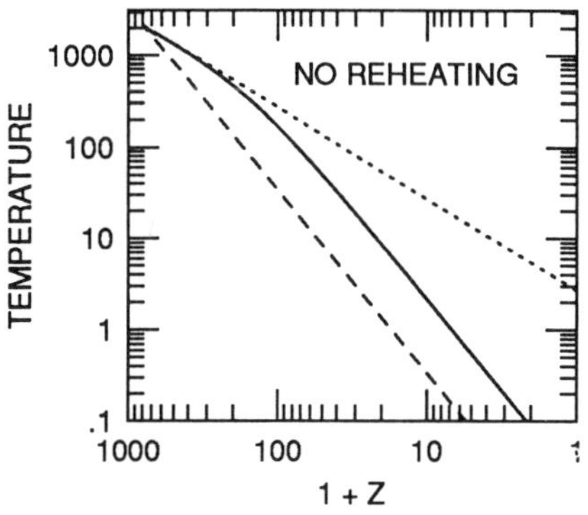

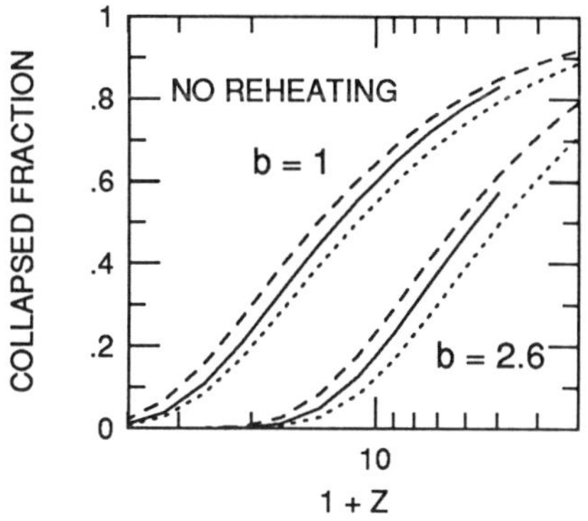

Fig. 1 (left) Evolution of gas temperature T in the post recombination IGM. Dashed curve is pure adiabatic evolution ($T \propto (1+z)^2$) from $T = T_{rad}$ at $z = 800$. Dotted curve is $T = T_{rad}$ ($T \propto (1+z)$). Solid curve is actual T.
Fig. 2 (right) Collapsed baryon fraction $f_c(z)$ versus redshift. Solid, dotted, and dashed curves show $f_c(z)$ for an IGM whose temperature evolution is given by the corresponding curves in Fig. 1. Curves are labelled with value assumed for bias factor b.

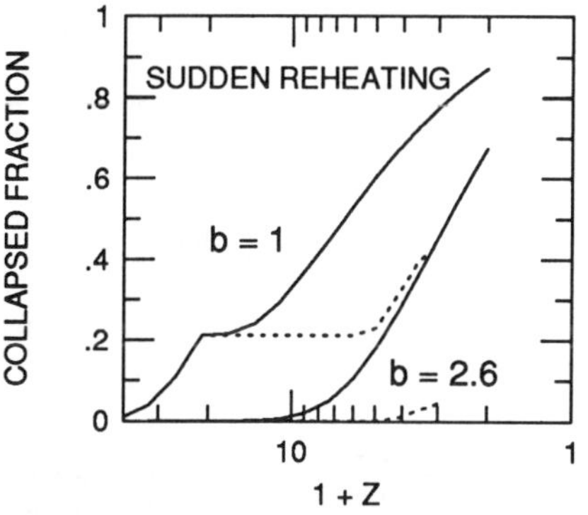

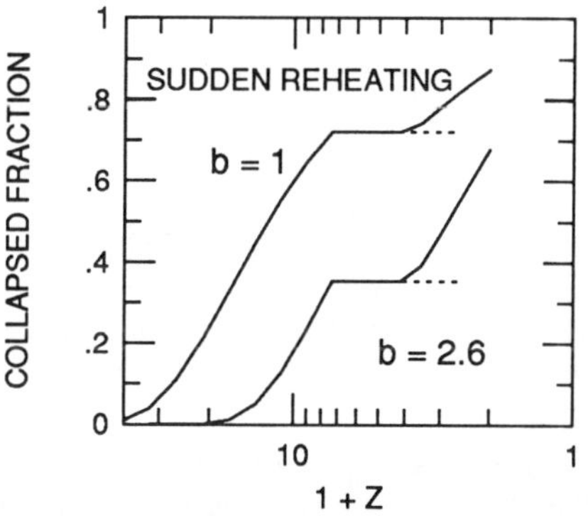

Fig. 3 (left) Collapsed baryon fraction $f_c(z)$ versus redshift for IGM suddenly reheated at $z = z_{reheat}$ to $T = T_{reheat}$ for all $z \leq z_{reheat}$, for $T_{reheat} = 10^4\ K$ (solid curves) and $10^6\ K$ (dotted curves). All curves assume $z_{reheat} = 15$ and T follows Fig. 1 (solid) for $z > z_{reheat}$. Curves labelled with bias parameter.
Fig. 4 (right) Same as Fig. 3, except $z_{reheat} = 5$.

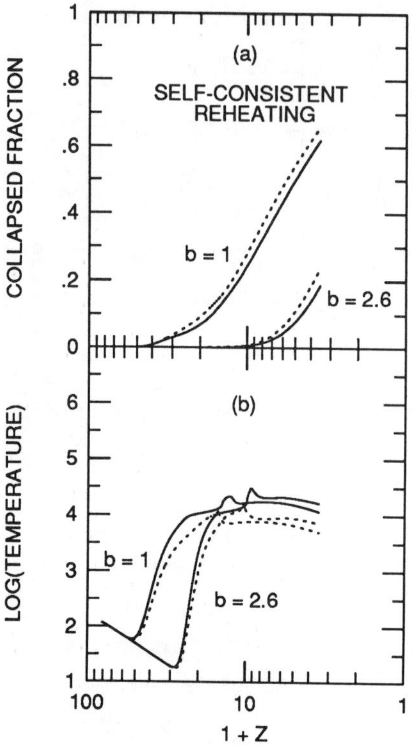

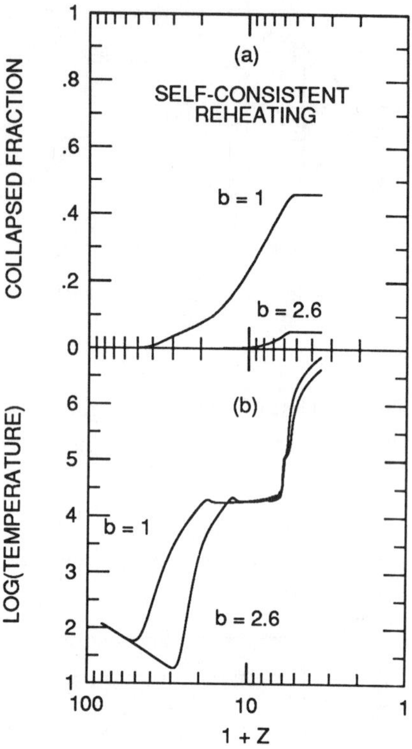

Fig. 5 (a) (top left) Collapsed baryon fraction versus redshift for self-consistent reheating case. The two sets of curves, for bias parameters 1 and 2.6, show the evolution of the collapsed baryon fraction in an IGM with photoionization sufficient to ensure that the GP Lyman α optical depth at $z = 4.1$ is less than 0.1 (the stronger constraint in this case). The solid curve assumes an AGN-like source spectrum with energy flux $\propto \nu^{-1.5}$ for $h\nu > 13.6$ eV, and the dotted curve assumes a stellar-type primeval galaxy source spectrum taken from a Bruzual galaxy synthesis model (assuming a Salpeter IMF). **(b)** (bottom left) IGM Temperature vs. redshift for the cases in Fig. 5 (a).

Fig. 6 (a) (top right) Collapsed baryon fraction versus redshift for self-consistent reheating case. The two curves, for bias parameters 1 and 2.6, show the evolution of the collapsed baryon fraction in an IGM with bulk heating sufficient to ensure that the GP Lyman α optical depth at $z = 4.7$ is less than 1 (the stronger constraint in this case). **(b)** (bottom right) IGM Temperature vs. redshift for the cases in Fig. 6 (a).

and $\tau_{GP} \leq 1.0$ at $z = 4.73$[8]. For the Ly α cloud opacity at $z > 5$, we used the empirical column density distribution and its evolution law observed at $z < 4$ to determine the cloud distribution out to $z = 5$, and for $z > 5$, extrapolated this by assuming constant cross sections per cloud and a total number of clouds per comoving volume $\propto f_c(z)$. For the photoionized cases, we considered two spectrum types for j_{ph}, an AGN-type with energy flux $\propto \nu^{-1.5}$ for $h\nu > 13.6\ eV$ and a stellar-type primeval galaxy spectrum taken from a Bruzual galaxy synthesis model (assuming a Salpeter IMF).

For the AGN cases, we find $\Omega_{IGM}/\Omega_b = 0.93(0.5)$ and $f_c\ \tilde{\zeta}_{ph} = 1780(590)$ at $z = 4.1$, which is well in excess of the observed quasar contribution at $z \gtrsim 3$, $(f_c\ \tilde{\zeta})_{obs} \lesssim 40(\Omega_b h^2/0.025)^{-1}$.[9,10] For the stellar cases, $\Omega_{IGM}/\Omega_b = 0.9(0.46)$ and $f_c\ \tilde{\zeta}_{ph} = 2190(670)$ at $z = 4.1$. Since the massive stars which dominate the ionizing photon emission also produce metals[2,3], this implies a metallicity mass fraction per collapsed baryon at $z = 4$ of 6×10^{-3} (6.9×10^{-4}), or $\approx 60\%$ (7%) of solar metallicity. For bulk heating, we find $\Omega_{IGM}/\Omega_b = 0.95(0.54)$ and $f_c\ \tilde{\zeta}_{th} = 1260(250)$ at $z = 4.1$. If this is due to supernovae from massive stars, then metal production also results[11]. Our bulk heating results then imply a metallicity per collapsed baryon at z = 4 of $1.5 \times 10^{-2}(5.5 \times 10^{-4})$, or $\approx 150\%$ (5.5%) of solar.

This work benefitted from the support of Robert A. Welch Foundation Grant F-1115, Texas Advanced Research Program Grant 4132, and NASA Grant NGT-50316. Numerical computations were performed on the UT Center for High Performance Computing Cray X-MP computer. A. B. is supported by a NATO Science Fellowship.

REFERENCES

1. Shapiro, P. R. and Giroux, M. L., 1989, in *The Epoch of Galaxy Formation*, eds. C. S. Frenk, R. S. Ellis, T. Shanks, A. F. Heavens, and J. A. Peacock (Boston: Kluwer Academic Publishers), pp. 153-161.
2. Giroux, M. L. and Shapiro, P. R., 1990a, in *Physical Processes in Fragmentation and Star Formation*, eds. R. Capuzzo-Dolcetta, C. Chiosi, and A. Di Fazio (Netherlands: Kluwer Academic Publishers), pp. 71-79.
3. Giroux, M. L. and Shapiro, P. R., 1990b, *Ap. J.*, to be submitted.
4. Babul, A., 1989, Ph. D. Dissertation. Princeton University.
5. Babul, A., 1990, *Ap. J.*, **349**, 429.
6. Davis, M., Efstathiou, G., Frenk, C., White, S., 1985, *Ap. J.*, **292**, 371.
7. Webb, J. K., Parnell, H. C., Carswell, R. F., McMahon, R. G., Irwin, M., Hazard, C., Ferlet, R., and Vidal-Madjar, A., 1988, *ESO Messenger*, **51**, 15.
8. Schneider, D. P., Schmidt, M., and Gunn, J. E., 1989, *Astron. J.*, **98**, 1951.
9. Koo, D. C. and Kron, R. G, 1988, *Ap. J.*, **325**, 92.
10. Turner, E. L., 1990, *Princeton Observatory Preprint No. 349.*
11. Miralda-Escude, G. and Ostriker, J. P., 1990, *Ap. J.*, **350**, 1.

DIAGNOSING THE REIONIZATION OF THE UNIVERSE: THE ABSORPTION SPECTRUM OF THE INTERGALACTIC MEDIUM AND LYMAN α CLOUDS

Mark L. Giroux and Paul R. Shapiro
University of Texas at Austin
Austin, TX 78712

ABSTRACT

We have calculated in detail the thermal and ionization evolution of a uniform intergalactic medium (IGM) composed of H and He undergoing "reionization," including the mean effect of gas clumps embedded in a smoothly distributed ambient gas. We have solved the rate equations for ionization and recombination, together with the equations of energy conservation, including the effects of cosmological expansion, radiative and Compton cooling, and the diffuse flux emitted by the gas, and radiative transfer. We have included the contribution to the continuum opacity of the universe due to the observed quasar absorption line clouds (QALC's). We have considered a variety of sources of photoionization, including quasars and primeval galaxies, as well as the possibility that hydrodynamical processes deposit thermal energy in the IGM.[1–3] We limit our discussion in what follows, to two questions:

(1) Can metal line ratios for Lyman limit quasar absorption line systems diagnose the metagalactic ionizing radiation background at $z \sim 3$?

The interesting suggestion has been made[4] that metal absorption line strengths for Lyman limit (LL) QALC's ($N_{HI} \gtrsim 10^{17}\ cm^{-2}$) at $z \sim 3$ can be compared with photoionized cloud predictions in order to diagnose the nature and origin of the metagalactic ultraviolet and/or soft x-ray background. Previous results have strongly favored an AGN-type, nonthermal source spectrum for this background over that emitted by a normal star-forming galaxy. The input spectra for these theoretical comparison models, however, were those of the sources in their own rest frame, unaffected by the intervening IGM and unevolved by cosmological expansion and redshift effects. We have reexamined this question by using our detailed thermal/ionization and radiative transfer calculations of the IGM to predict the proper cosmologically evolving metagalactic ionizing radiation field for each of several types of underlying source spectra. We find that a wider range of source types is acceptable than previously thought.

(2) Can the He II Gunn-Peterson effect be used to diagnose the reionization source and/or the nature of the Ly α forest clouds?

Our detailed IGM calculations predict the strength of the Gunn-Peterson (GP) He II Lyman α absorption troughs from the smoothly-distributed IGM, detectable by the HST, for a variety of assumptions about the nature of the Ly α absorbers and of the sources of ionization of the IGM and Ly α clouds, including a hydrodynamical source (e. g. shock-heating) for the former.

CALCULATION

For (1) above, we take $\Omega_{tot} = 1$ and an IGM with $\Omega_{IGM} h^2 = 0.01$ and $h = 0.75$, in which a constant space-averaged emissivity per comoving volume of ionizing radiation turns on at $z_{on} = 6$. We consider three possible emissivity spectra: (A) AGN-type sources with energy flux $\propto \nu^{-1.5}$ for $h\nu \geq 13.6$ eV; (B) stellar-type primeval galaxy sources, using a Bruzual galaxy synthesis model (with Salpeter IMF); and (C) combination of stellar and AGN sources with the ratio of (B) to (A) of ~ 4. The emissivity levels were chosen to marginally satisfy the requirement that the H GP Ly α optical depth $\tau_{GP} < 0.1$ at $z = 4.1$ and < 1 at $z = 4.7$

In general, we calculate the mean effect of the continuous H and He opacity of the Ly α clouds assuming the cloud number N_c follows $\partial^2 N_c/\partial N_{HI}\partial z = A(1+z)^\gamma N_{HI}^{-\beta}$. We consider two cases[5], intended to bracket the uncertainty in number of high N_{HI} clouds, Case L (Case H), for "Low N_{HI} cutoff" ("High N_{HI} cutoff"),

$$L \begin{cases} A = 2.0 \times 10^{10}, \beta = 1.7, \gamma = 2.4 & 10^{14} < N_{HI} < 7.9 \times 10^{16} cm^{-2}; \\ No\ clouds, & N_{HI} > 7.9 \times 10^{16} cm^{-2}, \end{cases}$$

$$H \begin{cases} A = 2.0 \times 10^{7}, \beta = 1.5, \gamma = 2.4 & 10^{14} < N_{HI} < 1.58 \times 10^{17} cm^{-2}; \\ A = 2.3 \times 10^{8}, \beta = 1.5, \gamma = 0.5 & 1.58 \times 10^{17} < N_{HI} < 10^{22} cm^{-2}. \end{cases}$$

To relate N_{HeI} and N_{HeII} to N_{HI}, we consider two limiting cloud models, fully neutral[6] "N" and highly photoionized[7] "P". For the latter, we take $N_{HeI} = 0$ and solve for the photoionization equilibrium value of N_{HeII} consistent with the metagalactic ionizing background.

The results for question (1) presented here are for Ly α cloud cases P and L only ("PL"). For case PL, emissivity type C, and the assumed $\Omega_{IGM}h^2$ value, the AGN fractional contribution corresponds to the ratio of the maximum *observed* quasar contribution[8,9] at $z \sim 4$ to the total required to satisfy the GP constraints described above.

We use the shape of the metagalactic ionizing radiation field at $z = 3$ thus calculated as an incident spectrum for our models of LL QALC's. These are constructed using Ferland's photoionization equilibrium code CLOUDY[10], assuming an HI column density $\log(N_{HI}) = 17.5$, and varying the incident ionization parameter Γ (i. e. photon-to-H-atom density ratio). In general, the metagalactic radiation spectral *shape* after IGM reionization "break-through" is relatively insensitive to $\Omega_{IGM}h^2$ for values as low as assumed here. Hence, we can interpret our variation of Γ either as a variation of the underlying emissivity level for fixed LL cloud internal density *or* as a variation of this density for fixed emissivity.

For (2) above, we calculate the IGM, Ly α cloud, and radiation field evolution as described above for (1), except that we also consider a "bulk heating" emissivity case (D) (i. e. thermal energy, instead of ionizing radiation) and we consider several of the Ly α cloud opacity cases described above.

RESULTS AND CONCLUSIONS

1) Absorption line ratios for Lyman Limit QALC's. Fig. 1 shows the metagalactic radiation fields contributed by the three types of sources discussed above, illustrating the strong effects of attenuation by the IGM and Ly α forest clouds. Figs. 2-4 show plots of the column densities of C II, C III, Si II, Si III, and Si IV with respect to C IV versus Γ, and the observed ratios[11] for the LL cloud with $logN_{HI} = 17.35 \pm 0.1$ at $z = 2.9676$ toward Q2126-1551 as an

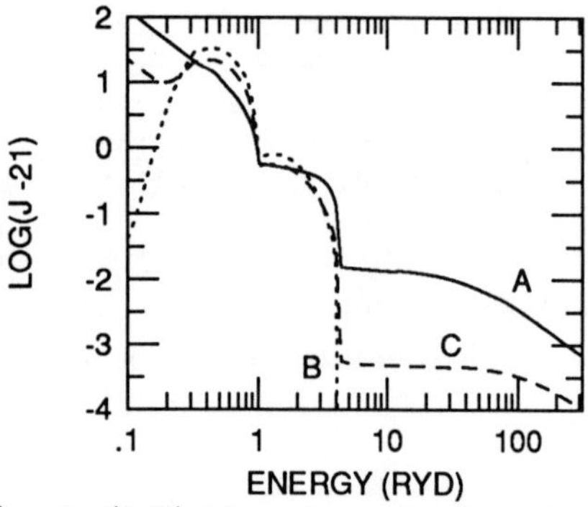

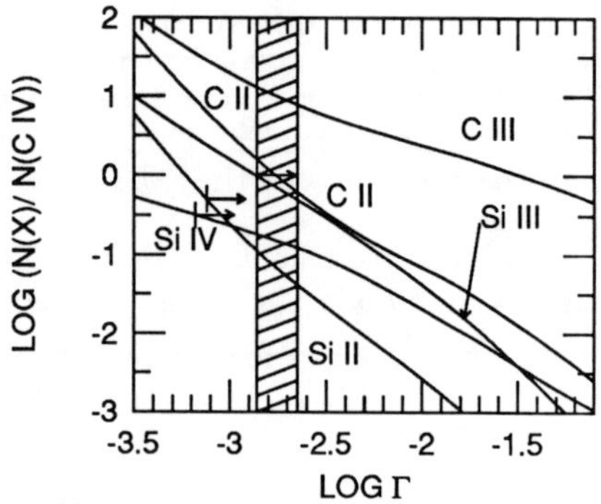

Fig. 1 (left) Mean intensity in units of 10^{-21} ($erg\ cm^{-2}\ s^{-1}\ Hz^{-1}\ ster^{-1}$) of the intergalactic radiation field at $z = 3$ as attenuated by the IGM for emissivity case A for AGN-like sources (solid curve labelled A), case B for stellar-type primeval galaxy sources (dotted curve labelled B), and case C, a mixture of cases A and B in ratio 1 : 4, the sum of a stellar-type spectrum and the maximum observed contribution from AGN's (dashed curve labelled C). All cases assume PL Lyman α cloud opacity.

Fig. 2 (right) Predicted quasar absorption line cloud (QALC) column densities of the labelled species relative to that of C IV versus Γ for a cloud at $z = 3$ exposed to spectrum A in Fig. 1. The ionization parameter $\Gamma = n_\gamma / n_H$ where n_γ and n_H are the number densities of ionizing photons in the incident radiation field and H atoms in the cloud, respectively. Comparison is made to observations of a LL QALC at $z \sim 3$ by ref. [11]. Observed upper limits are noted as arrows. (C III limit is ≤ 2.) The range of Γ compatible with the observations is hatched.

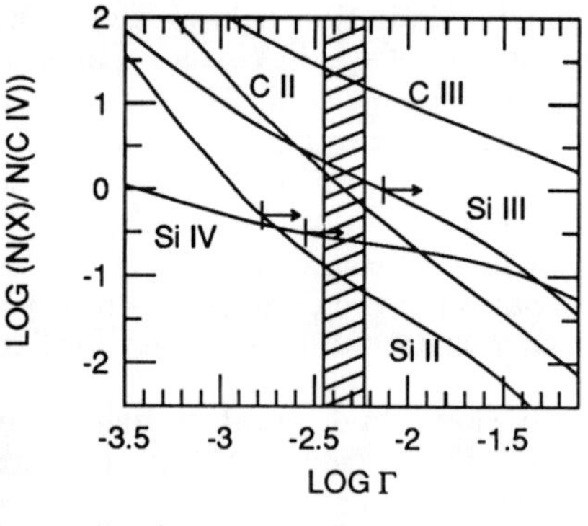

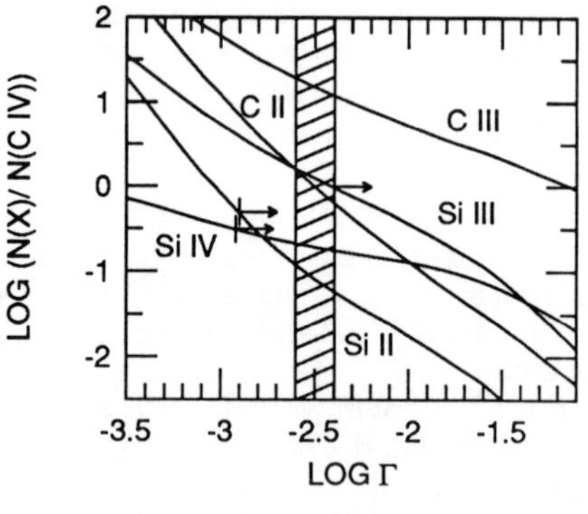

Fig. 3 (left) Same as Fig. 2, except cloud is exposed to radiation field B in Fig. 1.

Fig. 4 (right) Same as Fig. 2, except cloud is exposed to radiation field C in Fig. 1.

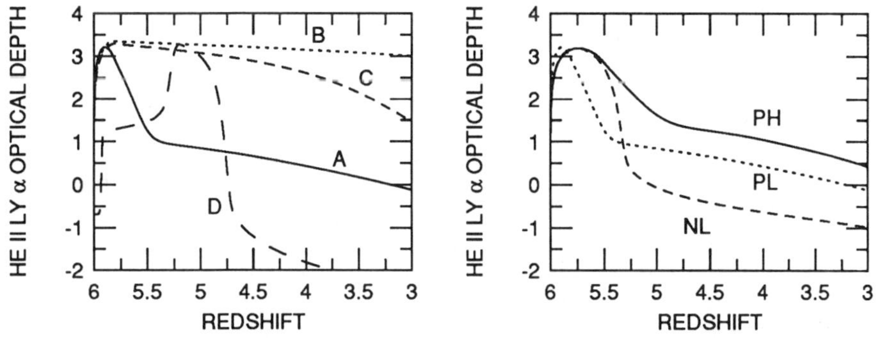

Fig. 5 (left) τ_{GP} (He II) at $304\mathring{A}(1+z)$ versus z for different ionization sources. Solid curve assumes photoionization by the AGN-type emissivity case A. Dotted curve assumes photoionization by the stellar-type emissivity case B. Short dashed curve assumes photoionization by the combination stellar/AGN emissivity case C. Long dashed curve assumes collisional ionization by bulk heating (D). All cases assume PL Lyman α cloud opacity.

Fig. 6 (right) τ_{GP} (He II) at $304\mathring{A}(1+z)$ versus z for different models for the contribution of the Lyman α clouds to the IGM opacity. In all cases photoionization by the AGN emissivity case A is assumed. Solid curve uses PH model for cloud opacity (i.e. cases "P" and "H"). Dotted curve uses PL model for cloud opacity. Dashed curve uses NL model for cloud opacity.

example for comparison. While the predicted Si III ratio for the stellar case is marginally incompatible with the observed constraints (Fig. 3), this is possibly not significant because the balance between Si III and Si IV is greatly affected by the transport of He II Ly α within the clouds, one of the largest uncertainties remaining in photoionization models [11,12], and a small depletion of Si relative to C would also help. For fixed Γ, N_{HI}, and N_{CIV}, the flux level of our calculated metagalactic radiation background determines the internal cloud density, cloud size, and metallicity. For the flux level which marginally satisfies the GP constraints above, the acceptable range in Γ corresponds to metallicities of order 1% solar and cloud sizes of 3-5, 9-20, and 7-20 kpc, for the AGN, stellar, and stellar/AGN sources, respectively, all compatible with known constraints on the cloud size. *Thus, the observed line ratios in such Lyman Limit QALC's are compatible with a stellar- or an AGN-dominated metagalactic ionizing flux.*

2) GP He II Ly α Optical depths. Fig. 5 shows the $\tau_{GP}(HeII)$ predicted at $304\mathring{A}(1+z)$ in the spectrum of a quasar at $z_{QSO} \geq z$, for the four sources of ionization and case PL only. The He II troughs for both the stellar and the stellar-dominated combination spectra are completely black, while for an AGN spectrum $\tau_{GP}(HeII) \gtrsim 1$ for $z \gtrsim 3$. For a collisionally ionized medium (i.e. thermal energy injection only) which marginally satisfies the H GP constraints above, however, $\tau_{GP}(HeII) \sim 0.5$ at $z = 4.73$, but $\tau_{GP}(HeII) < 0.02$ by $z = 4$.

Fig. 6 shows $\tau_{GP}(HeII)$ for AGN-type sources but with three models for the Lyman α cloud opacity. If AGN's ionize the IGM and $\Omega_{IGM} h^2 \not\lessdot 0.01$, then the observed absence of a pronounced $\tau_{GP}(HeII)$ would apparently favor the neutral cloud hypothesis for the Ly α forest ($\lesssim 10^{17} cm^{-2}$) and a significantly lower density of LL QALC's ($10^{17} < N_{HI} < 10^{20} cm^{-2}$) than our case H.

This work benefitted from the support of Robert A. Welch Foundation Grant F-1115, Texas Advanced Research Program Grant 4132, and NASA Grant NGT-50316. Numerical computations were performed on the UT Center for High Performance Computing Cray X-MP computer.

REFERENCES

1. Shapiro, P. R. and Giroux, M. L., 1989, in *The Epoch of Galaxy Formation*, eds. C. S. Frenk, R. S. Ellis, T. Shanks, A. F. Heavens, and J. A. Peacock (Boston: Kluwer Academic Publishers), pp. 153-161.
2. Giroux, M. L. and Shapiro, P. R., 1990a, in *Physical Processes in Fragmentation and Star Formation*, eds. R. Capuzzo-Dolcetta, C. Chiosi, and A. Di Fazio (Netherlands: Kluwer Academic Publishers), pp. 71-79.
3. Giroux, M. L. and Shapiro, P. R., 1990b, *Ap. J.*, to be submitted.
4. Steidel, C. C. and Sargent, W. L. W., 1989, *Ap. J. (Letters)*, **343**, L33.
5. Miralda-Escude, G. and Ostriker, J. P., 1990, *Ap. J.*, **350**, 1.
6. Hogan, C. J., 1987, *Ap. J.*, **316**, L59.
7. Ostriker, J. P. and Ikeuchi, S., 1983, *Ap. J. (Letters)*, **268**, L63.
8. Koo, D. C. and Kron, R. G, 1988, *Ap. J.*, **325**, 92.
9. Turner, E. L., 1990, *Princeton Observatory Preprint No. 349.*
10. Ferland, G. J., 1989, OSU Astronomy Dept. Internal Report, 90-001.
11. Steidel, C. C., 1990, *Ap. J. (Suppl.)*, **74**, 37.
12, Netzer, H., and Ferland, G. J., 1984, *P.A.S.P.*, **96**, 593.

THE HYDRODYNAMICS OF RELICT COSMOLOGICAL H II REGIONS: IMPACT ON THE IGM

Avery Meiksin and Piero Madau
Department of Physics and Astronomy, The Johns Hopkins University
and
Space Telescope Science Institute
3700 San Martin Drive, Baltimore MD 21218

ABSTRACT

Discrete sources of photoionizing radiation present at the time the intergalactic medium is largely neutral will create expanding H II regions. A source which turns off or dims quickly enough will not be able to maintain the advance of its ionization front indefinitely. The high pressure of the relict H II region will then drive a shock into the cold, neutral ambient medium. We discuss the hydrodynamical evolution of relict cosmological H II regions generated at redshifts $z \sim 5 - 15$. A thin shell of shocked material develops around the H II region. The peculiar velocity of the shell is typically $10 - 20\,\mathrm{km\,s^{-1}}$. The dense shell may cool and fragment into objects with baryonic masses as high as $10^6 - 10^8\,\mathrm{M_\odot}$ and column densities of order 10^{19} cm^{-2}. These values are independent of the nature of the photoionizing source. They are set only by the density of the IGM, the sound speed of 10^4 K gas, and the age of the shell. The shock propagation ceases once the filling factor of the H II regions reaches unity and the IGM is photoionized. The clouds formed may account for the Lyα forest seen in the spectra of high redshift QSOs. Gravitationally bound fragments may be candidates for blue-excess galaxies at high redshift.

1. INTRODUCTION

At epochs corresponding to $z \lesssim 1000$, the intergalactic medium (IGM) is expected to recombine and remain neutral until objects form that are capable of reionizing it. This appears to have been accomplished by $z \sim 5$ (Schneider, Schmidt, and Gunn 1989), in order to satisfy the Gunn-Peterson constraint on the amount of neutral hydrogen in the IGM. Discrete sources of photoionizing radiation present at the time the IGM is neutral will create expanding H II regions. Sources which die or fade quickly enough will not be able to support the advance of the ionization front indefinitely. Once the supply of ionizing photons ceases to reach the front, the high pressure of the H II region will drive a shock into the surrounding gas. We discuss the possible impact of such shocks on the IGM. A fuller account is provided in Madau and Meiksin (1990).

2. COSMOLOGICAL H II REGIONS

The evolution of an expanding cosmological ionization front generated by a point source of ionizing radiation is governed by the equation

$$4\pi r_I^2 n_{\rm H}\left(\frac{dr_I}{dt} - Hr_I\right) = S - \frac{4\pi}{3} r_I^3 n_{\rm H}^2 \alpha_B (1+\chi) \tag{1}$$

(Shapiro 1986), where r_I is the proper radius of the I-front, H is the Hubble constant, S is the number of ionizing photons emitted per second from the central source, $n_{\rm H}$ is the intergalactic number density of hydrogen, α_B is the total recombination coefficient to the excited states of hydrogen, and $\chi = 0.2$ corrects for helium. The right hand side of equation (1) vanishes at the Strömgren radius $r_{St} = \{3S/[4\pi n_H^2 \alpha_B(1+\chi)]\}^{1/3}$.

We generalize the constant luminosity result of Shapiro and Giroux (1987) and Donahue and Shull (1987) by taking the ionizing luminosity to decrease with redshift z as a power-law of the form $S = S_{in}(x/x_{in})^q$, where $x = 1+z$, $x_{in} = 1+z_{in}$, z_{in} is the redshift at which the source turns on, $\alpha = 3S_{in}/(4\pi n_0 H_0)$, and $\beta = n_0\alpha_B(1+\chi)/H_0$. Here $n_0 = (8.08\times10^{-6}\mathrm{cm}^{-3})\Omega_{IGM}h^2$ is the current number density of hydrogen, and Ω_{IGM} is the ratio of the IGM density to the closure density. The cosmological generalization of the Strömgren radius is $r_{St} = (\alpha/\beta)^{1/3}x^{q/3-2}/x_{in}^{q/3}$. For $r_I \lesssim 0.5r_{St}$, equation (1) can be integrated directly with good accuracy to obtain ($K = Ht$ for $\Omega = 0$ or 1),

$$r_I \simeq \left(\frac{\alpha}{q+3K/2-5/2}\right)^{1/3} x_{in}^{K/2-5/6}x^{-1}\left[1-\left(\frac{x}{x_{in}}\right)^{q+3K/2-5/2}\right]^{1/3}. \quad (2)$$

The Strömgren sphere of a rapidly dimming object ($q > 3$), shrinks in comoving coordinates: the density decrease associated with the Hubble expansion cannot balance the decrease in ionizing photons. Once the ionization front and the Strömgren radius cross, a shock is driven into the neutral gas of the IGM. For the case $\Omega = 1$ and $q = 4$, the comoving radius of the ionized bubble is $R_0 = (1+z_s)r_I(z_s) \simeq (23-29\,\mathrm{Mpc})h_{0.5}^{-1}(\Omega_{IGM}/0.1)^{-1/3}(S_{in}/10^{56}\mathrm{s}^{-1})^{1/3}$, where $h_{0.5} \equiv h/0.5$, z_s is the redshift at which the shock occurs, and the coefficient has been adjusted to match numerical integration including recombinations for x_{in} in the range 9-13. For $\Omega = 0$, R_0 is 20% larger. In Figure 1, the evolution of r_{St} and r_I is displayed for the case $q = 4, \Omega = 0$ (including recombinations).

3. THE STRUCTURE OF THE THIN SHELL

We wish to obtain an estimate for the mass which will be swept up by the shock driven into the IGM by the relict H II region, as well as estimates for the thickness and column densities of the swept up material. We do so by approximating the H II region as a uniform sphere of hot gas which expands into the surrounding medium due to its high overpressure. We treat the separation of the hot gas and the IGM as a discontinuity. The expanding gas occupies only a thin layer compared with the radius of the H II region. In this approximation, the expansion behaves as a 1-D cosmological shock tube. The flow admits an analytic solution, depicted in Figure 2.

The amount of shocked material in the shell depends on the initial sound speed in the H II region at the time of the shock. We determine the temperature assuming an IGM with cosmological abundances (He/H= 0.1), and including hydrogen and helium recombination radiation, thermal bremsstrahlung, and Compton cooling of electrons off the cosmic background radiation photons. We find that for $\Omega_{IGM}h^2 = 0.03$ today, at $z = 10$ the temperature of the H II region is 1.6×10^4 K, corresponding to an adiabatic sound speed of $20\,\mathrm{km\,s^{-1}}$. For $\Omega = 0$, we find that the mass swept up by the shock,

$$M_s \simeq (1.8\times10^{13}\,\mathrm{M_\odot})h_{0.5}\left(\frac{\Omega_{IGM}}{0.1}\right)\left(\frac{R_0}{30\,\mathrm{Mpc}}\right)^2\left(\frac{c_{4s}^{ad}}{20\,\mathrm{km\,s^{-1}}}\right)\left(1-\frac{x}{x_s}\right), \quad (3)$$

is concentrated in a thin shell of thickness

$$\Delta r_s \simeq (9.8\,\mathrm{kpc})h_{0.5}^{-1}\left(\frac{c_{4s}^{ad}}{20\,\mathrm{km\,s^{-1}}}\right)\left(\frac{x}{6}\right)^{-1}\left(1-\frac{x}{x_s}\right), \quad (4)$$

and hydrogen column density, $N_{\rm H} = 4n_{\rm H}\Delta r_s$, given by

$$N_{\rm H} \simeq (5.3\times10^{18}{\rm cm}^{-2})h_{0.5}\left(\frac{\Omega_{IGM}}{0.1}\right)\left(\frac{c_{4s}^{ad}}{20\,{\rm km\,s}^{-1}}\right)\left(\frac{x}{6}\right)^2\left(1-\frac{x}{x_s}\right). \qquad (5)$$

The results are comparable for $\Omega = 1$ and $5 < z < 10$.

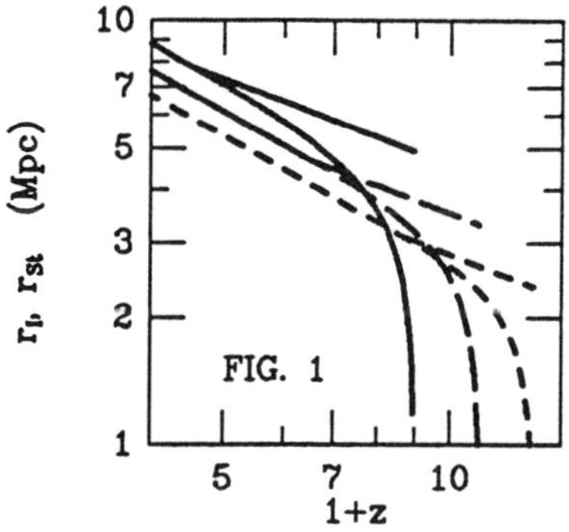

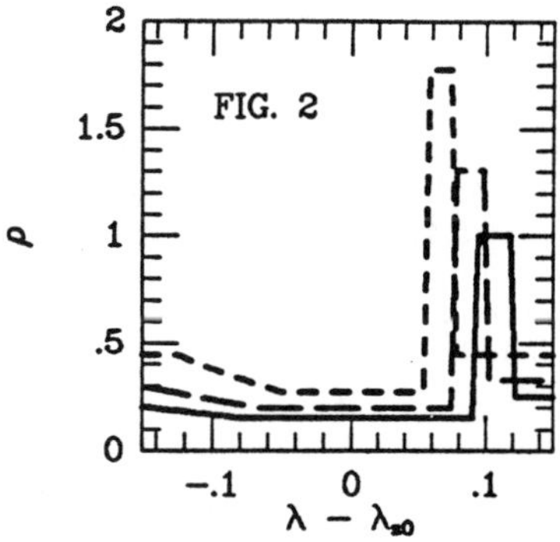

Figure 1: The evolution of the ionization front r_I and Strömgren radius r_{St} as a function of redshift for a radiation source with $S_{in} = 10^{56}$ s^{-1} and $q = 4$ ($\Omega = 0$, $\Omega_{IGM} = 0.1$, $h = 0.5$). The curves of slope $-2/3$ represent r_{St}, while the curves of varying slope represent r_I. When these radii cross, the central source can no longer maintain the advance of the I-front and a shock is driven into the IGM. The three families of curves correspond to initial redshifts of $z_{in} = 8$ (*solid line*), 10 (*long-dashed line*), and 12 (*short-dashed line*).

Figure 2: Evolution of physical gas density ρ as a function of comoving coordinate $\lambda - \lambda_{s0}$, where λ_{s0} is the comoving coordinate of the initial pressure discontinuity. The density is shown at the times $t/t_s = 1.5$, 1.75, and 2 for the case $\Omega = 1$, where t_s is the time of shock formation. A rarefaction wave moves into the driver gas while a shock front propagates into the IGM. Note that the density decreases uniformly throughout the flow structure like t^{-2}. The density normalization is arbitrary.

The overdense shell will cool as a result of the formation of H_2. The density in the shell may then become sufficiently high for self-gravitating unstable modes to grow. The characteristic unstable fragment mass is

$$M_c \simeq (1.7\times10^8\,{\rm M}_\odot)h_{0.5}^{-\frac{1}{2}}\left(\frac{\Omega_{IGM}}{0.1}\right)^3\left(\frac{c_{4s}^{ad}}{20\,{\rm km\,s}^{-1}}\right)^3\left(\frac{x}{6}\right)^2\left(1-\frac{x}{x_s}\right)^3, \qquad (6)$$

for $\Omega = 0$. It is a factor of 100 smaller for $\Omega = 1$. Irrespective of whether the shocked gas is gravitationally unstable, the cooling shell may be subject to hydrodynamical instabilities. We estimate a characteristic fragment mass of $\sim 10^7\,{\rm M}_\odot$.

4. ENVIRONMENTAL IMPACT

The results presented above depend on the specific assumption that the IGM is cold compared to the H II region. This might occur at redshifts $\gtrsim 5$, when the IGM is assumed to be still neutral. Whenever the timescale for full recombination is longer than the cosmological timescale, a simple relation can be derived between the (comoving) volume V of the H II bubble and the energy ΔE emitted in ionizing photons from the central source

$$\frac{V}{\Delta E} = \frac{56\pi G m_p}{15 H_0^2 \Omega_{IGM} e_I} = (2.3 \times 10^{17} \mathrm{cm}^3 \mathrm{erg}^{-1}) h_{0.5}^{-2} \left(\frac{\Omega_{IGM}}{0.1}\right)^{-1}, \tag{7}$$

where $e_I = 13.6$ eV, and we assumed a monochromatic ionizing emissivity.

Giant H II regions with comoving radii of 100 Mpc will be generated with an energy input of 5.4×10^{62}erg, characteristic of quasar sources. The region ionized by a young, star forming galaxy will be a few to several Mpc, depending on the amount of neutral hydrogen and dust present inside the galaxy that would absorb the radiation.

The fraction of the IGM which could be processed by the shocks is given by

$$\frac{\Omega_s}{\Omega_{IGM}} = 12 \frac{\Delta r_s}{r_B} Q_{\mathrm{HII}}, \tag{8}$$

where Ω_s pertains to the shocked material and $Q_{\mathrm{HII}} = (4/3)\pi r_B^3 n_B$ is the porosity parameter of the relict H II bubbles of proper radius r_B and number density n_B. For $z_s \sim 9$, $\Delta r_s \sim 4$ kpc at $z = 5$. Then $Q_{\mathrm{HII}}^{-1} \Omega_s / \Omega_{IGM} \sim 5 \times 10^{-2}$, 10^{-2}, and 3×10^{-3} for shells with comoving radii of 6, 30, and 100 Mpc, respectively.

Since gravitational modes grow more slowly than hydrodynamical ones, initially the fragments will not be gravitationally bound. It is then tempting to identify the resulting objects with Lyα absorption line systems. The total hydrogen column density, mass, and size of the fragments are comparable to the values inferred for the Lyα clouds (e.g., Foltz *et al.* 1984), assuming the clouds to be photoionized, while for galactic-type radiation sources and $Q_{\mathrm{HII}} \sim 1$, $\Omega_s \sim \Omega_{\mathrm{Ly}\alpha}$ is easily achieved.

Due to the presence of molecular hydrogen, the fragments will continue to cool and some might become bound objects. These may account for the faint blue-excess galaxies detected by Tyson (1988).

REFERENCES

Donahue, M. J., and Shull, J. M. 1987, *Ap. J. (Letters)* **323**, L13.

Foltz, C. B., Weymann, R. J., and Roser, H. 1984, *Ap. J. (Letters)* **281**, L1.

Madau, P. and Meiksin, A. 1990, submitted to *Ap. J.*

Schneider, D. P., Schmidt, M., and Gunn, J. E. 1989, *A. J.* **98**, 1507.

Shapiro, P. R. 1986, *P.A.S.P.*, **98**, 1014.

Shapiro, P. R., and Giroux, M. L. 1987, *Ap. J. (Letters)* **321**, L107.

Tyson, J. A. 1988, *A. J.* **96**, 1.

Bubbles in McVittie Spacetime

Paul Haines and J.D. Harris
Department of Physics and Astronomy
Dartmouth College
Hanover, NH, 03755

ABSTRACT

We present some results concerning vacuum-filled voids embedded in the McVittie metric.

The McVittie metric describes a spherically symmetric mass distribution embedded in an expanding spacetime. In the flat $k = 0$ case, it can be written

$$g = -\left(\frac{1-h}{1+h}\right)^2 dt^2 + S^2(1+h)^4(dr^2 + r^2 d\Omega^2),$$

where $S = S(t)$ and $h = m/2rS$.

The equation of motion for the bubble boundary in spherically symmetric cases has been derived by Lake [L84]:

$$\dot{R}^2 + 1 = \left(\frac{m_+ - m_-}{M}\right)^2 + \frac{m_+ + m_-}{R} + \frac{M^2}{4R^2}$$

$$= \left[\left(\frac{m_+ - m_-}{M} + \frac{M}{2R}\right)^2 + \frac{2m_-}{R}\right].$$

Here R is the bubble radius, $R = rS(1+h)^2$, and $m_+ = m + \frac{4}{3}\pi R^3 \rho_+$.

In the Newtonian limit, $1 \gg \dot{R}^2 - 2m_-/R$. So,

$$m \approx \frac{1}{2}M\dot{R}^2 - \frac{M^2}{2R} - \frac{Mm_-}{R} + M - \frac{4}{3}\pi R^3(\rho_+ - \rho_-),$$

and then m represents the total, conserved, energy. In this limit, 'm' gets parceled into the six Newtonian terms:

- $\frac{1}{2}MR^2$ is the kinetic energy of the shell;
- $-M^2/2R$ is the gravitational potential energy of attraction of the shell on itself;
- $-Mm_-/R$ is the gravitational potential energy of attraction of the mass inside the shell to the shell;
- M is the rest energy of the shell;
- $-\frac{4}{3}\pi R^3\rho_+$ is the mass removed from the inner region;
- $\frac{4}{3}\pi R^3\rho_-$ is the mass added to the inner region.

In the case where the bubble boundary surrounds a vacuum, described by Minkowski spacetime, $m_- = 0$; 'compensated' voids then have $M = \frac{4}{3}\pi R^3\rho_+$. A void can be 'overcompensated' by embedding it in the McVittie metric and allowing for $m \neq 0$. These voids will have gravitational effects in addition to lensing [T87].

We graph r/r_i for several bubbles, varying the initial shell velocity. Given large initial velocities, the bubble boundaries seem to asymptotically approach $r = t^\alpha$, $\alpha \approx 0.13$ [see B83]. The slower shells, however, begin to collapse, and some shells can appear to be co-moving or nearly co-moving. The nearest observed voids may be nearly co-moving[1].

References

[1] Lake, K., Phys. Rev D, **29**, 1861.

[2] Thompson, Keith L. and Ethan T. Vishniac, Ap.J. **313**, 517.

[3] Bertschinger, E., Ap.J. **268**, 17.

[1] private communication

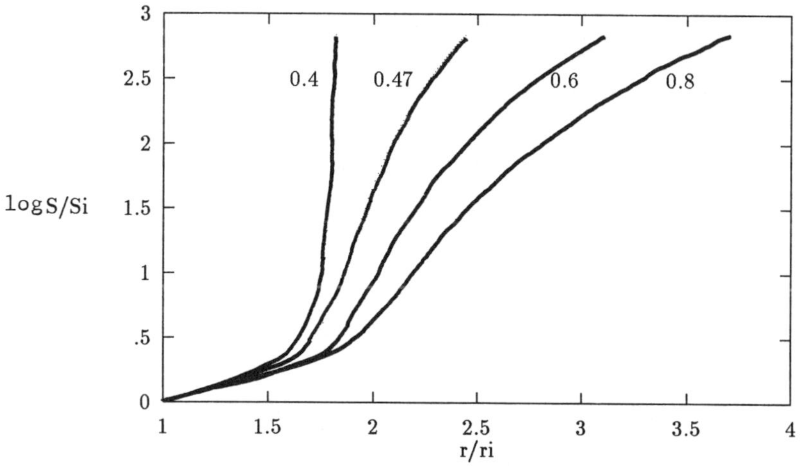

Figure 1: $m = 0$ bubble, with $v_i = 0.4, 0.47, 0.6, 0.8$.

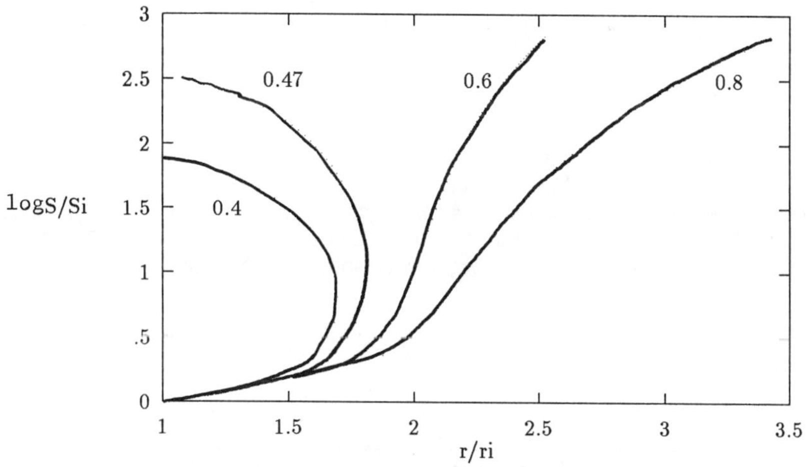

Figure 2: $m = 5e9$ bubble, with $v_i = 0.4, 0.47, 0.6, 0.8$.

INTEGRAL MAXIMA OF SCALE INVARIANT STRUCTURE

Peter von Gronefeld
Jet Propulsion Laboratory
Reston, VA 22091

An isotemporal model of cosmological structure in a strictly geometric sense, whether shaped primarily by a fragmenting or aggregating process, must coherently accommodate the apparent incongruence of the extreme isotropy of the cosmic microwave background radiation (CMBR) with the large scale anisotropy of scale invariant structure. The extreme isotropy of the CMBR also would indicate that, beyond some maximum scale, the distribution of matter transitions to homogeneity; however, this scale has not yet been established[1]. In defining the possible relative proportions of such a scale, considerations for an ideal model exhibiting these properties are discussed.

SCALE DIFFERENTIAL HOMOGENEITY

Transition to a homogeneous mass density from a stationary fractal[2] type distribution could be considered in two ways:

1. Beginning on the scale of the largest structures $\geq r$, where the largest scale correlation is statistically distributed evenly throughout the system (the statistic could either be relatively strong or weak, depending on the regularity of the structure); or
2. Beyond the scale of the largest structure $>r$, where the distribution of matter is no longer coherently associated with it.

The essential difference in these two interpretations is that in the former, the cosmos is organized into a matrix, symmetrically invariant about a maximum resolution scale; and in the latter, a single recursive type structure dominates, but is limited to some overall maximum scale (both of these being ideal cases).

MAXIMUM SUBMULTIPLE INTEGRAL

In either case, irrespective of Ω, the extreme isotropy of the CMBR can be considered as a spherical analogue that constrains the possible scale of anisotropic structure, which at a maximum must fit within this spherical volume. In principle, then, since an s dimensional[3] cosmological structure that scales in increments by a one-dimensional scaling factor λ occupies 3-space scaling in corresponding volumetric multiples by λ^3, it follows that

$$\frac{c}{\lambda H_o}$$

must be the maximum proportional scale possible for any asymmetric organization of matter as such, whether described by a maximum resolution cell or single form. This scale may be large enough to be detected by the Cosmic Background Explorer, and, with the value of the scaling factor λ, is the subject of further study.

1. P.H. Coleman, The Fractal Nature of the Galaxy Distribution, in *Fractals' Physical Origin and Properties*, ed. L. Pietronero (New York: Plenum Press 1989), p. 349.
2. B.B. Mandelbrot, *The Fractal Geometry of Nature* (San Francisco: Freeman 1983).
3. F. Hausdorff, Dimension und äußeres Mass, *Math. Annalen*, **79**, 157-179 (1919).

Primordial Magnetic Field and Density Fluctuation from Early Universe Phase Transition

Jin Wang
Department of Astrophysics
University of Illinois
Urbana, IL 61801

Abstract

In this paper, we consider a mechanism to generate primordial magnetic field and density fluctuation. We show that bulk viscosity being generated during the early universe phase transition might be the origin of the two poblems we mensioned above.

Primordial magnetic field and density fluctuation spectrum are two very interesting problems in early universe study. There are some mechanisms proposed. The big probelm for density fluctuation is that it needs fine tunning (such as in inflation theory [1]). Primordial turbulence model with non zero vorticity for generating primordial magnetic field [2] suffers from the fact that vortices decay during cosmic expansion [3].

In this paper we study the effect of early universe phase transition on the above two probelms. We show that primordial magnetic field can be gernerated from the viscosity caused by the early universe phase transition. Similar situation is dicussed for density fluctuation.

The basic mechanism for generating bulk viscosity is that when phase transition occurs [4], the Higgs fields get the vaccum expectation value, symmetry is spontaneously broken, also through gauge coupling gauge bosons get mass, through Yukawa coupling fermions get mass. Gauge bosons mass are usually large, so when they suddenly get mass, the motion changes from relativistic to non relativistic, It is this difference of velocity of non relativistic part of the motion of the particles and relativistic part of the motion of particles that generate the bulk viscosity. Fermions mass are usually small. They are mostly relativistic. So the viscosity force between fermions and charge bosons are balanced by the electric field between them. Since particles are moving and background is expanding, the electric field is moving too. The moving electric field will generate a magnetic field. Evolve this til recombination time, we can get some quantitative ideas about the values of this magnetic field. The effect of viscosity is to slow down particles so that it makes easier for the particles to stick together, therefore produce perturbation seed easier.

Let us make the above statement more quantitative. After simplification we get for the bulk viscosity coefficient [5], [6], [10]:

$$\zeta = \frac{R}{3\dot{R}} \left[\frac{5}{6} b T_{\text{II}}^{5/2} + \frac{d}{2} T_{\text{II}}^{3/2} \right] \qquad \ldots \quad (1)$$

For time of GUT phase transition $T \sim 10^{14}$ Gev . We have $H^{-1} = R/\dot{R} \sim 0.23 \times 10^{-35}$ s.

$$\zeta = 10^{62} \qquad \ldots \quad (2)$$

$\zeta \frac{\dot{R}}{R}$ is the viscosity force per unit volume , it is balanced by electric force between charge bosons and fermions.

$$\zeta \frac{\dot{R}}{R} = \frac{1}{2} E^2 \qquad \ldots \quad (3)$$

For GUT where $\zeta \sim 10^{62}$, $\dot{R}/R \sim 10^{35}$. We have

$$E = 10^{47} \qquad \ldots \quad (4)$$

For GUT, we have

$$v = 10^{-2} c$$

$$B = \frac{v}{c} \times E \sim 10^{45} \text{ Gauss} \qquad \ldots \quad (5)$$

If we assume flux conservation

$$B_{recombination}/B_i = \left(\frac{T_{rec}}{T_i}\right)^2 \qquad \ldots \quad (6)$$

For GUT we have $T_i = 10^{14}$ Gev, $T_{rec} \sim 10^3$ K .

$$B_{rec} = B_i \left(\frac{T_{rec}}{T_i} \right)^2 \sim 10^{-3} \text{ Gauss} \quad \ldots \quad (7)$$

From flux conservation:

$$B_{Intergalaxy} = B_{rec} \left(\rho_{IG}/\rho_{rec} \right)^{\frac{2}{3}} \sim 10^{-9} \text{ Gauss} \quad \ldots \quad (8)$$

$$B_{Galaxy} = 10^{-6} \text{ Gauss} \quad \ldots \quad (9)$$

This is consistent with cosmic magnetic field measurement, and we don't need dynamo effect to amplify magnetic field. Strittmatter [8] has shown that gravitational collapse of a cloud of mass M and density ρ will be haulted by a large scale magnetic field B unless

$$M \geq M_c \sim 3.6 \times 10^{-25} B^3/\rho^2 \, M_\odot \quad \ldots \quad (10)$$

where B is measured in Gauss and ρ is g/cm^3. Taking $B \sim B_{IG} \sim B_{rec} (\rho_{IG}/\rho_{rec})^{\frac{2}{3}}$ we find $M_c \sim 1.4 \times 10^7 M_\odot$, roughly the mass of a dwarf galaxy. In the instability theory gravitational collapse is possible as long as $M > M_{Jeans} \sim 10^5 M_\odot$, the Jean's mass is too small to be important in any theory of galaxy formation. If we consider microwave background radiation distrotion we have:

$$\frac{\delta\rho}{\rho} \sim 4 \frac{\delta T}{T} \quad \ldots \quad (11)$$

$$\frac{\delta T}{T} \sim \frac{1}{4} \frac{\delta\rho}{\rho} \sim \frac{1}{4} \frac{B^2/8\pi}{aT^4} \sim 10^{-5} \quad \ldots \quad (12)$$

It is acceptable since COBE now has just the low limit of 2*10 . Also using fluctuation dissiapation theorm [9] we can get an quantitative idea about the density perturbation caused by the bulk viscosity in both GUT and Electroweak phase transition.

$$\left(\frac{\delta\rho}{\rho}\right)^2 \sim \frac{1}{N} \left(\frac{hkT}{P+\epsilon} \right) \left(\frac{\Upsilon}{P+\epsilon} \right) \frac{\dot{R}}{R} \quad \ldots \quad (13)$$

For GUT we have

$$\frac{\delta\rho}{\rho} \sim \left(\frac{1}{10^{79}} \times \frac{10^{93}}{10^{95}} \right)^{\frac{1}{2}} \sim 10^{-41} \quad \ldots \quad (14)$$

We evolve this perturbation til recombination time.

$$\left(\frac{\delta\rho}{\rho}\right)_{rec} \sim \left(\frac{\delta\rho}{\rho}\right)_{init} \frac{t_{rec}}{t_{GUT}} \sim 10^{-41} \times 10^{47} \sim 10^6 \quad \ldots \quad (15)$$

It is too big for galaxy formation. The above argument is rather naive. If we want to know if galaxy will form, we have to compare the time scale of perturbation τ_p with collapse time scale $\tau_{collapse}$ (τ_p is perturbation time scale, it is related to gauge bosons decay rate once they get mass, this determines the decay time scale for the viscosity, therefore for the density perturbation) .

For GUT phase transition: $\tau_p \sim 10^{-34} s$

$$\tau_{collapse} \sim \frac{1}{\sqrt{G\rho}} \sim \frac{1}{\sqrt{7\times10^{-8}\times10^{73}}} \sim 10^{-32} s \quad \cdots \quad (16)$$

so we see that

$$\tau_p << \tau_{collapse}$$

The massive gauge bosons decay too fast or the density perturbation decay too fast that it doesn't have time to stick together and grow. So it really can't form a galaxy in this mechanism.

References:

[1]. Kolb, E. and Turner, M., "The Early Universe", (Addison and Wesley Publishing Company, 1990)

[2]. Harrison, E.R., M.N.R. AS. 147(1978)279

[3]. Rees, M., Q.J.R. Astr. Soc. 28(1977)197

[4]. Ross, G., "Grand Unified Theories", (Addison and Wesley Publishing Company, 1985).

[5]. Weinberg, S., Ap.J. 168(1971)175

[6]. Cheng, B., UIUC preprint, "Bulk Viscosity From GUT Phase Transition", (1990).

[7]. Eckart, C., Phys. Rev. 58(1940)919.

[8]. Strittmatter, P.A., M.N.R.A.S.,132(1966)359.

[9]. Zimdahl, W., Phys. Lett. A142(1989)229.

[10]. Wang, J., "Primordial Magnetic Field and Density Fluctuation from Early Universe Phase Transition", UIUC preprint.

(mostly) *Velocity-Dependent Evidence for Dark Matter*

EVIDENCE FOR DARK MATTER FROM ROTATION CURVES: TEN YEARS LATER

Vera C. Rubin, Department of Terrestrial Magnetism
Carnegie Institution of Washington, Washington D. C. 20015

ABSTRACT

During the past decade, observations have confirmed that rotation curves are generally flat or rising across the optical galaxies out to the limits of the neutral hydrogen. Regrettably, parameters of the dark halos can be evaluated only in terms of assumed mass distributions. However, it seems well established that dark matter forms an increasingly dominant component with decreasing galaxy luminosity. A major advance of the decade has been the observation that in some gas rich dwarf irregulars the mass in dark matter exceeds that in luminous matter at virtually all radii. Interior to the final measured point in DDO 154, 90% of the mass is dark, and integral $M/L \geq 75$.

INTRODUCTION

It is now over ten years since astronomers started asking questions about dark matter: Is it necessary?, Where is it?, How much is there?, What is it? In this talk, I shall emphasize recent attempts to answer the first three questions, based on observational evidence principally from rotation curves. With very few exceptions, rotation velocities within spirals are constant or rising to the limits of the observations. Hence only a lower limit to the mass of a galaxy is obtained from internal velocity measures. It is still correct, as Bosma (1978) stated, that the best rotation curves are composite: optical observations at small radial distances, and radio observations for HI gas extending beyond the optical image.

The usual assumptions in deducing a mass distribution from a velocity distribution are that Newtonian gravitational theory is valid, that a self gravitating disk of constant M/L dominates at small radial distances, that an isothermal halo dominates at large radial distances, and that the internal extinction is not a function of radius. An isothermal halo distribution is adopted, not because of its reality, but because it produces a flat rotation curve. Extinction surely decreases away from the center; a more realistic treatment will only enhance the observed increase of M/L with radius, a primary evidence for dark matter. Many arguments can be given to contradict recent statements (Valentijn 1990) that galaxies are opaque (although optical depths of unity may exist). Background galaxies detected through foreground galaxies (Hodge 1974; Gurwell & Hodge 1990) show that, for example, extinction reaches only 1.3 V-magnitudes at the center of the Small Magellanic Cloud. Evidence supports the following: spiral galaxy masses increase approximately linearly to the limits of the observations, while luminosities asymptotically reach a limiting luminosity. Hence cumulative M/L values increase with radius with a gradient that is in excess of any value which might be attributed to population variations. A dark halo is the major contributor to the mass.

With $H_o = 50$ km sec^{-1} Mpc^{-1}, a value of M/L (mass to blue luminosity ratio in solar units) of about 800 is necessary to close the universe. However, individual field spirals exhibit velocities and luminosities which predict M/L values generally less than 10 within the final measured point. Thus, internal dynamics of spirals require only a small percentage of the closure mass.

ESCAPE VELOCITIES

If the kinetic energy of a star at galaxy radius r exceeds the negative of its potential energy Φ at that r, the star will escape from the galaxy. For a point mass, $v_{esc} = (2GM/r)^{1/2} = 2^{1/2} v_{circ}$, where M is the galaxy mass interior to r. Because galaxies do not consist of central point masses, it is not possible to infer a total mass for a galaxy unless the complete rotation curve or the gravitational potential or the form of the mass distribution beyond r is known. For our own Galaxy, knowledge of the escape velocity at the position of the sun will not determine the total mass uniquely, due to the uncertainty in the exterior mass distribution. For rotation velocities assumed constant to r = 41 kpc, with v_{circ} = 220 km/s and v_{esc} = 500 km/s, the mass interior to $M(r = 41$ kpc$) = 5 \times 10^{11}$ $M_\odot$ (Binney and Tremaine 1987; see also Leonard and Tremaine 1990), with an uncertainty at least a factor of two.

Historically, Oort (1921) noted that the distribution of high velocity stars in the solar neighborhood showed an asymmetry in direction. In the direction of galactic rotation few stars had velocities greatly in excess of the solar rotation; Oort identified the observed velocity limit with the escape velocity from the Galaxy. A recent analysis of high velocity stars in the vicinity of the sun (Carney et al. 1988) shows a few stars with velocities approaching 500 km/s. This represents a lower limit to the escape velocity at the solar circle only if these stars are bound to the Galaxy, not bound to the Local Group, and are not the product of disruption of globular cluster binaries. Given difficulties in determining both the escape velocity and the uncertainty in mass distribution beyond the solar circle, it is unlikely that a mass for the Galaxy can be determined by this method independent of an assumed model.

HALO PARAMETERS

The rotation curve of the Sc galaxy NGC 3198 is everyone's favorite, extending flat to 11 disk optical scale lengths (van Albada et al. 1985). To evaluate parameters of the halo and their contribution to the observed velocities, we must first subtract in quadrature the contribution of the constant M/L disk. Limiting cases for so doing are shown in Figure 1 (Kormendy 1988). The upper figure illustrates the maximum possible disk (M/L=4.4); the lower shows the minimum disk (M/L=0). During the mid-1980's, it was customary to analyze rotation curves by adopting the maximum disk consistent with the inner rise of the rotation curve and with the disk surface brightness profile (Kent 1984, 1986, 1988). From these models, galaxies with a range of 100 in luminosity ($-21 < M_B < -16$) show a variation of only a few in central density (ρ_o=$10^{-3} M_\odot pc^{-3}$), and a ratio of dark matter to luminous matter $M_{dark}/M_{lum} \sim 1$ at the Holmberg radius (Carignan 1987). The halo mass so determined is a lower limit.

A major conceptual advance came when Athanassoula and her colleagues (1987) pointed out that a low mass halo would permit m=1 spiral modes to dominate, producing lopsided one-armed spirals. Adding mass to a halo inhibits m=1 modes until m=2 modes dominate, and so on. By requiring dominant m=2 modes, these authors showed that disk masses ~ one-half that of the maximum disk are required on the basis of present theories of spiral structure.

By combining the "no m=1 modes" condition for the disk with a modified isothermal mass distribution for the halo, Athanassoula et al. (1987) and Kormendy (1988, 1990) have deduced core radii, central densities, and velocity dispersions for dark halos and correlated these with Hubble types, optical luminosities, and maximum rotation velocities. It seems well established that dark

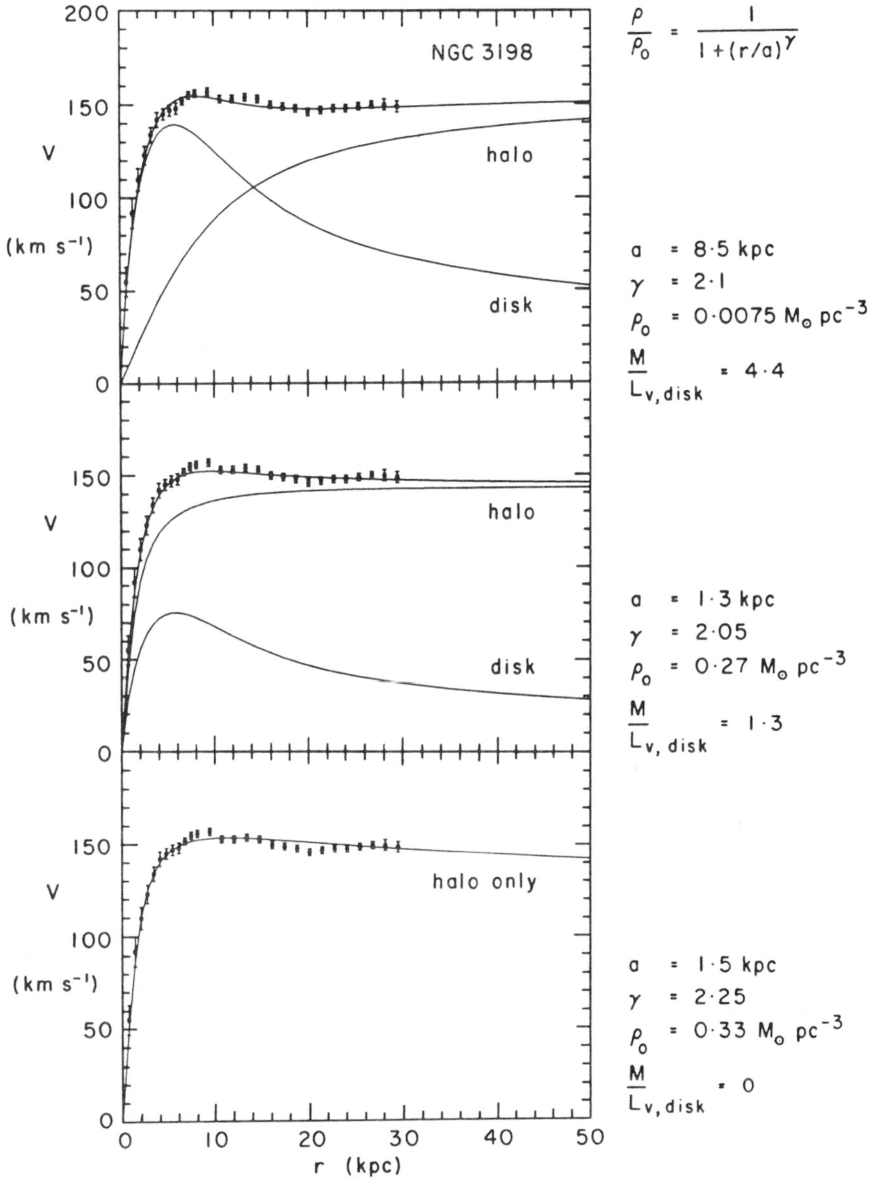

Fig. 1. Alternative decompositions of the rotation curve of NGC 3198 from Kormendy (1988). The observations (van Albada et al., 1985) are shown as points, along with the model disk and halo rotation curves and their sum in quadrature (curves). For the assumed form of the halo density distribution ρ, model parameters are given at right. Note the large variations in core radii a, central densities ρ_o, and M/L ratios among the three models.

matter becomes an increasing fraction of the total mass with decreasing galaxy luminosity. Smaller core radii, higher central densities, and higher central are less well established for low luminosity galaxies. Because the parameters are model dependent, the resulting correlations may be to some extent artifacts of the models. A valuable analysis by Lake and Feinswog (1989) shows how little the rotation velocities constrain the halo parameters. It is probably correct to conclude that values for parameters of the dark halos are imprecisely known.

THE INNER RISE OF THE ROTATION CURVE

For various reasons, both physical and observational, it may be unwise to use the inner rise of rotation curves to constrain the disk mass, as is done in maximum disk models. Photographic spectra were generally overexposed in the central regions, so as to detect the weak outer emission; innermost velocities from these spectra are poor. Even CCD spectra at low spatial resolution may not spatially resolve inner velocities. Echelle CCD spectra (Rubin and Graham 1987) reveal that some nearby galaxies have velocities which remain high at the innermost measured velocity; the entire velocity gradient occurs in the inner central pixel. Steep rises and sharp elbows in the rotation curves are unobservable at low (30″) resolution 21-cm HI observations. A comparison of optical and 21-cm velocities for Virgo galaxies (Rubin et al. 1989) reveals that 21-cm velocities are consistently lower than optical velocities at small radial distances; optical and 21-cm velocities match only at large radial distances.

Inner gas disks may not be in equilibrium and the predominant motions may not be circular. Gas disks can be tilted, warped, or twisted with respect to the plane defined by the disk; the Hα disk in M31 is twisted and appears much more face-on than the disk (Ciardullo et al. 1988). An even greater complexity arises if nuclear bulges are triaxial, with nuclear gas moving on elliptical orbits, as observed for NGC 4845 (Bertola et al. 1988). Hence velocities at small r deduced on the basis of assumed circular orbits and geometries of the outer disk may be incorrect.

NGC 2683 is a nearby large Sb galaxy which exhibits many interesting dynamical properties. It has a patchy dusty disk, seen projected against a nuclear bulge (Fig. 2). Measured optical velocities fall slightly from 220 to 200 km/s over the inner 2′. Virtually the entire 440 km/s nuclear velocity gradient occurs over the innermost pixel (30 pc). Velocities are double valued within the innermost arc minute, with some gas velocities negative with respect to the sense of rotation. We may be viewing the inner region through foreground disk material warped out of the plane.

VLA observations (Fig. 3; Casertano and van Gorkom 1991) detect HI gas to 700″, and show that velocities decrease beyond about 50″. Interior to this, velocities are distorted by the lower resolution. Even though the rotation curve falls, dark matter is required, for even a disk with M/L=8 predicts velocities lower than observed (Fig. 3). However, a falling rotation curve puts an *upper* limit to the halo mass, for too much halo produces too flat a rotation curve. For all of these reasons, NGC 2683 is an interesting galaxy which illustrates the need for high spatial resolution at small r, and the need for dark matter even with a falling rotation curve. We must learn why the dynamics of NGC 2683 are different from so many other field spirals.

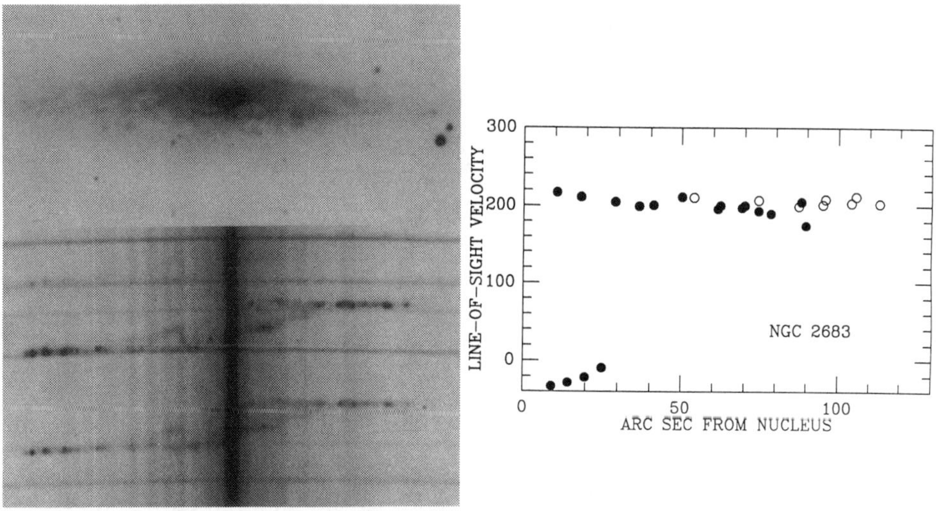

Fig. 2. (Left: image) NGC 2683 from a Kitt Peak 36″, Mould R filter + TI CCD frame; original scale 0.86″/pixel, 600 sec. (Left: spectrum) Hα and [NII] emission from major axis spectrum, inner 120″ only. KP 4-m; dispersion=0.33Å/pixel, scale=0.91″/pixel, 6000 sec. Note double valued velocities within 60″ of nucleus. (Right) Measured velocities; different symbols refer to different sides of major axis. Note that velocities remain nearly constant to within 10″ of nucleus.

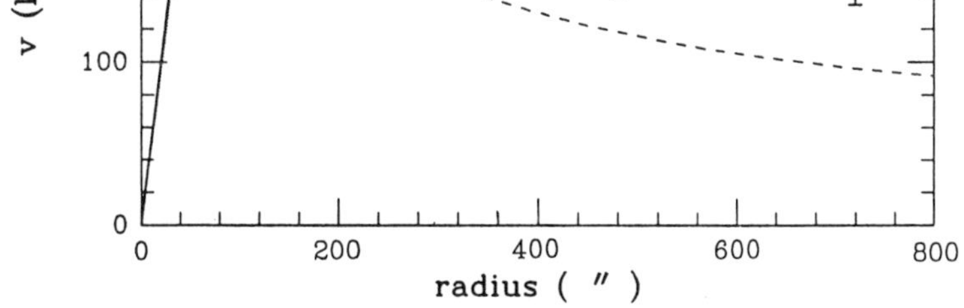

Fig. 3. HI velocities (Casertano and van Gorkom 1991) showing declining rotation curve in NGC 2683. The dashed line represents velocities for a disk with M/L=8; the solid line shows velocities predicted for disk+dark halo model. Horizontal line at small r indicates optical velocities. At large r, the rotation curve declines, but dark matter is still required.

DARK MATTER IN DWARF IRREGULARS

Very early in the study of the systematic properties of galaxy dynamics it was recognized that the galaxies of lowest luminosity might offer important clues concerning the distribution of dark matter. "Does the common form for the Sc rotation curves mean that all Sc galaxies, small or large, sit in generally similar potential wells, in which the optical galaxy defines only that region in which matter happens to be luminous? If we could observe beyond the optical image, especially for the smaller galaxies, would the velocities continue to rise?... Is the luminous matter only a minor component of the total galaxy mass?" (Rubin et al. 1980a).

Ten years later, a spectacular answer is available. There exist dwarf irregular galaxies which are very gas rich, and in which the dark mass exceeds the luminous mass by an enormous amount. The dwarf DDO 154, discovered by Krumm and Burstein (1984), has a ratio $M_{HI}/L_B \geq 5.4$, making it one of the most gas rich dwarfs known (Carignan and Freeman, 1988; Carignan and Beaulieu 1989). The HI has been detected extending 5 times the Holmberg radius (Fig. 4), and is dynamically more important than the stellar disk. The slow rise of velocity with radius precludes a high mass disk; the luminous disk is only a small fraction of the mass. Interior to the final measured point, 90% of the mass is dark, and integral $M/L \geq 75$. This is about 10 times that of a normal higher luminosity field spiral. Note however, that even with the enormous fraction of dark matter, the absolute amount is less than that for a higher luminosity spiral. The very high fraction of dark matter in low luminosity dwarf galaxies is one of the major discoveries of the past ten years. It leads naturally to the yet unanswered question "Is there a population of galaxies that is totally dark?"

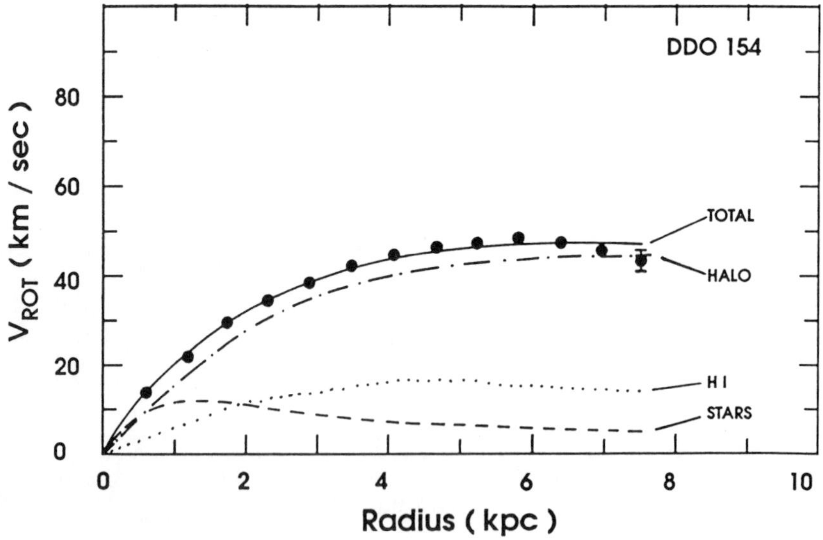

Fig. 4. Mass model for DDO 154 from Carignan and Beaulieu (1989). The stellar disk has $M/L_B = 1$, and a mass of 5.0 x $10^7 M_\odot$; the HI mass is 2.7 x $10^8 M_\odot$. The dark plus luminous mass to the last measured point is 3.8 x $10^9 M_\odot$.

ENVIRONMENTAL EFFECTS ON SPIRAL DYNAMICS

Fifty years ago, Hubble believed that a galaxy evolved in splendid isolation, living out its life constrained to the morphology it acquired at birth. Astronomers now recognize that a galaxy is a continuously evolving structure, which will acquire stars or lose stars through gravitational interactions, will acquire gas or lose gas through infall or galactic winds, and will be actively forming stars or quiescent depending upon its most recent history. Some evidence of this complex evolutionary history must be contained both in the present galaxy morphology and the galaxy dynamics.

Morphological differences between field and cluster spirals are observed. Galaxies in clusters are segregated by morphological type. Spirals in the Virgo cluster core have HI disks which are deficient, asymmetrical, and truncated in the direction of M87, yet the CO gas, denser and more difficult to strip, appears normal. Thus while morphological properties imply a complex evolutionary history for cluster spirals, dynamical history is harder to decipher.

For 8 Virgo spirals, Chincarini and de Souza (1985) conclude that rotation properties are normal. Yet from 15 spirals in 4 clusters, Whitmore et al. (1988) find that rotation curves for spirals within 1 Mpc of the cluster centers have rotation curves which are flat or falling; galaxies near the periphery or in the field have rotation curves which are flat or rising. NGC 7793, an Sd galaxy in the Sculptor group, also has a declining rotation curve (Carignan & Puche 1990). These tantalizing results need more evidence.

Galaxies in small compact groups live in an even more treacherous environment. Some of the galaxies in the Hickson (1982) compact groups (Rubin et al. 1991) are tidally interacting and distorted; some appear morphologically normal. Observed velocity patterns range from too peculiar to form rotation curves (1/3 of spiral sample), to abnormal (1/3), to normal (1/3); peculiar rotation curves are only loosely coupled with peculiar morphology. There is a surprisingly good correlation of luminosity with (log) maximum rotation velocity, i.e., the Tully-Fisher relation, particularly for galaxies with normal rotation curves.

For individual galaxies, values of integral M/L_B are low (median = 3.5), slightly lower than for field spirals of equivalent Hubble type. Historically, values of M/L_B ratios determined from the application of the virial theorem for compact groups are at least an order of magnitude higher than M/L_B ratios determined for individual group spirals from their rotation velocities. Hence the dark matter that produces the large group M/L_B is not clumped about each spiral, and is likely to be a common intergroup background.

In the dense cluster environment, it is expected that dark halos are modified, or may have not been permitted to form. Computer simulations (White 1978, Barnes 1989) suggest that galaxies or compact groups will merge into a single object within a few ($t \sim 10^8$ years) orbital periods. Hence dynamical records may reveal both evolutionary histories and the dark matter parameters. However, modifications due to the cluster environment appear subtle, so untangling these records has been difficult. That's what the future is for.

SYSTEMATIC PROPERTIES OF ROTATION CURVES

Now that rotation curves of spirals are accumulating at a fairly rapid rate, it will be possible to learn more about their systematic properties. Like the Sc's and Sb's studied earlier, the Sa's too show (Fig. 5; Rubin et al. 1985) an increase in radius and in rotation velocity as a function of absolute magnitude. Moreover, the forms of the rotation curves vary; low luminosity spirals reach their maximum velocities only near their outer radii, while high luminosity spi-

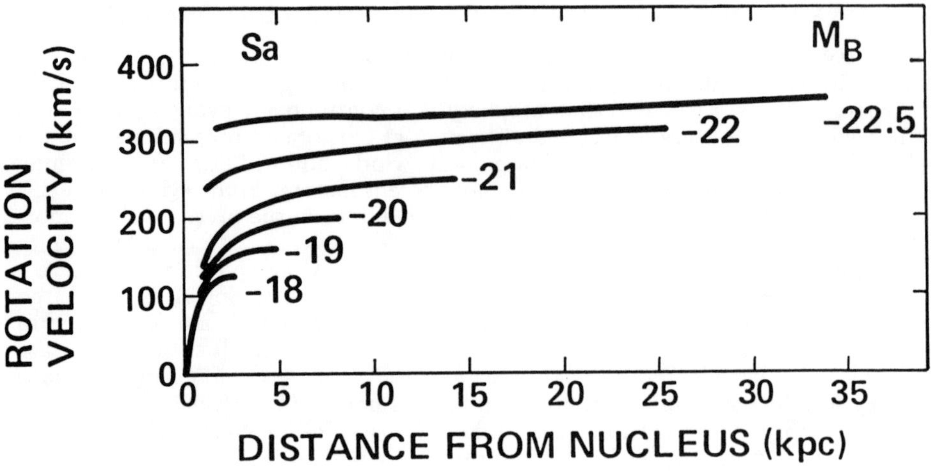

Fig. 5. Synthetic rotation curves (Rubin et al. 1985) for Sa galaxies formed from observations of 16 Sa with a wide range of luminosities. With increasing luminosity, galaxies get bigger, rotation velocities get higher, and the turnover occurs at a smaller fraction of the galaxy radius.

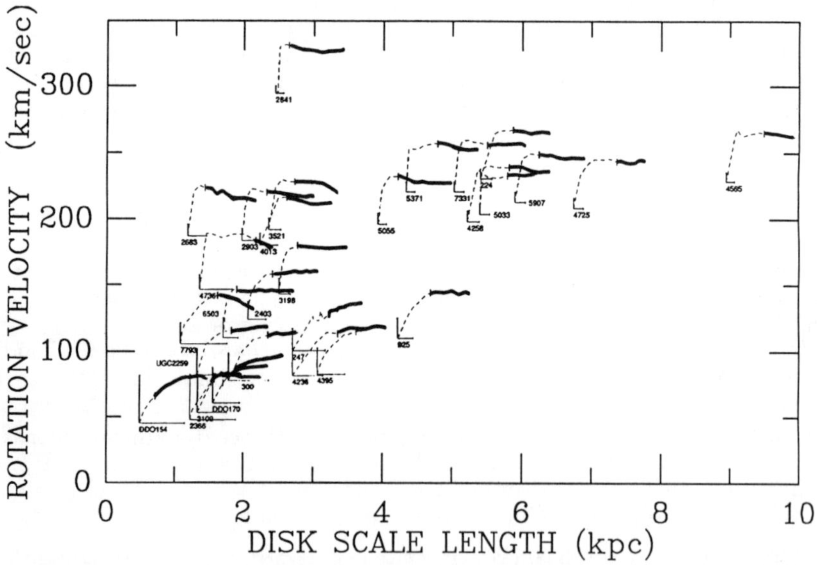

Fig. 6. Extended rotation curves for 28 spirals (Casertano and van Gorkom 1991) plotted (rescaled) as a function of disk scale length and maximum rotation velocity. The thicker line identifies that fraction of the rotation curve beyond 2/3 of the optical image. Note the falling rotation curves for the compact luminous (upper left) galaxies.

rals reach their maximum velocities just beyond their nuclei. For Sc galaxies, Rubin et al. (1980b) suggested that the radius where the velocity reaches 100 km/sec, in units of the optical radius, is an accurate luminosity discriminator. Early discussions of these systematic properties (Rubin 1983) led to a characterization of the mass forms of galaxies (Burstein and Rubin 1985) which has been controversial.

However, recent work by Distefano et al. (1990) has supported the earlier mass characterizations, suggesting that it may be profitable to continue similar analyses. Persic and Salucci have made numerous studies of systematic parameters of spiral dynamics; their recent parameterization (1990) of rotation curves nicely quantifies the systematic variations with luminosity. The most ambitious study is that of Casertano and van Gorkom (1991), which combines extended rotation curves with a luminosity parameter, the disk scale length, for 28 spirals. A plot of rotation curve forms on the Vmax, disk scale length plane (Fig. 6) offers a graphic look at the variation of forms of rotation curves as a function of these two parameters. These works are surely just the start of numerous compilations which will teach us more about the interplay of luminous matter and dark matter in spiral galaxies.

CONCLUSIONS

With the assumption that Newtonian gravitational theory is valid, dark matter is necessary to account for the flat or rising, and now even falling, rotation curves observed in spiral galaxies. The distribution and amount of the dark matter are not well constrained, although a few facts have emerged during the last decade. Dark matter becomes more dominant in galaxies of low luminosity, where it exceeds the luminous matter at virtually all radii. It is probable that totally dark galaxies exist. For almost no galaxy can we point to the "edge" of the mass. Thus rotation curves offer only lower limits to the masses and M/L values for spiral galaxies.

I thank Drs. Carignan, Kormendy, and Casertano for permission to reproduce figures from their papers.

REFERENCES

Athanassoula, E., Bosma, A., Papaioannou, S. 1987, *A&A* **179**, 23.
Barnes, J. E. 1989, *Nature* **338** ,123.
Bertola, F., Rubin, V. C., Zeilinger W. W. 1989, *ApJ Letts* **345**, L29.
Bosma, A. 1978, Ph.D. dissertation, Groningen.
Binney, J. & Tremaine, S. 1987, *Galactic Dynamics*, Princeton University Press.
Burstein, D. & Rubin, V. C. 1985, *ApJ* **297**, 423.
Carignan, C. 1987, in *Dark Matter in the Universe*, IAU Symp. 117 ed. Kormendy & Knapp (Reidel:Dordrecht), p. 135.
Carignan, C. & Beaulieu, S. 1989, *ApJ* **347**, 760.
Carignan, C. & Freeman, K. C. 1988, *ApJ Letts* **332**, L33.
Carignan, C. & Puche, D. 1990, *AJ* **100**, 394.
Carney, B. W., Latham, D. W., Laird, L. B. 1988, *AJ* **96**, 560.
Casertano, S. & van Gorkom, J. H. 1991, in press.
Chincarini, G. & de Souza, R. 1985, *A&A* **153**, 218.
Ciardullo, R., Rubin, V. C., Jacoby, G. H., Ford, H. C., Ford, W. K. 1988, *AJ* **95**, 438.
Distefano, A., Rampazzo, R., Chincarini, G., de Souza, R. 1990, *A&A*, in press.

Gurwell, M. & Hodge, P. 1990, *PASP* **102**, 849.
Hickson, P. 1982, *ApJ* **255**, 382.
Hodge. P. 1974, *ApJ* **192**, 27.
Kent, S.M. 1984, *Ap J Suppl* **56**, 105.
Kent, S.M. 1986, *Ap J* **91**, 1301.
Kent, S.M. 1988, *Ap J* **96**, 514.
Kormendy J. 1988, in *Origin, Structure, and Evolution of Galaxies*, ed. Fang Li Zhi (World Scientific: Singapore), p. 252.
Kormendy J. 1990, in *Evolution of the Universe of Galaxies*, ed. R.G. Kron. (ASP: San Francisco), p. 33.
Krumm, N. & Burstein, D. 1984, *AJ* **89**, 1319.
Lake, G. & Feinswog, L. 1989, *AJ* **98**, 166.
Leonard, P. J. T. & Tremaine, S. 1990, *ApJ* **353**, 486.
Oort, J. H. 1921, *BAN* **1**, 133.
Persic, M. & Salucci, P. 1990, *ApJ* in press.
Rubin V. C. 1983, in *Internal Kinematics and Dynamics of Galaxies, IAU Symp. 100*, ed. Athanassoula, Reidel:Dordrecht, p. 3.
Rubin V. C., & Graham, J. A. 1987, *ApJ Letts* **316**, L67.
Rubin, V. C., Burstein, D., Thonnard, N. 1980b *ApJ Letts* **242**, L149.
Rubin, V. C., Ford, W. K., Thonnard, N. 1980a *ApJ* **238**, 471.
Rubin, V. C., Hunter, D. A., Ford W. K. 1991, *ApJ Suppl* in press.
Rubin, V. C., Burstein, D., Ford, W. K., Thonnard N. 1985, *ApJ* **289**, 81.
Rubin, V. C., Kenney, J. D., Boss, A. P., Ford, W. K. 1989, *AJ* **98**, 1246.
Valentijn, E. A. 1990, *Nature* **346** 153.
van Albada, T. S., Bahcall, J. N., Begeman, K., Sancisi, R. 1985, *ApJ* **295** 305.
White, S. D. M. 1978, *MNRAS* **184**, 185.
Whitmore, B. C., Forbes, D. A., Rubin, V. C. 1988, *ApJ* **333**, 542.

Discussion

Casertano: Your point of the uncertainty of the disk mass derived from the inner rise of the rotation curve is well taken. However, in the case of NGC 2683 (Casertano and van Gorkom 1991), the decline in rotation velocity outside the optical disk offers some leverage on the disk mass. We find that the luminous mass must be at least *half* the maximum mass to explain the velocity decline. Unless, of course, the dark matter in NGC 2683 is truncated, which would make this galaxy truly unusual.

CONSTRAINTS ON PROPERTIES OF DARK HALOS

Keith M. Ashman
Space Telescope Science Institute
3700 San Martin Drive
Baltimore MD 21218

ABSTRACT

The requirement that dark galactic halos virialize by the present epoch gives a lower limit to the mean mass density within such objects. In conjunction with observations of HI rotation curves and the virial theorem, this constraint can be used to limit the radial extent and the mass of dark halos. We can use this constraint to calculate the maximum contribution to the density of the Universe made by such halos. Reversing the argument allows us to obtain an upper limit on the formation epoch of dark halos. We find that the halos of spirals formed more recently than a redshift of around 2 to 4 and contribute less than a tenth of the critical density required to close the Universe. Stronger limits are obtained by using estimates of the degree of dissipational gas collapse within dark halos along with observed disk scale-lengths.

I. THEORETICAL RELATIONS

We make the usual assumption that dark halos begin as density fluctuations which turn around from the Hubble flow at a redshift z_t and time t_t. At this epoch such fluctuations have an overdensity relative to the background *critical* density of about 5.5. Assuming that the subsequent collapse is dissipationless, the fluctuation virializes after a further time around $2t_t$. The virial radius of the halo R is about half that of the radius at maximum expansion.

For an $\Omega = 1$ model, these quantities are related to the halo mass M by

$$R \approx 170 h^{-0.67} \left(\frac{M}{10^{12} M_\odot} \right)^{0.33} (1+z_v)^{-1} \mathrm{kpc} \tag{1}$$

(*e.g.* Rees and Ostriker 1977) where z_v is the redshift of virialization and $h = H_0/100$ km s^{-1} Mpc^{-1}. Using the virial theorem in the form $V^2 \approx GM/R$ gives

$$(1+z_v) \approx 3.5 h^{-0.67} \left(\frac{M}{10^{12} M_\odot} \right)^{-0.67} \left(\frac{V}{300 \mathrm{kms}^{-1}} \right)^{2} \tag{2}$$

or, equivalently,

$$(1+z_v) \approx 2.1 h^{-0.67} \left(\frac{R}{100 \mathrm{kpc}} \right)^{-0.67} \left(\frac{V}{300 \mathrm{kms}^{-1}} \right)^{0.67} \tag{3}$$

The most important constraint that we obtain from the above equations is an upper limit on the virialization epoch of dark halos. This is achieved by substituting into equation (4) the observed asymptotic rotation velocity for V and the galactocentric distance of the last measured point of the rotation curve for R. Provided we restrict our attention to flat or falling rotation curves, this will overestimate z_v. This is because using the observed asymptotic value of V for the virial velocity corresponds to the assumption that the dark halo is a singular isothermal sphere. The presence of a core or a density fall-off that is more rapid than R^{-2} leads to a lower mean density and therefore a lower value of z_v.

Inverting equations (3) and (4) gives the mass and radial extent of dark halos in terms of the epoch of virialization. Explicitly, we find

$$M \approx 2.2 \times 10^{12} h^{-1} \left(\frac{V}{300\mathrm{kms}^{-1}}\right)^3 (1+z_v)^{-1.5} M_\odot \tag{4}$$

and

$$R \approx 330 h^{-1} \left(\frac{V}{300\mathrm{kms}^{-1}}\right) (1+z_v)^{-1.5} \mathrm{kpc} \tag{5}$$

Substituting $z_v = 0$ in equations (5) and (6) gives the maximum mass and radius of a virialized dark halo. We will denote these quantities by M_{max} and R_{max} respectively. The quantity R_{max} is not of great importance, although it does provide useful constraints on schemes involving very extended halos. The related quantity M_{max} is rather more useful since it gives an upper limit on the total mass-to-light ratio of a given galaxy. This can be converted into a constraint on the contribution to the density of the Universe made by such halos. This is addressed in the next Section.

There are two drawbacks to the technique described above. Firstly, it is apparent that not all halos are virializing at the present epoch. More realistically, one could use a value of z_v predicted by a particular cosmological model in equations (4) and (5) to calculate M and R. Secondly, the derived values of $(z_v)_{max}$ are really a reflection of how far each rotation curve has been measured. This makes it difficult to draw conclusions about a characteristic value of $(z_v)_{max}$ and the variation in z_v from one galaxy to another.

Fortunately, one can obtain an independent estimate of the halo radius which can be used in equation (3) to get a more reasonable value for z_v. The idea is to use the fact that the protogalactic gas and the dark halo are expected to extend to the same radius of maximum expansion before turning around from the Hubble flow. Under these conditions, Fall and Efstathiou (1980) have shown that the gas must collapse in radius by a factor F around 10 in order to explain the high angular momentum of spiral disks. This leads to the prediction that the disk scale-length α^{-1} and the halo radius should be related by the simple expression

$$F \approx \alpha R_h \tag{6}$$

Adopting $F = 10$ leads to a direct estimate of the halo radius. Using equations analagous to those given above then constrains the mass and virialization epoch.

We again obtain upper limits to these quantities, since the assumption of an isothermal sphere tends to overestimate mass and formation redshift. However, the uncertainty in the value of F means that the limits are less reliable than those obtained using the virial condition.

II. RESULTS FROM OBSERVED ROTATION CURVES

We are now in a position to obtain limits on individual galaxies from their measured rotation curves. In order to do this we have compiled a sample of over 20 spiral galaxies from the literature. A list of these galaxies is given in a separate paper (Ashman 1991). The only criterion for inclusion of a galaxy is that it has a reasonably flat HI rotation curve so that the assumptions of Section I are valid. Wherever possible we have converted the data of the original sources to a Hubble constant of $h = 0.75$. In some cases, the distances are based on indicators other than recessional velocity, so that some inconsistency may be introduced.

(a) The Epoch of Halo Virialization

The requirement that the halos of spirals have virialized leads to the result that most halos virialized more recently that a redshift of 4 and some virialized at redshifts less than 2. Using a collapse factor of 10 to obtain R_h for each galaxy supports relatively recent formation epochs. We find a mean value of $\langle z_v \rangle \approx 2.7$, although, as noted above, this may overestimate the true value. This number falls further if we demand that the value of R_h obtained from equation (6) exceeds the maximum observed radius of rotation curves, which it clearly must do, but which is not the case for all our galaxies. This suggests that a collapse factor exceeding 10 is appropriate for some of the galaxies.

(b) Halo masses and radii

Since the galaxies in our sample are chosen simply for having extended rotation curves it is unwise to draw general conclusions about typical halos. However, our results do indicate that the virialized halo of an L_* galaxy must have $R_h < 200$kpc and $M_h < 3\times10^{11} M_\odot$ for $h = 0.75$. The estimates for these quantities based on the observed disk scale-lengths are $R_h \approx 30$kpc and $M_h \approx 1\times10^{11} M_\odot$, assuming that $F = 10$.

(c) Mass-to-light ratios

The virial constraint suggests that for most galaxies $M/L < 200$ with some galaxies showing markedly lower values. This corresponds to a maximum contribution to the density of the Universe by such halos of $\Omega_h \approx 0.1$. The estimate obtained by assuming $F = 10$ gives a mean value of $\langle M/L \rangle \approx 24$ which implies $\langle \Omega_h \rangle \approx 0.01$. Both methods give higher values for two dwarf galaxies, DDO 154 and DDO 170, in agreement with mass models of these objects (Carignan and Freeman 1988; Carignan and Beaulieu 1989; Lake *et al.* 1990). This is also consistent with the reported trend of an increasing dark matter fraction in low-mass spirals (Persic and Salucci 1988; Casertano and van Gorkom 1990).

III. DISCUSSION

The above ideas are a useful way of constraining the unobservable properties of dark galactic halos. One particularly interesting result is that the parameters of the rotation curve of the HI cloud reported by Giovanelli and Haynes (1990) imply that the surrounding halo has not yet virialized. This suggests to us that the HI measurements are not indicative of a true rotation curve at all and that the HI cloud is unlikely to be a protogalaxy. A more probable scenario is that there are two gas-rich dwarfs orbiting each other. This interpretation is consistent with the observed HI distribution which shows two distinct peaks.

The limits on the mass-to-light ratios obtained above allow for the possibility that all the halo dark matter is baryonic, even if one accepts standard primordial nucleosynthesis calculations. These results also suggest that the halos of normal spirals cannot provide the dark matter densities characteristic of groups and clusters. Perhaps the dark halos of ellipiticals are significantly more massive than those around spirals. Another possibility, which has some observational support, is that dwarf galaxies have much higher dark matter fractions. Maybe the smallest galaxies are completely dark? A forthcoming paper addresses these issues in more detail.

This work was supported by an ESA postdoctoral fellowship.

REFERENCES

Ashman, K.M. 1991, *Ap.J.,* to be submitted.
Carignan, C., and Beaulieu, S. 1989, *Ap.J.,* **347**, 760.
Carignan, C., and Freeman, K.C. 1988, *Ap.J.,* **332**, L33.
Casertano, S., and van Gorkom, J.H. 1990, preprint; see also these proceedings.
Fall, S.M., and Efstathiou, G. 1980, *M.N.R.A.S.,* **193**, 189.
Giovanelli, R., and Haynes, M.P. 1989, *Ap.J.,* **346**, L5.
Lake, G., Schommer, R.A., and van Gorkom, J.H. 1990, *A.J.,* **99**, 547.
Persic, M., and Salucci, P. 1988, *M.N.R.A.S.,* **234**, 131.
Rees, M.J., and Ostriker, J.P. 1977, *M.N.R.A.S.,* **179**, 541.

DARK MATTER, MERGERS, AND MORPHOLOGY: CLUES FROM GALAXY PAIRS

Jane C. Charlton
Steward Obs., Univ. of Arizona, Tucson, AZ 85721
and
Edwin E. Salpeter
Space Sciences Bldg., Cornell Univ., Ithaca, NY 14853

ABSTRACT

We have examined the distributions of radial velocity differences Δv and projected separations r_p of pairs in low density regions of the CfA and SSRS redshift catalogs. Bound pairs exist even at relatively large projected separations (up to 1Mpc), indicating large amounts of dark matter around galaxies. The distribution of r_p for bound pairs is flat, with significant numbers of pairs at projected separations less than 50kpc. Pairs at small separations merge rapidly, but the distribution is replenished by those originally at larger separations that are being drawn toward merger by dynamical friction. As a result of this merger/replenishment process, at least 20% of galaxies in low density regions are the product of a merger; thus it seems that nearly all ellipticals and S0's in these regions are created in this way.

INTRODUCTION

The dynamics of galaxy pairs may serve as a probe of the distribution of dark matter around the galaxies. We have studied the properties of galaxies in the CfA (Huchra, *et al.* 1983) and SSRS (Southern Sky Redshift Survey) redshift catalogs. This paper is a brief summary of certain aspects of this work which is described in detail in Charlton and Salpeter (1991). Here we shall discuss an estimate of the galaxy merger rate in low density regions, and its consequences for the origin of morphology.

HISTOGRAMS OF PAIR PROPERTIES

Many previous pair studies have focused on isolated binaries, with samples having a bias toward small separations due to the use of an angular separation catalog. Magnitude or diameter limited redshift catalogs, such as CfA and SSRS, allow us to examine unbiased distributions in radial velocity difference Δv and projected separation r_p of pairs. Figures 1 and 2 display cuts through the two-dimensional distribution function of Δv and r_p, for pairs with primary members in low density regions of the redshift catalogs. A primary galaxy must be far enough from a catalog edge that a local density (number of galaxies within $\Delta v < 375$km/s and $r_p < 5$Mpc corrected for catalog incompleteness using the selection function) can be determined. If this local density is below the median, the primary is considered to be in a low density region and all pairs containing it will be included in the histograms. Only about half of the pairs displayed in the figures are isolated within 0.75Mpc in r_p and 300km/s in Δv of either pair member, and many of these will have fainter neighbors that are not included in the magnitude limited catalog. The lower, dotted, histograms include only the isolated pairs.

In Fig. 1a-b, we notice an excess of pairs with $\Delta v \lesssim 150$km/s, even for separations as large as 1Mpc. At least for separations smaller than ~ 0.5Mpc, the dispersion time is less than a Hubble time, so that any initial correlations between pairs would be destroyed unless the pairs were bound. This already suggests large amounts of dark matter in the vicinity of the galaxies. We focus on these "bound" pairs with $\Delta v < 150$km/s in Fig. 2a, where we examine their distribution in projected separation. This histogram is nearly flat, meaning there are roughly equal numbers of pairs separated by 0-100kpc and by 900-1000kpc, although the "area" of the latter bin is much larger than that of the former. (A random distribution of pairs would yield a distribution linearly increasing with r_p, as in Fig. 2b which presents only those pairs with relatively large Δv.) In fact, we find substantial numbers of pairs separated by less than 25kpc (still no feature or deviation from a flat distribution). We shall now examine some of the implications of these histograms.

DARK MATTER HALOS AND MERGERS

HI rotation curves of spiral galaxies have demonstrated that isothermal dark matter halos extend out to at least 30kpc in some cases (Rubin *et al.* 1985). However, if a pair of galaxies is separated by less than twice the halo radius dynamical friction of the overlapping dark matter halos will lead to merger within a few orbital periods. We can now rule out the simple model in which orbits are circular and all dark matter halos cut off abruptly at some small radius R_h (less than a few hundred kpc). In such a model we would expect all pairs separated by less than $2R_h$ to merge rapidly, depleting the r_p distribution at small values, while the distribution at larger separations remains unchanged. This is not in agreement with the observed distribution in Fig. 2a. We conclude that pairs at small separation must merge rapidly, however pairs originally at larger separations are destined to replenish the distribution (similar to an idea discussed in Ostriker and Turner 1979). This argument will only apply out to separations where the orbital periods are smaller than a Hubble time. In addition, any feature in the r_p distribution will be smoothed in the likely instance that different galaxies have different halo sizes.

Although the above argument (and the existence of bound pairs at relatively large r_p) indicates that there are large amounts of dark matter surrounding galaxies, we do not mean to imply that it is distributed in the form of spherical, isothermal halos about individual galaxies. In fact the lack of correla-

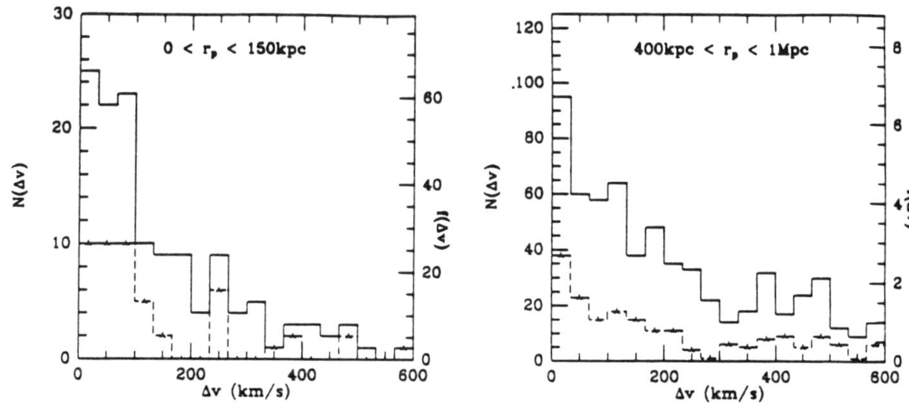

Figure 1: The histogram of radial velocity differences for pairs containing low density primaries. The scale on the left hand side is the actual double counted number labeled as $N(\Delta v)$, and the right hand side $f(\Delta v)$ is the normalised distribution obtained by dividing by the number of pairs expected in each bin if the mean number of galaxies in the 5Mpc density determination cylinder were spread out randomly over its volume. The lower, dotted histograms exclude all pairs that are not isolated within 0.75 Mpc and 300 km/s of either member.

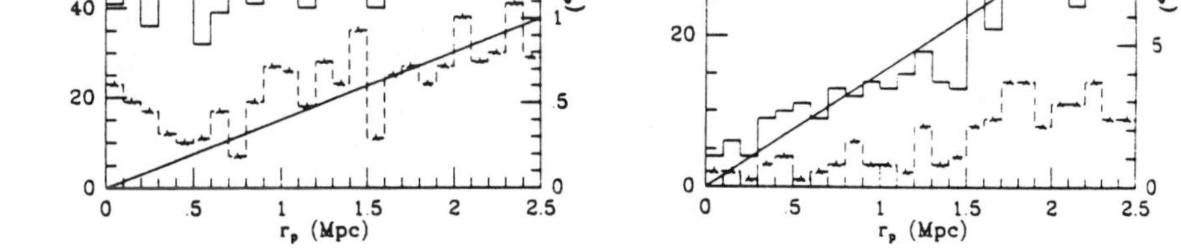

Figure 2: The distribution of projected separations r_p for pairs with low density primaries in various bins of Δv. The subset of pairs that are isolated within 0.75Mpc and 300km/s of both members are displayed in the dotted histograms. The left hand scale is the actual double counted number of galaxies $N(r_p)$, while the right scale $g(r_p)$ is normalised by the number expected in the right-most bin (2.4-2.5Mpc) if the mean number of galaxies in a density determination cylinder were spread uniformly over the volume. Solid lines represent this background level for all bins.

tion between Δv and both r_p and pair luminosity (as described in Charlton and Salpeter 1991) suggests that the pair dynamics are not governed by the masses of individual galaxies, but rather by the total mass of an irregular distribution of dark matter that surrounds them.

In the context of the merger/replenishment scheme, it is possible to estimate the galaxy merger rate among catalog pairs. Assuming a stationary distribution (constant current) of r_p, the number of mergers in the past Hubble time can be estimated by counting the pairs expected to merge within the next Hubble time. We define an "effective, orbit-averaged" rate of change of pair separation $\dot{R}$, which can be estimated as the typical orbital velocity divided by the number of decay orbits before merger N_{dk}. N-body simulations (White 1978; Carlberg 1982) show that a binary undergoes only a few orbits before merging, so that $N_{dk} < 5$. This yields the result that $\sim 10(5/N_{dk})\%$ of low density region galaxies have undergone a merger over the past Hubble time. In order to include all mergers of pairs with mass ratios smaller than four, we apply a correction factor (=2) to account for the less massive pair members that are often fainter than the catalog magnitude limit. Thus at least 20% of low density region primary galaxies are the product of a merger. The majority of the low density region galaxies are spirals, with 14% S0's and 5% ellipticals, and only a few irregular systems. In order to account for all the merger products, it seems that nearly all of the S0's and ellipticals in low density regions must have formed from mergers.

We have also examined the r_p histograms for pairs in the high density regions, and conclude that there are not many mergers occurring there at present. Thus the formation mechanism for E's and S0's in clusters and dense groups need not be the same as in low density regions. This leads to the prediction that E's and S0's in dense regions are, on the average, older than those in "the field". In the low density regions, we expect that half of the E's and S0's have formed over the past half a Hubble time, so that the fraction of these types should change as a function of redshift.

We gratefully acknowledge support from NSF grants AST87-14475 and AST86-57467 at Cornell Univ. and from AST88-22297 at Steward Obs. of the Univ. of Arizona.

REFERENCES

Carlberg, R. G. 1982, MNRAS, 199, 1159.

Charlton, J. C., and Salpeter, E. E. 1991, ApJ, submitted.

da Costa, L. N., *et al.* 1988, ApJ, 327, 544.

Huchra, J. P., Davis, M., Latham, D., and Tonry, J. 1983, ApJS, 52,89.

Ostriker, J. P., and Turner, E. L. 1979, ApJ, 234, 785.

Rubin, V. C., Burstein, D., Ford, W. K., and Thonnard, N. 1985, ApJ, 289,81.

White, S. D. M. 1978, MNRAS, 184, 185.

ULTRAMASSIVE DARK OBJECT IN THE LUMINOUS INFRARED GALAXY NGC 6240 ?

JONATHAN BLAND-HAWTHORN
Dept. of Space Physics and Astronomy, Rice University, TX 77251

ANDREW S. WILSON
Astronomy Program, University of Maryland, MD 20742

ABSTRACT

Fabry-Perot Hα observations of the superluminous infrared galaxy NGC 6240 have provided the first kinematic evidence that this system comprises at least two merging disks. One of the disks appears to be strongly associated with the main body of the 2μm continuum; the other disk displays remarkable kinematics and has no obvious counterpart in the continuum.

CONTEXT

During the past decade, observational evidence from all wavebands has shown that the chaotic appearance of NGC 6240 is the result of a galactic collision (Fosbury and Wall 1979). When IRAS uncovered the existence of a class of galaxies that emit the bulk of their luminosity at far-infrared wavelengths (Soifer *et al.* 1984), it was soon realized that NGC 6240 is one such object (Wright, Joseph and Meikle 1984). The origin of the enormous infrared luminosity ($\sim 6\times 10^{11}$ h_{75}^{-2} $L_\odot$)* remains highly controversial. Opinions differ on the relative roles of large-scale shocks (Harwit *et al.* 1987), massive star formation (Soifer *et al.* 1984), or a "buried" active nucleus (Becklin and Wynn-Williams 1986) in heating the dust. The weakness of the Pα line in relation to the infrared luminosity (Depoy, Becklin and Wynn-Williams 1986) suggests that the fraction of hot stars is considerably lower than in normal star formation regions within the Galaxy. The nature of the strong vibration-rotation H_2 lines from NGC 6240 has been widely discussed (e.g. Harwit *et al.* 1987; Draine and Woods 1990). Recent 2μm imagery reveals that the H_2 is extended and distinct from the two continuum nuclei which would seem to favor large-scale shocks as the excitation mechanism (Herbst *et al.* 1990). This inference is supported by recent simulations (Barnes and Hernquist 1990) which demonstrate that the observed nuclear gas concentration (Wang, Scoville and Sanders 1990) can only accumulate after dissipating its orbital energy through strong radiative cooling in shocks.

OBSERVATIONS

We have obtained a detailed Hα velocity map of the superluminous ($\sim 6\times 10^{11} L_\odot$), infrared galaxy NGC 6240 (z = 0.0245) with the Hawaii Imaging Fabry-Perot Interferometer (HIFI - Bland and Tully 1989). In Fig. 1, the kinematic data reveal two dynamical systems (disks 1 and 2) that exhibit radically different rotation and are closely spaced in velocity ($\approx$70 km s^{-1}) and position ($\approx 12''$, $\approx$5.7

* A distance of 100 Mpc is adopted throughout; h_{75} is Hubble's constant in units of 75 km s^{-1} Mpc^{-1}.

kpc). Disk 1 (P.A. = 45±5°) is roughly aligned with the major axis of the near-infrared continuum and exhibits flat rotation out to ~20″ in radius, centered on the double nucleus seen at optical, near-infrared and radio wavelengths. The rotation turns over at $r_{t1} \approx 7.2''$ with a peak-to-peak velocity amplitude $\approx 280/\sin i_1$ km s^{-1} where i_1 is the disk inclination. A preliminary analysis suggests $i_1 \simeq 70\pm7°$ although the effective beam of the filtered HIFI data (4″ FWHM) makes a reliable determination difficult.

The structure of disk 2 (P.A. = 155±3°) is much more remarkable. The rotation curve comprises an unresolved, or marginally resolved, central velocity gradient with a peak-to-peak amplitude of $\approx 800/\sin i_2$ km s^{-1} within $r_{t2} < 2.5''$, and a faster than Keplerian drop-off outside r_{t2}. The peak rotation implies a compact mass M_2 greater than $4.5\times10^{10} M_\odot/\sin^2 i_2$ within a radius <1.2 kpc. Presently, there are no observations of a luminous, compact counterpart to the core of disk 2 at radio, millimeter, infrared or x-ray wavelengths although our data show evidence for a weak optical continuum. We derive strict lower limits on the mass-to-light ratio from UBVRIzK photometry, in particular, $M_2/L_K >$ 3 ($M_\odot/L_\odot \sin^2 i_2$) and $M_2/L_R >$ 60 ($M_\odot/L_\odot \sin^2 i_2$).

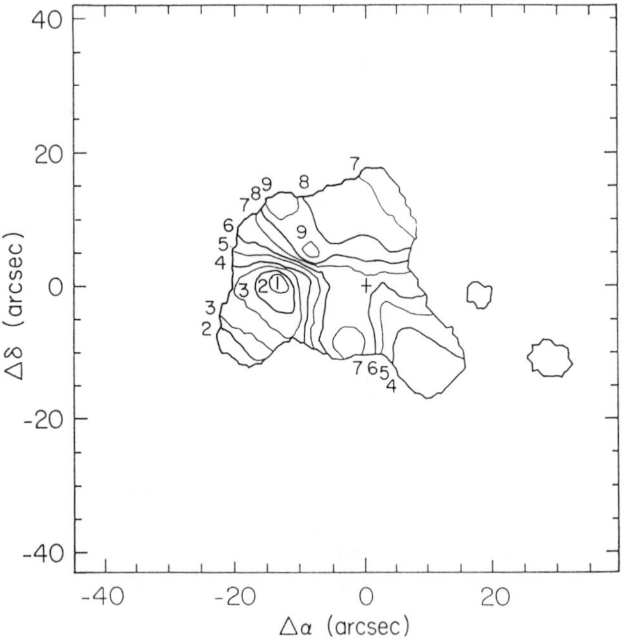

Figure 1. The Hα velocity field for an 85″×85″ field of view (north at top) sampled at 0.85″ increments. The outer boundary defines where the Hα surface brightness falls below 2 × 10^{-17} erg cm^{-2} s^{-1} arcsec^{-2}. The origin of disk 1, marked by a cross, coincides with the brighter peak of the compact, double nucleus; the core of disk 2 lies ≈12″ to the east. In order to increase the signal-to-noise ratio, the data have been convolved spatially with a 4″ FWHM gaussian beam. The spectra were sampled at 68 km s^{-1} increments over a 820 km s^{-1} velocity range. The velocity contours in km s^{-1} are: (1) 7120, (2) 7170, (3) 7220, (4) 7270, (5) 7320, (6) 7370 (systemic for disk 1), (7) 7420, (8) 7470, (9) 7520.

DISCUSSION

We are unable to find a luminous counterpart to the core of disk 2 in high resolution maps from the radio (Carral, Turner and Ho 1990), millimeter (Wang, Scoville and Sanders 1990), near-infrared (Herbst *et al.* 1990) or x-ray (Rieke 1988) bands. Conceivably, the large IRAS beam ($0.76' \times 4.6'$) could hide a bright, compact source at the center of disk 2, although the listed IRAS position is very close to the center of disk 1 and the core of disk 2 lies close to the 95% confidence position ellipse (Fullmer and Lonsdale 1989). Wang, Scoville and Sanders (1990) have observed NGC 6240 in the CO λ2.6mm line and find $\sim 1 \times 10^{10} M_\odot$ of molecular gas concentrated within a diameter of roughly 8″ at the center of disk 1; there is no evidence of CO emission at the core of disk 2. If the dust properties in NGC 6240 are similar to the Galaxy, then the visual extinction within a beam α in arcseconds is $A_V \sim (R/\alpha^2)(M_7/M_\odot)$ mag where M_7 is the mass of gas in units of $10^7 M_\odot$ and R is the selective extinction coefficient. If no more than $\approx 1 \times 10^9 M_\odot$ (3σ upper limit) could have escaped detection within the 7″ beam and we conspire to place all of this gas in front of the core of disk 2, the visual extinction could be as high as 12 mag although this would imply a value of A_V an order of magnitude higher towards the center of disk 1. Our estimate of A_V suggests, however, that the core of disk 2 could be significantly obscured at near-infrared wavelengths.

If the peak rotation is almost resolved at $r_{t2} \approx 2.5''$, the mass M_2 within this radius is $4.5 \times 10^{10} M_\odot / \sin^2 i_2$. If the isovelocity contours are matched to those of a planar disk in circular rotation, the derived inclination is $i_2 = 35° \pm 10°$. However, the confusion with disk 1 and the non-circular motions implied by the rapid fall in rotation invalidate this method for determining the inclination. Our estimate of M_2 assumes that the rotation curve reflects dynamical equilibrium inside r_{t2} and that the tidal radius of disk 2 extends beyond r_{t2}. The quadrupole moment from M_1 may account for the fast drop-off although this depends on the relative positions and the spin orientations of the disks with respect to their orbital angular momentum. Our strict upper limits on the K and R luminosities indicate that $M_2/L_K > 3\ (M_\odot/L_\odot \sin^2 i_2)$ and $M_2/L_R > 60\ (M_\odot/L_\odot \sin^2 i_2)$. The core of disk 2 is somewhat darker than the M/L_V limit for the nucleus of M87 (Sargent *et al.* 1978), and much more massive. A supermassive black hole coalescing with disk 1 is one possible interpretation; higher resolution data are needed to see whether the rotational velocity in disk 2 continues to increase towards smaller radii than our resolution. Other possibilities include compact, dark stellar systems (e.g. highly evolved or brown dwarf stars). If we assume virial equilibrium, the stellar velocity dispersion $< V_* >$ exceeds 800 km s^{-1} and gives a crossing time of $\tau_C \sim 3 \times 10^6$ yrs. For a homogeneous cluster of N_* stars near the thermonuclear threshold ($\approx 0.08 M_\odot$) with a dynamical cross-section σ_*, the relaxation time ($\tau_R \sim \tau_C N_*/8 \ln N_*$) and the coalescence time ($\tau_S \sim 1/N_* \sigma_* < V_* >$) are of order 10^{15-16} yrs. If the core condensed through strong tidal interactions with disk 1, Barnes and Hernquist (1990) have shown that only dissipative matter settles easily to the center, after shedding angular momentum through large-scale shocks. This would tend to rule out collisionless systems, in particular, stars or putative "dark matter". However, if the true core radius of the cluster is <100 pc and if the initial mass function is weighted strongly towards high mass stars, the relaxation times may be short enough in the core for matter to coalesce and form a massive black hole (Murphy, Cohn and Hut 1990).

The central engines of quasars are thought to be supermassive ($\geq 10^8 M_\odot$) black holes that radiate close to the Eddington limit. If one makes the usual association between radio-quiet quasars and powerful Seyferts (i.e. spirals), roughly a third of all local L_* galaxies should possess a supermassive black hole (Begelman, Blandford and Rees 1984). Thus, over half of all mergers now involve at least one spiral with a black hole. Stockton (1990) has reviewed observations supporting the notion that at least some quasars may be triggered by strong dynamical interactions. The light profile of the ultraluminous infrared galaxy Arp 220 obeys an $r^{1/4}$–law at $\lambda 2\mu$m (Wright *et al.* 1990) which presumably arises in the violent relaxation of the merging stellar systems (White 1979). If this is the ultimate fate of NGC 6240, then it may be that the dark core and the observed CO concentration will sink to the center, thus refueling the putative black hole and producing a short-lived quasar or a giant, radio elliptical galaxy. We expect high resolution observations with the NASA *Great Observatories* to determine the specific nature of the core of disk 2.

ACKNOWLEDGMENTS

JBH is partly funded through NSF grant AST 88-18900 that supports the HIFI "Nearby Active Galaxies" program.

REFERENCES

Barnes, J. E., and Hernquist, L. 1990, *Ap. J. (Letters)*, in press.

Becklin, E. E., and Wynn-Williams, C. G. 1986, in *Star Formation in Galaxies*, NASA CP 2466, Caltech, p.643.

Begelman, M. C., Blandford, R. D., and Rees, M. J. 1984, *Rev. Mod. Phys.*, **56** 255.

Bland, J., and Tully, R. B. 1989, *A. J.*, **98** 723.

Carral, P., Turner, J. L., and Ho, P. T. P. 1990, *Ap. J.*, **362** 434.

Depoy, D. L., Becklin, E. E., and Wynn-Williams, C. G. 1986, *Ap. J.*, **307** 116.

Draine, B. T., and Woods, D. T. 1990, *Ap. J.*, **363** 464.

Fosbury, R. A. E., and Wall, J. V. 1979, *M.N.R.A.S.*, **189** 79.

Fullmer, L., and Lonsdale, C. J. 1989, *Joint IRAS Sci. Work. Group*, JPL D-1932.

Harwit, M., Houck, J. R., Soifer, B. T., and Palumbo, G. G. C. 1987, *Ap. J.*, **315** 28.

Murphy, B. W., Cohn, H. N., and Hut, P. 1990, *M.N.R.A.S.*, **245** 335.

Rieke, G. 1988, *Ap. J. (Letters)*, **331** L5.

Sargent, W. L. W., Young, P. J., Boksenberg, A., Shortridge, K., Lynds, C. R., and Hartwick, F. D. A. 1978, *Ap. J.*, **221** 731.

Soifer, B. T., *et al.* 1984, *Ap. J. (Letters)*, **278** L71.

Stockton, A. 1990, in *Dynamics and Interactions of Galaxies*, ed. R. Wielen, Heidelberg, p. 440.

Wang, Z., Scoville, N. Z., and Sanders, D. B. 1991, *Ap. J.*, in press (Feb. 10 issue).

White, S. D. M. 1979, *M.N.R.A.S.*, **189** 831.

Wright, G. S., James, P. A., Joseph, R. D., and McLean, I. S. 1990, *Nature*, **344** 417.

Wright, G. S., Joseph, R. D., and Meikle, W. P. S. 1984, *Nature*, **309** 430.

SYSTEMATIC PROPERTIES OF DARK MATTER DISTRIBUTION

Stefano Casertano
University of Pittsburgh, Pittsburgh, PA 15260

J. H. van Gorkom
Columbia University, New York, NY 10027

The neutral hydrogen rotation curves of two galaxies observed at the VLA indicate that their rotation velocity declines outside the optical radius. These galaxies, NGC 2683 and NGC 3521, have maximum rotation velocities of about $200\,\mathrm{km\,s^{-1}}$, and the velocity decline is large, about 25%, between the optical radius and the last measured point (about two Holmberg radii). In order to explain the gravitational field of these galaxies, *some* dark matter is still needed, but much less than for galaxies with flat rotation curves.

We compare the optical and kinematic properties of these two galaxies with other systems with extended rotation curves. NGC 2683 and NGC 3521 turn out to be unusually small for their rotation velocity, with optical scale lengths of 1.2 and 2.4 kpc respectively ($H_0 = 75\,\mathrm{km\,s^{-1}\,Mpc^{-1}}$). Other galaxies with fast rotation velocity and small optical size also exhibit some degree of velocity decrease in the outer parts. The decline is absent or less pronounced in galaxies of larger optical size. On the other hand, dwarf galaxies have rotation curves that *rise* outside the optical region, and galaxies with intermediate rotation velocity ($\sim 150\,\mathrm{km\,s^{-1}}$) mostly have flat rotation curves.

These findings imply a systematic change in the properties of the dark matter. Dwarf galaxies must have a comparatively large amount of dark matter inside their luminous disk. For intermediate galaxies, dark and luminous matter are present in comparable amounts. In bright galaxies the luminous matter dominates, especially when it is concentrated in a small region (optical scale length < 3–4 kpc).

Our results are described in greater detail in a forthcoming publication (Casertano and van Gorkom, 1991).

REFERENCE

Casertano, S. and van Gorkom, J. H., 1991. *Astron. J.*, in press.

THE MISSING MASS IN CLUSTERS OF GALAXIES AND ELLIPTICAL GALAXIES

Richard F. Mushotzky
Laboratory for High Energy Astrophysics
Goddard Space Flight Center
Greenbelt Md 20771

ABSTRACT

We review the available data for the existence of dark matter in clusters of galaxies and elliptical galaxies . While the amount of dark matter in clusters is not well determined both the x-ray and optical data show that >50% of the total mass must be dark. There is in general fair agreement in the binding mass estimates between the x-ray and optical techniques but there is not detailed agreement on the form of the potential or the distribution of dark matter, The x-ray spectral and spatial observations of elliptical galaxies demonstrate that dark matter is also required in these objects and that it must be considerably more extended then the stellar distribution.

PROLOGUE

The dynamical mass of clusters of galaxies is one of the best determined parameters in the "field" of Dark Matter . The strong contrast between the amount of light detected in clusters, the "dynamically" determined mass and the mean mass to light ratio M/L for stellar systems is one of the best pieces of evidence for gravitationally "bound" dark matter.

In this talk I shall use H_0=50 because I prefer to believe in the ages of globular clusters rather then in the dynamical measures of H_0. Also varying the Hubble parameter will introduce more confusion. However please remember that $M_{dynamic}/L_{obs} \propto H_0$ and $M_{gas}/M_{dynamical} \propto H_0^{-3/2}$. Thus "small" M/L and large gas baryonic mass go with small values of H_0

THE PROBLEM

The earliest realization that there was "dark matter" in the universe was the determination of the mass of clusters by Zwicky (1933) and Smith (1936) and comparison with the observed total amount of light due to stars. Under the assumption that all the baryons were in stars and that use of a mass to light ratio could convert the observed amount of light to mass of stars they showed that there was considerably more gravitational binding mass than mass in stars.

SIMPLEST MASS CALCULATION

The "simplest" mass estimator assumes that the galaxies are bound in a potential well and that their velocity dispersion σ is due to the total mass of the system via the "virial" theorem. Using the simplest possible assumptions (isotropic orbits, the galaxy distribution traces the total matter distribution and an "isothermal" potential)

$$M_{\text{"virial"}} \sim 3 R_H \sigma^2/G \sim 7\times10^{14}\ \sigma_{1000} R_{Mpc}\ \mathcal{M}_{sun}$$

However to know whether this "measured" mass is greater than that expected from the total number of baryons in the cluster we need an estimate of this quantity. Before 1970 it was thought that most of the baryons are in stars which emit optical light. And therefore the efforts concentrated on determining the mass in stars.

Thus,traditionally, we needed 3 observational quantities: two to estimate the mass (1) σ, the velocity dispersion of the galaxies (2) the "effective" radius R_{Mpc} of the cluster. And one to estimate the mass in baryons (3)the total amount of light in the cluster. However, since we now know that most of the baryonic mass in clusters lies in the hot x-ray emitting gas one also requires (4) x-ray data to obtain the mass in this gas.

Recent very careful work (cf Zabludoff,Huchra, Geller 1990) has accurately determined the velocity dispersion of ~25 clusters (one must measure the velocity of more than 40 galaxies per system to determine cluster membership, the effect of foreground and background galaxies and estimate the effects due to sub-clustering).

For a large sample of clusters $<\sigma>$~800 km./sec . However this median value is probably biased due to selection effects at small velocity dispersions. Theoretical calculations (Frenk et al. 1990) indicate that the true distribution of σ should be more power law than gaussian in form.

To estimate the light we need to either : 1) count "all" the light from the galaxies 2) integrate up the luminosity function or 3) integrate the radial surface brightness profile.

There have been "heroic" efforts to count "all" the galaxies in nearby clusters but most systems have their total light estimated by use of #2 or 3.

Method #2 is the easiest to explain: It has been shown that most clusters luminosity function can be well approximated by a Schecter function e.g. a function of the form

$$N(L)dL=N_0(L/L_*)^{-\alpha} \exp(-L/L_*)d(L/L_*)$$

the data are fit to determine N_0- the normalization, α - the slope and L_*- the "characteristic" luminosity of a galaxy.

The total luminosity is then $L_{opt}=N_0\,\Gamma(2-\alpha)L_*$; where Γ is the gamma function. Since most clusters seem to have a narrow dispersion in α (~1.25) and L_* (corresponding to M_B~-21.6 in the B band) the "trick" is to determine N_0. For the rich clusters studied by Colles 1989 N_{mean}~94$\pm$51. An estimate of their total light is

$$L_{total} \sim 7.2\times10^{12}\, L_{sun}\, (N_0/N_{mean}) \quad \text{(using } M_B(sun)=5.41)$$

Method #3 The surface distribution of galaxies:

The galaxy distribution is well fit by a "King" profile

$$\Sigma_{gal} \sim \Sigma_0(1+y^2)^{-3/2}; \quad y=r/b$$

Thus one must determine the "core radius" b and central surface density Σ_0 and the integral amount of light is .

$$L(y)_{total} = \Sigma_0 b^2 \{\ln(y+(1+y^2)^{1/2})-(y/(1+y^2)^{1/2})\}.$$

Using typical values of Σ_0~50-100 galaxies per square core radius (Beers and Tonry 1986) and a mean L of ~6.4×10^{10} L_{sun} per galaxy and going out to ~5 core radii give a total luminosity of ~4-8×10^{12} not inconsistent with the results from integrating the luminosity function. However this result is slightly more uncertain due to the poorly determined core radii and the strong correlation between Σ_0 and b .

While the harmonic mean radius ,"the size" , of a cluster R_H is somewhat difficult to determine the best work on the subject (Godwin and Peach) find typical values of R_H~1-2 Mpc.

Thus an estimate of the mean mass to light ratio for clusters is

$$\frac{<M_{total}> \sim}{<L_{total}> \sim} \quad \frac{6.7\times10^{14}\,(\sigma/\sigma_{mean})\,(R_H/R_{Hmean})\,M_{sun}}{7.2\times10^{12}\,L_{sun}\,(N_0/N_{mean})} \sim 100$$

We show in figure 1 the M/L distribution for 17 well measured (more than 40 galaxies per cluster. Since the mean M/L for isolated galaxies is ~4-10 this indicates a considerable "excess" of binding mass to stellar baryons. However this simple estimate does not include uncertainties due to the large number of theoretical simplifications in the above analysis and observational problems in determining σ and R_H.

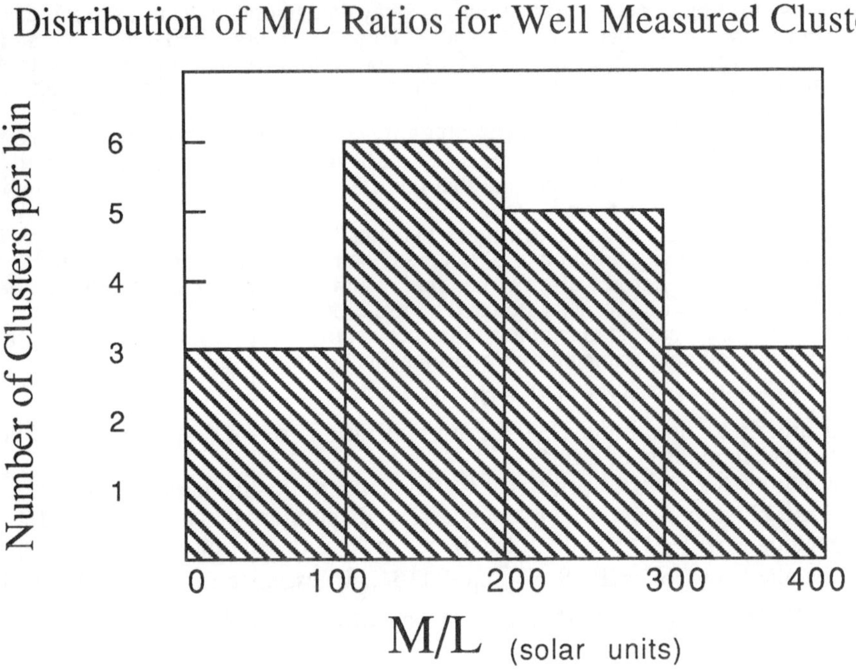

Figure 1: The distribution of mass to light for 17 clusters with more than 50 measured galaxy velocities

SIMPLE THEORY USING OPTICAL VELOCITY AND POSITIONAL DATA

As has been shown in detail by Binney and Strimple (1978), Merritt (1987), Richstone and Tremaine (1987) and The and White (1987) the accurate model independent, determination of the mass of a bound system from a relatively small (~300) sample of galaxy velocities is not possible. This is because the fundamental parameters of interest, the velocity and spatial distributions are always going to be poorly sampled and because the velocity distribution is a tensor only one of whose components can be directly measured.

However, under a wide range of assumptions about the form of the potential, the form of the velocity distribution, the ratio of the scale lengths of dark to visible matter etc, sensible **bounds** on the mass, inside physically interesting radii can be derived. If we define the measured galaxy surface density as $\Sigma(r)$,the line of sight velocity dispersion as $\sigma_p^2(r)$, the true galaxy space density $n(r)$, and the mass distribution $M(r)$ then we have two equations for the radial and tangential velocity dispersions and the mass distributions (Merritt 1987)

$$\Sigma(r)\sigma_p^2(R)=2\int (r\, n(r)/\sqrt{r^2-R^2})\, [\sigma_r^2-(R^2/r^2)(\sigma_r^2-\sigma_t^2)]dr$$

and the equation of stellar hydrodynamical equilibrium

$$d(n(r)\sigma_r^2)/dr \;+2n(r)/r(\sigma_r^2-\sigma_t^2)=Gn(r)M(r)/r^2$$

notice we have 3 unknowns $(\sigma_r^2, \sigma_t^2, M(r))$ where σ_r, σ_t are the radial and tangential velocity dispersion

3 knowable quantities $(\Sigma(r), \sigma_p^2(r), n(r))$

and 2 equations.

Often $\sigma_p^2(r)$ is not measured and just its average, V^2, is determined. Thus there is a "free" function to be determined. This is often assumed to be either the cluster potential or the velocity anisotropy $\beta(r)=\sigma(r)_r^2/\sigma(r)_t^2$. One must also reject "unphysical" solutions e.g. those with a negative mass or velocity values at any radius. As shown by Merritt, even with no *a priori* knowledge of the mass distribution the virial theorem puts a lower bound on the mass of the system of

$$M_{min} = \langle V^2 \rangle \langle r \rangle h^{-1}/G$$

which corresponds to the total mass of the system concentrated to a point at the center.

THE COMA CLUSTER

The Coma cluster is the best studied rich cluster. The minimum mass estimate for this systems corresponds to ~0.20 of the "virial" mass and gives $M(r<2\ \mathrm{Mpc})=3.9\times10^{14}M$ Using the light inside $2h^{-1}$ Mpc of $6\times10^{12}\ h^{-2}L_{sun}$ (The and White 1988) this sets a minimum **M/L~65**. Thus Coma is quite "optically dark".

The use of M/L ignores the baryons that the x-ray emitting gas has. Inside 2 Mpc the gas has a mass of $\sim1.5\times10^{14}\ M_{sun}$ about 3.5 times that in galaxies. **Thus the ratio of total mass to baryons inside 2 Mpc is >2.**

Thus under the most conservative possible assumptions there is still a need for dark matter in Coma

However, the observed velocity and galaxy position distribution is not consistent with a central point mass. More realistic models, consistent with the optical data, have the binding mass distributed on larger scales and give larger total mass estimates. A common assumption, is that optical light traces mass, e.g. that the galaxy and dark matter distributions are the same (and M/L is constant). Another possible assumption is that the x-ray light and dark matter e.g. the gas and the dark matter have the same distribution (the scale length of the gas is larger then that of the galaxies). If mass traces light and the galaxy orbits are isotropic then

$$M=(3\pi/2G)\sigma^2 R: \quad M=1\times10^{15}\sigma^2_{1000}R_{Mpc}\ M_{sun}$$

for Coma, the M/L derived in this way inside ~3 Mpc is ~150. The x-ray emitting gas has ~ 20% of the total mass and the stars ~5%. We have searched the literature for clusters with more than 50 galaxies available for determination of the velocity field and find that (fig 2) their M/L ratios are consistent with that for the Coma cluster. Frequently, lower quality data shows a higher value of M/L. Using Coma to scale the ratio of hot gas to stellar baryons one can show that the ratio of dark matter to baryons is ~3:1 for the average cluster inside ~3 Mpc. However this is model dependant. Merritt (1988) has examined the minimum mass consistent with the optical data under a wide variety of assumptions. He found that M_{min}~0.4 of the "viral, light traces mass " value and thus $M_{dark}/M_{baryonic}>2.5$.

The various models only diverge strongly at radii >1 Mpc and all are consistent within 10% at R~1 Mpc with $M_{Coma}=6\pm1\times10^{14}$ and $L_{Coma}=3.3\times10^{12}$ for M/L=180±30. The gas mass is 0.75×10^{14} and the stellar mass is ~ 0.23×10^{14} for a $M_{dark}/M_{baryonic}$ of ~ 6.

OTHER CLUSTERS

While the situation for Coma and ~10-15 other well measured clusters seems well understood the general situation is not clear. It is possible that the measured velocity dispersions are not true estimates of the actual velocities because of sub-clustering, mergers and projection effects (Zabludoff et al. 1990). It seems as if many clusters are composed of "lumps"(Dressler and Shectman 1989). If this is true the measured velocity dispersion is an overestimate of the true value and the cluster masses are overestimated. However the required factor of 3 overestimate in velocity dispersion to completely eliminate the need for dark matter is too large to be consistent with the data for all clusters.

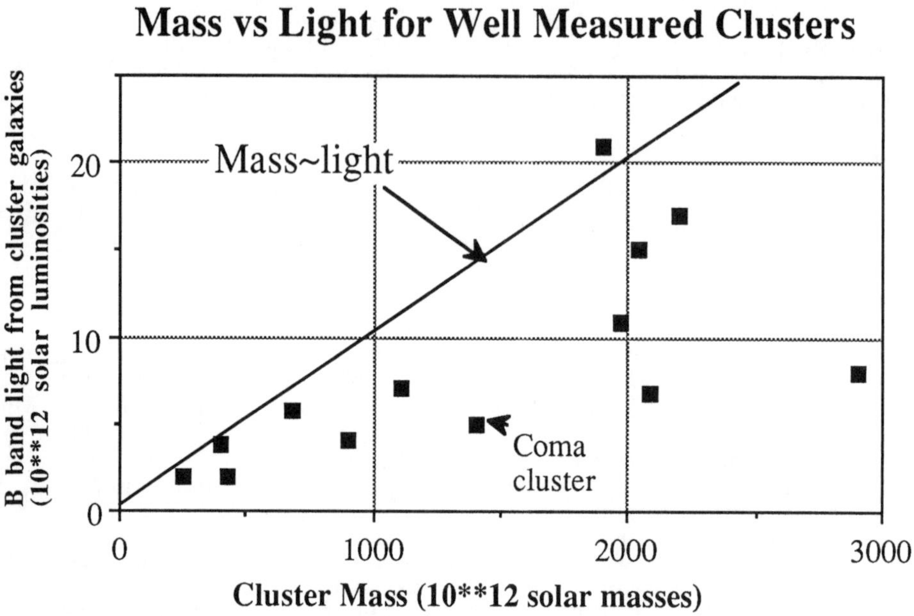

Figure 2 Distribution of mass determined from the virial theorm and light determined via integration of the luminosity function for well measured clusters

MASS DETERMINATIONS FROM X-RAY TEMPERATURE AND SURFACE BRIGHTNESS MEASUREMENTS

If the x-ray emitting gas which is "trapped" in the potential well of a cluster is in "rough" hydrostatic equilibrium then it is straight forward to determine the total binding mass of the cluster from the equation of hydrostatic equilibrium for a spherically symmetric system:

$$GM(r)=-(r^2/r_g)\ d/dr(r_g kT_g)$$

where r_g and T_g are respectively the gas density and temperature. Alternatively this may be written as

$$M(<r)=(kT_g(r)/G\mu m_H)[(d\log\rho_g/d\log)+(d\log T_g/d\log r)]\ r$$

The gas density distribution, ρ_g, is directly derivable from x-ray surface brightness profile as measured, for example by the Einstein observatory and Rosat for many 10's of clusters. The determination of the temperature profile for luminous, massive, hot x-ray clusters will have to wait until the flight of the U.S. Japanese mission Astro-D in

1993[1]. Notice that determination of the mass using this method does not require assumptions about the form of the potential or a particular scale length.

Jones and Forman (1984) have shown that the surface brightness distribution of most clusters can be well fit by a function of the form

$S(r)=S(0)(1+(r/a)^2)^{-3\beta+1/2}$ which gives a density distribution of

$\rho(r)_g=\rho(0)_g(1+(r/a)^2)^{-3\beta/2}$

the median value of $d\log \rho_g/d\log r$ for luminous clusters is ~-2 and thus the mass inside 2 Mpc, a typical size scale for the surface brightness and temperature measurements is

$M(<R)\sim\ 1.5\times10^{15}\ T_8\ (R/2Mpc)M_{sun}$

if the gas is isothermal where T_8 is the temperature in units of 10^8 degrees.
When it is possible,fig 3, direct comparison shows that the x-ray and optical mass estimates are of the same order for radii less then ~2 Mpc

At larger radii neither the optical velocity dispersions, the gas density gradient or the x-ray temperatures are well known. To make further progress the thermal structure of the gas must be assumed. If the equation of state is polytropic then the mass is (Cowie, Henriksen and Mushotzky 1988)

$M(<r)\sim1.5\times10^{15}\gamma(3\beta/2)T_8(r/a)^3/(1+(r/a)^2)^{-1/2}$

where γ is the polytropic index. For typical values of parameters the ratio of the mass of a polytropic gas to an isothermal one is

$M(poly)/M(iso)\sim1.25(r/a)^{-1/2}$

Thus is it only at large distances from the cluster center that the mass depends sensitively on the equation of state.
The x-ray spectral and imaging data are, at radii less then 2 Mpc, consistent with the simplest assumptions used to analyze the optical data e.g. mass traces light and isotropic orbits and confirm that clusters must have a substantial amount of "dark matter". Thus even though there is good reason to believe that the simplest assumptions are "wrong" it seems clear (Everad 1990) that their use to analyze both the optical and x-ray data does not result in errors in the mass of more than 50% inside 1 Mpc. However at larger radii where most of the dark matter is presumed to lie both the x-ray and optical data cannot strongly constrain the total mass.

COMBINED X-RAY AND OPTICAL FITS :ABELL 2256 AND ABELL 2670

The simplest solution used above of isothermal gas and galaxies, constant galaxy velocity dispersion with radius and isotropic galaxy orbits can be used to combine the equations of hydrostatic motion for the gas and galaxies together (Fabricant et al. 1989) to derive

$\mu\sigma/\kappa T(\rho)=[d(\log\rho/d\log r)]/[d\log\rho/d\log r]$

However this result is sometimes incompatible with the available data . In particular for the Perseus cluster, Abell 2256 and A2670 $\mu\sigma^2/kT_g(r)>1$--- the x-ray emitting gas is too cold or the galaxies are too hot In addition,the x-ray surface brightness data shows a shallower slope then the galaxy surface density e.g.

$[d(\log\rho_{gas}/d\log r)]/[d\log\rho_{galaxies}/d\log r]<1.$

[1] The temperature profiles of a few clusters, viz Perseus, Coma and Virgo have been roughly measured and within ~1 Mpc are more or less isothermal.

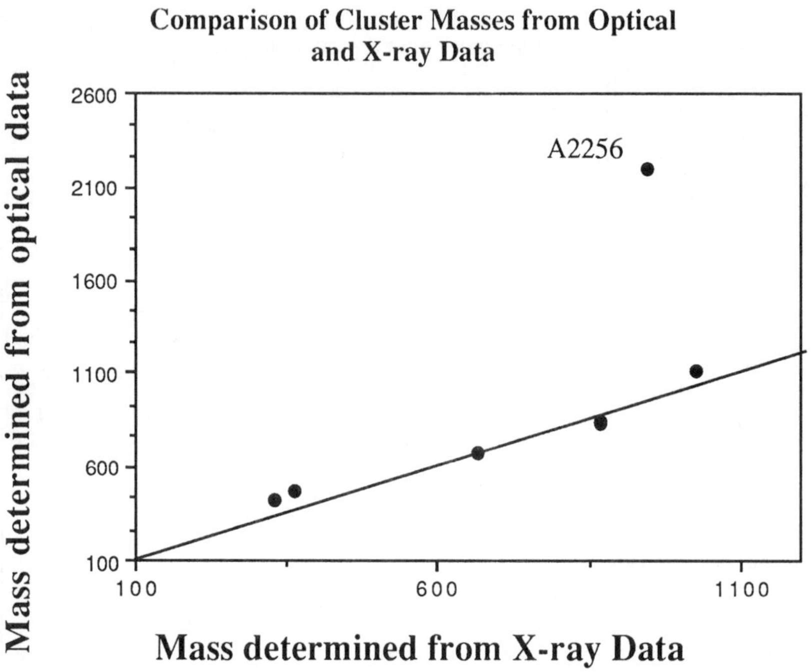

Figure 3: Comparison of masses derived from x-ray spectroscopic and imaging data with those from optical velocity dispersion and harmonic radii. Units of mass are 10^{12} solar masses. The line is equal x-ray and optical masses.

If mass follows light the predicted x-ray temperature

$$T(r)=1/\rho_{gas}\ [T_g(o)\rho_{gas}(0)-\mu/k\int(GM(r)/r^2)\ \rho_{gas}(r)dr$$

is a factor of 2 higher then the observed temperature.

We can try to accommodate this discrepancy by relaxing one (or more) of our assumptions

1) Let the velocity dispersion be anisotropic. This can lower the predicted temperature but not enough to match the observations.

2) Relax the assumption of mass following light:

A wide variety of models now become possible. However if we assume a polytropic equation of state for the gas we reduce the number of free parameters. This works but requires:

a)the gas temperature must increase with radius

b)the galaxy orbits must be radial at the outer parts of the cluster

c)the mass to light ratio increase with radius.

This is because the the gas temperature is less than the galaxy temperature even though the gas has a flatter density profile.

In conclusion the existence of "dark" (non optical, IR or x-ray emitting material) in clusters is irrefutable. Inside ~1 Mpc the derived values are stable with $M_{dark}/M_{baryonic}$ ~5. At radii 1<R<5 Mpc sensible statements can be made but the errors are large (factor of 3) . At larger radii we have little or no information.

WHERE IS THE DARK MATTER?

If we can determine the distribution of the dark matter we will obtain a clue to its physical nature: baryonic, hot dark matter or cold dark matter. Also biased galaxy formation theories require that the galaxies are more clustered than the mass. The distribution function of the orbits of the galaxies (isotropic, circular, radial) and the form of the temperature gradient can constrain the location of the dark matter. In particular it is of interest to know whether the dark matter is more (or less) centrally concentrated then the gas or the galaxies.

If, for ease of parameterization if we pick a polytropic "King" model, then the dark matter would be more concentrated than the galaxies if $(1-\gamma) < 0$ and more extended if >0. Spatially resolved x-ray spectral data would allow an unambiguous determination of this issue. If $\gamma > 1.2$ the galaxy orbits would have to be predominantly tangential.

The recent observation of sharp decrease in the orbital velocity with radius for A2670 (Sharples et al. 1989) indicate that the orbits of the galaxies in this cluster are radial in the outer regions and thus if the galaxies are in equilibrium that the dark matter is more extended then the galaxies (Merritt 1987) .
Thus the "jury is still out" on this issue.

ARE THERE PATTERNS IN THE DISTRIBUTION OF DARK MATTER?

At present we have very little information on the relative amount or distribution of dark matter in clusters. We do not know if more massive systems have a greater or less fraction of dark material nor do we know whether the ratio of dark material is related to the evolutionary condition of the cluster. There are some preliminary patterns from the x-ray imaging and spectroscopic data (Edge and Stewart 1990,White 1990) that indicate that more massive clusters have a greater fraction of their total mass in gas then less massive clusters. However it is not clcar (David et al. 1989) whether the baryonic fraction of the total mass is also increasing.

ELLIPTICAL GALAXIES :X-RAY CONSTRAINTS

The data for spiral galaxies seem to indicate that $M_{dark}/M_{baryonic}$ is ~ 5-10 "inside" the galaxies but the most recent optical results for elliptical galaxies (Binney et al. 1990) do not require dark matter in elliptical galaxies on scale lengths less than 5 kpc. The strongest evidence for dark matter in "isolated" ellipticals comes from the x-ray imaging and spectroscopic data.

The x-ray emission from massive elliptical galaxies is dominated by bremmstrahlung from hot gas which is in hydrostatic "equilibrium". Thus in principle the same techniques that can be used to determine the mass profile of clusters can be used for ellipticals However the physical situation is quite different from that in clusters.
1) The mass of the gas is much less than the mass in the stars
2) Over much of the volume of the galaxy the cooling time of the gas is considerably less than the Hubble time and thus the gas is cooling (in even the strongest cooling flow in a cluster the cooling time is less than the Hubble time over less than one core radius).
3) energy input from supernova is on the order of the energy due to gravitation (while in the past this may have been true for clusters in the present epoch the energy input from supernova is ~0.05 of the energy stored in the gas)
4) There exist good quality x-ray spectra for only 3 elliptical galaxies (there are ~50 for clusters).

Taken together this makes the analysis of the mass distribution more problematical. Be that as it may one can get a rough estimate of the mass :
The x-ray surface brightness profiles of Ellipticals are rather flat with
$d \log \rho / d\log r \sim -1.0\text{-}1.5$. If one assumes that the gas is "isothermal" (e.g. that all the variation in temperature is due to cooling) then

$$M(<r) \sim 3.4 \times 10^{12}\, T_{kev}(r/50\text{kpc}) M_{sun}$$

The Ginga satellite has determined the temperatures of three ellipticals (Awaki et al. 1990). Using these temperatures and estimates of the scale length R from Einstein imaging observations one derives the following simple mass estimates inside R(kpc)

Names	kT(kev)	R(kpc)	Mass($\times10^{12}M_{sun}$)
NGC4472	1.9±0.2	52	6.4
NGC4636	0.9±0.2	32	1.9
M87	2.2±0.2	100	15

Putting in "arbitrary" temperature profiles $T(r)\sim r^{\alpha}$ allows a considerably wider range of masses. If $-1/2< \alpha <1/2$ then the total mass varies between $2.3\text{-}4.5\times10^{12}\ T_{kev}\ (R/50kpc)M_{0}$ for the observed slope of the gas density .

This problem has been looked at in detail for M87 where there does exist some x-ray spectral imaging data (Fabricant and Gorenstein 1983, Huchra and Brodie 1988, Stewart et al. 1984). While the range of allowed models is considerable none of them are consistent with the "stellar" mass of M87 being solely responsible for binding the x-ray emitting gas and the "best" mass estimates are consistent with the simplest assumptions.

There are two other systems (NGC1399 (Killeen and Bicknell,1989) and NGC4472 (Loewenstein 1990) for which detailed analysis using the integrated x-ray spectra and the optical data taken together have been performed The case of NGC 4472 was studied in detail by M. Loewenstein in an attempt to discover what total mass distribution is consistent with the best data in both the optical and x-ray.The procedure he followed was to simultaneously solve both the equation of stellar hydrodynamics for the projected stellar velocity distribution and the equation of hydrostatic equilibrium for the hot gas temperature distribution. Stars and gas were assumed subject to a gravitational potential consisting of a stellar component derived from optical surface photometric and spectroscopic observations (Burstein et al. 1987; Franx, Illingworth, and Heckman 1989; Davies and Birkinshaw 1988), and a variable dark matter component parameterized by a scale length and total mass. The gas density distribution derived from the Einstein imaging data (Trinchieri, Fabbiano, and Canizares 1986) is utilized, and a range of assumptions about the behavior of the stellar velocity ellipsoid and the outer boundary gas pressure are made.

The figure 4 shows that on a plot of dark halo scale length vs. total mass-to-light (the dashed line is the assumed stellar mass-to light ratio), only in the small region marked "DC" (corresponding to relatively small dark matter scale lengths) would the dark matter significantly disturb the stellar dynamics inside the core. Figure 5 shows the stellar and halo mass distributions for the two models indicated by filled circles on the first figure -- the dark matter fails to become significant until nearly 10 kpc, well outside the radius where stellar velocity dispersions have been measured. A much more significant region in the first figure can be disallowed because gas in hydrostatic equilibrium in the corresponding total potential cannot have an emission weighted temperature high enough to be consistent with the Ginga derived temperature -- kT =1.5-2.3 keV (Awaki et al. 1990). This constrains the total mass-to-light ratio in NGC 4472 to be at least 30 (corresponding to a dark-to-light matter ratio of 3) --similar to the lower limit derived from considerations of convective stability by Fabian et al. (1986).

Conclusion: While the available x-ray data are rather poor they do seem to require a massive halo in several elliptical galaxies. If this is not an artifact of conditions in the Virgo cluster then M/L values of >20 are required for giant ellipticals. Since the stars have M/L~7 dark matter is required.

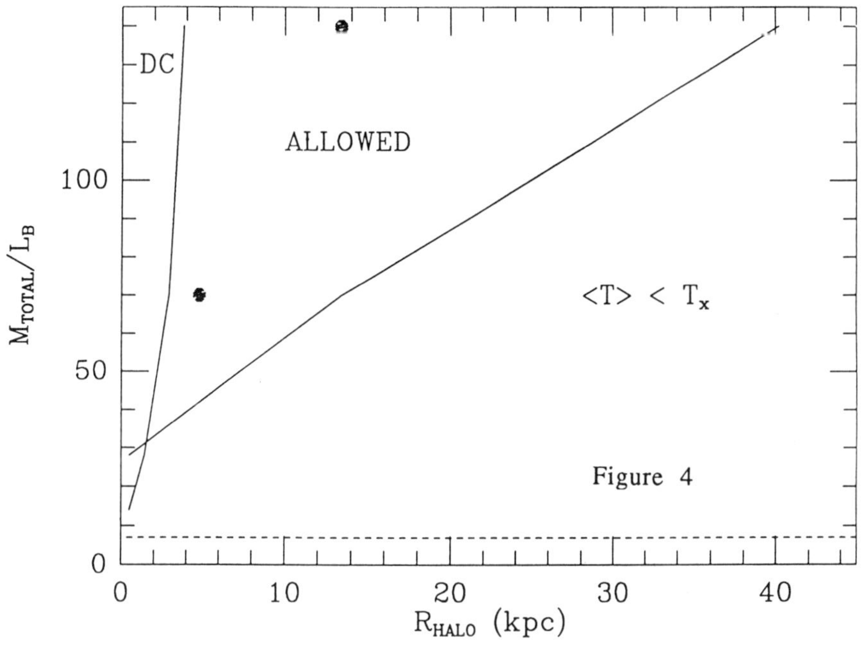
DC
ALLOWED
<T> < T_x
Figure 4
M_{TOTAL}/L_B
0
50
100
0
10
20
30
40
R_{HALO} (kpc)

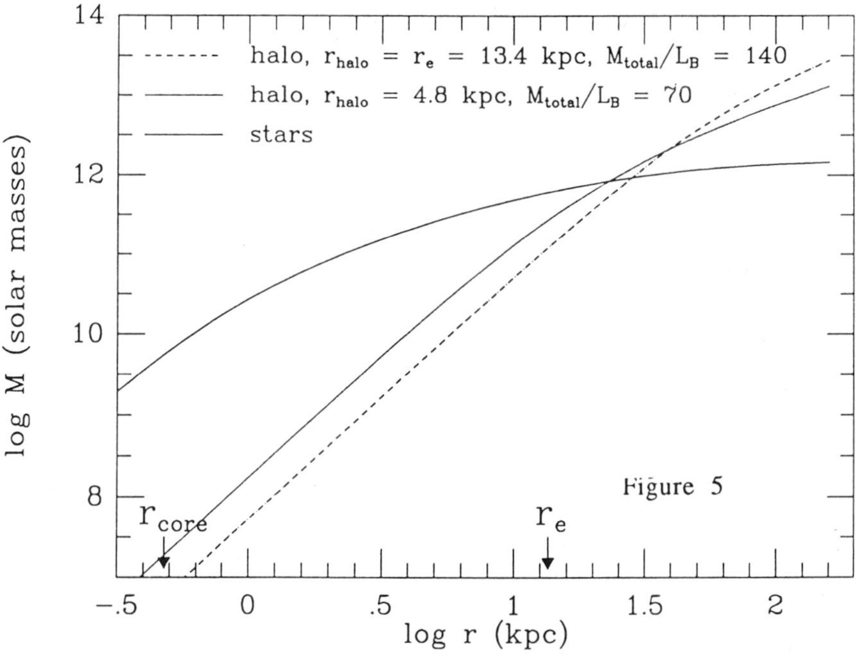
halo, $r_{halo} = r_e = 13.4$ kpc, $M_{total}/L_B = 140$
halo, $r_{halo} = 4.8$ kpc, $M_{total}/L_B = 70$
stars
Figure 5
r_{core}
r_e
log M (solar masses)
8
10
12
14
−.5
0
.5
1
1.5
2
log r (kpc)

References

Awaki , H., Koyama, K, Kunieda, H, Takano,S. Tawara,Y. and Ohashi,T. 1990 Ap.J. in press
Binney, J. and Strimple, O. 1978 M.N.R.A.S. 185,473
Binney, J. J. Davies, R.L., Illingworth, G. D. Ap.J.,361, P78
Beers T. and Tonry 1986 Ap.J. 300,557
Burstein, D. et al. 1987 Ap.J. Suppl 64,601
Colles M. 1989 M.N.R.A.S. 237,799
Cowie, L. L.,Henriksen, M. and Mushotzky, R. 1988 Ap.J.,317, P593
David, L. Jones,C., Arnaud, K. and Forman ,W. 1989 Ap.J.,356, P32
Davies, R. and Birkinshaw,, M. 1988Ap.J. Suppl 68,409
Dressler, A, Shectman, S. A. 1989 A.J.,95, 985
Edge, A. and Stewart, G.S. 1990 M.N.R.A.S. submitted
Evrard, A. 1990 Ap.J. in press
Fabian,A,Thomas,P,Fall,S, White,R.E. III 1986 M.N.R.A.S. 221,1049
Fabricant and Gorenstein, P.1983ApJ. 267, 535
Fabricant ,D., Kent, S. and Kurtz 1989
Franx, M, Illingworth, G Heckman, T Ap.J.,344, 613
Frenk, C.S. White, S. D. M., Efstathiou, G, Davis, M Ap.J.,351, P10 1990
Godwin and Peach M.N.R.A.S. 1989
Jones,C. and Forman, W. 1984 Ap.J. 276 38
Huchra, J. , Brodie, J, 1988 A.J.,93, 779
Killeen, N.E.B.and Bicknell, G.V.1989 Ap.J. 325,165
Merritt ,D. 1987 Ap.J. 313,121
Richstone, D. O.,Tremaine, S. 1987 Ap.J.,327, 82
Sharples,Ellis, and Gray M.N.R.A.S. 231,494
Smith , H 1936 Ap.J. 83,23
Stewart ,G. Fabian, A., Jones, C. and Forman, W. 1984 Ap.J. 285,1
The, L.S. and White ,S.D.M. 1987 A.J.,95, 15
Trinchieri, G.,Fabbiano,G., , and Canizares,C. 1986 Ap.J. 310,637
White, R. 1990 Ap. J. in press
Zabludoff,A. Geller, M and Huchra, 1990 Ap.J. Suppl.,74, 1
Zwicky, F. 1933 Helv Phys Acta 10

DISTRIBUTION OF DARK MATTER IN THE PERSEUS CLUSTER, AND MASS DISTRIBUTIONS IN THE COMA AND OPHIUCHUS CLUSTERS

C. J. Eyles, M. P. Watt, D. Bertram, M. J. Church, P. A. Knight,
T. J. Ponman, G. K. Skinner, and A. P. Willmore
School of Physics and Space Research
University of Birmingham
Edgbaston, Birmingham B15 2TT
ENGLAND

ABSTRACT

Data obtained with a coded mask X-ray spectral imaging telescope (XRT) built at Birmingham, England and flown on the Spacelab 2 mission constitute the first high quality spectral imaging observations of clusters of galaxies in the hard X-ray band (2.5-25keV).

The spectral imaging properties of the XRT allow the plasma temperature as a function of radius to be determined in larger, brighter clusters, and high quality data have been obtained for the Perseus, Coma and Ophiuchus clusters.

Since the observed spectral images are instrumentally smeared projections of the three dimensional distributions in the cluster, a model fitting procedure has been adopted to determine the true temperature and density distributions of the gas.

We have performed a detailed analysis of the properties of the intracluster gas in Perseus, and derived the distribution of the mass components in the cluster. Work is in progress on the Coma and Ophiuchus clusters.

INTRODUCTION

The XRT had hard X-ray spectral imaging capabilities, with a spatial resolution of 12′ FWHM, an energy range of 2.5–25keV and a field of view 6.4° square. The observations were made over the period 1985 July 29 – August 6; the total good quality observing time achieved on the clusters was 24600s for Perseus, 27800s for Coma and 3000s for Ophiuchus.

THEORY AND MODELLING PROCEDURE

Under the assumption that the X-ray emitting gas is in hydrostatic equilibrium with the cluster potential, the total gravitating mass within radius r of the cluster center is given by

$$M_{grav}(<r) = -\frac{kT_{gas}(r)}{G\mu m_p}\left[\frac{d\ln\rho_{gas}(r)}{d\ln r} + \frac{d\ln T_{gas}(r)}{d\ln r}\right] r.$$

Here $\rho_{gas}(r)$ and $T_{gas}(r)$ are the density and temperature of the X-ray emitting gas, which can be determined from our data.

The density profile was assumed to be of the form $\rho(r) = \rho_0(1+(r/a)^2)^{-\alpha_\rho}$ where ρ_0, a and α_ρ are free parameters.

Since temperature gradients have not been determined explicitly in clusters before, the choice of parameterization is rather unclear. We have used both a 'polytropic' model in which the gas density and pressure are related via a polytropic index which is a free parameter, and a power law temperature profile.

In fitting Perseus, we have also allowed for nuclear emission from the active galaxy NGC1275, by including a point source with a power-law spectrum, and we have taken fits to the Einstein IPC data to allow for the cooling flow.

RESULTS

We have assumed a Hubble constant of 50 km s^{-1} Mpc^{-1} ($h_{50} = 1$). Statistical errors are 1σ except where otherwise stated.

The cumulative gravitating mass as a function of radius is shown in Fig 1. The X-ray emitting gas makes up a significant fraction of the total cluster mass in each case, ~30% in Perseus, and ~15% in Coma and Ophiuchus at 1Mpc (Table 1.) For all three clusters, the gas is significantly more extended than a simple isothermal sphere model would suggest, and the radial surface brightness profiles (Fig 2) suggest that there is no break in the gas density profile out to the limits of the data implying that the gas fraction becomes increasingly significant out to this radius.

Because of the complexity of the error calculations required to determine the distribution of dark matter, we have only completed the analysis for Perseus at present (Fig 3, Eyles *et al* 1990, submitted to Ap. J.). This work shows that a scenario in which the dark matter is more centrally condensed than the galaxies is favoured, although a situation in which the dark matter is distributed in the same way as the galaxies is not excluded (Fig 4). Joint analysis of the XRT and IPC data is in progress which should reduce the error bounds in this plot.

Cluster	Total cluster mass ($M_\odot$)	Gas mass ($M_\odot$)
Perseus	3.7×10^{14}	1.0×10^{14}
Coma	5×10^{14}	0.8×10^{14}
Ophiuchus	7×10^{14}	1.1×10^{14}

Table 1: Values of total cluster mass and cluster gas mass within 1Mpc. The total cluster mass scales as h_{50}^{-1}, the gas mass scales as $h_{50}^{-5/2}$.

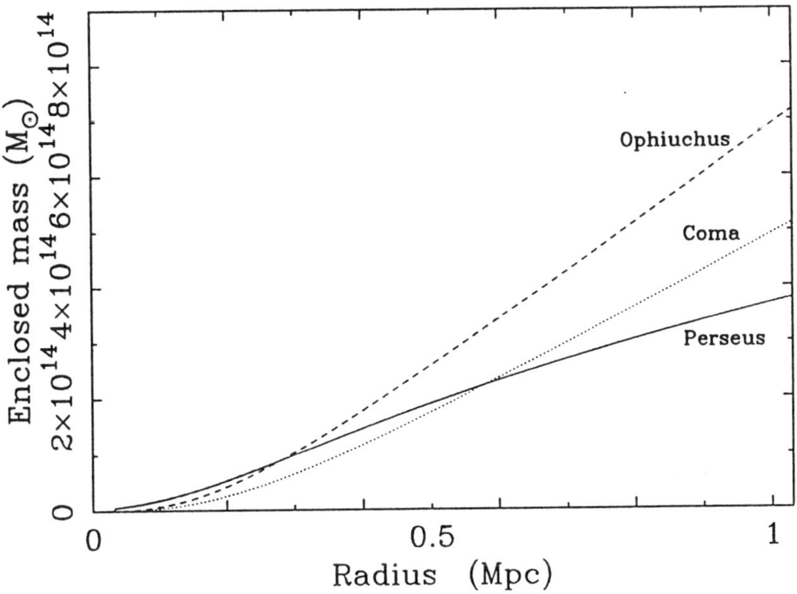

Fig 1. Total enclosed gravitating mass as a function of radius for the three clusters.

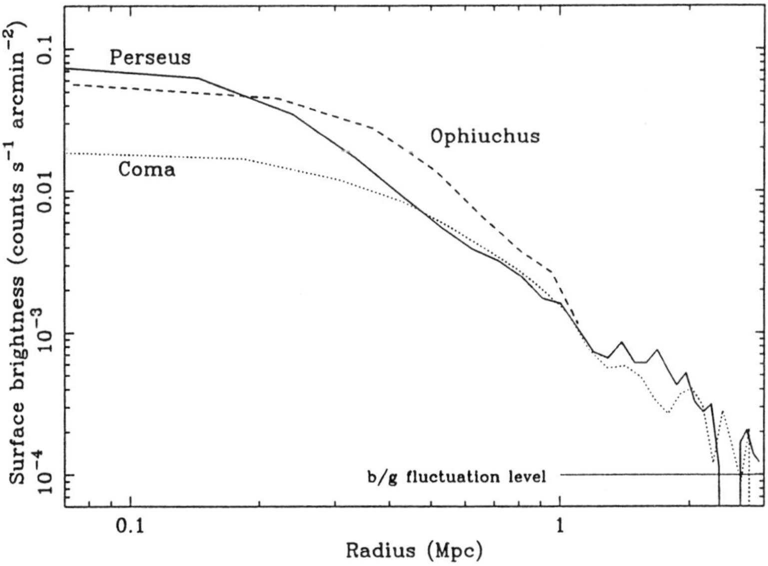

Fig 2. Surface brightness profiles of the three clusters. Significant emission is detected out to >2Mpc (>60′) in Perseus, and to ~ 2Mpc in Coma and 1Mpc in Ophiuchus.

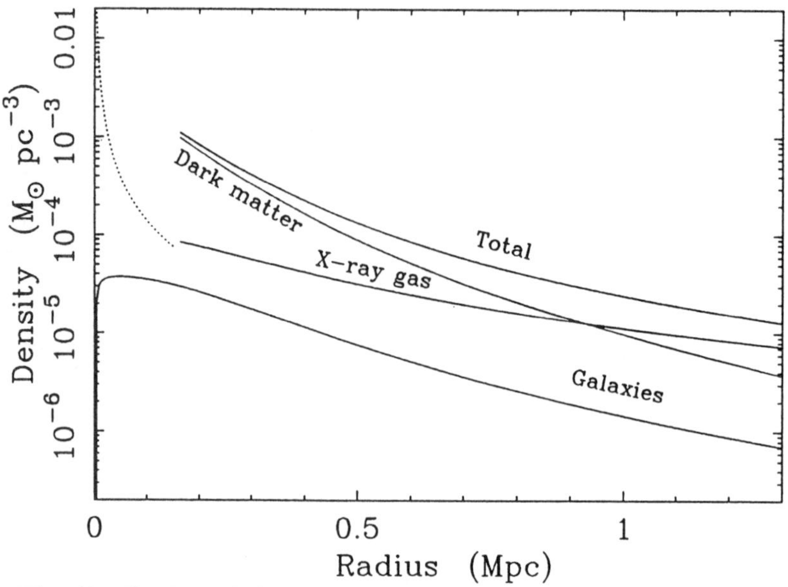

Fig 3. The distribution of the mass components in the Perseus cluster derived from the best fit temperature and density profiles to the data. The properties of the cooling flow (inside 5′) have been taken from Einstein IPC data, as the resolution of the present instrument is inadequate to map at this scale.

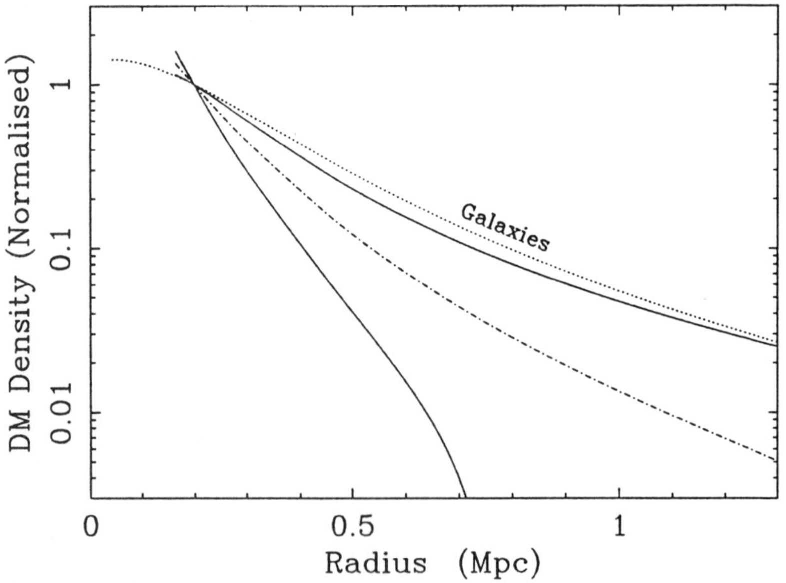

Fig 4. Relative distribution of dark matter and galaxies in the Perseus cluster. The band shows the permitted range of dark matter distributions, normalised to unity at 0.2Mpc, allowing for uncertainties in the cooling flow parameters, and the effects of statistical and systematic errors.

DARK MATTER IN THE HERCULES CLUSTER

Christina M. Bird and John M. Dickey
University of Minnesota
116 Church St. S.E., Minneapolis, MN 55455

ABSTRACT

We study the distribution of dark matter in the Hercules Cluster. Our dataset consists of 120 measurements of redshifts, 47 from 21-cm. observations and 73 from optical work. Using the entire dataset we derive a velocity dispersion vs. radius profile and a value for the total dynamical mass of the cluster of $1.6x10^{15}h^{-1}$ $M_\odot$. Corrections for sub-structure in the cluster are considered.

With the 21-cm. spectra we derive rotational velocities and masses for our radio detected galaxies. The sum of these masses is $6.2x10^{12}h^{-1}$ $M_\odot$. This sum is extrapolated assuming a range of mass-to-light ratios to estimate the cluster mass contained in the luminous galaxies. The total mass included INSIDE ALL the visible light in the cluster is 8 to $30x10^{13}h^{-1}$ $M_\odot$.

The ratio of the dynamical mass to the rotational mass is in the range from 5-25. This means that the bulk of the mass (80-95%) in A2151 is not contained within the luminous galaxies. The light in the cluster may or may not trace the mass, but most of the mass is apparently not where the light is.

INTRODUCTION

In this paper we study the gravitational potential in A2151, the Hercules Cluster. This cluster is very well observed, so many velocities for its galaxies exist in the literature. We have obtained new 21-cm. observations of faint (m_b=15.5-17.5) spirals at Arecibo to supplement this dataset. We use the combined sample to derive a velocity dispersion-radius profile. Following Heisler, Tremaine, and Bahcall (1985) we then derive a mass-radius profile. We call this the "projected mass." We also consider the effects of substructure on the projected mass determination.

The subset of our sample which is composed of 21-cm. observations of Hercules galaxies allows an estimate of the total mass which is contained **within** the HI radii of galaxies. The width of the HI line of an individual galaxy indicates its rotational velocity as far out as the neutral hydrogen disk extends, which is typically 50% further out from the center of the galaxy than the optical disk. This maximum rotational velocity can then be used to determine the total mass inside the outer edge of the HI disk. This includes all gravitational mass in both disk and halo components, not just the neutral hydrogen itself.

To obtain an estimate for the total mass of the galaxies in the cluster, to be compared to the projected mass (which is contained in the entire volume occupied by the cluster and not just the galaxies and their halos), the sum of the masses of the HI-detected galaxies must be augmented by an estimate of the mass inside the galaxies which we did not observe or detect at 21-cm. (mostly early type galaxies with little or no HI). Also we must correct for galaxies below the magnitude limit of our optical catalog. Then we can estimate the mass "directly observed" within the cluster. This we call the "galaxy-enclosed mass."

DYNAMICAL MASS

Using the complete set of redshift measurements for A2151 galaxies we determine the velocity dispersion profile for the core of the cluster. The galaxies are binned by projected radius from the cluster center (R.A.=$16^h03^m11^s$, dec=$17^0 55'55''$) in intervals of 400 arcsecs (=$0.1h^{-1}$ Mpc). We find an average recessional velocity for A2151 of 11,063 km sec^{-1}.

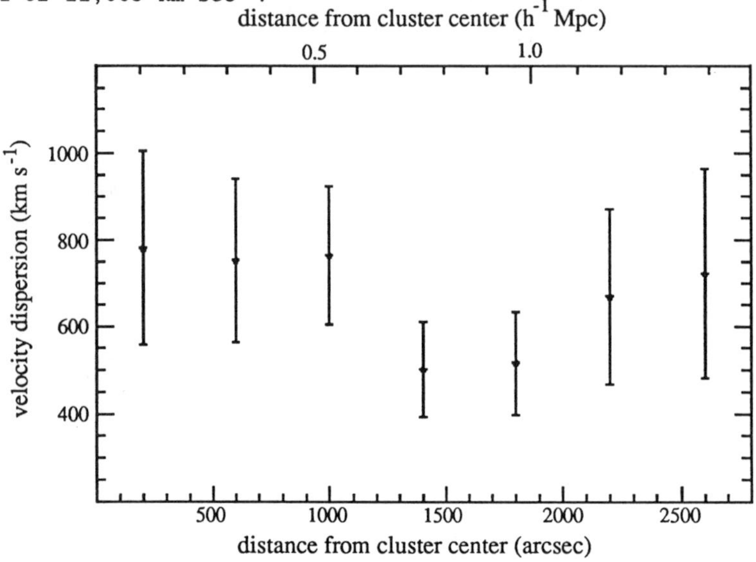

Figure 1. The velocity dispersion profile of A2151.

The HBT projected mass estimator for A2151 is $1.6x10^{15}h^{-1}$ $M_{\odot}$ within $1.5h^{-1}$ Mpc (projected) of the core, yielding a projected mass density of $1.9x10^{-26}h^2$ gm cm^{-3} $2.9x10^{-4}h^2$ $M_{\odot}$ pc^{-3}).

The presence of substructure in A2151 is the largest source of uncertainty in its mass determination. Substructure is an indication that a cluster has formed recently and has not evolved to a state of dynamical equilibrium. Since all dynamical methods of mass determination depend critically on the assumption of equilibrium, it is crucial to estimate its importance. We have explored this using the cumulative deviation test of Dressler and Shectman (1988b). For each galaxy in the cluster, a clump is formed which includes the galaxy and its n nearest neighbors. Then their average velocity and velocity dispersion are compared with the global values, and the resulting deviations for all galaxies are summed. This yields a cumulative deviation in A2151 which is 5σ over the expected value for a random distribution of velocities, a very strong signature of subclustering. This result is consistent with Dressler and Shectman's result (1988a,b). Preliminary studies of the Virgo Cluster, whose mass has been carefully determined using dynamical models which consider its clumpy structure, indicate that this level of subclustering may introduce a factor of 3-5 error in the dynamical mass determination.

ROTATIONAL MASS AND GALAXY-ENCLOSED MASS

21-cm. linewidths may be used to calculate the mass of a spiral galaxy within its HI radius by equating its rotational energy to its gravitational potential. Of the 47 galaxies in the radio sample, 26 have suitable inclinations to permit conversion of the linewidth to a rotational velocity and so calculate a mass. The sum of these masses is $6 \times 10^{12} h^{-1}\ M_\odot$.

To extend the rotational measurement to estimate the mass inside ALL visible galaxies, we use the Schecter luminosity function. This extrapolation depends on the assumption of a constant mass-to-light ratio for all the galaxies in the cluster. Ellipticals may have M/L's 5-10 times greater than spirals, hence we compute the total galaxy-enclosed mass using two possible extrapolations from the rotational mass measured for the 26 spirals above. The smaller estimate, $8.2 \times 10^{13} h^{-1}\ M_\odot$, comes from assuming a constant M/L for all light in the cluster. The larger estimate, $3 \times 10^{14} h^{-1}\ M_\odot$, comes from assuming that E's and S0's have a mass-to-light ratio 5 times that of gas-rich spirals.

The major uncertainties in the galaxy-enclosed mass are the values of the inclinations in the individual mass determination, the low signal-to-noise of many of the spectra (which decreases the reliability of the linewidths), and the variability of the mass-to-light ratio with galaxy type.

CONCLUSIONS

We can now begin to compare the distribution of dynamical mass in the entire volume of A2151 to the distribution of matter contained in the individual galaxies. The ratio of the projected mass to the galaxy-enclosed mass (which includes not only luminous matter in the galaxies, but also dark matter contained within the outer radius of the HI disks) is 5-24, which indicates that only about 5-20% of the total mass contained in the inner $1.5h^{-1}$ Mpc of the Hercules Cluster is enclosed within the radii of the HI disks. This may imply that, in the rich cluster environment, the halos of individual galaxies either never formed, or formed and were tidally stripped by interactions with the halos of other galaxies and the inter-cluster medium.

REFERENCES

Bird, C.M. and J.M. Dickey, in preparation.
Dressler, A. and S. Shectman. 1988a, A.J., 95, 284.
Dressler, A. and S. Shectman. 1988b, A.J., 95, 985.
Heisler, J., Tremaine, S., and J. Bahcall. 1985, Ap.J., 298, 8.

ACKNOWLEDGEMENTS

We are grateful to G.L. Hoffman for useful conversations about the situation in Virgo Cluster, and to B.M. Lewis and E.E. Salpeter, who collaborated on the observations. This work is supported in part by NSF grant #87-22990 to the University of Minnesota.

THE DYNAMICS OF THE RICH CLUSTER A2634

George Rhee[1], Jason Pinkney[1], Jack Burns[1],
J.M. Hill[2], W. Oegerle[3], P. Hintzen[4], and D. Batuski[5]
[1]Department of Astronomy, New Mexico State University, [2]Steward Observatory, University of Arizona, [3]Space Telescope Science Institute, [4]University of Nevada at Las Vegas, [5]University of Maine

ABSTRACT

We present results of a study of the dynamics of the rich cluster Abell 2634. Spectroscopic observations were obtained using the MX spectrometer on the Steward Observatory 2.3m telescope. The spectrometer utilizes 2" diameter silica fibers which are independently positioned in the telescope focal plane by microprocessor controlled mechanical arms. We have obtained $\sim$ 100 radial velocities of galaxies in this cluster with an accuracy of $\sim$ 100 km/s. This cluster is of particular interest since its cD galaxy (NGC 7720) has been identified with the wide-angle tail radio source 3C 465. The tail has a northwest extent of 130 Kpc. The observations of Scott, Robsertson and Tarenghi (1977) indicate that NGC 7720 has a peculiar velocity of $\sim$ 330 km/s with respect to the mean cluster velocity. We do not measure a significant peculiar motion for the cD galaxy. We also do not find evidence of substructure in this cluster or deviations from normality of the galaxy velocity distribution.

INTRODUCTION

We have made a study using the MX multifiber spectrograph of the cluster Abell 2634. This cluster is of particular interest as it contains a cD galaxy which is the prototype example of a wide angle tail (WAT) radio source (Eilek et al., 1984). The data presented here allow us to measure the motion of the cD galaxy relative to the cluster mean velocity. The cD peculiar motion amplitude is related to two key issues:

1) What is the formation mechanism of the cD? If these galaxies form during the collapse and virialization of compact groups which later merge to form a cluster, the cD may have an appreciable peculiar motion. If cD galaxies grow by cannibalism in the centers of rich clusters they should have no peculiar motion.

2) What physical process bends the jets in WATs? The jets may be bent by i)ram pressure of the intracluster gas, ii)buoyancy forces if the cD is not exactly at the cluster center, iii)a passive wake, iv) turbulent gas motions in the ICM. Measuring the cD velocity should yield clues as to whether WATs are produced

by the same processes as head-tail radio galaxies. A2634 is a Rood-Sastry type cD cluster; the cD is NGC 7720 also denoted 3C 465. The cD nucleus contains a weak emission line spectrum.

The X-ray distribution is roughly circular with local irregularities outside the inner few arc minutes (Eilek *et al.*,, 1984). In the central region the X-ray major axis coincides with the line of symmetry of the radio tails.

THE DATA

Galaxy positions were measured to 0.5" using the KPNO two axis grant measuring engine and KPNO glass copies of the PSS. The radial velocities were measured using the MX spectrometer on the 2.3m telescope of Steward Observatory. The MX uses 2" diameter optical fibers. These are positioned using microprocessor controlled mechanical arms. Spectra for 32 objects in a 45' field are obtained simultaneously, 24 additional fibers are used as sky fibers. Fibers from the MX are fed into the f/9 Cassegrain spectrograph. A UV flooded TI 800×800 CCD chip is used. A 300 groove/mm grating is used to record spectra in the 3800-6700Å range with 11Å resolution. The data were reduced using the IRAF Nessie package. Radial velocities were measured using M32 as a template by the standard cross correlation method.

We have obtained high S/N spectra for 94 galaxies. This is our MX data base to which we have added that data of Bothun and Schombert (1990). The histogram shows the velocity distribution (the cD velocity is marked by an arrow).

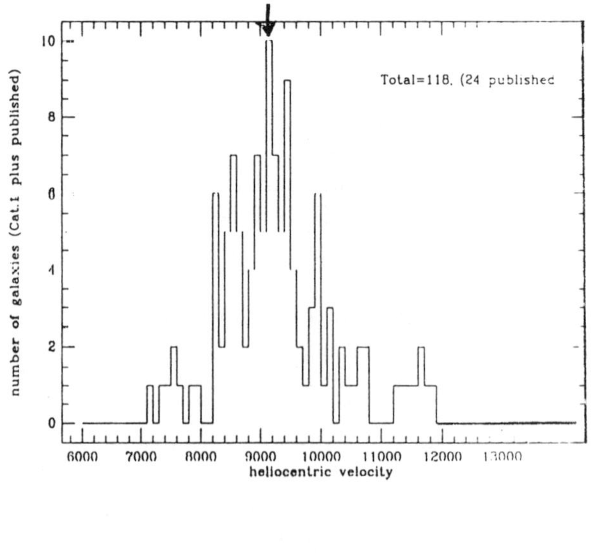

ANALYSIS

Our analysis addresses three questions. 1)Does the galaxy (x, y, v_z) dataset show evidence of substructure? 2)Does the cD galaxy have a measurable motion with respect to the cluster mean velocity? 3)Does the velocity distribution of the galaxies deviate significantly from a Gaussian distribution?

To test for substructure (clumping of galaxies) we have applied the Dressler and Schectman test (1988). This test enables one to search for local deviations from the mean sigma of the cluster. For each galaxy the ten nearest neighbors on the sky are selected. From this sample the local velocity mean and dispersion are calulated and compared to the cluster mean and dispersion. For each galaxy δ is evaluated: $\delta^2 = (11/\sigma^2)[\bar{v}_{local} - \bar{v})^2 + (\sigma_{local} - \sigma)^2]$. The cumulative deviation Δ is the sum of the δs for the N_v cluster members. To calibrate the statistic, Monte Carlo simulations were made by randomly shuffling the redshifts among the galaxies. Using this test we can rule out the presence of substructure at the 90% confidence level.

In order to measure the cD peculiar motion we must measure the mean velocity (μ) and velocity dispersion (σ), and the associated uncertainties in these numbers. From the study of Beers et al. (1990) we conclude that the biweight estimator is the best estimator of μ and σ for our cluster data. The biweight estimator uses part of the velocity distribution and is not influenced by fluctuations in the outer parts of the velocity distribution as are more conventional estimators.

The cD galaxy in A2634 has a radial velocity v_{cD} =9113±59 km/s. The cluster velocity dispersion is σ=976 ±82 km/s. The results for the cD peculiar motion are shown below. The first line lists the cluster velocity as measured using three estimators. The second line lists the cD peculiar motion caclulated using the three estimators.

Summary of Results for cD Motion (km/sec)

using:	mean	median	biweight
A2634	9376±111	9269±117	9267±103
cD-V_{pec}	263±133	156±140	154±133

We do not measure a significant peculiar motion for the cD galaxy. The above analysis suggests that previous large values for the peculiar motion were due to a)too small a sample of cluster redshifts b)use of non-robust and non-resistant location estimators (see also Gebhardt and Beers, 1990).

We have also applied 10 tests to the velocity distribution to test for non normality. These tests have been described by D'Agostino (1986). The 3σ clipping method of Yahil and Vidal (1977) was applied to the data to remove suspected field galaxies. The first five tests we used were the a, u, w, skewness and kurtosis tests. The a and w and kurtosis statistics are sensitive to a long tailed distribution. The u test tests for contamination by extreme values. None of the above tests produced evidence for non-normality with a significance higher than the

10% level. The second group of tests which we applied are based on empirical distribution function statistics. These tests involve measuring the discrepancy between the cumulative distribution function of the normal distribution function and the empirical distribution function. We have applied five tests based on this method: the Kolmogorov test, W^2, A^2, U^2, and the Kuiper test. None of the above tests produced evidence for non-normality with a significance higher than the 10% level. We conclude that the velocity histogram for A2634 does not deviate significantly from that expected for a normal distribution.

CONCLUSION

The goal of this project is to study the dynamics of the cluster A2634 with a view to finding the mechanism that gave rise to the wide angle tail radio galaxy at the cluster center. We list below the conclusions based on our analysis of the data. cD peculiar motion: We do not find evidence for a significant difference between the cD velocity and the cluster velocity. The amplitude that is measured is sensitive to the statistic used. It is important to use a robust and resistant statistic when measuring the cluster mean velocity. Substructure: We have applied the test developed by Dressler and Schectman to search for superposed clumps in redshift. We do not find evidence of such clumps in our data set. Non-normality: We have applied a series of statistical tests to see if the radial velocity distribution deviates significantly from a normal distribution. After 3σ clipping we do not find evidence of a non-normal velocity distribution.

The data presented here suggest that A2634 may be a cluster in virial equilibrium. The cD peculiar motion if present is small and cannot account for the bending of the radio jets either by a passive wake or by ram pressure of the intra-cluster medium. Future work will involve 1)studying other rich clusters to see if A2634 is typical of clusters containing wide-angle tail radio galaxies and 2)theoretical work to find the mechanism that bends the jets of the radio galaxy.

REFERENCES

Beers, T.C., Flynn, K., and Gebhardt, K., 1990, *Astron. J.*, **100**, 32

Bothun, G.D., and Schombert, J.M., 1990, *Astrophys. J.*, **360**, 436

D'Agostino, R.B., 1986, in *Goodness-of-Fit Techniques*, Eds. R.B. D'Agostino and M.A. Stephens (New York: Marcel Dekker), p. 367

Dressler, A., and Schectman, S.A., 1988, *Astron. J.*, **95**, 985

Eilek, J., Burns, J., O'Dea, C., and Owen, F., 1984, *Astrophys. J.*, **278**, 37

Gebhardt, K., and Beers, T.C., 1990, preprint

Scott, J.S., Robertson, J.W., and Tarenghi, M., 1977, *Astr. Ap.*, **59**, 23

Yahil, A., and Vidal, N.V., 1977, *Astrophys. J.*, **214**, 347

A SEARCH FOR DARK CLUSTERS IN REDSHIFT CATALOGS

John M. Dickey
University of Minnesota, Minneapolis, MN 55455

ABSTRACT

In this work I attempt to set limits on the abundance of "dark clusters", i.e., deep gravitational potential wells similar to those in clusters of galaxies but without the galaxies. Two methods based on statistics of galaxy velocities in redshift surveys are discussed. Both methods turn up some objects which look like groups but with large velocity dispersions (more than 1000 km s^{-1}). Some of these may be larger scale structures seen in projection, while others are compact groups which seem to be bound. Until the statistics of these objects are sorted out we cannot put a reliable upper limit on the density of cluster-like potential minima. Meanwhile, some of the objects which stand out in the analysis warrant further observational study.

INTRODUCTION

One of the most important observational advances of the last decade for studies of large scale structure is the redshift surveys done by the Center for Astrophysics group (e.g. Huchra et al., 1983, 1990) and the Southern Sky Redshift Survey (da Costa et al., 1987). As a tremendous favor to the community, the CfA group have made these and other redshift survey data available on magnetic tape (ZCAT). The kinematic picture of the nearby universe presented by these catalogs is compelling. In particular, clusters of galaxies stand out, in part because they have been preferentially studied at high redshifts; but even in the domains over which the surveys are complete the clusters make striking "finger" patterns, i.e. spurious radial features caused by the high random velocities of the member galaxies. [A similar effect is detected by Bahcall, Soneira and Burgett (1986) for cluster catalogs.] Clusters are particularly noticeable as fingers because they contain so many galaxies, but it is interesting to ask whether there are structures with similar finger shapes but many fewer galaxies. If not, can we set upper limits on their density in the local universe ? This is not an entirely academic question, since in the context of theories of biased galaxy formation one might expect local minima of the gravitational potential at early times to initiate galaxy formation. Deep minima evolve into clusters because galaxy formation is particularly efficient there, but shallower local minima would be much less efficient at forming galaxies, and would be barely traceable today.

Departures from the Hubble flow caused by the fluctuating gravitational potential are statistically understood in terms of the cosmic virial theorem (Peebles, 1980). That approach gives the expectation value of the mean square relative motion of pairs of galaxies vs. pair separation from the density autocorrelation function and the mean mass density, but it does not tell how to isolate specific regions of particular interest in a given catalog. Relative velocities of pairs of galaxies in the CfA and SSRS with various separations are considered by Charlton and Salpeter (1991), who find statistical evidence for large halos of dark matter, extending as much as 1 to 2 Mpc around typical galaxies. Specific larger scale concentrations of mass in the nearby universe can be traced by a technique introduced by Bertschinger and Dekel (1989) using galaxies with distances measured by the Dn-σ relation or the Fisher-Tully relation. For such a sample the measured velocity gives directly the departure from Hubble flow; assuming that the random velocity field is irrotational allows the derivation of the "velocity potential", whose gradient is the velocity field. This in turn can be related to the gravitational potential. In this study I am neither as ambitious as the latter approach nor as complete as the former. I study galaxies in the CfA and SSRS catalogs, most of which have no independent distances, but I do not restrict the analysis to statistics of pairs of galaxies. The objective is to set limits on the density of "dark clusters", or groups with anomalously large velocity dispersions suggestive of dark mass concentrations.

ANALYSIS

Two approaches to the problem are considered here. The first is based on gridding, which gives a "cube" of three dimensional pixels whose value is the number of galaxies (or the total light) in a volume element of R.A. - Dec. - velocity space. The cube is conveniently analysed and displayed using the AIPS package of spectral line software, which is fine for crude manipulation, but has fundamental limitations for statistics. The main difficulty is the observational distinction between angular position and velocity measurements, which makes it problematic to construct a Cartesian three dimensional coordinate system. In any case, to resolve the clusters requires such small grid cells that the vast majority are empty, so that Poisson fluctuations and the gross spatial correlation of galaxies dominate the appearance of the cube. Thus statistics of alignments of grid cells with high occupation are not convenient to use either to search for dark clusters or to put limits on their abundance.

The gridded data can at least be used to point up specific regions worthy of scrutiny. Figure 1 shows the velocity distribution of galaxies in grid points selected to have seven galaxies per square degree. Most of these are dominated by groups with

... k_v ...	i_α	j_δ	Notes
0000000000131000200000000000000000000000000000	10	31	
0000000000000000000000121110100000000000000000	39	29	
0000000000000000000000220100110000000000000000	46	50	
0000000002121000010000000000000000000000000000	50	21	near A1367
0022200000100000000000000000000000000000000000	51	26	
0001600000000000000000000000000000000000000000	63	56	
0000000001011110110000000000000000000000000000	64	28	near A1656(=Coma)
4000000000000000000000000010000000000000000101	83	36	
0000000000000000000000132000001000000000000000	113	25	
↑ 3750 velocity (step = 250 km s^{-1}) 15000 ↑			

FIGURE 1 :

Velocity distribution of galaxies in selected square degree cells which contain 7 galaxies each. This is an example of many pages designed to aid the search for things that would look like groups if we had no redshift information, but which are in fact spread more widely in velocity. Each line on the table corresponds to a one square degree cell (i,j) which has (in this case) 7 galaxies in the range $3750<cz<15000$ km s^{-1}. The velocity step between digits (k) along the rows is 250 km s^{-1}. Typically the galaxies are all in a few adjacent velocity cells, corresponding to a group, with perhaps one or two foreground or background galaxies. Occasionally the "group" spreads over a range of 2000 km s^{-1} or so. Some of these cases are near rich clusters, but others are isolated.

tight velocity distributions, often with one or two foreground or background galaxies. In some cases the galaxies are spread over a wide velocity range (2000 to 3000 km s^{-1}). It would be interesting to observe some such regions, for example to attempt distance determinations using the Fisher-Tully relation. However without independent distances we cannot interpret such a coincidence as a bound structure, since the spatial correlation of galaxies would predict many more such "coincidences" of galaxies than expected from Poisson statistics. Outside clusters galaxies are typically found in filamentary or sheet-like structures, so some of the structures picked out on figure 1 may be simply filaments seen nearly end-on (e.g. Einasto et al., 1984).

TREES AND ELLIPSOIDS

An objective approach to organizing galaxies into groups and clusters is to construct "trees" by joining all neighbors nearer than some search radius into a data structure which as no closed loops. I follow the "friends of friends" technique (e.g. Ramella, Geller, and Huchra, 1989, and references therein, Einasto, 1988 and

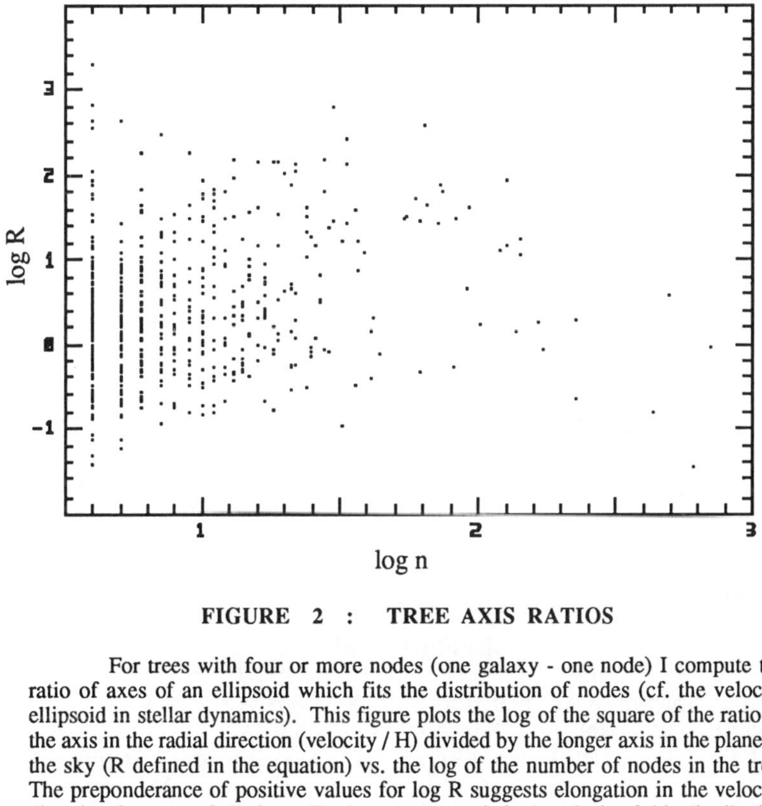

FIGURE 2 : TREE AXIS RATIOS

For trees with four or more nodes (one galaxy - one node) I compute the ratio of axes of an ellipsoid which fits the distribution of nodes (cf. the velocity ellipsoid in stellar dynamics). This figure plots the log of the square of the ratio of the axis in the radial direction (velocity / H) divided by the longer axis in the plane of the sky (R defined in the equation) vs. the log of the number of nodes in the tree. The preponderance of positive values for log R suggests elongation in the velocity direction for trees of all sizes. To do a proper statistical analysis of this distribution will take **further study**.

references therein), using a pillbox shape with radius 2 h^{-1} Mpc and velocity half height 200 km s^{-1} for the "bubble" within which to search for "friends". About 60% of the galaxies are assigned to trees in this case, although about 10% of these are in "trees" with only two members.

For each tree we can fit an ellipsoid by computing the mean square difference from the center of gravity in each coordinate (and the cross terms). The distribution of axis ratios is shown on figure 2, here defined as :

$$R = \frac{<(v - \bar{v})^2 H_o^{-2}>}{\sqrt{<(x - \bar{x})^2>^2 + <(y - \bar{y})^2>^2}}$$

where x and y are separations in the plane of the sky, and v is the separation in velocity (h cancels out of R). The apparent preponderance of trees with velocity dispersion much greater than spatial dispersion (log $R > 0$ on figure 2), even for trees with small numbers of elements, suggests that the trees found are typically elongated in the velocity direction more than in the spatial plane. This is to be expected, since groups as well as clusters are typically gravitationally bound. The next step is to study distributions of second moments like these for trees constructed with various size "bubbles" and to select various subsets of the catalog with particular completeness characteristics. This is underway.

DISCUSSION

The objective of this study is to set a limit on the abundance of "dark clusters" through analysis of the statistics of "fingers" in redshift surveys. Here "dark clusters", i.e. gravitational potential minima similar to those of clusters but without the galaxies, must have at least a few galaxies close enough to have their velocities perturbed from the Hubble flow in a noticeable pattern. The analysis so far would turn up groups with velocity dispersions of 2000 km s^{-1} or so. In fact, it does turn up some such objects, Arp 330 is an example. [Arp 330 has six galaxies in a chain of length 5' with velocities scattered from 7769 to 9005 km s^{-1}, with 10' away a galaxy at 9600 km s^{-1}. Bahcall, Harris, and Rood (1984) find X-ray emission which strongly suggests that Arp 330 is a bound group at the bottom of a deep gravitational potential well.] This group is typical of many objects which stand out in both the cube analysis and the tree analysis. Many of these may simply be large scale structures, filaments or sheets, in special orientations. Whether some may be dark clusters will require much further study; especially valuable would be X-ray observations and distance measurements for the individual galaxies.

REFERENCES

Bahcall, N. A., Harris, D.E., and Rood, H.J., 1984, *Ap. J. Lett.* **284**, L29.
Bahcall, N.A., Soneira, R.M., and Burgett,W.S., 1986, *Ap. J.* **311**, 15.
Bertschinger,E. and Dekel, A., 1989, *Ap. J. Lett.*, **336**, L5.
Charlton, J. C. and Salpeter, E. E., 1991, *Ap.J.*, in press (see article in this volume).
da Costa, L., Pellegrini, P., Sargent, W.L.W., Tonry, J., Davis, M., Meiksin, A., and Latham, D., 1987, *Ap. J.* **327**, 544.
Einasto, M., 1988, *Mon. Not. R. Astr. Soc.* **234**, 37.
Einasto, J., Klypin, A. A., Saar, E., and Shandarin, S.F., 1984, *Mon. Not. R. Astr. Soc.* **206**, 529.
Huchra, J., Davis, M., Latham, D., and Tonry, J., 1983, *Ap. J. Supp*, **52**, 89.
Huchra, J., Geller, M., de Lapparent, V., and Corwin, H., 1990, *Ap. J. Supp.* **72**, 433.
Peebles, P. J. E., 1980, **The Large Scale Structure of the Universe**, (Princeton : Princeton University Press), pp. 280 - 284.
Ramella, M, Geller, M.J., and Huchra, J.P., 1989, *Ap. J.* **344**, 57.

OBSERVATIONAL EVIDENCE FOR DARK MATTER FROM PECULIAR VELOCITIES IN THE LOCAL UNIVERSE

Ofer Lahav
Institute of Astronomy, Madingley Road, Cambridge CB3 0HA, UK

ABSTRACT

Deviations from the Hubble flow in the local Universe provide a direct measure of the underlying total mass distribution, assuming the gravitational instability picture. A comparison of the mass distribution deduced from the peculiar velocities with the galaxy distribution could tell us the amount of dark matter on large scales. We discuss the origin of motion of the Local Group with respect to the Cosmic Background Radiation and review the peculiar velocity field deduced from distances to hundreds of elliptical and spiral galaxies, including new results for the Shapley Supercluster. The density parameter Ω_0 deduced from the relation between the peculiar velocity and acceleration (as inferred from galaxies) suggests an open universe if light traces mass. However, it is still possible that there is enough matter to close the Universe if the galaxy distribution is more strongly clustered ('biased') than the mass distribution. We also consider the possibility of a non-zero cosmological constant Λ and 'anti-bias'.

1. INTRODUCTION

Peculiar velocities have become a major probe in Cosmology in recent years, together with the Background Radiations and the clustering of galaxies. Assuming the gravitational instability picture, they provide a direct probe of the underlying mass distribution. One can compare the measured peculiar velocities with those predicted from the distribution of luminous galaxies, or vice versa.

Given the rapid progress in this field it is important to revisit the basic questions: (i) Are the peculiar velocities real? (ii) Does light follow mass? (iii) Do we really test the gravitational instability picture? (iv) Do the observed peculiar velocities allow us to distinguish between different cosmological scenarios?

The outline of this paper is as follows: In §2 we discuss the origin of motion of the Local Group as deduced from the dipole anisotropy in the distribution of optical and infra-red galaxies. In §3 we review the observations of line-of-sight peculiar velocities of ellipticals and spirals. In §4 we discuss what can be inferred from the velocities in terms of cosmological parameters : Ω_0, Λ and biasing. In §5 we discuss the results and future work.

2. THE ORIGIN OF MOTION OF THE LOCAL GROUP

The best known peculiar velocity is that of the sun as deduced from the dipole anisotropy of the Cosmic Background Radiation (CBR). The COBE measurement of the dipole (e.g. Smoot in these Proceedings) confirmed the earlier measurements (e.g. Lubin & Villela 1986). The dipole in the temperature map indicates a motion of the sun at about 360 km/sec with respect to the CBR towards Galactic coordinates $l = 265°; b = 50°$. A similar dipole, although less obvious, appears in the X-ray Background (Shafer & Fabian 1983,

Boldt 1987). Here we make the standard assumption that the sun moves at 300 km/sec towards $l = 90°; b = 0°$ with respect to the Local Group. This gives a net motion of the Local Group with respect to the CBR of 600 km/sec towards $l = 268°; b = 27°$. This motion is shared by the other galaxies within 500 km/sec from us (the so called 'Coma-Sculptor' cloud, see e.g. Faber & Burstein 1988). It is also important to note that the motion of the Local Group has a significant component *perpendicular* to the Supergalactic plane ($SGZ \approx -370$ km/sec).

One way of exploring the cause of this motion is to study the dipole moment of the galaxy surface brightness distribution in a whole sky catalogue. Since both gravity and flux fall as the square of distance, the surface brightness dipole can measure the force acting on the Local Group due to the matter represented by galaxies in a flux-limited catalogue. The flux dipole **P** is calculated by summing up the unit vectors of galaxies $\hat{\mathbf{r}}_i$ weighted by their flux S_i:

$$\mathbf{P} = \frac{3}{4\pi} \sum_i S_i \, \hat{\mathbf{r}}_i. \tag{1}$$

In principle, this should give a model-independent estimate of the acceleration without knowing the distances to galaxies. However, there are a few problems in applying this method: (i) catalogues are flux-limited, so some modelling of the density field is required; (ii) one has to assume, somewhat arbitrarily, that the distribution of light at a certain wavelength is correlated with the underlying mass distribution; (iii) the sky coverage is not complete, mainly due to the Galactic plane.

In linear theory the acceleration **g** is parallel to the peculiar velocity **v** (Peebles 1980):

$$\mathbf{v} \propto f(\Omega_0, \lambda_0) \, \mathbf{g}, \tag{2}$$

where $f(\Omega_0, \lambda_0) \equiv (\frac{d(\ln \delta)}{d(\ln a)})_0 \approx \Omega_0^{0.6}$ measures the rate of growth of the fluctuation δ with respect to the scale factor a at the present epoch. It mainly depends on the density parameter Ω_0 and only very weakly on the cosmological constant $\lambda_0 \equiv \Lambda/(3H_0^2)$ (where H_0 is the Hubble constant). This relation can therefore be used to find Ω_0. It is worth clarifying (as emphasized by J. Ostriker) that eq. (2) is basically derived from the linearized continuity equation, so it does not test directly the gravitational instability picture. However, the derivation also assumes that the rate of growth of perturbations is independent of position, a property which gravity gives naturally. Moreover, it is interesting that values of the constant of proportionality between velocity and acceleration do give 'reasonable' values for Ω_0 when interpreted within the gravitational instability framework. So gravity is assumed and tested for consistency, rather than proved, by comparing the density and velocity fields using linear theory.

Another important point is that the acceleration **g** is deduced from the distribution of galaxies (e.g. from the dipole **P**), not from the distribution of matter. So if, for example, the fluctuation in galaxy density $\delta n/n$ is related to the fluctuation in matter density $\delta\rho/\rho$ by

$$\frac{\delta n}{n} = b \, \frac{\delta \rho}{\rho}, \tag{3}$$

where the b is the 'bias parameter', then one measures $\Omega_0^{0.6}/b$ by equation (2), not the 'true' Ω_0. The flux dipole method has been applied to several samples under the assumption that galaxies in a given sample trace the mass distribution perfectly. However, for example an optical sample of galaxies shows a stronger clustering than the Infra Red Astronomical Satellite (IRAS) sample, which only includes late-type galaxies. So it is clear that at least one of the samples (and most likely both) is not a perfect tracer of the mass distribution.

The flux dipole of galaxies was found (e.g. Yahil, Walker & Rowan-Robinson 1986) to lie within 20° − 30° of the CBR dipole. More recently the IRAS dipole has been deduced from a 2 Jy IRAS redshift survey (e.g. Strauss & Davis 1988; Yahil 1988 ; Strauss 1989) and from a One-in-Six 0.6 Jy survey (e.g. Kaiser in these Proceedings). Using the redshifts of galaxies they have calculated the force acting on *each* galaxy and corrected for its peculiar velocity in an iterative way.

Lahav (1987) has merged three optical diameter-limited catalogues: UGC (Nilson 1973), ESO (Lauberts 1982), and MCG (Vorontsov- Velyaminov & Arkipova 1963-8) in the 'missing strip' of 15°. The combined optical catalogue is shown in Figure 1.

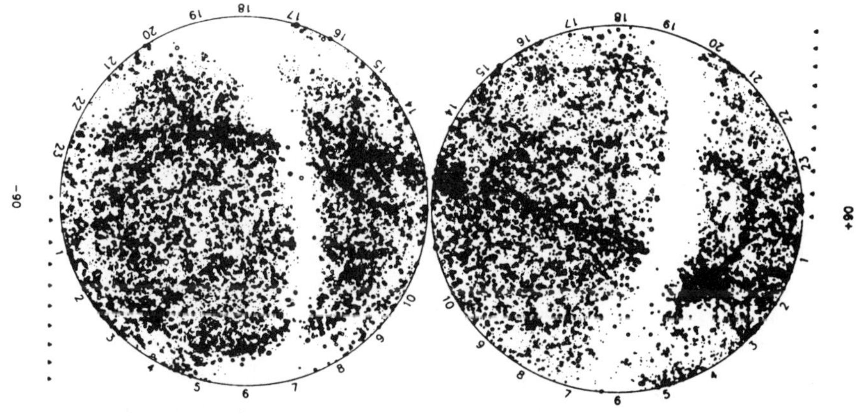

Figure 1 . An equal area projection (by Lynden-Bell & Lahav) of the 2 hemispheres in Equatorial coordinates of all galaxies in the UGC, ESO and MCG (15°) catalogues. The most prominent feature is the Supergalactic plane. The empty bands are due to the Galactic obscuration.

Lahav, Rowan-Robinson & Lynden-Bell (1988) and Lynden-Bell, Lahav & Burstein (1989) have revised the calculation of the optical dipole using a new calibration of the different catalogues, partial redshift information and a better correction for extinction near the Galactic plane. The characteristic depth of the combined catalogue is about 7000 km/sec.

The optical dipole is found to lie only 7° − 15° away from the CBR dipole. About 80 % of it is produced within a radius of 4000 km/sec. The major sources of the dipole are Virgo-Ursa Major, the Centaurus/Great-Attractor region (see nomenclature below) and the Local Void, which is pushing (together with a possible pull by Fornax and Eridanus) the Local Group *perpendicular* to the Supergalactic plane. However, we would like to emphasize that it is difficult to explain the motion of the Local Group (and probably of any other single galaxy) by simple models of two to three attractors or repulsors . The gravity acting on a galaxy is indeed an N-body problem.

It is interesting to look at the dipole using other tracers. For example Miyaji and Boldt (1990; also in these Proceedings) defined a sample of 70 low redshift AGNs from the HEAO1 A-2 whole-sky X-ray map. Although the sample is very sparse, it is interesting that they found a dipole which is within 30° of the CBR dipole and is mainly due to AGNs within 4500 km/sec. Another possible tracer is defined by samples of Abell clusters (Scaramella *et al.* 1989 and in these Proceedings; Plionis & Valdarnini 1990) and X-ray clusters (Lahav, Edge, Fabian & Putney 1989). The advantage of clusters is that they go deeper than the galaxy all-sky catalogues. However, the clusters only sample the high peaks of the density field, without showing its gradients. Another problem with the Abell sample is that it only includes clusters at distances larger than 5000 km/sec or so (e.g. Virgo is missing), so the dipole computed represents the force acting on a 5000 km/sec sphere, rather than on the Local Group. Optical clusters also tend to disappear faster than galaxies towards the Galactic plane.

Given a sample of a finite depth, it is always possible that there is a contribution to the acceleration due to matter outside the sample. An interesting 'geographical' example is the Shapley Supercluster (Shapley 1930, Scaramella *et al.* 1989, Lahav *et al.* 1989, Raychaudhury 1989) which is at the same direction of Centaurus/Great-Attractor, but at a distance of 14000 km/sec. Although this is an impressive supercluster, it is likely that the $1/R^2$ law of gravity attenuates its effect on the motion of the Local Group. Raychaudhury (1989) estimated the contribution of the Shapley Supercluster to be no more than 15 % of that of the optical dipole due to nearby galaxies.

One can also approach the problem in a statistical way, by assuming an *a priori* model and predicting the convergence and misalignment expected, given the condition that the Local Group is moving at 600 km/sec (Juszkiewicz, Vittorio & Wyse 1990; Lahav, Kaiser & Hoffman 1990; Davis, Strauss & Yahil 1990). For example, in a 'slightly biased' Cold Dark Matter model, 85% of the hypothetical observers will measure a misalignment less than 20° due to the matter out to 4000 km/sec, while in an Isocurvature Baryon model (with $\Omega_0 = 0.4$ and primordial spectral index of -1) only 8% will measure such a misalignment. However, the statistical approach is not necessarily conclusive, since we live after all in a special position! For example, it is possible that in a Universe with a large amount of power on large scales, the Local Group's acceleration is converging faster than expected due to a balance between the Centaurus/Great-Attractor and Perseus-Picses superclusters.

3. OBSERVATIONS OF PECULIAR VELOCITIES

It is possible to deduce the line-of-sight peculiar velocity of a galaxy by measuring its redshift (the easy part) and its distance (the more difficult part), i.e. $V_p = cz - H_0 r$. This way of mapping the peculiar velocity field was pioneered by Rubin *et al.* (1976) and in recent years several groups (e.g. Aaronson *et al.* 1986, Lynden-Bell *et al.* 1988 and Dressler in these Proceedings) have used various distance indicators to get more accurate peculiar velocities. The infrared Tully-Fisher (luminosity-line width) relation for spirals and the diameter-velocity dispersion ($D_n - \sigma$) relation for ellipticals give errors in distance of about 15% and 20% per galaxy, respectively. Another complication in deducing the peculiar velocity field is due to the so-called 'Malmquist bias'. For a review of the observational aspects see Burstein (1990). Given that the distance indicators are based on empirical relations it is questionable how universal they are (e.g. Silk 1989). The reality of the peculiar velocities can be tested by comparison to the galaxy distribution (e.g. the prediction of the 'Great-Attractor' from the velocity field by Lynden-Bell *et al.* 1988 was followed by an identification of a large concentration of galaxies in the optical catalogues - see their Figure 8). Another approach is to search for dependence on environment. No correlations were found with richness of clusters (Burstein *et al.* 1990) or with X-ray properties of clusters (Edge & Lahav, unpublished). Another approach is to compare independent distance indicators. Although a comparison of distances to spirals and ellipticals looks encouraging up to a 'zero-point' calibration (e.g. Bertschinger 1990), there are several alarming cases. A particular example is the cluster A2634 at $cz = 9000$ km/sec. Lucey *et al.* (1990) found from ellipticals and S0's that the cluster has a very large peculiar velocity, $V_p = -3400 \pm 600$ km/sec, while the Tully-Fisher relation for the cluster gives $V_p = +300$ km/sec. Although this might be due to the different assignment of ellipticals and spirals into groups, this case may indicate that environmental effects do take place.

Another related point is that the galaxies are not necessarily arbitrary test particles of the velocity field. Carlberg *et al.* (1990) found in simulations a 'velocity bias' which they attributed to dynamical friction.

Several groups are continuing to measure peculiar velocities of ellipticals and spirals. One of the $D_n - \sigma$ surveys is by Lucey (Durham), Lahav and Lynden-Bell (Cambridge), Terlevich (RGO) and Melnick and Infante (ESO). In this survey we have carried out fibre-spectroscopy (at the AAT) and V-band CCD photometry (at the ANU and Chile) for 15 clusters (with 10-20 galaxies in each) at the redshift range $3000 \leq cz \leq 14000$ km/sec.

One of the clusters in our sample is Centaurus. Its motion played a major role in leading the 'Seven Samurai' to propose a massive object *behind* Centaurus. There is continuing confusion over the terminology related to the Centaurus/Great-Attractor region. Unfortunately, astronomers are using their own different definitions for the 'Great-Attractor', to either prove or disprove its existence! Appendix A of Lynden-Bell, Lahav and Burstein (1989) gives a 'dictionary' to this region (see also Burstein *et al.* 1990). Here we refer to the 'Great-Attractor region' as the entire steradian on the sky at a depth of 2000-6000 km/sec, *including* Centaurus.

Whatever the definition for the Great-Attractor is, it is important to understand what is happening in Centaurus. Figure 2 shows the Hubble diagram for Centaurus ($12^h12^m < \alpha < 13^h20^m; -35° < \delta < -47°$) using available data from recent surveys (see Lucey *et al.* 1989 for details). The velocity field seems to be very complicated. On the one hand, if we look at the entire region as one unit we get a net outward motion of 800 km/sec, in accord with the 'Seven Samurai' (see also Burstein *et al.* 1990). On the other hand one can argue (Lucey *et al.* 1986) that Centaurus is made out of 2 clumps, at redshifts of 3000 km/sec (Cen30) and 4500 km/sec (Cen45), such that both are at about the same distance: Cen30 is nearly at rest with respect to the CBR, while Cen45 is falling towards Cen30 and away from us at 1500 km/sec. We find that with the present peculiar velocity data it is difficult to distinguish between these 'bulk motion' and 'local infall' models for Centaurus (for more discussion see Lucey *et al.* 1989). A correlation of the distribution of galaxies with the peculiar velocity in this region may shed some light on this problem.

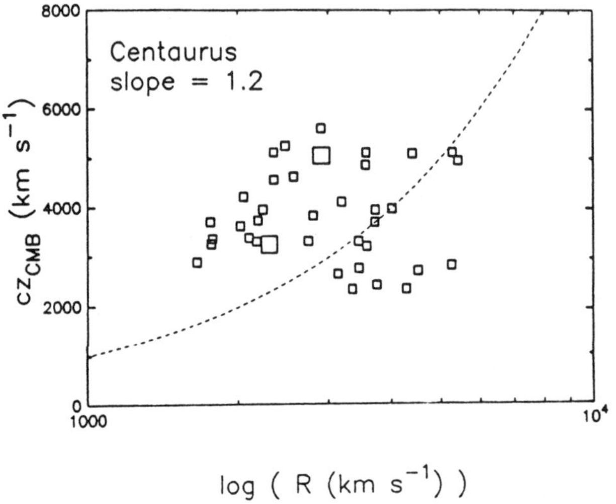

Figure 2 . The Hubble diagram, redshift *vs.* log(derived distance), for early-type galaxies in Centaurus (from Lucey *et al.* 1989). The distance is derived assuming a $D_n \propto \sigma^{1.2}$ relation. The Hubble line is marked by the dashed line. The two big squares indicate large ellipticals.

In the same direction of Centaurus we also measured the peculiar velocity of Shapley 8, a member of the Shapley Supercluster at 14000 km/sec. At such a large distance a peculiar velocity of 600 km/sec can only be detected by using at least $\{20\%/(600:14000)\}^2 = 25$ elliptical galaxies. Using only 20 ellipticals and S0's we found for this cluster $V_p = -300\pm^{1300}_{600}$ km/sec (where the error bars were deduced by using a 'bootstrap' analysis). So the cluster seems to be at rest with respect to the CBR, but the errors are fairly large (as expected). We find that 3 other clusters (A2063, D0107 and D0247) are also consistent with being at rest with respect to the CBR.

There is clearly a need for new techniques to deduce peculiar velocities at large distances. One promising method is the *kinematic* SZ effect (Sunyaev & Zeldovich 1980). This effect measures directly the peculiar velocity of a cloud of electrons from the anisotropy of the CBR, $\frac{\Delta T}{T} = -\frac{V_p}{c}\tau$, where τ is the optical depth. The intensity variation $\frac{\Delta I}{I}$ with frequency shows that the kinematic effect has a maximum at a frequency of 218 GHz (for the COBE CBR temperature of $2.735^o K$), exactly where the *thermal* SZ effect vanishes. In a band centred on this frequency one gets

$$(\frac{\Delta I}{I})_K = -5 \times 10^{-4}\ (V_p/1000\ \mathrm{km/sec})(\tau/0.01). \tag{4}$$

Therefore if clusters move at 2000-3000 km/sec they should be easily detected. Rephaeli and Lahav (1990) discuss some practical aspects and provide a list of cluster candidates to look for the effect. Another promising method is the fluctuations analysis (Tonry *et al.* 1989).

4. COSMOLOGICAL INFERENCE

What can we learn from the local peculiar velocities about Cosmology? One approach is to compare the velocities to the galaxy distribution and to deduce the combination of $\Omega_0^{0.6}/b$, assuming constant biasing. As emphasized in §2 there is obviously some sort of biasing on purely phenomenological grounds (e.g. IRAS vs. optical). It was also shown (Juszkiewicz *et al.* 1990, Lahav *et al.* 1990) that the deduction of Ω_0 is model-dependent and is valid only if not much power is expected on large scales. Despite all these problems, it is nevertheless interesting to see the values deduced. The optical dipole gives $\Omega_0^{0.6}/b \approx 0.4$ (Lynden-Bell, Lahav & Burstein 1989), while the IRAS one gives (Strauss 1989) $\Omega_0^{0.6}/b \approx 0.8$. A simple interpretation of this discrepancy is that optical galaxies are biased relative to the mass distribution in a different way than IRAS galaxies. This trend is supported by a direct comparison of the optical and IRAS correlation functions (Lahav, Nemiroff and Piran 1990; Babul & Postman 1990). An important step forward is provided by a comparison of the velocity field and galaxies at many points in space. Bertschinger & Dekel (1989) developed a reconstruction method in which the peculiar velocities are integrated along radial rays to give the 3-D potential (assuming a curl-free velocity field) and hence the 3-D density field. A comparison of their mass density field to the IRAS density field gives (Bertschinger 1990, Yahil 1990) $\Omega_0^{0.6}/b \approx 1.0 \pm 0.3$, somewhat higher but still consistent with the estimate from the IRAS dipole (see also Kaiser in these Proceedings for a similar estimate). A similar comparison with optical galaxies is also under the way (Bistolas, Dekel, Bertschinger & Lahav).

Another interesting question is whether such a comparison can teach us about the cosmological constant $\lambda_0 \equiv \Lambda/(3H_0^2)$. For a flat Universe ($\Omega_0 + \lambda_0 = 1$) Peebles (1984) has shown that $f(\Omega_0, \lambda_0)$ (see eq. 2) depends only weakly on λ_0. Lahav & Primack (1990) have generalized this calculation for any values of Ω_0, λ_0 and redshift (see Figure 3). They showed that indeed at the present epoch ($z = 0$) dynamical tests are insensitive to the value of λ_0. However, they pointed out that by carrying out dynamical tests at higher redshifts ($z \approx 0.5 - 2.0$) it is possible to decouple these two important cosmological parameters (subject to factors of biasing and evolution).

Another approach to analyze the velocity field is to confront the observations directly with the velocities expected in theoretical models for the primordial density fluctuations. The most commonly used and robust statistic is the

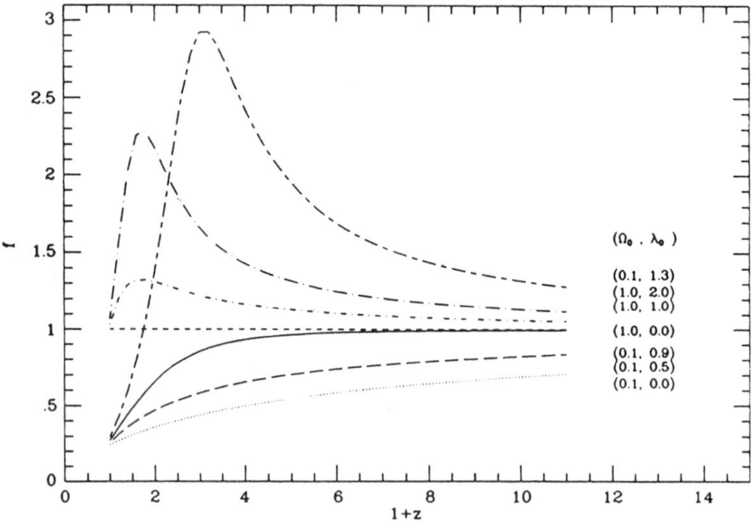

Figure 3 . The evolution of $f(\Omega_0, \lambda_0) \equiv (\frac{d(\ln \delta)}{d(\ln a)})$ with redshift for different values of the density parameter Ω_0 and the cosmological constant λ_0 (from Lahav & Primack 1990). Note that at $z = 0$, f is only sensitive to Ω_0, while at higher z it is becoming more sensitive to λ_0.

bulk flows statistic (e.g. Kaiser 1988). The measurement of the flow on a given scale R_* is compared with the rms expected in a hypothetical Universe specified by a given power-spectrum $P(k)$ and a 'window function' $W(R_*k)$ which describes the observational set-up:

$$\langle V_p^2(R_*)\rangle = H_0^2\Omega_0^{1.2} \int d^3k P(k) W(R_*k). \tag{5}$$

The freedom here is the normalization of the power spectrum which can be specified by the rms fluctuation σ_ρ on 800 km/sec sphere. Cold Dark Matter is consistent with the measured flows only for a high normalization (low 'biasing') of $\sigma_\rho \approx 0.6$ (e.g. Kaiser & Lahav 1989). Other statistics proposed recently are the velocity correlation function (Gorski *et al.* 1989; Groth *et al.* 1989) and the 'Mach number' (Ostriker & Suto 1989).

It is important to extend the analysis to the nonlinear regime, e.g. by N-body simulations, and to show what part of the present-day structure is due to gravity and what part is due to the initial conditions. Lilje & Lahav (1990) considered the infall around clusters in the mild non-linear regime using the idealized spherical infall model, and found that the final velocity profile 'remembers' the initial conditions for the density fluctuations only at distances larger than 1000 km/sec from the cluster centre.

5. DISCUSSION

The large galaxy and peculiar velocity samples offer the very exciting possibility of measuring the amount of dark matter on large scales. However, many of the deductions are model-dependent and require prior assumptions. If one prefers to adopt the 'paradigms' of inflation ($\Omega_0 + \lambda_0 = 1$), Big-Bang nucleosynthesis ($\Omega_B \approx 0.1$ for a reasonable choice of the Hubble constant) and some form of biasing (constant is clearly naive, but convenient for discussion) then two extreme possible pictures emerge. One is with $\Omega_B = 0.1$, large amount of non-baryonic matter ($\Omega_{NB} = 0.9$) and zero cosmological constant. In this picture (represented for example by the Cold Dark Matter and Hot Dark Matter models) the galaxies are clustered stronger than the underlying mass distribution. For example, if we adopt for the measurements $\Omega_0^{0.6}/b \approx 0.4$ for optical galaxies and $\Omega_0^{0.6}/b \approx 0.8$ for IRAS galaxies, then this model (in which $\Omega_0 = \Omega_B + \Omega_{NB} = 1.0$) gives $b_{opt} = 2.5$ and $b_{IRAS} = 1.3$.

An alternative (discussed by Peebles in these Proceedings) is to assume that all the matter in the universe is baryonic, but to add a cosmological constant of $\lambda_0 = 0.9$ in order to save inflation (i.e. zero curvature), so $\Omega_0 = \Omega_B$ and $\Omega_0 + \lambda_0 = 1$. However, such a model will suggest *anti-biasing*, i.e. that galaxies are less clustered than the matter. The above observational values for $\Omega_0^{0.6}/b$ will then suggest $b_{opt} = 0.6$ and $b_{IRAS} = 0.3$. This is not ruled out on the basis of astrophysical arguments. In fact we have evidence that IRAS galaxies under-represent the depth of the potential wells of clusters (as indicated by velocity dispersion and X-ray temperature). One problem with anti-biasing is that it will make the absence of anisotropies in the CBR even more embarrassing and will require new mechanisms to wash-out the fluctuations.

Regardless of any particular model, the next few years will hopefully see a detailed phenomenological comparison of the matter distribution deduced from the peculiar velocities and the galaxy distribution. This will allow us to see how 'biasing' varies with scale, where luminous galaxies prefer to reside and what the amount of dark matter is.

Acknowledgments. I thank my collaborators in the work presented here for their contribution and for many stimulating discussions. This work was supported in part by a research fellowship of St. Catharine's College and NSF grant no. AST-8802533.

REFERENCES

Aaronson, M., Bothun, G.D., Mould, J.R., Huchra, J., Schommer, R. & Cornell, M., 1986. *Ap. J.*, **302**, 536.

Babul, A. & Postman, M., 1990. *Ap. J.*, in press.

Bertschinger, E., 1990. in *Proceedings of the Moriond conference, March 1990.*

Bertschinger, E. & Dekel, A., 1989. *Ap. J.* (*Letters*), **336**, L5.

Boldt, E., *Phys. Rep.*, **146** (4), 215.

Burstein, D., 1990. *Rep. Prog. Phys.*, in press.

Burstein, D., Faber, S.M. & Dressler, A., 1990. *Ap. J.*, **354**, 18.

Carlberg, R.G., Couchman, H.M.P. & Thomas, P. A., 1990. *Ap. J.* (*Letters*), **352**, L29.

Davis, M., Strauss, M.A. & Yahil, A., 1990. *Ap. J.*, in press.

Faber, S.M. & Burstein, D., 1988. in *Large Scale Motions in the Universe*, Proceedings of the Vatican Study Week, eds. G. V. Coyne & V.C. Rubin, Princeton University Press,

Princeton.
Gorski, K., Davis, M., Strauss, M.A., White, S.D.M. & Yahil, A., 1989. *Ap. J.*, **344**, 1.
Groth, E.J., Juszkiewicz, R. & Ostriker, J.P., 1989. *Ap. J.*, **346**, 558.
Juszkiewicz, R., Vittorio, N. & Wyse, R., 1990. *Ap. J.*, **349**, 408.
Kaiser, N., 1988. *M.N.R.A.S.*, **231**, 149.
Kaiser, N. & Lahav, O., 1989. *M.N.R.A.S.*, **237**, 129.
Lahav, O., 1987. *M.N.R.A.S.*, **225**, 213.
Lahav, O., Rowan-Robinson, M. & Lynden-Bell, D., 1988. *M.N.R.A.S.*, **234**, 677.
Lahav, O., Edge, A.C., Fabian, A.C. & Putney A. ,1989. *M.N.R.A.S.*, **238**, 881.
Lahav, O., Nemiroff, R.J. & Piran, T., 1990. *Ap. J.*, **350**, 119.
Lahav, O., Kaiser, N. & Hoffman, Y., 1990. *Ap. J.*, **352**, 448.
Lahav, O. & Primack, J.R., 1990, in preparation.
Lauberts, A., 1982. *The ESO-Uppsala Survey of the ESO(B) Atlas*, European Southern Observatory.
Lilje, P.B. & Lahav, O., 1990. *Ap. J.*, submitted.
Lubin, P. & Villela, T., 1986. in *Galaxy Distances and Deviations from Universal Expansion*, pg. 169.,eds. B.F. Madore and R.B. Tully, Reidel, Dordrecht.
Lucey, J.R., Currie, M.J. & Dickens, R.J., 1986. *M.N.R.A.S.*, **222**, 427.
Lucey, J.R., Lahav, O., Lynden-Bell D., Terlevich, R. J. & Melnick, J., 1989. "Distances and peculiar velocities of clusters in Centaurus", in *Large Scale Structures and Motions is the Universe*, Rio workshop 1989, eds. D. Latham and N.L. Da Costa.
Lucey, J.R., Gray, P.M., Carter, D. & Terlevich, R. J. , 1990. *M.N.R.A.S.*, in press.
Lynden-Bell, D., Faber, S.M., Burstein, D., Davies, R.L., Dressler, A., Terlevich, R.J. & Wegner, G., 1988. *Ap. J.*, **326**, 19.
Lynden-Bell, D., Lahav, O. & Burstein, D., 1989. *M.N.R.A.S.*, **241**, 325.
Miyaji, T. & Boldt, E., 1990. *Ap. J.* (*Letters*), **353**, L3.
Nilson, P., 1973. *Uppsala General Catalogue of Galaxies*, Uppsala astr. Obs. ann., **6**.
Ostriker, J.P. & Suto, S., 1990. *Ap. J.*, **348**, 378.
Peebles, P.J.E., 1980. *The Large Scale Structure of The Universe*, Princeton University Press, Princeton.
Peebles, P.J.E., 1984. *Ap. J.*, **284**, 439.
Plionis, M. & Valdarnini, R., 1990. *M.N.R.A.S.*, in press.
Raychaudhury, S., 1989. *Nature*, **342**, 251.
Rephaeli, Y. & Lahav, O., 1990. *Ap. J.*, in press.
Rubin, V.C., Thonnard, N., Ford, W.K., Roberts, M.S., & Graham, J.A., 1976. *Astron. J.*, **81**, 719.
Scaramella, R., Baiesi-Pillastrini, G., Chincarini, G., Vettolani, G. & Zmorani, G., 1989. *Nature*, **338**, 562.
Shafer, R.A., & Fabian, A.C., 1983. in *Early Evolution of the Universe and Its Present Structure*, IAU Symp. 104, eds. G.O. Abell & G. Chincarini, Reidel, Dordrecht.
Shapley, H., 1930. *Harvard College Obs. Bull.*, **874**, 9.
Silk, J., 1989. *Ap. J.* (*Letters*), **345**, L1.
Strauss, M.A., 1989. Ph.D. thesis, University of California, Berkeley.
Strauss, M.A. & Davis, M., 1988. in *Large Scale Motions in the Universe*, Proceedings of the Vatican Study Week, eds. G. V. Coyne & V.C. Rubin, Princeton University Press, Princeton.
Sunyaev, R.A. & Zeldovich, Ya.B., 1980. *M.N.R.A.S.*, **190**, 413.
Tonry, J.L., Ajhar, E.A. & Luppino, G.A., 1989. *Ap. J.* (*Letters*), **346**, L57.
Vorontsov-Velyaminov, B.A., & Arkipova, A.A., 1963-1968. *Morphological Catalogue of Galaxies*, Moscow state University, Moscow.
Yahil, A. 1988. in *Large Scale Motions in the Universe*, Proceedings of the Vatican Study Week, eds. G. V. Coyne & V.C. Rubin, Princeton University Press, Princeton.
Yahil, A. 1990. in *Proceedings of the Moriond conference, March 1990.*
Yahil, A., Walker, D. & Rowan-Robinson, M., 1986. *Ap. J.*, **301**, L1.

THE X-RAY FLUX DIPOLE OF AGNS AS AN INDICATOR OF THE LOCAL GRAVITATIONAL FIELD

Takamitsu Miyaji[1,2], Keith Jahoda[1], and Elihu Boldt[1]
[1]Laboratory for High Energy Astrophysics, CODE 666,
NASA/Goddard Space Flight Center, Greenbelt, MD 20771
[2]Astronomy Program, University of Maryland,
College Park, MD 20742

ABSTRACT

We have investigated the dipole anisotropy of the sky X-ray flux from nearby AGNs to compare with the peculiar motion of the Local Group (LG) using a more extensive sample than that of our previous work (Miyaji and Boldt, 1990; hereafter MB). We have sampled 56 low redshift emission-line AGNs ($z<0.06$) from the HEAO 1 A-2 experiment with a flux limit lower than that of Piccinotti et al. (1982) which MB used and also with a lower galactic latitude cutoff ($|b|>15°$). Our sample shows a significant dipole anisotropy whose apex is only about 30° away from the direction of the Local Group's peculiar motion for the objects with $z<0.015$. The dipole growth vs redshift shows a sharp rise between $z=0.006$ and $z=0.015$; the amplitude of the dipole is 40±10% of the corresponding monopole at $z=0.015$. Using the shell model described in MB with a modification to include both the inner ($z=0.006$) and the outer ($z=0.015$) edges, we get a value of $b\Omega_0^{-0.6}=2.6\pm0.5$. The outer redshift cutoff is consistent with the results obtained from analyses using optical and IRAS galaxies. The present sample strengthens our previous conclusion that X-ray emission from AGNs traces the underlying mass distribution as strongly as optical and IR emission from galaxies.

INTRODUCTION

The dipole moment of the sky flux distribution from extragalactic objects is a convenient and simple measure describing a certain aspect of the inhomogeneity around us, especially as a tracer of the local gravitational field. According to linear perturbation theory (Peebles 1980) the motion of the Local Group (LG) with respect to the microwave background (towards $l=268°, b=27°$; hereafter the LG direction) is in the same direction as the local gravitational acceleration. Thus the dipole analysis gives a test for the notion that the luminous objects as traced by radiation reflects the underlying gravitational mass.

The flux (or [diameter]^2) dipoles of various tracers (IRAS galaxies,e.g. Yahil, Walker, and Rowan-Robinson 1986; Lahav, Rowan-Robinson, and Lynden-Bell 1988, hereafter LRL and optical galaxies, e.g. Lahav 1987; LRL; Lynden-Bell, Lahav and Burstein, 1989) have shown reasonable alignments with the LG direction. Our previous dipole analysis (MB) using X-ray fluxes from AGNs detected by the HEAO 1 A-2 experiment and catalogued by Piccinotti *et al.* (1982) has also shown this alignment. It is found that each tracer gives a different value of bias parameter b, i.e. $b\Omega^{-0.6}$~1 for IRAS galaxies and ~3 for optical galaxies and X-ray AGNs (see Lahav, this conference).

In this work, we have extended our sample of X-ray AGNs from the HEAO 1 A-2 experiment using a lower flux limit and lower galactic latitude cutoff in redoing the X-ray flux dipole analysis. The sample is almost fully identified with redshift values. Thus we can explore the behavior of the dipole in redshift space. The number of objects with the redshift range of particular interest ($z<0.02$) in this work is about twice that in MB and it may be particularly important to increase the number of objects in this counting statistics limited experiment.

THE SAMPLE

We have detected 56 AGNs ($z<0.06$) from the HEAO 1 A-2 experiment using less strict criteria and more sensitive methods than those of Piccinotti *et al.* Our new criteria and methods are: (1) A database with the 'Total Count Rate' has been used instead of R15 in Piccinotti et al. A two dimensional fitting program is used at positions of emission-line AGN identifications of LASS catalog by Remillard (private communication). (2) The objects with $\Delta\chi^2>4$ (change of χ^2 between the fits with and without the source, $\Delta\chi^2>30$ for Piccinotti *et al.*) have been treated as detected. With this criterion, about 2% of the detected sources at this marginal level are expected to be spurious. It is still much less than the errors associated with the counting statistics. A flux limit of 0.5 (Total Count Rate) is imposed to make a sample with a uniform sky sensitivity. (3) Sources with $|b|<15°$ ($|b|<20°$ in Piccinotti *et al.*) are excluded to avoid co nfusion problems with the galactic source.(4) An extremely bright source (Cen A, also the nearest one) is excluded.

RESULTS AND ANALYSIS

Fig. 1 shows the dipole growth versus redshift. Fig. 1 (a) shows some features of the growth of the dipole component along the LG direction: (1) Flat in $z\leq0.006$; (2) A steep rise in $0.006<z<0.015$; (3) A drop in $0.015\leq z<\sim0.04$ and a gradual flattening afterwards.

The X-ray flux dipoles for AGNs in various redshift ranges by direct vector sums are summarized in Table 1. The probability of a random sky distribution of sources with the fluxes of AGNs in the sample with $0<z\leq0.015$ would produce dipole to monopole ratio along a certain direction larger than observed (38%) (using a Monte Carlo simulation) is 0.3%.

The simple shell model of used in MB has been modified to include the effect of the inner edge of the dipole growth which can be seen in Fig.1 (a). Assuming a form:

$$\rho(\mathbf{r})= \rho_0 + \delta\rho \cos\theta \quad (r_1<r<r_2), \quad \rho(\mathbf{r})=\rho_0 \text{ (elsewhere)}$$

where ρ_0 and $\delta\rho$ are constants. Note that higher order terms do not contribute to the peculiar acceleration. Expressing the X-ray volume emissivity in the same form and denoting it by a subscript x, we can define the bias parameter as: $b=(\delta\rho_x/\rho_{0x})/(\delta\rho/\rho_0)$. Now expressing $\rho_0=\Omega_0\rho_c$ where Ω_0 is the cosmological density parameter and ρ_c is the critical density ($=3H_0^2/8\pi G$), following the similar procedure described in MB, we get:

$$|v|/c=\Omega_0^{-0.6}(z_2-z_1)/b \times D/M(z_1<z<z_2), \; (z_i=r_iH_0/c).$$

Assuming that the steep rise between $z=0.006$ and $z=0.015$ is solely responsible for the peculiar velocity of the LG, using $D/M(0.006<z<0.015)=57\pm12\%$, we get $b\Omega_0^{-0.6}=2.6\pm0.5$. If we ignore the inner edge, using $D/M(z<0.015)=38\pm11\%$, we get $b\Omega_0^{-0.6}=2.9\pm0.8$.

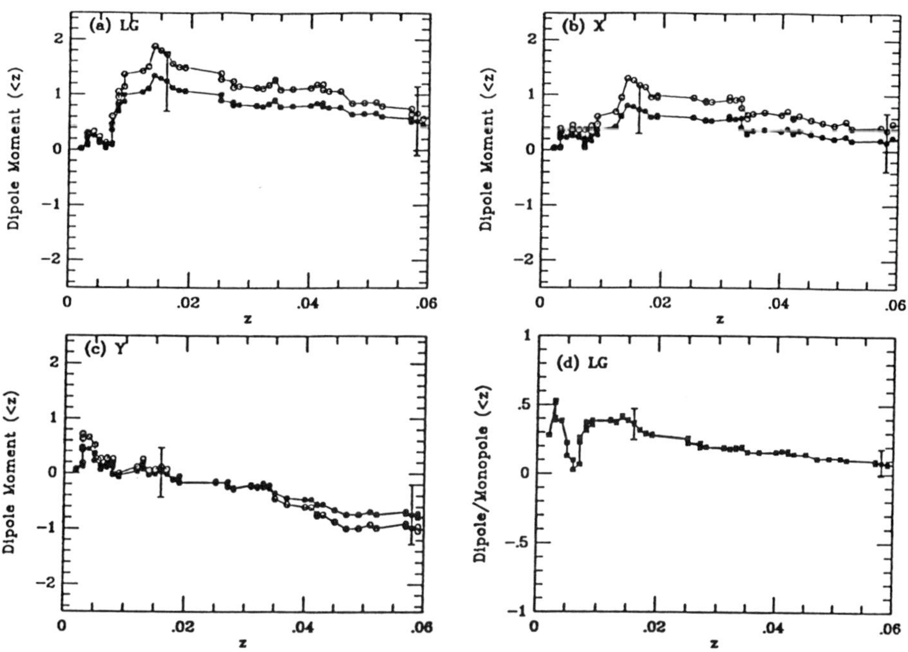

Fig. 1: (a)-(c) Dipole growth curves in three orthogonal directions:

$$D(\leq z)=(4\pi)^{-1}\sum_{\leq z} F_i \, \eta_i \cos\theta_i ,$$

where F_i is the flux of the object, η_i is the correction factor and θ_i is the angle from the axis. The growth curves with η_i =1 (no correction) are shown in filled symbols. The correction for the effect of hidden part in |b|<15° is considered by taking η_i =(the fraction of the ring θ_i from the axis for which |b|>15°)$^{-1}$(open symbols). The direction of the axes are: (a)-(LG):(l =268°, b=27°); (b)-(X): (l =22°, b=39°); (c)-(Y): (l =154°, b=39°). (d) Growth of the dipole to monopole ratio along the LG direction:

$$D/M(\leq z)=(\sum_{\leq z} F_i \, \eta_i \cos\theta_i)/(\sum_{\leq z} F_i \, \eta_i)$$

Table 1: The X-ray flux dipoles (Vector Sum)

Redshift Range z	*l*	*b*	Distance from the LG	D† along the LG	D/M along the LG
0 -- 0.015	299°	45°	31°	1.8±0.7	38±11%
0.006 - 0.015	292°	39°	24°	1.8±0.8	57±12%
0 -- 0.040	305°	17°	36°	1.1±0.8	16±10%
0 -- 0.060	319°	-9°	61°	0.6±0.9	7± 9%

†In (Total Count Rate•str^{-1}). Corrected for the hidden part (see caption of Fig. 1) only in this column.

DISCUSSION

The bias parameter from our sample is closer to the bias parameter for optical galaxies ($b_{opt}\ \Omega_0^{-0.6} \sim 3$) rather than IRAS galaxies ($b_{IRAS}\Omega_0^{-0.6} \sim 1$), confirming our previous result. Because most of our sample AGNs are Seyfert galaxies associated with disk populations, it may rather be an enhancement of the contrast of the same underlying population as the IRAS galaxies than the optical galaxies (with a significant contribution from ellipticals).This may have some implications on the environmental effect of the formation of AGNs. Calculating the cross correlation with IRAS and optical galaxies, autocorrelation of AGNs may be a reasonable next step.

We have to address the significance of the inner edge at z=0.006 (feature (1), see last section) and the drop in z>0.015 (feature (3)). Dipole analyses based on redshift surveys of IRAS galaxies (Yahil, 1988; Strauss and Davis, 1988) show no such features (see also Scaramella *et al.*, this conference). If (1) is real, it is possible that nearby AGNs in our sample (typically low X-ray luminosity AGNs) do not trace the underlying mass distribution as much as higher luminosity ones (typically represented by ones with z>0.006) or IRAS galaxies. (We leave open the reality of the feature (3). It is still possible that there are residual biases.)

It is possible that we are missing some contributions from the unsampled part of the sky, e.g. the main part of the 'GA' in $|b|<15°$. A rough estimation based on an enhancement of the X-ray background surface brightness by 4% in the GA region over the opposite side of the sky (Jahoda and Mushotzky, 1989) , assuming that the enhancement reflects the underlying mass with the same bias parameter as our AGN dipole, suggests that the GA could make as much contribution to the local gravitational acceleration as the AGN dipole presented here (MB).

The ROSAT all sky survey can give a much larger set of X-ray AGNs (although it may add different classes of objects from this work) and will enable us to do more extensive work on large scale structure traced by X-ray emission. This would not only be a supplement to the optical and IRAS works but would give us clues to understand the relationship between the environment and activities in galactic nuclei.

ACKNOWLEDGEMENTS

We thank Ron Remillard for prividing us with the unpublished identifications of the LASS X-ray sources. We also thank Ofer Lahav for stimulating discussions.

REFERENCES

Jahoda, K. and Mushotzky, R. 1989, *Ap. J,* **346**, 638.

Lahav, O., Rowan-Robinson, M., and Lynden-Bell, D.1988, *M. N. R. A. S.*, **234**, 677 (LRL)

Lynden-Bell, D., Lahav, O., and Burstein, D. 1989, *M. N. R. A. S.*, **241**, 325.

Peebles, P. J. E. 1980, *The Large Scale Structure of the Universe* (Princeton: Princeton University Press),§14.

Piccinotti, G. *et al.* 1982, *Ap. J.*, **253**, 485.

Miyaji, T. and Boldt, E. 1990, *Ap. J. (Letters)*, **353**, L3 (MB).

Strauss, M. A. and Davis, M. 1988, in *Large-Scale Motions in the Universe, A Vatican Study Week* ed.by Rubin and Coyne, (Princeton: Princeton University Press).

Yahil, A. 1988, in *Large-Scale Motions in the Universe, A Vatican Study Week* ed. by Rubin and Coyne, (Princeton: Princeton University Press).

Yahil, A., Walker, D., and Rowan-Robinson, M. 1986, *Ap. J. (Letters)*, **301**, L1.

Velocity-Independent Evidence for Dark Matter

DARK MATTER: GRAVITATIONAL LENS STUDIES

J. A. Tyson
AT&T Bell Labs
Murray Hill, NJ 07974

ABSTRACT

Studies of dark matter using gravitational lenses are reviewed, with emphasis on recent maps of dark matter in galaxy clusters. Although there are several QSOs which are multiply imaged by isolated intervening galaxies, distant galaxies are 4 million times more numerous than QSOs and thus can sample a lens at multiple positions. To a surface brightness of 29 B mag arcsec^{-2} there are about 100 background galaxies per square arcmin, which is sufficient to map statistically the dark matter distribution in a foreground galaxy cluster. Automated pattern recognition software generates a 2-d mass density map. There is evidence for a soft core of about 100 kpc radius for the dark matter in the rich clusters studied so far. These lens studies confirm the large mass in rich galaxy clusters which were implied by virial calculations using velocity dispersion. The dark matter distribution in a cluster is found to be similar to the distribution of the bright galaxies.

INTRODUCTION

It is likely that over 90% of the matter in the universe by its very nature cannot be seen directly. However, the gravitational influence of sufficiently large mass concentrations on passing photons can be detected. Gravitational lens distortions of background objects (galaxies or QSOs) can serve as a useful tool for the study of the total mass distribution of the foreground lens. These studies have the potential of going beyond being a complementary technique to dynamical studies: surveys for statistical lens distortions of background galaxies can reveal concentrations of dark matter where there may be no luminous matter. Naturally biased to objects which are luminous, astronomy traditionally represents the study of stars, galaxies, and clusters of galaxies. It is possible that non-luminous and massive objects inhabit the space between these stellar systems.

QSOs LENSED BY GALAXIES

In the thirty's Zwicky argued that galaxies would act as gravitational lenses, distorting and amplifying background objects [1]. Galaxy lenses offer an independent measure of the galaxy mass distribution. The light deflection is proportional to the square of the velocity dispersion in a (isothermal) lens, and is about 2 arcsec for a lens with 300 km sec^{-1} velocity dispersion. In 1973 Press and Gunn [2] showed how a cosmologically significant density of compact objects would create small arcs and split images of background objects. This important paper, perhaps a decade ahead of its time, was an inspiration to many. Not until 1979 was the first gravitational lens discovered: a distant QSO split into at least two images by the gravity of a foreground galaxy [3]. After a decade of searching 4681 QSOs for multiple images, nine unambiguous gravitationally lensed QSOs have been found. Several of the lensed QSOs found to date are caused by some isolated foreground galaxy dominated by its dark matter, and the data are consistent with galaxy masses of about 10^{12} solar masses, the number density of foreground galaxies, and the small numbers of background QSOs. The three best cases (2237, 1654, and 0142) yield values of M/r for the lensing galaxy around 3×10^{10} $M_{\odot}$ kpc^{-1}, typical of a heavy galaxy dominated by dark matter [4].

An important work by Turner, Ostriker, and Gott [5] in 1984 became the general theoretical reference for gravitationally lensed QSOs. The statistical evidence from lensed QSOs leads to an interesting conclusion: there is insufficient compact dark matter (black holes, for example) of mass equal to an average galaxy to close the universe. Multiple image lensed QSOs would have been more frequent were the universe closed by compact dark matter. This conclusion has been strengthened considerably by the systematic radio imaging (at higher resolution than usually obtainable optically) of a complete sample of radio QSOs [6]. Hewitt, et al. found that less than 40% of the closure density could be ascribed to any objects in the "galaxy" mass range 10^{11} to 10^{12} $M_{\odot}$.

With only one background object being lensed, no information on the mass distribution in the lens is available due to the undersampling of the lens plane. The small number of strongly lensed background QSOs (typical redshifts of 2) are due mostly to the relative scarcity of QSOs compared with galaxies at similar redshifts.

BACKGROUND GALAXIES

Deep CCD imaging surveys have revealed a population of faint blue resolved galaxies which uniformly cover the sky [7, 8]. At a surface brightness of 29 B mag $arcsec^{-2}$, the sky is over 15% covered with faint blue galaxies. This population has an unusually blue color near B-R = 0.3, which makes it particularly easy to distinguish from red cluster galaxies at lower redshift. Several lines of evidence suggest that most of these faint blue galaxies are distributed smoothly over a range of redshifts between 0.7 and 3 [9]. A lower redshift limit for this faint population can be derived from its response to a gravitational lens placed in front of it. A good test is provided by the dark matter lens in the clusters 3C295 (z=.46) and 0024+16 (z=.39). Roughly the same number of faint blue galaxies per square arcminute are seen behind these red clusters, but most of their images are stretched by the gravitational lens and are aligned orthogonal to the vector to the cluster center. We take this as evidence that most of the background galaxies are at redshifts larger than .7. Indeed, simulations fail to reproduce the observed distortions for these two high-redshift clusters for background mean redshift less than 0.9.

The observed ellipticity distribution of this population (after seeing deconvolution) is similar to that of nearby field galaxies. Some knowledge of the pre-seeing ellipticity distribution is required for accurate lens Monte Carlo simulations [10], discussed below. The dark matter investigations for galaxy clusters described here will use only clusters of redshift less than 0.3, so that uncertainties in the background source redshifts become relatively unimportant compared with noise and the ellipticity distribution of the background sources.

GALAXIES LENSED BY GALAXIES

Compared with galaxy-QSO lenses, the probability of galaxy-galaxy lensing is high. If all galaxies have massive dark halos sufficient to collectively close the Universe, then the systematic distortion of background galaxies within 10-20 arcseconds of these foreground galaxies would be detectable. Automated detection of galaxy-galaxy lensing has been used to set limits to the dark halo of individual foreground galaxies [11, 12]. 1200 foreground galaxies with background galaxies within 30 arcseconds were studied, out of a photographic sample of 50000 galaxies. No systematic background galaxy image gravitational lens distortion was found. That experiment was designed to test the hypothesis that all galaxies have massive dark halos, independent of absolute luminosity. Although suffering photographic nonlinearities and unknown background galaxy redshifts, those data were sufficient to rule out that hypothesis, for isothermal halos with greater than 200 km sec^{-1} circular velocity extending beyond 50 h^{-1} kpc radius. Distortions at radii larger than a few arcseconds are small, and the measurement of isolated galaxy mass distributions using this technique must wait for large area deep surveys using CCD mosaics.

GALAXY CLUSTERS AS LENSES

Evidence for gravitational lensing in galaxy clusters came with the discovery in 1988 of several bright blue "arcs" in compact rich galaxy clusters [13, 14]. A large blue arc occurs if a distant evolving galaxy falls near a caustic. Some theoretical work on the possibility of occasional bright arcs near caustics in lenses had been done by 1988 [15, 16, 17, 18]. However, these caustic projections of background objects are rare. The large mass associated with rich clusters of galaxies distorts background galaxy images over a large area of the sky. All the faint background galaxies within 1-2 arcminutes of the cluster will be measurably distorted by the shear. Foreground galaxy clusters at redshifts .2-.5 with radial velocity dispersions above 1000 km s^{-1} have sufficient mass density to significantly distort background galaxies of redshift greater than .4-1.

We have found systematic gravitational lens alignment of 10-60 faint background galaxy images, centered on foreground galaxy clusters of high velocity dispersion [19]. The background galaxy population is selected by its extreme blue color relative to the red cluster galaxies. At sensitivities achieved with CCDs on large telescopes, there are over 30 background galaxies $arcminute^{-2}$ anywhere in the sky, which is sufficient to statistically map the dark matter distribution in a foreground cluster. The 20-30 high redshift background galaxies within the central square arcminute of a foreground compact cluster of galaxies will be significantly lensed by the cluster mass distribution, distorting background galaxy images into faint arcs several arcsecond long. These are not the relatively rare caustic alignments: every background galaxy above some critical redshift will be seen as stretched along a circle centered on the lens. By going faint, the improved statistics of these distortions permits the construction of a map the dark matter distribution [19]. Such a dark matter map for the cluster Abell 1689 is shown in Figure 1, along with the orientation pattern of some of the brighter shear-distorted background blue galaxies. This new method is immune to the problems associated with unbound galaxy contamination and dynamical assumptions in cluster virial mass estimates.

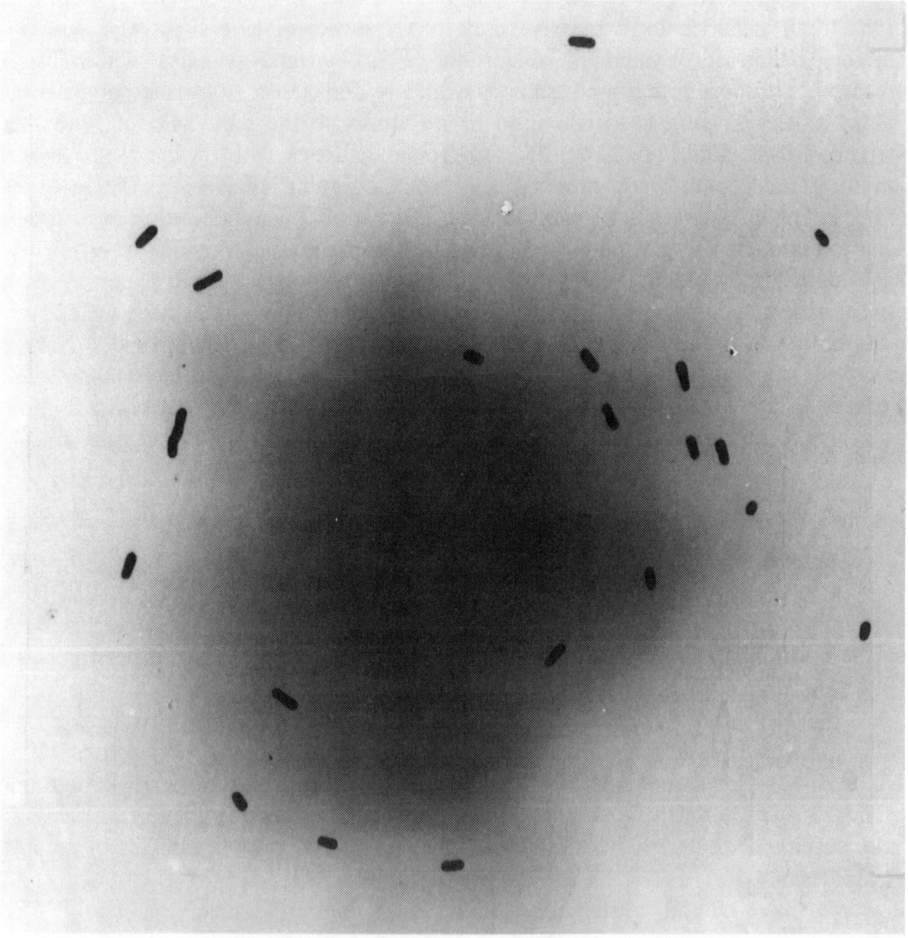

Fig. 1. Clumps of dark matter can form imperfect but powerful gravitational lenses, giving us an amplified and distorted view of the distant universe. This figure shows the alignment of some faint blue galaxies behind a foreground massive cluster of galaxies, and the dark matter map for this cluster, obtained from hundreds of these systematic lens distortions. The foreground cluster, A1689, has a redshift of 0.18.

We have measured multiple background galaxy gravitational lens distortions for 8 clusters of known high velocity dispersion and X-ray luminosity, in a sample of 23 clusters deep-imaged to date. In order to study the distribution of dark matter under a variety of states of cluster evolution the deep cluster imaging sample is distributed over a wide range of redshift and optical/X-ray properties. Each cluster is imaged to 29 B and 28 R magnitude arcsec^{-2}. This imaging survey of rich clusters is a collaboration between groups at Bell Labs (G. Bernstein, T. Tyson, R. Wenk), Institute for Advanced Study (P. Guhathakurta), Princeton University (J. Miralda-Escude, E. Turner), Toulouse Observatory (B. Fort, G. Soucail, Y. Mellier), and NOAO (F.Valdes). Of the 23 clusters imaged to 29 B mag arcsec^{-2} and 28 R mag arcsec^{-2} to date, 18 have Abell richness greater than 3, 5 are poor (for control), 8 have X-ray luminosities greater than 6x10^{44} erg sec^{-1}, 8 are optically compact (relaxed), 5 have redshift less than 0.2, and 7 have redshift greater than 0.4.

CLUSTER LENS DISTORTIONS

The effect of a cluster gravitational lens is to stretch the apparent size of the image tangent to a circle centered on the lens. The distorted images of background galaxies appear as mini-arcs, small parts of circles centered on the foreground cluster. A significant excess of faint blue background galaxies are aligned tangent to circles centered on the foreground cluster lens by the distortion of the gravitational lens. The cumulative systematic alignment of the numerous faint blue (background) galaxy ellipticities gives a strong gravitational signal. This stretch is measured using two scale lengths: the intensity weighted second moment of the galaxy image orthogonal and along the radius relative to the lens center. To detect the coherent stretching from the lens and cancel effects of noise and the intrinsic ellipticities of the galaxies, the scale lengths are averaged over the background population. From the two mean scale lengths we form a dimensionless quantity similar to an ellipticity, the difference of the lengths divided by the sum of the lengths, which we call the net tangential alignment. A random population of galaxies will give a net alignment of zero while a population of lensed galaxies will give a positive value.

Preprocessing and FOCAS automated detection, splitting and photometry reduced the multi-color images to a position-matched catalog [7]. This matched catalog may be filtered for objects with reproducible position, which are classed as galaxy, and their colors and second moments obtained. Those galaxies having B-R color bluer than the blue tail of the cluster color distribution have their positions, total magnitudes, and second moments sent to the lens orientation-correlation software *findlens*. To detect if a gravitational lens is present, the alignment statistic is computed over a grid of positions as candidate lens centers. The alignment statistic is computed using galaxies at various radii from the cluster center. All faint blue background galaxies with ellipticities > 0.2 within 300 kpc of that point (the radius is arbitrary and not critical) are included in the tangential and radial moment average. A real lens produces an alignment value which is a maximum at the true center. The alignment strengths at the lens centers are highly significant (> 99% likelihood) as determined by comparison to a large number of random fields created by randomizing the sample galaxy positions without changing their measured shapes. We study the alignment strength as a function of radius and compare it to the light distribution. Figure 2 shows, for the cluster A1689, the red luminosity profile compared with the alignment strength of faint blue galaxies, and the dark matter surface density.

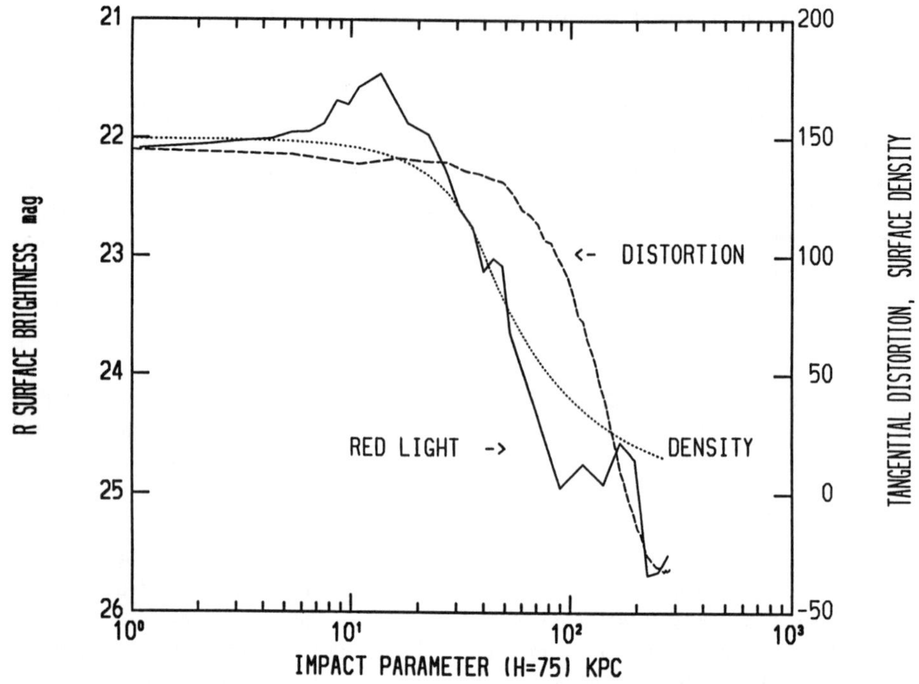

Fig. 2. Red luminosity profile in cluster A1689, compared with strength of tangential alignment of faint blue background galaxies. Observed log red flux (subtracting bright stars in foreground) vs log radius is plotted, together with the radial dependence of the net tangential alignment $a^2 - b^2 / a^2 + b^2$. The best fit surface density profile of the form [1.5 - .5 $(r/r_c)^2$], $r < r_c$, $[r_c/r]$, $r > r_c$, for a 67% peak density radius r_c = 40 h_{75}^{-1} kpc is also plotted as a dotted curve. The right scale is arbitrary units, normalizing all plots at zero radius.

The gravitational image distortions appear to extend radially at least as far as the cluster red light. The lens distortions are roughly correlated with cluster red light both in central position and radial extent. This implies that the dark matter distribution follows the smoothed cluster luminosity, with most of the mass interior to 100 h_{75}^{-1} kpc. A null result is found in eleven control fields, each with a surveyed area similar to that of our cluster data. Statistical significance of the lens alignments (7-10σ per cluster) and the integral distortion statistic is particularly high for the optically compact clusters we have examined.

CLUSTER LENS SIMULATIONS

We have simulated cluster lens distortions of background galaxies, by ray-tracing real images, as a function of cluster and DM parameters and compared them with the observations. The background galaxy fields are synthesized to have the same properties (number-magnitude counts, ellipticity distribution, galaxy intensity profiles) as found in the seeing-deconvolved deep surveys. These background galaxies are assigned redshifts in three model schemes: all at z=1, all at z=2, and equally at z=1 and 2. The cluster galaxies (found by their color, brightness, or redshifts) are assigned truncated isothermal mass distributions obtained from individual galaxy surface photometry to 28 B_J mag, K-corrections derived from the observed colors, and the Faber-Jackson relation σ_V = 220 dex [(20.2 - B_J)/10]. These cluster galaxies make a small contribution to the background galaxy image distortions beyond several arcseconds from the cluster galaxies. A soft core isothermal mass distribution is superposed on the cluster galaxies. The velocity dispersion (500-2000 km sec^{-1}) and core 67%-surface-density radius (25 to 100 h_{75}^{-1} kpc) are varied in different Monte Carlo simulations. After the distortion from all components of the lens, the image is convolved with the seeing profile, binned down to the CCD resolution, and noise is added.

To relate the observations to the simulations the net tangential alignment of background galaxies over the extent of the cluster is computed for both the observations and the simulations in exactly the same way (including FOCAS processing). This is particularly necessary because this distortion statistic is weakly dependent on the parent background galaxy ellipticity distribution. The alignment strength is found to be proportional to the radial velocity dispersion and weakly dependent on DM core size (R_{core} < 100 h_{75}^{-1}) and the background redshift (z_{bg}>1). The best agreement with the observed lens statistics for CL3C295, CL0024+16, and A1689 is obtained for 30 < R_{core} < 50 h_{75}^{-1} kpc, $\sigma_{V,DM}$ = 1100-1500 km sec^{-1}, for z_{bg}=2. Note these "DM" parameters refer to all mass, dark + luminous, *other* than that associated with the modeled 20-50 brighter cluster galaxies. These are three of the most optically compact clusters in our sample. For many of our clusters, the best fit DM parameters correspond to as large a mass, but distributed over a larger volume (larger core sizes).

Based on these initial clusters, we find that the DM distribution is similar to the smoothed red cluster light. This is consistent either with baryonic DM or baryonic - non-baryonic dissipative coupling. The scale of the sub-clumping of the dark matter within clusters of galaxies is also of interest. It appears that the DM cannot all be clumped on galaxy mass scales, otherwise the long arcs seen in several relaxed cluster lenses would be broken up on 1-2 arcsecond scales. We have also done an experiment to see how much of the DM can possibly be ascribed to cluster galaxies. We assigned a mass to each of several hundred cluster galaxies, from the Faber-Jackson relation, and then scaled all these galaxy masses up to very large value (σ_{gal} < 500 km sec^{-1}). The resulting number of blue rings around individual cluster galaxies was found to be much larger than observed, and none of these simulations showed arc patterns resembling those found in the real data. Little more can be said of the DM in clusters, based on the small number we have imaged. There appears to be a large scatter in the DM properties within our sample, generally correlated with optical compactness. A study of a large sample of galaxy cluster fields will provide a clearer picture of the forms of DM in clusters, and may constrain the nature of the dark matter.

The best fit lens data for A1689 imply a total velocity dispersion (dark + luminous matter) of about 1400 km sec^{-1}, somewhat smaller than the measured velocity dispersion of 68 galaxies assumed to be associated with the cluster (1800 km sec^{-1}). For the optically compact clusters we find the DM peak densities are about 1-2x10^5 times ρ_c, with 67% peak density core radii about 50 h_{75}^{-1} kpc. We have no data on the dark matter beyond 300 h^{-1} kpc. Larger CCDs are required in order to address the question of the possible outer cutoff of the mass. CCDs with 1024 or 2048 pixels per side are capable of addressing the outer cutoff problem, so data on this interesting question may be available soon.

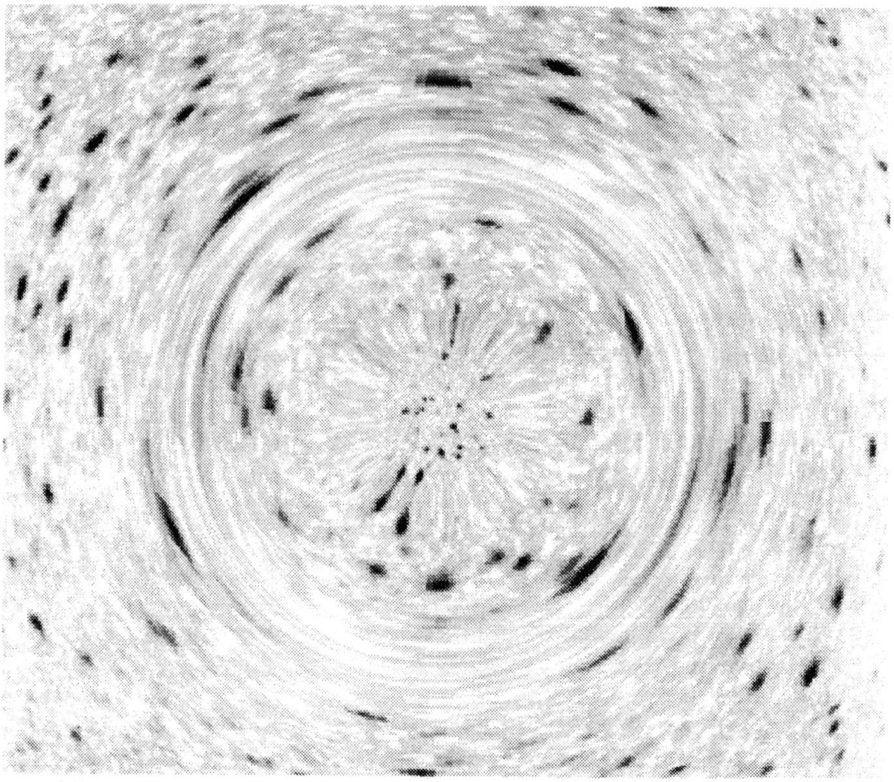

Fig. 3. A simulation of the lens in A1689. A synthetic background galaxy image, with all the statistical properties of seeing-deconvolved deep imaging observations is distorted with a 1750 km sec^{-1} isothermal mass distribution with a 67% core radius of 45 h^{-1} kpc. Faber- Jackson masses were used for the observed cluster galaxies. The mean properties of many of these Monte Carlo simulations are used to define the dark matter parameter space consistent with the cluster lens observations.

SUMMARY

The nature and dynamical history of dark matter will be constrained by the way it clumps. Prospects are good for a direct measure of the dark matter distribution on scales of stars (from "micro-lensing"), galaxies, clusters of galaxies, and eventually, inter-cluster scales. The first large-scale application of this technique has been to rich clusters of galaxies. Using the observed tell-tale pattern of background galaxy shapes, automated pattern recognition software generates a 2-d lens distortion map.

To date, we have studied 23 of these clusters to gather statistics on how the dark matter is distributed. The distribution of dark matter in rich galaxy clusters is found to be similar to the distribution of total luminosity (although on a smoother scale of 100 kpc). This is consistent with a baryonic origin for this dark matter, perhaps brown dwarf stars or primeval black holes. Currently, our data for clusters are also consistent with remnant neutrinos of about 10 ev rest mass.

These shear distortion studies measure mass internal to a radius, M/r. With the exception of the cluster 3C295, we find good agreement with dynamical measures of M/r for those few clusters for which both types of data exist. This suggests that, baring cases of accidental superposition of physically unrelated galaxies, the assumptions behind the application of the virial theorem to clusters may be generally valid. This large mass density in dark matter in clusters is not a prediction of current biased cold dark matter theory [20]. Blue rings are occasionally seen around individual bright cluster galaxies, offering the possibility of measurement of the relative mass in cluster galaxies and the diffuse DM mass component. Preliminary results suggest that cluster galaxies are no more massive than field galaxies, and that the dark matter in clusters is primarily in the diffuse component.

Larger scale applications of this dark matter mapping may eventually discover dark matter unrelated to galaxies or clusters of galaxies. Larger CCDs make such a large scale search for coherent alignment in the distant faint galaxies possible and dark matter on angular scales up to degrees can be studied in this way.

It is a pleasure to acknowledge my collaborators: G. Bernstein, B. Fort, R. Guhathakurta, Y. Mellier, J. Miralda-Escude, G. Soucail, D. Spergel, E. Turner, F. Valdes, and R. Wenk.

---

1. Zwicky, F. *Phys. Rev. Lett.* **51**, 290 (1937).
2. Press, W. and Gunn, J. E. *Astrophys. J.*, **185**, 397 (1973).
3. Walsh, D., Carswell, R. F., and Weymann, R. J. *Nature* **279**, 381-384 (1979).
4. Kochanek, C., preprint (1990).
5. Turner, E. L., Ostriker, J. P., and Gott, J. R., *Astrophys. J.* **284**, 1 (1984).
6. Hewett, J. N., Turner, E. L., Schneider, D. P., Burke, B. F., Langston, G. I., and Lawrence, C. R., *Nature*, **333**, 537 (1988).
7. Tyson, J. A., *Astron. J.*, **96**, 1 (1988).
8. Tyson, J. A. and Seitzer, P., *Astrophys. J.* , **335**, 552 (1988).
9. Guhathakurta, P., Tyson, J. A., and Majewski, S., *Astrophys. J.* , **357**, L19 (1990).
10. Miralda-Escude, J., *Astrophys. J.*, (1990), in press.
11. Tyson, J. A., Valdes, F., Jarvis, J. F., and Mills, A. P., *Astrophys. J.*, **281**, L59 (1984).
12. Tyson, J. A., in *Theory and Observational Limits in Cosmology*, W. R. Stoeger, ed. (Specola Vaticana), p. 441 (1987).
13. Soucail, G., Mellier, Y., Fort, B., Mathez, G., and Cailloux, M., *Astron. Astrophys.* **191**, L19 (1988).
14. Lynds, R., and Petrosian, V., *Astrophys. J.***336**, 1 (1989).
15. Blandford, R. D. and Narayan, R., *Astrophys. J.*, **310**, 568 (1986).
16. Paczynski, B. P., *Nature*, **325**, 572 (1987).
17. Blandford, R. D., Phinney, E. S., and Narayan, R., *Astrophys. J.*, **313**, 28 (1987).
18. Grossman, S. A. and Narayan, R., *Astropys. J.*, **324**, L37 (1988).
19. Tyson, J. A., Valdes, F., and Wenk, R. A., *Astrophys. J.*, **349**, L1 (1990).
20. Evrard, A. E., *Astrophys. J.*, **341**, L71 (1989).
21. Peebles, P. J. E., Daly, R. A., and Juszkiewicz, R., *Astrophys. J.*, **347**, 563 (1990).

A SEARCH FOR MASSIVE COMPACT HALO OBJECTS IN OUR GALAXY

presented by D. P. Bennett

in collaboration with C. Alcock, T. Axelrod, K. Cook, H. Park
Lawrence Livermore National Laboratory, Livermore, CA, USA

K. Griest, C. Stubbs
Center for Particle Astrophysics, U.C. Berkeley, Berkeley, CA, USA

K. Freeman, B. Peterson, P. Quinn, A. Rodgers
Mount Stromlo and Siding Spring Observatories, Canberra, Australia

November 14, 1990

Abstract

MAssive Compact Halo Objects such as brown dwarfs, Jupiters, and black holes are prime candidates to comprise the dark halo of our galaxy. Paczynski noted that these objects (dubbed MACHOs) can be detected via gravitational microlensing of stars in the Magellanic Clouds with the caveat that only about one in 10^6 stars will be lensed at any given time. Our group is currently involved in constructing a dedicated observing system at the Mount Stromlo Observatory in Australia. We will use a refurbished 1.27 meter telescope and an innovative two-color CCD camera with 3.4×10^7 pixels to monitor 10^6-10^7 stars in the Magellanic Clouds. During the first year of operation (1991-1992), we hope to detect (or rule out) objects in the mass range $0.001 M_\odot \leq M \leq 0.1 M_\odot$, and after five years, we hope to have covered the range $10^{-6} M_\odot < M \lesssim 100 M_\odot$.

I MICROLENSING BASICS

Gravitational microlensing refers to lensing in the case where there is significant amplification of the source, but the lensing angle is too small to be observed. Paczynski[1] has shown that the "optical depth" for microlensing by the dark halo of our galaxy is about 10^{-6}, so that at any given time about 1 star in 10^6 will be microlensed with amplification by a factor of 1.34 or larger. (The typical angular separation of the images is $\lesssim 0.001$ arc second, so they cannot be resolved.) The fundamental unit of length for a microlensing is the Einstein ring radius which is given by

$$R = 2\left(\frac{GMx(L-x)}{c^2 L}\right)^{1/2}$$

where L is the distance to the source star (in our case in the Magellanic Clouds) and x is the distance to the lensing object in the halo. If R is not too small, the source star can be assumed to be a point light source, and the microlensing amplification depends only on the dimensionless impact parameter[1-3] ($u \equiv b/R$):

$$A(u) = \frac{u^2+2}{u\sqrt{u^2+4}}$$

$u = 1$ corresponds to an amplification of 1.34.

The time scale of a microlensing event is given by the timescale for the lensing object to traverse a distance equal to the impact parameter b which yields

$$\Delta t = 7\,\text{days}\sqrt{\frac{M}{10^{-2}M_\odot}}$$

for a typical lensing object located at a distance of 10 kpc moving at a velocity of 200 km/sec.

The microlensing light curves will be distinguishable from the background of variable stars in several important ways: they are achromatic, time-symmetric, and non-repeating. These signatures should allow us to distinguish most (and hopefully all) of the variable stars from the microlensing events. In addition, the microlensing light curve is described by just 3 parameters: the maximum amplification, the time of the maximum amplification and the duration of the event. Clear evidence of this characteristic light curve will be required in order to distinguish microlensing by MACHOs from other phenomena such as variable stars.

There are several possible complications which might tend interfere with the characteristic microlensing light curve. First, the finite size of the source star adds another parameter to the microlensing lightcurve, but for halo objects with $M > 10^{-4}M_\odot$, most of the red giants in the Magellanic Clouds would appear as point-like sources. Another potential complication comes from double stars. These would add another parameter to the lightcurve fit and could make the microlensing events appear chromatic, but we have found that most double star light curves can be fit very well by a single star light curve. The only real problem would be the different light curves observed in the two color bands when a member of a red giant-main sequence pair is microlensed. Fortunately, red giants are bright only for a very short portion of their lifetime, so the probability that a red giant will be paired with a main sequence companion of comparable brightness is small. One might also worry that the microlensing of double stars might be more difficult to detect because apparent amplitude of the lightcurve will be reduced when only one member of the pair is microlensed. This is a very small effect, however, because the second member of the pair can also be lensed, and this possibility mostly compensates for the number of events lost due to the presence of double stars.

II THE TELESCOPE

We plan to carry out our microlensing search on the Great Melborne (50") Telescope at Mount Stromlo which is being refurbished for dedicated use in our microlensing search. This telescope is an "English" (two-pier, crossed axis) mounted f4.47 reflector of 1.27m aperture and 5.67m focal length. We will install our field corrector and double CCD camera package at the prime focus. The corrector package will reduce the effective focal length to 4.95m (f3.9) and provide a usable field of greater than 8.7cm = 60 arc minutes in diameter.

The main advantages of this telescope are the proximity to our collaborators in Australia and high degree of support from MSSSO. The Australian telescope will also be ready to observe starting in the 1991-1992 Magellanic cloud season, which is significantly earlier than a new telescope could be made ready. The main disadvantages of the telescope are the weather and seeing conditions. The telescope site is completely overcast about 35% of the nights and completely clear on about 35% of the nights. The remaining 30% of the nights have some useable time. The median seeing is about 1.6" FWHM. We have analyzed both of these difficulties, and found that they should not prevent us from detecting or ruling out the dark matter in the mass range that we plan to probe.

III THE CAMERA

The most innovative part of our experiment is our CCD camera. We plan to observe simultaneously in two different color bands. (The first generation camera will use color bands that are approximately equal to the Gunn g band and the Johnson R band.) This will be accomplished by splitting the beam with a glass cube dichroic element which will send the beams to separate 2×2 mosaics of 2k $\times$ 2k Ford Aerospace edge butable CCDs.

IV OBSERVATIONAL STRATEGY

Our precise observational strategy has been by no means decided yet, but the following seems to be the most reasonable strategy at present. We will survey stars in the LMC, the SMC and the galactic bulge, and we plan to work to a limiting magnitude of about $m < 19.5$ with 5 minute exposures (at full moon). We tentatively plan to work out to a radius of about 3 degrees from the center of the LMC and 2 degrees from the center of the SMC. (Both these numbers refer to the distance along the major axis.) The estimated star densities ($m < 19.5$) in these fields range from about 380,000 stars per square degree in the center of the LMC to about 85,000 stars per square degree 3 degrees away from the center of the LMC to 30,000 stars 2 degrees from the center of the SMC. An analysis of existing CCD frames of a region in the Magellanic Clouds taken in different seeing conditions using standard crowded field photometry software has indicated that we should be able to get accurate (better than 5%) photometry on 120,000 stars per square degree with 3" seeing and on $> 200,000$ stars per square degree with 1.5" seeing. This means that we should be able to get about 2,000,000 stars in a night with 3" seeing in the prime cloud season and 2,500,000 in a night with 1.5" seeing. During a new moon, we should be able to get about 3,200,000 stars on a good night.

We plan to concentrate initially on the mass range $10^{-3}M_\odot \leq M \leq 10^{-1}M_\odot$ which spans the range for Jupiters and brown dwarfs. The time scale for these events ranges from a couple days to a few weeks, so a reasonable strategy will be to observe as many different stars as possible each night. As many of the same stars as possible will be observed on successive nights so as to give daily sampling of the light curves for as many stars as possible.

In order to search for halo objects of larger mass (black holes?), it will be preferable to sample more stars less frequently. One relatively easy way to do this will be to co-add data from successive nights in order to get accurate photometry on fainter stars. In order to get data on the 10^7 stars which would be required to be able to detect microlensing by objects of more than $100M_\odot$ or more, we will probably have to take longer exposures in some fields that are further from the centers of the clouds.

Smaller mass dark matter candidates ($M \leq 10^{-4}M_\odot$) pose several different problems. First, the lensing light curves for red giants (which make up two thirds of the stars to $m < 19.5$) in the clouds become more complicated because the point-like source approximation begins to fail. For impact parameters comparable to the stellar radius, this results in a large *increase* in the amplification, but when the impact parameter is smaller than the stellar radius, the amplification is reduced. The other difficulty with these small mass objects is that they tend to have lightcurves which last about 24 hours which makes it difficult to obtain complete light curves. The detection of microlensing in this mass range will probably require the use of at least two different observing sites at different longitude.

V PREDICTED EVENT RATES

We have examined 9 years of detailed weather records to come up with the following event rates. The Halo was assumed to consist of a Halo density profile of the form[2]

$$\rho = \rho_0 \frac{a^2 + r_0^2}{a^2 + r^2}$$

where the core radius $a = 5\,\text{kpc}$ and the local Halo density $\rho_0 = 0.0079 M_\odot\,\text{pc}^{-3}$. The table shows the average number of events detected using 8 years of weather data from Mt. Stromlo for different masses of Halo objects. The two detection criteria used are (1) to require 2 points with $A > 1.2$ on each of the rising and falling parts of the light curve for a total of at least four points on the light curve with $A > 1.2$, and (2) at least 5 points on the light curve with at least 2 on the rising and falling portions of the curve and 1 point with $A > 2$. Note that each "point" on the light curve refers to 2 simultaneous images in different color bands.

Mass	4 pts. with $A > 1.2$	5 pts. with $A > 1.2$ & 1 with $A > 2$
$10^{-4} M_\odot$	2.3 events/yr.	0.8 events/yr.
$3 \times 10^{-4} M_\odot$	7.2 events/yr.	2.8 events/yr.
$10^{-3} M_\odot$	15.9 events/yr.	6.6 events/yr.
$3 \times 10^{-3} M_\odot$	23.7 events/yr.	10.8 events/yr.
$10^{-2} M_\odot$	24.2 events/yr.	11.4 events/yr.
$3 \times 10^{-2} M_\odot$	19.3 events/yr.	8.9 events/yr.
$10^{-1} M_\odot$	12.8 events/yr.	5.7 events/yr.

VI THE USE OF ADDITIONAL SITES

For Halo objects of $M \lesssim 10^{-4} M_\odot$, it is highly desirable to be able to track microlensing events with points evenly spaced at intervals of 2-10 hours. At most southern Hemisphere observing sites, the clouds can rarely be observed for much more than 8 hours in a single night, it may be necessary to use different telescopes at different longitudes to sample this mass range. One option that may become available is to collaborate with the French group[4] which will search for microlensing from a 40cm refracting telescope at ESO. Another option would be to develop an alert mechanism that would allow follow up observations of ongoing microlensing events at other telescopes.

REFERENCES

[1] B. Paczynski, Ap. J. **304**, 1 (1986).
[2] K. Griest, Ap. J., in press, (1990).
[3] R. Nemiroff, NRL preprint, (1990).
[4] A. Milsztajn, to appear in the proceedings of the 10th Moriond Workshop on new and exotic phenomena, Les Arcs, France, 1990.

DISCUSSION:

Nemiroff: What technical advances in the last four years make this experiment feasible now?

Bennett: It is not clear that this experiment couldn't have been performed four years ago when Paczynski first proposed it nor is it clear that it can be done now. The main difficulty with this experiment is the "background events" caused by variable stars, and it will only be when these background events are understood will it be clear what technology is necessary for this experiment. Our approach makes use of the recently developed large-format CCDs in order to do accurate photometry on a large number of stars, but the French approach is to obtain the photometry from Schmidt plates. If this is successful, then one might argue that this experiment could have been done four years ago.

PRESENT AND FUTURE LIMITS ON COMPACT DARK MATTER FROM GRAVITATIONAL LENSING

Robert J. Nemiroff
Naval Research Laboratory, Washington, DC 20375

ABSTRACT

If the dark matter is composed of black holes, stars, or any compact objects (COs) then certain gravitational lens effects are to be expected (Press and Gunn 1973). Non-detection places a limit on their mass and abundance. Current observational frontiers of candidate lensed sources are expanding in three ways: increased spatial resolution, increased sensitivity limits, and increased time coverage. Present limits exclude closure densities of compact objects ($\Omega_{co} < 1$) of masses greater than $10^{10}\ M_{\odot}$.

THE ANGULAR RESOLUTION FRONTIER

The most common lensing scenario expected with a compact lens is a dim counter-image near a bright QSO image (Nemiroff and Bistolas 1989). For lensing detection, the lens must fall inside a dynamic range detection volume but outside an angular resolution detection volume (Nemiroff 1989). To compute the probability of detectable lensing, the hypothesized number density of compact lenses is integrated over the allowed volume. If this probability predicts significantly more dim QSO counter-images than observed, the hypothesized scenario is ruled out. In the following analysis an $\Omega = 1$ universe will be assumed with no cosmological constant and $H_o = 50$ km sec^{-1} Mpc^{-1}.

The angular resolution limit θ of the telescope primarily determines the lowest lens mass M excluded such that

$$M \approx (\theta/2")^2\, 10^{12}\, M_{\odot}. \qquad (1)$$

Previous angular resolution limits in the optical were on the order of 2". This corresponds to a capability of detecting cosmological COs of order $10^{12}\ M_{\odot}$. Non-detection of a significant number of dim counter-images to QSOs by surveys such as Webster, Hewett, and Irwin (1988) and Tyson and Gullixson (1986) hence rule out a cosmological abundance of these objects.

The pioneering thesis of Hewitt (1986) set the first interesting limits on a cosmological abundance of compact lenses. This work in the radio with the VLA at an angular resolution of 0.4" (see also Burke 1990) excluded a cosmological density between 10^{11} and $10^{13}\ M_{\odot}$. VLBI currently has the capability to resolve images at the 0.001" level which make it capable of detecting a cosmological density of $10^{6}\ M_{\odot}$ objects. One problem is that most quasars inspected at this resolution show intrinsic structure.

The current best angular resolution search was done by Crampton, McClure, Fletcher, and Hutchings (1989, hereafter CMFH) where they were able to inspect distant QSO images for optical counterparts with 0.2" resolution. Subsequent analysis by Nemiroff (1990) excluded a closure density of cosmological compact objects of mass greater than $10^{9.9}\,M_{\odot}$.

Currently both the NTT and the HST have sub-arcsecond resolution capability in the optical and should soon reduce the maximum mass below $10^{9.5}\,M_{\odot}$. New image stabilizing equipment incorporating active mirrors (such as that described and used by CMFH) show much promise of allowing even greater angular resolution in the optical from ground based telescopes.

THE DYNAMIC RANGE FRONTIER

The highest detectable dynamic range between visible images, Δm, primarily determines the lowest cosmological lens density Ω_L excluded such that

$$\Omega_L \approx (10\,f)\,10^{-\Delta m/5}, \qquad (2)$$

where f is the fraction of inspected QSO images which show lens-induced multiple images, and $\Delta m \geq 2$.

Of the 8 currently accepted cases of gravitational lens induced multiple imaging of QSOs (as collected by Mellier, Fort, and Soucail 1990), 5 of the image groups are within 1 magnitude of each other and the largest dynamic range is 2.1 magnitudes. Yet there should be many more lens cases of higher dynamic range to be expected in a universe with a significant population of compact masses. Expanding sensitivity can therefore greatly decrease the allowed density of these compact objects in the universe.

Hewitt (1986) attained a dynamic range of 17 ($\Delta m \approx 3$) with VLA observations (A array), which detected a maximum lens fraction of $f = 0.32$. The corresponding upper limit on Ω_L is about 0.7. CMFH had a dynamic range that varied with angular distance from the bright QSO image. At 0.2" resolution their dynamic range was only $\Delta m = 0.5$, but at 0.6" resolution their dynamic range improved to $\Delta m = 4.0$. From their measurements it is possible to exclude $\Omega_L > 0.2$ (Nemiroff 1990).

Tyson and Gullixson (1986) record a QSO observation with dynamic range of $\Delta m = 4.6$. The recent deep photometry of Tyson and Seitzer (1988) appears to show an ability to record a much higher dynamic range, possibly on the order of $\Delta m \approx 10$. Were 100 QSOs inspected to this dynamic range without a detectable counter-image, the Ω_L could be constrained below 10^{-3}. Recently a proposed occluding telescope called the Circumstellar Imaging Telescope designed specifically for high dynamic range at high angular resolution was proposed Ftaclas and Terrile (1990) which could reduce Ω_L even further.

THE TIME COVERAGE FRONTIER

Most hypothesized lensing events with compact matter lenses are not expected to happen on a time scale so short as to inhibit our ability to measure it. A more practical problem is the amount of telescope time needed for adequate monitoring. For sufficient monitoring, the time between observations should be short compared to the duration of the lensing event and time between lensing events.

The currently most sought after transient CO lensing event is galactic halo microlensing of distant AGN (Gott 1981). Here the variability time scale for a low optical depth event is

$$t_{duration} \approx 200\, z_L\, \sigma_{250}^{-1} (M/\, M_{\odot})^{1/2} \text{years.} \tag{3}$$

where z_L is the lens redshift, and σ_{250} is the velocity dispersion of the microlensing galaxy in units of 250 km sec^{-1}.

The most promising microlensing observations are between macro-images of the same QSO (Vanderriest 1990, Irwin *et al.* 1989) and these specific QSOs are frequently monitored to determine the time delay between macro-images. There are many other single macro-image QSOs for which the probability of microlensing is high and would require monitoring only on a monthly basis to check for microlensing (Borgeest, Schramm, and Refsdal 1990). Most of these QSOs, however, have only been observed once. Such an observing program might only detect intrinsic variability.

It has also recently been suggested that stars in nearby galaxies could be inspected for lensing by potential COs in our Galaxy's halo (Paczynski 1986a, Alcock, Axelrod, and Park 1989, Griest 1990. Nemiroff 1991). The duration of these events would be (Paczynski 1986a)

$$t_{duration} \approx 0.2\, (M/\, M_{\odot})^{1/2} \text{ years,} \tag{4}$$

while the waiting time between high amplitude events would be (Nemiroff 1991)

$$t_{wait} \approx 0.14\, \frac{(M/\, M_{\odot})^{1/2} A}{N_6 \sigma_{250}^3 R_{20}^{1/3}} \text{ years.} \tag{5}$$

where A is the amplification, N_6 is the number of LMC source stars monitored in units of 10^6 stars, and R_{20} is the extent of our Galactic halo in units of 20 kpc. These observations have the capability of detecting compact objects between $10^{-8}\, M_{\odot}$ and $10\, M_{\odot}$

Another method of detecting a cosmological density of COs is the monitoring of distant supernovae (Schneider and Wagoner 1987). 5% of all supernovae discovered brighter than 22 magnitude would show an amplification of 10 or greater (Linder, Schneider, and Wagoner 1988). This method would have the capability of detecting COs in the stellar mass range and above: $M > 10^{-2}\, M_{\odot}$.

Frequent monitoring of the sky for cosmologically distant gamma ray bursts (Paczynski 1986b) also has the possibility of detecting COs with stellar mass. A time-singular burst would be expected to be lensed into a series of bursts of a bright burst followed by an echo. The mass of the CO that could be detected would be determined by the width of the temporal variation $t_{duration}$ such that

$$M \approx 100 \frac{t_{duration}}{10^{-3}\,\text{sec}} M_{\odot}. \tag{6}$$

This work was done while the author held a National Research Council - Naval Research Laboratory Research Associateship.

REFERENCES

Alcock, C., Axelrod, T., and Park, H. S. 1989, *Bull. Amer. Astron. Soc.*, **21**, 1149.

Borgeest, U., Schramm, K. J., and Refsdal, S. 1990, in Gravitational Lensing, eds: Y. Mellier, B. Fort, and G. Soucail, (Springer Verlag: Berlin), p. 203.

Burke, B. F. 1990, in Gravitational Lensing, eds: Y. Mellier, B. Fort, and G. Soucail, (Springer Verlag: Berlin), p. 127.

Crampton, D., McClure, R. D., Fletcher, J. M., and Hutchings, J. B. 1989, *Astron. J.*, **98**, 1188 (CMFH).

Ftaclas, C., and Terrile, R. J. 1990, *Bull. Amer. Astron. Soc.*, **22**, 757.

Gott, J. R. III, 1981, *Ap. J.*, **243**, 140.

Griest, K. 1990, preprint.

Hewitt, J. N. 1986, Ph D. Thesis, Massachusetts Institute of Technology.

Irwin, M. J., Webster, R. L., Hewett, P. C., Corrigan, R. T., and Jedrzejewski, R. I. 1989, *Astron. J.*, **98**, 1989.

Linder, E. V., and Schneider, P., and Wagoner, R. V. 1988, *Ap. J.*, **324**, 786.

Mellier, Y., Fort, B., and Soucail G., 1990, Gravitational Lensing, (Springer Verlag: Berlin), p. 311.

Nemiroff, R. J. 1989, *Ap. J.*, **341**, 579.

Nemiroff, R. J. 1990, *Phys. Rev. Lett.*, submitted.

Nemiroff, R. J. 1991, *Astron. Ap.*, in press.

Nemiroff, R. J., and Bistolas, V. G. 1989, *Ap. J.*, **358**, 5.

Paczynski, B. 1986a, *Ap. J.*, **304**, 1.

Paczynski, B. 1986b, *Ap. J.*, **308**, L43.

Press, W. H., and Gunn, J. E. 1973, *Ap. J.*, **185**, 397.

Schneider, P. and Wagoner, R. V. 1987, *Ap. J.*, **314**, 154.

Tyson, J. A., and Gullixson, C. A. 1986, *Science*, **233**, 1183.

Tyson, J. A. and Seitzer, P. 1988, *Ap. J.*, **335**, 552.

Vanderriest, C. 1990, in Gravitational Lensing, eds: Y. Mellier, B. Fort, and G. Soucail, (Springer Verlag: Berlin), p. 210.

Webster, R. L., Hewett, P. C., and Irwin, M. J. 1988, *Astron. J.*, **95**, 19.

THE DISTORTION OF DISTANT GALAXY IMAGES BY LARGE SCALE STRUCTURE

R. D. Blandford and A. B. Saust
Caltech, Pasadena, CA 91106.

T. G. Brainerd and J. V. Villumsen
The Ohio State University, Columbus, OH 43210.

ABSTRACT

Inhomogeneity in the large scale distribution of mass in the universe on scales $\sim 100h^{-1}$ Mpc can induce a coherent shear distortion in the images of background galaxies. This distortion may be measurable over patches of the sky a few degrees in size. Simulations of giant voids, great attractors and walls as randomly located structures, expanding with the universe, indicate that the mean correlated ellipticity of the images of high redshift ($z \gtrsim 2$) galaxies is roughly $0.3\Omega_{LSS}$, where Ω_{LSS} is the density parameter of the inhomogeneously distributed mass on scales $\sim$ 100Mpc. The correlation length is roughly 1°. Simulations of biased cold and hot dark matter cosmogonies produce correlated ellipticities ~ 0.03 with correlation lengths $\sim 0.5°, 0.6°$, respectively. The faint blue galaxies discovered by Tyson and collaborators have a surface density of $\sim 3 \times 10^5$ per square degree and should provide an ideal set of background sources for measuring this effect.

1. GALAXY IMAGE DISTORTION

Recent observations of arcs and mini-arcs in rich clusters have reinforced the view that there is a large population of blue galaxies at redshifts $1 \lesssim z \lesssim 3$, (*e.g.* Soucail *et al.* 1987, Tyson, Valdes and Wenk 1990 and references therein.) Galaxy counts in deep fields (*e.g.* Tyson 1988) suggest that there are $\sim 3 \times 10^5$ such galaxies per square degree down to 29^m. It is therefore practical to contemplate using CCD's to measure the shapes of $\gtrsim 10^6$ of these galaxies and mapping their mean distortion on the sky. This will provide a means of quantifying the amount of large scale structure in the mass distribution, in contrast to that already reported in the distribution of local galaxies (*e.g.* Geller and Huchra 1989, Broadhurst *et al.* 1990).

We measure the mean distortion of a galaxy image using the image polarization, defined by the average

$$P = < \epsilon \cdot e^{2i\phi} >$$

where ϵ is the observed ellipticity of an individual galaxy and ϕ is the associated position angle. When a ray congruence passes by a large density inhomogeneity, a small contribution to P will result. Polarization induced by unrelated masses will add stochastically.

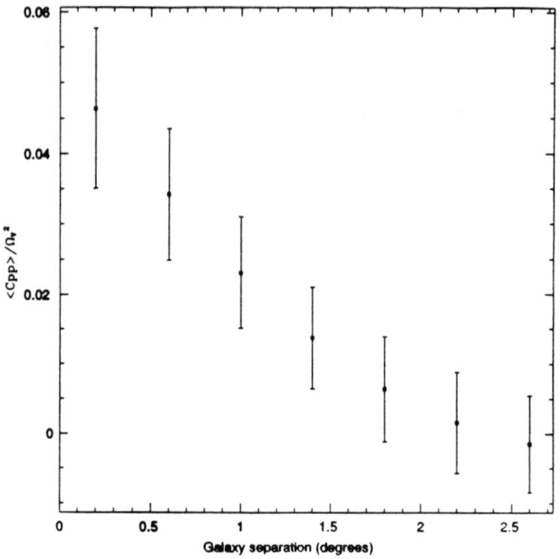

Fig. 1. Two point correlation function for the image polarization of galaxies at a redshift $z = 3$ in a simple model universe containing spherical voids of radius $60h^{-1}$Mpc and a filling factor of 0.67. The mean fractional underdensity in the voids is Ω_V.

2. MODELS OF INDIVIDUAL DEFLECTORS

We model individual deflectors as points, spheres or disks. The largest distortion is created by point masses which produce a mean two point polarization correlation function which can be computed analytically in an Einstein-De Sitter cosmology.

$$C_{PP} = < P(\theta')P^*(\theta' + \theta) > = \frac{16\Omega_M M[(1+z)^{1/2} - 1]}{\theta^2}$$

where Ω_M is the fraction of the critical density contained in points of mass M. However, the variance about this mean is considerable on account of the strong shear caused by occasional close encounters.

Spherical inhomogeneities of negative mass can be used to model voids. If we distribute these randomly per unit comoving volume throughout the universe with size and filling factor matched to those observed, then we can compute the polarization correlation function. This is given roughly by

$$C_{PP} \sim 8 \times 10^{-3} |\Omega_V|$$

for image separations $\theta \lesssim 1°$, where Ω_V is the density parameter for the voids. Therefore, if the mean void density deficits are ~ 0.1, then the mean correlated ellipticity will be roughly 2-3 per cent.

Conversely, we can attempt a simple kinematic model of a wall as a uniform density, circular disk and randomly distribute these structures through the universe. We find that image distortions of similar magnitude to those found in the void models are created.

3. POWER SPECTRUM OF DENSITY FLUCTUATIONS

A complementary description of large scale structure postulates a power spectrum of density fluctuations, $P(k)$ in conventional notation. Again restricting attention to an Einstein De-Sitter universe, we find that we can compute the image polarization coefficient to be

$$C_{PP}(\theta) = 288\pi^2[1-(1+z)^{-1/2}]^3 \int_0^\infty dk k P(k) \int_0^1 ds s^2 (1-s)^2 J_0(st)$$

where $t = 2k[1-(1+z)^{-1/2}]\theta$ and units where $c = H_0 = 1$ are used.

Using N-body simulations (see Villumsen 1989), we have adopted standard cold and hot dark matter cosmogonies in a universe with $h = 1/2$ and a bias factor $b = 2$. We find that if the source galaxies are at redshifts $z \sim 3$, then

$$C_{PP}(0) \sim 8 \times 10^{-4}(\mathrm{HDM})\ ;\ C_{PP}(0) \sim 6 \times 10^{-4}(\mathrm{CDM})$$

The correlation lengths (defined to be the point at which $\int_0^\theta 2\pi\theta' C_{PP}(\theta')d\theta'$ reaches half its asymptotic value) are $\sim 0.5°$ for CDM and $\sim 0.6°$ for HDM. In general the integral for the polarization correlation function is dominated by the peak of the density fluctuation power spectrum $P(k)$. Small scale correlations associated with small groups and clusters should not contaminate the signal from large scale structure provided that we average over large enough angular scales and avoid nearby rich clusters.

OBSERVATIONAL PROSPECTS

We have shown that the images of distant galaxies whose intrinsic shapes we presume to be isotropically oriented on the sky ought to exhibit a ~ 3 percent correlated ellipticity in either an Einstein- De Sitter model universe with a ten per cent mass density contrast in voids or walls or a standard hot or cold dark matter universe. The effect is significantly larger than originally estimated (Blandford 1990). CCD's with fields of view as large as $50' \times 50'$ will soon be available (Tyson, private communication) and it should be possible to record the shapes of the images of $\sim 10^5$ galaxies with a single frame, The associated random errors in the measured correlated ellipticity should then be ~ 0.003, far smaller than the signals proposed here. These observations will be subject to a variety of systematic errors, including atmospheric refraction, CCD anisotropies, telescope optics and trailing. If these are not too serious a problem, then mapping galaxy distortion on the sky promises to provide a powerful probe of large scale structure in the contemporary universe in much the same way as

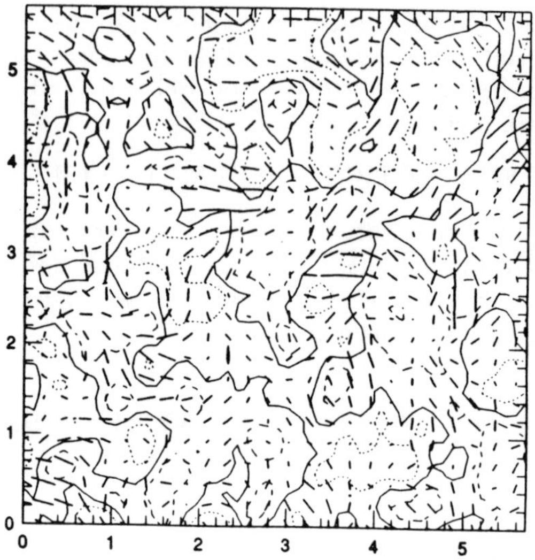

Fig. 2. Contours of constant magnification as computed by ray tracing in a model cold dark matter universe (see text), with superposed image polarizations, measured by the lengths and orientiations of the lines.

the measurement of microwave background fluctuations once promised to be a probe of inhomogeneities at the time of recombination. A fuller treatment will appear shortly.

ACKNOWLEDGEMENTS

We acknowledge financial support from the National Science Foundation under grants AST 86-15325, 89-15326 (RDB), and 89-21001 (JVV), the Carlsberg Foundation and the Danish Research Academy (ABS). The N-body simulations were performed at the Ohio Supercomputer Center.

REFERENCES

Blandford, R. D.. 1990, *Q. J. R. Astr. Soc.*, **31**, 305.

Broadhurst, T. J., Ellis, R. S., Koo, D. C. and Szalay, A. S.. 1990, *Nature*, **343**, 726.

Geller, M. J. and Huchra, J.. 1989, *Science*, 246, 897.

Soucail, G., Fort, B., Mellier, Y. and Picat, J.. 1987, *Astr. Astrophys.*, **172**, L14.

Tyson, T.. 1988, *Astr. J.*, **96**, 1.

Tyson, J. Valdes, F. and Wenk, R.. 1990, *Astrophys. J. (Lett.)*, **349**, L1.

Villumsen, J. V.. 1989, *Astrophys. J. Suppl.*, **71**, 407.

THE MASS OF THE GALAXY LENSING Q0957+561: LIMITS ON H_0

George Rhee
Dept. of Astronomy, New Mexico State University, Las Cruces, New Mexico,

ABSTRACT

The double quasar Q0957+561 A B is believed to be a gravitational lens image formed by a galaxy at z=0.36 (Young *et al.,* 1979). The object is a single QSO at z=1.41 with a double image produced by the intervening galaxy. Deep images taken by Young *et al*, reveal a rich cluster of galaxies whose brightest member is nearly superposed on Q0957+561B. For the past eight years Vanderriest *et al.*, 1989, have been monitoring the flux from the images A and B of the lensed quasar. A time delay between the two images has been obtained with a highly significant correlation peak. Falco Gorenstein and Shapiro (1985) have presented a two component model for the lens where the components approximate the bending due to the elliptical galaxy plus a cluster. Their analysis shows that the one dimensional velocity dispersion of the elliptical galaxy and the time delay can be used to constrain the value of H_0. I have measured the velocity dispersion of the galaxy lensing Q0957+561 and report this measurement and the corresponding bounds on H_0.

INTRODUCTION

The first example of a lensed quasar (Q0957+561 AB) was discovered in 1979[1]. This quasar is believed to be a gravitational lens image formed by a galaxy at a redshift of 0.36[2]. The object is a single quasar at z=1.41 with a double image produced by the intervening galaxy. Deep images[2] reveal a cluster of galaxies whose brightest member (the deflector) is nearly superposed on Q0957+561B. Multichannel scans of the quasar show a contribution from the galaxy and the location of the 4000Å break in the galaxy spectrum indicating a redshift of 0.36. Images taken in 0.3 to 0.5 arcsecond seeing[3] confirm that the galaxy is 1.0±0.03 arcseconds north and 0.24 ± 0.09 arcseconds east of the southern quasar image(B). The system has been imaged in the infrared (K-band) [11], where it is found that the brightness profile of the galaxy matches that of a cD (giant elliptical with an extended halo) galaxy. The galaxy has an apparent magnitude of 18.49 ± 0.03 in Gunn r, a core radius of 0".24 ± 0".09, and an ellipticity of 0.13 ± 0.01.

THE TIME DELAY

From 1980 onwards photometric monitoring of the A and B images was carried out by two groups[5,6]. The quoted time delays are 415 ± 20 days[5] and 404 ± 10 days[6]. The time delay is such that $\Delta\tau_{BA}$ is positive i.e. image A varies first. These values are based on data acquired independently on different telescopes by different observers and their consistency is strong evidence that the effect is real. Both groups also suggest that the variation with time of the true brightness variation between the two images is indicative of microlensing.

MODELS

Simple models have been developed that acount for the known properties of the lens system[6,7]. The matter doing the lensing is modelled by three components; 1) the bright galaxy, 2) a compact nucleus and 3) the cluster. The brightest galaxy is represented as a circularly symmetric smooth King[8] type surface density profile whose parameters are the core radius and velocity dispersion. The compact nucleus has one parameter its mass, and the cluster is modelled by a smooth distribution of dark matter with a given surface density. The compact nucleus is required to account for the relative magnification of the A and B images. The model has fewer parameters than the observations yet accounts well for all the observations. Using this model[7], one can obtain bounds on the Hubble constant. This is possible because the time delay is proportional to the product of the distance to the deflector and the distance to the qso divided by the distance from the deflector to the qso. Since each distance scales as H_0^{-1}, the time delay is inversely proportional to H_0. The full expression in terms of observable parameters is: $H_0 = (90 \pm 10)(\sigma_v/390km/s)^2(\Delta\tau_{BA}/1yr)^{-1}$ km/s/Mpc. $\Delta\tau_{BA}$ is the difference in the light propagation times from the source to the observer, via images A and B and σ_v is the velocity dispersion of the brightest galaxy.

THE PRESENT OBSERVATIONS

Motivated by the unambiguous measurement of $\Delta\tau_{BA}$ mentioned above, I set out to measure the velocity dispersion σ_v of the lensing galaxy and the redshifts of the brightest galaxies in the putative cluster. Measuring the velocity of the brightest galaxy is rendered non trivial by the fact that the galaxy spectrum is seriously contaminated by quasar light. In fact the "sky background" is dominated by the continuum emission from the quasar. Observations were carried out on the night of February 19/20 1990 using the 4.2-m William Herschel Telescope at the Observatorio del Roque de los Muchachos. The ISIS triple spectrograph[10] was used at the f/11 Cassegrain focus of the William Herschel Telescope. The R600R grating was used with effective blaze at 7000Å and a resolution of $\sim$ 1Å. The wavelength coverage for observing the cD galaxy was 6600-7400Å, chosen so as to include the Mgb triplet at 5174Å redshifted at z=0.36.

A 1" slit oriented in the north-south direction was centered on Q0957+561B. Five 2400s integrations were made with this setup. The five CCD frames were median filtered a wavelength calibration was made using Helium-Argon arc exposure, the spectra were corrected for misalignment (the dispersion axis did not coincide exactly with the y-axis on the CCD), and the data were sky subtracted. The reduction was done using the IRAF reduction package. With such a setup the profile of Q0957+561B is asymetric with the northern wing of the profile being brighter than the southern wing due to the contribution of the cD galaxy light. The columns of the northern wing of the quasar profile were summed so as to give a spectrum which contained the maximum signal to noise for the galaxy. The galaxy redshift was measured and the spectrum was deredshifted. The lines of the Mg1b triplet are clearly visible in figure 1. The redshift and velocity dispersion were measured using the cross correlation technique. The template spectrum used was a spectrum of the standard star HD207229. This method yields a line of sight velocity dispersion of 303 ± 50 km/s for the galaxy. Using the simple model described above and assumimg the time delay between the two quasar images to be 415 days one obtains the value of H_0 to be 50 km/s/Mpc $\pm$ 17 km/s/Mpc.

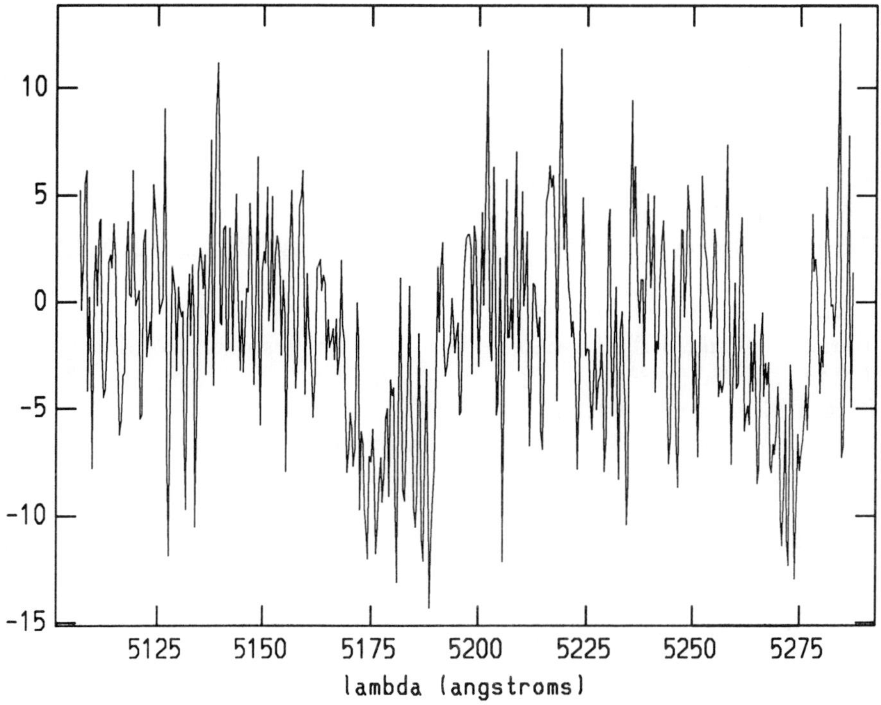

1. The galaxy spectrum continuum, subtracted and shifted to z=0.

DISCUSSION

A key unknown in this problem is the mass of the cluster and its contribution to the bending. During the same observing run I intended to measure the velocity dispersion of the cluster using 30 galaxies. Due to poor weather only six redshifts of cluster members were obtained. These data yield an rms line of sight velocity dispersion[12] of 530 km/s. Redshifts for the galaxies were measured using a standard cross correlation technique[13], the template used was M32[14]). The uncertainties in this number or large and $\sim$ 30 redshifts of cluster members will be required to assess the contribution of the cluster gravitational potential to the lensing.

To conclude I have obtained an estimate of the value of the Hubble constant by measuring the velocity dispersion of the galaxy lensing Q0957+561. Using a simple model and the known time delay between the A and B images of the quasar I obtain a value for H_0 of 50 ± 17 km/s/Mpc.

REFERENCES

1. Walsh, D., Carswell, R.F., & Weymann, R.J. *Nature* bf 279, 381- 384 (1979).
2. Young, P., Gunn, J.E., Kristian, J., Oke., J.B., & Westphal, J.A. *Astrophys. J.* **204**, 507-520 (1980).
3. Stockton, A. *Astrophys. J. Lett.* **242**, L141-L144 (1980).
4. Vanderriest, C., Schneider, J., Herpe, G., Chevreton, M., Moles, M. & Wlerich, G. in *Gravitational Lenses* (eds. Moran J.M., Hewitt, J.N. & Lo, K.Y.) (Dordrecht: Reidel, 1989).
5. Schild, R.E. *Astrophys. J.* (submitted).
6. Falco, E., Gorenstein, M. & Shapiro, I. *Astrophys. J. Lett.* **289**, L1-L4 (1985).
7. Falco, E., Gorenstein, M. & Shapiro, I. *Astrophys. J.* (1990), (in press).
8. Young, P., Gunn, J.E., Kristian, J., Oke., J.B., & Westphal, J.A. *Astrophys. J.* **244**, 736-755 (1980).
9. King, I.R. *Astron. J.* **67**, 471-485 (1962).
10. Unger, S.W., Brinks, E., Laing, R.A., Tritton, K.P. & Gray, P.M. La Palma User's Manual (Royal Greenwich Observatory) (1988).
11. Gallais, P., Rouan, D., Lacombe, F. & Tiphene, D. *Astron. Astrophys.* **232**, 16-18 (1990).
12. Danese, L., De Zotti, G. & di Tullio, G. *Astron. Astrophys.* **82**, 322-327 (1980).
13. Tonry, J., & Davis, M. *Astron. J.* **84**, 1511-1523 (1979)
14. Keel, W.C. *Astrophys. J.* **269**, 466-486 (1983).

Phase Space Distribution of Halo Particles and Detection of WIMPS

Dongsu Ryu
NASA/Fermilab Astrophysics Center, Fermilab, Batavia, IL 60510
and Princeton University Observatory, Princeton, NJ 08544
and
Armando Pérez
NASA/Fermilab Astrophysics Center, Fermilab, Batavia, IL 60510
and Departamento de Física Teórica, Universidad de Valencia, Spain

ABSTRACT

We calculate the velocity distribution of halo particles using realistic models for the density profiles of the Galactic disk, spheroid, and halo components. We get the resulting velocity distribution, which is different from the Maxwellian distribution of isothermal halo. We investigate the consequences concerning the time averaged count rate and the amplitude of the annual modulation in WIMPS detection experiments. Compared to the previous estimate based on the Maxwellian distribution, the time averaged count rate is up to 40% large. On the other hand, the amplitude of the annual modulation is up to a factor of two large.

I. INTRODUCTION

It is known that there are massive dark halos around spiral galaxies including our Galaxy, even though the nature of the halo material is unknown. Strong evidence for the existence of massive halos comes from the rotation curves of spiral galaxies which remain relatively flat out to 2 – 3 Holmberg radii[1]. Previously, several authors considered the possibility that the halos of spiral galaxies are composed of some form of baryonic matter, especially low-mass stars or Jupiter-like objects, or stellar remnants such as neutron stars and white dwarfs[2]. Although such baryonic halos are not very plausible, their possibility can not be completely ruled out. Also, the possibility of baryonic black holes with low mass $m_{BH} \lesssim 10^6 M_\odot$ as halo matter is a viable hypothesis. On the other hand, particle physics models suggest a host of non-baryonic dark matter candidates for halo matter[3]. These include Dirac or Majorana neutrinos and various particles predicted by supersymmetry, such as photinos, Higgsinos, or scalar neutrinos. The constraint that these particles do not overclose the Universe ($\Omega_{\rm DM} \lesssim 2$) requires particle mass $m_{\rm DM} \gtrsim 2$Gev (an exception is the axion).

Here, we assume that the Galactic halo is composed of WIMPS and analyze their expected signals in some detection experiments. Previously, several authors considered the same problem, by assuming that the Galactic halo is isothermal and the velocity distribution of halo particles is Maxwellian[4]. Such assumption was based, not so much on any hard evidences, but rather on simplicity and the lack of more compelling models. We try to improve the situation by considering realistic models for the mass distribution of the Galaxy.

II. PHASE SPACE DISTRIBUTION OF HALO PARTICLES

We assume the Galaxy as a system with spherical mass distribution. If the velocity dispersion tensor is isotropic, the velocity distribution of halo particles can be calculated from the Eddington's formula[5],

$$f(\varepsilon) = \frac{1}{\sqrt{8}\pi^2}\frac{d^2}{d\varepsilon^2}\int_0^{\varepsilon}\frac{\rho_{\rm H}(\Psi_{\rm T})}{\sqrt{\varepsilon - \Psi_{\rm T}}}d\Psi_{\rm T},$$

where $\rho_{\rm H}$ is the *halo* density, $\Psi_{\rm T}$ is the negative *total* potential, and $\varepsilon = \Psi_{\rm T} - 1/2v^2$. Similarly, for a system with anisotropic velocity tensor, the Osipkov-Merritt model[5] can be used to calculate the velocity distribution.

For $\rho_{\rm H}$ and $\Psi_{\rm T}$, we use the models suggested by Caldwell and Ostriker[6]. In their models, three coexisting mass components for the model density distribution were adopted. They are an exponential disk with an exponential central gap ($\rho_{\rm D} = \rho^{\rm o}_{\rm D}[\exp(-r/r_{\rm D}) - \exp(-r/r_{\rm G})]$), a Hubble law spheroid ($\rho_{\rm S} = \rho^{\rm o}_{\rm S}(r/r_{\rm S})^{-3}$), and a pseudo-isothermal halo ($\rho_{\rm H} = \rho^{\rm o}_{\rm H}/(r^2 + r^2_{\rm H})$). The model parameters and the solar galactocentric position were chosen by fitting a set of Galactic observational constraints in a maximum likelihood sense.

Figure 1 shows the resulting velocity distribution of halo particles in the Galactic rest frame. The velocity has been normalized, so that $v = 1$ corresponds to 160km/s. The distribution function has been normalized, so that $4\pi\int fv^2dv = 1$. The density distributions from the 8 different models of Caldwell and Ostriker[6] have been used. (Here, Models A75 and B75 have the tidal radius at 75kpc, and so on.) For comparison, the Maxwellian velocity distribution of isothermal halo is also shown. The rotational speed of the local standard of rest with respect to the Galactic center has been set to be 220km/s[6,7] in the Maxwellian distribution. Because of large mass inside the solar position which is mainly contributed from the disk and spheroid components, the peak in the velocity distribution shifts into a high velocity compared to the peak of the Maxwellian distribution. The velocity distribution is more or less symmetric around the peak. Since a boundary is imposed in the Galaxy models, the velocity distribution is truncated at the Galactic escape velocity and its high velocity tail is modified. The shape of the high velocity tail depends on the assumed shape and width of the transition zone around the boundary. For our calculations, an exponential boundary with continuous density and density derivative has been used.

III. WIMP DETECTION

The estimation of the expected signals in WIMPS detection experiments involves two steps: 1) converting the velocity distribution in the Galactic rest frame to the velocity distribution in the Earth's frame, and 2) counting the number of WIMPS registered in a specific detector with threshold energy ϵ. In the first step, we set the net speed of the Sun with respect to the Galactic rest frame to be 230km/s and the orbital speed of the Earth around the Sun to be 30km/s with the angle between the axis of the orbit and the velocity vector of the Sun 31°[7]. In the second step, following Freese, Frieman, and Gould[4], we calculate the time averaged count rate and the amplitude of the annual modulation in experiments measuring nuclear recoiling due to the scattering of WIMPS.

Figure 2 shows the dimensionless measures of the resulting expected signals. Here, A^2 is the dimensionless threshold energy defined as $3(M_{\rm W} +$

$M_N)^2/(4M_W^2 M_N v_n^2)\epsilon$ where M_W is the mass of WIMPS, M_N is the mass of nuclei in the detector, and v_n is the normalization velocity (160km/s). The time averaged count rate is $S(\epsilon) \propto \gamma(\epsilon)\rho_H(R_o)\bar{v}$ and the amplitude of the annual modulation is $S_m(\epsilon) \propto \beta(\epsilon)\rho_H(R_o)\bar{v}$, where $\rho_H(R_o)$ is the local halo density and $\bar{v}$ is the average velocity of WIMPS in the Galactic rest frame. (For the detailed discussions and the discussions about the modulation significance, refer Freese, Frieman, and Gould[4]). For given $\rho_H(R_o)$ and $\bar{v}$, compared to the previous estimate based on the Maxwellian velocity distribution, the time averaged count rate is slightly large (up to 40%). On the other hand, the amplitude of the annual modulation is significantly large (up to a factor of two) and its peak occurs in a high threshold energy. The effects of the boundary are important only when the threshold energy of detectors is much larger than the maximum energy deposited by a typical WIMP.

IV. DISCUSSION

By assuming the Galaxy as a spherical system with an isotropic velocity distribution, we have ignored the effects of the non-sphericity of the mass distribution and the anisotropy of the velocity distribution. When we allow the anisotropic velocity distribution, the average radial velocity increases while the average tangential velocity decreases. The effects of the velocity anisotropy on WIMPS detection experiments will be discussed in a subsequent publication.

The biggest uncertainty of the astrophysical aspect in WIMPS detection experiments comes from a factor of 2 uncertainty in the local halo density. Except that, the estimated astrophysical uncertainty is less than $\sim 50\%$, when a moderate degree of anisotropy is assumed in the velocity distribution. Of course, we expect there are uncertainties from the view points of particle or nuclear physics which may exceed the astrophysical uncertainties.

ACKNOWLEDGMENTS

We would like to thank Joshua A. Frieman, David N. Spergel, and Michael S. Turner for discussions. This work was supported in part by the DOE and by the NASA at Fermilab through grant NAGW-1340.

REFERENCES

[1]See *e.g.* N. Krumm and E. E. Salpeter, *Astron. J.* **84**, 1138 (1979).

[2]D. J. Hegyi and K. A. Olive, *Phys. Lett.* **126B**, 28 (1983); D. J. Hegyi and K. A. Olive, *Astrophys. J.* **303**, 56 (1986); D. Ryu, K. A. Olive, and J. Silk, *Astrophys. J.* **353**, 81 (1990).

[3]For a review see J. Primack, D. Seckel, and B. Sadoulet, *Ann. Rev. Nucl. Part. Sci.* **38**, 751 (1988).

[4]For instance see A. K. Drukier, K. Freese, and D. N. Spergel, *Phys. Rev. D* **33**, 3495 (1986); K. Freese, J. Frieman, and A. Gould, *Phys. Rev. D* **37**, 3388 (1988).

[5]See J. Binney and S. Tremaine, *Galactic Dynamics* (Princeton University Press, Princeton, 1987), Chapter 4.

[6]J. A. R. Caldwell and J. P. Ostriker, *Astrophys. J.* **251**, 61 (1990).

[7]See *e.g.* C. W. Allen, *Astrophysical Quantities* (The Athlone Press, London, 1973); K. Lang, *Astrophysical Formulae* (Springer, Berlin, 1980); D. Mihalas and J. Binney, *Galactic Astronomy* (Freeman, San Francisco, 1981); F. J. Kerr and D.Lynden-Bell, *Mon. Not. R. Astron. Soc.* **221**, 1023 (1986).

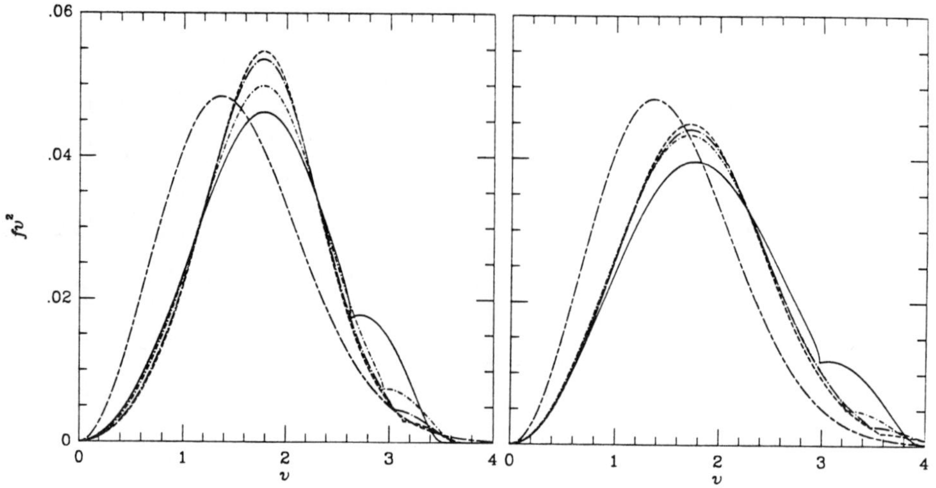

FIGURE 1. Velocity distributions of halo particles in the Galactic rest frame for (a) Models A75 (solid line), A150 (dot-short dashed line), A225 (dot-long dashed line), and A300 (short dashed line), and (b) Models B75 (solid line), B150 (dot-short dashed line), B225 (dot-long dashed line), and B300 (short dashed line) of Caldwell and Ostriker[6]. For comparison, the Maxwellian velocity distribution (short dash-long dashed line) for isothermal halo is also plotted.

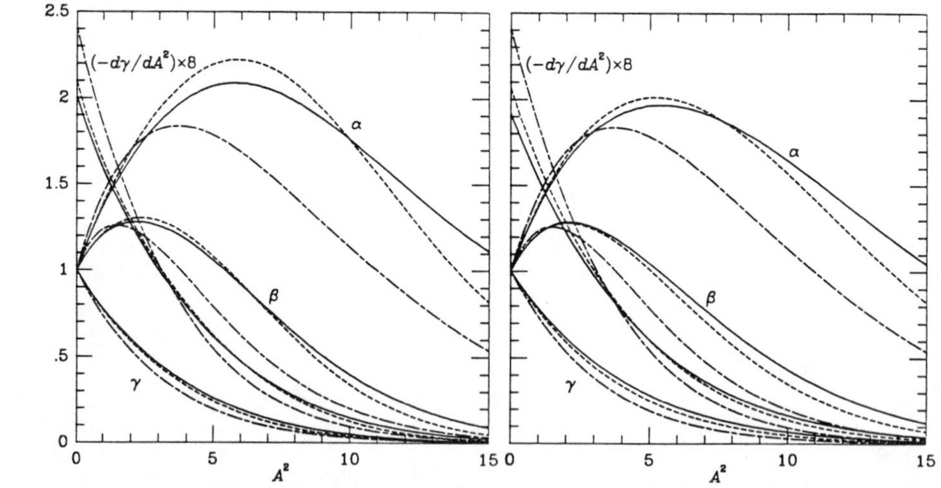

Figure 2. Dimensionless measures of the time averaged count rate and the modulation amplitude for (a) Models A75 (solid line) and A300 (short dashed line) and (b) Models B75 (solid line) and B(300) (short dashed line). For comparison the dimensionless measures for isothermal halo (short dash-long dashed line) are also plotted.

TYPE Ia SUPERNOVAE: CONSTRAINTS ON BARYONIC DARK MATTER

Tammy A. Smecker
Rosemary F.G. Wyse
Dept. of Physics and Astronomy, The Johns Hopkins University
Baltimore, MD USA 21218

ABSTRACT

We examine models in which the dark matter of galaxy halos is in the form of white dwarfs (WDs). Major constraints, which have been previously neglected, result from the occurrence of Type Ia supernovae, which are the products of merging WDs in binary systems whose orbits decay due to the emission of gravitational wave radiation. The lack of observed supernovae in the outer regions of galaxy halos limits the fraction of binaries to $\ll 1\%$ in extant models of WD dominated halos, for standard Type Ia parameters. This binary fraction is much smaller than that believed typical for luminous stars, necessitating a further peculiarity in the star-formation process during formation of the dark halo.

1. INTRODUCTION

The baryonic content of the Universe is constrained by primordial nucleosynthesis; homogeneous models place limits on the contribution of baryons to the critical density as (Olive *et al.* 1990)

$$0.01 \leq \Omega \leq 0.063$$

for values of the normalized Hubble parameter $0.5 \leq h \leq 1$. That BDM must exist can be seen from consideration of the contribution of stars to Ω; a reasonable estimate of which is

$$\Omega_{lum} \lesssim 0.005.$$

Clearly a large fraction of baryons are *not* in visible stars.

A number of suggestions for the repository of BDM have been discussed. Remnants of massive stars are plausible candidates for BDM (*c.f.* Larson 1986). Restricting the IMF of dark halo precursors is attractive in order to minimize the chemical enrichment which can contaminate subsequent stellar populations. In particular, WDs might be favorable dark matter candidates, as suggested by Ryu, Olive and Silk (1990).

Intermediate-mass stars ($2 - 8\,M_\odot$) produce less enrichment during their quiescent evolution to WDs than do higher mass stars which explode in Type II supernovae, but the carbon and helium enrichment from stellar winds and planetary nebulae formation is non-negligible for $10^{12}\,M_\odot$ of WD remnants. In addition, no previous consideration has been given to metal production in Type Ia supernovae or the observable consequence of these explosions which arise from the merging of a binary system of WDs precipitated by orbital energy lost by gravitational wave radiation. The timescale for a binary to merge can be longer than a Hubble time, so we can expect these explosions to occur even after a stellar population has faded beyond detection by any other means.

There is every reason to expect a significant population of binaries among the precursor stars of remnant-dominated dark halos – the binary fraction of stars in the solar neighborhood is estimated to be from 50% to ∼ 100 %. Consider IMFs restricted to $2 - 8\,M_\odot$ stars; only those above $\sim 4\,M_\odot$ will contribute to the SN Ia rate, but the fraction of these stars in this IMF is a factor of 30 above that in a solar-neighborhood IMF with a lower cutoff of $0.1\,M_\odot$ (Miller and Scalo 1979). Also, the relative masses of dark to luminous matter is approximately a factor of 10. Therefore, we constrain the binary fraction of WD dark halos with the observed lack of SN Ia in dark halos.

2. THE SUPERNOVA RATE CONSTRAINT

We calculate the SN Ia rate expected from a population of intermediate-mass stars, where the origin of SN Ia is the merging of pairs of carbon-oxygen WDs, as described in Iben and Tutukov (1987). We identify sets of initial masses and separations which result in mergers leading to one object of total masses greater than $1.4\,M_\odot$ – these we identify as SN Ia. The major parameter of the SN Ia model is the efficiency of the common envelope interaction in reducing the orbital separation of the stars, α, for which we follow common practice and adopt $\alpha = 1$. However, it should be noted that the derived rates are unfortunately sensitive to this parameter, and some numerical studies suggest it may be lower (*c.f.* Livio 1989).

The SN Ia rate arising from a stellar population formed with a star-formation rate, $\psi(t)$, of which p% of stars are born in binaries, is given by

$$R(t) = \frac{p}{200} \times C \times \int dM \int d\log A\ \Phi(M)\,\psi(t')$$

where the integrals are taken over all masses and separations yielding SN Ia at time t; t' is the time at which the binary was born, and is related to t by

$$t' = t - \tau_{M2} - \tau_{GR}$$

where τ_{M2} is the lifetime of the secondary, and τ_{GR} is the timescale for gravitational wave radiation to merge the WDs.

We assume binaries are of equal component masses, although relaxing this assumption does not significantly alter our conclusions. Initial binary separations are assumed to be drawn from a flat distribution in $d(\log A)$ over the range $\log(A/R_\odot) = 1 - 5$ giving the normalization factor $C = 0.25$.

Stars are created in the range $M_{lower} \leq M \leq M_{upper}$ with an IMF which is assumed to be constant in time as given by

$$\Phi(M) \propto M^{-(1+x)}$$

where $\Phi(M)dM$ is the fractional number of stars in the interval, $M \rightarrow M + dM$. We adopt the Miller-Scalo (1979) slope of $x = 1.5$, although the value of the slope has little effect on the results since the range of masses is very limited.

An exponential star-formation rate with e-folding time, τ, as given in the table below, is used. We truncate the star-formation rate at 2 Gyr to distinguish the WD precursor population from the luminous galaxy, which is assumed to form subsequently.

In BDM models star formation can occur at high redshifts ($z \lesssim 100$) when structure goes nonlinear, thus the age of the WD dark halos are approximately the age of the Universe. We require that our models produce a $10^{12}\,M_\odot$ WD dark halo at the present epoch of ~ 17 Gyr. The initial mass of the gas cloud must be somewhat larger to take account of the WDs lost in SN Ia – SN Ia are expected leave no remnants since the the entire WD is disrupted, and only the nucleosynthetic yields ejected (Thielemann *et al.* 1986). Normalization of the models then requires that we adopt a standard value for the binary fraction; we adopt 20% for illustrative purposes.

Stars which evolve into halo WDs must recycle their ejecta in order to prevent contamination of the luminous components of galaxies. Therefore, the WD halo models rapidly evolve to high metallicity. We compute the chemical enrichment of our models using non-instantaneous recycling in a closed box system. Since the SN Ia rates are sensitive to the initial metallicity of the stars, we combine low metallicity stellar models (Tornambé and Castellani 1986), and solar metallicity models (Iben and Tutukov 1987), in forming the predicted SN Ia rate.

The following table gives a summary of the model parameters and the derived value of the SN Ia rate from a $10^{12}\,M_\odot$ WD halo at the present epoch (17 Gyr). Figure 1 presents the SN Ia rates for our three models. Details can be found in Smecker and Wyse (1991).

Model	M_{lower} ($M_\odot$)	M_{upper} ($M_\odot$)	τ (Gyr)	R(17 Gyr) (# SN Ia/yr)
1	2	8	2	0.647
2	4	6	1	0.243
3	5	5	0.6	0.187

3. COMPARISON WITH OBSERVATIONS

If 20% of the WDs were born in binaries then we predict the present SN Ia rate for a dark halo would be greater than 30 times the SN Ia rate observed for an average elliptical galaxy (van den Bergh *et al.* 1987).

Crane, Tammann, and Woltjer (1977) estimate the intergalactic supernovae rate by searching the Palomar plates of the central 3° of the Coma cluster. This region was investigated by blinking the plates, with the regions between galaxies checked as carefully as luminous areas. With an effective search of 15 yr, they reported ten supernovae in Coma cluster members, with *no* supernova occurring outside of a galaxy's Holmberg radius. The number of galaxies surveyed was ~ 1000, thus if each had a $10^{12}\,M_\odot$ dark halo, an upper limit to the supernova rate is obtained by assuming that one supernova actually occurred, leading to

$$R \leq 6 \times 10^{-5} \quad \text{SN Ia/yr}/10^{12}\,M_\odot \text{ dark halo.}$$

Thus scaling the models linearly in p, the maximum percentage of dark halo WDs in binaries is 0.002%, 0.005%, and 0.006% for Models 1, 2, and 3 respectively. Our derived SN Ia rates suggest that there can be effectively *no* WDs remaining in binary systems at the present epoch.

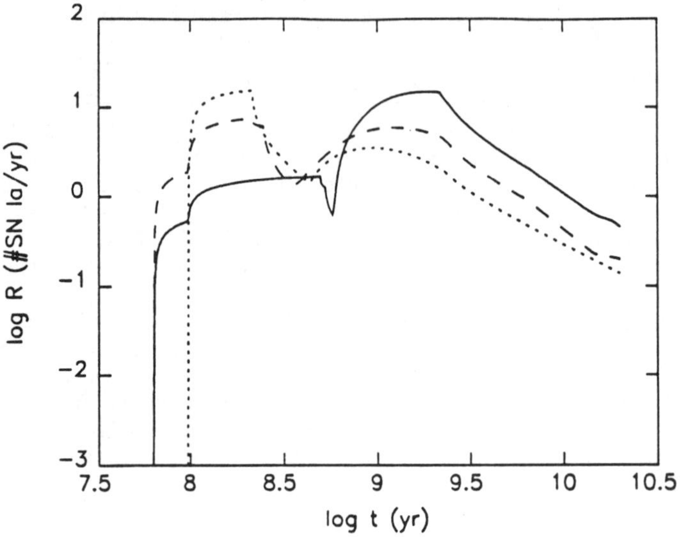

Fig. 1. — The SN Ia rates for WD halo Models 1 (solid), 2 (dashed), and 3 (dotted).

4. CONCLUSIONS

Our work emphasises that although a crucial parameter, α, of double-degenerate merging is debated, a large region of parameter space provides long-lived tracers in the form of SN Ia. Thus the rate of SN Ia presently observed can constrain stellar populations whose stars have aged beyond detection. The predicted SN Ia rates for WD dark halos greatly exceed observed constraints if $\alpha \sim 1$, unless the percentage of stars born in binaries is $\ll 1\%$, thus further necessitating different physical processes operative in creating stars which become dark than that which has created today's luminous stars.

T.A.S. wishes to acknowledge support from the Davis Fellowship of the Department of Physics and Astronomy, The Johns Hopkins University.

REFERENCES

Crane, P., Tammann, G. A., and Woltjer, L. 1977, *Nature*, **265**, 124.
Iben, I., Jr., and Tutukov, A. V. 1987, *Ap. J.*, **313**, 727.
Larson, R. 1986, *M.N.R.A.S.*, **218**, 409.
Livio, M. 1989, *Space Sci. Rev.*, **50**, 299.
Miller, G. E. and Scalo, J. H. 1979, *Ap. J. Suppl.*, **41**, 513.
Olive, K. A. *et al.* 1990. *Phys. Letters*, in press.
Ryu, D., Olive, K. A., and Silk, J. 1990, *Ap. J.*, **353**, 81.
Smecker, T. A., and Wyse, R. F. G. 1991. *Ap. J.*, in press.
Tornambé. A., and Chieffi, A. 1986, *M.N.R.A.S.*, **220**, 529.
van den Bergh, S., McClure, R. D., and Evans, R. 1987, *Ap. J.*, **323**, 44.

Detecting Halo Weakly Interacting Massive Particles with a Cryogenic Phonon Detector

N. Wang, P.D. Barnes, A. Cummings, J. Emes, Y. Giraud-Héraud[¶], E.E. Haller[†], A. Lange, R. Ross, B. Sadoulet, T. Shutt, C. Stubbs
Center for Particle Astrophysics and Department of Physics, University of California, and Physics Division, Lawrence Berkeley Laboratory
[¶]Laboratoire de Physique Corpusculaire, Collège de France, 75231 Paris, France
[†]Center for Particle Astrophysics and Department of Materials Science and Mineral Engineering, University of California, and
Materials and Chemical Sciences Division, Lawrence Berkeley Laboratory, Berkeley, CA 94720

ABSTRACT

If dark matter in the halo is made of weakly interacting massive particles(WIMPs), it may be detected with a cryogenic phonon detector. Such a detector must have low energy threshold, high energy resolution, and, more importantly, good background rejection capability. It has been proposed that a good background rejection ratio can be obtained by a simultaneous measurement of phonons and ionization. To test this experimentally, we have built a 60 g Ge detector and have measured simultaneous phonon and ionization signals, each with 4 keV energy resolution. We present a simple estimate of the rejection ratios, and WIMP event rates that may be achieved by such a detector.

DETECTING WEAKLY INTERACTING MASSIVE PARTICLES

There is little doubt about the existence of dark matter, but little is known about its nature. If we assume $\Omega=1$ and standard big bang nucleosynthesis, we need both baryonic and nonbaryonic dark matter. Among the various nonbaryonic candidates, cold dark matter, which includes weakly interacting massive particles (WIMPs) and axions, is favored because it provides an attractive explanation for galaxy formation[1].

WIMPs scattering elastically from a nucleus[2] can be detected using a phonon detector which consists of a target crystal with phonon sensors. The recoiling nucleus generates phonons and charge carriers. The latter may be detected as an ionization signal if the target is a semiconductor. To probe the interesting phase space of various WIMP models, the detector has to have an energy threshold of a few keV and a crystal mass of around 100 g. Phonons generated by an interaction are expected to first be quasi-ballistic with energies of about 1 meV, and to subsequently down-convert to thermal energies. To detect the thermalized phonons, it is essential to decrease the heat capacity of the crystal. For large crystal mass, this implies an operating temperature of about 10^{-2} K. The detection of ballistic phonons requires a fast phonon sensor and a phonon-transparent interface between the target crystal and the phonon sensors. Several groups are working on various detection methods[3].

There are two major challenges in a dark matter experiment: low energy threshold and good background rejection. Low mass WIMPs have a higher rate and a smaller average energy deposition than high mass WIMPs[3]. The recent results from LEP and hadron colliders seem to favor high mass WIMPs (>20 GeV). In this case, the requirement for energy threshold is reduced and background rejection becomes more crucial. Background signals can be caused by cosmic rays, fast neutrons, and internal and external radioactive decays. Cosmic rays can be easily identified and greatly reduced by going deep underground. Fast neutrons and external decay products can be suppressed by proper shielding. Internal radioactive decay is very difficult to eliminate because the signal originates inside the detector. To reduce these signals, we need to use intrinsically pure materials and minimize their exposure to cosmic rays. In addition, wewould like to distinguish them from actual dark matter events. It has been proposed that we can reject internal background by simultaneous measurement of phonon and ionization signals. This is because WIMP interactions cause nuclear recoil, while most of the background interactions cause electron recoil. Nuclear recoil generates a higher ratio of phonon to ionization energy than that generated by electron recoil (Fig. 1); consequently, the simultaneous measurement of phonon and ionization signals could provide a powerful background rejection capability.

In this paper, we describe results obtained with a 60 g Ge detector operating near 30 mK. We demonstrate simultaneous measurement of phonon and ionization energies. Using a Monte Carlo simulation, we estimate the background rejection capability of this detector.

CURRENT STATUS OF THE DEVELOPMENT

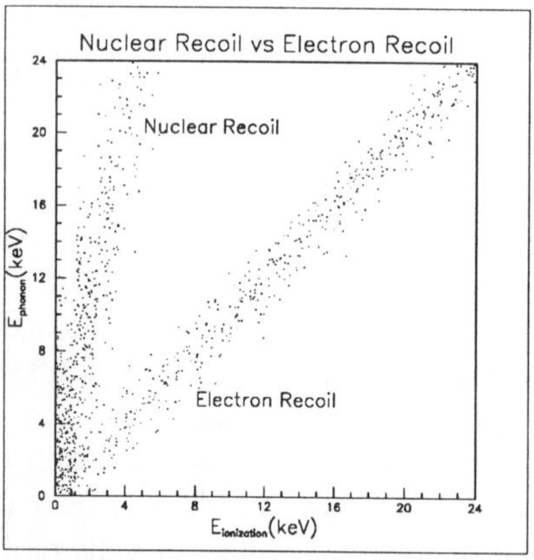

Fig. 1. A scatter plot of phonon energy vs ionization energy from a Monte Carlo simulation. $\sigma_{ph}=\sigma_{ion}$ =0.5 keV, Vbias=0.4 Volt, energy distribution is flat for electron recoil, and exponential for nuclear recoil with an average energy of 6.8 keV.

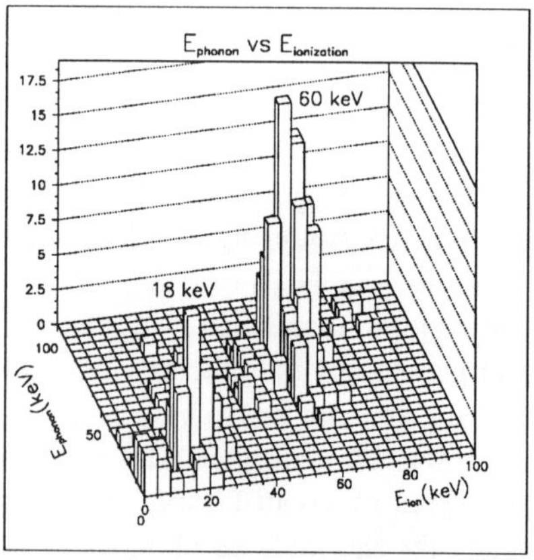

Fig. 2 A histogram of phonon vs ionization energy measured when the detector was irradiated with 18 and 60 keV photons.

We have made a 60 g Ge detector with six phonon sensors attached[4]. The phonon sensors are made of neutron transmutation doped Ge[5], and the detector is fabricated so that we can measure simultaneous phonon and ionization signals. We have observed 18 and 60 keV photons from ^{241}Am in both phonons and ionization. We show in Fig. 2 a histogram of phonon vs. ionization energy. The data were taken at 30 mK and the bias voltage across the crystal was 0.4 V. The signal sizes are as expected and are normalized so that a pulse corresponding to a 60 keV photon has an energy of 60 keV in both phonons and ionization. Energy resolutions are shown in Fig. 3 and Fig. 4.

POTENTIAL BACKGROUND REJECTION CAPABILITY

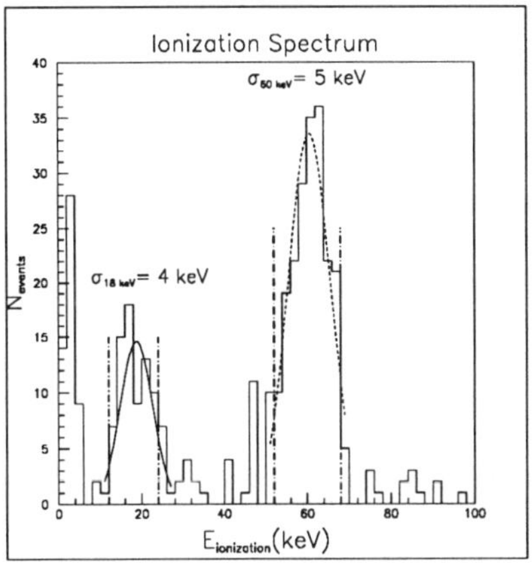

Fig. 3. An ionization spectrum. A gaussian fit to the two peaks gives an energy resolution of 4 keV at 18 keV and 5 keV at 60 keV.

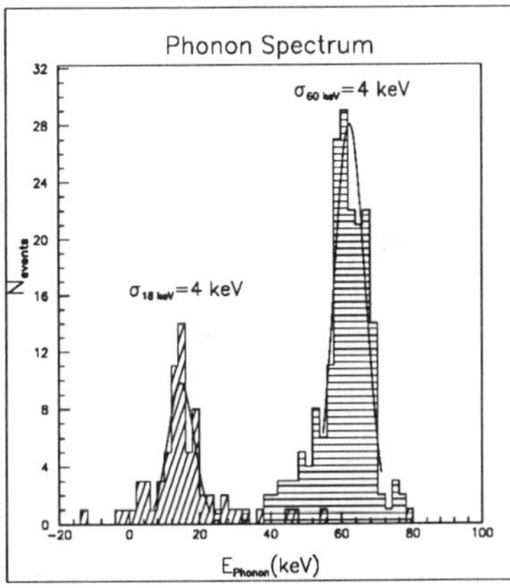

Fig. 4. A phonon spectrum corresponding to the ionization events within the two vertical lines around the two peaks in Fig. 3. A gaussian fit to both peaks gives an energy resolution of 4 keV at both 18 keV and 60 keV.

We present here a simple Monte Carlo estimate of the rejection ratio. We assume that one third of the total energy of background events goes into ionization. For a dark matter event, the fraction of the energy in ionization is parametrized by Lindhart's formula[6]. We also assume that the differential background rate is flat over energy, and the differential WIMP rate is exponential[3], with the exponent determined by the mass of the WIMP. At a given energy, the detected energies are described by a gaussian with the width given by the resolution of the detector, σ. We count all events from a threshold at 3σ to energies up to five times the average energy of the exponential distribution. Finally, we reject events whose ionization energy is larger than 1/6 of the total energy.

With the above assumptions, taking a conservative background rate of 10 event / kg/day/keV, and a counting time of 30 kg-day, we obtain results shown in Table 1. The resolution σ is assumed to be equal for both phonon and ionization signals. The mass of the WIMP is denoted by M_{DM}. The rejection ratio R is defined as ratio of the number of unrejected to total number of background events, and N_{bkgd} is the number of unrejected background events. The WIMP rate (events/kg/day) which can be detected with a 95%

confidence level is given by R_{DM}.

Table 1

σ(keV)	M_{DM}(GeV)	R	N_{bkgd}	R_{DM} (event/kg/day)
5	50	3.5×10^{-5}	2.7×10^{-1}	.5
5	100	6×10^{-6}	1.4×10^{-1}	.2
2	20	3×10^{-4}	1.1×10^{-2}	3
2	50	7×10^{-6}	7×10^{-2}	.2
0.5	20	10^{-5}	2×10^{-2}	.3

DISCUSSION AND CONCLUSION

From Table 1, we see that the predicted rejection capability is very good, provided that the energy distribution is gaussian. With a 30 kg-day counting time, in order to be sensitive to 1 event/kg/day at a WIMP mass of 20 GeV, the detector resolution has to be better than 2 keV. Since in addition to the gaussian peak, the energy distribution usually has a tail, we must also take this into account to obtain a more realistic rejection ratio. Nevertheless, the simple Monte Carlo simulation is very encouraging.

In conclusion, we have built a 60 g Ge detector with simultaneous measurement of phonons and ionization, each with 4 keV energy resolution. A preliminary estimate indicates that such a detector has a very good background rejection capability. We are currently preparing a test to demonstrate experimentally the effect shown in Fig. 1. It is not unreasonable to hope for a dark matter experiment within two years.

This work is supported by the Center for Particle Astrophysics, a National Science and Technology Center operated by the University of California at Berkeley under Cooperative Agreement no. ADT-8809616.

REFERENCE

[1] G.R. Blumenthal, S. M. Faber, J.R. Primack and M.J. Rees, *Nature*, vol.311 p.517, 1984.

[2] L.M. Krauss et al., Ap. J., vol. 299, p.1001,1985., and B. Sadoulet, "Prospects for Detecting Dark Matter Particles by Elastic Scattering", in *13th Texas Symposium on Relativistic Astrophysics*, Chicago, Ill., Dec. 14-19 1986, p.260.

[3] J.R. Primack, D. Seckel, and B. Sadoulet, "Detection of Cosmic Dark Matter", *Ann.Rev.Nucl.Part.Sci.*, vol.38, p.571, 1988.

[4] A. Cummings, N. Wang, T. Shutt, P. Barnes, J. Emes, Y. Giraud-Héraud, E. E. Haller, A. Lange, J. Rich, R. Ross, B. Sadoulet, G. Smith, C. Stubbs, "Performance of a 60 g Cryogenic Ge Detector", to be published in *IEEE Trans. in Nucl. Sci.*, 1990.

[5] N. Wang, F.C. Wellstood, B. Sadoulet, E. E. Haller, and J. Beeman, "Electrical and Thermal Properties of Neutron Transmutation Doped Ge at 20 mK", *Phys. Rev. B*, vol.41 no.6, p.3761, 1990.

[6] G.Gerbier et al., "Measurement of the Ionization of Slow Silicon Nuclei in Silicon for the Calibration of a Silicon Dark Matter Detector", CEN Saclay preprint, DPHPE.

DEVELOPMENT OF 3-D UV AND SOFT X-RAY DETECTORS BASED ON ENERGY SENSITIVE SUPERCONDUCTING ELEMENTS

A. K. Drukier
Applied Research Corporation, 8201 Corporate Drive, Suite 920, Landover, MD 20785

The understanding of the very early Universe and missing mass problem is truly fundamental. A development of a new class of cryogenic particle/radiation detectors is essential for:

a) searches for cold dark matter candidates(1);

b) better estimate of amount of dark matter in galaxies and clusters of galaxies.

Currently, the soft X-ray images of galaxies and clusters of galaxies are used to study the hot gas distribution and thus to evaluate the shapes of gravitational potential, ergo the amount of dark matter. These studies, however, are severely limited due to low energy resolution of existing X-ray imagers ($dE/E \approx 5$ - 10% for CCD's or gas detectors at $E \approx 1$ keV). Furthermore, the fluxes of X-ray from distant/extended objects are quite small and use of UV should be considered.

The possibility of obtaining simultaneous high quality information on photon arrival time, 2-D spatial resolution, and photon energy is a highly desirable feature for a UV detector. Existing UV detectors can be divided into either dispersive imaging spectrometers or 2D imagers, throwing away either spatial or spectral information. When 2-D imagers are equipped with UV-filters, the energy resolution is not better than 10% and the photon flux is cut by at least one order of magnitude. Both methods leads to considerable, (often orders of magnitude) losses in photon statistics and significantly limits the possibility to spectrally resolve images of weak/distant extended UV sources. The optimal detectors would measure for each individual photon the time of arrival, position and photon energy (t, X, Y, E). This class of detectors is commonly called 3D detectors. Their advantages have been clearly recognized in soft X-ray astronomy. Previously, it was believed that no 3-D detector would be possible in the UV spectral range but progress in superconducting detector technology is opening new possibilities. A general argument based on the statistics suggest that

$$dE/E(FWHM) \approx 1/\sqrt{N} \approx (E_{uv}/\varepsilon_{gap})^{-1/2}$$

where N is number of created quanta, E_{uv} is photon energy and ε_{gap} is energy necessary to create a measurable quantum, either a secondary photon or photoelectron. It can be seen that for $E_{uv} \approx 10$ eV, the dE/E (FWHM) $\approx 1\%$ requires $\varepsilon_{gap} \leq 10^{-3}$eV. Thus essentially gapless detectors have to be used; to permit a good signal-to-noise ratio $\varepsilon_{gap} >> kT$, i.e., cryogenic detectors seem necessary. We propose the development of superconducting UV-detectors, in which the absorption of a UV-photon leads to a small temperature increase and thus to breaking of Cooper pairs; in this case ε_{gap} may be as small as a few microelectronvolts.

Non-dispersive energy sensitive detectors are usually operated in a proportional mode, i.e., the number of secondary quanta generated is proportional to the energy of incident photons. The number of detectable quanta (visible photons in scintillators, ionization electrons in gas detectors, conduction band electrons in semiconductors) is usually small and the statistical fluctuations of these secondaries limit the energy resolution. The several generations of radiation detectors listed below are characterized by a decreasing energy gap, ε, and thus a better energy resolution.

Scintillators:	$\varepsilon \approx 30$ eV	$(\delta E/E)_x \approx 50$-75%	$(\delta E/E)_{uv}$ = none
Gas detectors:	$\varepsilon \approx 10$ eV	$(\delta E/E)_x \approx 10$-20%	$(\delta E/E)_{uv}$ = none
Semiconductors:	$\varepsilon \approx 1$ eV	$(\delta E/E)_x \approx$ 2-4%	$(\delta E/E)_{uv}$ = 100%
Superconductors:	$\varepsilon \approx 10^{-5}$ eV to 10^{-3} eV	$(\delta E/E)_x \approx 5\times10^{-3}$	$(\delta E/E)_{uv}$ = 1-5%

A typical energy resolutions, $\delta E/E$ (FWHM), at 5 keV, and 10 eV are indicated in the table above. The energy resolution for superconductors is based on estimated practical limitations rather than statistics, the fundamental limit is $(\delta E/E) \approx 10^{-3}$. Future UV and X-ray astronomy instrumentation will benefit greatly from improvements in both imaging capability and energy resolution. Therefore, studies of the potential applications of superconducting devices are clearly warranted.

Superconducting detectors are sensitive to the breaking of bound pairs of electrons (Cooper pairs). The typical binding energy is: $\varepsilon_{gap} \approx \varepsilon_o\,(1-t^2)$ with $t = T_o/T_c$ and $\varepsilon_o \approx kT_c$ where T_o and T_c are the operating and critical temperature respectively. We are considering low-temperature superconductors with a T_c typically several °K or lower. Materials such as Sn, In, Pb and Nb have transition temperatures accessible in liquid helium cryostats and have ε_o's of several milli-electronvolts. In other materials, such as Be, Ir, Hf, and W, ε_o can be as small as several micro-electronvolts.

Significant progress has been made in the development of small volume cryogenic devices for X-ray detectors. One promising class of cryogenic detectors are crystal bolometers investigated by the Wisconsin/NASA collaboration.[2] They recently have obtained $\delta E = 10$ eV for $E = 6$ keV using a small silicon bolometer with an implanted thermistor. Thus, even at the present time, at X-ray range, bolometric detectors have much better energy resolution than the best semiconducting detectors. Unfortunately, the crystal bolometers have to be operated at very low temperature ($\leq$50 mK), and the readout of these devices requires 4 wires per pixel, which limits their application as imagers. Superconducting tunnel junctions[3] have also been developed into very sensitive soft X-ray detectors with an energy resolution of about 40 eV at 6 keV. Interestingly, a one dimensional X-ray detector featuring energy resolution of about 50 eV and excellent spatial resolution was developed by the group at TU, Munich. The practical limitations appear to be the small size of tunnel junctions, low production yield and general fragility, e.g. sensitivity to vibrations which complicates space-borne applications.

In collaboration with groups at NRL and UBC, Vancouver, we have investigated techniques that demonstrate potential for larger focal plane detectors, while also providing excellent energy and positional resolution. Here we discuss a class of devices, Superconducting Granular Detectors (SGDs), in which the detection medium is an ordered array of superconducting spherical granules[4]. The SGD depends on the change of state, from superconducting to normal, of the grain in which the photon was absorbed. For example, an energy deposition of 10 eV in a 3- micron Sn sphere at a temperature of 0.1 K will result in a temperature rise of about 10 millikelvin. If the granules are biased just below the superconducting transition temperature (say $\delta T = 1$ mK) an energy loss of less than 10 eV will result in the granule becoming <u>non</u>-superconducting.

We believe that within the next 5 years, the detectors with the following properties can be developed:

Table 1. SGD UV Detector	UV	X-ray
Surface (cm^2)	1 x 1	3 x 3
Granularity (pixels)	1028 x 1028	2048 x 2048
Spatial Resolution (FWHM) (μm)	$\leq$10	~20
QDE (%)	50 - 80	$\geq$50%
Timing (nsec)	10	10
dE/E (FWHM)	1-5%	5×10^{-3}
Detector noise (photons/min)	$\leq$1.	$\leq$0.1
Maximum count rate (photons/sec)	$\geq 10^8$	$\geq 10^8$
Radiation Hardness (MRad)	$\geq$100	$\geq$100

Recently significant progress has been made in the development of SGDs. Areas of improvements are:

1. Granule quality: Arrays of spheres (128 x 128) with excellent uniformity have been produced such that a temperature increase of $\delta T/T$ <<1% leads to change of state of all grains,[(5)]

2. Readout electronics: a very good signal/noise has been achieved for detection of the transition in very small volume granules ($V \sim 10 \mu m^3$),[(6)]

3. Observation that in large collections of grains the order in which the grains change state when heated is highly reproducible[(7)].

References

1) Drukier, A. K., Stodolsky, L., Phys. Rev. D30, 2296, (1984).
Goodman, M. W., Witten, E., Phys. Rev., D31, 3059, (1985).
Drukier, A. K., Freese, K., Spergal, D., Phys. Rev. D33, 3495, (1986).
Avignone, F. T., et al., Phys. Rev. D., 35, 2752, (1987).
Ahlen, S., et al., Physics Letters B, 195, 603, (1987).
Primack, J. R., Seckel, D., Sadoulet, B., Ann Rev. Particles, 38, 751, (1989).

2) Moseley, S. H., J. C. Mather, and D. McCammon, J. Appl. Phys., 56, 1257, (1984).
McCammon, D., Mosley, S. H. Mather, J. C., and Mushotsky, R. F., J. Appl. Phys. 56, 1263, (1984).
McCammon, Juda, M., Zhang, J., Holt, S. S., Kelly, R. L., Moseley, S. H. and Szymkowiak, A. E., Proceedings LT 18, Kyoto 1987.

3) Twerenbold, D., Europhysics Letters 1, 209, (1986).
Twerenbold, D., Phys. Rev. B34, 7748, (1986).
Twerenbold, D., and Zehnder, A., J. Appl. Phys., 61, 1, (1987).
Gare, P., Engelhardt, R., Peacock, A., Twerenbold, D., Lumley, J., and Someth, R. E.
Twerenbold, D., Nucl. Instr. and Meth. A273, 575, (1988).
Rothmund, W., and Zehnder, A., in "Superconductive Particle Detectors," ed. A. Barone, World Scientific, Singapore, (1988).

Kraus, H., Peterreins, Th., Probst, F., von Feilitzsch, F., Mossbauer, R. L., Zacek, V., and Umlauf, E., Europhysics Letters, 1, 161, (1956).
Kraus, H., von Feilitzsch, F., Jochum, J., Mossbauer, R. L., Peterreins, Th., and Probst, F., submitted to Phys. Lett. B, (1989).
Kraus, H., et al., Proc. 3rd Workshop on Low Temperature Detectors..., Gran Sasso, Sept. 1989.
4) A. K. Drukier et al., Lett. Nuovo Cimento 14, 300 (1975).
A. K. Drukier, NIM 154, 83 (1978).
A. K. Drukier et al., NIM 173, 259 (1980).
D. Hueber, C. Valette, G. Waysand, Cryogenics 387 (1981).
A. K. Drukier, NIM 201, 77 (1982).
A. K. Drukier, L. Stodolsky, Phys. Rev. D30, p. 2295-2309 (1984).
K. Pretzel et al., Proc. 1st Workshop on Low Temperature Detectors..., Ringberg Castle, Tegernsee, May 1987; publisher: Springer Verlag.
L. Gonzales-Mestres, D. Perret-Gallix, Proc. 2nd Workshop on Low Temperature Detectors..., Annency-Le-Vieux, May 1988, p. 297-320; publisher: Editions Frontieres.
5) M. LeGros, A. Kotlicki, A. da Silva, B. G. Turrell, A. K. Drukier, in "Proc. 3rd Workshop on Low Temperature Detectors..., Gran Sasso, Sept. 1989, p. 417, published by Editions Frontieres.
B. G. Turrell, M. LeGros, A. da Silva, A. Kotlicki, and A. K. Drukier, NIM, A289, p. 512, (1990).
M. LeGros, A. Kotlicki, A. da Silva, B. G. Turrell, A. K. Drukier, App. Phys. Lett., 56, (22) May 1990.
M. LeGros, G. Meagher, A. Kotlicki, B. G. Turrell, and A. K. Drukier, "A Position Sensitivity of Planar Array of Superheated Superconducting Grains," to be published in Proc. Int. Conference on Position Sensitive Detectors, London, August 1990.
6) A. da Silva, M. LeGros, B. G. Turrell, A. Kotlicki, A. K. Drukier, Proc. 2nd Workshop on Low Temperature Detectors..., Annency-Le-Vieux, May 1988, p. 417-424, publisher: Editions Frontieres.
A. K. Drukier, W. L. Waldon, J. of Superconductivity, 2, 463, (1989).
A. de Bellefon et al., Proc. 2nd Workshop for Low Temperature Detectors..., Annency-Le-Vieux, May 1988, P. 425-439; publisher: Editions Frontiers.
7) A. K. Drukier et al., "Superconducting Granular Detectors: State-of-the-Art," p. 1-50, to be published in the Journal of Superconductivity.

Dark Matter Candidates and Alternatives

Astronomical Limits on the Nature of the Dark Matter

David N. Spergel
Princeton University Observatory, Princeton, NJ 08544 USA

Abstract. Astronomical observations can constrain, and potentially detect, the nature of the dark matter. Baryonic dark matter or black holes can be detected through their microlensing of stars in nearby galaxies. Dwarf galaxies serve as a laboratory for searches for dark matter. WIMPs can be potentially detected through annihilation in stars. The current limits on axions from supernovae and cosmology may rule out these particles as dark matter candidates.

INTRODUCTION

The flat galactic rotation curves argue that either Newtonian gravity fails at weak acceleration or that most of the mass of our galaxy (and other galaxies) is in non-luminous form. (See Trimble 1988 for a review of the astrophysical evidence for dark matter).

Candidates for the dark matter range in mass form micro-eV axions to 10^6 solar mass black holes. I will review several of the most popular dark matter candidates: baryons, black holes, neutrinos, WIMPs and axions and discuss the possibility of detecting these candidates either in the laboratory or at the observatory. (See Primack, Sadoulet and Secket 1989 for a review of laboratory search techniques).

BARYONS

Baryons, ordinary protons and neutrons, make up most of the mass of the Earth and the Sun. They are the most obvious dark matter candidate.

These baryons can not be bound into luminous stars, nor can they be in either hot or cold gas. Cold gas can be detected through hyperfine and molecular lines. Hot gas is detectable through Oxygen ultra-violet lines and through X-ray thermal bremstrahlung. Current observations constrain the gas mass to be much less than the dynamical mass.

Baryons can escape detection only if they bound together into clumps, either as comets, bound by atomic forces, or as planets or very low mass stars, bound by gravitational forces. Low mass objects may be detected through gravitational microlensing. These objects may pass between us and a distant star in a neighboring galaxy (e.g. Andromeda or the Lesser Magellanic Cloud(LMC)). When the extra-galactic star passes through the Einstein ring radius of the halo object, its image is focused, increasing its apparent luminosity. Paczynski (1986) has proposed monitoring stars in a nearby galaxy. A program of monitoring one million stars for several years could detect several lensing events. This technique could detect halo objects more massive than Jupiter regardless of their composition. Bennett's talk describes recent progress by the Livermore group in developing a detected microlensing telescope.

NEUTRINOS

Neutrinos, known non-baryonic matter, are a natural dark matter candidate. In the hot big bang, neutrinos are produced in roughly equal numbers as photons. These weakly interacting particles decouple from the thermal bath of electrons, positrons, muons and photons while there are still relativistic. Each neutrino species contributes to the closure density of the universe:

$$\Omega_{\nu_i} \simeq \left(\frac{m_{\nu_i}}{30\mathrm{eV}}\right)\left(\frac{H_0}{50\,\mathrm{kms^{-1}Mpc^{-1}}}\right)^{-2},$$

where H_0 is the Hubble constant. In an $\Omega = 1$ universe, requiring that the age of the universe exceed 10 Gyr implies that $H_0 < 65$ km s^{-1} Mpc^{-1} and $\Sigma m_{\nu_i} < 50$ eV. The lack of correlation between the arrival times and the energies of the electron anti-neutrinos detected from SN 1987a place a 16 eV upper limit on the electron neutrino mass (Spergel and Bahcall 1988) . Current experimental limits on the τ and μ neutrino mass do not rule out cosmologically interesting masses.

Neutrinos have fallen from favor as astrophysically popular dark matter candidates. In N-body simulation of galaxy formation in a neutrino dominated universe with adiabatic primordial density fluctuations, clusters form before galaxies. Observations suggest that galaxies formed before clusters. Neutrino-dominated universes have structures like massive X-ray emitting gas haloes that are not observed in our universe (Hut and White 1987). If the initial fluctuations that generate galaxy formation do not come from inflation, but rather from topological defects, then neutrinos are more attractive dark matter candidates.

THE SOLAR NEUTRINO PROBLEM AND THE NEUTRINO REVIVAL

As Arnon Dar described in his paper in these proceedings, there has been a recent revival of the neutrino as dark matter motivated mainly by the recent results from the Soviet American Gallium Experiment (SAGE).

The most popular explanation of the solar neutrino problem was proposed by Mikheyev, Smirnov and Wolfenstein (MSW). They suggested that neutrinos have mass and the flavor eigenstates (e, μ, τ) have small mixing angles. They noted that if either the μ or τ mass exceeded the electron mass by $\sim 10^{-3}eV$, then the electron neutrino would be resonantly converted into another neutrino species on its way out of the Sun. The resonance occurs because in the center of the Sun, the interactions between the electron neutrino and the plasma give the electron neutrino an effective mass, much like the effective mass that plasma gives to a photon, which leads to the plasma cutoff at low frequencies.. Thus, in the center of the Sun, the electron neutrino flavor state is more massive than the μ neutrino flavor state. While in the vacuum, the μ neutrino is more massive. Along the path from the center of the Sun to the earth, the electron neutrino is converted into the μ neutrino (MSW effect),

The MSW effect is consistent with both Davis Chlorine experiment and the Kamiokande (water) experiment, which measure the flux of ^{8}B neutrinos (PPIII chain). These experiments see only 40% of the predicted flux. There are currently two Gallium experiments underway, SAGE (Soviet American Gallium Experiment) and GALLEX (German/French). The Gallium experiment is sensitive to low energy PPI neutrinos. SAGE recently reported that they saw 0 neutrinos (the standard solar model predicts 21). The MSW hypothesis is consist with converting nearly all of the PPI e neutrino to μ neutrinos. I refer the interested reader John Bahcall's excellent recent book on the solar neutrino problem.

What does this mean for dark matter? Particle physicists have suggested the "see-saw" mechanism as an explanation of small neutrino masses. In some GUT models, the neutrino mass matrix looks like:

$$\begin{vmatrix} 0 & m_q \\ m_q & M \end{vmatrix}$$

where m_q is the mass of a quark and M is a large mass scale. Diagonalizing this matrix yields a neutrino mass of m_q^2/M This model predicts that the neutrino masses scale as the quark masses. If the μ neutrino mass is $\sim 3 \times 10^{-3}$ eV, then the tau neutrino mass is $\sim (m_{top}/m_{Charm})^2 \sim 10$ eV, where I have assumed that the tau mass is about 140 GeV. Accurate measurements of the shape of the Z^0 resonance imply that $m_t = 137 \pm 35 \pm 20$ GeV (OPAL collaboration 1990, Abe 1990). This suggest that the τ neutrino mass may lie in a cosmological interesting range, which has fueled the recent neutrino renaissance.

DWARF GALAXIES AND NEUTRINOS

Dwarf galaxies, because of their weak internal accelerations and shallow gravitational wells, are excellent laboratories for testing alternative theories of gravity and for constraining candidates for the galactic dark matter. Lake (1989) emphasized the discrepancies between MOND and rotation curves of dwarf irregular galaxies (but also, see Milgrom 1989). In this section, I describe some recent work by Ortwin Gerhard and myself (Gerhard and Spergel 1990) on the constraints that the presence of significant central densities of dark matter in Draco and Ursa Minor place on neutrinos and on other modifications of Newtonian gravity.

Both Draco and Ursa Minor appear to contain large amounts of dark matter. In the absence of dark matter, the observed distribution of stars would imply velocity dispersions of $\sigma \simeq 2\,\mathrm{km\,s^{-1}}$ in Draco and $\sigma \simeq 1\,\mathrm{km\,s^{-1}}$ in Ursa Minor. However, measurements of individual stellar velocities find $\sigma \simeq 10\,\mathrm{km\,s^{-1}}$ in both cases (Aaronson 1983, Aaronson

& Olszewski 1988). Aaronson & Olszewski (1988) argue that long-term monitoring of these dwarfs has reduced the possibility that contamination of the sample by binaries has spuriously raised these dispersions. The mass-to-light ratios seemingly implied for Draco and Ursa Minor by these data are $\zeta \sim 20 - 100$, although the precise values depend on the assumed core radii of the dark material relative to those for the stars, and the possible anisotropy of the stellar orbits.

If the spatial distribution of dark matter and luminous stars is similar, the central dark matter densities needed to explain the kinematic data are $\rho_D \sim 0.5 - 1\,\mathrm{M_\odot/\,pc^3}$ (Kormendy 1987); much larger than in any bright galaxy with good rotation curve data. The effects of both a different spatial extent (core radius) for the dark and luminous matter, and of an anisotropic velocity dispersion tensor for the stars were considered by Lake (1990) and Pryor & Kormendy (1990). These authors compared a set of two-component models to the published star count profiles and the observed velocity dispersions. A model independent lower limit on the halo density can be obtained from the virial theorem (Merritt 1987); applying this result Pryor & Kormendy (1990) found lower limits of around $\rho_D \sim 0.07 - 0.15\,\mathrm{M_\odot/\,pc^3}$ in Draco and Ursa Minor. These values will be somewhat reduced if the velocity distribution is strongly radial and the observed stars are mostly near the center.

It is easy to predict qualitatively the effect on the dark matter density introduced by a large halo core radius or by anisotropic stellar velocities. As long as the velocity distribution of the luminous matter is nearly isotropic, its observed core radius and velocity dispersion directly determine the dark matter density via $(G\rho_D)^{-1/2} \sim 2r_c/\sqrt{3}\sigma_0$, or $\rho_D \sim 1\,\mathrm{M_\odot/\,pc^3}$. Thus if the stars have the expected mass-to-light ratio of ~ 2, their contribution to the density is small, and the halo density must be high. There is no constraint on the halo core radius from the star count profile in this case, as one may invert an Abel equation to determine the energy distribution of the stars that produces the observed profile in an infinite constant density halo which dominates the potential. The constraints claimed by Pryor & Kormendy are due their specific choice of Michie-King models.

Anisotropy, however, can substantially modify the result. Consider an anisotropic distribution of test stars in a constant density halo core. If all stars were on *exactly* radial orbits, their space density profile would be singular. For a *nearly* radial orbit distribution the core radius of the test stars is set by the distribution of central impact parameters rather than by any scale referring to the potential itself. The velocity dispersion, of course, is still characteristic of the total mass distribution probed by the orbits. Thus in the anisotropic case, the core radius and velocity dispersion of the luminous matter do not immediately determine the central density, and one must resort to comparing dynamical models to the radial star count profile. Again, because of the enormous freedom of an arbitrary distribution function $f(E, L)$ for the stars, it is unlikely that any constraint on the halo core radius can be set, but Pryor & Kormendy (1990) in their extensive modelling found a lower limit to the halo densities in both Draco and Ursa Minor of $\simeq 0.05 - 0.1\,\mathrm{M_\odot/\,pc^3}$. This is consistent with the virial theorem result.

Can the dense halos in Draco and Ursa Minor be made of massive neutrinos? The present neutrino phase-space density is constrained by the collisionless Boltzmann equation to be less or equal to the neutrino phase-space density at the time of neutrino decoupling (Tremaine and Gunn 1979). This places a lower limit to the mass of the neutrinos that can make up an isothermal spherical halo. When this argument was applied to the dwarf spheroidals by Faber & Lin (1983), the difficulty proved to be the unknown halo core radius. Formally, a halo with infinite core radius has zero phase-space density.

With the lower limit to the density now known to be $0.05 - 0.10\,\mathrm{M_\odot/\,pc^3}$, we are now in a better position to use the phase space argument for constraining the contribution of neutrinos to the observed dark matter in Draco and Ursa Minor. Using King's formula we can convert the conventional neutrino mass constraint for an isothermal sphere model to

$$m_\nu \gtrsim 170\mathrm{eV} \left(\frac{1\,\mathrm{kpc}}{r_D}\right)^{3/4} \left(\frac{0.1\,\mathrm{M_\odot/\,pc^3}}{\rho_D}\right)^{1/8} g_\nu^{-1/4}, \tag{1}$$

where g_ν is the product of the number of neutrino spin states and number of neutrino species of this mass, and r_D and ρ_D are the core radius and core density of the dark matter halo, assumed in this case to consist of neutrinos. Notice that in this form the constraint is not very sensitive to the halo density. Requiring a cosmologically reasonable mass for the dominant neutrino, $m_\nu = 30\mathrm{eV}$, and using (12) and the inferred densities, we obtain halo core radii for Draco and Ursa Minor of $r_D = 10\,\mathrm{kpc}$, or *core* masses of $4 \times 10^{11}\,\mathrm{M_\odot}$. The missing mass in Draco and Ursa Minor can be composed of neutrinos, only if the neutrino halo core radius is 100 times larger than the stellar core radius and if the global mass-to-light ratios exceeds 2×10^6. Objects of such enormous masses would have spiralled into and merged with the Milky Way long ago, and models of the Galaxy do not allow for this much dark matter at such a high density. As noted earlier, if the neutrinos in the halo are on radial orbits, then King's formula can overestimate the core mass for given r_D and velocities, (Richstone and Tremaine 1984), or here: overestimate the phase-space density at fixed space density.

We can turn the argument around and ask what limits can be set on m_ν by demanding that Draco and Ursa Minor have not undergone significant dynamical friction against the Galactic halo during the past Hubble time. Even if the Milky Way's halo goes out only to somewhat less than their galactocentric distance, we still expect significant friction due to resonance coupling; cf. Weinberg (1986). One objection to this argument might be that the two dwarfs could have just spiralled to their present distances from further out in the last $10^{10}\,\mathrm{yr}$. But to avoid assuming special initial conditions, one would have to assume that our galaxy once contained several other such massive satellites, which started at smaller radii and have spiralled into the inner Galaxy, where they would have heated the disk. Based on the small velocity dispersions of old disk stars, Toth and Ostriker (1990) find that the Galaxy can at most have accreted a few percent of its mass during the age of the disk— the accretion of just one 10^{10} solar mass object violates their limit.

The friction time for an object of mass M on a circular orbit in an isothermal halo with $v_c = 220\,\mathrm{km\,s^{-1}}$ is

$$t_{fric} = \frac{1.0 \times 10^{10}\,\mathrm{yr}}{\ln \Lambda} \left(\frac{r}{60\,\mathrm{kpc}}\right)^2 \left(\frac{v_c}{220\,\mathrm{km\,s^{-1}}}\right) \left(\frac{2 \times 10^{10}\,\mathrm{M_\odot}}{M}\right). \tag{2}$$

(Binney & Tremaine 1987). t_{fric} is essentially unchanged if the object moves on non-circular orbits or if the Galactic halo is slowly rotating. Using distances for Draco and Ursa Minor of $r = 75\,\mathrm{kpc}$ and $r = 60\,\mathrm{kpc}$, respectively, and a value of $\ln \Lambda = 3$, the requirement that $t_{fric} \lesssim 10^{10}\,\mathrm{yr}$ then leads to mass constraints of $M_{\mathrm{Draco}} \lesssim 1 \times 10^{10}\,\mathrm{M_\odot}$ and $M_{\mathrm{UMi}} \lesssim 8 \times 10^9\,\mathrm{M_\odot}$, and thus to neutrino mass constraints of $m_\nu \gtrsim 77\mathrm{eV}$ and $m_\nu \gtrsim 81\mathrm{eV}$ for a dark matter density of $\rho_D = 0.1\,\mathrm{M_\odot/\,pc^3}$. The total masses obtained for the two dwarfs in this way are still very large; yet the neutrino mass limits derived from them are uncomfortable in view of both current measurements for the electron neutrino mass, and - given that we know the total number of neutrinos in the Universe - the limits set by the age of the Universe on the mass of any type of neutrino.

Thus to the extent that the velocity dispersion measurements in these two dwarf spheroidals are reliable, neutrinos are unlikely to be the constituents of the halos of Draco and Ursa Minor.

ANISOTROPIC HALOS AND LIMITS ON NEUTRINOS

Several authors have argued that the Tremaine and Gunn neutrino limit is overly conservative and can be violated if the neutrino has an extremely radial velocity distribution function (Ralston 1989, Ralston and Smith 1990). These authors, however, have constructed extremely anistropic distribution functions which are likely to be unstable to radial perturbations. This type of instability would significantly lower the central phase space density.

Ortwin Gerhard and I have been investigating radially anisotropic distribution functions. We have found that functions with low central phase space density are often violently unstable. We have been unable to construct a *stable* distribution function that reduces the central phase space density by more than 30%.

Neutrinos, of course, do not "know" how to arrange themselves so to achieve the maximum phase space density. Their distribution function is determined by initial conditions. In a recent paper, Villumsen and van Dalen (1990) followed the evolution of neutrinos around a seed mass. They found that the neutrino distribution function did not differ significantly from an isothermal sphere. Their conclusions imply that the Tremaine-Gunn limit is a reasonably solid constraint on neutrino properties.

AXIONS

One of the outstanding problems in the standard model is understanding the lack of CP violation in the strong interaction. An elegant solution to this problem is the existence of a new U(1) symmetry (the Peccei-Quinn symmetry) that is spontaneously broken. The pseudo-goldstone boson associated with this broken symmetry is called the axion.

The axion acquires its mass through non-perturbative QCD anomaly effects— in its original incarnation, the axion mass was ~MeV. This particle was ruled out by several laboratory experiments. Theorists revived the axion in a new form: the "invisible axion", an extremely weakly interacting low mass particle (Kim 1979, Shifman, Vainstein and Zahkorov 1980, Dine, Fischler and Srednicki 1981). If the scale of the Peccei-Quinn symmetry breaking is $\sim 10^{12}$ GeV, then this particle can make up the missing mass Even the invisible axion can be detected: Axion couple to photons through the axial anomaly and stars can exploit this $a-\gamma-\gamma$ coupling as a alternative cooling channel. Red giants do not cool use this cooling mechanism— this constrains the axion mass to be less than 10^{-2} eV (Dearborn and Raffelt 1987). The best limits on the Peccei-Quinn symmetry breaking scale comes from SN1987A: since we did see neutrinos, we know that axions did not carry off the lion's share of the thermal energy of the hot neutron star. This argues that the P-Q scale is larger than $\sim 10^{10}$ GeV (Turner 1988).

Davis and Shellard (1989) have argued that most of the energy density in axions was produced not at the QCD phase transition, but rather by axion strings. These global strings radiate axions, which can close the universe if the P-Q scale is near 10^{10} GeV. Harari and Sikivie (1987) have argued that global string radiation is suppressed by backreaction and that this may weaken the limits. However, recent work by Dabholkar and Quashnock (1990) shows that backreaction is unimportant. Their work closes the axion window and implies that only inflation can save the axion.

WIMPs

There are several hypothetical particles that have been labeled as WIMPs (Weakly Interacting Massive Particles): heavy neutrinos, supersymmetric relics and solar cosmions.

Heavy neutrinos have the oldest pedigree of the WIMP candidates. They arise in 4 generation models: if the heaviest neutrino is stable and its mass is $\sim$ 2 GeV, then it can comprise the missing mass. WIMP searches, combined with accelerator limits have succeeded in ruling out these particles as dark matter candidates.

Supersymmetry offers an attractive dark matter candidate. In supersymmetry, there is a new conserved quantum number: R- parity. This implies that there will be a lowest mass particle with charge R=1 and that charge conservation will imply that it is stable. In most theories, this lightest supersymmetric particle (LSP) is a linear combination of photino, higgsino and zino interaction eigenstates. The photino is the fermionoic supersymmetric partner of the photon. The higgsino is the partner of the Higgs and the zino is the partner of the Z-boson. In minimal supersymmetry models, there is a large range of parameter space within which the LSP can close the universe (Ellis 1984).

Solar cosmions are particles proposed to solve the solar neutrino problem. These particles would provide an alternative means of energy transport in the solar core. This will lower the Sun's central temperature and decrease the flux of solar neutrinos (Steigman 1978, Spergel and Press 1985, Press and Spergel 1985, Faulkner and Gilliland 1985). Several models have been proposed for the solar cosmion. Many of these proposed candidates, because of their relatively large cross-section, are already ruled out by laboratory and accelerator searches.

WIMPs are potentially detectable through their direct interactions in the laboratory. If WIMPs comprise the halo, then their number density is ~ 0.1 cm^{-3} and their flux is $\sim 10^7$ cm^{-2} s^{-1}. A tiny fraction of this incident flux will collide with a nucleus in an experiment and deposit ~ 1 keV of kinetic energy. The experimental challenge is detecting this rare event.

WIMPs can also be detected through their annihilation products: anti-protons, gamma rays and high energy neutrinos (Silk and Srednicki 1984). The predicted fluxes of annihilation products of halo WIMPs are below background from cosmic ray collisions (Stecker and Tylka 1989). New satellites, such as the Gamma Ray Observatory, with improved spectral resolution may be able to detect the gamma-ray lines from WIMP annihilation.

A more promising place to look for dark matter is in the Sun. Halo WIMPs scatter off of nuclei in the Sun, lose kinetic energy and become bound (Press and Spergel 1985). The WIMP number density will accumulate until balanced by WIMP annihilation. The high energy neutrinos produced by WIMP annihilation are potentially detectable by the IMB, Kamiokande and Frejus detectors (Silk, Olive and Srednicki 1985; Gaisser, Steigman and Tilav 1986; Ng, Olive and Srednicki 1987; Ritz and Seckel 1988).

Accelerator experiments have placed important limits on WIMPs (see Larry Krauss' talk in these proceedings). Together, with direct laboratory searches, these experiments rule out fourth generation neutrinos and significantly constrain other WIMP candidates.

ACKNOWLEDGEMENTS

I would like to thank my collaborator, Ortwin Gerhard, for many useful discussion and would like to acknowledge support by NSF grant AST88-58145 (PYI) and an Alfred P. Sloan Fellowship.

References

Aaronson, M. 1983, *Ap.J. (Lett.)*, **266**, L11.

Aaronson,M. & Olszewski, E.W. 1986, Astron. J., **92**, 580.

Aaronson,M.& Olszewski, E.W. 1988, in *IAU Symposium 130: Large Scale Stucture of the Universe*,ed. J.Audouze, M.-C. Pelletan & A. Szalay (Dordrecht:Kluwer Academic Press)

Abe, F. *et al.* 1990, *Phys. Rev. Lett.*, **64**, 1491.

Binney, J.J. & Tremaine, S. 1987, *Galactic Dynamics*, (Princeton: Princeton University Press).

Dabholkar, A. and Quashnock, J.M. 1990, *Nucl.Phys.*, **B333** 815.

Davis, R.L. 1986, *Phys. Lett.* **189B**, 225.

Davis, R.L. and Shellard, E.P.S. 1990, to appear in *Nucl. Phys.*

Dearborn,D.S.P. and Raffelt, G., *Phys. Rev.*, **D 36** ,2211.

Dine, M. Fischler,W. and Srednicki,M. 1981,*Phys. Lett* 104B, 199.

Ellis, J. *et al.*, Nucl. Phys. **B238** (1984) 453.

Faber, S.M. & Lin, D.N.C. 1983, *Ap. J.*, **266**, L21.

Faulkner, J. & Gilliland, R.L. 1985, *Ap. J.*, **299**, 994.

Gaisser,T.K. Steigman, G. and Tilav, S. 1986, *Phys. Rev. D* **34** 2206.

Gerhard, O.E. & Spergel, D.N. 1990, submitted to *Ap.J.*

Harari, D. and Sikivie, P. 1987, *Phys. Lett.*, **195B**, 1868.

Hut, P. and White, S.D.M. 1987, *Nature* **310**, 637.

Kim, J.E., 1979, *Phys. Rev. Lett*, **43**, 103.

King, I.R. 1966, *Astron. J.*, **71**, 64.

Kormendy, J. 1987. in *IAU Symposium No. 117, Dark Matter in the Universe*, J. Kormendy and G.R. Knapp, eds. (Dordrecht: Reidel).

Lake, G. 1989, *Ap. J. (Letters)*, **345**, 17.

Lake, G. 1990, *M.N.R.A.S.*, **244**, 701.

Ng, K. Olive ,K.A.and Srednicki,M. 1987, *Phys. Lett.* **188B** *138.*

OPAL collaboration 1990, CERN preprint EP/90-81(1990)

Paczynski, B.,*Ap.J.* **304** (1986) 1.

Paltoglou, G. & Freeman, K.C. 1987, in *IAU Symposium No. 127, Structure and Dynamics of Elliptical Galaxies*, ed. T. de Zeeuw (Dordrecht: Reidel).

Press, W.H. and Spergel, D.N. 1985, *Ap.J.*,**296**,663.

Primack, J.R., Seckel,D. and Sadoulet, B, 1989, *Ann. Rev. of Nuc. and Particle Science*, **38**, 751.

Pryor, T. & Kormendy, J. 1990, *Astron. J.*, **100**, 127.

Pryor, C., McClure, R.D., Fletcher, J.M. and Hesser, J.E. 1988. in *IAU Symposium No. 126, Globular Cluster Systems in Galaxies*, J.E. Grindlay and A.G.D. Philip, eds. (Dordrecht: Reidel).

Ralston, J.P. and Smith, L.L. 1990, to appear in *Ap.J.*

Ralston, J.P. 1989, *Phys. Rev. Lett.*, **63**, 1038.

Richstone, D.O. & Tremaine, S. 1984, *Ap.J.*, **286**, 27.

Ritz, S. and Seckel,D. 1988, *Nucl. Phys.* **B304** 877.

Shifman, M.A. Vainshtein, A.I. and Zakharov,V.I. 1980, *Nucl. Phys.*,**B166** (1980) 493.

Silk, J., Olive, K.A. and Srednicki, M. 1985*Phys. Rev. Lett.***55** 257.

Silk ,J. and Srednicki, M. 1984, *Phys. Rev. Lett.* **53**, 624.

Spergel, D.N. and Press, W.H 1985, *Ap. J.***296**, 679.

Spergel, D.N. and Bahcall, J.N. 1988,*Phys. Lett.* **200B (1988)**, 366.

Steigman, G. Sarazin, C. Quintana, P. and Faulkner, J. 1978 *Astron. J.* **83**, 1050.
Stecker, F.W. and Tylka, A.J. 1989,*Ap.J. (Lett)***336**,L51.
Toth, G. & Ostriker, J.P. 1990, private communication.
Trimble, V., 1987 *Ann. Rev. Astron. and Astrophys.* **25**, 425.
Tremaine, S. & Gunn, J.E. 1978, *Phys. Rev. Lett.*, **42**, 407.
Turner, M.S., 1988, *Phys. Rev. Lett.,* **60**, 1797.
Villumsen, J. and Van Dalen, A., 1990, Ohio State University preprint.
Weinberg, M.D. 1986, *Ap.J.*, 300, 93.

DARK MATTER THEORY: CONSTRAINTS FROM PARTICLE PHYSICS

Lawrence M. Krauss
Center for Theoretical Physics and Department of Astronomy
Sloane Laboratory, Yale University,
New Haven ,CT 06511

ABSTRACT

I briefly review recent results from particle theory and experiment which are relevant to the question of what might make up galactic halo dark matter, and more generally what might contribute a closure density in dark matter. I talk in turn about Cold Dark Matter candidates, Baryons, and Hot Dark Matter (i.e. light neutrinos).

INTRODUCTION

There has been a lot of discussion at this meeting about the confrontation between cold dark matter models and observations of large scale structure. Indeed, "cold dark matter-bashing" has become a favorite theme. Nevertheless, we must remember that if we are to dispense with cold dark matter, we also dispense with essentially the only model which comes even close to explaining how structure might form in a way which is compatible with the growing number of constraints from observational cosmology. It therefore becomes increasingly important to examine how constraints from particle physics are evolving to limit both exotic non-baryonic, and baryonic dark matter in the universe. This is the purpose of my brief review. The first, and largest section of this review will deal with cold dark matter, and in particular with the new constraints on WIMPs from LEP. I will then review how these results impact on the possibility that the galactic halo dark matter is baryonic, and will conclude with a few remarks on the recent resurgence of interest in tau neutrinos as hot dark matter.

1. COLD DARK MATTER: ON LEP, UNDERGROUND DETECTORS AND STAR WARS MAGNETS

(a) LEP and WIMPS: The emblem of this conference is the spectrum of the cosmic microwave background as measured by the COBE satellite. This was an extremely important result, giving a single curve which limited many possible energy release mechanisms in the pre- and post- recombination universe, and equally many exotic scenarios for the formation of large scale structure. Curiously, at the same time, an experiment on earth provided a curve of similar importance for particle physics. Like the microwave background spectrum it was a curve whose shape could have been predicted more than 20 years ago. Also, like the microwave background spectrum, a single comprehensive determination of its overall shape has provided incredibly powerful limits on much exotica beyond the standard model. I refer of course to the measurement at LEP and the SLC of the shape of the Z boson resonance, via Z production and subsequent decay into hadrons. This resonance is restricted to have a Breit-Wigner form, given by three free parameters, the Z mass, its total width Γ, and the peak value of the cross section $\sigma^0{}_h$, as shown in the figure below:

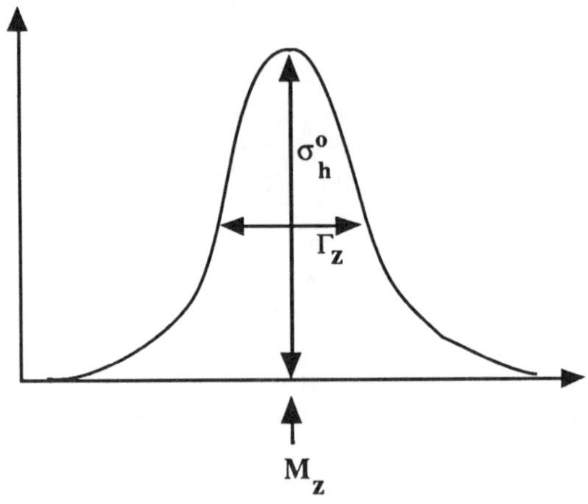

The total decay width of the Z boson can be *defined* to be $\Gamma_z \equiv N_\nu \Gamma_\nu + 3\Gamma_e + \Gamma_h$, where Γ_e is the partial width into electrons, Γ_h is the width into hadrons, and Γ_ν is the partial width into massless neutrinos, as calculated in the standard model [1]. This defines the "number of neutrinos", $N_\nu \equiv \frac{1}{\Gamma_\nu}\left(\Gamma_z - 3\Gamma_e - \Gamma_h\right)$. One's naive expectation might be that to extract a limit on the number of neutrinos one would first fit the measurements to the Breit-Wigner form (1), determine a best fit value for Γ_z and then use (3). However, because the integrated cross section for Z production is a constant, fixed by the strength of the weak interactions, this width and peak height are related, so that $\sigma_h^o = \frac{12\pi\Gamma_e\Gamma_h}{M_z^2\Gamma_z^2}$. Since the peak height depends *quadratically* on the total width, while an extra neutrino species adds 6% to Γ_z, it changes the peak height by 12%! Thus, while we often say that a limit on the number of neutrino species was extracted from the width of the Z peak, it is actually extracted from the height of the peak. The present experimental upper limit is $N_\nu < 3.2$ (2σ).

Be that as it may, the important point is that N_ν means more than just light neutrinos! It codifies all our ignorance about any new invisible decays of the Z. In particular, as long as new particles are less than 1/2 the Z mass, they can contribute to N, with their contribution being increasingly suppressed by phase space factors as their mass approaches 1/2 M_Z. What makes all of this interesting for dark matter, is that essentially all WIMP (Weakly Interacting Massive Particle) candidates for cold dark matter fall in this category. For example, the LEP limit on N_ν implies that any new neutrino must have a mass greater than about 35-40 GeV. For neutrinos whose remnant abundance is fixed by their annihilations in the early universe, this implies that their contribution to Ω today must be <<1 (unless their mass is in a narrow range around 1 TeV), ruling them out as WIMPS. In fact, this result can be strengthened by complementary terrestrial direct search experiments for WIMPS, which suggests that m<10 GeV for neutrino WIMPS [2], independent of whether their remnant abundance is determined by their annihilations in the early universe or not. Combining these results, one finds that **heavy neutrinos are categorically ruled out as WIMP candidates for Cold Dark Matter.**

The Z decay results go further. The chief other WIMP candidate is the lightest supersymmetric partner of ordinary matter, which in the minimal models was expected to have a mass in the O(1-100) GeV range. Here too Z decay gives new limits. The same combination of arguments which rule out a normal neutrino WIMP **rule out a "sneutrino"**, the SUSY partner of the neutrino [1]. Finally, the lone remaining SUSY WIMP in the minimal models is a linear combination of the SUSY partners of the photon, Z, and Higgs particles in the standard model. This "neutralino" has sufficient flexibility so that it is not entirely ruled out on the basis of the Z decay limits alone. However, the new limits, when combined with a new lower bound on the Higgs particle of about 30 GeV from LEP, and limits on SUSY particles from the Tevatron, suggest that the neutralino must be heavier than about 15-20 GeV [1,3]. This is an important result because: (a) a search of the available model space then suggests that this particle, if it exists, and is to have a sufficient remnant abundance to make up the dark matter, is likely to be even heavier than the W and Z particles, and is likely > O(100) GeV[4]; and (b) this mass range is in the opposite direction of previous efforts to extend the range to which direct detectors were sensitive, and moreover, the properties of such a WIMP are such as to make it much more difficult to detect directly in underground detectors [5]. The event rate at such detectors is predicted to be between 3-4 orders of magnitude smaller than that due to the heavy neutrinos which they presently constrain. Direct searches, which I believe are of vital importance, are now more challenging in light of LEP.

Finally, LEP not only limits these realistic dark matter candidates, it also limits more exotic possibilities. "Cosmions", a far-fetched solution of the solar neutrino problem [6], are now almost completely ruled out by combining the LEP measurements with new direct search limits [1,7].

(b) Axions: Axions remain, for me at least, the most attractive Cold Dark Matter candidates. They too are being squeezed however. The observation of neutrinos from SN 1987a decreased the upper bound on their mass by about an order of magnitude, to about 10^{-3} eV [8], leaving apparently about 2 orders of magnitude before they get into other cosmological troubles. In the last few years theorists have been arguing that maybe the lower bound might have to be raised, further constraining cosmic axions. Rick Davis and collaborators have argued that if the cosmic phase transition responsible for axions occurs after inflation, then the lower bound on the axion mass is raised, so that the available range for cosmic dark axions is restricted to almost vanish [9]. More recently, it has been suggested that if this phase transition happens before inflation, resulting density fluctuations from inhomogeneities in the cosmic axion field are unacceptable unless the previous lower bound of 10^{-5} is also raised somewhat [10].

Of course, all the cosmological arguments associated with axion lower bounds are kind of slippery at best. The proof of the pudding will be in the tasting. Here too, new developments are taking place. A collaboration from Florida, Chicago and Livermore may have convinced the Livermore labs to allow them to use a monstrous magnet to perform an axion conversion experiment which would be sensitive to cosmic axions in almost all of the range which is currently of interest for dark matter. Stay tuned.

2. BARYON DARK MATTER

It is by now very well known that the number of light neutrinos strongly impacts upon cosmology via primordial big bang nucleosynthesis. Traditionally, cosmologists have used the quantitative agreement between the predicted and the observationally inferred primordial abundances of light elements to limit the number of light neutrinos. Now that this number has been well established experimentally via the Z width, one can hope to use it to further constrain other aspects of big bang cosmology. In particular, with the number of neutrinos known (*assuming* no other light states which can affect the expansion rate during helium production, but which do not couple to the Z), the uncertainty in the predicted abundance of ^{4}He is the smallest of all the light elements. The chief remaining uncertainty is in the neutron half life, which is now less than .5%. Below I display the BBN predictions for the fraction by mass of ^{4}He (Y_p), for 3 and 4 neutrinos as a function of the baryon to photon ratio η, including the 2σ uncertainty due to our uncertainty in the BBN reaction rates [11]. Also shown for 3 neutrinos is the reduced 2σ uncertainty with the most recent neutron half life measurement included.

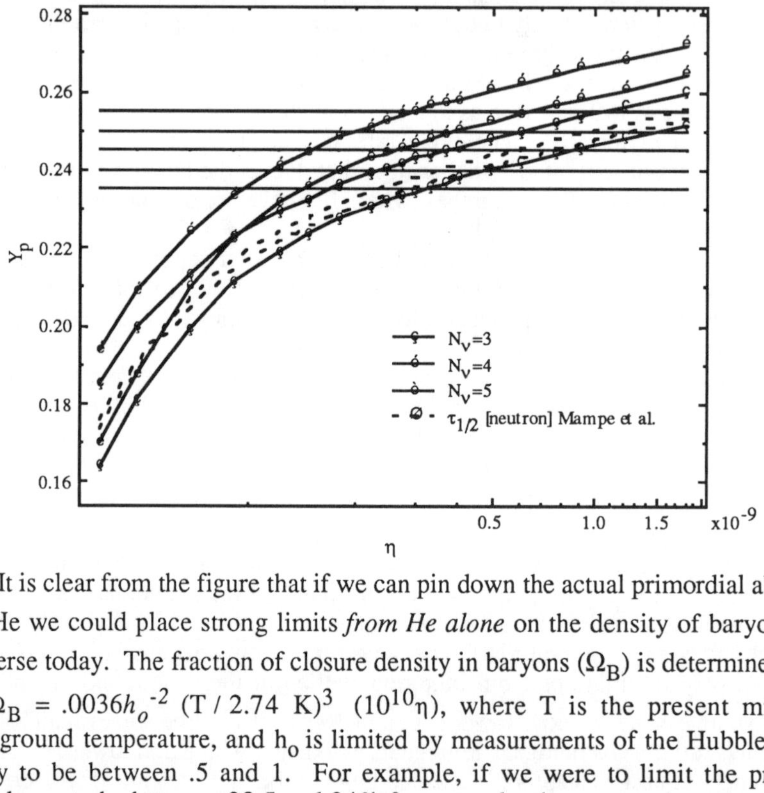

It is clear from the figure that if we can pin down the actual primordial abundance of ^{4}He we could place strong limits *from He alone* on the density of baryons in the universe today. The fraction of closure density in baryons (Ω_B) is determined from η as $\Omega_B = .0036 h_o^{-2}$ (T / 2.74 K)^3 ($10^{10}\eta$), where T is the present microwave background temperature, and h_0 is limited by measurements of the Hubble constant today to be between .5 and 1. For example, if we were to limit the primordial abundance to be between 23.5 and 24% for example, then, *assuming at most 3 light neutrino equivalents,* η would be constrained to be between 2.5-4.8 x 10^{-10}. This in turn would limit Ω_B to be between 0.01 and 0.09. The lower bound (obtained for the extreme value h_0 =1, which leads to conflicts with limits on the age of the universe) already implies that some dark matter must be baryonic. The upper limit is in marginal disagreement with the possibility that all dark matter inferred by virial estimates is baryonic. It has already been claimed that an upper limit of 24% on ^{4}He is suggested by the present data. If this remains true, then ^5He will give the strongest upper limit on the baryon density today. In fact, it is worth noting that the best fit limit of 23% gives such a tight constraint as to provide an upper bound which is lower than the lower bound obtained from arguments about D and ^3He abundances from BBN for 3 neutrino species. Either the tau neutrino is too massive to contribute to the BBN constraint, or the He abundance estimates from observation must be raised slightly, or perhaps this is a signal for some new physics....

Be that as it may, the fact that some of the dark matter is strongly suggested to be baryonic, combined with the negative results from LEP, have led some particle experimentalists and theorists to follow up on an idea of Paczynski's for looking for baryonic dark matter in compact halo objects by microlensing [12]. Two groups, one U.S. and one French, are presently launching efforts to give a definitive answer to whether the halo is baryonic or not within the next year.

3. LIGHT NEUTRINOS

The question of whether the tau neutrino might have a mass in the cosmologically interesting range of 10-30 eV has recently been raised because of possible solutions to the solar neutrino problem which involve a mass for the muon neutrino near 10^{-3} eV. To this I want to comment: (a) the results of the Chlorine and Kamiokande experiments do not point unambiguously in this direction. Certain interpretations of those results are in fact in disagreement with this inference, (b) The SAGE experiment has reported preliminary limits which have been grossly overinterpreted. The SAGE result is consistent with the standard solar model prediction, just as it is consistent with a measurement of 0 SNU's. Any quoted best fit is suspect at this point. All that is model independent is the upper limit of about 135 SNU's. Thus, no inferences about a neutrino based solution to the solar neutrino problem should yet be drawn, until the experiment has a clear result. Nevertheless, what if the tau neutrino really does have a cosmologically interesting mass? The only way to probe this at present may be through observations of neutrinos from the next galactic supernova. Preliminary estimates we have performed [15] suggest that such a limit might be possible in the next generation Super-Kamiokande detector, although it will not be possible from current detectors.

CONCLUSION

Recent results from particle theory and experiment are narrowing down the possibilities in the search for what makes up the dominant matter in the universe. Combined with important observations of large scale structure, the 90's may be the decade when this question is finally resolved. However, this will require some experiment without a null result or an upper limit...

The research of the author is supported in part by a Presidential Young Investigator Award from the NSF, and by a grant from the DOE. Thanks to P. Romanelli and G. Canali for proofreading the draft of this manuscript.

REFERENCES

1. For a further discussion of these limits, and the techniques used to derive them, along with references to experiments and theory, see L.M. Krauss, Phys. Rev. Lett. **64**, 999 (1990); L.M. Krauss, *The Implications of Z Width Measurements for the Search for Dark Matter,* YCTP-P9-90, to appear, Proceedings, 25th Moriond Conference, Editions Frontières (1989), J. Tran Than Van ed.
2. D. Caldwell et al, Phys. Rev. Lett. **61**, 510 (1988); S. Ahlen et al, Phys. Lett. **B195**, 603 (1987) For a description of a more recent experiment with Si, which is now reporting limits in the range of 5 GeV, see B. Sadoulet et al, Astrophys. J. Lett. **324**, 75 (1988) and D. Caldwell et al, Phys. Rev. Lett. **65**, 1305 (1990)
3. J. Ellis, D. Nanopoulos, L. Roszkowski, D. Schramm, CERN preprint CERN-TH-5705/90, Phys. Lett. B, in press; L. M. Krauss, *Proceedings of Tokyo IUPAP meeting on Big Bang Nucleosynthesis and the Early Universe*, to appear.
4. M. Srednicki and K. Olive, Minnesota pub. UMN-TH-801/89; K. Griest, M. Kamionkowski, M. Turner, Fermilab preprint Pub-89/239-A; see also J. Ellis et al, reference 3.
5. i.e. A.F. Pacheco D. Strottman, Phys. Rev. **D40**, 2131 (1989); J. Engel, P. Vogel, Phys. Rev. **D40**, 3132 (1989); and most recently, F. Iachello, L.M. Krauss, and G. Maino, Yale preprint YCTP-P14-89 , Phys. Lett B, in press
6. L. M. Krauss, Harvard preprint HUTP-85/A008a (1985); W. Press, D. Spergel, Astrophys. J. **296**, 673 (1985); J. Faulkner, R. Gilliland, Astrophys. J. **299**, 994 (1985) ; D. Spergel, W. Press, Astrophys. J. **294**, 663 (1985)
7. D. Caldwell et al, Phys. Rev. Lett. **65**, 1305 (1990)
8. i.e. see, G. C. Raffelt and D. Seckel, Phys. Rev. Lett. **60**, 1793 (1988); M. S. Turner, Phys. Rev. Lett. **60**, 1797 (1988); A. Burrows, R.P. Brinkman, M.S. Turner, Phys Rev. **D38**, 1020 (1989)
9. R. L. Davis, Phys. Lett **B180**, 225 (1986)
10. F. Wilczek, private communication
11. This comes from L. M. Krauss, P. Romanelli, Astrophys. J. **358**, 47 (1990).
12. B. Paczynski, Astrophys. J. **304**, 1 (1986)

THE COSMIC NEUTRINO BACKGROUND

Arnon Dar[#1]

Laboratory For High Energy Astrophysics, NASA

Goddard Space Flight Center, Greenbelt, MD 20771.

Abstract: The cosmic neutrino background is expected to consist of relic neutrinos from the Big Bang, of neutrinos produced during nuclear burning in stars, of neutrinos released by gravitational stellar collapse and of neutrinos prodced by cosmic ray interactions with matter and radiation in the interstellar and intergalactic medium. Formation of baryonic dark matter in the early Universe, matter-antimatter annihilation in a baryonic symmetric Universe and dark matter annihilation could have also contributed significantly to the cosmic neutrino background. The purpose of this paper is to review the properties of these cosmic neutrino backgrounds, the indirect evidence for their existence and the prospects for their detection in view of the dramatic progress in neutrino physics and astronomy that took place last year.

Neutrinos, 1990. In 1990 we have learned new fundamental facts about neutrinos: The measurements of the shape of the Z^0 resonance in e^+e^- collisions at LEP (the large electron positron collider at CERN, the European Center For Nuclear Research) showed (Fernandez 1990) that there are only three generations of light neutrinos, ν_e , ν_μ and ν_τ . Precision tests at LEP and other particle physics laboratories have also shown that their interactions are well described by the minimal

---

#1 **Permanent address: Department of Physics and Space Research Institute, Israel Institute of Technology - Technion, Haifa 32000, Israel.**

standard electroweak theory of Glashow (1961), Weinberg (1967) and Salam (1968), where neutrinos are usually assumed to be massless (and therefore also unmixed), although no established symmetry requires this. In fact, massive neutrinos are expected in most extensions of the minimal model. The many attempts to determine neutrino masses and mixings in terrestrial experiments and from astrophysical and cosmological observations, so far have yielded only the upper bounds (see for instance Dar 1990) which are summarized in Table 1. However, if the recent results reported by the Kamiokande II solar neutrino experiment (Hirata et al. 1990) and by SAGE - the Soviet American Gallium experiment (Gavrin 1990) are correct, they imply neutrino properties beyond the standard model. Moreover, the Mikheyev-Smirnov-Wolfenstein (MSW) solution to the solar neutrino problem (see Bahcall 1989 and references therein) which was first exposed by the Homestake Cl experiment (Davis 1968; Davis 1988), fully explains also the results reported by Kamiokande II and the prelimenary results reported by SAGE, if ν_e is mixed with ν_μ (or with ν_τ) with a mixing angle $sin^2 2\theta \sim 4 \times 10^{-2}$ and mass difference $m^2_{\nu_\mu} - m^2_{\nu_e} \sim 10^{-6}\ eV^2$ (Bahcall 1990, Dar and Nussinov 1990). This MSW solution will be tested within few years with the new generation of solar neutrino experiments that will be able to detect the "missing" solar ν_e's as ν_μ's or ν_τ's. The MSW solution and the "seesaw" mechanism (Yanagida 1978; Gell-Mann, Ramond and Slansky 1979), which is the only scheme presently known for naturally generating small, non-vanishing, neutrino masses, can be combined to yield the individual neutrino masses: The seesaw mechanism relates the masses of the neutrinos to the masses of corresponding Dirac fermions. If the latter are identified with the "up quarks", then

$$m_{\nu_e} : m_{\nu_\mu} : m_{\nu_\tau} \approx m_u^2 : m_c^2 : m_t^2 \approx 0.005^2 : 1.5^2 : 135^2\ , \tag{1}$$

where we used the recent "world" estimate for the mass of the top quark (Fernandez 1990) $m_t = (135\pm20)\ GeV$. With the MSW solution the seesaw mechanism suggests (Dar and Nussinov 1990) $m_{\nu_e} \sim 10^{-8} eV$, $m_{\nu_\mu} \sim 10^{-3} eV$, $m_{\nu_\tau} \sim 10\ eV$. Below we shall point out the implications of these estimates for the cosmological neutrino background. In particular a 10–30 eV tau neutrino can account for most of the dark matter in the Universe and play an important role as hot dark matter in galaxy and large scale structure formations.

1. Relic Neutrinos From The Big Bang

The standard hot Big-Bang cosmological model (see for instance Weinberg 1972) predicts that the present Universe consists mainly of photons, neutrinos and small traces of baryonic matter. The temperature of the relic neutrinos, T_ν, is related to that of the Cosmic Microwave Background (CMB), T_γ, through $T_\gamma \approx (11/4)^{1/3} T_\nu$ (the CMB temperature is higher because the neutrinos decoupled from matter before electron-positron annihilation reheated the photons). Using the observed CMB temperature (Mather 1990; Gush 1990), $T_\gamma \approx 2.735^0 K$, the standard model thus predicts a temperature $T_\nu \approx 1.953^0 K$ and a number density $n_{\nu_i} \approx (3/11) n_\gamma \approx 113\ cm^{-3}$ for each neutrino flavour. For massless neutrinos the corresponding flux is quite large, $\phi_{\nu_i} = c n_{\nu_i}/4 \approx 8.5 \times 10^{11}\ cm^{-2} s^{-1} sr^{-1}$, but the standard electroweak cross sections and the energy deposition of such neutrinos in matter are extremely small, $\sigma \sim G_F^2 E_\nu^2 \sim 2.5 \times 10^{-61}\ cm^{-2}$ and $\delta E \sim E_\nu^2/2m_e \sim 10^{-12}\ eV$, respectively, and cannot be measured. Some authors (e.g. Opher 1974, 1982; Lewis 1980) suggested that the mechanical pressure due to coherent scattering of massless relic neutrinos can be measured. However, it was shown by Langacker et al. (1983) and by Cabibbo and Maiani (1982) that the net momentum transfer vanishes to first order in the Fermi constant G_F and is not measurable by terrestrial detectors. The situation is not much different for massive neutrinos: Distortion of the endpoint spectrum in beta decay due to Pauli blocking by relic neutrinos (Weinberg 1962; Irvine and Humphries 1983), mechanical pressure due to coherent scattering of neutrinos in solids (Cabibbo and Maiani 1982; Shvartsman et al. 1983; Smith and Lewin 1983, 1984, 1985; Lewin and Smith 1984; Smith 1984), spin precession effects on polarized electrons (Stodolsky 1975, Cambell and O'Donell 1982), induced currents in superconductors (Opher 1982, 1984; Lewin and Smith 1984; Smith and Lewin 1987) and Bremsstrhalung in coherent scattering in detectors (Loeb and Starkman 1990) are also extremely small and/or masked by various backgrounds and practically non detectable. However, while direct detection appears to be hopeless, indirect evidence for a cosmological background of relic neutrinos from the Big-Bang may be obtained from primordial nucleosynthesis, from the observed cosmological energy density, from dark matter observations and from large scale structure observations.

1.1 Big Bang Nucleosynthesis. Relic neutrinos play a major role in the standard Big Bang nucleosynthesis (SBBN) theory (Alpher and Herman 1950; Alpher, Follin and Herman 1953; Peebles 1966; Wagoner, Fowler and Hoyle 1967; Wagoner 1969; Wagoner 1973; Yang et al. 1984. For reviews see for instance Schramm and Wagoner 1977; Boesgard and Steigman 1985). The theory has recently received a powerful support from the high statistics results of LEP on Z^0 production in e^+e^- collisions which show that there are only three generations ($N_\nu = 2.99 \pm 0.09$) of light neutrinos (Fernandez 1990) as had been anticipated before from the SBBN theory (for a recent review see Denegri, Sadoulet and Spiro 1990). The success of the SBBN theory can be considered as indirect evidence for relic neutrinos. Moreover, recent precise measurements of the neutron lifetime (Mampe et al. 1989; Byrne et al. 1990) makes the SBBN predictions of the primordial abundances of the light elements depending only on a single unknown paramete, $\eta = n_b/n_\gamma$, the baryon to photon ratio in the Universe. A recent comparison (Olive et al. 1990) between the predicted primordial abundances and those deduced from observations shows that theory is consistent with observations if $\eta \approx 3 \times 10^{-10}$. Since $n_\gamma \approx 414\ cm^{-3}$ it follows that $n_b \approx 1.2 \times 10^{-7} cm^{-3}$ and $\rho_b \approx 0.12\ KeV\ cm^{-3}$, i.e., $\Omega_b \equiv \rho_b/\rho_c \approx 1.1 \times 10^{-2} h^{-2}$, where $\rho_c \equiv 3H^2/8\pi G \approx 1.88 \times 10^{-29} h^2 g\ cm^{-3}$ is the critical energy density and h is the Hubble constant, H, in units of 100 $km\ s^{-1}\ Mpc^{-1}$. Independent evidence from observations for this value of the cosmic baryonic density could provide further evidence for the existence of the cosmic neutrino background.

Neutrinos lighter than 10 eV cannot account for the dark matter in the halos of spiral and dwarf galaxies (Tremaine and Gunn 1979). The only remaining candidates in the standard particle physics model for dark matter in galactic halos are baryons. Interestingly, if the galactic dark matter is mainly hadronic then its cosmological mass density is in good agreement with that deduced from SBBN and it can account for the missing baryonic matter, i.e. for the difference between the total luminous baryonic matter and that inferred from SBBN theory: The measured luminosity density in the Universe (see for instance Kolb and Turner 1990 and references therein) is $L \approx 2.4 \times 10^8 h L_\odot/Mpc^3$. The mass to light ratio in the solar neighbourhood is $M/L \sim 4M_\odot/L_\odot$. If this ratio is universal then the density of luminous matter in the Universe is $\Omega_L \approx 3.4 \times 10^{-3} h^{-1}$. The mass to light ratio of spiral galaxies deduced from rotational curves and escape velocities, and recently also from gravitational lensing (Langstom 1990) is $< M/L >_G \approx 16hM_\odot/L_\odot$. Consequently, $\Omega_{Halo} \approx < M/L >_G L/\rho_c \approx 1.4 \times 10^{-2}$, in excellent agreement with

$\Omega_b \approx 1.1 \times 10^{-2}h^{-2}$ deduced (Olive et al. 1990) from SBBN, especially if one uses the recent precise estimate of Jacoby et al. (1990), $h = 0.87 \pm 0.12$.

1.2 The Cosmological Energy Density. If the present Universe consists only of baryonic matter with a mass density that was deduced from SBBN, and of photons and neutrinos with masses that are suggested by the MSW solution to the solar neutrino problem and the seesaw mechanism, then its energy density is given by

$$\rho \approx n_\gamma(\eta m_p + 2.70T_\gamma + (3/11)\Sigma_i m_{\nu_i}) \approx 1.3\ KeV\ cm^{-3}\ , \tag{2}$$

corresponding to $\Omega \equiv \rho/\rho_c \approx 0.12h^{-2}$. In particular, with the recent precise estimate $h \approx 0.87 \pm 0.12$ the predicted cosmological energy density is $\Omega \approx 0.15$. This value is indeed in good agreement with the best estimated value, $0.1 < \Omega < 0.2$, that was deduced from the dynamics of clusters of galaxies (see for instance Peebles 1986 and references therein).

1.3 Relic Neutrinos And Dark Matter. If the dark matter in galactic halos consists of massive neutrinos its density is limited by the maximum phase space allowed by the self consistent solution of the Vlasov equation for collisionless self gravitating neutrinos (Tremaine and Gunn 1979). For sufficiently massive neutrinos the distribution function of relic neutrinos becomes Maxwellian in the frame of the galaxy. The demand that the local phase space density is constrained by the original cosmic value $(2\pi)^{-3}$ per flavour then yields the bound (Lake 1989, Ralston 1989)

$$m_\nu^4 \geq \frac{\rho(r)}{2}\left(\frac{2\pi}{\beta^2}\right)^{3/2}\ , \tag{3}$$

where $\beta =< \beta_x^2 >=< \beta_y^2 >=< \beta_z^2 >$ is the one dimensional local velocity dispersion. If the neutrinos have the masses predicted by the seesaw mechanism and the MSW solution to the solar neutrino problem, i.e., ($m_\nu \leq 10\ eV$), they cannot obey the above lower bound and they cannot form the halos of dwarf and spiral galaxies. For instance, the dark matter density in the solar neighborhood is $\rho_D \sim 4 \times 10^8\ eV\ cm^{-3}$ and the local velocity dispersion is $\sim 220\ km\ s^{-1}$, requiring that $m_\nu > 24\ eV$.

Within the standard particle physics model the remaining candidates for dark matter particles are baryons. Thus, in the above particular scenario where neutrino masses are lighter than 10 eV the dark matter in galactic halos is likely to be hadronic (e.g., brown dwarfs, white dwarfs, neutron stars and black holes) and as we saw its cosmological averaged density is consistent with the cosmological baryonic mass density that was derived from Big-Bang nucleosynthesis (Olive et al. 1990). However, it is not clear what mechanism can produce and eject selectively such dark matter candidates into the galactic halo. Moreover, baryonic dark matter implies declining rotational curves for field galaxies beyond some finite radius, for which there is not yet any convincing evidence. On top of that, a large population of neutron stars and black holes would imply over enrichment of the interstellar space with heavy elements, which is not observed (for a detailed discussion of the observational constraints on baryonic dark matter see for instance Carr 1985; 1990).

Since the seesaw estimate for the neutrino masses is only a rough estimate and it is based on rather prelimenary results of the Soviet American Gallium experiment, a tau neutrino mass in the range 10 – 30 eV is not excluded. Such a mass can account for the dark matter in galactic halos (Lake 1989; Ralston 1989). In that case galactic halos extend to much larger radii than those usually explored by HI rotation curves ($r_{HI} \leq 30\ kpc$) and they contain much more dark matter than estimated above, i.e., $\Omega_G \gg 1\%$. Such halos may also contain the missing baryonic dark matter inferred from SBBN theory. Conversely, SBBN theory may be only an idealized model of a more complicated reality and may not yield the correct baryonic density. Moreover, the primordial abundances that were extracted from observations and were used to infer the cosmological baryonic density (Olive 1990) may be inaccurate and actually there is no "missing baryonic matter problem". Dark matter in groups and clusters of galaxies may consist of relic tau neutrinos with mass in the range 10 – 30 eV which yield a cosmological energy density in the range $0.1 < \Omega < 0.3$. Such dark matter could have a distribution smoother than the distribution of luminous matter in rich clusters of galaxies, as inferred by Tyson et al. (1990) from gravitational lensing of faint blue galaxies by rich clusters of galaxies. However, it is not clear yet whether the sharp decline of the dark matter distributions near the edge of the clusters, where the red light intensity decreases sharply, can be reproduced by dissipationless neutrino dark matter (Tyson 1990).

1.4 Relic Neutrinos And Large Scale Structure. The current "standard model" of galaxy formation is based on linear adiabatic perturbations and cold dark matter (Blumenthal et al. 1984; Davis et al. 1985). It gives encouraging results on the scale of galaxies and clusters. However, recent observations indicate that there is much more structure on large scales than expected from the cold dark matter model. Clusters are seen to congregate into superclusters forming filaments of length larger than $50h^{-1}$ Mpc, and sheets of length and width of $50h^{-1}$ Mpc and thickness $5h^{-1}$ Mpc lying around voids of diameter 60 $h^{-1}Mpc$ (see these proceedings). Moreover, the standard particle physics model with light neutrinos does not offer any candidates for cold dark matter and there is no evidence for the existence of any particles beyond those of the standard particle model which might have served as such candidates. Relic neutrinos can serve as hot dark matter but, a model with adiabatic perturbations and hot dark matter predicts that galaxies should be younger than clusters in conflict with the data, and anisotropies of the CMB which are barely compatible with the present observational upper bounds. However, it has recently been shown that with non adiabatic seed perturbations the above problems of the hot dark matter model disappear and that dissipationless clustering of neutrinos in cosmic-string induced wakes can generate significant large scale structure in the Universe consistent with observations (Perivolaropoulos, Brandenberger and Stebbins 1989).

2. Neutrinos From Nuclear Burning In Stars

The general success of stellar evolution theory suggests that hydrogen fusion via the pp and CNO nuclear cycles powers stars like the sun. Baryon number, charge and lepton flavour conservation then require that the net reaction is

$$4p + 2e^- \longrightarrow {}^4He + 2\nu_e + Q \ , \tag{4}$$

and energy conservation requires that $Q \approx 26.78\ MeV$. Consequently, for stars that derive their energy from fusion of Hydrogen into Helium, their neutrino luminosity is given by

$$L_\nu = \frac{2L_\odot}{Q - 2\bar{E}_\nu}\frac{L_*}{L_\odot} \approx 1.86 \times 10^{38}\frac{L_*}{L_\odot}\ \nu\ s^{-1} \ , \tag{5}$$

where $L_\odot \approx 3.86 \times 10^{33}\ erg\ s^{-1}$ is the solar luminosity, and $\bar{E}_\nu$ ($\sim 0.27\ MeV$), is the average energy of solar neutrinos. The cosmic luminosity density is $L_B \approx 2.4 \times 10^8 h L_\odot\ Mpc^{-3}$. Consequently, if the luminosity density of the Universe is mainly due to hydrogen fusion, and if it is largely time independent over the Hubble expansion time, $t_0 \approx H^{-1} \approx 9.8h^{-1}\ GY$, then the cosmic neutrino density due to hydrogen fusion is given approximately by

$$n_\nu(pp) \approx t_0 L_B(L_\nu/L_\odot) \approx 6.2 \times 10^{-9}\ cm^{-3} \tag{6}$$

and the corresponding cosmic ν_e flux is $\phi_{\nu-e} \sim 50\ cm^{-2}s^{-1}sr^{-1}$. The average energy of these neutrinos is approximately 3/5 of that of the solar neutrinos due to redshift effects. Any possible signal of this isotropic neutrino background in underground detectors is masked completely by the solar neutrino flux ($\sim 6 \times 10^{10}\ cm^{-2}s^{-1}$), by Earth radioactivity (Krauss, Glashow and Schramm 1984) and by the cosmic ray and the radioactive backgrounds inside the detectors.

3. Neutrinos From Baryonic Dark Matter Production

If a significant amount of baryonic dark matter resides in neutron stars and black holes, e.g., remnants of a population of pregalactic massive stars (see for instance Carr 1985, 1990), their formation should have produced a relic neutrino background. An upper bound on the flux of relic neutrinos from past stellar collapses can be obtained as follows: In gravitational stellar collapse which produces a neutron star with a "canonical mass" of $\sim 1.4 \times M_\odot$ about 10% of the gravitational mass ($\sim 3 \times 10^{53}\ ergs$) of the neutron star is released by neutrino emission ($\sim 10^{57}$ neutrinos with average energy of $\sim 15\ MeV$). Since neutrino emission in gravitational stellar collapse into black holes is relatively much smaller than that in neutron star formation, the maximum flux of relic neutrinos from past stellar collapses would have been produced if all the baryonic dark matter consisted of neutron stars. The SBBN theory suggests (Olive et al. 1990) that $\rho_b \approx 0.12\ KeV\ cm^{-3}$. Consequently, the relic neutrino number density is bounded by $n_\nu < 0.1\rho_b/15\ MeV \approx 8 \times 10^{-7}\ cm^{-3}$, and the flux per flavour is bounded by $\phi_{\nu_i} \approx \phi_{\bar{\nu}_i} < 10^3\ cm^{-3}s^{-1}sr{-1}$. These relic neutrinos, in particular the $\bar{\nu}_e$'s, may have detectable energies, i.e. energies above 5 MeV where the background from terrestrial radioactivity and atmospheric neutrinos is much smaller (Krauss et al, 1984), if their energies (temperature) have not been red shifted by a factor much smaller than $1/(1+z) \sim 1/2$. The 90% C.L. limit on a thermal $\bar{\nu}_i$ background which was reported by Kamiokande II (W. Zhang et al. 1988), $\phi_{\bar{\nu}_i} \leq 3.6 \times 10^3\ cm^{-3}s^{-1}sr{-1}$ if the present temperature is above 2 MeV (i.e., if $< E_\nu > \approx 3.15T_\nu \geq 6.3\ MeV$), does not constrain yet a neutrino background from baryonic dark matter formation.

4. Neutrinos From Matter-Antimatter Annihilation

The overwhelming preponderance of matter over antimatter in the cosmic radiation provides convincing evidence that our galaxy is composed entirely of matter. Observations of extragalactic γ - ray background also indicate that if matter and antimatter exist in equal amounts, they are most probably separated into large domains (Stecker 1978 and references therein). Matter - antimatter annihilation that took place at the domain - interface regions produced mainly π^0's and $\pi^\pm$'s that have decayed, respectively, into photons and neutrinos. Thus matter - antimatter annihilation produced similiar cosmic backgrounds of of γ–rays and neutrinos. Furtheremore, if the difference between the cosmic baryonic density that was inferred from SBBN theory and the presently observed density of luminous matter is because most of the matter and antimatter have annihilated each other then the number density of the relic Mev neutrinos and γ–rays should have been $\sim 2n_b \approx 1.2 \times 10^{-7}\ cm^{-3}$, and the corresponding flux should have been

$\phi \sim 10^3\ cm^{-2}s^{-1}sr^{-1}$. The observed γ–ray background is much smaller, in conflict with the above assumption.

5. High Energy Cosmic Neutrino Background

The high energy ($E_\nu > 1\ GeV$) cosmic neutrino background is produced by cosmic ray interactions with matter and radiation at the sources and in the interstellar and intergalactic space. This background and the prospects for its detection with underground and underwater neutrino telescopes will be discussed in the following talk by Cline (1990). More details can be found, for instance, in recent reviews by Learned (1990) and Suzuki (1990) and will not be reproduced here because of lack of space.

Acknowledgements: This research was done while the author held a NRC senior research associateship at NASA Goddard Space Flight Center, which he gratefully acknowledges.

References

Alpher, R.A., Follin, J.W. and Herman, R.C., 1953; Phys. Rev. **92**, 1347.

Alpher, R.A. and Herman, R.C., 1950; Rev. Mod. Phys. **22**, 153.

Bahcall, J.N. and Davis jr., R., 1978, Science, **191**, 264.

Blumenthal, G. et al., 1984; Nature **311**, 517.

Boesgard, M. and Steigman, G., 1985; Ann. Rev. Astron. Astrophys. **23**, 319.

Cabibbo, N. and Maiani, L., 1982; **114b**, 115.

Campbell, A. and O'Donnell, 1982; Phys. Rev. **D26**, 148.

Carr, B.J., 1985; in "Observational And Theorertical Aspects Of Relativistic Astrophysics And Cosmology" (edts. J.L. Sanz and L. Goicoeches, World Scientific

1985) pp. 1-75.

Carr, B.J., 1990; Comments Astrophys. **14**, 257.

Cline, D., 1990; these proceedings.

Denegri, D., Sadoulet, B. and Spiro, M., 1990; Rev. Mod. Phys. **62**, 1.

Dar, A., 1990; Science (in press).

Dar, A. and Nussinov, S., 1990; Submitted to Nature.

Davis, M. et al., 1985; Ap. J. **292**, 371.

Davis jr., R. et al., 1968; Phys. Rev. Lett. **20**, 1205

Davis jr. R. et al., 1989; "Neutrino 88", eds. J. Schneps et al. (World Scientific 1989).

Fernandez, E., 1990; "Neutrino 90", Proceedings of the 14th Int. Conf. On Neutrino Physics and Astrophysics. Editor K. Winter, (North Holland), in press.

Gavrin, V.N., 1990; "Neutrino 90" (in press).

Gell-Mann, M., Ramond, P. and Slansky, R., 1979; "Supersymmetry", p. 315, edited by P. Van Nieuwenhuizen and D.Z. Freedman (North Holland 1979).

Glashow, S., 1961; Nucl. Phys. **22**, 579.

Gush, H.P., 1990; These proceedings.

Hirata, K.S. et al., 1990; Phys. Rev. Lett. **65**, 1297.

Jacoby, G.H. et al., 1990; Ap. J. **356**, 332.

Krauss, L.M., Glashow, S.L. and Schramm, D.N., 1984; Nature **310**, 191.

Kolb, E.W. and Turner, M.S., 1990; "The Early Universe" (Addison-Wesley Pub. Com. 1990).

Lake, G., 1989; Ap. J. **98**, 1253.

Langacker, P. et al., Phys. Rev. **27**, 1228.

Langstom, A. et al., 1990; Nature **344**, 4.

Learned, J.G., 1990; Proceedings Of The Venice Workshop On Neutrino Telescopes, February 1990.

Lewin, J.D. and Smith, P.F., 1984; Ap. J. (Lett) **24**, L59.

Lewis, R.R., 1980; Phys. Rev. **27**, 1228.

Mather, J., 1990; These proceedings.

Olive, K.A. et al., 1990; Phys. Lett. **236**, 454.

Opher, R., 1974; Astr. Ap. **37**, 135.

Opher, R., 1982; Astr. Ap. **108**, 1.

Opher, R., 1984; Ap. J. **282**, 398.

Peebles, P.J.E., 1966; Ap. J. **146**, 542.

Peebles, P.J.E., 1986; Nature **321**, 1.

Perivolaropoulos, L., Brandenberger , R.H. and Stebbins, A., 1989; Brown University Preprint HET–711 (July 1989).

Ralston, J.P., 1989; Phys. Rev. Lett. **63**, 1038.

Salam, A., 1968; "Elementary Particle Theory", ed. N. Svartholm (Almqvist and Wiksells, Stockholm 1968) p. 367.

Schramm, D.N. and Wagoner, R.V., 1977; Ann. Rev. Nucl. Part. Sci. **27**, 37.

Shvartsman V.F. et al., 1983; JETP Lett. **36**, 277.

Smith, P.F., 1984; Nuov. Cim. **83A**, 263.

Smith, P.F. and Lewin, J.D., 1983; Phys. Lett, **127B**, 185.

Smith, P.F. and Lewin, J.D., 1984; Acta Phys. Polonica **15**, 1201.

Smith, P.F. and Lewin, J.D., 1985; Acta Phys. Polonica **16**, 137.

Smith, P.F. and Lewin, J.D., 1987; Ap. J. **318**, 738.

Stecker, F.W., 1978; Nature **273**, 493.

Suzuki, A., 1990; Proceedings of The Venice Workshop On Neutrino Telescopes,

February 1990.

Stodolsky, L., 1975; Phys. Rev. Lett. **34**, 110.

Tremaine, S. and Gunn, J.E., 1979; Phys. Rev. Lett. **42**, 407.

Tyson, T., 1990; These proceedings.

Tyson, T. et al,. 1990; Ap. J. (Letters) , L .

Wagoner, R.V., 1969; Ap. J. Suppl. **162**, 247.

Wagoner, R.V., 1973; Ap. J. **179**, 343.

Wagoner, R.V., Fowler, W.A. and Hoyle, F., 1967; Ap. J. **148**, 3.

Weinberg, S., 1967; Phys. Rev. Lett. **19**, 1264.

Weinberg, S., 1972; "Gravitation And Cosmology" (John Wiley & Sons, Inc. 1972).

Yanagida, T., 1978; Prog. Theor. Phys. **B315**, 66.

Yang, J. et al., 1984; Ap. J. **281**, 493.

Zhang, W. et al., 1988; Phys. Rev. Lett. **61**, 385.

TABLE I: PRESENT LIMITS ON NEUTRINO MASSES

	LABORATORY	ASTROPHYSICS	COSMOLOGY
m_{ν_e}	<9.6 eV	<16 eV	<40 eV+
m_{ν_μ}	<250 KeV	<15 KeV*	<40 eV+
m_{ν_τ}	<35 MeV	<15 KeV*	<40 eV+

* Dirac Neutrinos +Stable Neutrinos (τ >20 Gy)

TECHNIQUES FOR DETECTION OF COSMIC DIFFUSE NEUTRINOS AND ANTI-NEUTRINOS

DAVID B. CLINE
Departments of Physics & Astronomy, University of California,
Los Angeles, CA 90024-1547

ABSTRACT

We first indicate the energy regions where diffuse neutrinos or anti-neutrinos are expected. Techniques for the detection of these neutrinos are described. We focus on the prospects for detecting neutrinos or anti-neutrinos in the window between the atmospheric and solar neutrino flux, from 10 − 30 MeV. We also describe methods to detect neutrinos in the 100 MeV to GeV region using a lunar neutrino telescope. A brief comment is made concerning the detection of the 1.9° relic neutrinos.

INTRODUCTION

The detection of diffuse microwave background photons provided a revolution in Astrophysics and Cosmology. The detection of a primordial neutrino background would likely be as important. Unfortunately this detection is going to be extremely difficult.

On the other hand, so little is known about the low energy neutrinos that the flux could possibly be much larger than is presently expected. This would be a major discovery.

There are many possible sources of diffuse neutrinos. For example, from processes like Wimp annihilation in the early universe to exploding black holes. The low energy gamma flux is attenuated by electro-magnetic processes in the universe; the neutrino flux is not. Thus, the ν-flux could be much larger than the observed γ-flux.

In this report we will do two things. First, we will discuss the windows for detection of diffuse neutrinos. Second, we will comment on a possible detection technique. In the process several possible detectors will be discussed, including a Lunar Base Detector or telescope.

TABLE 1

NEUTRINO DETECTION TECHNIQUES

$E_\nu \sim 10^{-3}$ eV – Coherent Interactions with many nuclei
individual σ $\left[(e^+ e \to \nu_e + e)\ \sigma \sim 10^{-62}\ \mathrm{cm}^2\right]$
(Relic ν)
Coherent $\sigma \propto N^2$, $N =$ nuclei in sample
but detection very difficult!

$E_\nu \sim$ KeV - MeV – Neutrino Processes

$$\sigma \sim 10^{-45} - 10^{-44}\ \mathrm{cm}^2 \begin{cases} - & \nu_e + \text{Nucleus} \to N' + e \\ & (\text{eg. } \nu_e + C\ell \to Ar + e - \text{Homestake}) \\ - & \nu_e + e \to \nu_e + e \\ & \nu_{\mu,\tau} + e \to \nu_{\mu,\tau} + e \end{cases}$$

(Solar Neutrinos)
Coherent $\nu_x + N \to \nu_x + N$

E_ν 10 – 100 MeV
Supernova ν detectors – Diffuse $\nu,\bar{\nu}$ in window

$\bar{\nu}_e$: $\bar{\nu}_e + p \to e^+ + n\ \sigma \sim 10^{-40}\ \mathrm{cm}^2$
ν_e : $\nu_e + N \to e^- + N'$
ν_x : $\nu_x + N \to \nu_x + N'_{\to n,\gamma}$

$E \gtrsim 100$ MeV – GeV

$$\left.\begin{array}{l} \nu_e + N \to e + x \\ \nu_\mu + N \to \mu + x \\ \nu_x + N \to \nu_x + x \end{array}\right\} \sigma \sim 10^{-38}\ \mathrm{cm}^2$$

E >> GeV – Large Water Detector – DUMAND for example

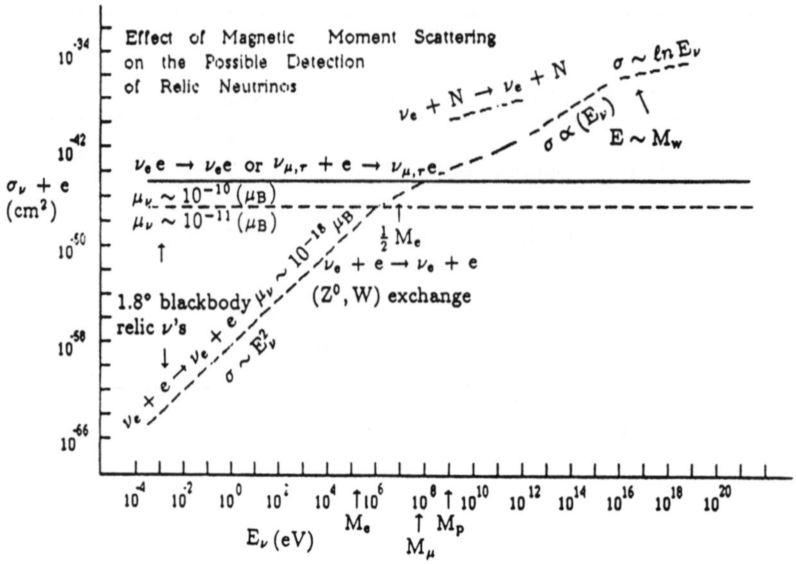

Figure 1

1. NEUTRINO DETECTION TECHNIQUES

Neutrino detection is very difficult, especially when the fluxes are small and at low neutrino energy. The detection of solar neutrinos is a case in point of this fact. Nevertheless, detection of new sources of neutrinos in nature is a promising field.

A summary of the different neutrino reactions that can be used, as well as some cross sections, is given in Table 1. In the very low energies where relic neutrinos are likely important, coherent effects of one sort or another may be used. In this region kilograms or multi-ton detectors are useful.

At the highest energy the cross sections become large but the expected flux from various sources is expected to be very low. Thus, a very large detector with target mass of $10^6 - 10^8$ tons is required. Figure 1 shows the neutrino cross section as a function of energy. Note: the existence of a neutrino magnetic moment of $\sim 10^{-10} - 10^{-11}\ \mu_B$ could give a relatively large enhancement of the very low energy cross section.[1]

2. DETECTION OF DIFFUSE NEUTRINOS AND ANTI-NEUTRINOS ON EARTH

The major problem of detecting neutrinos in the few MeV to multi TeV energy region on earth is the background due to radioactivity in the earth and the atmospheric flux of cosmic ray neutrinos. In addition, nuclear reacters provide a large background below 10 MeV anywhere on Earth.

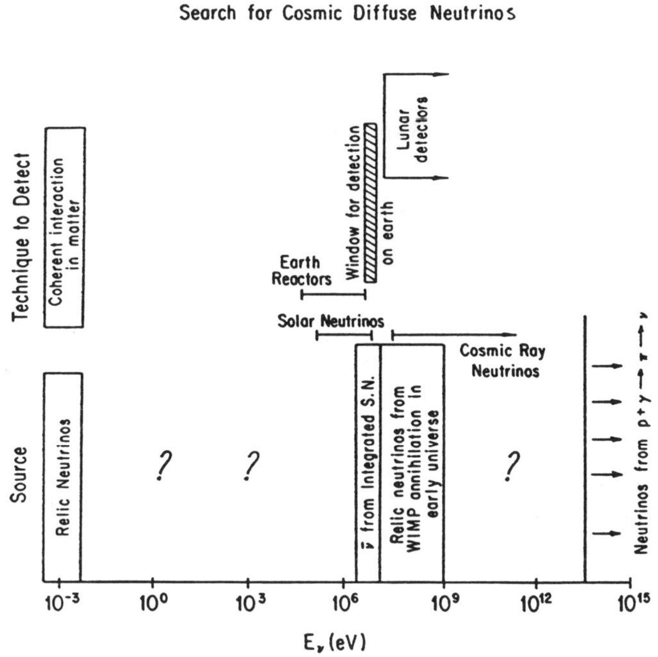

Figure 2

Figure 2 shows some possible sources for diffuse neutrinos. The most interesting source of anti-neutrinos is due to the integrated total of all previous supernova.[2] Fortunately, there is some hope to detect such neutrinos since there is a window in the background between $\sim 10 - 30$ MeV, as shown in Figure 3.

In Table 2 an outline of a possible detection technique is given. A 10^5 ton detector could be used to detect the currently expected flux of $\bar{\nu}_e$ from all previous SuperNova in this window.

There is no other comparable window for diffuse neutrino detection on Earth unless the flux is greater than that from the atmosphere (which is very unlikely). Also, at very high energy the Earth will become opaque and only downward neutrinos can be used. This is a definite disadvantage.

3. DETECTION OF DIFFUSE NEUTRINOS WITH A LUNAR TELESCOPE

To detect diffuse neutrinos above 30 MeV will likely require a non-terrestrial detector. The Moon is the most likely site for such a detector.

A small group of people at UCLA (D. Chrisman, Th. Müller and I) have looked into the question of whether a massive neutrino telescope can be constructed using a very low mass sensor that would be transported from the

TABLE 2

A TERRESTRIAL DETECTOR TO SEARCH FOR RELIC $\bar{\nu}$ IN THE WINDOW (10 – 50 MeV)

Reaction: $\bar{\nu}_e + p \rightarrow e^+ + n$

use material with $H_2O \cdots$ (water, clear ice, etc.)

(Require a **very** low Radioactivity Background and is a Coincidence Technique)

– use Simple neutron counters and coarse Č detectors (PMT's) 20 MeV $\bar{\nu}$

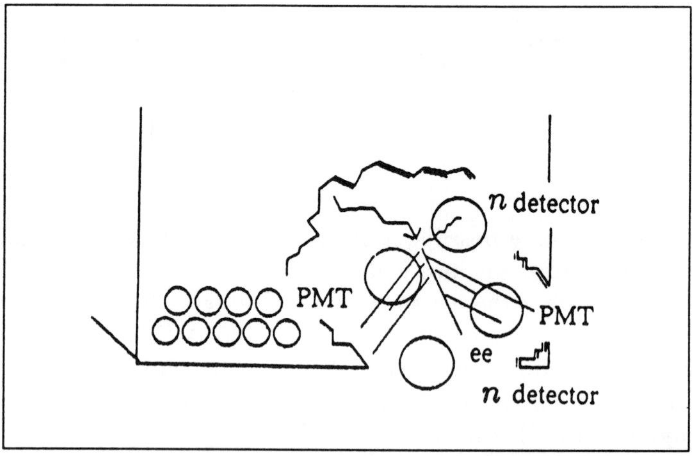

10^5 tons / 1 Year of Search

$$\Longrightarrow F_{\bar{\nu}} \leq 5 \times 10^{-2}\,\text{cm}^{-2}\,\text{sec}^{-1}$$

LIMIT OR DETECTION

TABLE 3

ORDER OF MAGNITUDE EFFECTS FOR ν EXPERIMENTS

EXPERIMENTS AT REACTORS	$\bar{\nu}_e e \to \bar{\nu}_e e$	$\sigma \sim 10^{-44}\,\mathrm{cm}^2$ $F_{\bar{\nu}} \sim 10^{13}\,/\,\mathrm{cm}^2\,/\,\mathrm{sec}$

$$\begin{aligned} \text{RATE} &\sim \sigma \cdot F \cdot N \\ &\sim 10^{-44} \times 10^{13} \times N \simeq N \cdot 10^{-31} \\ N &\sim 10^{30} \quad \text{for 1 TON} \\ \text{RATE} &\sim 10^{-1}\,\mathrm{sec}^{-1} \implies 10^{6}\,/\,\mathrm{year} \end{aligned}$$

RELIC NEUTRINOS	$\nu_e + e \to \nu_e + e$ $E_e \simeq 10^{-4}\,\mathrm{eV}$	$\sigma \sim 10^{-62}\,\mathrm{cm}^2$ $F_\nu \sim 10^{12}/\mathrm{cm}^2/\mathrm{sec}$

$$\begin{aligned} \text{RATE} &\sim 10^{-62} \times 10^{12} \cdot N = 10^{-50} \cdot N \\ \text{For RATE} &\sim 10^{3}\,/\,\text{year } N = 10^{44} \quad [\sim\sim 10^{14}\ \text{TONS}] \end{aligned}$$

TWO WAYS TO IMPROVE $\sigma \to 10^{-44}\,\mathrm{cm}^2$ MAGNETIC MOMENT

OR

COHERENT EFFECT

$$R \sim \sigma \cdot F \cdot N^2 \simeq 10^{-50}\,N^2$$

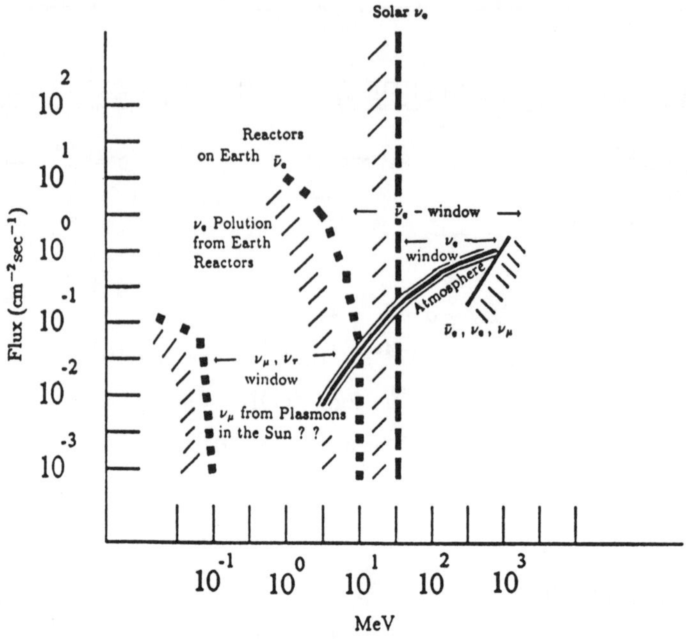

Figure 3

Earth.[3] The flux of neutrinos on the Moon is not zero due to interactions in the rock that give μ, π, K and charm particle decays. Thus, we have decided the best way to search for neutrinos is to look for downward going neutrinos with a proton veto shield above the detector (see Figure 4a).[3] With the shield removed this same detector can be used to reach for gamma ray sources or high energy diffuse gamma rays.

The initial lunar telescope should be ~ 1000 tons of lunar concrete. In our design the mass of the sensor could be $\lesssim 1$ ton. (Most neutrino detectors have a much larger sensor / mass ratio of $\lesssim [10^{-2} - 10^{-1}]$.) A schematic of the detector tubes is shown in Figure 4a. The tubes would be inflated by the detector gas on the lunar base (Figure 4b). This detector could be used to search for diffuse neutrinos from ~ 100 MeV to $\sim$multi GeV.

4. COMMENTS ON THE DETECTION OF RELIC NEUTRINOS

Table 3 provides an order of magnitude comparison of the rates of relic neutrinos and reactor experiments. Without the help of coherent effects it would take 10^{14} tons of material to detect the relic neutrinos.

The use of coherent effects is very difficult due to the very small recoil momentum.[4,5] We have looked at a recent proposal to use the process and

$$\nu_e + e \rightarrow \nu_e + e + \gamma$$

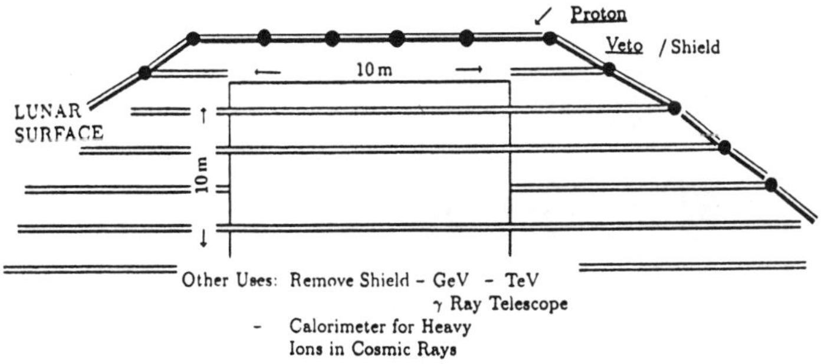

Figure 4a

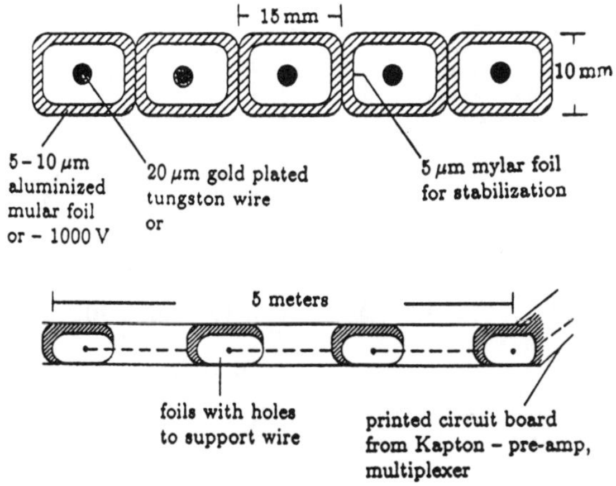

Figure 4b

are unable to confirm the estimated rates (we find rates that are $\sim 10^{-10}$ lower).[6,7] There is no viable technique known to us to detect relic neutrinos. However, novel techniques might be used to search for a flux that is several orders of magnitude larger than expected.

5. SUMMARY OF TECHNIQUES TO DETECT COSMIC DIFFUSE NEUTRINOS

In Figure 5 we present the projected neutrino and anti-neutrino flux levels that can be searched for using some of the techniques discussed in this paper. The only near term hope would be to construct a very large detector to search in the 10 − 30 MeV window. On a longer time scale a lunar neutrino telescope could open the whole field of cosmic diffuse neutrinos, provided there are sources in nature.

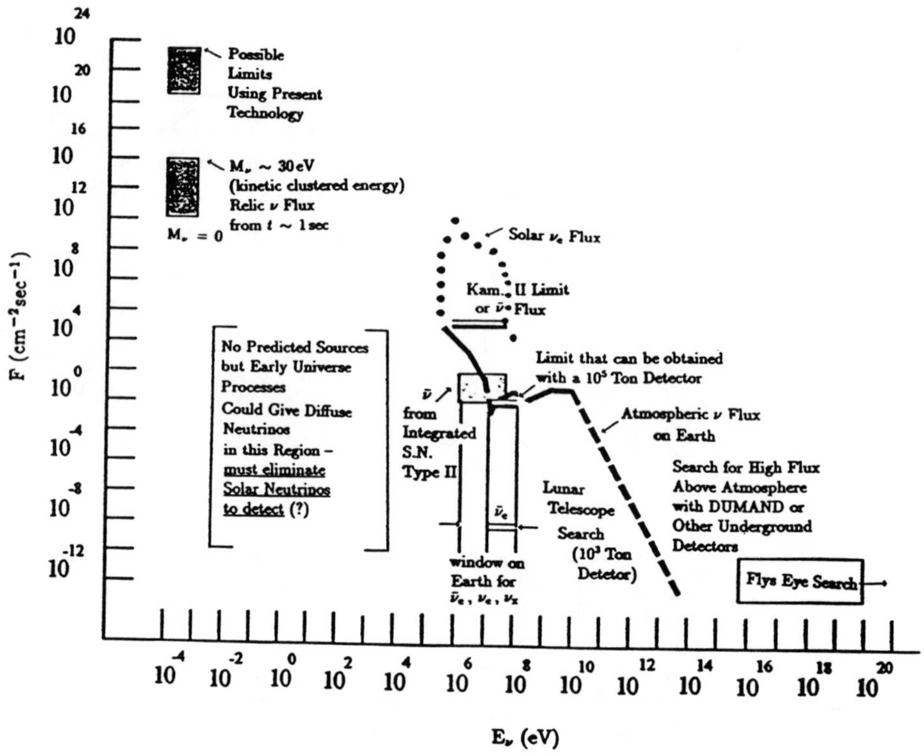

Figure 5

REFERENCES

[1] D. Cline, Proceedings of the 1986 Telemark Meeting

[2] See the Kamiokande II paper: Phys.Rev.Lett., 61, 385 (1988); and the enclosed references.

[3] D. Chrisman, D. Cline, Th. Müller, UCLA report (1989) in preparation.

[4] See for example, P. Langacker, J. Leveille and J. Sheiman, Phys.Rev. D, 1228 (1983) – these authors showed that the relic neutrino effect goes like G^2, not like G, as had been previously believed.

[5] P. Smith and J. Lewan, Phys.Lett., 127B, 185 (1983).

[6] A. Loub and G. Starkman, "A Detector for Cosmic Neutrino Background", Inst.for Adv.Studies preprint (1990).

[7] D. Cline and W.-P. Hong, "Relic Neutrino Detection by Phonon Detector", UCLA internal note (1990).

THE NEW DISCUSSION OF A NEUTRINO MASS AND ISSUES IN THE FORMATION OF LARGE-SCALE STRUCTURE

Adrian L. Melott
Department of Physics and Astronomy
Univeristy of Kansas
Lawrence, Kansas 66045 U.S.A.

INTRODUCTION

A number of persons here have emphasized new evidence which suggests lepton flavor nonconservation and with it a neutrino mass. There is a strong feeling in the particle physics community that we should again give strong consideration to the idea that neutrinos are the dark matter. This has been a controversial area in cosmology, with most persons concluding that the process of structure formation is grossly incompatible with our universe. I wish to argue that the discrepancy is not overwhelming, and probably not greater than that which exists with Cold Dark Matter. Given the primitive state of all simulations, we probably should remain open to this and other nonstandard ideas about the dark matter and its interaction with the galaxy distribution seen today.

CELL STRUCTURE

In 1980, there was a considerable flurry of interest in neutrinos of mass of order 30 eV, based on some experimental work, which has since been called into question. (Rather than give a comprehensive review, I refer the reader to the 1989 review paper of Shandarin and Zel'dovich). This led to a series of numerical simulations. The formation of structure in neutrino-dominated universe with adiabatic perturbations is greatly affected by the free-streaming which damps away perturbations up to some characteristic scale – tens of Megaparsecs for our $\sim$ 30 eV neutrino. It is the prime example of Hot Dark Matter (HDM).

The first milestone in understanding the development of structure in such a universe was the cell structure found by Doroshkevich *et al.* 1980). The interconnected flattened structure surrounding nearly empty voids must count as a *prediction*, since only hints existed in the data then. The numerical result was a surprise to the investigators, but is now well-understood. The existence of such structures has been definitely confirmed in recent years (*e.g.* de Lapparent *et al.* 1990). It has also been shown that other models, such as Cold Dark Matter (CDM) can give use to such structures, at least with a little "fiddling" in the form of biasing (*e.g.* Weinberg and Gunn 1991). However, the most straightforward, assumption-free prediction of these structures is provided by assuming an HDM dominated universe. This prediction was made before the crucial observations were made.

CORRELATION AMPLITUDE

The next stage in the study of HDM models was the important discovery of a problem by Frenk, White, and Davis (1983). Although the problem has been described in many ways, it is best understood as described there: an excessive autocorrelation amplitude. When one examines simulations and assumes that galaxies can only form in regions which have undergone collapse, the resulting distribution is too strongly clustered as compared to the observed distribution of galaxies. The amplitude is too high by about a factor of two, this result is only weakly dependent on the assumed redshift of galaxy formation.

This discrepancy is real and is the most serious problem with the standard gravitational clustering scenario for the neutrino–dominated universe. Yet it is of the same approximate amplitude as problems with Cold Dark Matter (CDM) which have been "fixed up" by biasing. It is important to note that in simulations we calculate a mass–mass correlation, while the observed quantity is a number–count based galaxy–galaxy correlation function. Thus a modest variation in mass per galaxy could change the slope and amplitude. There is some substantial evidence that galaxies in rich clusters have sunk (by dynamical friction) to the bottoms of those potential wells. As a part of this, galaxies in cluster cores seem to have merged, which lowers the number count without changing the mass. Also, the dark mater is left behind as a diffuse halo around the cluster. It does not appear in conventional estimates of the cluster mass. (However, see Tyson, these proceedings.) Such effects could easily introduce the kind of amplitude change needed to reconcile simulations with observations.

Another caveat that must be applied is the simulation method. The clustering of collisionless particles is a reasonable good approximation for theories with a lot of power on small scales in the intial conditions, because much of the cooling can take place before larger scale fluctuations grow to make superclusters. In HDM, large "pancakes" of gas and dark matter form. The process of galaxy formation requires following the cooling of this gas, which in turn means high–resolution hydrodynamics coupled to a collisionless system for the dark matter. This is only now becoming possible, and the results will be unreliable for some time. We have no detailed understanding of how and where the galaxies will form in the full nonlinear problems.

It seems premature to exclude HDM on the basis of this discrepancy factor of two, when our understanding of the galaxy formation process is so poor.

NONLINEARITY AND THE ONSET OF GALAXY FORMATION

A related problem concerns the era of galaxy formation. This is widely assumed to begin when the RMS density fluctuation, extrapolated from linear theory, approaches unity. This can be a very bad approximation. Phase correlations, resulting in mode coupling are extremely important in HDM (*e.g.* Scherrer *et al.* 1991). The distribution of densities becomes skewed (non Gaussian) sufficiently to be noticed even in relatively coarse simulations when $(\delta_\rho/\rho)_{RMS} = 0.1$ (Centrella *et al.* 1988). The formation of collapsed regions is quite strong by the time this reaches 0.2. If we allow that QSO's are rare bright objects, then their existence at redshifts even up to 5 or more can be accommodated with a relatively low "linear theory" fluctuation amplitude. The normalization in Centrella *et al.* (1988) would imply $Z_{NL} \sim 0$. This result, which includes the nonlinear effects, is very different from what is normally assumed for predicating microwave background fluctuations (*e.g.* Bond *et al.* 1990). In those predictions, Z_{NL} (where $\delta_\rho/\rho = 1$) is assumed to be the *beginning* of galaxy formation, leading to much too high a normalization and the claim that HDM predicts too high an amplitude. Attention to nonlinear effects reduces the amplitude by about a factor of five from that generally assumed. Our normalization *does* imply strong recent evolution in the galaxy luminosity function, but that seems to be just what is observed (Maddox *et al.* 1990).

LARGE SCALE STREAMING VELOCITIES

The next major objection to HDM arose from Kaiser (1983). He argued that HDM universes were ruled out, because they predicted large–scale streaming velocities on supercluster scales which were thought not to exist. However, in the years following strong evidence arose that there in fact *are* large–scale streaming motions in our universe.

It would seem that an objective response to this observation would be to count this prediction (made when the observations seemed to indicate the opposite) as strong evidence supporting that theory. Instead, the general response has been to try to "patch up" CDM, which was originally preferred because it predicted low velocities.

Based on the above, I believe that the situation is muddled. Neither CDM or HDM theory can be said to agree extremely well with data, but both do far better than a priori model should be expected to, unless it is getting close to the truth.

TOPOLOGY OF LARGE–SCALE STRUCTURE

Another constraint emerges from the study of the topology of superclusters and voids (Melott 1990). Numerical models indicate that both HDM and CDM structures are partially spongelike, with interconnecting superclusters and voids. The CDM model has a weak shift in the direction of clumpiness, or isolated clusters, which is stronger with decreasing bias. The HDM model predicts a weak shift in the direction of "bubbliness," or isolated voids. Analysis of the observations to data cannot definitively rule out one or the other, but must be counted as agreeing better with CDM. When better data become available, we believe a stronger result can be obtained. The new CfA survey (de Lapparent *et al.* 1990) should be especially useful for this.

CONCLUSIONS

Taken together, there does not seem to be a clear preference for HDM or CDM. CDM numerical models produce too low a correlation amplitude. If this is adjusted by biasing, then large–scale streaming velocities are far too small. Both models can produce structures that visually resemble the observations. HDM easily incorporates and in fact was shown to predict both large voids and substantial large–scale streaming velocities well in advance of observations. If nonlinear effects are included, some galaxy formation can begin early enough in either theory to account for the existence of QSOs, even with the so–called "nonlinear era" beginning at $Z_{NL} = 0$. Nonlinear effects are important in HDM theories when the fluctuations amplitude is below unity, and must be taken into account. The microwave background fluctuation from the Sachs–Wolfe effect are a few times 10^{-6}, and not ruled out by current observations.

Given the state of observational data, and the relative crudeness of our predictions, it seems that both HDM and CDM with adiabatic perturbations have problems. Theories based on cosmic strings, global texture, etc. are often so poorly specified as to represent a moving target. We must remain open to a variety of possibilities, including some that may require considering entirely new approaches to galaxy formation.

ACKNOWLEDGEMENTS

I am grateful for research support from NASA (NAGW–1288 and NAGW–1793) and NSF (AST-8911327).

REFERENCES

Bond, J.R., Efstathiou, G., Lubin, P.M. and Meinhold, P.R. 1990 *Phys. Rev. Lett.* in press.

Centrella, J.M., Gallagher, J.S., Melott, A.L. and Bushouse, H.A. 1988 *Ap.J.* **333**, 24.

de Lapparent, V., Geller, M. and Huccha, J. (1990) *Ap.J.* in press and references therein.

Doroshkevich, A.G., Kotok, E.V., Novikov, I.D., Polyudov, A.N., Shandarin, S.F. and Sigov, Yu S., 1980 *M.N.R.A.S.* **192**, 321.

Dresler, A. 1991 this book.

Kaiser, N. 1982 Ph.D. thesis, University of Cambridge.

Kaiser, N. 1983 *Ap.J. Lett.* **273**, L17.

Maddox, S.J., Sutherland, W.J., Efstathiou, G., Loveday, J. and Peterson, B.A. 1990 *M.N.R.A.S.* **247** 1p.

Melott, A.L. 1990 *Physics Reports* **193**, 1.

Scherrer, R.J., Melott, A.L., and Shandarin, S.F. *Ap. J.* 1991 in preparation.

Shandarin, S.F. and Zel'dovich, Ya.B. 1989 *Rev. Mod. Phys.* **61**, 1895.

Weinberg, D.H. and Gunn, J.E. 1990 *M.N.R.A.S.* in press.

Zabotin, N.A. and Naselski, P. 1982 *Astron. Zh.* **59**, 447.

COSMIC STRINGS, NEUTRINOS AND THE LARGE-SCALE STRUCTURE OF THE UNIVERSE

Robert H. Brandenberger
Department of Physics
Brown University
Providence, RI 02912, USA

ABSTRACT

A cosmological model in which the energy density perturbations are seeded by cosmic strings and in which the dark matter is hot (e.g. neutrinos with a mass of $25h_{50}^2 eV$) produces large-scale structure dominated by planar overdensities with mean radius and separation determined by the comoving horizon at the time of equal matter and radiation (about $40h_{50}^{-2} Mpc$). These planar overdensities are the consequence of velocity perturbations in the wake of long rapidly moving strings. As a consequence of the dark matter being hot, clustering about cosmic string loops is suppressed, leading to little nonlinear structure between the wakes.

1. INTRODUCTION

This is a summary of recent work performed in collaboration with Leandros Perivolaropoulos and Albert Stebbins[1,2]. The motivation comes from recent observations which indicate that there is more large-scale structure than predicted in the most popular theoretical model of structure formation. In particular, there is evidence for a peak in the power of large-scale structure on a length scale of between 40 and 100 Mpc. The structures on this scale seem to be planar or filamentary. Some of these results have been summarized at this meeting by Neta Bahcall, Alan Dressler, Margaret Geller, Nick Kaiser and Alex Szalay.

An obvious question for theoretical cosmologists is to determine whether there are theories of structure formation which produce both a distinguished length scale for large-scale structure and planar topology. The "standard" cold dark matter model[3)] predicts a change in slope of the power spectrum at a scale of about 50 Mpc, but no maximum. At least in the linear regime (which should give good results on large scales), there is no mechanism which produces planar structures.

It is not surprising that a scale of the order 50 Mpc should be special in all models since it corresponds to the comoving horizon at t_{eq}, the time of equal matter and radiation. However, it is very interesting that in a model in which cosmic strings seed the density perturbations and in which the dark matter is hot (e.g. $25h_{50}^2 eV$ neutrinos) there is a mechanism which produces planar perturbations, and that the perturbations with a length scale given by the comoving horizon at equal matter and radiation are both the most numerous and the most prominent (i.e. thickest).[1,2)] The mechanism responsible for forming the large-scale structure in the cosmic string model is accretion onto "wakes"[4,5], velocity perturbations which form in the wake of long straight moving strings[6].

Recently, global textures have been suggested as the origin of structure in the Universe[7]. This model also has the comoving horizon at t_{eq} as a scale distinguished in a similar way as in the cosmic string model. However, the predicted topology on large scales differs.

It is important to note that in the context of models in which nonadiabatic seeds (e.g. cosmic strings) seed structures, hot dark matter (and neutrinos in particular) can be resurrected as a viable dark matter candidate[8]. I will explain the reason in the third part of this short review.

2. COSMIC STRING WAKES

Cosmic string wakes are caused by the nontrivial structure of space near a long straight string. Space is locally flat, but globally conical in the "plane" perpendicular to the string[9]. The deficit angle is $8\pi G\mu$, where μ is the mass per unit length in the string.

A long straight string moving with relativistic speed v through the background matter and radiation bath will induce a velocity perturbation in its wake. The easiest way to see this is to consider the frame in which the string is at rest and the particles are moving. Their trajectories will be bent behind the string, i.e. they will receive a velocity perturbation $8\pi G\mu v$ towards the plane in the wake of the string. This initial velocity perturbation will form a planar density perturbation (a "wake") which will grow by gravitational instability. The planar dimensions of the wake are given by the curvature radius of the long string, and by the distance the string will move in one Hubble expansion time. The corresponding comoving distances are both of the order $tz(t)$, where t is the time of formation, and $z(t)$ is the corresponding redshift.

Wakes created before t_{eq} dissipate as a consequence of radiation pressure. It is now easy to argue heuristically that the comoving horizon at t_{eq} will be the distinguished scale of the wake network. The density contrast in wakes created after t_{eq} increases as $t^{2/3}$, the usual linear growth rate. The initial density contrast is independent of the time of formation. Hence, the wakes formed at the earliest time, i.e. at t_{eq}, have the largest density contrast today. They are also the most numerous, since according to the scaling solution for the cosmic string network[10], the number of long string segments per horizon volume is constant. An improved argument to show that the earliest wakes are the most prominent will be sketched below.

3. RESURRECTING HOT DARK MATTER

Free streaming is the crucial effect which distinguishes hot dark matter from cold dark matter. Particles which have large thermal velocities at t_{eq}, the time when structure formation starts, cannot be bound on small scales until very late times.

The minimal comoving length on which dark matter with velocity $v(t)$ can be bound at time t is

$$\lambda_J(t) = v(t)z(t)t \ ,$$

the free streaming length. For $t > t_{eq}$, $\lambda_J(t)$ decreases as $t^{-1/3}$. This leads to a significant difference in the fortunes of hot dark matter for structure formation between models based on primordial adiabatic perturbations (e.g. the "standard" cold dark matter model) and theories in which nonadiabatic seeds (e.g. cosmic strings) form the structures.

In models with primordial adiabatic perturbations of dark matter, perturbations on a scale smaller than the free streaming length at t_{eq} are erased. Since the velocity of $25h_{50}^2 eV$ neutrinos at t_{eq} is about $v_{eq} \simeq 0.1$, no structures on scales smaller than a few Mpc survive. This is very hard to reconcile with observations. In contrast, in models with nonadiabatic seeds, neutrinos can cluster to form small scale structures as long as the seeds survive sufficiently long. Growth is not prevented but only delayed. Hence, many of the objections against hot dark matter disappear when considering models with nonadiabatic seeds.

4. COSMIC STRINGS AND STRUCTURE FORMATION

Cosmic string evolution simulations[11)] show that the network of infinite strings satisfies a scaling solution which means that it looks statistically the same at all times when scaled to the horizon distance t. Hence, at all times $t > t_{eq}$, there will be a couple of long string segments crossing each horizon volume, plus a distribution of small loops. Loops can in principle seed structures by spherical accretion whereas the long string segments will lead to large-scale structure formation by the wake mechanism sketched above.

In models with hot dark matter, accretion onto small cosmic string loops is suppressed by free streaming. Hence, almost all of the mass of the Universe which is bound in nonlinear structures is formed in wakes. According to the recent cosmic string simulations[11)] there are very few large loops which could seed structures in the regions between the wakes. Hence, the cosmic string model with hot dark matter predicts that the large-scale structure should be dominated by a network of wakes with planar dimensions and mean separation of the order of the comoving horizon at t_{eq}, with voids between them. A significant percentage of the dark matter has not clustered at all. Thus, in principle this model can reconcile a flat Universe with observations which indicate $\Omega < 1$ in clusters.

The thickness of the wakes can be calculated in various ways. The simplest way is to use a modified Zel'dovich approximation[1)] and to follow the time dependence of the physical height $h(t)$ of a dark matter particle above a given wake. The hot nature of dark matter can be taken into account by letting the gravitational attraction force start only at the time when $h(t)$ becomes equal to the physical free streaming length. We can then calculate the time when $h(t)$ starts to decrease (from then on the dark matter particle is bound to the wake). We find that a region of comoving thickness

$$v_{eq} t_{eq} z(t_{eq}) \simeq 4Mpc$$

starts to be bound at the same time. Demanding that this time be before a redshift of 1 leads to a lower bound on $G\mu$, $G\mu > 5 \cdot 10^{-7}$, which is compatible with the known upper bounds from millisecond pulsar timing. An improved calculation of the thickness of wakes uses a full phase space analysis[2)].

5. DISCUSSION AND SUMMARY

The cosmic string model of structure formation with hot dark matter leads to the prediction that the large-scale structure of the Universe should be dominated by a network of planar wakes with planar dimensions and mean separation given by the comoving horizon at t_{eq}. The exact length is determined by the level of the scaling solution for the cosmic string network, i.e. the number of long string segments crossing each horizon volume. According to the latest simulations, the length is unfortunately rather on the low side to be compatible with the scale which emerges from the recent observations. It is unclear whether this is a fundamental problem for the cosmic string model or merely an artifact of incomplete simulations.

The model has other nice features: the regions between the wakes contain hardly any nonlinear structures - they are voids. In addition, a large fraction of the dark matter is still smooth, thus making a flat Universe compatible with the low value of Ω determined by dynamical measurements in clusters. Again, however, the exact fraction of the dark matter which remains smooth depends on numerical results from the cosmic string evolution simulations.

An open problem is the small scale structure in this model. Determining the answer to this question will have to include hydrodynamical and nonlinear gravitational effects. Work on these issues is in progress.

I thank Leandros Perivolaropoulos and Albert Stebbins for an enjoyable collaboration, and the Alfred P. Sloan Foundation for a fellowship. This research has been supported in part by DOE grant DE-AC02-76ER03130, Tasks A & K.

REFERENCES

1. R. Brandenberger, L. Perivolaropoulos and A. Stebbins, *Int. J. of Modern Phys.* **A5**, 1633 (1990).

2. L. Perivolaropoulos, R. Brandenberger and A. Stebbins, *Phys. Rev.* **D41**, 1764 (1990).

3. G. Blumenthal, S. Faber, J. Primack and M. Rees, *Nature* **311**, 517 (1984).

4. T. Vachaspati, *Phys. Rev. Lett.* **57**, 1655 (1986).

5. A. Stebbins, S. Veeraraghavan, R. Brandenberger, J. Silk and N. Turok, *Ap. J.* **322**, 1 (1987).

6. J. Silk and A. Vilenkin, *Phys. Rev. Lett.* **53**, 1700 (1984).

7. N. Turok, *Phys. Rev. Lett.* **63**, 2625 (1989).

8. R. Brandenberger, N. Kaiser, D. Schramm and N. Turok, *Phys. Rev. Lett.* **59**, 2371 (1987).

9. A. Vilenkin, *Phys. Rev.* **D23**, 852 (1981).

10. A. Vilenkin, *Phys. Rep.* **121**, 263 (1985).

11. A. Albrecht and N. Turok, *Phys. Rev. Lett.* **54**, 1868 (1985); D. Bennett and F. Bouchet, *Phys. Rev. Lett.* **60**, 257 (1988); A. Albrecht and N. Turok, *Phys. Rev.* **D40**, 973 (1989); B. Allen and E.P.S. Shellard, *Phys. Rev. Lett.* **64**, 119 (1990).

Cosmic Strings in the Sun?

Jin Wang
Department of Astrophysics
University of Illinois
Urbana, Il 61801

Abstract

We apply Chudnovsky and Vilenkin's idea [1] of possible existence of cosmic strings in the Sun. We did the stellar evolution caculation for one solar mass star with solar age (that is Sun). Stellar evolution with cosmic strings at solar age gives us radius and luminocity of the star which are in contradiction with observation.

References:

1. Chudnovsky, E., Vilenkin, A., Phys. Rev. Lett. 61(1988)1043.

DARK MATTER OR NEW PHYSICS?

Demosthenes Kazanas
Laboratory for High Energy Astrophysics
NASA/Goddard Space Flight Center, Greenbelt, MD 20771

and

Philip D. Mannheim
Department of Physics
University of Connecticut, Storrs, CT 06269

ABSTRACT

The deviation of galactic rotation curves from their expected Keplerian fall-off with distance as based on the observed distribution of light, has prompted the suggestion, by a small number of authors, that this phenomenon signifies a drastic departure of certain laws of physics at large scales from their smaller scale forms. A number of proposals is reviewed with particular attention being given to a locally scale invariant theory of gravity proposed by the authors. This latter theory and several of its consequences and comparisons to observation are presented.

INTRODUCTION

Several of the articles in these proceedings have dealt with the question of the need for dark matter to account for the dynamics of galaxies and clusters of galaxies. Such a desire for dark matter is not without precedent in the history of astronomy. In the middle of last century observations indicating that the motion of the planet Uranus deviated from that expected on the basis of the influence of the then known planets led to the supposition of the existence of a then unobserved planet (Neptune), with calculable orbit elements, which was subsequently discovered. Further, at the beginning of the twentieth century it was found that the perihelion of the planet Mercury exhibited an annoying, unexplained but nonetheless persistent precession. Though this feature could be accounted for in terms of Newtonian gravity by assuming that the Sun had a sufficiently large enough quadrupole moment, as is well known the resolution of this issue came only with the revision of Newtonian gravity and its replacement by General Relativity. A significant contribution to the acceptance of this latter interpretation came from the fact that it was a by-product of a new principle of physics, namely that of general covariance. Thus in only one of the two cases discussed above did the answer to the problem turn out to be the presence of "dark" matter.

With the above in mind, it should not be surprising that the issue of galactic rotation curves has prompted explanations other than that of dark matter, and encouraged speculation that a revision of the Newtonian law of force might be in order for large scale motions in the Universe. However, since the Newtonian potential is an approximation to the g_{oo} component of the Schwarzschild metric at large distances, abandoning the Newtonian force also implies abandoning General Relativity, a course that not too many people are willing to take, thus accounting for the small number of alternative gravitational proposals presented over the years. However, the data are sufficiently striking, and the dark matter alternatives sufficiently numerous and unconstrained, that some physicists and astrophysicists have dared to pursue this (professionally) dangerous path. For the most part (see 1 and 2 below) such attempts have been rather *ad hoc* motivated by the search for a resolution of the issue of dark matter in galactic dynamics. However, to quote from Binney and Tremaine (1987), "*If a new theory of gravity is required, it will ultimately be accepted because of its beauty and unifying properties rather than because it eliminates the need for dark matter.*". Such, in fact, is the character of the last alternative that will be discussed here which will occupy most of the space allocated for this review.

THE ALTERNATIVES

1. Modification of the Newtonian Potential

The earliest reference we are aware of to an alternative to Newtonian gravity as a means of accounting for the discrepancy between the observed and the theoretical galactic rotation curves is that of Finzi (1963). He observed that this discrepancy would be lifted if the gravitational force were to fall off more slowly than $1/r^2$ at large distances. He in fact suggested as a trial possibility the form

$$F = \frac{km_1m_2}{\rho^2}\left(\frac{\rho}{r}\right)^{3/2}, \quad V = -2\frac{km_1m_2}{\rho}\left(\frac{\rho}{r}\right)^{1/2}, \quad (\rho \ll r) \tag{1}$$

where ρ is a length scale he required to be of the order of 0.5 *kpc*. Finzi pointed out that such a modification would solve a number of problems ranging form the dynamics of clusters of galaxies to estimates of element production in our own galaxy. He acknowledged however that such a modification would not be compatible with General Relativity.

More recently Sanders (1984), motivated by theories attempting to unify gravity with the other fundamental interactions, suggested that besides the massless graviton a massive field might exist and be an additional carrier of the gravitational force. This field, of coupling constant β and range r_o, would then add a Yukawa-type term to the total gravitational potential resulting in

$$V = -\frac{GM}{(1-\beta)r}(1-\beta e^{-r/r_o}), \qquad (\beta, r_o > 0). \tag{2}$$

choice of the parameters β and r_o one can then ensure that the galactic rotation curves are "flat" in the observed range of parameters.

2. Modification of Newtonian Dynamics (MOND)

An alternative approach to answering the question of galactic dynamics is that employed by Milgrom (1983). Instead of changing the law of gravity Milgrom has opted to generalize Newton's second law suggesting that the inertial force should be not simply proportional to the acceleration but a more general function of it which only reduces to Newton's 2nd law in the limit of large accelerations. In particular he suggested the form

$$\boldsymbol{F} = \boldsymbol{a}\mu(a/a_o)m_g = \boldsymbol{a}\mu(x)m_g \tag{3}$$

$$\mu(x >> 1) \simeq 1 \ , \qquad \mu(x << 1) \simeq x$$

to replace $\boldsymbol{F} = \boldsymbol{a}m_g$ as the law providing the acceleration $\boldsymbol{a}$ of an object of gravitational mass m_g moving in an arbitrary force field $\boldsymbol{F}$. The parameter a_o is a characteristic acceleration, which he determined from observation and found to be of order $cH_o \simeq 5 \times 10^{-8}$ cm s^{-2} (H_o is the Hubble constant), thus giving a Machian flavor to his approach by relating an inertial parameter to cosmological quantities.

This theory (or rather scenario) can account for the flat galactic rotation curves in a straightforward fashion: Substituting the centrifugal acceleration, $a = V^2/r$, into eq. (3) (in the $a/a_o \ll 1$ regime) and setting it equal to the Newtonian force, GMm_g/r^2, one obtains $V^4 \simeq GMa_o$, independent of the distance. In addition, one also obtains a relation between the mass and velocity dispersion for galaxies similar to that of the Tully-Fisher relation.

Milgrom's approach is successful in dealing with the observations, at least at a first approximation; it does contain however an arbitrary function, $\mu(x)$, and an arbitrary parameter, a_o, which Milgrom hoped would eventually be provided by a more fundamental theory which is presently lacking.

3. Conformal Weyl Gravity

The general reluctance of the community to seriously consider any of the above alternatives has been based mainly on the fact that all of them appear *ad hoc* devised solely for the purpose of accounting for the dynamics of galaxies. In addition, it is thought that their acceptance would likely force us to abandon General Relativity and along with it the principle of general covariance.

Considering general covariance however, it should be pointed out that this principle alone is not in fact sufficient to uniquely determine the equations of the gravitational field; rather, any theory whose action is a general coordinate scalar will obey this principle. Einstein himself was well aware of this inherent freedom and he imposed an additional constraint, namely that the resulting equations be no higher than second order, in order to uniquely fix the gravitational field will obey this principle. Einstein himself was well aware of this inherent freedom, and he imposed an additional constraint, namely that the resulting equations be no higher than second order, in order to uniquely fix the gravitational field equations. This latter demand then also fixes the action to simply be linear in the Ricci scalar, $R^{\alpha}{}_{\alpha}$, leading to the well known Hilbert-Einstein action of General Relativity

$$I_{HE} = (1/16\pi G)\int d^4x(-g)^{1/2}R^{\alpha}{}_{\alpha} \ . \tag{4}$$

Since the action has to be dimensionless, a constant of dimension $(length)^{-2}$ has to be introduced into the theory, and Newton's constant precisely serves this purpose as can be seen from eq. (4), with its associated length scale being the so-called Planck length $l_P \propto G^{1/2}$.

Given the action of eq. (4), gravity then stands alone among the modern theories of fundamental interactions in that it is the only one with a dimensionful coupling constant in its action (a property which, incidentally, is directly responsible for the non-renormalizability of the quantized theory). The other fundamental interactions all have actions with dimensionless coupling constants and derive all their masses from dynamics, i.e. from spontaneous symmetry breaking. Moreover, Lorentz invariance alone (the flat space analog of general covariance) does not suffice to uniquely fix the form of the fundamental actions (a situation similar to that discussed above concerning gravity); so these theories then appeal to ***principles of local invariance***, the so-called gauge invariances, to precisely restrict the available freedom and uniquely fix the relevant actions.

Given the above considerations, and motivated by the way that local gauge invariance uniquely restricts the form of allowed actions of Lorentz invariant theories in flat space-time, Mannheim (1990) and Mannheim and Kazanas (1989) have considered the analogous principle of local scale (conformal) invariance to restrict the form of the gravitational action, thus abandoning Einstein's demand of second order equations, while still retaining his principle of general covariance. This latter principle, which demands that the laws of physics be invariant under any and all local stretchings of the geometry $g_{\mu\nu}(x) \rightarrow \Omega^2(x)g_{\mu\nu}(x)$ leads indeed to a unique action for the gravitational field which has the form

$$I_W = -\alpha\int d^4x(-g)^{1/2}C_{\lambda\mu\nu\kappa}C^{\lambda\mu\nu\kappa} \tag{5}$$

where $C_{\lambda\mu\nu\kappa}$ is the conformal or Weyl tensor and α is a universal dimensionless coupling constant.

The equations of motion for the gravitational field are derived from the action of eq. (5) by the standard methods of functional variation with respect to $g_{\mu\nu}$, and lead to equations of the form $4\alpha W_{\mu\nu} = T_{\mu\nu}$ where $W_{\mu\nu}$ is to replace the familiar Einstein tensor and $T_{\mu\nu}$ is the stress energy tensor. These equations are the analog of the Einstein equations in this alternative theory of gravity, hereafter referred to as *Weyl gravity*. Due to their complicated form they will not be presented here; they can be found for instance in Mannheim & Kazanas (1989). We will nonetheless discuss some of their general features. In general they are fourth order, non-linear equations for the metric coefficients $g_{\mu\nu}$. They consequently do not reduce to those of Einstein in any obvious limit and hence this theory does not trivially reduce to that of Einstein gravity, which immediately raises the question of its viability. This issue is discussed more extensively elsewhere in these proceedings (Mannheim and Kazanas 1990); we only point out here that in order that this theory not contradict experiment it suffices that its ***solutions*** (but not necessarily its equations) reduce to those of General Relativity. Indeed, inspection of the equations indicates that they involve powers of the Ricci tensor $R_{\mu\nu}$ and its second derivatives. Noting that the vacuum ($T_{\mu\nu} = 0$) Einstein equations reduce to $R_{\mu\nu} = 0$, any vacuum solution of Einstein gravity would then also be a vacuum solution to the equations of Weyl gravity, thus directly and immediately guaranteeing that the theory would pass the classic static tests of General Relativity. On the other hand, the vacuum equations of Weyl gravity, $W_{\mu\nu} = 0$, may be realized in ways other than simply by having the Ricci tensor vanish, thus leading to solutions of Weyl gravity more general than those of Einstein gravity (even while being compatible with them). To explore the situation fully however one needs to obtain exact solutions to the fourth order theory.

Since most of our intuition with the astrophysical aspects of General Relativity has been developed from the study of the static, spherically symmetric, Schwarzschild problem, Mannheim & Kazanas (1989) set out to investigate the same problem within Weyl gravity. The most general static, spherically symmetric line element can be written as

$$ds^2 = -B(r)dt^2 + A(r)dr^2 + r^2 d\Omega \tag{6}$$

where $A(r)$, $B(r)$ are arbitrary functions of the radial coordinate r. The conformal symmetry then allows one to set $B(r) = 1/A(r)$, and with this simplification the rr-component of the gravitational tensor is then found to take the exact non-perturbative form

$$\begin{aligned} W^{rr}/B(r) &= B'B'''/6 - B''^2/12 - (BB''' - B'B'')/3r \\ &-(BB'' + B'^2)/3r^2 + 2BB'/3r^3 - B^2/3r^4 + 1/3r^4 \,. \end{aligned} \tag{7}$$

In the vacuum, the above expression is set equal to zero yielding an equation for $B(r)$, whose complete and exact solution is given by (Mannheim & Kazanas 1989)

$$-g_{00} = 1/g_{rr} = B(r) = 1 - \beta(2 - 3\beta\gamma)/r - 3\beta\gamma + \gamma r - kr^2 \tag{8}$$

where β, γ, and k are three appropriate dimensionful integration constants.

It is apparent from the form of eq. (8) that the above solution reduces to that of Schwarzschild when $\gamma = 0, \quad k = 0$, as it should according to the above discussion. Interestingly, while the solution to the same problem in Einstein gravity gives an asymptotically flat solution, Weyl gravity provides us with a solution which is naturally not asymptotically flat; i.e. the theory even in its vacuum solution does address the issue of the large scale structure of the Universe. When $\gamma = 0, \quad k \neq 0$ the solution reduces to another well known solution of General Relativity, namely the Schwarzschild-de Sitter one. Interestingly, the latter solution while a vacuum solution in Weyl gravity, can only be obtained in Einstein gravity in the presence of a non-vanishing cosmological constant. As was pointed out by Mannheim (1990), a de Sitter geometry can in fact be obtained in Weyl gravity without the use of a cosmological constant. This may at first appear strange. However, it suffices to note that the de Sitter metric is a purely geometric concept; its association with the cosmological constant comes only through the dynamics of the Einstein equations. Since they have been replaced by those of Weyl gravity this association simply does not hold anymore. Concerning the question of the cosmological constant, discussed in this meeting in the context of both inflationary models and galaxy formation (see the contributions by Schramm and Peebles respectively), a few comments are in order since Weyl gravity also addresses these issues. Specifically, since conformal invariance does not allow the presence of any mass scales in the theory, a cosmological term with dimension $(length)^{-4}$ must be identically zero. Indeed, in the conformal theory the trace of the stress energy tensor vanishes identically, $(T^{\mu}{}_{\mu} = 0)$, thus forbidding any contribution to $T^{\mu\nu}$ from a cosmological term. As for the inflationary scenario, it has been devised to account for the discrepancy between the natural gravitational (Planck) scale and the scales associated with the observed Universe (in particular between the Planck time $t_P \propto G^{1/2}$ and the age of the Universe $t_U \simeq 1/H_o$). Since Newton's constant (and the associated Planck scale) has no place in Weyl gravity, such a discrepancy is meaningless to start with and no such account appears necessary anymore.

Because the terms associated with the parameter γ are not obviously identifiable with any known geometry, in order to gain insight into their nature we resort to a limiting procedure similar to that used above. Setting, for example, $k = 0, \quad \beta = 0$ (or considering sufficiently large distances such that the β dependent terms can be ignored), one can write the resulting metric in isotropic coordinates via the coordinate transformation

$$\rho = \frac{4r}{2(1+\gamma r)^{1/2} + 2 + \gamma r} \qquad \text{and} \qquad \tau = \int R(t)dt$$

with the metric then taking the form

$$ds^2 = \frac{1}{R^2(\tau)} \frac{[1-\rho^2\gamma^2/16]^2}{[(1-\gamma\rho/4)^2]^2} \left[-d\tau^2 + \frac{R^2(\tau)}{[1-\rho^2\gamma^2/16]^2}(d\rho^2 + \rho^2 d\Omega) \right] \tag{9}$$

so that in fact it becomes conformal to a Robertson-Walker (RW) metric with arbitrary scale factor $R(\tau)$ and 3-space curvature $K = -\gamma^2/4$. Then, in analogy with the form of the metric with $\gamma = 0$, (i.e. the one where $B(r) = 1-2\beta/r-kr^2$ which represents the geometry of a static spherically symmetric object embedded in a de Sitter background), we see that the metric with $\gamma \neq 0$, (which interpolates smoothly between the Schwarzschild and the RW geometries), seems to represent the geometry of a spherical object embedded in a space conformal to RW with 3 space curvature $K = -\gamma^2/4$. (It should be noted that no such type of solution is known within General Relativity). Since $\gamma^2/4$ is the curvature of the background space, it seems reasonable to associate γ with the inverse Hubble length $R_H = c/H_o$, thus obtaining an estimate $\gamma \simeq H_o/c = 5 \times 10^{-29}(H_o/50)\ cm^{-1}$, H_o being the value of the Hubble constant in units of 50 $km\ s^{-1}\ Mpc^{-1}$. With this interpretation for γ and the associated estimate of its magnitude one can then examine its possible observational consequences.

Because of the association of the parameter β with the Schwarzschild radius, its numerical value is $2\beta \simeq 3 \times 10^5\ M_o\ cm$, M_o being the mass of the object in solar masses. This value then, along with the estimated magnitude of γ, implies for objects with masses substantially smaller than the mass of the Universe that $\beta\gamma \ll 1$, (so that the $\beta\gamma$ term can then be neglected) to yield for the metric the approximate form $B(r) \simeq 1 - 2\beta/r + \gamma r$. The Newtonian term thus dominates at small distances and provides dynamics in accordance with observation, with the effects of the additional linear term becoming important only at larger distances. On the basis of these arguments alone one would then expect substantial deviations (i.e. of $O(1)$) from Newtonian dynamics at distances

$$r \simeq (2\beta/\gamma)^{1/2} \simeq 7.7 \times 10^{16}\ M_o^{1/2}(H_o/50)^{-1/2}\ \ cm. \tag{10}$$

For a solar mass this distance is well outside the solar system and hence inaccessible to observation; however for the mass of a galaxy, $M_o \simeq 10^{11}$, this distance is $\simeq 2.4 \times 10^{22}\ (H_o/50)^{-1/2}\ cm$ or $\simeq 8\ kpc$, i.e. comparable to the size of the galaxy. This is also roughly the characteristic distance at which the galactic rotation curves deviate from the values expected on the basis of the observed luminous matter and Newtonian dynamics (Rubin, these proceedings). It is rather striking that such an effect could in principle be "predicted" by theory.

The above reasoning then suggests that the deviation of galactic rotation curves from the behavior expected on the basis of the Newton-Einstein theory is due to the curvature of the cosmological background in which the galaxies are embedded. At first sight it appears surprising that cosmological effects could become important at galactic length scales; this is because our intuition is based on the Schwarzschild-de Sitter solution (i.e. the one with $\gamma = 0,\ k \neq 0$), the only previously known solution representing a source embedded in a non-flat background. In this case, setting $k \simeq R_H^{-2}$, one finds that for a galaxy the cosmological kr^2 term would compete with the Newtonian $1/r$ one at a distance $r \simeq 1.4 \times 10^{24}[M_{11}(R_H/10^{28})^{-2}]^{1/3}$, i.e. at a distance much larger than a typical galactic size.

In fact, the relation of eq. (10) suggests that embedding effects may be significant (i.e. of order one) not only galactically but also on a multitude of scales which depend on the masses of the objects considered. Though a precise theory has yet to be developed, one could argue qualitatively that background curvature effects might prevent a perturbation from collapsing to a bound object if its size is larger than that given by eq. (10). This argument would then provide a scale for the formation of structure by gravity and would imply a mass-radius relation for objects in virial equilibrium given precisely by eq. (10) (Incidentally, stars are not expected to obey such a relation because of their having radiated away much of their internal energy in collapsing to much smaller sizes).

Given that the Newtonian potential ϕ is related to the metric via $g_{oo} \simeq 1 + 2\phi/c^2$, and given that the virial velocity projected along the line of sight v_r can be expressed in terms of ϕ through $v_r \simeq c(\phi/3)^{1/2}$ (for an isotropic velocity distribution), if the above arguments are indeed correct and if virially supported objects do obey the relation given by eq. (10), there should also exist equivalent radius-velocity-dispersion (along the line of sight) and mass-velocity-dispersion relations of the form

$$v_r \simeq c(\gamma r/6)^{1/2} \simeq 1.5 \times 10^5 \ (r/pc)^{1/2} \ (H_o/50)^{1/2} \ \ cm \ s^{-1} \tag{11}$$

and

$$v_r \simeq 2.4 \times 10^4 M_o^{1/4} (H_o/50)^{1/4} \ \ cm \ s^{-1} \ . \tag{12}$$

In fact, observationally there does exist a relation essentially identical to eq. (11) both in slope and normalization for molecular clouds (Solomon et al. 1987; see also figure 2); with the equivalent observed mass-velocity-dispersion relation of eq. (12) being known as the Tully-Fisher or Faber-Jackson relation. In figure 2 we have also plotted along with eq. (11) (solid line), the same relation as derived for molecular clouds (dashed line, Solomon et al. 1987), two extreme points of a similar IR survey of galaxies (Rubin et al. 1985), as well as the point corresponding to the Coma cluster of galaxies (Mushotzky, private communication). As is apparent, the objects of all these diverse classes obey *the same* (both in slope and normalization) Tully-Fisher relation, namely that given by eq. (11) which is in fact derived from the mass-radius relation of eq. (10), shown in figure 1. In these days of a major effort toward the understanding of the formation of galaxies and their clustering, eqs. (10) and (11) and the associated observations suggest that our ideas on structure formation may have to be expanded to incorporate globular clusters and molecular clouds, benign and insignificant as these objects may appear on cosmic scales. A further testimony to the importance of eq. (10) is its implication that all objects obeying it (including the Universe) should have the same surface density, a fact which, while not generally appreciated, is observationally supported and has been noticed to hold for globular clusters and galaxies (Fall and Rees 1977).

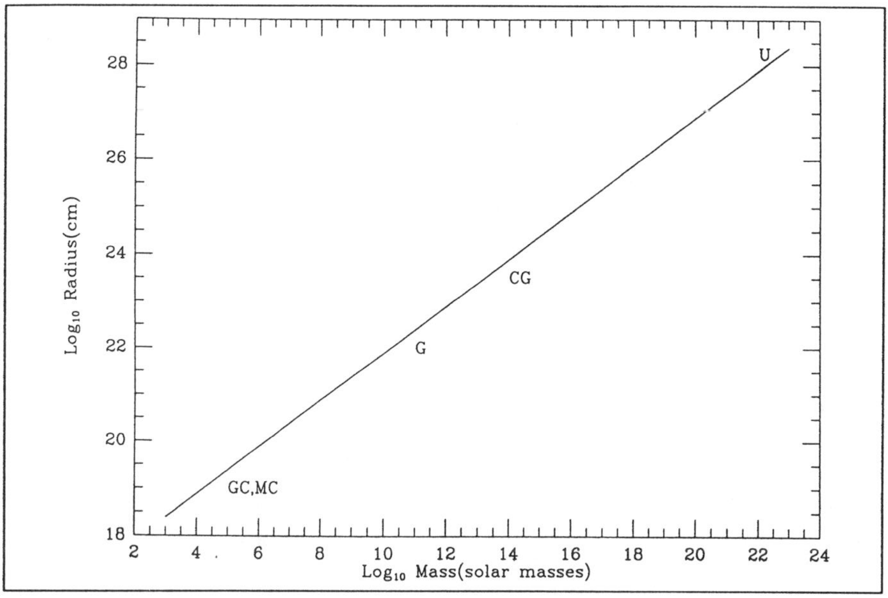

Figure 1. The mass-radius relation of eq. (10). The symbols stand for: MC molecular clouds, GC globular clusters, G galaxies, CG clusters of galaxies, U the Universe.

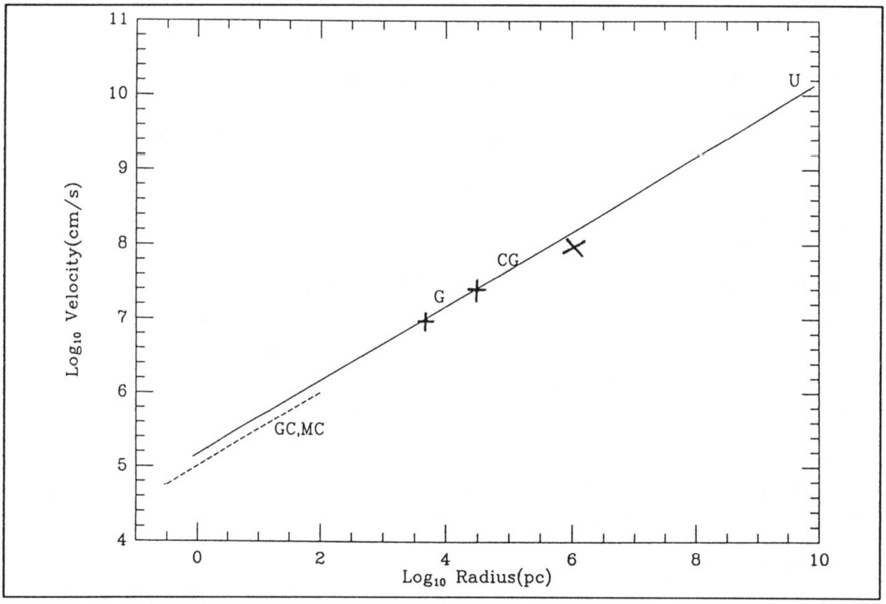

Figure 2. The radius - velocity dispersion relation. The symbols are as in fig. 1. The dashed line is the relation derived for molecular clouds (Solomon et al. 1987). The crosses are from IR data for galaxies (Rubin et al. 1985) and the × is the Coma cluster.

With respect to the relation of eq. (10) we would like to add that when extrapolated to the mass of the electron, $m_e \simeq 10^{-60}\ M_o$, it yields a radius roughly equal to the classical electron radius $r_0 \simeq 10^{-13}$ cm. The reason for this coincidence (reminiscent of Dirac's large number hypothesis) is not at this point apparent, though it is suspected that it may be related to the special Newtonian-like $1/r$ structure of the point charge (Reissner-Nordstrom type) solution in Weyl gravity. (For further details see the contribution of Mannheim and Kazanas presented in these proceedings).

In conclusion, we would like to point out that numerical simulations (Christodoulou 1990) of disks with a potential per particle similar to that of eq. (8) have indicated that such disks are stable against non-axisymmetric perturbations and bar formation. This fact is of particular importance since stability arguments (Ostriker and Peebles 1973) provided the original motivation for the consideration of "dark matter" in galaxies. With the use of the metric of eq. (8) we can also make (heuristic) contact with Milgrom's parameter a_o: taking the derivative of g_{oo} with respect to r (times c^2 to take care of the units) one obtains besides the Newtonian acceleration $2\beta c^2/r^2$, the additional acceleration $\gamma c^2 \simeq c^2/R_H = cH_o \simeq a_o$, so that Milgrom's a_o parameter is in some sense simulating the effects of our linear γ term, with a_o then perhaps being an effective parameter which should physically be associated with the the dynamics of embedding of an object in the proper cosmological background.

The work of one of us (P. D. M.) has been supported in part by the Department of Energy under grant no. DE - AC02 -79ER10336.A.

REFERENCES

Binney, J. and S. Tremaine, 1987, *Galactic Dynamics*, Princeton, p. 637.
Christodoulou, D., 1990, *Ap. J.*, in press.
Fall, S. M. and M. J. Rees, 1977, *M. N. R. A. S.* **181**, 37.
Finzi, A., 1963, *M.N.R.A.S.* **127**, 21,
Mannheim, P. D. and D. Kazanas, 1989, *Ap. J.* **342**, 635.
Mannheim, P. D., 1990, *Gen. Rel. and Gravit.* **22**, 289.
Mannheim, P. D. and D. Kazanas, 1990, these proceedings.
Milgrom, M., 1983, *Ap. J.* 270, 365, 371, 384.
Ostriker, J. P. and P. J. E. Peebles, 1973, *Ap. J.* **186**, 467.
Rubin, V. C. et. al., 1985, *Ap. J.* **289**, 81.
Sanders, R. H., 1984, *Astron. Ap.* **136**, L21.
Solomon, P. M. et. al., 1987, *Ap. J.* **319**, 730.

Discussion

NEMIROFF: Why doesn't M 31 disrupt our Galaxy's rotation curve?
KAZANAS: The perturbation of our galaxy by M 31 is due to tidal effects; these effects, calculated on the basis of eq. (8), fall as a function of distance (even while the potential itself grows).

CURRENT STATUS OF CONFORMAL WEYL GRAVITY

Philip D. Mannheim
Department of Physics
University of Connecticut, Storrs, CT 06269
and
Demosthenes Kazanas
Laboratory for High Energy Astrophysics
NASA/Goddard Space Flight Center, Greenbelt, MD 20771

Abstract

Fourth order conformal Weyl gravity has recently been advanced by Mannheim and Kazanas (for details and bibliography see Kazanas's invited talk in the proceedings of this workshop) as a candidate alternative theory of gravity to the standard second order Einstein theory. In this short presentation we review its current status, discussing what has been achieved and what open questions remain.

Conformal gravity is the theory of gravity which as well as being general covariant is also based on the requirement of invariance of the geometry under any and all local conformal stretchings of the form $g_{\mu\nu}(x) \to \Omega^2(x) g_{\mu\nu}(x)$ and leads to the unique gravitational action $I_W = -\alpha \int d^4x(-g)^{1/2} C_{\lambda\mu\nu\kappa} C^{\lambda\mu\nu\kappa}$ where $C_{\lambda\mu\nu\kappa}$ is the conformal Weyl tensor and α is a purely dimensionless coupling constant. This action is to replace the Einstein-Hilbert action of the standard theory. While the conformal theory is a somewhat intractable nonlinear fourth order theory nonetheless we have found some quite remarkable exact nonperturbative solutions to it. In particular the vacuum metric exterior to a static, spherically symmetric matter distribution is found[1] to be of the form

$$-g_{00} = 1/g_{rr} = 1 - \beta(2 - 3\beta\gamma)/r - 3\beta\gamma + \gamma r - kr^2 \tag{1}$$

where β, γ and k are three integration constants. The metric thus recovers the standard Schwarzschild - de Sitter solution when the linear γr term is small (the de Sitter geometry is a vacuum solution in the fourth order theory with the conformal symmetry actually forbidding the presence of any cosmological term[2]), and in this limit the theory meets all the standard static tests of Einstein gravity. However, as the distance scale increases the theory starts to deviate from Newton-Einstein. In particular on galactic scales the linear term prevents the rotational velocities of stars from undergoing a Keplerian fall as a function of distance from the center of a galaxy, and thus opens up the possibility that galactic dynamics can be explainable by an alternate theory of gravity rather than by the presence of so far unobserved non-luminous dark matter. For the moment this is only a purely gross phenomenological expectation of the model, and the theory awaits a detailed calculation using a more realistic model of a galaxy than a static spherically symmetric one to see if it really can fit rotational curves.

Beyond the need for dark matter in galaxies the standard Newton-Einstein theory also requires yet more dark matter in clusters of galaxies than that contained within the individual galaxies within the cluster. This fact also accords well qualitatively with the conformal theory since the linear potential term leads to bigger and bigger deviations from the standard theory on bigger and bigger distance scales. An actual possible comparative test of the relative merits of the conformal theory and the standard dark matter scenario is in principle provided by galactic gravitational lensing, and it may even prove to be definitive once the detailed picture of gravitational lensing is calculated (a somewhat non-trivial issue since the relevant geometry is not asymptotically flat according to Eq. (1)) in the fourth order theory.

Since the conformal theory is a relativistic field theory of gravity it necessarily contains retardation and radiation reaction, so the orbit of a binary pulsar is expected to decay in the fourth order theory just as it does in the second order one. However the details still remain to be worked out to see whether or not the conformal theory can also quantitatively meet this test, this being the only non-static test of the Einstein theory that is currently available. (It is not that the conformal theory does not meet this test it is simply that the calculations have yet to be done with the binary pulsar question being the only currently available observational challenge to the fourth order theory on stellar scales).

A further astrophysical challenge to the theory is to formulate a cosmological model since a Robertson-Walker metric is a vacuum solution to the conformal theory. While this is a difficulty, it is not necessarily an insuperable one since the implied local vanishing of the energy-momentum tensor can be achievable non-trivially by a cancellation between the energy density of the matter field and the energy density of the gravitational field itself. (An explicit illustration of how this can occur is given in the dynamically induced de Sitter case discussed in Ref. (2)).

In addition to the exact fourth order analog of the Schwarzschild metric, we have also found exact fourth order analogs of the exterior Reissner-Nordstrom, Kerr and Kerr-Newman solutions. Beyond their interest simply as exact solutions, they are also interesting in that just like the analog exterior Schwarzschild metric given above they are all qualitatively different from their second order counterparts. Of particular interest is the Reissner-Nordstrom case. Here we find that the modification to the metric is not in the form of the $1/r^2$ term familiar from the second order theory, but rather it is in the form of an additional $1/r$ term, a term whose form is thus just like that of the standard Newton term itself. This feature may thus enable us to reopen the old idea of a possible electromagnetic-gravitational connection.

Beyond looking for exact solutions to the theory and identifying the urgent challenges to the theory given above, we have also had to address theoretical objections raised in the literature in the past which had been used to argue against the viability of fourth order theories. Some of the criticisms (the presence of ghosts and anomalies in the associated quantum theory) relate to the perturbative structure of the theory when quantized around flat spacetime. Now that we have the exact non-perturbative solution of Eq. (1) given above we see that solutions to the theory are far from flat, so the wisdom obtained from standard perturbation theory may thus simply not be relevant to the structure of the full theory.

A more serious concern is the connection of the exterior solution of Eq. (1) above to its associated interior solution, and arguments have been made in the literature (again based on perturbation theory however) that the exterior and interior solutions are not compatible with each other, or with Newton's law of gravity, or with positivity of the matter distribution. To address these issues we have studied[3] some new exact properties of the interior solution. Specifically we have found that in the presence of matter the metric coefficient $B(r)$ $(= -g_{00}(r))$ remarkably obeys the exact non-perturbative fourth order Poisson type equation

$$B'''' + 4B'''/r = (Br)''''/r = \nabla^4 B = f(r) \tag{2}$$

where the function $f(r)$ is related to the energy-momentum tensor according to

$$f(r) = 3({T^0}_0 - {T^r}_r)/4\alpha B(r) \tag{3}$$

For a spherically symmetric source $f(r)$, Eq. (2) may be integrated completely analytically using the Green's function $-|\mathbf{r} - \mathbf{r}'|/8\pi$ associated with the fourth order Laplacian operator ∇^4 to yield the exact exterior solution

$$B(r > R) = -\frac{1}{6}\int_0^R dr' f(r')[3r'^2 r + \frac{r'^4}{r}] \tag{4}$$

We thus see that our exterior vacuum solution of Eq. (1) above is indeed compatible with the interior solution, and that by matching Eqs. (1) and (4) at the radius R of the matter distribution we can now express the integration parameters β and γ of Eq. (1) as moments of the interior matter distribution. As we can also see, the coefficient of the Newtonian $1/r$ term of Eq. (1) is now expressed in terms of elements of the energy-momentum tensor and the fundamental coupling constant α of the initial conformal action. Thus through the coupling constant α gravity couples universally to matter in our theory as should of course be the case in any gravitational theory.

As an initially unexpected bonus we see that even though the Green's function of the fourth order theory is linear in the distance r, after integration we find that we recover not only the linear potential term but also the $1/r$ Newtonian term in Eq. (4) even though the second order Laplacian operator ∇^2 is not present anywhere in the fourth order theory. Thus while a second order Poisson equation is sufficient to generate a $1/r$ potential, we see that is not necessary, with Newton's law of gravity being valid in the fourth order theory also. While the solution of Eq. (4) shows that the Newtonian term will be obtainable for any spherically symmetric matter distribution we note that its strength is related to the fourth moment of $f(r)$ rather than to the second one as is the case in the familiar second order Einstein theory. Since this fourth moment would vanish for a delta function source, we see that in order to obtain a Newtonian potential in the fourth order theory the source must be extended rather than pointlike. While this violates our standard second order intuition (but not any observational information incidentally), it is not all that surprising since our experience with dynamical mass generation in the other fundamental interactions indicates that we should generally anyway expect elementary particles to actually be extended soliton or bag-like objects rather than pointlike ones, with the only new feature here being that curvature now also plays a role in their generation.

Since the solutions in our theory are extended according to Eq. (4) there is some concern as to whether such extended objects are compatible with positivity of the source distribution $f(r)$. To this end we have constructed an explicit though only illustrative candidate source which is positive definite everywhere. Its specific form is given by

$$f(r) = -2p\delta(r)/r^2 - (3q/2)[\nabla^2 - (r^2/12)\nabla^4][\delta(r)/r^2] \tag{5}$$

where p and q are new intrinsic parameters which characterize the source. In terms of this source we find that Eqs. (1) and (4) yield

$$\beta(2 - 3\beta\gamma) = q \quad , \quad \gamma = p \tag{6}$$

which shows that the strengths of the linear and Newtonian potential terms in our solution are in principle independent in the absence of more detailed dynamical information. While the source function given in Eq. (5) is not of course mandated as the only possible source appropriate to the theory, its very existence serves as a counter-example to the claim that in the fourth order theory Newton's law is incompatible with positivity of the matter distribution. On the contrary, our analysis shows that Newton's law is perfectly viable in the fourth order theory, and that it is only our second order intuition that needs modifying.

The work of one of us (P. D. M.) has been supported in part by the Department of Energy under grant no. DE-ACO2-79ER10366.A.

References

1. Mannheim P. D. and D. Kazanas, Exact vacuum solution to conformal Weyl gravity and galactic rotation curves, Astrophysical Journal **342**, 635 (1989).

2. Mannheim P. D., Conformal cosmology with no cosmological constant, General Relativity and Gravitation **22**, 289 (1990).

3. Mannheim P. D. and D. Kazanas, Exact interior structure of fourth order conformal Weyl gravity and the Newtonian limit, NASA preprint, October, 1990, to be published.

WEYL GRAVITY, ASTROPHYSICAL OBSERVATIONS, AND LABORATORY MEASUREMENTS

Kent S. Wood and Robert J. Nemiroff
Naval Research Laboratory, Washington, DC 20375

ABSTRACT

Observational aspects of Weyl gravity (Mannheim and Kazanas 1989, MK) are investigated. We find the limits that current laboratory tests and ranging on the moon and satellites in low earth orbit place on the γ parameter. These results rule out a constant γ equal to the reciprocal of the Hubble constant (MK). On galactic scales, discrepancies exist between Weyl gravity predictions and galactic rotation curves.

BACKGROUND

The exterior line element for a non-rotating uncharged point mass in Weyl gravitation (MK) is

$$B(r) = 1 - \frac{\beta(2-3\beta\gamma)}{r} - 3\beta\gamma + \gamma r - kr^2. \tag{1}$$

where r is distance from the point mass and β and γ are integration constants. This contrasts with the analogous line element in General Relativity (GR) of $B^{GR}(r) = 1 - R_S/r$ where R_S is the Schwarzschild radius of the point mass. It is plausible to assign $R_S \approx 2\beta$.

In the weak field limit ($r >> R_S$) and for masses much less than $c^2/2\gamma G$ (well satisfied by galaxies, the solar system, and the terrestrial laboratory) the effective gravitational potential is, by analogy with the Schwarzschild solution in GR,

$$\phi \approx -\frac{GM}{r} + \frac{\gamma c^2 r}{2}, \tag{2}$$

where G is the gravitational constant, M is the mass and c is the speed of light. The equivalent expected velocity for circular rotation is, from $g = \nabla\phi = v_{circ}^2/r$

$$v_{circ} = \sqrt{\frac{GM}{r} + \frac{\gamma c^2 r}{2}}. \tag{3}$$

One effect of this potential is to produce a constant acceleration equal to $\gamma c^2/2$ at distances much larger than a critical distance R_{crit} where

$$R_{crit} \equiv \sqrt{R_S/\gamma}. \tag{4}$$

LABORATORY AND SATELLITE LIMITS ON γ

There is no confirmed report of a deviation from Newtonian gravity in the laboratory and there are sensitive upper limits. Chan, Moody, and Paik (1982) have tested for deviations from the Newtonian gravitational potential in the laboratory by using a 1600 kg mass and verified the potential at an average distance of 2.3 meters ,

$$\phi(r) = -(GM/r)(1 + \alpha e^{-\mu r}), \tag{5}$$

is identical to Newtonian such that $\alpha = 0.024 \pm 0.036$ at $\mu = 1$ meter. This limits the Weyl parameter γ such that $\gamma < 10^{-28.6}$ cm^{-1}.

Hoskins, Newman, Spero, and Schultz (1985) provide a more constraining null result on a still shorter scale. A fractional limit in the deviation of the potential of 8 x 10^{-4} was found for a mass of 7 kg and radius $\approx$ 2 cm on a scale of 105 cm. This corresponds to an upper limit on the γ parameter in this case such that $\gamma < 10^{-31.1}$ cm^{-1}.

The mass of a body can be computed from the orbit of satellites circling it. If two satellites orbit at different radii, two measured central masses can be compared. Weyl gravity would predict a slight discrepancy between these two measurements, on order

$$\frac{\Delta GM}{GM} = \frac{r_1^2 - r_2^2}{R_{crit}^2}, \tag{6}$$

where r_1 is the orbital radius of the outer satellite and r_2 is the orbital radius of the inner satellite. The GM of the earth has been measured for man-made satellites and compared with GM measured by lunar ranging (Ries *et al.* 1989). A limit on γ for the earth can then be calculated such that $\gamma < 10^{-28.9}$ cm^{-1}. In principle it is possible to derive a similar limit for ΔGM for the sun from Mars ranging data.

These measurements exclude the version of Weyl gravity where γ is a universal constant equal to the reciprocal of the Hubble constant ($\gamma = 10^{-28}$ cm^{-1}, MK, see also Wood and Nemiroff, 1991).

GALACTIC ROTATION CURVES FITS

Arguably the most significant aspect of Weyl gravity is that it can explain galactic rotation curves without recourse to dark matter. To study this issue we have compared the predictions of Weyl gravity to an extremely extended rotation curve: UGC 2885. We have tried to fit this rotation curve with the velocity dependence given by (3). No combination of β and γ will provide even a remotely satisfactory fit as measured by χ^2. (Typical deviations are 50 km/sec while typical uncertainties are $\pm$ 3 km/sec.

It is possible to create a rotation curve from Weyl gravity that is approximately flat over a factor of $\approx$ 4 in radius, using β and γ as free parameters.

The rotation curve of UGC 2885, however, is approximately flat over a factor of ≈ 20 in radius (Rubin, Ford, and Thonnard 1980). A good fit with the Weyl functional form for a point mass is not possible. Calculating the effects of a distributed rather than a point mass requires knowing how to combine the gravitational attractions arising from many point masses. This has not yet been specified for Weyl gravity, hence the only rigorous comparison one can make is the one that has been presented. Given such a prescription for one would presumably apply it to the luminous matter, not invoking any dark matter, in order to retain the main advantage of the Weyl theory.

The figure shown depicts the UGC 2885 rotation curve as measured by Rubin, Ford, and Thonnard (1980). Fits to this rotation curve with the Weyl potential were found assuming a central mass point and β and γ as free parameters. The best fit curve is for $\gamma = 1.1 \times 10^{-29}$ cm^{-1} and $\beta = 9.5 \times 10^{15}$ cm (equivalent Newtonian rest mass of $6.3 \times 10^{10}\ M_{\odot}$), for which $\chi^2 = 9500$, an unacceptably bad fit. A truly flat rotation curve of 260 km sec^{-1} has $\chi^2 = 1300$.

Weyl gravity predicts an upturn in velocity at large radii, eventually rising as $r^{1/2}$. There are galaxies for which a rising rotation curve has been measured. It also appears quite difficult, however, to explain the large variety of rotation curves with the single functional form of Eq. (3).

This work was supported by the Office of Naval Research and was done while RJN held a National Research Council - Naval Research Laboratory Research Associateship.

SUMMARY

- Laboratory measurements constrain $\gamma < 10^{-31.1}$ for a masses on the order of kilograms of grams and radii on the order of centimeters.

- Satellite and moon ranging observations constrain $\gamma < 10^{-28.9}$ for the earth.

- A point mass under Weyl gravity does not provide a good fit to the rotation curve of UGC 2885 for any (γ, β).

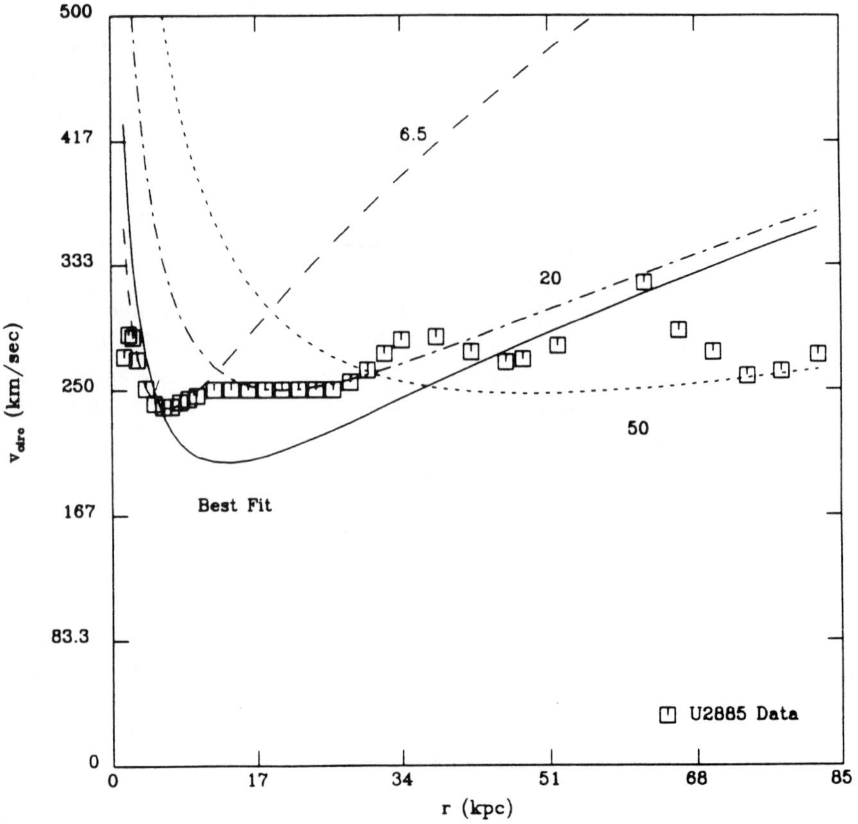

FIGURE CAPTION

The measured rotation curve of UGC 2885 (Rubin, Ford, and Thonnard 1980) and four fits to it with the Weyl potential of a central point mass. The number labelling the dashed curves shows the distance in kpc at which the given curve becomes flat. The best fit curve was determined by a least squares fit to the UGC 2885 data in the complete (γ, β) parameter space.

REFERENCES

Chan, H. A., Moody, M. V., and Paik, H. J. 1982, *Phys. Rev. Lett.*, **49**, 1745.

Hoskins, J. K., Newman, R. D., Spero, R., Schultz, J. 1985, *Phys. Rev. D.*, **32**, 3084.

Mannheim, P. D., and Kazanas, D. 1989, *Ap. J.*, **342**, 635.

Ries, J. C., Eanas, R. J., Huang, C., Schultz, B. E., Shum, C. K., Tapley, B. D., Watkins, M. M., and Yuan, D. N. 1989, *Geophys. Res. Lett.*, **16**, 271.

Rubin, V. C, Ford, W. K., and Thonnard, N. 1980, *Ap. J.*, **238**, 471.

Wood, K. S., and Nemiroff, R. J. 1991, *Ap. J.*, **368**, in press.

Summary

Overviews

IN THEIR OWN WORDS: EXTRACTS FROM THE FIRST FIFTY YEARS

Virginia Trimble
Astronomy Program, University of Maryland, College Park MD 20742
and
Department of Physics, University of California, Irvine CA 92717

ABSTRACT

Many of the topics on the program of this meeting have been discussed in nearly their present form for more than 40 years. These include dark matter; gravitational lensing; very large scale structure and streaming (scale, amplitude, topology, origin); galaxy formation; and the microwave background and cosmological nucleosynthesis. Collected here are a number of extracts from early papers on these subjects, most of them by very well-known astronomers, for comparison, if the reader wishes, with the versions he carries in his memory.

I. GROUND RULES

History of science is an independent discipline, with its own methods, literature, PhD programs, and points of view (not always focussed on who was first to ask a scientific question, get the answer, or persuade his colleagues of it). I claim no expertise in this subject; and the present paper is an exercise in the folklore of astronomy, not its history. Thus each section stops roughly when someone has published what we now regard as the right answer. The starting points are determined by a friends-of-friends algorithm. That is, all the authors directly quoted are people I have met either first or second hand. This is not as restrictive as it sounds. If you have had the pleasure of meeting Jan Oort, then you know Kapteyn and de Sitter at one remove. Similarly, William McCrea "introduces" you to Milne and Jeans, Peter Bergmann to Einstein, and so forth. The purpose of this restriction is to stay close enough to our our times and paradigms for words to carry essentially their current meanings. Finally, the exigencies of oral presentation dictated a focus on papers in English or English translation. Where possible, the references give the page on which the quoted remark occurs, not the first page.

II. DARK MATTER

Velocity data over a wide range of length scales have now persuaded most astronomers that dark matter (stuff that contributes to gravitational potential without radiating its fair share of light) dominates the dynamics of galaxies, clusters, and the universe, without telling us what it is made of. A discussion of such dark matter formed part of Kapteyn's (1922) first attempt at a theory of the structure and dynamics of the galaxy. He derived mass in the galactic plane from the distribution of star numbers and velocities perpendicular to it and concluded "As matters stand at present, this mass estimating the mass of dark matter in the universe. As matters stand at present it appears at once that this mass cannot be excessive. If it were otherwise, the average mass [per star] as derived from binary stars would have been very much lower than what has been found for the effective mass." In the same year and using much the same data, Jeans (1922) decided that "there must be about three dark stars in the universe to every bright star." Their "universe" was, of course, our "galaxy," as Oort (1932) knew a decade later when, with improved data and theory on "the force exerted by the stellar system in the direction perpendicular to the galactic plane" he found that "one unit of light corresponds to 1.8 solar masses." These determinations bracket the modern values.

On the scale of individual galaxies, Babcock (1939) was first to follow a rotation curve (M31) far enough out to see "the ratio of mass to luminosity, in solar units, is about 50" (17 for the modern distance scale). He wrote (private communication 1988) that the reception of his conclusions, both at his graduate institution and at an AAS meeting, was such as to persuade him to a career well away from extragalactic astronomy. His published thesis makes no mention of the two following results which also indicate large masses for galaxies.

Holmberg (1937) pioneered in taking binary galaxies seriously as bound systems and said, "if the orbits are assumed to be circular we now arrive at the following value: Average mass of the galaxies = $1.0 \cdot 10^{11} \odot$" (which is 5×10^{11} $M_\odot$ for H = 100 km/s/Mpc).

On still larger scales, Zwicky (1933) collected enough redshifts for galaxies in the Coma cluster to estimate M/L = 400 (old distance scale) and to opine "dass dunkle Materie in sehr viel grösserer Dichte vorhanden ist als leuchtende Materie." Sinclair Smith (1936), using a spectrograph of his own design and construction, brought the total number of Virgo redshifts up to 32 and found "the mass per nebula is $2 \times 10^{11} \odot$." (which is 10^{12} $M_\odot$ for H = 100). Notice that, when all are expressed in a consistent distance scale, the binary galaxy value nestles comfortably between the M31 and the cluster numbers.

A sort of renaissance of dark matter occurred at a 1961 conference in Santa Barbara, where Abell (1961) presented "evidence for the existence of second-order clusters" and said "typical masses for these systems are estimated at from 10^{16} to 10^{17} solar masses." At this same meeting, Von Hoerner (1961) presented what may have been the first n-body simulation of a cluster. He concluded that "deviations from the time averages in the virial theorem could be no larger than this [a factor of two]." His n was 16!. Alternatives to dark matter discussed there included unbound expanding clusters (advocated by Ambartsumyan) and substructure, interlopers, and observational errors (Holmberg). The first alternative in the form of "a law of gravitation that implies a much strunger attraction at the long distance than that predicted by the law of Newton" came two years later (Finzi 1963).

The end of the folklore period in this subject came with a pair of papers that summarized the existing data as "the mass of spiral galaxies increases almost linearly with radius to nearly 1 Mpc" (Ostriker et al. 1974) and "the mass-luminosity ratio rises to $1 \approx 100$

$1 \gtrsim 120$ for elliptical galaxies" (Einasto et al. 1974). In passing, we note (a) an early cosmological limit to the mass of the muon neutrino of 400 eV/c^2 (Gershtein & Zeldovich 1966) and (b) the coining of the name WIMP (Steigman & Turner 1985).

III. GRAVITATIONAL LENSES

Gravitational lensing would presumably result from Newtonian gravitation bending the paths of Newtonian light particles; but the first discussions of it as an astronomical phenomenon in fact came in the wake of general relativity and the detection in 1919 of the deflection of starlight by the sun. I have found some difficulty in disentangling the early history and so begin in the middle. Zwicky (1937ab) pointed out "that extragalactic nebulae offer a much better chance than stars for the observation of gravitational lens effects" and that "observations on the deflection of light around nebulae may provide the most direct determination of nebular masses." He was right on both counts, in that the only lensing we have unambiguously seen is that of one galaxy (or quasar) by another galaxy or cluster and that often the implied lens mass is quite high. And it took only about 50 years from promise to fulfillment!

Zwicky cites the (fairly) well-known Einstein (1936) paper (it addresses lensing of and by stars only), but says that he had the idea of "image formation through the action of gravitational fields" from V.K. Zworykin, to whom it had been suggested by R.W. Mandl. Einstein's attribution is "Some time ago, R.W. Mandl paid me a visit and asked me to publish the results of a little calculation which I had made at his request." But there are two confounding earlier items. First, the earliest lens paper I have found ("Über eine Mögliche Form fiktiver Doppelstern", Chwolson 1924) fills only about a quarter of its page and is followed immediately by an even shorter note on the index of refraction of an electron gas by --- A. Einstein. And it is hard to believe that he didn't at least glance at that page some time and so see the Chwolson note. Second, Zwicky (1937b) footnotes that "Dr. G. Strömberg of the Mt. Wilson Observatory kindly informs me that the idea of stars as gravitational lenses is really an old one. Among others, E.B. Frost, late director of the Yerkes Observatory, as early as 1923 outlined a program for the search of such lens effects among stars." Only on a late rereading did I finally turn over the A.N. page with Chwolson and Einstein, thinking that surely the last short paper thereon would be of no interest. But it bears the name Leigh Page, father of our colleague Thornton L.P. The world is a small and sometimes confusing place! (And the grandson is Leigh Page II.)

IV. VERY LARGE SCALE STRUCTURE AND STREAMING

Recognition that the spiral nebulae are concentrated in a supergalactic plane on the sky predates understanding of their nature and is usually credited to the younger Herschel (not a friend of a friend) The hypothesis of local mediocrity having become fashionable only rather recently, it seems fair to date the discovery of superclusters to 1934 when Shapley (1934) found the second one, saying "if it were not for its great distance...the twin supergalaxy in Hercules...would be equally interesting [to Virgo]. Future studies of this region of the sky may show that in fainter members the twins are one, and show also that they are but concentrations in an exceedingly rich region, a metagalactic cloud."

The very existence of superclustering long remained in dispute, particularly by Zwicky, who very early reported "that practially all nebulae must be thought of as being grouped in clusters" (Zwicky 1937c), but maintained to the end that "the concept of secondary clustering, in our opinion, thus may be discarded as superfluous" (Zwicky & Rudnicki 1963). And the author remembers George Abell in the mid 60's offering Herzog "political asylum" at UCLA when he had presumed to find such structure in the Zwicky, Herzog & Wild Catalogues. But in fairness one should say that Zwicky's unit of inhomogeneity was the "cluster cell" whose "indicative diameter...based on a redshift constant of 100 km/sec per million parsecs is...41 million parsecs," (Zwicky & Rudnicki 1963), more or less the size we now attribute to superclusters, as they themselves note.

The amplitude and range of large-scale deviations from uniform Hubble flow have remained controversial down to our own time. Whatever the eventual outcome, both credit and gratitude in large quantities belong to those who first took the issue seriously. Among the foremost is Vera Rubin. Cognoscenti will think first of the Rubin-Ford effect (Rubin et al. 1976). But still more prescient is Rubin (1951) which reported a peculiar motion of our galaxy of 179 km/s toward l = 105°, b = +39° relative to galaxies with velocities up to 1000 km/s and "a peculiar motion of our galaxy of 373 km/sec toward l = 160°, b = +22°" relative to galaxies with velocities from 1000 to 4000 km/s.

Finally, one cannot close this section with mentioning Gerard de Vaucouleurs, an early and firm proponent of superclustering and also of what we would now call Virgocentric infall, in the form of "a differential expansion effect explainable by the hypothesis that the expansion rate is small or negligibe in the denser central regions and increases asymptotically outwards in regions of decreasing density... around a center located in the Virgo cluster" (de Vaucouleurs 1958).

V. GALAXY FORMATION

Since we do not know the answer to how galaxies formed (or, rather, many people know the answer, but they know different ones) this section has no coherent focus, though I think the line between "history" and "current events" can be drawn fairly cleanly when Eggen, Lynden-Bell & Sandage (1962) presented "Evidence from the Motions of Old Stars that the Galaxy Collapsed."

The basic idea goes back at least to Jeans (1929) who said (crediting Kant and Laplace) that "the picture to which we are naturally led is that of matter scattered uniformly through space. It is satisfactory to find that if this matter were in the gaseous state, its molecules moving with reasonable velocities of thermal agitation, then the next stage in the evolution of the universe would be the formation of distinct aggregations having masses comparable with those of the spiral nebulae" via what he called the gravitational instability and we now call the Jeans instability. That this still occurs in an expanding universe is not self-evident, but was demonstrated by Lifshitz (1946) and Bonnor (1957) who went on to say that "the condensation process is too slow to account for the formation of the nebulae," which is true if you start with only statistical fluctuations.

Von Weizsäcker (1943, 1951; Heisenberg & von Weizsäcker 1948) discussed formation of stars (with planets) and galaxies together and said that "galaxies seem to have been formed by a competition between expansion and turbulence" and that "irregular nebulae must be young, spirals intermediate, elliptical galaxies genetically old," (though many of his contemporaries envisioned ellipticals gradually spinning up to spirals as revealed in the terminology "early" and "late" type galaxies). Hoyle's (1953) concept, now called heirarchical fragmentation, is closely related and predicted "that galaxies are likely to form, not as single objects, but in clusters."

Gamow returned several times to the difficult problem of making galaxies in a hot, expanding universe, proposing first that "the type of expansion necessary for the formation of nebulae indicates that space is infinite and unlimitedly expanding" (Gamow & Teller 1939), somewhat later that "shadow casting" or "mock gravity" can "effectively increase the attraction between various parts of the expanding gas and could be made responsible for the formation of rudimentary condensations" (Gamow 1949), and that "the primordial gas uniformly filling the universe during the early stages of its expansion must have been broken up into separate giant gas clouds (protogalaxies) during the transition epoch...from the predominantly radiative into the predominantly material state...when the density of gas and the mass density of radiation were comparable" (Gamow 1954).

Bonnor (1957) being absolutely right that statistical fluctuations cannot grow into galaxies in the available time, the origin of the primordial fluctuations has received much attention. Just before the discovery of the 3K background radiation, Sakharov (1965, whom we can all claim as an acquaintance once-removed through M.M. Shapiro, whose 75th birthday is being celebrated at a conference immediately following this one) suggested "that the initial inhomogeneities arise as a result of quantum fluctuations of cold baryon-lepton matter at densities of the order of 10^{98} baryons/cm^3" that is, at the Planck time. The first fluctuations to separate out in his model are the globular clusters, so it is an early "bottom-up" scenario.

VI. THE MICROWAVE BACKGROUND AND COSMOLOGICAL NUCLEOSYNTHESIS

Alpher and Herman (1988) have recently published an account of their own participation in the development of this subject and should be regarded as the primary source. The "prediscovery" observation of excited interstellar CN by W.S. Adams is widely known. McKellar (1941) reported that it implied "for the region where the CN absorption takes place, the 'rotational' temperature T = 2.3K." He remarked that the value is quite insensitive to the exact line ratio, while Herzberg (1951) quotes the result as "a rotational temperature of 2.3^{o}K...which has of course only a very restricted meaning." (This was one of our favorite phrases in graduate school just after Penzias and Wilson's ApJ letter appeared).

The famous Alpher, Bethe, Gamow (1948) paper really does exist. It occupies rather less than two pages in the April 1, 1948 issue of Phys. Rev. Nucleosynthesis is described as successive neutron captures, beginning with deuterium, "just above the upper fringe of the stable elements" followed by beta-decays, so that "individual abundances of various nuclear species must depend not so much on their intrinsic stabilities (mass defects) as on the values of their neutron capture cross sections." No mention is made of pervasive radiation or of a present temperature. This first appears in the November 13, 1948 Nature (Alpher & Herman 1948) as "the temperature in the universe at the present time is found to be about 5°K." But this is still based on a universe consisting initially of pure neutrons and building all the elements by successive captures thereof. Details of the calculation appear in Alpher and Herman (1949). It is arguably significant that, in 1949, Gamow assured an aspiring graduate student with a background in microwave spectroscopy that he knew of no interesting thesis problems in that field.

Hayashi (1950) recognized that at sufficiently high temperature the n/p ratio was determined by rates of electron and positron captures and emissions. He found that the ratio should be n/p = 1/4 at the beginning of nucleosynthesis and that "the hydrogen-helium abundance ratio (in numbers) resulting from the initial n-p ratio 1:4 become 6:1" (Y = 0.4). He also hoped that the presence of free protons would help to bridge the A = 5 and 8 gaps.

When American and Soviet theorists began thinking seriously about the early universe again in the 60's, it was in the wake of a paper by Ohm (1961) that reported a 2390 MHz atmospheric temperature very close to what was expected theoretically (I think it came from extrapolating csc θ dependence to csc θ = 0 and so was bound to). This was widely misinterpreted. But a couple of pages lated in the article comes a discussion of the contribution of the horn antenna to the system temperature, 2±1°K. And "this estimate is based on the temperature 'not otherwise accounted for' in a previous experiment; it is somewhat larger than the calculated temperature expected from back lobes measured on a similar antenna." This is evidently where the background is hiding.

Taking the Ohm result to imply an upper limit of 1K to the cosmic temperature, Zel'dovich (1962, 1963) for instance concluded that a hot big bang was ruled out because "such a density of radiation is obviously in contradiction with both radio-astronomical observations and the indirect data of cosmic-ray theory." (This latter is still a bit of a problem.) He turned his attention (though not that of all his colleagues) to a cold big bang, consisting initially of only protons, electrons, and neutrinos. Doroshkevich and Novikov (1964) had just published a prediction of the "spectrum of the metagalaxy" from 10^9 to 10^{15} Hz, and Dicke and his colleagues were about to do so, when

Penzias and Wilson (1965) collected the data they reported as "A Measurement of Excess Antenna Temperature at 4080 Mc/s," bringing this subject, and, really, all of cosmology, into the modern era.

VII. OTHER TOPICS

There are obviously many other interesting aspects of cosmology and its history that are not addressed here. Early discussions of an expanding universe vs. tired light; the catalytical effects of steady state; and the continuing saga of the distance scale come immediately to mind as topics treatable under the friends-of-friends algorithm. They will have to wait for another meeting!

ACKNOWLEDGEMENTS

This presentation is, in principle, based entirely on published papers. But of course my own thinking has been heavily influenced by seminars, private conversations, and other modes of communication (and this is reflected in a few of the points made above). Inevitably I recall only the tip of the iceberg in thanking for their various insights George O. Abell, Horace W. Babcock, George Gamow, James E. Gunn, Vera C. Rubin, Joseph Weber, Olin C. Wilson, Yakov B. Zel'dovich, and Fritz Zwicky.

REFERENCES

Abell, G.O. 1961. AJ 66, 580
Alpher, R.A., H. Bethe, & G. Gamow 1948. Phys. Rev. 73, 803
Alpher, R.A. & R. Herman 1948. Nature 162, 775
..... 1949. Phys. Rev. 75, 1092
..... 1988. Physics Today, August p. 24, and November p. 157
Babcock, H.W. 1939. Lick Observatory Bulletin 19, 41 (No. 498)
Bonnor, W.B. 1956. MNRAS 117, 104
Chwolson, O. 1924. A.N. 221, 329
Doroshkevich, A.G. & I.D. Novikov 1964. Dokl. Akad. Nauk USSR 154, 809 (in Russian)
Eggen, O.J., D. Lynden-Bell, & A.R. Sandage 1962. ApJ 136, 748
Einasto, J., A. Kaasik & E. Saar 1974. Nature 250, 310
Einstein, A. 1924. A.N. 221, 329
..... 1936. Science 84, 506
Finzi, A. 1963. MNRAS 127, 21
Gamow, G. 1949. Rev. Mod. Phys. 21, 372
..... 1954. Proc. NAS 40, 480
Gamow, G. & E. Teller 1939. Phys. Rev. 55, 654
Gershtein, S.S. & Ya. B. Zeldovich 1966. JETP Lett. 4, 174 (p. 120 in Russian)
Hayashi, C. 1950. Prog. Theor. Phys. 5, 224 & 235
Herzberg, G. 1950. Spectra of Diatomic Molecules (Van Nostrand) p.496
Holmberg, E. 1937. Lund Annals 6, 102
Hoyle, F. 1953. ApJ 118, 513
von Hoerner, S. 1961. AJ 66, 580

Heisenberg, W. & C.F. von Weizsäcker 1948. Zs. f. Phys. 125, 290
Jeans, J.H. 1922. MNRAS 82, 130
Jeans, J.H. 1929. Astronomy and Cosmogony (2nd ed. Cambridge University Press) Sect. 322
Kapteyn, J.C. 1922. ApJ 55, 302
Lifshitz, E.M. 1946. JETP 16, 587 (in Russian)
McKellar, A. 1941. Publ. DAO 7, 251
Ohm, E.A. 1961. Bell Syst. Tech. J. 40, 1073
Oort, J.H. 1932. BAN 6, 249
Ostriker, J.P., P.J.E. Peebles & A. Yahil 1974. ApJ 193, L1
Penzias, A.A. & R.E. Wilson 1965. ApJ 142, 420
Rubin, V.C. 1951. AJ 56, 47-48
Rubin, V.C. et al. 1976. AJ 81, 687 & 719
Shapley, H. 1934. MNRAS 94, 815
Smith, S. 1936. ApJ 83, 23
Steigman, G. & M.S. Turner 1985. Nuclear Phys. B253, 375
Sakharov, A.D. 1966. JETP 49, 345 (in Russian)
de Vaucouleurs, G. 1958. AJ 63, 253
von Weizsäcker, C.F. 1943. Zs. f. Ap. 22, 319
..... 1951. ApJ 114, 165
Zeldovich, Ya.B. 1962. JETP 43, 1561 (in Russian)
..... 1963. Uspk. Akad. Nauk USSR (Physics) 6, 475 (translation)
Zwicky, F. 1933. Helvetica Physica Acta 6, 110
..... 1937ab. Phys. Rev. 51, 290 & 679
..... 1937c. ApJ 86, 218
Zwicky, F. & K. Rudnicki 1963. ApJ 137, 718

OF BITS, FLOPs, AND FREE PARAMETERS

Robert J. Nemiroff (with apologies to Dr. Seuss)

New data had come to Skyville and all the funded residence rushed off to see it at the town meeting. The meeting hall buzzed with news of the new data and the technological inventions which made it possible. "They've got a Floating Frozen Microwave Mapper," someone overheard. "Did you hear about the Tree-Pronged Bit Swapper?" whispers repeated. Skyville was bustling with excitement.

Finally the meeting was called to order and the observers got up to speak about the new data. "10^{12} bits" the observers said, showing some of them to a hushed audience. The audience was impressed: this was much more data than last year, and more impressive. Some of the data showed fascinating features like great groups and winding walls of galaxies, while other data showed featureless photon fields and because of this were just as interesting. It seemed more of the sky was explored during the last year than in all the history of Skyville combined.

Next it was the computationalists turn to speak about the improved computer calculations completed over the last year. "10^{16} FLOPs" they said, letting the audience see some in action with a million mass movie. Again the audience was impressed, and again it was more than ever in the entire history of Skyville.

Next the theorists got up to speak on the current state of Skyville's understanding of the sky. "3 free parameters" the theorists said. An unsettling murmur spread over the audience. Skyville was not impressed. This was also more than last year, and more free parameters usually meant less understanding. Skyville yearned to understand the sky. Their hope now was that this new data would help increase understanding.

As the meeting continued Skyville learned more and more about the Sky they loved so much. But a complete understanding without free parameters eluded everyone. Try as they might, no one in Skyville could postulate a comprehensive theory to explain the data. It was an exciting time, but because of this it was also a time with an undercurrent of frustration.

Skyvillians made many comments to each other during the meeting as they strived to understand the sky. Here's what some of them said:

"Hoyle, Hubble, toil and trouble . . ."

"Frankly, my dear, I don't believe CDM."

"Read my MIPS."

They also sang of old and changing times:

"High H_o, high H_o, Hubble's constant we soon will know . . ."

"The answer, my friend, is string dominated HDM . . ."

"Ω_B, you're not half the value you used to be . . ."

"All in all it's just another indication of Walls . . ."

As the meeting drew to a close the theorists again spoke about the likely understanding of the Sky in the next year, in light of the new data. "4 free parameters" they said sorrily. It seemed all of Skyville groaned collectively. And what was worse, many of the free parameters were either newly added, not very popular, or not well understood. Many of the theorists were dismayed, but others took this as a challenge. Some thought that it was this very *lack* of understanding that drove Skyville to its present frenzy.

Next the computationalist spoke about the future of Skyville's ability to compute the Sky. They were preceded by rousing rumors of even more improved computational ability. Surely this would help. "A Max VAX Universe Tracker," one computationalist was heard to say. "A Rapid Reading Correlation Catcher," whispers repeated. The computationalists called for order and spoke confidently about their expectations in the next year: "10^{17} FLOPs", they forecasted. Skyville's excitement increased.

But the most impressive rumors were of new observational technology which would allow even newer, even more vast data sets. "A Quantum-Quick Quasar Quaffer," repeated one Skyvillian to another. "A Tinsel Tiled Galaxy Grabber," "A New SNU ν Snarfer," "A Flying Refocused Photon Finder," rumored others. A hush descended as the observers finally spoke about what they expected in next year: "10^{13} bits," they promised proudly. Skyville roared in approval as the meeting ended. With this pronouncement, exciting times in Skyville were sure to continue.

SUMMARY OF AFTER THE FIRST THREE MINUTES

J. Richard Bond
CIAR Cosmology Program
Canadian Institute for Theoretical Astrophysics
60 St. George, Toronto, ON M5S 1A1

ABSTRACT

Readers of the New York Times over the past few years have variously learned that: the Big Bang is dead; dark matter isn't cold; the Universe is bubbly, spongy, periodic, replete with Great Voids, Attractors and Walls. And every six months or so obituaries appear for the CDM model of cosmic structure formation. The observational evidence for the Hot Big Bang model, for dark matter, for the large scale structure of the Universe as revealed by galaxy surveys and for the very large scale structure as constrained by microwave background observations was thoroughly explored in *After The First Three Minutes*. From this meeting we learned that: The Hot Big Bang model is healthier than ever. There is evidence from IRAS surveys that $\Omega \gtrsim 0.5$, but from Big Bang nucleosynthesis that $\Omega_B \lesssim 0.07$, suggesting a great deal of nonbaryonic dark matter exists, which cannot be vacuum energy (nonzero Λ). There are fewer viable candidates for cold dark matter (CDM) after the LEP results from CERN, but searches are still promising. Hot dark matter in the form of light massive neutrinos got a boost from the MSW solution of the solar neutrino problem. The standard 'biased' CDM model, the 'minimal' inflation model with adiabatic Gaussian scale invariant fluctuations and cold dark matter, probably should have been 'unbiased' in its primordial amplitude all along — although some galaxy types and their relative velocities would be biased. The indications for extra density fluctuation power on large scales over that of the 'minimal' CDM model continue to firm up. The emerging dilemma is what can give the observed large scale structure, and yet such CMB isotropy on slightly larger scales. The key clue may reside in cosmological background radiation. All wavebands are being experimentally mined to unearth it. Energy injection models with hot gas or dust are already under pressure in the sub-mm. The recurring invocation of the cosmological constant to solve our cosmological problems causes theorists to either shudder from the extreme fine tuning required or hope that resolution of the greatest unsolved mystery in physics will clarify our cosmology.

1. INTRODUCTION

After The First Three Minutes brought together all of the experimental results of relevance for people like me who wish for a 'first principles' theory for how cosmic structure arose. That is, all of observational cosmology was covered in Maryland, a field exploding with new data and the promise of much much more to come in the not very distant future. All of this data and the dilemma that large scale structures pose for the minimal theories that dominated the last decade (*e.g.*, CDM) have turned many 'first principles' theorists into phenomenologists, a trend which Jim Peebles felt would dominate the physical cosmology scene in the nineties, with *ab initio* theorists back at the drawing board trying to make sense out of the phenomenology. Let us hope that galaxy formation will not be akin to the weather, but that the patterns of clustering and flow and of CMB anisotropy will offer a glimpse of conditions in the early universe. Phenomenology did dominate the discussions in Maryland. The best example was the density field maps constructed from radial velocity field and redshift data. Although there is already some sophistication in these endeavours, map-making is only in its infancy, but with a very bright future, requiring intensive collaboration between observers and theorists to understand our cosmic data base.

To evolve a universe from the beginning, or even to construct a phenomenological model which includes evolution, the global parameters of our observable patch of the Universe are needed (§2), such as $H_0 \equiv 100\text{h}\,\text{km s}^{-1}$, Ω, Ω_B, Ω_{dm}, Λ, *etc.* But it is the information about fluctuations from these global averages gleaned from large scale galaxy clustering and streaming (§3) and microwave background anisotropies (§4) which hold the promise of guiding a successful first principles theory. The figure given at the end of this paper illustrates how a huge effort using vastly different techniques is brought to bear on testing complete models of structure formation such as the standard CDM theory, or in constructing a phenomenological spectrum of fluctuations. The figure shows the CDM spectrum of density fluctuations, and indicates which band in comoving spatial wavelength the various observations summarized in §3 and §4 probe; below this I relegate various speakers to the wavenumber band their talks concentrated on. A more complete discussion is given in §6, where the fate of the CDM model is addressed. The issue of energy generation, how much there can have been, and whether it can induce much structure formation is described in §5.

Too Many Biasing Factors

Since different people mean different things by the CDM model, in spite of its characterization as being the standard by which all other theories are judged — or were judged if rumours of its death are true (§6) — I shall spend the rest of this section describing what is really meant, and in particular attempt to clarify the many biasing factors that were used in the course of the meeting. Of course the CDM model is considered to be the minimal model consistent with inflation and so has $\Omega \approx 1$, $\text{h} \lesssim 0.5$ for age reasons, $\Omega_B \lesssim 0.064$ for primordial nucleosynthesis reasons, and $\Omega_{cdm} \approx 1 - \Omega_B$ by Occam's razor that precludes us from adding many dark matter species. The density perturbations at the end of the inflationary epoch are Gaussian and scale-invariant, following naturally from a quantum mechanical mechanism for the generation of fluctuations in a weakly coupled scalar field. Finally there is an overall fluctuation amplitude, characterized by one number, the *amplitude* 'biasing factor', b_ρ.

It is perhaps unfortunate that this quantity, whose purpose was only to define the overall level of fluctuation power, has come to be known as the biasing factor, thereby mixing it up with visions of the complex gas dissipative processes that will define the overall clustering of galaxies of different types relative to the mass distribution. Gas and radiative processes can and do generate *astrophysical* biasing of objects such as galaxies. The simplest version of this supposes that on large scales the galaxy biasing factor b_g, the ratio of the relative fluctuation in the number density of galaxies to the relative fluctuation in the smoothed large scale mass density, is a scale-independent constant. Although this is often assumed in phenomenological work, for example for relating large scale flows to the density field estimated from redshift surveys, it is certainly suspect. The streaming and IRAS redshift survey data allow limits to be placed upon a combination of Ω and the b_g for IRAS galaxies. However, even if b_g is constant, we expect it to vary from galaxy-type to galaxy-type, and that it may bear no relation to b_ρ at all. Biasing factors for clusters, for groups, *etc.* would also exist and be very different than IRAS b_g's.

With this confusion between b_ρ and b_g, b_ρ is better thought of physically as a local measure of the initial amplitude of the gravitational potential fluctuations. For the scale invariant spectra that one gets in the simplest and most robust inflation models, the gravitational potential fluctuations are scale-independent, and related to b_ρ by $\approx 2\times10^{-5}\, b_\rho^{-1}$ (in units of c^2 per log of scale). The constancy in scale is not something *ab initio* theorists will give up lightly, for otherwise primordial physics is required to have a characteristic scale just among the length scales we can access with our observations. This is an old argument, invented long before inflation was, by Harrison, Peebles and Zeldovich, and thus the no-scale spectrum bears their name. Because of the b_ρ confusion, its inverse, labelled σ_8 (the *rms* mass density fluctuations in spheres of radius $8\,\mathrm{h}^{-1}$Mpc as determined in linear perturbation theory) is often used. We have seen both notations at this meeting.

The trend of b_ρ with time has been downward, from $b_\rho \approx 2.5$ used in much of the Davis, Efstathiou, Frenk and White work, all the way down to $b_\rho \sim 1$, which is now preferred by the data and by the numerical models, as we saw at this meeting. However, we also saw that the data prefer a different shape as well as overall amplitude, with more large scale power than CDM. Cosmic microwave background anisotropy constraints described at this meeting can also be expressed in terms of b_ρ for models like CDM. The values are nearing the b_ρ suggested by the large scale structure data; and extra power makes the confrontation worse.

2. GLOBAL PARAMETERS

To specify a cosmological model we need to know global parameters as well as the nature of the fluctuations. Some are now known fairly well, such as the relic photon temperature T_γ, the baryon abundance parameter $\Omega_B\mathrm{h}^2$, the number of light relic neutrinos N_ν, but we still have much room to maneuver in the space of theories, through uncertainties in H_0, total Ω, the types of dark matter, the value of the cosmological constant, and in whether Λ or some of the dark matter is decaying.

T_γ

COBE's FIRAS and COBRA, the UBC rocket experiment described by Mark Halpern, give among the world's best experimentally measured blackbody curves. The current limits on the temperature, 2.735±0.06K from FIRAS, using 9 minutes of data from December 1989, and 2.736 ± 0.017K from COBRA, using 5 minutes of data from January 1990, should tighten further with the improved modeling of the FIRAS thermometers now underway. The dipole spectrum, subtracting a 60° patch in the north from one in the south, with 0.1% errors, displays the promise of limits to come. John Mather and the rest of the COBE team have given NASA a much needed great unblemished success, and have astounded the astro world with their early results, and whet our appetites for the main course to come. But we of the frozen north lead the applause for the spectacular outcome of Herb Gush's decade long struggle in Vancouver.

N_ν

The number of light neutrinos, strongly implied by the standard primordial nucleosynthesis story as retold by Dave Schramm, has been definitively determined by CERN and SLAC. As of August 1990, $N_\nu = 2.98 \pm 0.06$, from 4 experiments at LEP.

With T_γ and a little weak interaction theory, we get T_ν, and thus the number density of relic neutrinos. Alas, the possibility of directly observing this sea, *e.g.*, by using very soft photon emission accompanying neutrino scattering as was recently advocated by Avi Loeb and Glen Starkman, was discounted by both Dave Cline and Arnon Dar.

Ω_B

With advances in neutron lifetime measurements and determination of the light neutrino number, the number of relic photons per baryon at the first three minutes is more tightly constrained than ever; together with the sharpening of T_γ, this translates into a more restrictive upper bound on Ω_B of $0.064\,(2\mathrm{h})^{-2}$. If the combination of He^3+D does not decline during stellar evolution, there is also a lower bound, $\Omega_B \gtrsim 0.038\,(2\mathrm{h})^{-2}$, as Dave Schramm emphasized, which should be compared with ~ 0.007 in luminous baryons, suggesting at least some baryonic dark matter exists.

But if some, why not all? If the $\Omega \gtrsim 0.5$ estimates of §3 are correct, a non-baryonic form is needed, unless nucleosynthesis is more complex. As Dave described, most workers now seem to agree that the inhomogeneous nucleosynthesis that is a possible byproduct of the quark-hadron phase transition still has a strongly constrained Ω_B, $\lesssim 0.4$, with $\Omega_B \lesssim 0.1$ likely and $\Omega_B \approx 1$ quite unlikely; and the most plausible and largest region of parameter space gives inhomogeneity on a scale that does not perturb the standard tale very much at all. However, the powerful voice of Willy Fowler invites us to be cautious at outright rejection of an inflationary $\Omega_B \approx 1$ $H_0 = 50$ cosmology with dark matter in the form of collapsed stellar remnants: our theoretical imaginings of $t < 3$ minutes may be too staid; and, for that matter, of $t > 3$ minutes.

H_0

Jacoby's recent planetary nebula value of $H_0 = 87 \pm 12\,\mathrm{km\ s^{-1}\,Mpc^{-1}}$ was reported, highlighting the high H_0 trend of late and the huge age problem looming for those theorists who are enamoured with inflation. The ways out are to assume globular cluster ages are dramatically overestimated, which seems unlikely to the extent that is required, or adopt an open Universe, anathema to inflationary cosmology, or adopt one with a non-zero cosmological constant.

Λ

The particle physics case against Λ was emphasized by Dave Schramm: whatever known particle physics scale we refer it to, the current constraints on Λ in those units are tiny; this is usually expressed in Planck energy density units ($m_P^4 \sim 10^{90}\mathrm{gm\,cm^{-3}}$): $\Lambda/(8\pi G) \lesssim 10^{-120} m_P^4$. Interpreting Λ as a measure of the energy density of the 'vacuum', we expect it to undergo many renormalizations in value during the cooling of the universe (before the first 3 minutes): it is not protected from this within current particle theories. Thus, it is not only hard to understand why the effective Λ is so small now, but even if it is zero we have no idea why it should be so.

The Λ problem is the biggest unsolved problem of early Universe physics, and, arguably, of all physics. Yet inflation relies on a large effective Λ generated by a scalar field to drive the accelerating expansion. We therefore should have no confidence whatsoever that, when the Λ problem is finally solved, our inflation picture and the density fluctuation spectrum it generates will emerge unscathed. We should take heart that the only Λ-experiments that can be performed are cosmological, and strongly connect we of the $t > 3$ minutes community with the latest particle theory and quantum gravity thoughts on Λ.

$\Lambda \neq 0$ models have become the rage this year among data-oriented cosmologists. They can solve the age problem and still have $H_0 \sim 80$. Fluctuations grow more than in the equivalent open universe, which helps to lower microwave background anisotropies, although this is still the Achilles' heel of these models. The simplest way to match the Maddox and Efstathiou APM survey angular correlation function is found by taking the CDM spectrum and varying Ω_{cdm}h until the best fit is obtained. Ω_{cdm} can be interpreted as the total Ω, so the universe is open, or a cosmological constant of magnitude $\Omega_\Lambda = 1 - \Omega_{cdm}$ can be introduced in a flat universe model. The result is $\Omega_{cdm} \sim 0.2$ with h ~ 0.8. However, other models with extra power can also fit the data. $\Lambda \neq 0$ has also been used to explain the high counts of faint blue galaxies at $z \sim 1/2$, although, more recently, K band counts have been shown by Len Cowie and Simon Lilly to be compatible with a $\Lambda \approx 0$ flat cosmology, and not to give agreement with the large Λ models; and K band counts are much less susceptible to evolutionary effects.

One might hope that a dynamical solution exists to the $\Lambda \approx 0$ problem, not a static one; *e.g.,* decaying Λ, as advocated by Peebles. Still, we are apparently stuck with a great coincidence if it is just now dynamically important, so that we are entering a steady state deSitter phase, a second inflation phase from which there are no prospects for escape as there must have been from our first inflation phase (if indeed there was one).

The coincidence required for open universes — that we are just now entering into curvature energy dominated expansion — is about as great, and unlikely small numbers are often invoked (especially by Dicke and Peebles, although not recently by Jim) to make cosmologists uncomfortable with open models as well as with $\Lambda \neq 0$ models. Anthropic arguments that restrict the values of Ω and Λ to ensure humans exist (*e.g.*, requiring a matter-dominated epoch to have reasonable perturbation growth) make most physicists cringe, but I rather like them.

Peebles 'Phenomenological' Isocurvature Model: Small Ω_B (~ 0.03) and large cosmological constant are the cornerstones of Jim's provocative baryonic model, crafted to provide both luminous matter and clustered dark matter from just the Ω_B of standard primordial nucleosynthesis and an $\Omega_\Lambda = 1 - \Omega_B$ large enough to make the universe nearly flat so the beauty of inflation can still be invoked. I suspect Jim is not very serious about the extreme numbers in the incarnation of the model unveiled at this meeting, designed to maximally perturb the ab initio theorists in the audience, but chose it as a framework to display Jim's feeling that galaxy formation occurred very early, at $z \sim 20$, with sufficient energy input to reionize the universe to erase tell-tale primary CMB anisotropies on small and intermediate scales, and that the form of the fluctuations may preclude measurable effects at the large angles probed by COBE's DMR. A requirement for Jim is to have fluctuations different from the usual scale invariant ones that simple inflation gives, but why not? There is a busy little industry of theorists (including me) trying to fashion viable non-Gaussian non-scale-invariant models. PS: I don't think Jim's model can generate the amplitude of large scale flows nor does it explain the high values of Ω inferred from the two IRAS redshift surveys. It might also run into problems with large secondary anisotropies of the CMB (§4).

Dark Matter

The most exciting possibility for the dark matter once again is the venerable light massive neutrino, linked as it may be to the solution of the solar neutrino problem via the MSW effect, and discussed by many people at the meeting. The 20 year average count rate for Davis's chlorine experiment is 27 % of the standard solar model, and for Kamiokande 46 % of it, while SAGE's gallium has not yet detected solar neutrinos, although the Soviets caution that the background subtractions are large; soon the Italian GALLEX experiment will help to firm up the story. The oscillation parameter for the MSW effect to fit this data is $m_{\nu\mu}^2 - m_{\nu e}^2 \sim (10^{-3}\,\mathrm{ev})^2$. In conjunction with the theoretical seesaw mechanism for neutrino mass generation (ν mass $\propto$ (quark or lepton mass)^2/heavy particle mass), the neutrino mass hierarchy would be something like $m_{\nu e} \sim 10^{-6}$ ev, $m_{\nu\mu} \sim 10^{-3}$ ev, and $m_{\nu\tau} \sim 10^{-1} - 10^2$ ev, the range depending upon whether quarks or leptons are used in the seesaw mass formula. Recalling that $\Omega_{\nu\tau} \approx (100\mathrm{h}^{-2} ev)/m_{\nu\tau}$, open season is declared for rethinking hot dark matter models of the Universe. The standard inflation model with scale invariant fluctuations and only one type of dark matter, HDM, seems to me too sick to cure (although Dr. Adrian Melott is trying), with the greatest ill being CMB anisotropy limits. With cosmic strings or other seeds, it is more palatable, and some would say even pretty — but not me.

Most people's best bet for a cold dark matter candidate is the lightest supersymmetric particle (LSP). The LEP results from CERN on the Z_0-boson lifetime give surprisingly strong constraints on the properties of generic massive cold dark matter candidates with interactions that are weak or weaker than weak, as Larry Krauss discussed. For example, a fermion LSP such as the photino must be above the surprisingly high 20 GeV. Sneutrinos are ruled out, the door has been closed to cosmions (dark matter candidates that get trapped in stars, in particular the sun, and do interesting stellar heat transport), and there can be no 4th generation neutrino with mass below about 40 GeV. Cold dark matter detectors now need to be bigger to be hopeful of success in a narrowing (but not discouragingly so) parameter space of possibilities. The LSP is still the main CDM contender.

The axion remains a viable CDM candidate. The generic phenomenon is a spatially coherent scalar field which oscillates rapidly in time. Decaying cosmological constants are of the same ilk, except the field slowly slides down a potential rather than rapidly oscillating in one. Thus to me Λ is just another type of dark matter, the 'hottest' type of all in the sense that it doesn't cluster, and yet the 'coldest' type of all in terms of its absolute immobility. All possibilities in between are imaginable, and most have been invoked by one theorist or another; *e.g.,* slow oscillations on cosmological timescales.

Baryonic dark matter can be constrained by background light limits. Bernard Carr showed how the COBE data constrains the 'black hole remnants as dark matter' hypothesis, although there are still wavebands to hide the predictable energy release (§5). About the least interesting type of dark matter from the observational and theoretical point of view is the Jupiter or brown dwarf solution. MACHO microlensing to the rescue?

Vera Rubin summarized the evidence for dark matter from galaxy observations, including her own heroic work on spirals. There is an uncomfortable amount of freedom in fitting rotation curves, even with all of the advances in modeling using spiral arm stability criteria by Athanassoula over the past few years, which leaves halo masses, and thus the Ω stored in halo dark matter, relatively uncertain. Vera emphasized the significance of the beautiful Carignan and Freeman work on dark matter in dwarf irregular galaxies which Dave Spergel used to rule out light massive neutrinos in their halos from the Tremaine-Gunn limit. Spergel suggested that relaxation of the TG limit by invoking radial orbits, as advocated by Ralston recently, leads to dynamically unstable models, so cannot be used to support neutrinos as dark matter in dwarfs.

Jane Charlton used galaxy pair observations to lend weight to the classical hierarchy story that large amounts of dark matter are in irregular blobs about parent galaxies which have already experienced many mergers.

Richard Mushotsky displayed all of the difficulties one encounters in modeling the dark mass in clusters from X-ray observations, but also envisioned a golden era when ASTRO-D begins to deliver data in the 95-96 period. ROSAT's relatively low energy cutoff restricts its utility in this. Richard highlighted the discrepency between the maximum temperature for X-ray clusters and the upper range for velocity dispersions of cluster galaxies, indicating the latter are generally (and not surprisingly) too high. New results using UHURU, HEAO-1 and Ariel-V data were displayed in a poster by Alaister Edge *et al.* showing cluster temperatures above 10 kev are very rare. The trend to lower velocities is good news for CDM-like models, which always had problems with the large values.

For ellipticals, the X-ray observations show one has to go a long way out to see the dark mass, explaining why there has been no evidence for it in the optical, and that only a lower bound for M/L is obtainable.

Everyone's imagination was fired by Tony Tyson's exciting new results on the Einstein-ring lensing of faint blue galaxies (with redshifts between 3 and 0.8) by clusters. From the spectacular beginnings of the great arc discovered by Lynds and Petrosian and the Toulouse group in A370, the subject has developed into a powerful probe of dark matter in clusters. Tony reported that he has 23 clusters imaged to 29th magnitude per square degree at $z \sim 0.2 - 0.4$. Good rings are found in poorer central clusters, not in rich clusters, where instead there are lots of little bits of arc. He passes the images through a complicated but apparently robust inversion process (itself a computational tour de force) to give a map of the dark matter. So far only regions near cluster centers have been probed, because of CCD size limitations. Within the central 300 kpc, he gets $M_{DM}/M_* \sim (10-25)$. Although Tony finds some evidence that optical velocity dispersions differ from his dark matter ones (in A1689), most seem to give reasonable agreement. He also finds tiny core radii for the dark matter, about 5 times smaller than the typical X-ray gas core radii of $\sim 300\,(2\mathrm{h})^{-1}$kpc. We have to wait for bigger CCDs to get the crucial maps farther out in the clusters, but the future looks very promising indeed.

3. FLUCTUATIONS AND LARGE SCALE STRUCTURE

Redshift Surveys

Margaret Geller told us that CfA2 now has 11500 galaxies with redshifts, and will soon finish the full 15500. The most spectacular inhabitant is of course the Coma pancake, now known as the Great Wall. It appears to be a few linked pancake-like entities, possibly enhanced in part by the 'palm of god' effect (redshift space distortion accentuating a collapsed appearance). It may be hinting at extra large scale power over CDM, although it is hard to say with only one example. As Margaret emphasizes, most redshift surveys to date have found structures on scales up to the survey size, although with the 1 in 6 IRAS survey we might at last be getting beyond this limitation. The pencil beam surveys were interpreted by Alex Szalay as giving evidence for the pervasity of the Great Wall phenomenon. Knowing Alex's long and close relationship with Zeldovich, I am sure he is pleased to see his work firming up that great analytic idea – first unveiled in 1970 – of the cellular nature of gravitational instability in a cold medium. I understand that Zeldovich thought his efforts on large scale structure to be his most profound work; and let us recall how much the man did for all of astrophysics.

The basic Zeldovich pancake, filament, cluster and void picture has nothing to do with the specifics of Gaussian or non-Gaussian initial perturbations, or degree of biasing. Pancakes will appear up to a model-dependent coherence scale, even with hierarchical clustering models such as CDM. Expressions for the relevant characteristic scales were sketched by Sergei Shandarin. Changbom Park's CDM simulations, among others, show that the Szalay *et al.* result is not even too far off what CDM can give, even without extra power fixes.

The pencil beams must be opened up in angle for us to be sure that short distance clumping is not prejudicing the visual impression of great pancakes or filaments, fooling us with ambient groups and poor clusters piercing the beams. The issue is not whether these short distance clustering effects are there, but only how significant they are. As Alex and Nick Kaiser have both discussed, a nice way to understand the issue is to think in wavenumber space. A long narrow pencil transforms to an oblate region near the origin in k-space (a k-space pancake). The transverse flat part extends its reach into high wavenumbers where there is a large known amount of clustering (the $r^{-1.8}$ power law). To what extent does this reduce the sensitivity to probe extra large scale power in the longitudinal wave direction? The cure of course is to rein in the transverse dimensions by having broader beams, which is what Szalay *et al.* are doing.

As it is, Alex showed us much more data than the original BEKS (Broadhurst, Ellis, Koo and Szalay) Nature paper had; in particular, 400 more at the South Galactic Pole so that it now has more data than the North. It was the North which gave the strongest visual impression of periodicity in BEKS. Did I mention periodicity? Pencils in other directions do not strongly indicate periodicity.

Roger Windhorst's ultra deep angular correlation function down to 26th magnitude shows the expected decreasing amplitude with depth, which, however, goes up at the faintest levels, presumably since galaxies in denser environments are picked out.

Nick Kaiser discussed the QDOT redshift survey (the key letter is the T for Toronto, the rest refer to places somewhere in England) which randomly sampled 1 in 6 IRAS galaxies, the strategy being to avoid wasting a lot of time measuring all the galaxies in groups and clusters since only a representative few are needed to obtain the two point correlation function, if that is the object of one's desire. Consequently the QDOT survey went a factor of 2.5 deeper than the Strauss, Davis, Yahil *et al.* IRAS survey, out to a maximum redshift ~ 0.1, which is still substantially less than the pencil beam probes. Although IRAS galaxies have the virtue of allowing an almost-all-sky survey, I am somewhat worried that the physics needed to make an IRAS galaxy could give a depth-dependent bias that appears as large scale structure.

Their data gives a compelling case for large supercluster entities being pervasive, even though rich cluster ellipticals are not included in the survey. Nick derived a galaxy-galaxy power spectrum, a nontrivial exercise for complicated selection functions, which shows more power at large scales than the CDM spectrum does. The extra power is compatible with that inferred from the angular galaxy correlation function of the APM survey. Does this mean that the primordial mass density spectrum has extra power as well? This is the simplest interpretation and provides further strong indications of problems with the minimal CDM model. However, beware of non-local biasing.

Large Scale Flows

What is the Great Attractor anyway? Alan Dressler said it was the entire Hydra-Cen, Pavo-Indus overdense region. His Las Companas z survey in the GA direction shows two peaks in the density with overdensity in galaxies above unity, at 3000 and 5000 km s^{-1}. He presented evidence for backflow onto the GA at larger distances, deemed essential for demonstrating that the GA is a distinct beast. There are, however, the usual worries about Malmquist bias associated with distance errors mimicking backflow. Most theorists prefer to think of superposed long wavelength flows that constructively interfere at some places, destructively at others, as measured by the velocity-velocity correlation function. The v-field then probes the ups and downs of mass within a few velocity coherence scales. A catalogue of the overdense and underdense beasts is useful but we should allow the beasts to fatten or slim down or change shape as the view the data affords us continues to improve. With Dressler's loose definition of the GA, there is no doubt that it lives on.

The very large scale entities seen in redshift surveys might force us to contend with even more powerful flows. The Shapley concentration of Abell clusters that Raychaudhury and Scaramella found also in the GA direction also contributes to our pull, but at no more than the 15% level. Jeff Willick's indication of a Perseus-Pisces flow in the GA direction as well is more worrisome since the coherence scale for the velocity would be very large indeed if true; still, I shall rest easy until more data is forthcoming. Nick Kaiser points to Hercules, which looms large in the QDOT survey, as possibly being a candidate for an embarrassingly (for CDM) large flow field. More accurate distance indicators are needed to map such deep regions, but I was encouraged by Alan Dressler's optimism. Tully-Fisher for spirals has an accuracy of 10-15%, $D_n - \sigma$ for ellipticals has an accuracy of 15-20%, but these might be improved with further selections and John Tonry's method of getting distance from stellar fluctuations in galaxies could possibly provide a distance indicator with as little as a 5% error.

When the spiral and elliptical data is processed through the Bertschinger-Dekel POTENT computer code to deliver a local gravitational potential map from the radial velocity measurements (assuming quite reasonably that the flow is curl-free flow on large scales), a consistent picture emerges in which the GA is clearly seen, as the largest density bump with a relative overdensity of 1.2 ± 0.4 (smoothed on a $\sim 12\,h^{-1}$Mpc scale). A significant and necessary development in this reconstruction game is the error estimation, which Ed and Avishai Dekel have done. Nick Kaiser and Albert Stebbins have an alternative approach to density reconstruction, but are not yet quoting errors.

Comparing v-fields and z-surveys: Ω and b_g

The estimates of Ω from IRAS and optical dipoles and large scale streaming assume the linear growth law for velocities derived from gravitational instability and linear biasing. That is, $b_g \equiv (\delta n_g/n_g)/(\delta\rho/\rho)$ on large scales, and it is a constant, and the velocities are proportional to $b_g^{-1}\Omega^{0.6}$. Thus the Ω issue is inextricably bound up with the biasing issue, and the large scale biasing factor may well vary from one galaxy type to another. Ofer Lahav showed that IRAS and optical dipoles align within $\sim 10^o$ of the CMB and XRB dipoles, and gets $\Omega/b_g^{5/3} \approx 0.2$–$0.3$ for the optical (with worries about patching very different optical catalogues together) and 0.5–0.8 for IRAS. Even if one adopts $b_g = 1$, much higher values of Ω than conventional methods give are indicated. These estimates and the two IRAS redshift survey estimates ($\Omega \gtrsim 0.5$ for unbiased universes) have arguable assumptions built in, but they are on a much firmer foundation than the old (now obsolete) virgocentric flow results ever were. If we choose to suppose that Ω in attracting mass is unity, then the bias factor estimates are $b_g \sim 2$ for optical galaxies and ~ 1.2 for IRAS galaxies. Of course, the galaxy biasing factors do not necessarily give us a measure of the amplitude of the primordial fluctuations (as measured by yet another biasing factor, b_ρ). Nonetheless, within the context of CDM models, $b_\rho \sim 1$ which people once strongly rejected is now preferred by this large scale data. Redshift survey data alone cannot determine b_ρ or b_g.

I was encouraged that the combination of the Strauss *et al.* IRAS survey with the flow field and the QDOT survey with the flow field seem to give roughly consistent stories (using different methods). The data also lend weight to the hypothesis that the flow is driven by gravitational instability. What we would like to have is a quantitative measure of the extent to which nongravitational forces contribute to the flow. Kaiser showed the beginnings of how this might be done by comparing the mismatch between the flow field and the redshift survey field.

Velocity Biasing from *N*-Body Simulations

Ed Bertschinger gave us a very nice review of N-body methods and dazzled us with a beautiful high precision movie of a CDM simulation made with Jim Gelb. CDM certainly makes a pretty universe, exhibiting to my eyes at least all the features we want, although possibly not to a large enough scale. Ed showed truly vast PM simulations of CDM and scale free models with 256^3 particles and a 384^3 mesh: but we want more please, to higher and higher resolution. All simulations now, no matter what the resolution or initial condition setup, reveal a startling network structure, at least for spectra redder than white noise, that seems to be more filamentary than pancake-like. Pancakes and filaments are a result of relatively coherent large scale flows which reflect the initial conditions and which carry along with the flow nonlinear sublumps that collapsed at earlier stages in the clustering hierarchy. The nonlinear tidal forces generated in the collapses themselves are apparently not sufficient to overcome the memory of the flow in the initial conditions, unlike the situation for Poisson initial conditions, which aren't filamentary. Although I think Jim Peebles remains unconvinced that this is so, most of the community seems satisfied.

Is the future in trees? Warren and Zurek displayed calculations in a poster that ran a tree code with 10^6 particles on the hypercube at Caltech. Impressive.

However, for now P^3M and even PM codes are still generating spectacular results. With the inclusion of spatially adaptive methods, now underway, they might find a useful place in the cosmological toolbox of the future. For example, there is Jens Villumsen's hierarchical PM method; and now Hugh Couchman's mesh-refined P^3M method which Ed and Jim Gelb have adopted, so that their calculations do not get stuck in the devastatingly slow particle-particle direct force summation part of P^3M, but can use PM (Fourier) methods to estimate the force.

A very important recent result from the simulations has been the recognition by Ray Carlberg and Hugh Couchman, and confirmed by Ed, that velocity biasing occurs, in which the pair velocities of galaxies in the simulation are significantly smaller than those for the dark matter particles. Ed gets pair velocities of only 250 km s^{-1} for an amplitude biasing factor $b_\rho = 1$, which matches with observations, and much smaller values for $b_\rho = 2.5$, the old preferred value of Davis, Efstathiou, Frenk and White. Thus an 'unbiased' cold dark matter model is rapidly becoming the standard CDM model. One issue that must be resolved is whether the inferred $\Omega \sim 0.2$ from mass-to-light ratio in clusters can arise in an $\Omega = 1$ Universe with $b_\rho = 1$. Why not? That doesn't mean $b_g = 1$ for cluster galaxies. Also there is now N-body evidence that dynamical friction at earlier stages in the hierarchy may result in the luminous matter being dynamically segregated from the mass towards the cluster core, although it is as yet unclear whether this effect is large enough to allow an unbiased scenario.

Ed also presented evidence that the Gaussian density peak analytic approach of Bardeen *et al.* (which was much used in the eighties and still is, for example in Park's simulations) is flawed because of overmerging. Although I am one of the *et al.* I completely agree: this is the cloud-in-cloud problem, which a number of us have been trying to solve with a multiscale peaks algorithm.

Clusters and Large Scale Structure

Clustering of the deepest virialized potential wells — related to, but not synonomous with, Abell's rich clusters — should provide an excellent probe of large scale structure. Margaret Geller's recent deep cluster survey gives 20 h^{-1}Mpc for the correlation length, to be compared with earlier author-dependent values ranging from 14–25 h^{-1}Mpc for shallower surveys. However, she was quick to qualify her result with a list of complaints about the Abell catalogue, pointing out: the missing clusters that Abell did not get, the extra clusters that he put in, the smallness of the samples with measured redshifts and the ones with bad redshifts. Neta Bahcall argued in favour of the catalogue, especially pointing to its success in identifying X-ray clusters, and the robustness of the correlation functions derived for different surveys. Although rich clusters do give an indication of extra power, we know that all errors we can think of will go in the direction of making more large scale power and more super-duper-clusters than a bias-free catalogue would give, so proclamations of the death of theories based on ξ_{cc} are premature.

Fortunately, ROSAT will soon come to the rescue; it can detect clusters to at least $z = 0.3$, according to Alaister Edge. In conjunction with the optical we shall have close to an ideal cluster finding algorithm more directly matched to the deep relaxed potential wells we theorists are really interested in, as measured by the X-ray temperature (which ROSAT cannot do well). ROSAT will finish its all sky survey in Jan 1991 and go into pointed mode at higher resolution. Edge *et al.* (§2) find strong evolution in the highest luminosity X-ray clusters, which will have important implications for the ROSAT cluster sample, and for probes of the fluctuation amplitude in the cluster regime of the density perturbation spectrum. Although they claim their results favour a b_ρ closer to 2 than 1 for CDM type models, there are many assumptions that go into estimating the theoretical mass function so I do not as yet regard this as evidence against the $b_\rho \sim 1$ value suggested by other large scale data.

4. FLUCTUATIONS AND CMB ANISOTROPY

Large Angle Anisotropies

The dipole data presented from FIRAS and DMR and Cottingham's 1.6 cm Princeton balloon experiment agree with the previous balloon and RELICT 1 result. DMR now gets 3.3 ± 0.2 mK. The errors should narrow as a longer timeline of DMR data is analyzed. Once the dipole was subtracted, the maps George Smoot showed revealed nary a morsel for the hungry theorists.

Have we converged on an agreed-upon notation for quoting quadrupole limits? I have changed my mind on this and now agree with the Soviets, DMR and Cottingham that $(\Delta T/T)_{rms}$ for the multipole is better since it is more directly comparable to $\Delta T/T$ limits one gets for other measures, such as the *rms* fluctuation in a DMR beam. The quadrupole limit from both DMR and Cottingham is then 3×10^{-5} at the 95% confidence limit, about the same as the RELICT-1 limit. The quadrupole by itself is not of much interest. We want the amplitude, the spectrum, and the statistics of the primordial fluctuations, none of which the quadrupole alone determines well. In particular, we could have a very small quadrupole and yet a healthy amplitude. The homogeneous anisotropic Bianchi models popular in the 70s which give interesting quadrupoles but not much else are of little theoretical interest now.

For inflation-based theories, the multipoles are Gaussian-distributed. Bob Schaeffer expressed scepticism that DMR could prove the scale invariance for the *rms* amplitudes predicted by these theories. Nicola Vittorio expressed scepticism that DMR could prove that the fluctuations are Gaussian. The reason for both is that we get to observe only one realization from an ensemble. What DMR should be able to do is to get into the $b_\rho \gtrsim 2$ territory *assuming* scale invariance, and could, by relative likelihoods, distinguish among power laws. However, the per pixel sensitivity of DMR is such that to get to high b_ρ a sizeable number of pixels will be required, of various frequencies, so the results will be sensitive to how good the Galaxy model is. I am optimistic. However, we learned from Cottingham that anisotropy detected in his 3° beam map is not well fit by the current Galaxy model, so much work is required.

Steve Meyer showed first results from the MIT 3.8^o beam balloon experiment. The sensitivity they have achieved looks excellent — and they have more data to add. Like DMR it has 4 channels, from 1.8 mm to 0.44 mm. So far the limits given assume a Gaussian correlation function, which no models predict, so it is difficult to compare with theories. They get $\Delta T/T \lesssim 4 \times 10^{-5}$ for radiation coherence scales above their beam size. We look forward to their results when the scale invariant form is adopted. COBE has also not analyzed their data this way. Cottingham has, and I infer $b_\rho \gtrsim 0.2$, which is below the corresponding limit from RELICT 1, $b_\rho \gtrsim 0.4$. However, the DMR single pixel *rms* of $\sim 4 \times 10^{-5}$ gives the slightly better result, $b_\rho > 0.33$ at the 95% confidence limit for the $\Omega_B = 0.1$, $\Omega_{tot} = 1$ standard CDM model; Meyer's experiment currently gives about the same limit.

A poster by Neil Turok and collaborators predicted that textures lead to a non-Gaussian pattern of anisotropies on the sky, but only on large scales, with about 10 hot and cold spots with amplitudes above 10^{-4}, of size about 8°, conveniently close to DMR's beam size. Textures are higher order field defects generated in an early universe phase transition which act as seeds for structure formation. The particle theory is reasonably well motivated, a rarity in first principles theories. However, since George Smoot reported no effects above $\sim 10^{-4}$, except for the dipole, we might soon see DMR's first victim.

Intermediate and Small Angle Anisotropies

Mark Fisher gave us a progress report on MAX, the UCB/UCSB mm-wave balloon collaboration, using multiple frequencies and beams $\sim 30'$. The data analysis is in progress, but MAX promises eventually to yield sub-10^{-5} amplitudes. And the scale is about the best to probe the adiabatic CDM and HDM models, if there were no early reionization to erase small angle anisotropies. Nicola Vittorio described his analysis of the UCSB South Pole experiment (Peter Meinhold and Phil Lubin), which probes the same angular scale as MAX, and showed constraints with $b_\rho \gtrsim 0.4 - 0.6$ for CDM models, in agreement with work that George Efstathiou and I did on this data set, using Bayesian statistical methods rather than the frequentist methods used by Nicola. The rapidly improving limits on primary CMB anisotropies and the increasing complication of the experiments are driving we theorists to adopt ever more sophisticated statistical methods to ensure that we do not prematurely sound the death knell of a given theory.

Steve Myers reviewed small angle experiments and in particular unveiled his 96 field 'ring' data set using Owens Valley. The 7-field OVRO experiment gives similar constraints to the UCSB South Pole experiment for the standard CDM model with $\Omega_B = 0.1$ and no early ionization, and combined they give $b_\rho \gtrsim 0.9$. This is getting perilously close to the $b_g \sim 1 \pm 0.3$ and $b_g \sim 1.15 \pm 0.2$ inferred from comparing IRAS galaxy redshift surveys with the spiral and elliptical flow fields, by Bertschinger and Kaiser, respectively, but some caution is useful in that the IRAS galaxy biasing b_g is not necessarily close to the primordial spectrum amplitude measure b_ρ. Myers' constraints on models from the 96-field ring are not competitive with the 7-field ones, mainly because fluctuations are detected, probably faint radio sources, since deep VLA searches of a few of the fields have identified 3. This source problem plagues centimeter wavelength anisotropy experiments, and emphasizes that an analysis method is needed to simultaneously constrain source models and primordial perturbations. Multiple frequency experiments will be a great help in source decontamination.

Kris Gorski, Roman Juskiewicz and Ed Bertschinger adopted phenomenological rather than *ab initio* approaches, using velocity field information to predict anisotropy levels, assuming the anisotropies are entirely given by the Sachs-Wolfe effect for flat cold dark matter universes: *i.e.,* that the anisotropies are one-third of the gravitational potential fluctuations. Although on the scales where this has been used this assumption is somewhat suspect, it provides a useful forecast of the level at which primary anisotropies might appear: they all get around 10^{-5}. Kaiser took the power spectrum determined from QDOT and estimated values just below 10^{-5} assuming Sachs-Wolfe only. None of these estimates hold if there was early reionization. I am very much in favour of this approach, but we need to find better estimation (or constraint) procedures that jointly analyze CMB and large scale structure data, including their errors, such as a Bayesian approach with a parameterized fluctuation spectrum.

Very Small Angle Secondary Anisotropies

Roman Juskiewicz and Jim Peebles estimated the anisotropies generated by the nonlinear flows of plasma at high redshift, after matter has collapsed and been heated, and inspired by Jim's picture of $z \sim 20$ galaxy formation. For that model, these secondary effects are best probed on protogalactic scales of a few arcseconds, where the amplitude could be as large as $\sim 10^{-4}$. Sub-mm limits on scales $\sim 10''$ are only somewhat above this. Bruce Partridge reported that VLA limits have improved significantly: 2 fields chosen to be free of obvious radio sources have $\Delta T/T < (7-8) \times 10^{-5}$ on $10''$ and $\Delta T/T < (1-2) \times 10^{-5}$ on $60''$. I think these will prove very useful for constraining early reionization models when the nonlinear effects are better calculated, especially if the flow at that time is tied via gravitational instability to the current observed velocity flow. If the medium is dusty, a reasonable sub-mm background with detectable fluctuations might also be expected, as discussed by Bernard Carr. Whether the ionization can ensure that these are wiped out will undoubtedly be model sensitive and require much more theoretical work.

Thus the case looks promising for fluctuations of some sort to appear at the sensitivity levels the experiments are nearing. We know, however, that sources will abound at that level as well, so we will all be kept very busy trying to separate the cosmic from the mundane.

5. COSMOLOGICAL ENERGY INJECTION

The XRB, UVB and Compton Cooling

To follow the evolution of structure, we need to know where and when gas dynamics driven by non-gravitational energy release operated, and over what scale. Clues should be found in radiation backgrounds in a variety of wavebands. Bernard Carr and Avi Loeb discussed the limits on this energy release from COBE's FIRAS and UBC's COBRA. Energy released at $z \gtrsim 10^{6.4}$ will likely be reprocessed into a slightly hotter Planck spectrum, so we cannot tell what happened before this period from radiation backgrounds. Energy released between $z \sim 10^{6.4}$ and $z \sim 10^{4.6}$ leaves a spectrum which is approximately a Bose-Einstein distribution of nonzero chemical potential, which is at most $|\mu_\gamma|/T_\gamma < 0.008$ from COBRA and FIRAS, and we expect improvements. Below $\sim 10^{4.6}$, the energy imprints a classic Compton cooling distortion on the CMB, characterized by y, currently below 10^{-3} from COBE and UBC, but COBE should be able to probe to the few times 10^{-5} range. Avi showed that the hot IGM hypothesis for the origin of the X-ray background is now strongly ruled out if the medium is uniform, and the relative *rms* density contrast would have to exceed ~ 20 for a clumpy medium in pressure equilibrium to give it. There are other problems associated with a hot IGM origin for the XRB as well.

3D hydrodynamics is now at the level where it can be applied to the cosmological evolution of model universes such as the CDM and HDM models, as Loeb illustrated with the Cen, Ostriker and Jameson (1990) calculations. Although estimates of X-ray emission and y-distortions from the high temperature gas arising from pancake, filament and cluster shocking have been around almost as long as the pancake model has, detailed calculations can now be made. The results are very sensitive to biasing factor; for models with $b_\rho \lesssim 1$, FIRAS might be able to see the average Compton cooling distortion and the fluctuations should be accessible to anisotropy experiments. What is not feasible is to get the hard XRB from these sources, because the highest temperatures should not realistically exceed those in clusters. However, X-rays in clusters and ambient gas as cosmological diagnostics look more important than ever as GINGA, ROSAT, BBXRT, *etc.* deliver their data.

Bremsstrahlung from hot gas would also manifest itself by a distortion of the CMB at wavelengths longer than those that FIRAS probes. There has been a new round of spectrum measurements at these wavelengths, as Dave Wilkinson and Al Kogut discussed, which should eventually be able to constrain the bremsstrahlung component, once the systematics of these difficult measurements are better understood.

If the hard XRB cannot come from a hot IGM or from gravitational collapse, what gives rise to it? Daryl Leiter discussed the answer that I like: Precursor Active Galaxies (PAGs) radiating at $z \gtrsim 4$ with a higher number density than Seyferts, more than ~ 5000 per square degree, which satisfies the XRB anisotropy limits. Leiter supposes that the key evolutionary quantity, the compactness parameter (the luminosity-to-radius ratio), evolves from a high value (compact) to a low one so the X-ray emission evolves from hard to soft.

Paul Shapiro presented a puzzle that PAGs may perhaps solve. Even for the $z = 4.1$ quasar, there is no Gunn-Peterson trough to the blue of the QSO's Lyman α emission, indicating a sizeable UV background exists at $z \sim 4$ (although it is still well off current background light constraints) and the IGM is highly ionized. Other sources besides observed quasars are definitely required, and if the light comes from normal stars, a large amount of metal production will accompany the ionizing radiation emission. Paul infers solar metallicity in the collapsed gas for CDM with biasing factor $b_\rho = 2.5$ and one-tenth solar metallicity for $b_\rho = 1$.

PAGs can heat and ionize the IGM enough to satisfy Gunn-Peterson, with no metal restrictions, and give a spectrum harder than the thermal spectrum of a nuclear burning star so the loss of the ionizing radiation through redshift effects is not as large. (Very Massive Stars, which give black hole remnants as well, have the same virtues, but PAGs are less of an extrapolation from things that we know about). Richard Griffiths, pointing to the $L_X - L_{IR}$ correlation for starbursts from Einstein data, suggested X-ray sources in starbursts might be able to give the XRB. They can certainly give an interesting IRB.

The sub-mmB, FIRB and NIRB

Far infrared and submillimeter backgrounds used to be easy places to hide significant energy injection, but this is no longer true with COBE. Bernard Carr showed that the idea of dark matter as black hole remnants from Very Massive Objects (stars above $\sim 200\,M_\odot$) forming at $z \sim 300 - 30$ is difficult to reconcile with the FIRAS and COBRA limits, assuming the radiation is degraded by dust, although it is not yet impossible.

FIRAS has the sensitivity to see normal galaxies and starbursts as well, assuming even moderate evolution, as Carr, Steven Eales and Boqi Wang discussed. Thus, we may be stuck with a foreground subtraction (if a welcome one) to get at the waste energy from $z \gtrsim 5$, which is surely the holy grail for the FIRAS and DIRBE teams.

DIRBE is now the benchmark by which previous experiments are judged. IRAS fares reasonably well, except at 60 μm, and at 100 μm, where it is a factor of two off. Matsumoto *et al.* 's near IR rocket results have similar amplitudes, but a different shape. I was impressed with the agreement with ZIP, the vintage 1980-81 AirForce zodiacal light satellite experiment covering 2-30 μm, which I had not heard of before this meeting.

Going from the beautiful DIRBE maps to estimates of isotropic backgrounds is a formidable task for Mike Hauser and his team. Until DIRBE is ready to deliver processed data to the waiting theoretical horde, we must treat the current observations by COBE as upper bounds on backgrounds from the sub-mm to the near IR, in spite of prior claims of 100 μm and 2.2 μm cosmic background detections.

The radiation from VMOs would lie in the near infrared if no dust degradation occurred, and be accessible at DIRBE's sensitivity level if they leave black hole remnants. The complicated business of foreground subtraction implies this model for the dark matter is still safe. Plausible energy injection by more ordinary pregalactic stars, decaying particles and primeval galaxies is also accessible. Mike Hauser noted that, although helium expired in late September, good quality data can be obtained for quite a while as the dewar slowly heats, especially in the near IR. Thus the DIRBE problem is not to be sensitivity at all, but how to view the high redshift universe amid the cosmologically boring that probably dominates the observations.

Anisotropy measurements, touted as a method for distinguishing models of sub-mm energy emission seen by DIRBE or FIRAS, runs into a severe problem with the confusion unleashed by CIRRUS clouds, especially for COBE resolutions, although hopes for a bright future using sub-mm telescopes were expressed by Carr, Eales and Wang.

Background radiations can drive large dynamical actions. Avery Meiksen showed how the UV background could result in shell formation and fragmentation of cosmological HII regions. Bernard Carr pointed out that the FIRAS data rules out the mock gravity scenario of radiation pressure pushing dust around enough to have generated a substantial amount of cosmic structure.

The gamma ray background reviewed by Loeb is not at the moment of direct significance for cosmology, although GRO might reveal an exciting high redshift origin of the diffuse background. We could in principle learn about energy injection from the neutrino background that arises from collapsing stars, associated with normal supernovae or black hole collapses of VMOs, but I came away pessimistic: back side of the moon indeed.

6. A REQUIEM FOR CDM?

Hierarchical Fluctuation Spectra

As mentioned in §1, the accompanying figure illustrates how different observations discussed at this meeting can combine to cover most of the spectral bands in (comoving) wavenumber k of the density fluctuation spectrum. Below the figure, various speakers in the meeting are organized into the waveband region their work is mostly probing. This illustrates the wide range of topics covered at this meeting and, even so, there were some speakers that I couldn't easily classify this way.

The corresponding wavelength is $\lambda = 2\pi/k$ and a rough mass estimate is $M = \bar{\rho}(\lambda/2)^3 \approx 10^{13}\Omega\mathrm{h}\,(k^{-1}/(\mathrm{h}^{-1}\mathrm{Mpc}))^3$. (This diagram is constructed explicitly for a flat $\Omega = 1$ universe.)

Three power spectra are also shown. The $b_\rho = 2.5$ standard h=0.5 $\Omega = 1$ CDM spectrum was used in the original work of Davis, Efstathiou, Frenk and Davis. The CDM model with $b_\rho = 1$ is supported by the new CDM simulations described by Bertschinger that include velocity biasing. In 84-86 work I often used $b_\rho = 1.7$; from 86 on I have favoured 1.4. Thus *the* CDM model has meant different things to different people. Still, the possibility of $b_\rho = 1$ CDM was a surprise to me.

The spectrum which gives rise to a $\xi = (r/r_0)^{-\gamma}$ correlation function, with $\gamma = 1.8$ and $r_o = 5.4\,\mathrm{h}^{-1}\mathrm{Mpc}$, the conventional values of Davis and Peebles derived from the original CfA1 redshift survey, is shown for comparison. The CDM spectra shown have no nonlinear effects included, which act to steepen the spectrum on sub-cluster scales so it more closely matches the $\gamma = 1.8$ one.

The spectrum is defined as $d\sigma_\rho/d\ln k$, the square root of the power in the density fluctuations per logarithm of the wavenumber. For an $\Omega = 1$ universe it roughly gives $1 + z$, where z is the redshift when the *rms* fluctuations on scale

k reach unity. Thus with $b_\rho = 1$ cluster-scale amplitudes are nearly unity. For a given fluctuation spectrum, the pattern with maximum entropy (maximally random) is a Gaussian one. Non-Gaussian fluctuations are still characterized by a spectrum, but much else is needed to characterize the degree of non-random localization of the power. Non-Gaussian fluctuations could arise from the primordial generation process, be induced through biasing (*e.g.,* by gas-dynamical processes), or be induced through nonlinear gravitational dynamics. Indeed dynamics can completely determine the spectrum below that of cluster scales, so we cannot claim to have a measurement of the primordial power spectrum on intermediate scales from galaxy clustering.

CDM is an example of a hierarchical scenario with the formation of objects of increasing mass being associated with progressively longer (spatial) wavelength bands of the spectrum, as indicated by the sequence first stars, dwarf galaxies, galaxies, groups and clusters, which form at decreasing redshifts. Of course the final collapsed objects are much smaller than the waves whose constructive interference gave rise to them. Although the clustering of cold dark matter will be hierarchical, the gas clustering is likely to be quite different, on subgalactic scales at least. One example is photoionization from a UV background discussed by Shapiro and Meiksen, which effectively filters the power spectrum of gas on scales below the 'gals' scale. Given Gunn Peterson, this is probably unavoidable, and gives rise to a nice model for Lyman alpha clouds as forming proto-dwarf galaxies with gas outflow. Structure formation as well as destruction may accompany energy injection, however (Meiksen).

The Wavenumber Space View of Experiments

The regions of wavenumber space that are well probed by the observations of the clustering of galaxies and of clusters are indicated by the lines under ξ_{cc} (the cluster-cluster correlation function), ξ_{cg} (the cluster-galaxy correlation function) and ξ_{gg} (the galaxy-galaxy correlation function). Up until recently, ξ_{gg} as determined from redshift surveys didn't probe much beyond $10\,\mathrm{h}^{-1}\mathrm{Mpc}$, if that. However, the deep angular catalogues, both Lick and the APM survey, probe larger scales. In particular, the Maddox and Efstathiou result on the angular correlation function w_{gg} probes out to about $15\,\mathrm{h}^{-1}\mathrm{Mpc}$ in k^{-1}, maybe up to 20. That is why it is such a powerful result.

The large scale texture (LST) as revealed by Great Walls goes out to only $\sim 15\,\mathrm{h}^{-1}\mathrm{Mpc}$, or maybe $20\,\mathrm{h}^{-1}\mathrm{Mpc}$ in k^{-1} if the $120\,\mathrm{h}^{-1}\mathrm{Mpc}$ wavelength that is indicated by the pencil beam survey is a characteristic scale. Large scale streaming velocities (LSSV) probe scales from that associated with the Local Group up to about $15\text{-}25\,\mathrm{h}^{-1}\mathrm{Mpc}$, although the current data subtends a region which is too small to estimate the spectrum out to the k shown.

Nick Kaiser's estimate for the QDOT galaxy-galaxy power spectrum, the Fourier transform of ξ_{gg}, goes out to about 20-25 $\mathrm{h}^{-1}\mathrm{Mpc}$. The extension of the CDM power spectrum indicated by the $\gamma = 1.8$ dotted curve is roughly the extra power inferred. The vintage 86-87 large scale structure problems led to a phenomenological spectrum (yes me too) we called CDM+X, which looks like

QDOT's extension. It was invoked to explain the high values for the cluster-cluster correlation function, large scale velocities and also the angular correlation function of galaxies (when the Maddox and Efstathiou results came in).

Thus, with varying belief levels in the intervening years, X has been around to stimulate theorists to write papers exploring the parameter space that could give rise to it: it could arise with more than one type of dark matter, non-zero Λ, open universes, or with funny inflation, none of which are compelling stories. What we also emphasized at the time was that X almost inevitably leads to hefty intermediate angle CMB anisotropies, and although around that time there was evidence from Tenerife that they existed, we now have stronger constraints than ever, especially if recombination is standard (*e.g.,* Vittorio's talk).

The microwave background anisotropy probes are characterized in the figure by the labelled filters which are functions of the (comoving) wavenumber k. (The filters act on a spectrum for $\Delta T/T$ not on the one shown for $\delta\rho/\rho$, but the bands of k probed are as shown.) On the largest scale is COBE's DMR, with its 7° beam. Shifting it over by a factor of two to higher k gives the filter for Cottingham's Princeton and Steve Meyer's MIT balloon experiments. Shifting by a factor of 10 gives the DIRBE beam; this takes it over as far as the Lubin and Meinhold South Pole experiment, which is the second filter shown, basically that of a two beam difference experiment. The filter for the MAX experiment described by Fisher is the same. If the Universe is reionized very early, then the width of the photon decoupling region around the point where the universe passes through optical depth unity is the NR vertical straight line, so that a larger beam experiment than MAX would be better for probing the theory. The OVRO region probed is indicated by the third filter. If recombination is standard the width of the last scattering surface is indicated by the SR line. Thus OVRO is a very poor probe of universes with early reionization, and only catches the tail of the SR power. Steve Myers' 96 field experiment has the same filter.

The IRAM sub-mm telescope filter is typical for sub-mm experiments and probes small scales well below the SR line, hence only tests for secondary anisotropies that arise after matter has collapsed. It is, however, too small to be ideal for probing the hot X-ray emitting gas discussed by Loeb, but is reasonable for protogalaxy flows discussed by Peebles and Juskiewicz and primeval and normal infrared emitting galaxies discussed by Carr, Eales and Wang. Thus it is expected to probe nonlinear parts of the spectrum as well as linear parts: patterns in these secondary backgrounds offer a potentially rich window to $z > 5$ structure.

If b_g is not a function of position — which would induce large scale power — then w_{gg} and QDOT give powerful indications of extra large scale power over that in the scale invariant CDM spectrum. In that sense, the theory is dead. For me, however, the definitive death notice for CDM will be delivered by those intrepid microwave background explorers whose drive towards higher precision has led us to the brink of CDM predictions, even with early reionization. And only the most outrageous nonlocal biasing could be invoked to avoid the anisotropy limits, unlike the galaxy clustering case.

Anisotropy predictions from flows and clustering show how near we should be to seeing primordial perturbations in the CMB; failure to do so will lead to a true crisis, the potential loss of the gravitational instability framework, but without much extra energy to power non-gravitational forces allowed by COBE. Thus *After the First Three Minutes* foresaw a wonderful next few years with CMB data-bases, especially from COBE, combining with large scale structure data bases, to definitively judge the first principles theories and guide the construction of phenomenological models.

With apologies to the collaborators and colleagues of the speakers, I have chosen to summarize only what was said at the meeting and refer only to who said it. The full credit listings should be found within the other contributions to these proceedings.

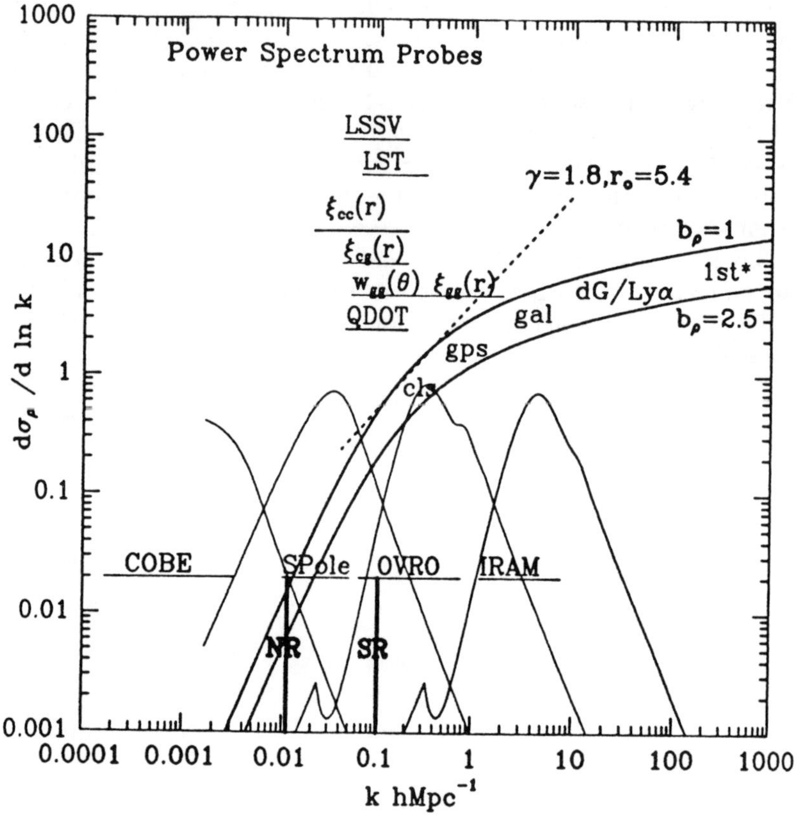

Smoot Myers Shapiro
Cottingham Edge Meiksen
Meyer Loeb Carr
Schaeffer Mushotsky
Hauser Rubin
Fisher Charlton
Juskiewicz
Dressler
Lahav
Geller
Bahcall
Szalay
Windhorst
Vittorio Kaiser
Gorksi Peebles
Bertschinger

Workshop Epilog

Charlcs L. Bennett
NASA Goddard Space Flight Center, Code 685
Laboratory for Astronomy and Solar Physics
Greenbelt, MD 20771

It was a great pleasure for me to have helped organize this workshop. I am grateful to all the contributors for their talks and for providing the written contributions contained in these Proceedings. I am also grateful to Steve Holt for his single-minded determination to make this workshop a reality and a success.

> *What we call science today consists of a haphazard heap of information, united by nothing.* Tolstoy in What is Religion?

The words of Tolstoy, above, provide a fairly accurate description of this workshop. It was the intention of the workshop organizers to place a heavy emphasis on the new observational results in cosmology. In the last few years, with the rapid advance of technology, observational astrophysics and cosmology have flourished. A wealth of new data and understanding have been attained only recently. Our understanding is far from complete, but our current and near-future observational capabilities assure continuing rapid advances in cosmology. At the workshop we heard reports on new observational information, but our theoretical frameworks have failed to unite our pictures of the early and current universe in a simple and compelling way.

> *The aim of science is to seek the simplest explanation of complex facts. We are apt to fall into the error of thinking that the facts are simple because simplicity is the goal of our quest. The guiding motto in the life of every natural philosopher should be, "Seek Simplicity and distrust it."*
> Alfred North Whitehead in Concept of Nature

Jim Peebles set the tone of the workshop by recalling that cosmological models before the 1980s were largely based on semi-empirical interpretations of observational data, but in the 1980s the theoretical community turned towards *ab initio* theories which would explain the observables. Jim Peebles concludes, "But, to put it bluntly, the *ab initio* approaches have not yet given us anything that is believable according to the standard rules of evidence in physical science." He predicts a return to semi-empirical modeling with the great new data influxes expected in the decade to come.

Dick Bond, in his outstanding summary of the workshop, has defended the other side of the theoretical community where he has rededicated himself to seek *ab initio* explanations for the observables.

Dave Schramm set the boundary conditions for this workshop by outlining the Big Bang model of the early universe. The Big Bang model has strong support from: (1) the observed expansion of the galaxies, (2) the nucleosynthesis calculations which successfully predict the observed light element abundances, and (3) the temperature, spectrum, and isotropy of the cosmic microwave background (CMB) radiation.

Since the pioneering work of Slipher and Hubble over half a century ago, almost all agree that we live in an expanding universe, but the value of the Hubble constant is known only roughly. Much work but little progress has come in attempting to sort out discrepant measurements of the Hubble constant. Observations and analyses peg the value at 50 km/sec/Mpc $<H_o<$ 100 km/sec/Mpc, almost always with error bars much smaller than the discrepancies between independent measurement techniques. The obvious conclusion is that systematic error bars are almost never overestimated!

Light element nucleosynthesis is lucidly summarized in this volume by Dave Schramm. While original work in the field concentrated on ^{4}He, the additional light element abundances of ^{2}H, ^{3}He, and ^{7}Li are now successfully predicted also. The abundances of these latter species depend on the total baryon density of the Universe. In short, the light element abundances are consistent with the Big Bang model, but only if the baryon density in the universe is $\Omega_b \sim 0.065 \pm 0.045$, a range which decisively excludes $\Omega_b=1.0$.

The CMB radiation, discovered in 1965 by Penzias and Wilson in a single frequency measurement described simply as "isotropic" is now being measured with astonishing precision over decades of wavelengths. The COBE FIRAS experiment has made a precise determination that the CMB radiation is described by a blackbody of temperature 2.735±0.060 K, as presented by John Mather in this volume. This measurement is confirmed by the measurement of Halpern, Gush, and Wishnow (also in this volume) which, somewhat coincidentally, resulted in a blackbody temperature of 2.736±0.017 K. There are no indications of extra energy injections in the early universe which would have resulted in a distortion of the spectrum. These measurements lend strong support for a delightfully simple picture of the early universe - a very clean Big Bang scenario.

George Smoot announced preliminary new CMB large angular scale anisotropy results from the COBE DMR experiment at this conference. We also heard from other groups stoically carrying out difficult balloon measurements but with the major new advantage of bolometric detection. The new sensitivities are wonderful. As pointed out by Dave Wilkinson, the $\Delta T/T$ business is no longer limited by sensitivity, but only by the all-important systematic errors of each measurement. These anisotropy measurements also point to a simple isotropic Big Bang model.

The Big Bang model is a great success in the best classical spirit of physical models; existing measurements were naturally explained within a theoretical framework which made definite predictions, and those predictions were beautifully borne out by further measurement.

While the Big Bang model is a great success, many questions remain, especially when attempting to match the currently observed universe with the Big Bang initial conditions. Coupling the Big Bang prediction of $\Omega_b \sim 0.065 \pm 0.045$ with the assumptions that $\Omega_o=1$ and $\Lambda_o=0$ leads to the immediate conclusion that either most of the matter in the universe is nonbaryonic, or that $\Lambda_o \neq 0$, or both.

It is often stated that inflation requires $\Omega_o=1$, but it actually requires $\Omega_o+\Lambda_o/3H_o^2 = 1$ (although this could be modified with additional assumptions which would further complicate the model). The assumption that $\Lambda_o=0$ is the reason for the great interest in possible nonbaryonic dark matter in the universe. Inflation effectively uses $\Lambda \neq 0$ to drive the inflation, the idea being that Λ_o is now negligibly small. Since $\Omega_b \sim 0.06$ from nucleosynthesis agrees well with galaxy (including dark halo) dynamics, to claim the universe is critically closed (Ω_0=1, Λ_0=0) is to claim that 94% of the mass in the universe is nonbaryonic dark matter and does not participate in galaxy dynamics.

Dark matter is known to exist in the universe. Evidence for dark matter is clear from galaxy rotation curves, gravitational lensing systems, dynamics of clusters of galaxies, and peculiar motions in the local universe, as reported in this volume by Vera Rubin, Tony Tyson, Rich Mushotzky, and Ofer Lahav, respectively, as well as several of the other contributions. From an observational standpoint, there is not evidence for enough dark matter to make Ω_0=1, although Nick Kaiser's analysis of IRAS galaxies in the QDOT survey arrives at Ω_0=0.8±0.3.

There is no doubt that a substantial amount of baryonic dark matter exists, but there is also little evidence that baryonic matter makes up the full content of Ω_0=1. Making Ω_0=1 with baryonic matter would conflict with the Big Bang nucleosynthesis results, so while the nature of the baryonic dark matter is an important and interesting question, the prime cosmological question centers on the nonbaryonic dark matter candidates. One such candidate is a massive neutrino. The nucleosynthesis models put a limit on the number of neutrino generations, $N_\nu \leq 4$. As reported in several of the papers in these proceedings, recent particle accelerator measurements put N_ν=2.98 ± 0.06, i.e. there are three neutrino generations: ν_e, ν_μ, and ν_τ. Experimental limits on the masses of the ν and τ neutrinos do not yet rule out cosmologically significant masses, but neutrinos may have difficulty serving as seeds for galaxy formation. Axions and Weakly Interacting Massive Particles ("WIMPs") of various kinds are also possibilities for explaining the nature of the nonbaryonic dark matter, but their possible energies are seriously constrained, as discussed by Dave Spergel and Larry Krauss.

Many forms of dark matter were in equilibrium with the CMB radiation until decoupling, forcing the constraint on dark matter theories that the amplitude of angular fluctuations in the CMB radiation must be kept very small while maintaining the large-scale power seen in galaxy correlations.

In summary, there exists a substantial amount of baryonic dark matter in the universe, but an even greater amount of nonbaryonic dark matter would be needed to reach the inflation prediction of Ω_0=1 if Λ_0=0. We therefore reach the conclusion that (1) most of the mass in the universe is in a form heretofore unknown, or (2) inflation requires substantial modification, or (3) $\Lambda_0 \neq 0$, or (4) Big Bang nucleosynthesis is wrong. While many may not like any of these options, at least one must be true.

The quantity and nature of the dark matter in the universe plays a critical role in galaxy formation and the ability to form large-scale structures in the universe. Surprising evidence for a characteristic length scale of ~120 Mpc has been found in the new pencil beam surveys, as described by Alex Szalay. The nearer of the excess concentrations are seen to correspond with known clusters, as described by Neta Bahcall.

Earlier, the discovery of the large-scale velocity flows came as a surprise as well. The fact that each major new observation leads to a struggle for the models makes it clear that the models, far from being able to predict, are barely able to accommodate. The theories developed to explain the diverse observations of large streaming velocities, large-scale power in the distribution of luminous object in the sky, but a lack of power as determined by measurements of only upper limits on CMB anisotropies center on the existence of cold dark matter, sometimes combined with hot dark matter, and fairly free range of the cosmological parameters H_0 and Ω_0. To date these theories have not been predictive. Indeed, they have required successive modifications (referred to in this conference as "squirming") to survive in the face of new measurements.

Many new measurements will soon be upon us. I summarize here several of the ones I think will make a significant impact in the next few years.

There is no doubt that improved $\Delta T/T$ experiments will continue. On large angular scales, the COBE DMR experiment continues to aquire data, and analysis efforts continue toward values of $\Delta T/T$ approximately an order of magnitude below current limits should potential systematic error limits and Galaxy subtraction be cooperative. On medium and small angular scales there will continue to be progress as reported by Boughn, Meyer, Myers, Fischer, and others. With some of these same instruments, more definitive measurements of the Sunyaev-Zeldovich effect should be forthcoming using the new bolometric detection sensitivities. More accurate measurements of the low frequency spectrum of the CMB would be very desirable, but will probably need to be done aboard balloons or even a space based platform.

Mike Hauser reported that the COBE DIRBE experiment is conducting the first all sky sensitive search for a remnant glow from the earliest generation of luminous objects. The technique is similar to that of Penzias and Wilson: account for everything you see, subtract it away, and see if anything is left. (Of course, Mike Hauser is going through the process with somewhat less cosmological serendipity than Penzias and Wilson). Either a finite detection of an infrared background or the lack of such a background will be a serious constraint on theories.

The COBE FIRAS experiment is also well on its way to producing better limits on the pressure integral along line of sight (the "y-parameter"); having already ruled out a smooth hot intergalactic medium, the experiment will measure or place stringent limits of energy releases in the early universe.

The blackbody nature of the CMB, with no spectral deviations, immediately tells us that there is no hot intergalactic medium, so the X-ray background is not primarily from a diffuse intergalactic medium, but future X-ray observations will be very important for cosmology. As we heard at this conference from Edge *et al.*, the ROSAT mission will be providing a list of clusters of galaxies an order of magnitude larger than the workhorse Abell list. The implications of this for large scale structure measurements, as well as many other measurements, should not be underestimated.

The gravitational lens work reported by Tony Tyson will produce powerful new results in the measurements of M/L ratios in the lensing systems.

The very question of whether there is a solar neutrino problem will soon be known definitively from ongoing experiments. If the problem remains, then either our standard stellar models are in error or the neutrino must have mass. This is obviously very important, although it should be recognized that the current limits on such a neutrino mass may already prevent the neutrino from being a dominant mass in the universe, as discussed by Dar.

Given these and other ongoing projects which promise important new data, I express my optimism that in the near future we will understand our universe considerably better than we do today.

> *the place where optimism flourishes is the lunatic asylum*
>
> H. Ellis in The Dance of Life

> *To be ignorant of one's ignorance is the malady of the ignorant.*
>
> A. B. Alcott in Table Talk

CREDITS

Best Feature-Length Motion Picture	Ed Bertschinger N-Bodies in Motion (NC-17)
Best Picture	Mike Hauser The Milky Way Galaxy
Best Musical	Rainer Weiss COBEians set to music
Best Comedy	Ned Wright "One advantage of the Steady-State Theory is that if the Universe is optically thick then we wouldn't see these pesky quasars whose redshifts we can't explain"
Longest Focal-Length Lens	Tony Tyson
Best Defense of Cold Dark Matter	Dick "Bayesian" Bond "It's not dead...is not...is not...is not...!"
Best Director (& Schedule Tyrant)	Steve Holt
Best Documentary	Virginia Trimble
Lifetime Achievement Award	Vera Rubin "I've been telling you for 15 years there is large-velocity streaming and dark matter"
Longest Rabbi Joke	Arnon Dar

Appendix A:

Conference Programme

AGENDA
AFTER THE FIRST THREE MINUTES
An Astrophysics Workshop
University of Maryland
Center of Adult Education
College Park, Maryland

14 - 17 OCTOBER 1990

Sunday, 14 October

1800 - 2100	Registration & Welcoming Reception	

Monday, 15 October

0800 - 1700	Registration	
0800 - 0900	Continental Breakfast	

1. OPENING SESSION (S. Holt, Chair)

0900 - 0910	Opening Remarks:	S. Holt
0910 - 0950	The First Three Minutes:	D. Schramm
0950 - 1030	General Introduction:	J. Peebles
1030 - 1100	BREAK	

2. COSMIC BACKGROUNDS I - CMB SPECTRUM
(R. Partridge, Chair)

1100 - 1130	COBE Measurements:	J. Mather
1130 - 1200	Ground Based Measurements:	D. Wilkinson
1200 - 1207	Low Frequency Measurements:	A. Kogut
1207 - 1217	Recent Rocket Measurements:	M. Halpern
1217 - 1230	Open Discussion	
1230 - 1400	LUNCH	

3. COSMIC BACKGROUNDS II - CMB ANISOTROPIES
(C. Bennett, Chair)

1400 - 1430	COBE Measurements:	G. Smoot
1430 - 1450	Overview and Small Scale Measurements:	S. Myers
1450 - 1500	Medium-Scale Measurements:	M. Fischer
1500 - 1505	Large-Scale Low Frequency Measurements:	D. Cottingham
1505 - 1510	Large-Scale High Frequency Measurements:	S. Meyer
1510 - 1530	Open Discussion	

AGENDA

1530 - 1600	BREAK	

4. COSMIC BACKGROUNDS III - CMB IMPLICATIONS (K. Gorski, Chair)

1600 - 1630	CMB Implications:	N. Vittorio
1630 - 1700	CMB Implications:	R. Juszkiewicz
1700 - 1730	Open Discussion	

5. POSTER SESSION

1730 - 2100	Unstructured Poster Session with Light Refreshments	

Tuesday, 16 October

0800 - 1700	Registration	
0800 - 0830	Continental Breakfast	

6. COSMIC BACKGROUNDS IV - THE CIB AND GALAXY FORMATION (E. L. Wright, Chair)

0830 - 0900	COBE and other Observations:	M. Hauser
0900 - 0930	Implications:	B. Carr
0930 - 0940	Integrated Far-IR Background from Galaxies:	B. Wang
0940 - 0950	Sub-mm Background Expected from Normal Galaxies:	S. Eales
0950 - 1000	Diagnosing the Reionization of the Universe:	P. Shapiro
1000 - 1015	Open Discussion	
1015 - 1045	BREAK	

7. COSMIC BACKGROUNDS V - X-RAYS, GAMMA-RAYS AND NEUTRINOS (E. Boldt, Chair)

1045 - 1115	The Diffuse X And Gamma Ray Backgrounds:	A. Loeb
1115 - 1125	Discrete Sources and the IGM:	D. Leiter
1125 - 1140	Open Discussion of X-Ray and Gamma-Ray Backgrounds	
1140 - 1210	The Cosmic Neutrino Background:	A. Dar
1210 - 1220	Neutrino Detection Techniques:	D. Cline
1220 - 1230	Open Discussion of Neutrino Background	
1230 - 1400	LUNCH	

AGENDA

8. LARGE SCALE STRUCTURE I - OBSERVATIONS
(N. Bahcall, Chair)

1400 - 1430	Local Surveys:	A. Dressler
1430 - 1500	Deep Surveys:	M. Geller
1500 - 1530	Deep Surveys:	A. Szalay
1530 - 1535	Galaxy Correlation Function to 26 mag:	R. Windhorst
1535 - 1540	ROSAT All-Sky Survey Results for Clusters:	A. Edge
1540 - 1600	Open Discussion	
1600 - 1630	BREAK	

9. LARGE SCALE STRUCTURE II - THEORY
(S. White, Chair)

1630 - 1700	Formation Of Large Scale Structure:	N. Kaiser
1700 - 1730	N Body Simulations:	E. Bertschinger
1730 - 1740	Cosmological HII Regions and the IGM:	A. Meiksin
1740 - 1800	Open Discussion	
1800 - 1900	RECEPTION - *"No Host" Bar*	
1900 - 2200	BANQUET	
	After-Dinner Speaker:	R. Weiss

Wednesday, 17 October

0800 - 0830	Continental Breakfast	

10. DARK MATTER I - OBSERVATIONAL EVIDENCE I
(D. Burstein, Chair)

0830 - 0900	Rotation Curves and Escape Velocities:	V. Rubin
0900 - 0930	Dynamics and X- Ray Observations:	R. Mushotzky
0930 - 0937	Galaxy Pairs in Redshift Catalogs:	J. Charlton
0937 - 1000	Open Discussion	
1000 - 1030	BREAK	

11. DARK MATTER II - OBSERVATIONAL EVIDENCE II
(R. Blandford, Chair)

1030 - 1100	From Peculiar Velocity Field of the Local Universe:	O. Lahav
1100 - 1130	From Gravitational Lensing:	T. Tyson
1130 - 1140	Compact Matter Limits from Gravitational Lensing:	R. Nemiroff
1140 - 1200	Open Discussion	

AGENDA

1200 - 1330	LUNCH	

12. DARK MATTER III - THEORY (L. Krauss, Chair)

1330 - 1355	Constraints from Astrophysics:	D. Spergel
1355 - 1405	Constraints from Particle Physics:	L. Krauss
1405 - 1430	Alternative Interpretation:	D. Kazanas
1430 - 1435	Halo Phase and Detection of WIMPs:	D. Ryu
1435 - 1440	MACHO Project	D. Bennett
1440 - 1500	Open Discussion	
1500 - 1530	BREAK	

13. CLOSING SESSION (S. Holt, Chair)

1530 - 1545	The First Fifty Years of the First Three Minutes:	V. Trimble
1545 - 1630	Summary Talk:	R. Bond

* * * * * * * *

<u>LOCATIONS:</u>

Registration...	Main Consourse
Welcoming Reception...	Chesapeake Room
SESSIONS 1 - 9...	Auditorium
SESSIONS 10 - 13...	Volunteer Fire Fighters Room (VFF)
Breaks...	Main Concourse
Lunches...	Founders Room
POSTER SESSION...	Main Concourse
Banquet...	Founders Room

Appendix B:

List of Attendees

AFTER THE FIRST THREE MINUTES
The University of Maryland, Center of Adult Education
15 - 17 October 1990

LIST OF PARTICIPANTS

Name	Affiliation
Adams, Jim	Naval Research Laboratory
Aizenman, Morris L.	National Science Foundation
Alsop, David	University of California/Berkeley
Ashman, Keith	Space Telescope Science Institute
Bahcall, Neta	Princeton University
Barnes, William J.	Massachusetts Institute of Technology
Barthelmy, Scott	USRA/NASA Goddard Space Flight Center
Beatty, James	Boston University
Bennett, Charles	NASA/Goddard Space Flight Center
Bennett, David	Inst. of Geophysics & Planetary Physics, LLNL
Bernstein, Gary	AT & T Bell Labs.
Berriman, Graham	GSC/NASA Goddard Space Flight Center
Bertschinger, Edmund	Massachusetts Institute of Technology
Bhavsar, Suketu	University of Kentucky
Bird, Christina	University of Minnesota
Blanchard, Paul	BDM International
Blandford, Roger	California Institute of Technology
Blitz, Leo	University of Maryland
Boggess, Nancy	NASA/Goddard Space Flight Center
Boldt, Elihu	NASA/Goddard Space Flight Center
Bond, Richard	Canadian Inst. of Theoretical Physics
Boughn, Stephen	Haverford College

Brandenberger, Robert	Brown University
Burstein, David	Arizona State University
Cameron, Robert	USRA/Naval Research Laboratory
Carr, B. J.	Queen Mary College
Casertano, Stefano	University of Pittsburgh
Casey, Sean	NRC/NASA Goddard Space Flight Center
Charlton, Jane C.	Steward Observatory/University of AZ
Cheng, Ed	NASA/Goddard Space Flight Center
Chiu, Hong-Yee	NASA/Goddard Space Flight Center
Clapp, Andre	University of California/Berkeley
Cline, David	UCLA
Cline, Thomas	NASA/Goddard Space Flight Center
Cottingham, Dave	University of California/Berkeley
Crotts, Arlin	NASA/Goddard Space Flight Center
Daly, Ruth	Princeton University
Danks, Anthony	ST Systems Corporation
Dar, Arnon	Technion/NASA GSFC
De Amici, Giovanni	University of California, Berkeley
Dickey, John M.	University of Minnesota
Dressler, Alan	The Observatories of CIW
Drukier, A. K.	ARC/USC
Dubin, Maurice	NASA/Goddard Space Flight Center
Dwek, Eli	NASA/Goddard Space Flight Center
Eales, Stephen	University of Toronto
Ebeling, Harald	Max-Planck Institut fur Extraterr. Physik
Edge, Alastair	Univ. of Cambridge & MPE, Garching

Ekers, Ron	Australia Telescope National Facility
Eplee, Eugene	General Science Corp.
Fagg, Lawrence	Catholic University of America
Fall, Michael	Space Telescope Science Institute
Feldman, Hume A.	CITA, University of Toronto
Felten, James	USRA/NASA Goddard Space Flight Center
Fichtel, Carl	NASA/Goddard Space Flight Center
Fischer, Marc	University of California/Berkeley
Fixsen, Dale	USRA/NASA Goddard Space Flight Center
Freedman, Immanuel	ST Systems Corporation
Freudenreich, Henry	ST Systems Corporation
Frost, Kenneth J.	NASA/Goddard Space Flight Center
Gelb, James M.	Massachusetts Institute of Technology
Geller, Margaret	Center for Astrophysics
Giroux, Mark	University of Texas
Gorski, Krzystof	Los Alamos National Laboratory
Griffiths, Richard	Space Telescope Science Institute
Guhathakurta, P.	Institute for Advanced Study
Gulkis, Samuel	Jet Propulsion Laboratory
Gunderson, Joshua	University of California/Santa Barbara
Gursky, Herbert	Naval Research Lab/Space Science Div.
Haines, Paul	Dartmouth College
Haken, Michael	University of Maryland
Halpern, Mark	University of British Columbia
Hartman, Robert	NASA/Goddard Space Flight Center
Hauser, Michael	NASA/Goddard Space Flight Center

Hawthorn, J. Bland	Rice University
Henning, Patricia	University of Maryland
Henriksen, Mark	Applied Research Corporation
Herreid, Leon	ST Systems Corporation
Hinshaw, Gary	USRA/NASA Goddard Space Flight Center
Holt, Stephen S.	NASA/Goddard Space Flight Center
Inman, Casey	Massachusetts Institute of Technology
Jackson, Peter	ST Systems Corporation
Jahoda, Keith	NASA/Goddard Space Flight Center
Janssen, Michael	Jet Propulsion Laboratory
Jones, Frank	NASA/Goddard Space Flight Center
Juszkiewicz, Roman	Institute for Advanced Studies
Kafatos, Menas	George Mason University
Kaiser, Nick	University of Toronto, CITA
Kazanas, Demosthenes	NASA/Goddard Space Flight Center
Keegstra, Phillip B.	ST Systems Corporation
Kelley, Richard	NASA/Goddard Space Flight Center
Kelsall, Thomas	NASA/Goddard Space Flight Center
Kerr, Frank J.	USRA/University of Maryland
Kinzer, Robert	Naval Research Laboratory
Kniffen, Don	NASA/Goddard Space Flight Center
Kogut, Al	NRC/NASA Goddard Space Flight Center
Kondo, Yoji	NASA/Goddard Space Flight Center
Kontizas, Evan	LASP/NASA Goddard Space Flight Center
Krauss, Lawrence	Yale University
Krolik, Julian	Johns Hopkins University

Krombel, Keith E.	USRA/Naval Research Laboratory
Kurfess, James	Naval Research Laboratory
Lahav, Ofer	Cambridge University, Inst. of Astronomy
Landy, Stephen	Johns Hopkins University
Lange, Andrew	University of California, Berkeley
Leiter, Darryl	FSTC - Science Division
Levin, Janna	Massachusetts Institute of Technology
Lui, Chuan Sheng	University of Maryland
Loeb, Avi	Princeton University
McDonald, Frank	University of Maryland
Madau, Piero	Space Telescope Science Institute
Mannheim, Philip	University of Connecticut
Mangus, John D.	NASA/Goddard Space Flight Center
Maran, Stephen P.	NASA/Goddard Space Flight Center
Marks, Dennis	Valdosta State College
Marshall, Frank	NASA/Goddard Space Flight Center
Mather, John C.	NASA/Goddard Space Flight Center
Mattox, John	NASA/Goddard Space Flight Center
Mead, Jaylee	NASA/Goddard Space Flight Center
Meiksin, Avery	Space Telescope Science Institute
Melott, Adrian	University of Kansas
Meyer, Stephan S.	Massachusetts Institute of Technology
Michalitsianos, Andy G.	NASA/Goddard Space Flight Center
Mithell, Kenneth	GSC/NASA Goddard Space Flight Center
Miyaji, Takamitsu	University of Maryland
Morikawa, Masahiro	Brown University

Moseley, Harvey	NASA/Goddard Space Flight Center
Mushotzky, Richard	NASA/Goddard Space Flight Center
Myers, Steven T.	University of Toronto, CITA
Nemiroff, Robert	NRC/Naval Research Laboratory
Neupert, Werner	NASA/Goddard Space Flight Center
Norris, Jay	NASA/Goddard Space Flight Center
Obenauer, John	Sachs-Freeman Associates @ NRL
Odenwald, Sten F.	SFA Inc./Naval Research Laboratory
Ormes, Jonathan	NASA/Goddard Space Flight Center
Ozernoy, Leonid	Los Alamos National Laboratory
Partridge, Robert B.	Haverford College
Peebles, P. J. E.	Princeton University
Pemmaraju, Ammiraju	Stevens Institute of Technology
Peterson, Ivars	Science News
Peterson, Jeffrey	Princeton University
Read, Shirley M.	ST Systems Corporation
Rhee, George	New Mexico State University
Rhie, Sun Hong	Lawrence Livermore Nat'l Lab.
Rickard, Lee J.	Naval Research Laboratory
Riegler, Guenter	NASA Headquarters
Rokke, Laurie	ST Systems Corporation
Rosen, Larry	ST Systems Corporation
Rosenblatt, Edward	University of California/La Jolla
Rubin, Vera C.	Carnegie Institution of Washington
Ruhl, John	Princeton University
Ryu, Dongsu	Princeton University Observatory

Santana, Jairo	ST Systems Corporation
Scaramella, Roberto	Osservatorio Astronomico di Roma
Schaefer, Robert K.	Bartol Research Institute
Schlegel, Eric	USRA/NASA Goddard Space Flight Center
Schramm, David	University of Chicago
Schreier, Ethan J.	Space Telescope Science Institute
Scott, Courtney	ST Systems Corporation
Shafer, Rick	NASA/Goddard Space Flight Center
Shandarin, Sergei	Univ. of Kansas/Inst. for Physical Problems
Shapiro, Maurice M.	University of Maryland
Shapiro, Paul	University of Texas
Shaya, Edward	University of Maryland
Shuman, Fred	ST Systems Corporation
Signore, Monique G.	Ecole Normale Superieure
Silverberg, Robert	NASA/Goddard Space Flight Center
Sinha, Ramesh	USRA/TIRF, Bombay
Skard, John Arthur	ST Systems Corporation
Smecker, Tammy A.	Johns Hopkins University
Smith, William (Chris)	Princeton University
Smith, Howard A.	Smithsonian Institution
Smoot, George	University of California/Berkeley
Snyder, William	Naval Research Laboratory
Soker, Noam	Center for Astrophysics
Spiesman, William	GSC/NASA Goddard Space Flight Center
Spergel, David	Princeton University Observatory
Staggs, Suzanne	Princeton University

Stecher, Ted	NASA/Goddard Space Flight Center
Stecker, Floyd	NASA/Goddard Space Flight Center
Stemwedel, Sally M.	Applied Research Corporation
Stiller, Bertram	(Retired)
Stone, Robert G.	GSRC/NASA Goddard Space Flight Center
Streitmatter, Robert	NASA/Goddard Space Flight Center
Swank, Jean	NASA/Goddard Space Flight Center
Szalay, Alex	Eotvos University, Budapest
Szapudi, Istvan	Space Telescope Science Institute
Tewari, Krishna	ST Systems Corporation
Timbie, Peter	Brown University
Toller, Gary	GSC/NASA Goddard Space Flight Center
Torres, Sergio	ST Systems Corporation
Trasco, John	University of Maryland
Traschen, Jennie	University of Massachusetts
Trimble, Virginia	University of Maryland
Trower, W. Peter	Virginia Tech.
Truran, Jim	University of Illinois
Turner, Kenneth	National Science Foundation
Tyson, Tony	AT & T Bell Labs.
Verter, Fran	USRA/NASA Goddard Space Flight Center
Vittorio, Nicola	University Dell'Aquila
von Gronefeld, Peter	Jet Propulsion Laboratory
Waldrop, Mitchell	Science
Wang, Boqi	University of California, Berkeley
Wang, Jin	University of Illinois

Wang, Ning	UC/Berkeley
Warren, Michael S.	Los Alamos National Laboratory
Watt, Martin	Birmingham University
Weiland, Janet	GSC/NASA Goddard Space Flight Center
Weiss, Ray	Massachusetts Institute of Technology
White, Simon	University of Arizona
Wilkinson, David	Princeton University
Wilson, Andrew S.	University of Maryland
Windhorst, Rogier A.	Arizona State University
Wolff, Michael T.	USRA/Naval Research Laboratory
Woo, Ching-Hung	University of Maryland
Wright, Edward L.	UCLA
Zurek, W. H.	Los Alamos National Laboratory

Appendix C:

Physical Constants

TABLE OF PHYSICAL CONSTANTS

CONSTANT	SYMBOL	MKS	CGS	OTHER
speed of light	c	$3.00 \cdot 10^{8}$ m/s	$3.00 \cdot 10^{10}$ cm/s	(2.997925)
electron charge	e	$1.60 \cdot 10^{-19}$ coul	$4.80 \cdot 10^{-10}$ esu	
Planck constant	h	$6.63 \cdot 10^{-34}$ J•s	$6.63 \cdot 10^{-27}$ erg•s	
	$\hbar$	$1.05 \cdot 10^{-34}$ J•s	$1.05 \cdot 10^{-27}$ erg•s	
	hc	$1.99 \cdot 10^{-25}$ J•m	$1.99 \cdot 10^{-16}$ erg•cm	
	$\hbar c$	$3.15 \cdot 10^{-26}$ J•m	$3.15 \cdot 10^{-17}$ erg•cm	200 MeV•fm
Boltzmann constant	k	$1.38 \cdot 10^{-23}$ J/K	$1.38 \cdot 10^{-16}$ erg/K	$8.6 \cdot 10^{-5}$ eV/K
	k/h	$2.08 \cdot 10^{10}$ s^{-1}/K	$2.08 \cdot 10^{10}$ s^{-1}/K	
	k/hc	67.5 m^{-1}/K	0.675 cm^{-1}/K	
Gravitational constant	G	$6.67 \cdot 10^{-11}$ $N \cdot m^2/kg^2$	$6.67 \cdot 10^{-8}$ $dy \cdot cm^2/gm^2$	
Gas constant	R	8.314 J/K•mole	$8.31 \cdot 10^{7}$ erg/K•mole	
Avagadro's number (= R/k)	N	$6.02 \cdot 10^{26}$ amu/kg	$6.02 \cdot 10^{23}$ amu/kg	$6 \cdot 10^{23}$ molecules/mole
electron mass	m_e	$9.11 \cdot 10^{-31}$ kg	$9.11 \cdot 10^{-28}$ gm	0.51 MeV
proton mass	M_p	$1.67 \cdot 10^{-27}$ kg	$1.67 \cdot 10^{-24}$ gm	938 MeV
neutron mass	M_n	$1.67 \cdot 10^{-27}$ kg	$1.67 \cdot 10^{-24}$ gm	939 MeV
pion mass (=270•m_e)	m_π	$2.46 \cdot 10^{-28}$ kg	$2.46 \cdot 10^{-25}$ gm	140 MeV
muon mass (=207•m_e)	m_μ	$1.89 \cdot 10^{-28}$ kg	$1.89 \cdot 10^{-25}$ gm	106 MeV
classical elect radius (=e^2/mc^2)	r_c	$2.82 \cdot 10^{-15}$ m	$2.82 \cdot 10^{-13}$ cm	
Compton wavelength (=h/mc)	λ_c	$2.43 \cdot 10^{-12}$ m	$2.43 \cdot 10^{-10}$ cm	0.02 Å

Thomson cross-section	σ_T	$6.65 \cdot 10^{-29}$	m^2	$6.65 \cdot 10^{-25}$	cm^2	
Planck length ($=\sqrt{\hbar G/c^3}$)	ℓ_{Pl}	$1.61 \cdot 10^{-35}$	m	$1.61 \cdot 10^{-33}$	cm	
Planck time ($=\sqrt{\hbar G/c^5}$)	t_{Pl}	$5.39 \cdot 10^{-44}$	s	$5.39 \cdot 10^{-44}$	s	
Planck density ($=c^5/\hbar G^2$)	ρ_{Pl}	$5.16 \cdot 10^{96}$	kg/m^3	$5.16 \cdot 10^{93}$	gm/cm^3	
Bohr radius ($=\hbar^2/me^2$)	r_B	$0.53 \cdot 10^{-10}$	m	$0.53 \cdot 10^{-8}$	cm	0.5 Å
Fine structure constant ($=e^2/\hbar c$)	α	$7.30 \cdot 10^{-3}$		$7.30 \cdot 10^{-3}$		1/137
Bohr magneton ($=e\hbar/2m_e c$)	μ_B	$9.27 \cdot 10^{-24}$	J/T	$9.27 \cdot 10^{-21}$	erg/gauss	
Nuclear magneton ($=e\hbar/2M_p c$)	μ_N	$5.05 \cdot 10^{-27}$	J/T	$5.05 \cdot 10^{-24}$	erg/guass	
Permittivity of vacuum	ε_o	$8.85 \cdot 10^{-12}$	fd/m			$1/4\pi\varepsilon_o = 9.0 \cdot 10^9$
Permeability in vacuum	μ_o	$4\pi \cdot 10^{-7}$	Hen/m			
Stefan-Boltzmann constant	σ	$5.67 \cdot 10^{-8}$	$W/m^2 \cdot K^4$	$5.67 \cdot 10^{-5}$	$erg/s \cdot cm^2 \cdot K^4$	
Rydberg ($=m_e e^4/2\hbar^2$)	R_∞	$2.18 \cdot 10^{-18}$	J	$2.18 \cdot 10^{-11}$	erg	13.6 eV
1 amu		$1.66 \cdot 10^{-27}$	kg	$1.66 \cdot 10^{-24}$	gm	931.5 MeV
1 calorie		4.19	J	$4.19 \cdot 10^{7}$	erg	
1 year		$3.16 \cdot 10^{7}$	s	$3.16 \cdot 10^{7}$	s	
1 atmosphere		$1.01 \cdot 10^{5}$	N/m^2	$1.01 \cdot 10^{6}$	$dyne/cm^2$	$14.2\ lbs/in^2$
		$1.01 \cdot 10^{5}$	Pascal			760 Torr
1 eV		$1.6 \cdot 10^{-19}$	J	$1.6 \cdot 10^{-12}$	erg	11,605 K
		$1.24 \cdot 10^{-6}$	m	$1.24 \cdot 10^{-4}$	cm	
		1	Tesla	10^4	gauss	

ASTROPHYSICAL CONSTANTS

CONSTANT	SYMBOL	MKS	CGS	OTHER
atronomical unit	AU	$1.50 \cdot 10^{11}$ m	$1.50 \cdot 10^{13}$ cm	
	AU/year			4.74 km/s
parsec	pc	$3.09 \cdot 10^{16}$ m	$3.09 \cdot 10^{18}$ cm	3.26 LY
solar mass	$M_\odot$	$1.99 \cdot 10^{24}$ kg	$1.99 \cdot 10^{33}$ gm	
solar luminosity	$L_\odot$	$3.90 \cdot 10^{26}$ J/s	$3.90 \cdot 10^{33}$ erg/s	
solar effective temperature	$T_{eff\odot}$	5780 K	5780 K	
solar radius	$R_\odot$	$6.96 \cdot 10^{8}$ m	$6.96 \cdot 10^{10}$ cm	
Earth radius	$R_\oplus$	$6.38 \cdot 10^{6}$ m	$6.38 \cdot 10^{8}$ cm	
Earth mass	$M_\oplus$	$5.98 \cdot 10^{24}$ kg	$5.98 \cdot 10^{27}$ gm	
Earth density	$\rho_\oplus$	5520 kg/m^3	5.52 gm/cm^3	
Jansky	Jy	$1.0 \cdot 10^{-26}$ $W/m^2 \cdot Hz$	$1 \cdot 10^{-23}$ $erg/s \cdot cm^2 \cdot Hz$	
Hubble constant	H_o	$3.24h \cdot 10^{-18}$ s^{-1}	$3.24h \cdot 10^{-18}$ s^{-1}	$100h$ km/s•Mpc
critical density ($=3H_o^2/8\pi G$)	ρ_o	$1.88h^2 \cdot 10^{-26}$ kg/cm^3	$1.88h^2 \cdot 10^{-29}$ gm/cm^3	
plasma frequency				$8.98\sqrt{n_e(cm^{-3})}$ kHz/gauss
radian				$57.29578° = 206{,}265''$
CMB photon density	n_γ	$4.15 \cdot 10^{5}$ m^{-3}	415 cm^{-3}	

Author Index

Subject Index

AIP Conference Proceedings

		L.C. Number	ISBN
No. 159	Applications of Radio-Frequency Power to Plasma: Seventh Topical Conference (Kissimmee, FL, 1987)	87-71812	0-88318-359-5
No. 160	Advances in Laser Science–II (Seattle, WA, 1986)	87-71962	0-88318-360-9
No. 161	Electron Scattering in Nuclear and Particle Science: In Commemoration of the 35th Anniversary of the Lyman-Hanson-Scott Experiment (Urbana, IL, 1986)	87-72403	0-88318-361-7
No. 162	Few-Body Systems and Multiparticle Dynamics (Crystal City, VA, 1987)	87-72594	0-88318-362-5
No. 163	Pion–Nucleus Physics: Future Directions and New Facilities at LAMPF (Los Alamos, NM, 1987)	87-72961	0-88318-363-3
No. 164	Nuclei Far from Stability: Fifth International Conference (Rosseau Lake, ON, 1987)	87-73214	0-88318-364-1
No. 165	Thin Film Processing and Characterization of High-Temperature Superconductors (Anaheim, CA, 1987)	87-73420	0-88318-365-X
No. 166	Photovoltaic Safety (Denver, CO, 1988)	88-42854	0-88318-366-8
No. 167	Deposition and Growth: Limits for Microelectronics (Anaheim, CA, 1987)	88-71432	0-88318-367-6
No. 168	Atomic Processes in Plasmas (Santa Fe, NM, 1987)	88-71273	0-88318-368-4
No. 169	Modern Physics in America: A Michelson-Morley Centennial Symposium (Cleveland, OH, 1987)	88-71348	0-88318-369-2
No. 170	Nuclear Spectroscopy of Astrophysical Sources (Washington, DC, 1987)	88-71625	0-88318-370-6
No. 171	Vacuum Design of Advanced and Compact Synchrotron Light Sources (Upton, NY, 1988)	88-71824	0-88318-371-4
No. 172	Advances in Laser Science–III: Proceedings of the International Laser Science Conference (Atlantic City, NJ, 1987)	88-71879	0-88318-372-2
No. 173	Cooperative Networks in Physics Education (Oaxtepec, Mexico, 1987)	88-72091	0-88318-373-0
No. 174	Radio Wave Scattering in the Interstellar Medium (San Diego, CA, 1988)	88-72092	0-88318-374-9
No. 175	Non-neutral Plasma Physics (Washington, DC, 1988)	88-72275	0-88318-375-7
No. 176	Intersections Between Particle Land Nuclear Physics (Third International Conference) (Rockport, ME, 1988)	88-62535	0-88318-376-5
No. 177	Linear Accelerator and Beam Optics Codes (La Jolla, CA, 1988)	88-46074	0-88318-377-3

No. 178	Nuclear Arms Technologies in the 1990s (Washington, DC, 1988)	88-83262	0-88318-378-1
No. 179	The Michelson Era in American Science: 1870–1930 (Cleveland, OH, 1987)	88-83369	0-88318-379-X
No. 180	Frontiers in Science: International Symposium (Urbana, IL, 1987)	88-83526	0-88318-380-3
No. 181	Muon-Catalyzed Fusion (Sanibel Island, FL, 1988)	88-83636	0-88318-381-1
No. 182	High T_c Superconducting Thin Films, Devices, and Application (Atlanta, GA, 1988)	88-03947	0-88318-382-X
No. 183	Cosmic Abundances of Matter (Minneapolis, MN, 1988)	89-80147	0-88318-383-8
No. 184	Physics of Particle Accelerators (Ithaca, NY, 1988)	89-83575	0-88318-384-6
No. 185	Glueballs, Hybrids, and Exotic Hadrons (Upton, NY, 1988)	89-83513	0-88318-385-4
No. 186	High-Energy Radiation Background in Space (Sanibel Island, FL, 1987)	89-83833	0-88318-386-2
No. 187	High-Energy Spin Physics (Minneapolis, MN, 1988)	89-83948	0-88318-387-0
No. 188	International Symposium on Electron Beam Ion Sources and their Applications (Upton, NY, 1988)	89-84343	0-88318-388-9
No. 189	Relativistic, Quantum Electrodynamic, and Weak Interaction Effects in Atoms (Santa Barbara, CA, 1988)	89-84431	0-88318-389-7
No. 190	Radio-frequency Power in Plasmas (Irvine, CA, 1989)	89-45805	0-88318-397-8
No. 191	Advances in Laser Science–IV (Atlanta, GA, 1988)	89-85595	0-88318-391-9
No. 192	Vacuum Mechatronics (First International Workshop) (Santa Barbara, CA, 1989)	89-45905	0-88318-394-3
No. 193	Advanced Accelerator Concepts (Lake Arrowhead, CA, 1989)	89-45914	0-88318-393-5
No. 194	Quantum Fluids and Solids—1989 (Gainesville, FL, 1989)	89-81079	0-88318-395-1
No. 195	Dense *Z*-Pinches (Laguna Beach, CA, 1989)	89-46212	0-88318-396-X
No. 196	Heavy Quark Physics (Ithaca, NY, 1989)	89-81583	0-88318-644-6
No. 197	Drops and Bubbles (Monterey, CA, 1988)	89-46360	0-88318-392-7
No. 198	Astrophysics in Antarctica (Newark, DE, 1989)	89-46421	0-88318-398-6
No. 199	Surface Conditioning of Vacuum Systems (Los Angeles, CA, 1989)	89-82542	0-88318-756-6
No. 200	High T_c Superconducting Thin Films: Processing, Characterization, and Applications (Boston, MA, 1989)	90-80006	0-88318-759-0

No. 201	QED Stucture Functions (Ann Arbor, MI, 1989)	90-80229	0-88318-671-3
No. 202	NASA Workshop on Physics From a Lunar Base (Stanford, CA, 1989)	90-55073	0-88318-646-2
No. 203	Particle Astrophysics: The NASA Cosmic Ray Program for the 1990s and Beyond (Greenbelt, MD, 1989)	90-55077	0-88318-763-9
No. 204	Aspects of Electron–Molecule Scattering and Photoionization (New Haven, CT, 1989)	90-55175	0-88318-764-7
No. 205	The Physics of Electronic and Atomic Collisions (XVI International Conference) (New York, NY, 1989)	90-53183	0-88318-390-0
No. 206	Atomic Processes in Plasmas (Gaithersburg, MD, 1989)	90-55265	0-88318-769-8
No. 207	Astrophysics from the Moon (Annapolis, MD, 1990)	90-55582	0-88318-770-1
No. 208	Current Topics in Shock Waves (Bethlehem, PA, 1989)	90-55617	0-88318-776-0
No. 209	Computing for High Luminosity and High Intensity Facilities (Santa Fe, NM, 1990)	90-55634	0-88318-786-8
No. 210	Production and Neutralization of Negative Ions and Beams (Brookhaven, NY, 1990)	90-55316	0-88318-786-8
No. 211	High-Energy Astrophysics in the 21st Century (Taos, NM, 1989)	90-55644	0-88318-803-1
No. 212	Accelerator Instrumentation (Brookhaven, NY, 1989)	90-55838	0-88318-645-4
No. 213	Frontiers in Condensed Matter Theory (New York, NY, 1989)	90-6421	0-88318-771-X 0-88318-772-8 (pbk.)
No. 214	Beam Dynamics Issues of High-Luminosity Asymmetric Collider Rings (Berkeley, CA, 1990)	90-55857	0-88318-767-1
No. 215	X-Ray and Inner-Shell Processes (Knoxville, TN, 1990)	90-84700	0-88318-790-6
No. 216	Spectral Line Shapes, Vol. 6 (Austin, TX, 1990)	90-06278	0-88318-791-4
No. 217	Space Nuclear Power Systems (Albuquerque, NM, 1991)	90-56220	0-88318-838-4
No. 218	Positron Beams for Solids and Surfaces (London, Canada, 1990)	90-56407	0-88318-842-2
No. 219	Superconductivity and Its Applications (Buffalo, NY, 1990)	91-55020	0-88318-835-X
No. 220	High Energy Gamma-Ray Astronomy (Ann Arbor, MI, 1990)	91-70876	0-88318-812-0
No. 221	Particle Production Near Threshold (Nashville, IN, 1990)	91-55134	0-88318-829-5
No. 222	After the First Three Minutes (College Park, MD, 1990)	91-55214	0-88318-828-7